SYSTEMS & CONTROL ENCYCLOPEDIA

Theory, Technology, Applications

VOLUME 8
INDEXES

Editor-in-Chief

Madan G Singh

University of Manchester
Institute of Science and Technology,
Manchester, UK

PERGAMON PRESS

OXFORD · NEW YORK · BEIJING · FRANKFURT
SÃO PAULO · SYDNEY · TOKYO · TORONTO

U.K.	Pergamon Press, Headington Hill Hall, Oxford, OX3 0BW, England
U.S.A.	Pergamon Press Inc., Maxwell House, Fairview Park, Elmsford, New York 10523, U.S.A.
PEOPLE'S REPUBLIC OF CHINA	Pergamon Press, Qianmen Hotel, Beijing, People's Republic of China
FEDERAL REPUBLIC OF GERMANY	Pergamon Press, Hammerweg 6, D-6242 Kronberg, Federal Republic of Germany
BRAZIL	Pergamon Editora, Rua Eça de Queiros, 346, CEP 04011, São Paulo, Brazil
AUSTRALIA	Pergamon Press Australia, P.O. Box 544, Potts Point, N.S.W. 2011, Australia
JAPAN	Pergamon Press, 8th Floor, Matsuoka Central Building, 1–7–1 Nishishinjuku, Shinjuku-ku, Tokyo 160, Japan
CANADA	Pergamon Press Canada, Suite 104, 150 Consumers Road, Willowdale, Ontario M2J 1P9, Canada

---

First edition 1987

Library of Congress Cataloging in Publication Data
Main entry under title:
Systems and control encyclopedia.
Includes bibliographies.
1. System analysis—Dictionaries.
2. Control theory—Dictionaries.
3. Systems engineering—Dictionaries.
I. Singh, Madan G.
QA402.S968 1987 003'.03'21 86-15085

British Library Cataloguing in Publication Data
Systems and control encyclopedia: theory, technology, applications.
1. Control theory—Dictionaries I. Singh, Madan G.
003'.03'21 QA402

ISBN 0–08–028709–3 (set)

Printed in Great Britain by A. Wheaton & Co. Ltd, Exeter
Indexes computer-generated and composed
by Information Services Division,
BPCC Graphics Ltd, Derby

CONTENTS

Honorary Editorial Advisory Board	vi
Subject Editors	vii
Foreword	Volume 1
Preface	Volume 1
Introduction	Volume 1
Guide to Use of the Encyclopedia	Volume 1
Alphabetical Entries	
A–Com	Volume 1
Con–E	Volume 2
F–H	Volume 3
I–L	Volume 4
M–O	Volume 5
P–Sim	Volume 6
Sin–Z	Volume 7
Systematic Outline of the Encyclopedia	5193
List of Contributors	5225
Author Citation Index	5275
Subject Index	5423
Information Sources in Systems and Control	5575
List of Acronyms and Abbreviations	5587

HONORARY EDITORIAL ADVISORY BOARD

SUBJECT EDITORS

Applications of Control to Aerospace and Aeronautics

M Pélegrin
Centre d'Etudes et de Recherches de Toulouse
Toulouse, France

Applied Robotics, Control Applications to Manufacturing Industries, Flexible Manufacturing and Ergonomics

P Drazan
University of Wales
Institute of Science and Technology
Cardiff, UK

Building Systems

D J Fisk
Department of the Environment
London, UK

Classical Control, CAD of Control Systems, Multivariable Control

N Munro
University of Manchester
Institute of Science and Technology
Manchester, UK

Database Management, Theory and Applications of Distributed Computing for Control

E Gelenbe
Université Paris-Sud
Paris, France

Distributed Parameter Systems, Discrete Systems

J L Lions
Centre National d'Etudes Spatiales
Paris, France

Environmental Measurements, Measurements in Chemistry

Y Sawaragi
Japan Institute of Systems Research
Kyoto, Japan

Environmental Systems and Agricultural Systems

E Halfon
National Water Research Institute
Burlington, Ontario, Canada

Hierarchical Control, Decentralized Control, Model Simplification for Complex Systems Team Theory, Robustness and Sensitivity

A Titli
Centre National de la Recherche Scientifique
Toulouse, France

History of Systems and Control

S Bennett
University of Sheffield
Sheffield, UK

Management Systems, Educational Systems, Health Care Systems, Social Effects of Automation

A P Sage
George Mason University
Fairfax, Virginia, USA

Measurements in the Biological Area

P A Payne
University of Manchester
Institute of Science and Technology
Manchester, UK

Measurements in the Control Area including Intelligent Instrumentation

M G Mylroi
University of Bradford
Bradford, UK

Modelling and Control of Biological Systems, Biomedical Engineering

D A Linkens
University of Sheffield
Sheffield, UK

Pattern Recognition, Fuzzy Systems

M M Gupta
University of Saskatchewan
Saskatoon, Saskatchewan, Canada

Physical Systems Modelling, Identification, Estimation Theory and Applications, Signal Processing

T Kailath
Stanford University
Stanford, California, USA

L Ljung
Linköping University
Linköping, Sweden

Process Control Applications

A J Niemi

R Ylinen
Technical Research Centre of Finland
Espoo, Finland

Realization Theory for Linear and Nonlinear Systems

C Lobry
Nice, France

M Fliess
Centre National de la Recherche Scientifique
Gif-sur-Yvette, France

Y Rouchaleau
Ecole des Mines de Paris
Valbonne, France

Self-Tuning Regulators and Adaptive Control, Stochastic Systems

D W Clarke
University of Oxford
Oxford, UK

Simulation Techniques, Model Simplification

B P Zeigler
University of Arizona
Tucson, Arizona, USA

Stability Theory, General Systems Theory

G J Klir
State University of New York at Binghamton
Binghamton, New York, USA

Static Optimization, Dynamic Optimization, Optimal Control, Game Theory

A Bensoussan
Institut National de Recherche en
Informatique et en Automatique
Le Chesnay, France

Technology of Final Control Elements

J P Elloy
Ecole Nationale Supérieure de Mécanique
Nantes, France

Transportation Systems, Communication Systems

A J Fossard
Ecole Nationale Supérieure de l'Aeronautique et de l'Espace
Toulouse, France

Utility Systems and Energy Systems

A Brameller
University of Manchester
Institute of Science and Technology
Manchester, UK

SYSTEMATIC OUTLINE OF THE ENCYCLOPEDIA

The Systematic Outline of the Encyclopedia, which is supplementary to the Subject Index, groups the contents of the Encyclopedia by article title under the main subject headings used in commissioning the Encyclopedia. These headings, which are summarized in the Classified List of Contents below, are further subdivided into appropriate subheadings as each subject dictates. The reader is thus presented with a general overview of the contents of the Encyclopedia. Some articles inevitably relate equally well to more than one heading. Rather than make an arbitrary decision as to which section they belong, each article has been listed wherever appropriate.

Article titles can be located by two methods: (a) by referring to the Classified List of Contents below and then looking up the appropriate Subject Area in the main part of the Analytical List of Contents, or (b) by consulting the Alphabetical List of Contents given overleaf.

The publisher wishes to thank Professor Singh for help in overseeing the collation of the Systematic Outline which was prepared by Danny Dicks, Debbie Puleston and other members of the editorial staff of the Encyclopedia from the original commissioning plans of the Subject Editors.

CLASSIFIED LIST OF CONTENTS

1. THEORY

A. Modelling and Simulation

(i) Physical Systems Modelling
(ii) Identification
(iii) Estimation Theory
(iv) Signal Processing
(v) Simulation Techniques
(vi) Model Simplification

B. Systems and Control Theory

(i) Fuzzy Systems
(ii) Pattern Recognition
(iii) General Systems Theory
(iv) Stability Theory
(v) Realization Theory for Linear and Nonlinear Systems
(vi) Classical Control
(vii) Computer-Aided Design of Control Systems
(viii) Multivariable Control
(ix) Distributed Parameter Systems
(x) Discrete Systems
(xi) Self-Tuning Regulators and Adaptive Control

C. Optimization and Operations Research

(i) Static Optimization
(ii) Dynamic Optimization
(iii) Optimal Control
(iv) Game Theory
(v) Stochastic Systems
(vi) Robustness and Sensitivity

D. Complex Systems Theory

(i) Hierarchical Optimization and Control
(ii) Decentralized Control
(iii) Model Simplification for Complex Systems
(iv) Singular Perturbations
(v) Stability of Interconnected Systems
(vi) Team Theory

2. TECHNOLOGY

(i) Technology of Actuators
(ii) Special Actuators Required in Communications, Power Generation, Nuclear Industry, etc.
(iii) Measurements in the Control Area
(iv) Intelligent Instrumentation
(v) Measurements in the Biological Area
(vi) Environmental Measurements
(vii) Measurements in Chemistry
(viii) Database Management
(ix) Distributed Control

3. APPLICATIONS

A. Technological Applications

(i) Process Control in Chemical Industries (Paper, Cement, Sugar, etc.)
(ii) Control of Power Generation and Distribution
(iii) Control in Aeronautics and Space
(iv) Robotics

B. Semitechnological Applications

(i) Road Traffic Control
(ii) Air Traffic Control
(iii) Other Transportation Systems
(iv) Manufacturing Systems
(v) Building Systems
(vi) Utility Systems (Gas, Electricity, Water)
(vii) Communication Systems
(viii) Management Systems
(ix) Health Care Systems
(x) Social Effects of Automation

C. Nontechnological Applications

(i) Modelling and Control of Biological Systems
(ii) Biomedical Engineering
(iii) Environmental Systems
(iv) Agricultural Systems

4. HISTORY

(i) History of Theoretical Techniques
(ii) History of Applications
(iii) History of Organizations

ALPHABETICAL LIST OF CONTENTS

This list enables the reader to locate all the headings and subheadings in the Systematic Outline. Major subdivisions of the Outline are indicated by capitals.

Actuator Technology 2.i
Adaptive Control and Self-Tuning Regulators ... 1.B.xi
Aerospace Control 3.A.iii
Agricultural Systems 3.C.iv
Air Traffic Control 3.B.ii
APPLICATIONS 3
Automation, Social Effects 3.B.x
Biological Measurements 2.v
Biological Systems Modelling and Control 3.C.i
Biomedical Engineering 3.C.ii
Building Systems 3.B.v
Chemical Industries Process Control 3.A.i
Chemical Measurements 2.vii
Classical Control 1.B.vi
Communication Systems 3.B.vii
COMPLEX SYSTEMS THEORY 1.D
Computer-Aided Design of Control Systems 1.B.vii
CONTROL AND SYSTEMS THEORY 1.B
Control in Aeronautics and Space 3.A.iii
Control of Power Generation and Distribution ... 3.A.ii
Database Management 2.viii
Decentralized Control 1.D.ii
Design of Control Systems, Computer-Aided 1.B.vii
Discrete Systems 1.B.x
Distributed Control 2.ix
Distributed Parameter Systems 1.B.ix
Dynamic Optimization 1.C.ii
Electricity Systems 3.B.vi
Environmental Measurements 2.vi
Environmental Systems 3.C.iii
Estimation Theory 1.A.iii
Fuzzy Systems 1.B.i
Game Theory 1.C.iv
Gas Systems 3.B.vi
General Systems Theory 1.B.iii
Health Care Systems 3.B.ix
Hierarchical Optimization and Control 1.D.i
HISTORY 4
Identification 1.A.ii
Intelligent Instrumentation 2.iv
Linear Systems Realization Theory 1.B.v
Management Systems 3.B.viii
Manufacturing Systems 3.B.iv
Measurements in Chemistry 2.vii
Measurements in the Biological Area 2.v
Measurements in the Control Area 2.iii
Model Simplification 1.A.vi
Model Simplification for Complex Systems 1.D.iii
Modelling and Control of Biological Systems 3.C.i
MODELLING AND SIMULATION 1.A
Modelling of Physical Systems 1.A.i
Multivariable Control 1.B.viii
Nonlinear Systems Realization Theory 1.B.v
NONTECHNOLOGICAL APPLICATIONS 3.C
OPERATIONS RESEARCH AND OPTIMIZATION 1.C
Optimal Control 1.C.iii
OPTIMIZATION AND OPERATIONS RESEARCH 1.C
Pattern Recognition 1.B.ii
Physical Systems Modelling 1.A.i
Power Generation and Distribution Control 3.A.ii
Process Control in Chemical Industries (Paper, Cement, Sugar etc.) 3.A.i
Realization Theory for Linear and Nonlinear Systems 1.B.v
Road Traffic Control 3.B.i
Robotics 3.A.iv
Robustness and Sensitivity 1.C.vi
Self-Tuning Regulators and Adaptive Control ... 1.B.xi
SEMITECHNOLOGICAL APPLICATIONS ... 3.B
Sensitivity and Robustness 1.C.vi
Signal Processing 1.A.iv
SIMULATION AND MODELLING 1.A
Simulation Techniques 1.A.v
Singular Perturbations 1.D.iv
Social Effects of Automation 3.B.x
Special Actuators Required in Communications, Power Generation, Nuclear Industry etc. 2.ii
Stability of Interconnected Systems 1.D.v
Stability Theory 1.B.iv
Static Optimization 1.C.i
Stochastic Systems 1.C.v
SYSTEMS AND CONTROL THEORY 1.B
Team Theory 1.D.vi
TECHNOLOGICAL APPLICATIONS 3.A
TECHNOLOGY 2
Technology of Actuators 2.i
THEORY 1
Transportation Systems 3.B.i-ii
Utility Systems(Gas, Electricity, Water) 3.B.vi
Water Systems 3.B.vi

ANALYTICAL LIST OF CONTENTS

1. THEORY

A. Modelling and Simulation

(i) Physical Systems Modelling

Dynamic-Systems Modelling: Basic Principles and Lumped-Parameter Systems
Dynamic-Systems Modelling: Distributed-Parameter Models and Discretization
Ordinary Differential Equations
Stationary Time Series and Their Spectra

(ii) Identification

Estimation or Identification
Extended Least Squares Method
Fuzzy Identification of Systems
Fuzzy Systems: Identification
Fuzzy Systems: Simultaneous Identification and Control
Generalized Least Squares Method
Identification: Asymptotic Theory
Identification: Basic Problem
Identification: Correlation Methods
Identification: Experiment Design
Identification: Frequency-Domain Methods
Identification: History
Identification: Instrumental Variable Techniques
Identification: Least Squares Method
Identification: Maximum Likelihood Method
Identification: Model Structure Determination
Identification: Practical Aspects
Identification: Pseudo-Random Signal Method
Identification: Recursive Methods
Identification: Time-Domain Methods
Identification: Transient- and Frequency-Response Methods
Parameter Estimation: Basic Problem
Parameter Estimation: Fuzzy Logic
Validation of Identified Models
Validation of Simulation Models: General Approach
Validation of Simulation Models: Statistical Approach

(iii) Estimation Theory

(a) Nonlinear Estimation Theory

Extended Kalman Filter
Nonlinear Estimation
Nonlinear Estimation Theory: Asymptotic Analysis Methods
Nonlinear Estimation Theory: Implications for Stochastic Control
Nonlinear Filtering and Quantum Physics
Nonlinear Filtering: Computational Aspects
Nonlinear Filtering: Iterated Ito Integral Expansions

(b) Linear Estimation with Given Second-Order Statistics

Block Matrix Formulae
Constant-Parameter Systems: Fast Algorithms
Discrete-Time Processes: Innovations
Distributed Parameter Systems: Estimation and Control
Kalman Filter Formulae: Variations
Kalman Filtering: Applications
Kalman Filtering: State-Space Models for Discrete-Time Processes
Matrix Triangularization: Givens Transformations
Matrix Triangularization: Householder Transformations
Parameter Estimation: Basic Problem
Parameter Estimation: Fuzzy Logic
Random Number Generation
Smoothing Algorithms
Smoothing Algorithms, Two-Filter
State-Space Modelling: Square Root Algorithms
State-Space Modelling: State-Space Transformation

(iv) Signal Processing

Adaptive Signal-Processing Algorithms: An Introduction
Adaptive Signal-Processing Algorithms: Convergence and Tracking
Adaptive Signal-Processing Algorithms: On-Line Detection of Abrupt Changes
Adaptive Signal-Processing Algorithms: Rare Events and Large Deviations

See also: *Measurements in the Control Area* (*2.iii*)

(*v*) *Simulation Techniques*

(a) Simulation Methodology: General

Simulation in Automatic Control
Simulation Methodology and Model Manipulation
Simulation Methodology and Systems Control and Management
Simulation Methodology: Top-Down Approach
Validation of Simulation Models: General Approach
Validation of Simulation Models: Statistical Approach

(b) Simulation Modelling Systems Formalisms

Model Specification Interfaces
Models as System-Theoretic Specifications
Simulation Models: Taxonomy
Simulation Taxonomy

(c) Simulation Modelling Formalisms

Formalism
Simulation Modelling Formalism, Activity-Based
Simulation Modelling Formalism: Arithmetic Relators
Simulation Modelling Formalism: Bond Graphs
Simulation Modelling Formalism: Cellular Automata
Simulation Modelling Formalism, Discrete Arithmetic-Based
Simulation Modelling Formalism, Event-Based
Simulation Modelling Formalism: Extended Petri-Net Graphs
Simulation Modelling Formalism: Heterarchical Systems
Simulation Modelling Formalism: Hierarchical Decomposition
Simulation Modelling Formalism: Markov Chains
Simulation Modelling Formalism: Ordinary Differential Equations
Simulation Modelling Formalism: Partial Differential Equations
Simulation Modelling Formalism, Process-Based
Simulation Modelling Formalism: Systems Dynamics
Simulation Modelling Formalism: Transaction-Flow Techniques
Simulation Models of Autopoiesis: Variable Structure

(d) Simulation Model Management

Databases for Simulation
Simulation: Model Base Organization and Utilization
Simulation Model Management Objectives and Requirements

(e) Model Behavior Generation/Simulation Techniques

Global and Regional Models
Model Behavior: Type Taxonomy, Generation and Processing Techniques
Moving-Boundary Models: Numerical Solution
Ordinary Differential Equation Models: Numerical Integration of Initial-Value Problems
Ordinary Differential Equation Models: Symbolic Manipulation
Partial Differential Equation Models: Numerical Solution
Random Number Generation
Random Variate Generation
Sensitivity Analysis and Simulation Experimentation
Simulation and Optimization Interfacing
Simulation Objectives: Experimental Frames and Validity
Simulation: Time-Advance Methods
Stochastic Simulation: Experiment Design
Stochastic Simulation: Initial Transient Techniques
Stochastic Simulation: Variance-Reduction Techniques
Structure/Parameter Identification by Simulation Experimentation

(f) Simulation Software and Computer Systems

Cellular Space Models: Simulation Methods
Combined Discrete and Continuous Models: Firmware
Computer Architecture
Computer Memory Hierarchy
Computer Systems Simulation Models
Discrete-Event Simulation: Microprocessor Architectures
Distributed Computer Systems
Distributed Model Simulation Software
Distributed System Models: Distributed Simulation
Emulation and Microprogramming
Flowcharting in Computing

Hybrid Analog–Digital Computers
Interactive Simulation Model and Program Generation
Interactive Terminals
Large-Scale Software Systems: Reliability
Microprocessors
Microprogramming
Multiprogramming
Operating Systems (Computers)
Parallelism and Concurrency in Computers
Process-Based Models: Simulation Software
Program Compilers
Program Interpreters
Programming Languages
Programming Methodology
Semantics in Computing
Simulation and Model Oriented Languages: Taxonomy
Simulation Modelling Support Environments
Simulation of Activities-Based Models: Software
Simulation Program Development: Structured Design Methodology
Simulation Software, Event-Based
Simulation Study Credibility
Simulators for Training
Simulators, Real-Time
Software Modules
Synchronization in Computing
Syntax in Computing
Systems Software
Transactions-Flow Models Simulation Software
Utility Programs

(g) Simulation Behavior Display and Analysis

Character Representation
Simulation and Graphics
Simulation Models Symbolic Processing: Taxonomy
Simulation Output: Statistical Analysis

(h) Simulation: Industrial Applications

Chemical Engineering: Simulation
Industrial Engineering and Operations Research: Simulation
Mechanical System Simulation

See also: *Management Systems (3.A.viii); Database Management (2.viii)*

(*vi*) *Model Simplification*

Linear Continuous Model Manipulation Software
Model Behavior: Regression Metamodel Summarization
Model Structure Simplification
Simulation Methodology and Model Manipulation
Simulation Models: Documentation
System Model Specifications: Manipulation and Simplification

See also: *Model Simplification for Complex Systems (1.D.iii)*

B. Systems and Control Theory

(*i*) *Fuzzy Systems*

(a) Fuzzy Set Theory

Eigen Fuzzy Sets
Extension Principle
Fuzzy Cartesian Product
Fuzzy Connectives (Intersection and Union): General Class
Fuzzy Mapping on Fuzzy Sets
Fuzzy Mapping on Ordinary Sets
Fuzzy Membership Evaluation
Fuzzy Set Theory: An Introduction
Fuzzy Set Theory: Possibilities and Probabilities
Fuzzy Set Theory: Review
Fuzzy Sets: α-Cut
Fuzzy Sets and Systems: Fundamentals
Fuzzy Singletons
Modus Ponens: Generalization
Modus Tollens: Generalization
Ordinary Mapping on Fuzzy Sets
Random Sets in Fuzzy Set Theory
Vagueness in Scientific Theories

(b) Fuzzy Databases and Fuzzy Calculus

Fuzzy Algebra
Fuzzy Databases
Fuzzy Databases: Retrieval Processing
Fuzzy Events
Fuzzy Integrals
Fuzzy Numbers
Fuzzy Numbers: Applications
Fuzzy Random Variables
Language Databases

(c) Possibility Theory, Fuzzy Statistics and Fuzzy Measures

Entropy Measures in Fuzzy Set Theory
Fuzzy Causes: Probability
Fuzzy Events: Probability
Fuzzy Measures: Classical Approach
Fuzzy Measures: Fuzzy Integral Approach
Fuzzy Probability
Fuzzy Set Theory: Possibilities and Probabilities
Fuzzy σ-Fields in Fuzzy Measure Theory
Measures of Fuzziness: A Survey
Measures of Fuzziness: Conditions
Normal Fuzzy Measures and Fuzzy Random Measures
Parameter Estimation: Fuzzy Logic
Plausibility Measures
Possibility Distribution Functions
Possibility Distributions
Possibility Measures
Probabilistic Sets
Probabilistic Sets: Distribution Function
Subjective Entropy

(d) Fuzzy Logic Controllers and Applications of Fuzzy Control

Adaptive Fuzzy Automation
Adaptive Fuzzy Control System
Adaptive Fuzzy Controller
Control Rules: Competition
Control Rules: Generation
Crisp Controller
Direct Current Series Motor: Fuzzy Control
Direct Current Series Motor: Fuzzy Model
Expert Fuzzy Controller
Fuzzy Control Algorithm Completeness
Fuzzy Control Language
Fuzzy Control Processes: Convergence
Fuzzy Control Systems: Linguistic Analysis
Fuzzy Controllers: Reproducibility Property
Fuzzy Logic Controller
Fuzzy Logic Controller: Algebraic Analysis
Fuzzy Logic Controller: Good-Mapping Property
Fuzzy Logic Controller: Implementation of Parameters
Fuzzy Logic Controller: Interactivity
Fuzzy Logic Controller: Self-Improvement
Fuzzy Logic Controller: Self-Organization
Fuzzy Logic Controller: Sensitivity
Fuzzy Logic Controller: Stability
Fuzzy Logic Controller: Structure
Fuzzy Łukasiewicz Controller
Industrial Processes: Fuzzy Logic Controllers
Industrial Production: Fuzzy Control Systems
Learning Fuzzy Controller
Linguistic Controller
Linguistic Variables
Look-Up Tables (Decision Tables)
Manual Process Control

(e) Fuzzy Logic and Inferences

Analogical Inference
Approximate Reasoning (Fuzzy Reasoning)
Fuzzy Algorithms
Fuzzy Conditional Inference
Fuzzy Conditional Inference: Compositional Rules
Fuzzy Implication Operators
Fuzzy Implication Operators: A Survey
Fuzzy Logic
Fuzzy Łukasiewicz Logic
Fuzzy Reasoning Methods
Fuzzy Switching Functions
Fuzzy Switching Functions: Algebra
Many-Valued Logics
Many-Valued Logics, Algebraic
Many-Valued Logics: Applications
Multidimensional Fuzzy Implication
Neural Fuzzy Learning Strategies

(f) Fuzzy Systems and Decision Making

Brains as Fuzzy Systems
Brains: Creative Functions
Communication Theory: Fuzzy Detection and Estimation
Composite Fuzzy Relational Equations: Solution Algorithms
Composite Fuzzy Relations
Composite Fuzzy Relations and Relational Equations
Computer-Assisted Medical Diagnostic Systems: Fuzzy Methods
Computer Security Systems: Fuzzy Logics
Computer Security Systems: Possibility Theory
Conflict Resolution: Fuzzy-Ideal Concepts
Defuzzification Operators

Education Concepts: Fuzzy Aspects
Fuzzification Operators
Fuzzy Automata
Fuzzy Automata: Identification
Fuzzy Catastrophes
Fuzzy Discretization
Fuzzy Identification of Systems
Fuzzy Learning Systems
Fuzzy Linear Programming
Fuzzy Restrictions
Fuzzy Systems: Analysis and Synthesis
Fuzzy Systems, Dynamical
Fuzzy Systems: Identification
Identification of Fuzzy Sets with Random Sets
Incompatibility Principle
Optimal Control of Fuzzy Systems
Optimal Fuzzy Target Control
Pragmatic Fuzzy Systems
Prediction in Fuzzy Systems
Relational Assignment Equations
Stabilization Control of Fuzzy Systems
Stochastic Systems in Fuzzy Environments: Control
Systems Management

(g) Fuzzy Ranking, Modelling and Optimization

Fuzzy Cardinals
Fuzzy Mathematical Programming Problem: Resolution
Fuzzy Multicriteria Modelling
Fuzzy Ranking
Fuzzy Sets: Ordering by Fuzzy Entropy
Multicriteria Decision Problem
Operations Research: Fuzzy Set Theory
Optimal Control of Fuzzy Systems
Optimal Fuzzy Target Control
Optimization: Fuzzy Set Theory
Optimization in a Fuzzy Environment
System Performance: Fuzzy Methods

(h) Linguistic Approximations and Expert Systems

Aggregate Planning Using Linguistic Variables
Biofeedback
Brain Activity During Language Perception
Computer-Assisted Medical Diagnostic Systems: Fuzzy Methods
Computerized Electrocardiogram Diagnosis: Fuzzy Approach
Electrocardiogram Diagnosis: Fuzzy Approach
Expert Fuzzy Controller
Expert Systems
Expert Systems in Medical Reasoning
Fuzzy Control Systems: Linguistic Analysis
Information Systems
Linguistic Approximation
Linguistic Controller
Linguistic Hedges
Linguistic Models
Linguistic Probabilities
Linguistic Variables
Structural Damage Assessment: Fuzzy Approach
Subjectivity: Human Communication
Verbal Reports
Verbalization

(i) Reliability Theory and Applications

Reliability of Control
Reliability of Systems
Reliability of Systems: Basic Theory
Reliability of Systems: Specific Applications

(j) Random and Pseudo-Random Signals

Binary Random Signals and Random Signals
Identification: Pseudo-Random Signal Method
Maximum-Length Sequences
Power System Component Identification: Pseudo-Random Signal Method

(ii) Pattern Recognition

Character Recognition
Cluster Analysis, Nonmetric
Discriminant Analysis
Feature Selection: Fuzzy Classification
Fuzzy Partitions
Fuzzy Transformations
Industrial Vision Systems
Industrial Vision Systems for Inspection and Manufacturing
Industrial Vision Systems: Segmentation of Gray-Scale Images
Pattern Recognition
Pattern Recognition: Decision-Theoretic Approaches
Pattern Recognition: Syntactic Methods
Scene Analysis, Hierarchical
Thinning Transformations

(iii) General Systems Theory

(a) Taxonomy of Systems

Adaptive Systems
Anticipatory Systems
Autopoiesis
Behavior Systems
Closed Systems
Data Systems
Directed Systems
Dynamic Systems
Dynamic Systems: A Survey
Effective Systems
Efficient Systems
Epistemological Hierarchy of Systems
General Systems
Goal-Oriented Systems
Learning Systems
Lindenmayer Systems
Linear Systems: General Aspects
Metasystems
Neutral Systems
Petri Nets
Self-Organizing Systems
Self-Reproducing Systems
Source Systems
State-Transition Systems
Structure Systems
Systems
Value Systems

(b) Information Aspects of Systems

Cybernetics
Cybernetics: A Survey
Cybernetics: History
Entropy Minimax Theory
Expert Systems
Expert Systems in Medical Reasoning
Information Laws of Systems
Information Theory
Information Transmission
Law of Requisite Variety
Maximum-Entropy Principle
Possibilistic Information
Pragmatic Information
Semantic Information
Shannon Entropy
U-Uncertainty

(c) Methodological Issues of Systems

Design of Systems
General Systems Problem Solver
Interpretive Structure Modelling
Metamethodology of Systems
Monte Carlo Methods
Multicriteria Decision Making
Q-Analysis
Reconstructability Analysis
Set-Covering Problems

(d) Special Theories

Catastrophe Theory
Information Theory
Linear Systems: General Aspects
Praxiology
Similarity
Stability Theory

(e) Philosophical Issues of Systems

Autology
Causality
Explanation
Holism
Prediction
Reductionism
Retrodiction
Semiotics
Teleology

(f) Computational Aspects of Systems

Artificial Intelligence
Computability
Computational Complexity
Decidability
Turing Machines

(g) General Systems Concepts

Antisymmetric Relations
Chaos
Complexity of Systems
Emergent Properties
Environment of a System
Equivalence Relations
General Systems
Generalized Dependence
Homomorphism of Systems
Intelligence Amplification
Isomorphic Systems
Linear Orderings
Partial Orderings
Partition of a Set
Past Determinancy and Finite Memory
Projections
Quasiorderings
Relational Join
Relations, Mathematical
Reliability of Systems
Set Covers
Simplical Complexes
Symmetric Relations
Systems Interactions
Total Relations

Transitivity of Relations

See also: *Management Systems (3.B.viii)*

(iv) Stability Theory

Connective Stability
Domains of Attraction
Lyapunov Equation
Nonlinear Feedback Systems: Stability Conditions
Nonlinear Systems Stability: Vector Norm Approach
Stability: Dissipativeness Concept
Stability of Values and Pulses
Stability Theory
Structural Stability
Structural Stability of Control Systems

See also: *Classical Control (1.B.vi); Stability of Interconnected Systems (1.D.v)*

(v) Realization Theory for Linear and Nonlinear Systems

(a) Linear Realization Theory

Abstract Realization Theory
Approximate Realization Theory: Comparison of Optimal Methods
Balanced Realization of a Linear Stationary Dynamical System
Internal Model Principle
Internal Model Principle: Output Tracking and Disturbance Rejection
Linear Multivariable Systems: Zeros
Linear Passive Network Synthesis
Linear Systems: Robustness
Linear Systems: Stability
Partial Minimal Realization
Pole Assignment by State Feedback and Luenberger Observer
Realization Theory: Gauss–Markov Processes
Realization Theory: Hankel Matrix Approach
Realization Theory: Matrix Fraction Approach
Realization Theory, Multivariate
Realization Theory: Numerical Aspects

(b) Nonlinear Realization Theory

Distillation Columns: Bilinear Models
Functional Expansions
Input–Output Maps of Nonlinear Discrete-Time Systems
Noncommuting Variables: Finite Automata
Noncommuting Variables: Rational Generating Series
Noncommuting Variables: Realization of Bilinear Systems
Nonlinear Discrete-Time Systems: Algebraic Theory
Nonlinear Systems: Approximation by Simple State-Space Methods
Nonlinear Systems: Local Realization Theory
Power Plants: Nonlinear Modelling
Realization of Finite Volterra Series
Realization Theory, Nonlinear
Universal Inputs

(c) System Structure

Affine Systems: Controllability
Attitude Control of Rigid Spacecraft
Extension Techniques
Input–Output and Disturbance Decoupling
Lie Brackets
Lie Groups: Controllability
Matrix Groups: Action on State Spaces and Control
Multivalued Differential Equations in Control Systems
Nonlinear Decoupling
Nonlinear Observability
Nonlinear System Theory: Geometric Methods
Real Analyticity
Structural Stability of Control Systems

See also: *Classical Control (1.B.vi); Multivariable Control (1.B.viii); Stability Theory (1.B.iv)*

(vi) Classical Control

(a) Linear Systems

Laplace Transforms and Bode Diagrams
Nyquist Stability Criterion
PID Controllers
PID Controllers, Digital

(b) Nonlinear Systems

Describing-Function Reliability
Nonlinear Systems: A Survey
Nonlinear Systems: Lyapunov Method
Nonlinear Systems: Lyapunov Stability
Sinusoidal Input Describing Function: Evaluation and Properties

(c) Sampled-Data Systems

z-Transform Method

See also: *Realization Theory for Linear and Nonlinear Systems (1.B.v); Multivariable Control (1.B.viii)*

(vii) *Computer-Aided Design of Control Systems*

Computer-Aided Design: Environments for Control Design
Computer-Aided Design: Multivariable Systems
Nonlinear Systems: Computer-Aided Design

(viii) *Multivariable Control*

(a) System Representations

Smith–McMillan Canonical Forms for Rational Matrices
State-Space Modelling: State-Space Transformation

(b) Analysis

Linear Multivariable Systems Controllability and Observability: Numerical Aspects
Linear Multivariable Systems: Zeros
Multivariable Systems: Functional Analysis
Polynomial Matrices in Control Theory
Realization Theory: Hankel Matrix Approach
Realization Theory: Matrix Fraction Approach

(c) Synthesis Techniques

Linear Multivariable Control Systems Synthesis: Geometric Approach
Multivariable Systems: Minimal Design Problem
Pole Assignment: A Review of Methods
Pole Assignment: Polynomial Matrices
Pole Assignment: The General Problem

(d) Design Techniques

Characteristic-Locus and Multivariable-Root-Locus Design Methods
Characteristic-Locus Method: Design Study
Inverse Nyquist Array Design Method
Inverse Nyquist Array Design Method: Applications
Inverse Nyquist Array Design Method: Dominance, Optimization and Sharing
Inverse Nyquist Array Design Method: Multivariable Control of a Milling Circuit
Linear Quadratic Gaussian Multivariable Control Designs: Robustness
Multivariable Systems: Method of Inequalities
Multivariable Systems: Robustness in Design
Robust Controller Design

(e) Nonlinear Systems

Nonlinear Systems: Circle Criterion Extensions

See also: *Classical Control (1.B.vi); Realization Theory for Linear and Nonlinear Systems (1.B.v); Optimal Control (1.C.iii); Stochastic Systems (1.C.v); Robustness and Sensitivity (1.C.vi); Self-Tuning Regulators and Adaptive Control (1.B.xi); Complex Systems Theory (1.D)*

(ix) *Distributed Parameter Systems*

(a) Distributed Parameter Systems and Their Control

Control Equipment Location Problems
Distributed Elliptic Systems: Optimal Design
Distributed Parameter Systems
Distributed Parameter Systems: Estimation and Control
Distributed Parameter Systems: Identification
Distributed Parameter Systems: Self-Tuning Control
Distributed Systems with Delay
Free Surfaces: Control
Integrodifferential Systems
Linear Hyperbolic Systems: Control Theory
Linear Systems: Control Theory
Optimal Control: Gradient Related Numerical Solution of Optimal Control Problems
Optimal Control of Multistate Systems
Parabolic Systems
Sensor and Control Location Problems
Stochastic Distributed Parameter Systems: Linear Estimation and Control

(b) Methods for Analysis

Calculus of Variations
Control in Coefficients
Duality Methods
Eigenfunctions and Eigenvalues: Control
Finite Difference Method
Finite Element Method
Fourier Series: Fast Fourier Transforms
Fourier Transforms
Isoperimetric Problems
Laplace Transforms
Multigrid Methods
Semigroup Theory
Spectral Methods
Transport Equations
Variational Inequalities
Variational Inequalities: Control

See also: *Optimization and Operations Research (1.C)*

(x) *Discrete Systems*

Discrete and Combinatorial Optimization: Introduction
Discrete-Time Control
Nonlinear Discrete-Time Systems: Algebraic Theory
Optimal and Suboptimal Stochastic Control: Discrete-Time Systems

See also: *Distributed Parameter Systems (1.B.ix); Classical Control (1.B.vi); Optimization and Operations Research (1.C)*

(xi) *Self-Tuning Regulators and Adaptive Control*

(a) General Adaptive Control Theory

Adaptive and Stochastic Control: Weighted Minimum Variance Method
Adaptive Control: Fundamental Problems
Adaptive Control: Positive-Real Condition
Adaptive Controllers: Convergence Analysis
Adaptive Controllers: Design Packages
Adaptive Controllers: Supervision Level
Adaptive Systems: Hyperstability and Positivity
Adaptive Systems: Lyapunov Synthesis
Dual Control
Learning Systems
Martingales and Semimartingales
Minimum-Phase Systems: Adaptive Controllers and Design
Non-Minimum-Phase Systems: Adaptive Control
Predictive Control, Fast-Model
Predictor Models
Self-Organizing Control Systems
Variable-Structure Systems: Adaptive Control

(b) Model Reference Adaptive Control Theory

Adaptive Systems: Model Reference Approach
Discrete Model Reference Adaptive Systems: Design
Model Reference Adaptive Control
Model Reference Adaptive Control: Microcomputer Implementation
Model Reference Adaptive Control Theory
Model Reference Adaptive System Direct and Indirect Designs
Model Reference Adaptive Systems: Augmented Error Signals
Model Reference Adaptive Systems: Error Models
Model Reference Adaptive Systems Methods: Comparative Studies
Model Reference Adaptive Systems: Model Updating
Self-Tuning and Model Reference Adaptive Systems: Interrelation

(c) Adaptive Control Applications

Adaptive Control: Industrial Applications
Cement Industry: Adaptive Control
Concrete Mixing: Cautious Adaptive Control
Distillation Columns: Adaptive Control
Engine Speed: Self-Tuning Control
Glass Industry: Adaptive Control
Heating Systems: Adaptive Control
Hydroelectricity: Self-Selecting Control
Load-Frequency Regulation: Adaptive Control
Nuclear Power Plants: Adaptive Control
Papermaking: Adaptive Control
pH Regulation
Radio Telescopes: Adaptive Control
Satellites: Adaptive Control
Ship Positioning: Adaptive Control
Ship Steering: Model Reference Adaptive Control

Titanium Dioxide Kilns: Adaptive Control

(d) Self-Tuning Regulators

Distributed Parameter Systems: Self-Tuning Control
Explicit Self-Tuners: Convergence Analysis
Multivariable and Multirate Self-Tuning Controllers
Multivariable Systems: Self-Tuning Control Design
Nonlinear Parameter-Adaptive Control Algorithms
Parameter-Adaptive Algorithms: Feedback System
Parameter-Adaptive Controllers: Applications
Pole-Assignment Self-Tuning
Predictive Control, Fast-Model
Predictor Models
Self-Tuners as Data Predictors
Self-Tuners Design: Linear Quadratic Gaussian Theory
Self-Tuners Design: Predictive Control Theory
Self-Tuning and Extremal Control Systems
Self-Tuning and Model Reference Adaptive Systems: Interrelation
Self-Tuning Control, Feedforward
Self-Tuning Control: General Aspects
Self-Tuning Control, Hybrid
Self-Tuning Control: Models
Self-Tuning Controller Implementation
Self-Tuning Controllers: Input–Output Analysis
Self-Tuning On–Off Control
Self-Tuning Proportional-Integral-Derivative Controllers: Pole Assignment Approach

(e) Stochastic Systems and Recursive Algorithms

Gaussian Random Variables
Optimal and Suboptimal Stochastic Control: Discrete-Time Systems
Optimal Stochastic Control: General Aspects
Optimal Stochastic Control: Numerical Methods
Optimization Under Uncertainty
Recursive Algorithms: Variable Forgetting Factors
Recursive Stochastic Algorithms: Ordinary Differential Equation Method
Stochastic Control: Introduction
Stochastic Integrals and Stochastic Calculus

See also: *Stochastic Systems (1.C.v); Process Control in Chemical Industries (3.A.i)*

C. Optimization and Operations Research

(*i*) *Static Optimization*

Static Optimization: Introduction

(a) Linear Programming

Decomposition Methods
Linear Programming: An Introduction
Linear Programming: Duality
Simplex Method

(b) Nonlinear Programming

Nonlinear Programming: A Survey

(c) Variational Methods

Calculus of Variations
Differential Inclusions and Viability Theory
Variational Inequalities
Variational Inequalities: Control
Variational Methods: Introduction

(d) Sensitivity in Static Optimization

Optimization: Sensitivity Analysis

(e) Numerical Algorithms for Optimization

Constrained Optimization: Computational Methods
Constrained Optimization: Feasible-Direction Methods
Constrained Optimization: Lagrange-Multiplier Methods
Constrained Optimization: Lagrangian Methods Based on Sequential Quadratic Programming
Constrained Optimization: Penalty Methods
Linearly Constrained Optimization: Numerical Methods
Optimization of Functionals by Algorithms: Introduction
Unconstrained Optimization: Basic Descent Method

Unconstrained Optimization: Conjugate Gradient Methods
Unconstrained Optimization: Newton and Quasi-Newton Methods

(f) Nondifferential Optimization

Optimization, Nondifferentiable

(g) Homotopy

Homotopy

(h) Combinatorial Optimization

Combinatorial Optimization: Independence Systems
Combinatorial Optimization Problems: Approximative Algorithms
Combinatorial Optimization Problems: Branch-and-Bound Approach
Discrete and Combinatorial Optimization: Introduction
Polyhedral Combinatorics and Cutting-Plane Methods

See also: *Distributed Parameter Systems (1.B.ix)*

(*ii*) *Dynamic Optimization*

Differential Equations: An Introduction
Dynamic Optimization: Introduction
Ordinary Differential Equations

(*iii*) *Optimal Control*

(a) Fundamentals

Bang–Bang Principle
Local Controllability
Maximum Principle
Optimal Control: Perturbation Methods
Optimal Control Theory: Introduction
Path-Following Methods
Singular Control: Higher-Order Conditions
Transversality

(b) Dynamic Programming

Dynamic Programming: Introduction
Hamilton–Jacobi–Bellman Equation
Hamilton–Jacobi–Bellman Equation: Numerical Methods
Optimality Principle

(c) Sensitivity Analysis in Optimal Control

Sensitivity Analysis

(d) Feedback

Irregular Optimal Feedback and Generalized Trajectories
Optimal Control: Number of Switchings
Optimal Control, Open-Loop and Closed-Loop
Optimal Feedback: Linear Quadratic Problem
Optimal Feedback: Linear Time-Optimal Control Problem
Regular Synthesis and Piecewise Analytic Feedback

(e) Linear Quadratic Problems

Linear Quadratic Regulator Problem
Optimal Control: Riccati Equations
Riccati Equation and Conjugate Points

(f) Hereditary Systems

Hereditary Systems
Hereditary Systems Control: Introduction
Optimal Control: Riccati Equations

(g) Discrete-Time Control Problems

Discrete-Time Control

(h) Inventory Control Problems

Backward Dynamic Programming
Forward Dynamic Programming
Impulse Control
Inventory Control: Introduction
Optimal Control: *s,S* Policy and *K* Convexity
Optimal Stopping
Planning Horizons Problem: Fundamentals
Production Smoothing

(i) Numerical Algorithms in Optimal Control

Optimal Control: Gradient Related Numerical Solution of Optimal Control Problems
Optimal Control of Multistate Systems
Optimal Control: Singular Arcs

(j) General Mathematical Concepts and Techniques

Brownian Motion
Closed Sets
Compact Sets
Formal Languages
Polyhedra
Semicontinuous Functions
Stationary Points

See also: *Distributed Parameter Systems (1.B.ix); General Systems Theory (1.B.iii)*

(*iv*) *Game Theory*

(a) General Game Theory

Competitive Equilibrium
Cooperative Game Theory
Games, Fuzzy
Games, Noncooperative
Games, Repeated
Games with Infinitely Many Players
Minimax Theory
Nash Equilibrium
Pareto Optimum
Social Choice Theory

(b) Differential Games

Differential Games, Closed-Loop
Differential Games: Introduction
Differential Games: Isaacs Equation
Differential Games, Linear Quadratic
Differential Games, Open-Loop
Dynamic Duopoly Theory
Incentives
Stochastic Differential Games
Stochastic Games and the Problem of Information
Stochastic Incentive Problems
Team Theory and Information Structure

See also: *Team Theory (1.D.vi); Management Systems (3.B.viii)*

(*v*) *Stochastic Systems*

Certainty Equivalence Principle
Gaussian Random Variables
Girsanov Transformation
Impulse Control
Optimal and Suboptimal Stochastic Control: Discrete-Time Systems
Optimal Stochastic Control: General Aspects
Optimal Stochastic Control: Numerical Methods
Optimal Stopping
Optimization Under Uncertainty
Recursive Algorithms: Variable Forgetting Factors
Recursive Stochastic Algorithms: Ordinary Differential Equation Method
Separation Principle
Stochastic Control: Introduction
Stochastic Differential Games
Stochastic Hamilton–Jacobi–Bellman Equation
Stochastic Integrals and Stochastic Calculus
Stochastic Maximum Principle

See also: *Self-Tuning Regulators and Adaptive Control (1.B.xi); Decentralized Control (1.D.ii)*

(*vi*) *Robustness and Sensitivity*

Linear Systems: Robustness
Multivariable Systems: Robustness in Design
Optimization: Sensitivity Analysis
Robust Controller Design
Robust Controller Design, Decentralized
Sensitivity Analysis

See also: *Decentralized Control (1.D.ii)*

D. Complex Systems Theory

(*i*) *Hierarchical Optimization and Control*

(a) Interaction Analysis Decomposition

Data Analysis and Interaction Evaluation
Decomposition and Structural Properties
Interaction Analysis: Informational Approach
Structural Analysis: Graph Representation

(b) Numerical Aspects of the Decomposition Approach

Auxiliary Problem Principle and Decomposition of Optimization Parameters
Decomposition–Coordination: A New Method
Distributed Asynchronous Fixed-Point Algorithms: General Convergence Theory
Large-Scale Systems: Linear Programming
Multistage Decompostion Algorithms
Overlapping Coordination, Hierarchical

(c) Control

Combined Systems Optimization and Parameter Estimation
Complex Systems Control: Tracking Methods

Complex Systems Methodology
Costate Coordination Method
Decentralized Fixed Modes
Dynamic Programming: Spatial Decomposition
Dynamic Systems: On-Line Cordination
Goal Coordination Method
Hierarchical Control: Partition
Mixed Method of Coordination
Model Coordination Method
Multilayer Control: Vertical Decomposition
Optimal Control: Decomposition and Coordination Using Asynchronous Iteration
Overlapping Decentralized Control
Reliability of Control
Robust Controller Design, Decentralized
Steady-State Systems: On-Line Coordination
Structure-Constrained Control: Parametric Optimization
Structure-Constrained Robust Controller Design

See also: *Classical Control (1.B.vi); Multivariable Control (1.B.viii); Optimization and Operations Research (1.C)*

(*ii*) *Decentralized Control*

(a) Algebraic and Geometric Approaches

Decentralized Control: Geometric Approach
Large Linear Systems: Algebraic Approach to Decentralized Control
Linear Quadruples ($A,B;C,D$): Geometric Approach

(b) Large Stochastic Systems

Decentralized Observers: Design and Evaluation Methods
Decentralized Stochastic Control
Interconnected Systems: Decentralized Control
Large-Scale Interconnected Systems: Well Posedness
Large-Scale Systems: Multilevel Filter
Large-Scale Systems: Observers
Large-Scale Systems: Parameter Estimation
Large Stochastic Systems
Large Stochastic Systems: Learning Automata

See also: *Stochastic Systems (1.C.v)*

(*iii*) *Model Simplification for Complex Systems*

Aggregation and Suboptimal Control
Aggregation, Chained
Model Simplification: Frequency-Domain Approach

See also: *Model Simplification (1.A.vi)*

(*iv*) *Singular Perturbations*

Multi-Timescale Systems
Multi-Timescale Systems: Dynamical Location
Singular Perturbations: Boundary-Layer Problem
Singular Perturbations: Discrete Version
Singular Perturbations in Automatic Control
Singularly Perturbed Systems: Output Feedback Control
State Feedback Control: Two-Stage Design by Slow–Fast Decomposition

(*v*) *Stability of Interconnected Systems*

Decentralized Control: Pole Assignment
Decentralized Stabilization
Decentralized Stabilization: A Review of Methods
Decentralized Stabilization: Block Gerschgorin Theorem
Interconnected Dynamical Systems: Stability
Interconnected Systems: Lyapunov Approach
Large-Scale Interconnected Systems: Input–Output Stability
Stability: Dissipativeness Concept
Structural Stability

See also: *Stability Theory (1.B.iv)*

(*vi*) *Team Theory*

Active Goal-Seeking Systems
Complex Systems Management: Game Theory
Dynamic Multistage Processes: Multiobjective Control
Multiattribute Utility Theory
Multilevel Systems: Stackelberg Strategies
Multiobjective Decision
Multiobjective Optimization
Surrogate Worth Trade-Off Method

Team Theory and Information Structure

See also: *Management Systems (3.B.viii); Optimization and Operations Research (1.C)*

2. TECHNOLOGY

(i) *Technology of Actuators*

Control Valve Sizing
Control Valve Technology
Control Valves: Gas Applications
Motor Drives, Alternating Current
Motor Drives, Direct Current
Servovalves, Electrohydraulic
Stepping Motors
Variable Speed Induction Machines: Numerical Control

(ii) *Special Actuators Required in Communications, Power Generation, Nuclear Industry, etc.*

Digital Communications: Satellite Systems
Industrial Gas Equipment Control and Safety Systems
Nuclear Reactors
Switching Networks: General Structures

(iii) *Measurements in the Control Area*

(a) Electrical Measurements

Amplifiers for Instrumentation
Electric Voltage Measurement: Basic Principles
Electric Voltage Measurement: Commercial Systems
Electrical Current Measurement
Electrical Measurement: Bridge Circuits
Electrical Noise: Principles and Measurement
Frequency Measurement

(b) Instrument Systems

Amplifiers for Instrumentation
Electrical Measurement: Bridge Circuits
Errors and Uncertainty in Measurements and Instruments
Instruments and Instrument Systems: Concepts and Principles
Instruments and Instrument Systems: Design Principles
Microprocessors in Instruments
Microprocessors in Instruments: Design Case Study

(c) Process Industry Measurement

Flow Measurement: Mass Rate
Flow Measurement, Quantity
Flow Measurement: Two-Phase Systems
Flow Measurement: Volumetric Rate
Force and Weight Measurement
Level Indication for Intermediate Storage
Level Measurement for Bulk Liquid Storage
Level Measurement: Liquids and Bulk Material
pH Measurement
Pressure Measurement
Speed and Acceleration Measurement
Temperature Measurement: General Aspects
Temperature Measurement: Types of Thermometer
Temperature Measurement: Use and Installation of Equipment
Viscosity Measurement

(d) Optical Fibers

Optical Fiber Interferometric Sensors
Optical Fiber Sensors: General Aspects

(e) Signals Characteristics

Electrical Noise: Principles and Measurement
Frequency Measurement
Harmonics
Signal Analysis: Fundamentals
Signal Processing: Analog
Signal Processing: Digital
Signal Transmission

(f) Analytical Techniques

Chromatography
Gas Chromatography
Inductively Coupled Plasma Emission Spectrometry
Liquid Chromatography
Mass Spectrometry
Nuclear Magnetic Resonance
Oxygen Concentration Determination in Gases
Oxygen Concentration Determination in Liquids
pH Measurement

Spectroscopy: Fundamentals and Applications

See also: *Process Control in Chemical Industries (3.A.i)*

(*iv*) *Intelligent Instrumentation*

Microprocessors in Instruments
Microprocessors in Instruments: Design Case Study

(*v*) *Measurements in the Biological Area*

(a) Mammalian Systems

Auditory Perception
Biological Measurement: Blood Gas Analysis
Biological Measurement: Blood Mechanical Properties
Biological Measurement: Dielectric Properties of Soft Tissue
Biological Measurement: Electrical Characteristics of the Heart
Biological Measurement: Lung Function
Biological Measurement: Myoelectric Activity of the Gut
Computerized Electrocardiogram Diagnosis: Fuzzy Approach
Electrocardiogram Diagnosis: Fuzzy Approach
Electrocardiography, His Bundle
Electroencephalography
Gait Analysis: Measurements
Human Kidneys
Lung Modelling: Measurement Aspects
Speech
Temperature Measurement, Physiological
Visual System: Psychophysics
Visual System: Structure and Function

(b) Biological Imaging Techniques

Absorption Spectroscopy in Biology
Dental Measurements: Radiography
Electron Microprobe Analysis
Nuclear Magnetic Resonance Imaging in Biology
Radioisotopic Imaging in Biology
Scanning Electron Microscopy in Biology
Thermal Imaging in Biology
Transmission Electron Microscopy in Biology
Ultrasound Imaging in Biology
X-Ray Imaging in Biology

(c) Telemetry

Telemetry, Underwater

(d) Dental Measurements

Dental Measurements: Characteristics of Jaws and Teeth
Dental Measurements: Clinical Equipment and Devices
Dental Measurements: Clinical Research
Dental Measurements: Clinical Visual and Tactile Methods
Dental Measurements: Radiography
Dental Measurements: Research into Stomatognathic Physiology

(e) Specialized Measurement Techniques

Coulter Counters
Stereological Analysis of Heterogeneous Solids in Biology

See also: *Modelling and Control of Biological Systems (3.C.i); Biomedical Engineering (3.C.ii)*

(*vi*) *Environmental Measurements*

(a) Measurements of Atmosphere

Air Pollution: Aerosol Chemical Analysis
Air Pollution: Aerosol Particle Size
Air Pollution: Automatic Analysis of Hydrocarbons
Air Pollution: Automatic Analyzers
Air Pollution: Chemiluminescence Monitoring Methods
Air Pollution: Monitoring Systems
Air Pollution: Photochemical Smog Formation
Air Pollution: Wind Tunnel Simulation
Environmental Data Statistical Analysis: Nonphysical Modelling Approaches
Environmental Data Statistical Analysis: Physical Modelling Approach
Environmental Measurement: Automatic and Continuous Analysis
Environmental Measurement: Odorous Compounds in Air
Environmental Measurement: Optical Lasers for Long-Range Wind Velocity Determination
Laser Spectroscopy in Atmospheric Analysis
Meteorological Measurement Techniques
Remote Sensing: Laser Beam Analysis of Atmospheric Conditions

Spline Interpolation in Air Pollution Pattern Analysis

(b) Measurements of Hydrosphere

Environmental Measurement: Distribution of Chemical Elements in the Ocean
Environmental Measurement: Water Quality Monitoring
Landsat Imaging: Removal of Atmospheric Effects
Remote Sensing: Water Quality
Water Pollution: Chemical Analysis
Water Pollution: Fingerprinting of Oil
Water Pollution Indices: Algal Growth Potential Determination
Water Pollution Indices: Biochemical Oxygen Demand
Water Pollution Indices: Chemical Oxygen Demand
Water Pollution Indices: Dissolved Oxygen
Water Pollution Indices: Total Organic Carbon
Water Pollution Indices: Total Oxygen Demand
Water Pollution: Inorganic Nitrogen Compound Determination
Water Pollution: Organic Nitrogen Compound Determination
Water Pollution: Trace Element Determination
Water Pollution: Trace Organic Compound Determination

(c) Measurements of Biosphere

Environmental Measurement: Image Instrumentation for Evaluating Pollution Effects on Plants

(d) Measurements of Social Environment

Environmental Evaluation of Green Spaces
Environmental Systems Analysis for Policy Making
Environmental Systems Analysis for Policy Making: A Support System
Waterfront Appraisal

(e) Noise and Vibration

Environmental Measurement: Noise

(f) Chemical Measurements

Activation Analysis: Environmental Trace Metals
Air Pollution: Aerosol Chemical Analysis
Air Pollution: Automatic Analysis of Hydrocarbons
Air Pollution: Chemiluminescence Monitoring Methods
Biodegradability Determination
Gas Chromatography in Environmental Analysis
Gas Chromatography–Mass Spectrometry in Environmental Analysis
Gas Chromatography–Mass Spectrometry in Environmental Chemicals Analysis
Inductively Coupled Plasma Emission Spectrometry
Infrared Spectrometry in Environmental Analysis
Ion-Selective Electrodes in Environmental Analysis
Photoacoustic Spectrometry in Environmental Analysis
Water Pollution: Chemical Analysis
Water Pollution: Fingerprinting of Oil
Water Pollution Indices: Algal Growth Potential Determination
Water Pollution Indices: Biochemical Oxygen Demand
Water Pollution Indices: Chemical Oxygen Demand
Water Pollution Indices: Dissolved Oxygen
Water Pollution Indices: Total Organic Carbon
Water Pollution Indices: Total Oxygen Demand
Water Pollution: Trace Element Determination
Water Pollution: Trace Organic Compound Determination

(g) Physical Measurements

Environmental Measurements: Optical Lasers for Long-Range Wind Velocity Determination
Meteorological Measurement Techniques

(h) Environmental Management

Environmental Systems Analysis for Policy Making

(i) Simulation Models

Air Pollution: Wind Tunnel Simulation

(j) Surveillance Systems

Air Pollution: Monitoring Systems
Environmental Data: Allocation Problems in Information Collecting Networks
Environmental Measurement: Water Quality Monitoring
Spline Interpolation in Air Pollution Pattern Analysis

(k) Remote Sensing

Earth–Atmosphere System: Identification of Optical Parameters
Ground Albedo Mapping: Invariant Embedding
Landsat Imaging: Removal of Atmospheric Effects
Laser Spectroscopy in Atmospheric Analysis
Remote Sensing: Classification of Multispectral Images
Remote Sensing: Laser Beam Anaylsis of Atmospheric Conditions
Remote Sensing: Satellites
Remote Sensing: Water Quality

(l) Databases and Data Processing

Environmental-Data Compression
Environmental-Data Smoothing
Environmental Data Statistical Analysis: Nonphysical Modelling Approaches
Environmental Data Statistical Analysis: Physical Modelling Approach
Environmental Databases
Environmental Databases: INFOTERRA
Multivariate Analysis

(*vii*) *Measurements in Chemistry*

Chromatography
Gas Chromatography
Inductively Coupled Plasma Emission Spectrometry
Liquid Chromatography
Mass Spectrometry
Nuclear Magnetic Resonance
Oxygen Concentration Determination in Gases
Oxygen Concentration Determination in Liquids
pH Measurement
Spectroscopy: Fundamentals and Applications

See also: *Process Control in Chemical Industries (3.A.i)*

(*viii*) *Database Management*

Database Concurrency Control Theory
Database Management Systems: Introduction
Distributed Database Systems: Concurrency Control
Distributed Database Systems: Efficiency
Distributed Database Systems: Failure Recovery Procedures
Distributed Systems: Synchronization and Interprocess Communication

See also: *Simulation Techniques* (*1.A.v*)

(*ix*) *Distributed Control*

(a) Distributed Systems

Distributed Computer Systems: An Introduction
Distributed Computer Systems: Reliability
Distributed Data Processing Systems and Decentralized Control
Distributed Database Systems: Reliability
Distributed Systems: Synchronization and Interprocess Communication

(b) Computer Networks

Distributed Systems: Synchronization and Interprocess Communication
Packet-Switching Networks

(c) System Performance

Computer Systems: Performance Measurement
Performance Modelling Methodology in Computing

3. APPLICATIONS

A. Technological Applications

(*i*) *Process Control in Chemical Industries (Paper, Cement, Sugar, etc.)*

(a) Control in Industries

Adaptive Control: Industrial Applications
Cement Industry: Adaptive Control
Concrete Mixing: Cautious Adaptive Control
Food Industry: Process Control
Glass Industry: Adaptive Control
Glass Industry: Process Control
Inorganic Chemicals Industry: Process Control
Mineral Processing and Cement Industries: Process Control
Nonferrous Metallurgical Industry: Process Control
Papermaking: Adaptive Control
Petrochemical Industry: Process Control
Pharmaceutical Industry: Process Control
Polymer Industry: Process Control
Production Control in Process Industries
Pulp and Paper Industry: Process Control
Steel Industry: Hierarchical Control
Sugar Industry: Process Control

(b) Control of Components and Parameters

Chemical Reactor Control
Consistency, Density and Viscosity: Control
Distillation Column Control
Distillation Columns: Adaptive Control
Drying and Dewatering Control
Electrolysis Control
Engine Speed: Self-Tuning Control
Evaporation Control
Fermentation Control
Fermenter Control
Fluid Flow Control
Heat Exchanger Control
Heating Systems: Adaptive Control
pH and Chemical Concentration Control
pH Regulation
Pressure Control
Rolling-Mill Control
Separation Process Control
Steam Generator Control
Titanium Dioxide Kilns: Adaptive Control

(c) Process Measurement

Flow Measurement: Mass Rate
Flow Measurement, Quantity
Flow Measurement: Two-Phase Systems
Flow Measurement: Volumetric Rate
Force and Weight Measurement
Level Indication for Intermediate Storage
Level Measurement for Bulk Liquid Storage
Level Measurement: Liquids and Bulk Material
pH Measurement
Pressure Measurement
Speed and Acceleration Measurement
Temperature Measurement: General Aspects
Temperature Measurement: Types of Thermometer
Temperature Measurement: Use and Installation of Equipment
Viscosity Measurement

See also: *Measurements in the Control Area* (*2.iii*); *Measurements in Chemistry* (*2.vii*)

(*ii*) *Control of Power Generation and Distribution*

Electric Generators: Load and Frequency Control
Electric Generators, Rotating
Electric Generators: Unit Commitment
Gas Turbines in Power Generation
Hydroelectricity: Self-Selecting Control
Load-Frequency Regulation: Adaptive Control
Nuclear Energy
Nuclear Power Plants: Adaptive Control
Nuclear Reactors
Power System Operation
Power System Planning
Power System Protection
Power System Reliability
Power System Transients
Power Systems
Power Systems: Automatic Generation Control
Power Systems: Economic Dispatching
Power Systems: High-Voltage DC Transmission

Power Systems: Integrated Software for Control
Power Systems: Load Forecasting
Power Systems: Maintenance Scheduling
Power Systems: Microprocessor Monitoring and Control
Power Systems: On-Line Control
Power Systems: State Estimation
Power Systems: Steady-State Power Flow Modelling
Reactive Power Control: Loss Minimization
Steam Boiler Control
Steam Thermal Units

See also: *Utility Systems* (*3.B.vi*)

(*iii*) *Control in Aeronautics and Space*

(a) General Aspects and Background

Aerospace Control
Aerospace Systems and Control
Civil Aircraft: Long-Term Future
Combat Aircraft: Long-Term Future
Flight-Crew Workload
Fuel Conservation: Air Transport
Helicopters
Helicopters: Long-Term Future

(b) Flight Mechanics and Flight Control

Active Control Technology in Aircraft
Active Control Technology in Aircraft: General Aspects
Aircraft Artificial Control Load (Artificial Feel Spring)
Aircraft Automatic Landing in Bad Weather Conditions
Aircraft Automatic Landing Systems
Aircraft Dynamics: State Equations
Aircraft State Equations: Singular Perturbations
Airship Control and Guidance
Angle of Attack
Attitude Control
Autopilot Modes
Autopilots
Autopilots: Feedback Structure
Control-Configured Vehicles
Flexibility of Aerospace Vehicles
Flight Envelope
Flight Management Systems
Fly-by-Wire Systems
Full-Authority Control in Aerospace Vehicles
Helicopters
Helicopters: Flight Control Assistance
Missile Guidance and Control
Wing Load Alleviation
Wing Load Alleviation Modelling

(c) Propulsion Systems

Aircraft Engine Control: General Aspects
Aircraft Engine Control: Physical Principles
Aircraft Engine Control: System Design
Aircraft Engines: Survey of Systems
Helicopters: Turbine-Engine Governing Systems

(d) Navigation and Landing Systems

Air Traffic Control Near Airports
Air Traffic Control Systems
Aircraft Automatic Landing in Bad Weather Conditions
Aircraft Automatic Landing Systems
Data Links in Aerospace
Four-Dimensional Navigation
Fuel Conservation: Air Transport
Homing of Aerospace Vehicles
Inertial Systems
Inertial Systems: Selection and Integration
Visual Flying Rules and Instrument Flying Rules

(e) Instrumentation and Simulation

Aerospace Simulators
Air Traffic Simulators: The CAUTRA Subsystem
Aircraft Instrumentation: Development of Cathode Ray Tube Technology
Head-Up Displays and Trajectory Control
Simulation of Aerospace Systems
Simulators for Training
Wing Load Alleviation Modelling

(f) Space Vehicles

Drag-Free Satellites
Flexible Spacecraft: Computer-Oriented Control Analysis
Launch Vehicles, Multistage
Launch Vehicles: Problems
Missile Guidance and Control
Satellites: Adaptive Control
Solar Power Satellites
Space-Shuttle Control Systems Reliability: Redundant Processing
Space-Vehicle Trajectories: Optimization
Telemetry and Data Processing in Space Communication

(g) Reliability and Safety

Aircraft Automatic Landing in Bad Weather Conditions
Aircraft Sensor Failure: Detection by Analytic Redundancy
Space-Shuttle Control Systems Reliablilty: Redundant Processing

See also: *Air Traffic Control* (*3.B.ii*); *Simulation Techniques* (*1.A.v*)

(*iv*) *Robotics*

(a) Robot Sensing

Automatic Modelling of Three-Dimensional Objects
Computer Vision by Three-Dimensional Rangefinding
Industrial Robots: Laser-Beam Position Reference
Industrial Vision Systems
Industrial Vision Systems: Segmentation of Gray-Scale Images
Industrial Vision Systems, Three-Dimensional
Position Sensing in Three Dimensions: Laser Tracking
Robots: Tactile and Force Sensing and Control
Robots: Tactile Sensors

(b) Robot Control

Industrial Robots: An Optimal Control System
Industrial Robots: Conventional Controller Design
Industrial Robots: Manipulator Control
Robot Arms: Dynamic Control
Robots: Coordination Control
Robots: Teach Function
Robots, Walking

(c) Robot Mechanisms

End Effectors
Industrial Robot Anatomy
Robot Arms
Robots: Mechanical Structure

(d) Robot Servosystems

Articulated Robot Arms: Electrohydraulic Servosystem Design
Robot Arms, Flexible

(e) Robot Systems

Industrial Robot Arms
Industrial Robots
Industrial Robots: Application to Automatic Adjusting
Industrial Robots: Modular Systems
Robots
Robots, Walking
Robots, Snakelike

(f) Languages and Operating Systems

Industrial Robots: Error Recovery
Industrial Robots: Programming Languages

(g) Robot Applications

Factory Automation: Case Study in Production Line Management
Industrial Robots: Applications
Industrial Robots: Design Case Studies
Industrial Robots in Garment Manufacture
Industrial Robots in the Aerospace Industry

(h) Automatic Inspection

Industrial Vision Systems for Inspection and Manufacturing

(i) Flexible Manufacture

Flexible Manufacturing Systems
Flexible Manufacturing Systems: LMAC System for Control Language Generation
Flexible Manufacturing Systems: Modelling and Simulation
Industrial Production Systems: Robot Integration

(j) Social Aspects

Automation: Social Effects
Personnel Aspects of Automation
Robotics: Social Aspects

See also: *Social Effects of Automation* (*3.B.x*)

B. Semitechnological Applications

(*i*) *Road Traffic Control*

(a) Urban Traffic Control and General Aspects

Automotive Electronics
Bus-Route Control
Fuel Conservation: Road Transport

Road Information and Electronic Guidance Systems
Road Information: Automatic Diversion Control
Road Information: Communication
Traffic Control: Intelligent Systems
Traffic Detectors and Recorders

(b) Freeway Traffic Control

Freeway Traffic Incidents: Automatic Detection
Freeway Traffic: Modelling, Surveillance and Multilayer Control
Traffic Corridor Control: General Priciples
Traffic Corridor Control: Ramp Assignment
Traffic Corridor Control: Ramp Metering

(c) Modelling

Freeway Traffic: Modelling, Surveillance and Multilayer Control
Traffic Simulation: Hydrodynamic Theory and Shock Waves
Traffic Simulation: Macroscopic Models
Traffic Simulation: Microscopic Models
Traffic Simulation: Predictive Models
Traffic Variables
Transportation Modelling

(*ii*) *Air Traffic Control*

Air Traffic Control Near Airports
Air Traffic Control Systems
Air Traffic Simulators: The CAUTRA Subsystem
Aircraft Automatic Landing in Bad Weather Conditions
Aircraft Automatic Landing Systems
Data Links in Aerospace
Visual Flying Rules and Instrument Flying Rules

See also: *Control in Aeronautics and Space* (*3.A.iii*)

(*iii*) *Other Transportation Systems*

(a) Sea Transport

Fuel Conservation: Sea Transport
Navigation Control
Ship Control and Automation
Ship Positioning: Adaptive Control
Ship Propulsion Control
Ship Steering: Model Reference Adaptive Control
Vessel Traffic Monitoring Systems

(b) Rail Transport

Automated Guideway Transit Systems and Personal Rapid Transit Systems
Automatic Train Control: Line Supervision and Control
Automatic Train Control: Organization in Subways
Automatic Train Control: Reliability and Safety
Automatic Train Control: Train Driving
Automatic Train Control: Train Protection

(c) Goods Transport Management

Harbor Terminal Management
Warehouse Automation

(*iv*) *Manufacturing Systems*

Flexible Manufacturing Systems
Flexible Manufacturing Systems: LMAC System for Control Language Generation
Flexible Manufacturing Systems: Modelling and Simulation
Industrial Production Systems: Robot Integration

See also: *Robotics* (*3.A.iv*)

(*v*) *Building Systems*

(a) Air Conditioning

Air Conditioning
Air Conditioning, Air Floor
Air Conditioning Control Systems
Air Conditioning, Dual Duct
Air Conditioning, Fan Coil
Air Conditioning, Induction Unit
Air Conditioning, Reversible Cycle Heat Pump
Air Conditioning Systems
Air Conditioning, Variable Air Volume
Economizer Control
Reset Control in Air Conditioning
Reset Ratio in Air Conditioning

(b) Electrical Load Control

Critical Time Override
Duty Cycling
Minimum On-Time

(c) Heating and Ventilation Systems

Anticipating Control

Building Management Systems
Deadband Control of Temperature
Degree Day
Demand Control
Domestic Heating
Domestic Heating: Boiler and Furnace Efficiency
Domestic Heating: Boilers and Furnaces
Domestic Heating: Temperature Control Operation
Domestic Heating: Temperature Control Strategies
Flame Safeguard Control
Heat Pumps
Heating, Ventilation and Cooling Systems: Controller Tuning
Humidistats
Humidity Control Systems
Hydronic Systems
Pump Characteristics: Control Applications
Thermostats, Multistage

(d) Heating and Ventilation System Simulation

Heating, Ventilation and Cooling Systems: Actuator Models
Heating, Ventilation and Cooling Systems: Analog Simulation
Heating, Ventilation and Cooling Systems: Control Process Interactions
Heating, Ventilation and Cooling Systems: Digital Simulation

(e) Human Comfort Requirements

Clothing Insulation
Comfort Control: Modelling
Comfort Equations
Comfort Indices
Comfort Indices: Standard Effective Temperature
Globe Thermometers
Metabolic Rate
Net Radiometers
Thermal Comfort
Thermal Environment
Thermal Sensation
Thermoregulation: General Aspects

(f) Lighting

Light Switching
Lighting Control
Lighting Control: Applications

(g) Vertical Transportation

Elevator Control Systems
Elevator Management
Elevator Systems: Computer-Aided Design and Simulation Techniques
Elevator Systems: Computer Control
Elevator Systems: Design
Elevator Systems: Performance Measurement
Elevator Systems: User Requirements
Vertical Transport: History

(*vi*) *Utility Systems* (*Gas, Electricity, Water*)

(a) Power Systems

Hydroelectricity
Hydroelectricity: Production Planning
Power System Operation
Power System Planning
Power System Protection
Power System Reliability
Power System Transients
Power Systems
Power Systems: Automatic Generation Control
Power Systems: Economic Dispatching
Power Systems: High-Voltage DC Transmission
Power Systems: Integrated Software for Control
Power Systems: Load Forecasting
Power Systems: Maintenance Scheduling
Power Systems: Microprocessor Monitoring and Control
Power Systems: On-Line Control
Power Systems: State Estimation
Power Systems: Steady-State Power Flow Modelling
Reactive Power Control: Loss Minimization

(b) Gas Systems

Control Valves: Gas Applications
Domestic Gas Equipment Control and Safety Systems
Fuel-Gas Interchangeability
Gas Distribution Systems
Gas Distribution Systems: Balancing Supply and Demand
Gas Distribution Systems: Steady-State Analysis
Gas Leakage Control
Gas, Manufactured
Gas Metering
Gas, Natural
Gas Storage Systems
Gas Supply Systems: Control and Telemetry

Gas Supply Systems: Load Forecasting
Gas Transmission Systems
Gas Transmission Systems: Supervision and Control
Gas Transmission Systems: Transient Analysis
Gas Utility System Planning
Industrial Gas Equipment Control and Safety Systems

(c) Water Distribution Systems

Water Distribution Systems
Water Distribution Systems: Planning
Water Distribution Systems: Steady-State Analysis
Water Networks: Extended-Period Simulation
Water Resources: Systems Engineering
Water Sources and Supply Systems
Water Supply Leakage Control
Water Supply Pumps: Optimal Use
Water Supply Pumps: Optimization
Water Supply Systems: Demand Forecasting
Water Supply Systems: Suppression and Control of Transient Phenomena

(d) Other Energy Systems

Energy Sources
Energy Sources, Renewable
Fossil-Fuel Energy
Hydroelectricity
Hydroelectricity: Production Planning
Nuclear Energy
Nuclear Power Plants: Adaptive Control
Nuclear Reactors
Solar Power Satellites

See also: *Control of Power Generation and Distribution (3.A.ii)*

(vii) Communication Systems

Digital Communications: Fundamentals
Digital Communications: Satellite Systems
Switching Networks: General Structures
Switching Networks: Learning Algorithms for Control
Switching Networks: Long-Term Planning
Switching Networks: Modelling
Switching Networks: Quality Parameters
Switching Networks: Real-Time Management
Switching Networks: Short-Term Planning
Switching Networks: Traffic Routing and Real-Time Control
Telecommunication Traffic: Variables and Measurement

(viii) Management Systems

(a) Behavioral Factors in Systems Design

Activity
Bounded Rationality
Cognition
Collective Enquiry
Coupling
Descriptive Methods or Results
Evaluation
Evaluation Design: Systems and Models Approach
Group Decision Making and Voting
Human–Computer Interaction
Human Factors Engineering
Human Information Processing
Human Judgment and Decision Rules
Interacting Groups
Interpretive Structure Modelling
Lotteries
Needs
Options
Organization Charts
Subjective Probability
Surveys
Systems Management: Cognitive Styles
Systems Management: Conflict Analysis

(b) Knowledge-Based Systems Design

Decision Support Systems
Delphi Method
Diagnostic Inference
Expert Systems
Frames
Fuzziness
Inference
Inference Engines
Knowledge Bases
Knowledge Representation
Metarules
Production Rules
Scripts

(c) Decision Making

Alternatives
Constraints
Critical Activity
Critical Paths
Decision Analysis

Decision Support Systems
Decomposition
Delphi Method
Descriptive Methods or Results
Diagnostic Inference
Difference Equations, Ordinary
Differential Equations
Digraphs
Directed Relations (Directed Graphs)
Discrete Variables
Dominance
Dynamic Programming
Fault Trees
Feedback Loops
Forecasts
Game Theory
Goal-Driven Approaches
Group Decision Making and Voting
Heuristics
Human Judgment and Decision Rules
Hypotheses
Impact Analysis and Hierarchical Inference
Inference
Information Requirements Determination
Input–Output Analysis
Inputs
Iteration
Linear Programming: Fundamentals
Management: Game Theory
Markov Decision Processes
Markov Processes
Multiattribute Utility Theory in Decision Analysis
Multicriteria Decision Making
Multicriteria Decision Problem
Multiobjective Decision
Multiple Objective Optimization: Pareto Optimality
Nodes
Nonlinear Programming
Null Hypothesis
Objectives
Optima
Options
Outputs
Parameters
Performance Index
Sensitivity
Stochastic Processes
Stochastic Programming
Strategies
Suboptimal Solutions
Time Horizons
Trade Off
Tree Structures
Trees

(d) Modelling and Simulation

Bayes Rule
Certainty Equivalents
Closed Systems
Constraints
Correlation
Decomposition
Delphi Method
Deterministic Systems
Difference Equations, Ordinary
Differential Equations
Digraphs
Directed Relations (Directed Graphs)
Discrete Variables
Dominance
Dynamic Programming
Dynamic Systems
Edges
Endogenous Variables
Estimation or Identification
Extrapolation
Feedback Loops
Graph Theory
Impact Analysis and Hierarchical Inference
Independence
Independent Variables
Inputs
Iteration
Level of Significance
Likelihood Ratio
Markov Decision Processes
Markov Processes
Models
Monte Carlo Methods
Multidimensional Scaling
Nondirected Relations
Null Hypothesis
Outputs
Parameters
Partition
Queuing Models
Random Variables
Regression
Samples
Sensitivity
Spatial Allocation: Gravity Models
Stochastic Processes
Stochastic Programming
Structured Modelling
Technological Forecasting
Time Horizons
Transients

Transition Probability
Transitivity
Variables

(e) Operations Research

Algorithms
Bayes Rule
Constraints
Correlation
Cost–Benefit and Cost–Effectiveness Analysis
Cost–Benefit Ratios
Critical Activity
Critical Paths
Databases
Deterministic Systems
Dynamic Systems
Estimation or Identification
Extrapolation
Forecasts
Impact Analysis and Hierarchical Inference
Inputs
Latest Completion
Latest Start
Life-Cycle Costing
Linear Programming: Fundamentals
Markov Decision Processes
Nonlinear Programming
Operations Research
Optima
Queuing Models
Random Variables
Server Occupancy
Service Time
Slack
Spatial Allocation: Gravity Models
Trade Off

(f) Multiobjective Analysis

Attribute Measures
Attributes
Efficient Frontiers
Efficient Systems
Hierarchy
Multiattribute Utility Theory
Multiattribute Utility Theory in Decision Analysis
Multicriteria Decision Making
Multicriteria Decision Problem
Multiobjective Decision
Multiple Objective Optimization: Pareto Optimality
Objectives
Objectives Trees
Performance Index
Preferential Independence
Prescriptive Methods or Results
Revealed Preference Theory
Satisficing
Scenarios
Strategies
Structured Modelling
Subjective Probability
Utility
Utility Measurements
Utility Theory
Voting Paradox

(g) Economic Systems

Economic Equilibrium
Economic Models
Empirical Results
Endogenous Variables
Exogenous Variables
Input–Output Analysis
Leontief Inverse
Market Clearing Equations
Means–Ends Analysis
Microeconomic Systems Analysis
Needs
Net Present Value
Normative Methods
Perfectly Competitive Economy
Production Functions
Public Goods
Rationality
Revealed Preference Theory
Risk Aversion
Satisficing
Shadow Prices
Substitution in Economics
Tatonnement
Theory of the Firm
Utility
Utility Measurements
Utility Theory
Welfare Functions
Welfare or Normative Economics

(h) Methodological and Philosophical Concerns

Axiology
Cardinals
Contextual Relation
Cost–Benefit and Cost–Effectiveness Analysis
Cost–Benefit Ratios
Cybernetics
Data
Deduction
Design and Evaluation of Systems

3.B.viii

...nt of Systems and
...nt: Systems Engineering
...Programming
...Systems
...cal Results
...uation
...aluation Design: Systems and Models Approach
Events
Externality
Floats
Hierarchy
Holistic Methods
Human Factors Engineering
Human Information Processing
Interaction Matrices
Interactive Approaches
Large-Scale Systems: Development
Life-Cycle Costing
Management: Hypothesis Testing
Methodology
Network Methods
Office Automation
Open Questions
Ordinals
Pattern Recognition
Planning Horizons
Project Management: Network Models
Reliability of Systems: Statistical Design and Control
Risk Aversion
State Variables
System Dynamics Modelling
System Theory
Systemic Processes
Systems
Systems Analysis
Systems Engineering
Systems Knowledge: Philosophical Perspectives
Systems Management
Systems Management: Cognitive Styles
Systems Management: Conflict Analysis
Systems Methodology
Systems Science
Systems Synthesis
Top-Down Methods
Trade Off
Transactions Tables
Validation
Validation of Management Models
Value Systems
Verification

(i) Systems Management Applications

Communication, Command and Control Information Systems
Large-Scale Systems: Development
Lotteries
Management: Game Theory
Management: Hypothesis Testing
Population Control
Program Evaluation and Review Technique
Systems Acquisition and Procurement
Validation of Management Models
Water Resources: Systems Engineering
Workshop Dynamic Modelling

See also: *Optimization and Operations Research* (*1.C*); *Complex Systems Theory* (*1.D*); *General Systems Theory* (*1.B.iii*); *Fuzzy Systems* (*1.B.i*)

(*ix*) *Health Care Systems*

Computer-Assisted Medical Diagnostic Systems: Fuzzy Methods
Expert Systems in Medical Reasoning

See also: *Modelling and Control of Biological Systems* (*3.C.i*); *Biomedical Engineering* (*3.C.ii*)

(*x*) *Social Effects of Automation*

Automation: Social Effects
Personnel Aspects of Automation
Robotics: Social Aspects
Technological Innovation: Issues and Perspectives

See also: *Robotics* (*3.A.iv*)

C. Nontechnological Applications

(*i*) *Modelling and Control of Biological Systems*

(a) Anesthesia

Anesthesia: Feedback Control of Depth
Anesthesia: Inhalational Agent Identification and Control
Demand Analgesia: Modelling and Control of Postoperative Pain

(b) Biochemical Control

Biochemical Oscillators: Dynamic Behavior

Biochemical Systems Modelling: Switching Behavior
Chaotic Behavior in Physiological and Chemical Systems

(c) Biotechnology

Fermentation Control
Fermentation Measurement

(d) Cardiovascular System

Blood Pressure Hypertension Treatment: Feedback Control
Blood Pressure Postoperative Treatment: Adaptive Control
Cardiac Electrical Activity Modelling
Cardiovascular System: Baroreflexes Model
Cardiovascular System: Hemodynamic Model
Coronary Blood Flow Modelling

(e) Metabolic Systems

Cellular Protein Synthesis Control
Glucose Regulation: Adaptive Control
Glucose Regulation: Modelling
Glucose Regulation: Modelling and Control in Diabetics
Lipoprotein Dynamics and Control
Protein Metabolism: Dynamics and Control
Tracers and Metabolic Dynamics

(f) Nerve, Muscle and Brain

Biofeedback
Biological Motor Systems: Nervous Control
Gait Analysis and Locomotion Control
Intercranial Pressure Regulation by Feedback
Muscle Relaxation Control
Muscle-Spindle Modelling and Identification
Muscles: Dynamic Models
Muscles: Nervous Control

(g) Pharmacology and Pharmacokinetics

Drug Effect Prediction *In Vivo*
Ferrokinetic Modelling in the Body
Hepatobiliary Kinetics Modelling
Hormonal Regulation
Ketone Body Metabolism Modelling
Kinetic Data Diagnosis in Biological Modelling
Pharmacokinetics: Nonlinear Modelling

(h) Respiration

Lung Modelling
Respiratory System: Control During Exercise
Respiratory System: Modelling and Control of External Respiration
Respiratory System: Modelling and Identification
Respiratory System: Optimal Control

(i) Smooth Muscle

Digestive Tract: Gastric Electrical Activity Modelling
Digestive Tract: Large Intestine Electrical Activity Modelling
Digestive Tract: Small Intestine Electrical Activity Modelling
Ureter Electrical Activity Modelling

(j) Theoretical and Experimental

Biomedical Systems Modelling: Experiment Design
Biomedical Systems Modelling: Structural Identifiability
Educational Computer Models in Medicine
Identification Errors in Biomedicine
Mode Analysis: Coupled Nonlinear Oscillators

(k) Therapeutic Modelling

Tumor-Disease Treatment: Simulation

(l) Visual System

Eye Movement: Slow-Phase Caloric Nystagmus Modelling
Eye Movement: Vestibular System Modelling and Control
Image Filtering by the Retinocortical Visual Pathways
Visual System: Nonlinear Modelling

(m) Whole Body Behavior

Circadian Rhythms: Modelling
Hunger and Thirst: Motivational Control Mechanisms
Thermoregulation: Modelling
Thermoregulation: Physiological Control

See also: *Measurements in the Biological Area* (*2.v*)

(*ii*) *Biomedical Engineering*

Artificial Kidneys

3.C.ii

…dneys, Portable and
…e
…ungs
…Respiratory Systems
…anics of Bone
…echanics of Connective Tissue
…mechanics of Skin
…iomechanics of Soft Tissue

…) Environmental Systems

(a) Catastrophe

Catastrophe Theory in Ecosystem Analysis

(b) Control

Abiotic Control Mechanisms in Terrestrial and Freshwater Environments
Population Control
Toxic Chemicals Assessment: Simulation Modelling

(c) Ecosystem Models

Algal Growth Modelling: Case Study
Ecosphere
Ecosystem Diagraphs: Decomposition into Elementary Components
Ecosystem Networks: Measures of Structure
Ecosystems
Fate Modelling of Organic Chemicals: Case Study
Flow Network Ascendency and Self-Organization in Living Systems
Lake Ecosystem Modelling
Lake Ecosystem Modelling: Input–Output Models
Nitrification Modelling in Lotic Systems
Population Modelling
Radioecological Modelling
Radioecological Modelling in Agricultural Food Chains
Stochastic Modelling in Ecology
Toxic Substances Modelling: Fugacity Concepts

(d) Energy

Energy Systems in Ecology

(e) Environmental Data

Agricultural Soils: Fate of Toxic Substances
Aluminum and the Fate of Nutrients and Toxic Substances in Terrestrial and Freshwater Environments
Ecological Diversity and Information
Environmental Problem Solving: Systems Concepts
Fuzzy Set Theory in Ecology
Herbicides in Agroecosystems
Limnology: Dispersion of Toxic Substances

(f) Model Analysis in Ecological Modelling

Bilinear Equations in Ecology
Catastrophe Theory in Ecosystem Analysis
Directed Graphs: Stability Conditions
Ecological Modelling
Ecological Modelling, Hierarchical
Ecological Modelling: Law of the Minimum
Ecological Modelling: Michaelis–Menten Models
Ecological Population Modelling: Leslie's Equations
Ecological Population Modelling: Lotka–Volterra Equations
Ecological Population Modelling: Volterra's Equations
Ecosystem Compartment Modelling
Environ Theory and Analysis
Environmental Risk Analysis
Error Analysis and Sensitivity Analysis in Ecology
Error Analysis, First-Order
Flow Network Ascendency and Self-Organization in Living Systems
Lake Ecosystem Modelling: Uncertainty
Maximum Power Principle in Ecology
Population Growth, Logistic
Population in Ecology
Sensitivity Analysis and Simulation Experimentation
Size Dependence in Ecology
Systems Ecology

(g) Model Development in Ecological Modelling

Ecological Modelling: Aggregation Errors
Ecological Modelling: Bond Graphs
Ecological Modelling: New Perspectives
Ecological Modelling: Temperature Model Formulation
Grammars in Ecology
Lake Ecosystem Modelling: Input–Output Models

Limnology: One- and Two-Box Models of Lake Response Time
Stochastic Modelling in Ecology
Tide Modelling
Top-Down Modelling in Ecology
Trophic Structures Modelling
Water Quality Modelling: Case Study

(h) Model Validation in Ecological Modelling

Agricultural Modelling: Field Applicability
Gnotobiotic Systems
Validation of Phytoplankton Models
Validation of Simulation Models: Philosophy and Statistical Methods of Confirmation

(i) Nutrient Cycling and Eutrophication Models

Carbon-Cycle Modelling
Food Chains
Lake Ecosystem Modelling
Lake Ecosystem Modelling: Input–Output Models
Nitrification Modelling in Lotic Systems
Nitrogen Cycle Modelling: Nominal and Perturbed Dynamics in Watersheds
Nutrient Cycling
Primary Productivity
Stochastic Modelling in Ecology
Trophic State Criteria
Trophic Structures Modelling

(j) Reduction

Ecological Modelling: Aggregation Errors
Model-Order Estimation in Ecological Modelling

(k) Sensitivity Analysis

Ecological Modelling: Aggregation Errors
Environ Theory and Analysis
Error Analysis and Sensitivity Analysis in Ecology
Error Analysis, First-Order
Lake Ecosystem Modelling: Uncertainty
Parameter Estimation in Ecological Modelling
Sensitivity Analysis in Ecological Simulation Models
Sensitivity Techniques in Input–Output Flow Analysis

(l) Simulation

Carbon-Cycle Modelling

(m) Stability and Perturbation

Abiotic Control Mechanisms in Terrestrial and Freshwater Environments
Catastrophe Theory in Ecosystem Analysis
Chemical Stress in Agriculture: Modelling
Complex Ecosystems Stability
Directed Graphs: Stability Conditions
Ecological Disturbance Theory
Ecological Disturbances
Ecosystem Modelling: Qualitative Stability
Ecosystems, Disturbed
Environmental Risk Analysis
Error Analysis and Sensitivity Analysis in Ecology
Error Analysis, First-Order
Perturbation Techniques in Ecosystem Modelling
Sensitivity Analysis in Ecological Simulation Models

(n) Stochastic Models

Dynamic Eutrophication Modelling: Variance Estimates
Ecosystems, Disturbed
Error Analysis and Sensitivity Analysis in Ecology
Error Analysis, First-Order
Forecast Evaluation in Ecological Modelling
Hydrologic Forecasting
Markov Chains in Ecological Modelling
Stochastic Modelling in Ecology
Tides as Time Series
Water Bodies Time Scale Estimation: Mass Concepts

(o) Toxic Substances Models

Agricultural Soils: Fate of Toxic Substances
Aluminum and the Fate of Nutrients and Toxic Substances in Terrestrial and Freshwater Environments
Chemical Stress in Agriculture: Modelling
Ecological Effects of Toxic Substances
Fate Modelling of Organic Chemicals: Case Study

Limnology: Dispersion of Toxic Substances
Radioecological Modelling
Toxic Chemicals Assessment: Simulation Modelling
Toxic Substances Modelling: Fugacity Concepts

(*iv*) *Agricultural Systems*

Agricultural Modelling: Field Applicability
Agricultural Soils: Fate of Toxic Substances
Chemical Stress in Agriculture: Modelling
Fate Modelling of Organic Chemicals: Case Study
Herbicides in Agroecosystems
Toxic Chemicals Assessment: Simulation Modelling

4. HISTORY

(*i*) *History of Theoretical Techniques*

Cybernetics: History
Feedback Control: History
Frequency-Response Techniques: History
Identification: History
Nonlinear Systems: History
Optimal Control Theory: Developments 1950–1962
Optimal Relay and Saturating Control Systems: History
Sampled-Data Systems: History
State-Space Techniques: History
Systems Concepts: History

(*ii*) *History of Applications*

Automatic Flight Control Systems: History
Cybernetics: History
Electrical Control in the Nineteenth Century
Process Control: History
Ship Stabilization: History
Speed Regulation: History
Vertical Transport: History

(*iii*) *History of Organizations*

International Federation of Automatic Control: History
International Federation for Systems Research: History

LIST OF CONTRIBUTORS

Contributors are listed in alphabetical order together with their affiliations. Titles of articles which they have authored follow in alphabetical order, along with the respective volume and page numbers. Where articles are co-authored, this has been indicated by an asterisk following the article title.

Aarna, O.
(Tallinn Technical University, Tallinn, USSR)
Inorganic Chemicals Industry: Process Control **4**: 2511–7

Abignoli, M.
(CRAN, Nancy, France)
Stepping Motors * **7**: 4593–8

Aczél, J.
(University of Waterloo, Waterloo, Ontario, Canada)
Information Theory * **4**: 2504–8

Adlassnig, K.-B.
(University of Vienna, Vienna, Austria)
Computer-Assisted Medical Diagnostic Systems: Fuzzy Methods * **1**: 736–40

Aeyels, D.
(University of Ghent, Ghent, Belgium)
Nonlinear Observability **5**: 3336–41

Agarwal, G. C.
(University of Illinois at Chicago, Chicago, Illinois, USA)
Biological Motor Systems: Nervous Control * **1**: 462–6

Ahrens, U.
(Opto Elektronik, Heilbronn, Stuttgart, FRG)
Robots: Tactile Sensors * **6**: 4118–23

Aiga, I.
(National Institute for Environmental Studies, Ibaraki, Japan)
Environmental Measurement: Image Instrumentation for Evaluating Pollution Effects on Plants * **2**: 1516–22

Akagi, S.
(Osaka University, Osaka, Japan)
Fuel Conservation: Sea Transport **3**: 1761–6

Akimoto, H.
(National Institute for Environmental Studies, Ibaraki, Japan)
Air Pollution: Photochemical Smog Formation **1**: 185–8

Albisser, A. M.
(Hospital for Sick Children, Toronto, Ontario, Canada)
Glucose Regulation: Modelling and Control in Diabetics * **3**: 2025–32

Allan, R. N.
(UMIST, Manchester, UK)
Power System Reliability **6**: 3790–5

Allen, P. S.
(University of Alberta, Edmonton, Alberta, Canada)
Nuclear Magnetic Resonance Imaging in Biology **5**: 3421–5

Allidina, A. Y.
(ICI, Northwich, UK)
Self-Tuning Control, Feedforward **6**: 4187–9

Amouroux, M.
(University of Perpignan, Perpignan, France)
Control Equipment Location Problems * **2**: 801–8
Sensor and Control Location Problems * **6**: 4238–45

Andrew, A.
(Viable Systems, Chillaton, UK)
Cybernetics: A Survey **2**: 867–9
Self-Organizing Systems **6**: 4165–8

Andrews, R. W.
(University of Michigan, Ann Arbor, Michigan, USA)
Simulation Output: Statistical Analysis * **6**: 4393–6

Aoki, Y.
(National Institute for Environmental Studies, Ibaraki, Japan)
Environmental Evaluation of Green Spaces **2**: 1506–7
Waterfront Appraisal **7**: 5161–2

Arcelli, C.
(Institute of Cybernetics, Arco Felice, Italy)
Thinning Transformations **7**: 4881–3

Arigoni, A. O.
(University of Bologna, Bologna, Italy)
Fuzzy Causes: Probability **3**: 1788–91
Language Databases **4**: 2655–9

Arimoto, S.
(Osaka University, Osaka, Japan)
Robots, Walking **6**: 4123–8

Armstrong, P. J.
(Queen's University of Belfast, Belfast, UK)
Industrial Robots in the Aerospace Industry **4**: 2449–53

Artola, M.
(University of Bordeaux I, Talence, France)
Distributed Systems with Delay **2**: 1209–11
Fourier Series: Fast Fourier Transforms **3**: 1719–23
Fourier Transforms **3**: 1723–8
Laplace Transforms **4**: 2659–64
Semigroup Theory **6**: 4216–23

Asbjørnsen, O. A.
(University of Houston, Houston, Texas, USA)
Chemical Reactor Control **1**: 585–9
Electrolysis Control **2**: 1413–6

Åström, K. J.
(Lund Institute of Technology, Lund, Sweden)
Adaptive Control: Fundamental Problems **1**: 45–7
Papermaking: Adaptive Control **6**: 3590–1

Athans, M.
(MIT, Cambridge, Massachusetts, USA)
Linear Quadratic Gaussian Multivariable Control Designs: Robustness * **4**: 2842–6

Atherton, D. P.
(University of Sussex, Brighton, UK)
Nonlinear Systems: A Survey **5**: 3366–73
Nonlinear Systems: History **5**: 3383
Sinusoidal Input Describing Function: Evaluation and Properties **7**: 4443–52

Atkins, M. S.
(University of British Columbia, Vancouver, British Columbia, Canada)
Simulation Modelling Formalism: Transaction-Flow Techniques **6**: 4368–70
Transactions-Flow Models Simulation Software **7**: 4933–5

Aubin, J.-P.
(University of Paris IX, Paris, France)
Differential Inclusions and Viability Theory * **2**: 1023–33
Games, Fuzzy **3**: 1913–7
Minimax Theory **5**: 2994–8

Auer, M. T.
(Michigan Technological University, Houghton, Michigan, USA)
Algal Growth Modelling: Case Study * **1**: 257–62

Auslender, A.
(University of Clermont II, Aubière, France)
Nonlinear Programming: A Survey **5**: 3352–62

Aylmer, S. F.
(Petromin Shell Refinery Company, Industrial City, Saudi Arabia)
Gas Distribution Systems: Steady-State Analysis **3**: 1947–54

Azorín, F.
(University of Madrid, Madrid, Spain)
Discriminant Analysis **2**: 1102–7

Baba, M. L.
(Wayne State University, Detroit, Michigan, USA)
Technological Innovation: Issues and Perspectives **7**: 4828–33

Babary, J. P.
(LAAS-CNRS, Toulouse, France)
Control Equipment Location Problems * **2**: 801–8
Sensor and Control Location Problems * **6**: 4238–45

Baccelli, F.
(INRIA, Le Chesnay, France)
Brownian Motion **1**: 511–2
Distributed Database Systems: Performance Analysis of Consistency Preserving Algorithms * **2**: 1165–9

Bacciotti, A.
(Turin Polytechnic, Turin, Italy)
Multivalued Differential Equations in Control Systems * **5**: 3180–4

Baheti, R. S.
(General Electric Company, Schenectady, New York, USA)
Kalman Filtering: Applications **4**: 2618–22

Balaskovic, P.
(CNRS, Paris, France)
Airship Control and Guidance **1**: 252–7

Balci, O.
(Virginia Polytechnic Institute and State University, Blacksburg, Virginia, USA)
Simulation Model Management Objectives and Requirements * **6**: 4328–33

Banatre, J. P.
(IRISA, Rennes, France)
Distributed Database Systems: Reliability **2**: 1169–74

Bandle, C.
(University of Basel, Basel, Switzerland)
Isoperimetric Problems **4**: 2609–14

Bandler, W.
(Florida State University, Tallahassee, Florida, USA)
Computer Security Systems: Fuzzy Logics * **1**: 741–3

Fuzzy Implication Operators: A Survey * **3**: 1806–10
Relations, Mathematical * **6**: 4000–8

Baptistella, L. F. B.
(Telebras CPGD, Campinas, Brazil)
Multiobjective Optimization **5**: 3152–6

Bar-Shalom, Y.
(University of Connecticut, Storrs, Connecticut, USA)
Dual Control **2**: 1253–7

Barbenel, J. C.
(University of Strathclyde, Glasgow, UK)
Biological Measurement: Blood Mechanical Properties **1**: 440–5
Biomechanics of Connective Tissue **1**: 474–8
Biomechanics of Skin **1**: 478–82

Barbu, V.
(University of Iaşi, Iaşi, Romania)
Distributed Parameter Systems **2**: 1182–6
Variational Inequalities **7**: 5031–6
Variational Inequalities: Control **7**: 5036–41

Barel, N.
(Rhineland–Westphalia Technical University, Aachen, FRG)
Discrete-Event Simulation: Microprocessor Architectures **2**: 1083–7

Barker, H. A.
(University College, Swansea, UK)
Signal Processing: Analog **6**: 4281–4
Signal Processing: Digital **6**: 4284–9
Signal Transmission **6**: 4289–94

Barnes, G. R.
(RAF Institute of Aviation Medicine, Farnborough, UK)
Eye Movement: Vestibular System Modelling and Control **2**: 1591–6

Barnett, S.
(University of Bradford, Bradford, UK)
Nonlinear Systems: Lyapunov Stability **5**: 3398–402

Barney, G. C.
(UMIST, Manchester, UK)
Elevator Control Systems **2**: 1420–4
Elevator Management **2**: 1424–7
Elevator Systems: Computer-Aided Design and Simulation Techniques **2**: 1427–31
Elevator Systems: Computer Control **2**: 1431–4
Elevator Systems: Design **2**: 1435–41
Elevator Systems: Performance Measurement **2**: 1441–5
Elevator Systems: User Requirements **2**: 1446–50
Vertical Transport: History **7**: 5047–51

Barnthouse, L. W.
(Oak Ridge National Laboratory, Oak Ridge, Tennessee, USA)
Gaussian Random Variables * **3**: 1998–9

Barraud, A. Y.
(ENSIEG, Saint-Martin-d'Hères, France)
Realization Theory: Numerical Aspects **6**: 3977–83

Başar, T.
(University of Illinois, Urbana, Illinois, USA)
Deterministic Incentives * **2**: 993–5
Multilevel Systems: Stackelberg Strategies **5**: 3145–9

Batchelor, B. G.
(UWIST, Cardiff, UK)
Industrial Vision Systems for Inspection and Manufacturing **4**: 2473–9

Bates, J. F.
(University of Wales College of Medicine, Cardiff, UK)
Dental Measurements: Characteristics of Jaws and Teeth **2**: 963–5
Dental Measurements: Clinical Equipment and Devices **2**: 965–7
Dental Measurements: Clinical Research **2**: 968–9
Dental Measurements: Clinical, Visual and Tactile Methods **2**: 969–73
Dental Measurements: Radiography **2**: 973–6
Dental Measurements: Research into Stomatognathic Physiology **2**: 976–8

Bayoumi, M. M.
(Queen's University, Kingston, Ontario, Canada)
Multivariable Systems: Self-Tuning Control Design **5**: 3229–32

Beatty, P. C. W.
(Withington Hospital, Manchester, UK)
Artificial Respiratory Systems **1**: 315–9

Beck, M. S.
(UMIST, Manchester, UK)
Flow Measurement: Two-Phase Systems * **3**: 1685–94

Becker, P. W.
(Technical University of Denmark, Lyngby, Denmark)
Reliability of Systems **6**: 4011–5

Bel, G.
(CERT/DERA, Toulouse, France)
Flexible Manufacturing Systems: Modelling and Simulation * **3**: 1642–9

Bélanger, P.
(McGill University, Montreal, Quebec, Canada)
Adaptive Control: Industrial Applications **1**: 47–51
Titanium Dioxide Kilns: Adaptive Control **7**: 4892–4

Belforte, G.
(Turin Polytechnic, Turin, Italy)

Identification Errors in Biomedicine * **4**: 2253–7

Bell, D. J.
(UMIST, Manchester, UK)

Optimal Control: Singular Arcs **5**: 3506–8

Bellville, J. W.
(UCLA, Los Angeles, California, USA)

Respiratory System: Modelling and Identification * **6**: 4055–62

Benenati, F. E.
(US Environmental Protection Agency, Washington, DC, USA)

Agricultural Modelling: Field Applicability * **1**: 137–42
Chemical Stress in Agriculture: Modelling * **1**: 589–92
Toxic Chemicals Assessment: Simulation Modelling * **7**: 4895–7

Bennett, S.
(University of Sheffield, Sheffield, UK)

Feedback Control: History **3**: 1604–7
Ship Stabilization: History **6**: 4271–5
Speed Regulation: History **7**: 4513–7

Benoît, A.
(European Organisation for the Safety of Air Navigation, Brussels, Belgium)

Fuel Conservation: Air Transport * **3**: 1749–56

Bensoussan, A.
(INRIA, Le Chesnay, France)

Backward Dynamic Programming **1**: 407–9
Certainty Equivalence Principle **1**: 564–5
Differential Equations: An Introduction **2**: 1003–4
Dynamic Optimization: Introduction **2**: 1276–8
Forward Dynamic Programming **3**: 1713–4
Girsanov Transformation **3**: 2007–8
Impulse Control **4**: 2339–43
Inventory Control: Introduction **4**: 2583
Nonlinear Estimation Theory: Implications for Stochastic Control **5**: 3311–6
Optimal Control: Perturbation Methods **5**: 3497–502
Optimal Control: s,S-Policy and K-Convexity **5**: 3509–10
Optimal Stopping **5**: 3531–5
Ordinary Differential Equations **5**: 3560–3
Planning Horizons Problem: Fundamentals **6**: 3703–7
Production Smoothing **6**: 3890–2
Separation Principle **6**: 4245–8
Static Optimization: Introduction **7**: 4554–5
Stochastic Control: Introduction **7**: 4611–3
Stochastic Differential Games **7**: 4613–7
Stochastic Maximum Principle **7**: 4638–44

Benton, R.
(Honeywell, Minneapolis, Minnesota, USA)

Heating, Ventilation and Cooling Systems: Analog Simulation **3**: 2099–105
Heating, Ventilation and Cooling Systems: Digital Simulation **3**: 2113–7

Benveniste, A.
(IRISA, Rennes, France)

Adaptive Signal-Processing Algorithms: An Introduction **1**: 73–5
Adaptive Signal-Processing Algorithms: Convergence and Tracking **1**: 75–81
Adaptive Signal-Processing Algorithms: On-Line Detection of Abrupt Changes **1**: 81–6
Adaptive Signal-Processing Algorithms: Rare Events and Large Deviations **1**: 86–91

Bergman, R. N.
(University of Southern California, Los Angeles, California, USA)

Glucose Regulation: Modelling * **3**: 2021–5

Berkovitz, L. D.
(Purdue University, West Lafayette, Indiana, USA)

Optimal Control Theory: Developments 1950–1962 **5**: 3510–3

Bernadou, M.
(INRIA, Le Chesnay, France)

Finite Difference Method * **3**: 1623–7
Finite Element Method * **3**: 1627–32

Bernhard, P.
(INRIA, Sophia Antipolis, France)

Differential Games, Closed-Loop **2**: 1004–9
Differential Games: Introduction **2**: 1009–10
Differential Games: Isaacs Equation **2**: 1010–7
Differential Games, Linear Quadratic **2**: 1017–20
Differential Games, Open-Loop **2**: 1020–3
Linear Quadratic Regulator Problem **4**: 2846–9
Riccati Equations and Conjugate Points **6**: 4066–70

Bernstein, P. A.
(Wang Institute, Tyngsboro, Massachusetts, USA)

Distributed Database Systems: Concurrency Control * **2**: 1143–55

Bernussou, J.
(LAAS-CNRS, Toulouse, France)

Structure-Constrained Control: Parametric Optimization * **7**: 4678–85
Switching Networks: Real-Time Management * **7**: 4733–41

Bertrand, P.
(CNRS-ESE, Gif sur Yvette, France)

Aggregation and Suboptimal Control * **1**: 126–33
Decomposition–Coordination: A New Method * **2**: 952–6

Bertsekas, D. P.
(Massachusetts Institute of Technology, Cambridge, Massachusetts, USA)
Distributed Asynchronous Fixed-Point Algorithms: General Convergence Theory **2**: 1121–5

Bianchini, R. M.
(University of Florence, Florence, Italy)
Multivalued Differential Equations in Control Systems * **5**: 3180–4

Bickford, J. P.
(UMIST, Manchester, UK)
Power System Transients **6**: 3802–6

Billingsley, J.
(Portsmouth Polytechnic, Portsmouth, UK)
Industrial Robots: Application to Automatic Adjusting **4**: 2412–5
Predictive Control, Fast-Model **6**: 3866–70

Billinton, R.
(University of Saskatchewan, Saskatoon, Saskatchewan, Canada)
Reliability of Systems: Basic Theory **6**: 4015–20
Reliability of Systems: Specific Applications **6**: 4020–6

Binder, Z.
(ENSIEG, Saint-Martin-d'Hères, France)
Complex Systems Control: Tracking Methods * **1**: 675–82

Birta, L. G.
(University of Ottawa, Ottawa, Ontario, Canada)
Simulation and Optimization Interfacing **6**: 4306–10

Black, M. M.
(University of Sheffield, Sheffield, UK)
Biomechanics of Soft Tissue * **1**: 482–6

Blanchard, B. S.
(Virginia Polytechnic Institute and State University, Blacksburg, Virginia, USA)
Development of Systems and Equipment: Systems Engineering **2**: 995–1003
Life-Cycle Costing **4**: 2760–4

Blanchard, N.
(Claude Bernard University of Lyons I, Lyons, France)
Fuzzy Cardinals **3**: 1782–6

Blankenship, G. L.
(University of Maryland, College Park, Maryland, USA)
Nonlinear Estimation Theory: Asymptotic Analysis Methods **5**: 3303–11

Blume, C.
(University of Karlsruhe, Karlsruhe, FRG)
Industrial Assembly Robots: Programming Languages and Systems * **4**: 2347–51

Boesefeldt, J.
(Heusch/Boesefeldt, Aachen, FRG)
Road Information and Electronic Guidance Systems **6**: 4075–9

Bohlin, T.
(Royal Institute of Technology, Stockholm, Sweden)
Identification: Practical Aspects **4**: 2301–7
Validation of Identified Models **7**: 4996–5001

Boissonnat, J. D.
(INRIA, Le Chesnay, France)
Automatic Modelling of Three-Dimensional Objects **1**: 343–7

Bona, B.
(Turin Polytechnic, Turin, Italy)
Kinetic Data Diagnosis in Biological Modelling * **4**: 2631–6

Bonfield, W.
(Queen Mary College, London, UK)
Biomechanics of Bone **1**: 468–74

Bonnard, B.
(ENSIEG, Saint-Martin-d'Hères, France)
Attitude Control of Rigid Spacecraft **1**: 319–23

Bonne, U.
(Honeywell, Bloomington, Minnesota, USA)
Domestic Heating **2**: 1214–26
Domestic Heating: Boiler and Furnace Efficiency **2**: 1226–7
Domestic Heating: Boilers and Furnaces **2**: 1227–33
Domestic Heating: Temperature Control Operation **2**: 1234–9
Domestic Heating: Temperature Control Strategies **2**: 1239–41

Booker, G. A.
(East Worcestershire Waterworks Company, Bromsgrove, UK)
Water Supply Pumps: Optimal Use **7**: 5147–50

Boothby, W. M.
(Washington University, St Louis, Missouri, USA)
Matrix Groups: Action on State Spaces and Control Theory **5**: 2944–9

Borne, P.
(Institut Industriel du Nord, Villeneuve d'Ascq, France)
List of Acronyms and Abbreviations **8**: 5587–611
Laplace Transforms and Bode Diagrams **4**: 2664–5
Lyapunov Equation **4**: 2904

Multi-Timescale Systems: Dynamical Location * **5**: 3176–80
Nonlinear Systems Stability: Vector Norm Approach **5**: 3402–6
PID Controllers * **6**: 3699–701
PID Controllers, Digital * **6**: 3701–3
Singular Perturbations: Boundary-Layer Problem * **7**: 4425–9
State-Space Modelling: State-Space Transformation * **7**: 4540–7

Bortolan, G.
(LADSEB-CNR, Padua, Italy)
Computerized Electrocardiogram Diagnosis: Fuzzy Approach * **1**: 760

Bosch, P. P. J. van den
(Delft University of Technology, Delft, The Netherlands)
Model Reference Adaptive Systems: Model Updating **5**: 3070–1
Satellites: Adaptive Control **6**: 4146–8

Bose, B. K.
(General Electric Company, Schenectady, New York, USA)
Motor Drives, Alternating Current **5**: 3089–92

Bose, N. K.
(Pennsylvania State University, University Park, Pennsylvania, USA)
Multidimensional Systems Theory: Applications **5**: 3127–31
Multidimensional Systems Theory: Stability **5**: 3131–5
Realization Theory, Multivariate **6**: 3969–72

Bosserman, R. W.
(University of Louisville, Louisville, Kentucky, USA)
Catastrophe Theory in Ecosystem Analysis **1**: 550–2
Ecosystem Digraphs: Decomposition into Elementary Components **2**: 1345–8
Ecosystem Networks: Measures of Structure **2**: 1352–6
Fuzzy Set Theory in Ecology **3**: 1858–62
Sensitivity Techniques in Input–Output Flow Analysis **6**: 4235–8

Bouchon, B.
(University of Paris VI, Paris, France)
Fuzzy Partitions **3**: 1835–8

Bouillot, J. C.
(Centre National d'Etudes Spatiales, Evry, France)
Launch Vehicles, Multistage **4**: 2723–7
Launch Vehicles: Problems **4**: 2728–30

Bourdon, P.
(Electricité de France, Clamart, France)
Power Plants: Nonlinear Modelling * **6**: 3772–7

Bousquet, M.
(ENSAE, Toulouse, France)
Digital Communications: Satellite Systems * **2**: 1051–7

Bowler, A. L.
(University of Salford, Salford, UK)
Control Valves: Gas Applications **2**: 837–40
Gas Distribution Systems **3**: 1939–43
Gas Distribution Systems: Balancing Supply and Demand **3**: 1943–7
Gas Leakage Control **3**: 1954–8
Gas Storage Systems * **3**: 1969–75
Gas Supply Systems: Control and Telemetry **3**: 1975–9
Gas Supply Systems: Load Forecasting **3**: 1979–82
Gas Transmission Systems **3**: 1982–5
Gas Utility System Planning **3**: 1995–8

Bradley, J. M.
(North West Water Authority, Manchester, UK)
Water Distribution Systems: Planning **7**: 5099–102

Brans, J. P.
(Free University of Brussels, Brussels, Belgium)
Multiobjective Decision * **5**: 3150–2

Bransby, M. L.
(Central Electricity Generating Board, Harrogate, UK)
Steam Boiler Control **7**: 4572–7

Breddermann, R.
(Rhineland–Westphalia Technical University, Aachen, FRG)
Self-Tuning On–Off Control * **6**: 4209–11

Brewer, J. W.
(University of California, Davis, California, USA)
Ecological Modelling: Bond Graphs **2**: 1326–9
Ecological Populations: Optimal Impulsive Control **2**: 1335–7
Population Modelling **6**: 3756–62

Bristol, E. H.
(The Foxboro Company, Foxboro, Massachusetts, USA)
Process Control: History **6**: 3879–84

Brockhaus, R.
(Technical University of Braunschweig, Braunschweig, FRG)
Aircraft Dynamics: State Equations **1**: 213–20
Autopilots: Feedback Structure **1**: 388–93

Brown, B. H.
(Royal Hallamshire Hospital, Sheffield, UK)
Biological Measurement: Myoelectric Activity of the Gut **1**: 459–62

Brown, T. W.
(British Aerospace, Filton, UK)
Aircraft Engine Control: Physical Principles **1**: 226–31

Brunovský, P.
(Comenius University, Bratislava, Czechoslovakia)
Irregular Optimal Feedback and Generalized Trajectories **4**: 2608–9
Optimal Control: Number of Switchings **5**: 3487–8
Optimal Control, Open-Loop and Closed-Loop **5**: 3496–7
Optimal Feedback: Linear Quadratic Problem **5**: 3513–4
Optimal Feedback: Linear Time-Optimal Control Problem **5**: 3514–5
Regular Synthesis and Piecewise Analytic Feedback **6**: 3998–4000

Bryant, G. F.
(Imperial College of Science and Technology, London, UK)
Multivariable Systems: Optimization in Dominance Design **5**: 3220–4

Bryant, R. M.
(IBM Thomas J. Watson Research Center, Yorktown Heights, New York, USA)
Simulation Modelling Formalism, Event-Based **6**: 4346–8
Simulation Software, Event-Based **6**: 4403–6

Buckland, E. C.
(B.S. Instruments, Rustington, UK)
Pressure Measurement **6**: 3876–7

Buckles, B. P.
(University of Texas, Arlington, Texas, USA)
Fuzzy Databases * **3**: 1795–7

Bullinger, H. J.
(Fraunhofer Institute, Stuttgart, FRG)
Personnel Aspects of Automation * **6**: 3659–64

Bullingham, R. E. S.
(Syntex Research, Maidenhead, UK)
Demand Analgesia: Modelling and Control of Postoperative Pain * **2**: 960–3

Bunge, M.
(McGill University, Montreal, Quebec, Canada)
Causality **1**: 552–6
Holism **3**: 2163–4
Reductionism **6**: 3997–8

Burkov, V. N.
(USSR Academy of Sciences, Moscow, USSR)
Active Goal-Seeking Systems **1**: 31–7

Burns, N. D.
(Unilever Research, Port Sunlight, UK)
Industrial Robots: Design Case Studies **4**: 2440–4

Burris, H. R.
(Neotechnic Industries, Redondo Beach, California, USA)
Emulation and Microprogramming **2**: 1450–1

Cabriere, J.
(Avions Marcel Dassault–Breguet Aviation, Saint-Cloud, France)
Combat Aircraft: Long-Term Future * **1**: 618–23

Cale, W. G.
(University of Texas at Dallas, Richardson, Texas, USA)
Ecological Modelling: Aggregation Errors **2**: 1324–6

Calvet, J. L.
(LAAS-CNRS, Toulouse, France)
Hierarchical Control: Partition **3**: 2153–63

Camoin, B. G.
(BULL, Paris, France)
Switching Networks: Quality Parameters **7**: 4732–3
Telecommunication Traffic: Variables and Measurements **7**: 4833–5

Canale, R. P.
(Michigan Technological University, Houghton, Michigan, USA)
Algal Growth Modelling: Case Study * **1**: 257–62

Carpentier, J.
(Electricité de France, Paris, France)
Reactive Power Control: Loss Minimization **6**: 3950–4

Carson, E. R.
(City University, London, UK)
Cellular Protein Synthesis Control * **1**: 556–9
Lipoprotein Dynamics and Control * **4**: 2884–7
Protein Metabolism: Dynamics and Control * **6**: 3905–9

Cartwright, D. E.
(Institute of Oceanographic Sciences, Birkenhead, UK)
Tide Modelling **7**: 4883–9
Tides as Time Series **7**: 4889–92

Carver, M. B.
(Chalk River Nuclear Laboratories, Chalk River, Ontario, Canada)
Distributed Model Simulation Software **2**: 1180–1

Cassier, A.
(Aérospatiale, Marignane, France)
Helicopters **3**: 2118–23
Helicopters: Turbine-Engine Governing Systems **3**: 2137–9

Castel, C.
(CERT/DERA, Toulouse, France)
Switching Networks: Modelling * **7**: 4726–32

Casti, J.
(IIASA, Laxenburg, Austria)
Catastrophe Theory **1**: 547–50
Chaos **1**: 565–7
Connective Stability **2**: 776–7
Domains of Attraction **2**: 1211–2
Linear Systems: General Aspects **4**: 2861–6
Q-Analysis **6**: 3919–22
Stability of Values and Pulses **7**: 4522–3
Stability Theory **7**: 4523–9

Castoldi, C.
(Camel Robots, Milan, Italy)
Factory Automation: Case Study in Production Line Management **3**: 1597–600

Cavaillé, J. B.
(CERT/DERA, Toulouse, France)
Flexible Manufacturing Systems: Modelling and Simulation * **3**: 1642–9

Cekirge, H. M.
(King Fahd University of Petroleum and Minerals, Dhahran, Saudi Arabia)
Moving-Boundary Models: Numerical Solution **5**: 3097–9

Cellier, F. E.
(University of Arizona, Tucson, Arizona, USA)
Ordinary Differential Equation Models: Numerical Integration of Initial-Value Problems **5**: 3555–9
Simulation Modelling Formalism: Ordinary Differential Equations **6**: 4356–60

Cellina, A.
(University of Paris IX, Paris, France)
Differential Inclusions and Viability Theory * **2**: 1023–33

Chalabi, Z.
(Rothamsted Experimental Station, Harpenden, UK)
Eye Movement: Slow-Phase Caloric Nystagmus Modelling **2**: 1586–91

Chang, P. K.
(Purdue University, West Lafayette, Indiana, USA)
Industrial Robots: An Optimal Control System * **4**: 2392–412

Chankong, V.
(Case Western Reserve University, Cleveland, Ohio, USA)
Multiple Objective Optimization: Pareto Optimality * **5**: 3156–65
Surrogate Worth Trade-Off Method * **7**: 4703–9

Chapman, P. K.
(Arthur D. Little, Cambridge, Massachusetts, USA)
Solar Power Satellites * **7**: 4472–6

Char, B. W.
(University of Waterloo, Waterloo, Ontario, Canada)
Ordinary Differential Equation Models: Symbolic Manipulation **5**: 3559–60

Chatterji, B. N.
(Indian Institute of Technology, Kharagpur, India)
Character Recognition **1**: 570–5

Chaudhuri, B. B.
(Indian Statistical Institute, Calcutta, India)
Fuzzy Membership Evaluation **3**: 1823–5

Chavent, G.
(INRIA, Le Chesnay, France)
Distributed Parameter Systems: Identification **2**: 1193–7

Chemouil, P.
(CNET, Issy les Moulineaux, France)
Large-Scale Systems: Parameter Estimation * **4**: 2702–6
Singularly Perturbed Systems: Output Feedback Control **7**: 4439–43
Switching Networks: Traffic Routing and Real-Time Control * **7**: 4749–51

Cheng Shaozhong
(Chengdu University of Science and Technology, Chengdu, People's Republic of China)
Fuzzy Probability * **3**: 1839–43
Normal Fuzzy Measures and Fuzzy Random Measures * **5**: 3410–3

Chilcoat, R. T.
(The BOC Group, Murray Hill, New Jersey, USA)
Anesthesia: Feedforward Control of Depth **1**: 281–4

Chow, L. R.
(Tamkang University, Taipei, Taiwan)
Information Systems **4**: 2502–4

Christensen, R. A.
(Entropy Ltd., Lincoln, Massachusetts, USA)
Entropy Minimax Theory **2**: 1473–9

Ciarlet, P. G.
(INRIA, Le Chesnay, France)
Finite Difference Method * **3**: 1623–7

Finite Element Method * **3**: 1627–32

Clarke, D. W.
(University of Oxford, Oxford, UK)
Minimum Variance Control **5**: 3004–6
Predictor Models **6**: 3870–1
Self-Tuners Design: Predictive Control Theory **6**: 4176–80
Self-Tuning Control: Models **6**: 4197–9
Self-Tuning Controller Implementation **6**: 4199–204

Clymer, A. B.
(Clymer Technology, Ocean, New Jersey, USA)
Simulators for Training **6**: 4415–7

Cobelli, C.
(University of Padua, Padua, Italy)
Glucose Regulation: Modelling * **3**: 2021–5
Ketone Body Metabolism Modelling * **4**: 2628–31

Cohen, G.
(ENSMP, Fontainebleau, France)
Auxiliary Problem Principle and Decomposition of Optimization Problems **1**: 400–6

Cohen, S.
(INRETS, Arceuil, France)
Traffic Simulation: Hydrodynamic Theory and Shock Waves **7**: 4920–3
Traffic Variables **7**: 4930–3

Coleman, T. G.
(University of Mississippi, Jackson, Mississippi, USA)
Biological Systems Simulation * **1**: 466–8

Collier, P. I.
(UWIST, Cardiff, UK)
Harbor Terminal Management **3**: 2072–6

Collins, C. D.
(New York State Museum, Albany, New York, USA)
Validation of Phytoplankton Models **7**: 5005–8

Conant, R. C.
(University of Illinois, Chicago, Illinois, USA)
Information Laws of Systems **4**: 2492–7
Intelligence Amplification **4**: 2542–3

Concepcion, A. I.
(Wayne State University, Detroit, Michigan, USA)
Distributed System Models: Distributed Simulation **2**: 1199–201

Connor, N. E.
(University of Salford, Salford, UK)
Gas, Manufactured **3**: 1958–61
Gas Metering **3**: 1962–5
Gas Storage Systems * **3**: 1969–75

Cook, P. A.
(UMIST, Manchester, UK)
Nonlinear Feedback Systems: Stability Conditions **5**: 3316–9
Nyquist Stability Criterion **5**: 3440–1

Corbeel, D.
(Institut Industriel du Nord, Villeneuve d'Ascq, France)
Petri Nets in Manufacturing * **6**: 3670–3

Cornet, B.
(Center for Operations Research, Louvain-La-Neuve, Belgium)
Optimization: Sensitivity Analysis **5**: 3547–51

Cory, B. J.
(Imperial College of Science and Technology, London, UK)
Power Systems **6**: 3806–9

Cottam, M.
(UWIST, Cardiff, UK)
Robot Operating System Design Using Syntax Graphs * **6**: 4089–93

Coulbeck, B.
(Leicester Polytechnic, Leicester, UK)
Water Networks: Extended-Period Simulation **7**: 5106–9
Water Supply Pumps: Optimization **7**: 5150–3

Courtney, J. M.
(University of Strathclyde, Glasgow, UK)
Artificial Kidneys, Portable and Wearable **1**: 310–2

Cramp, D. G.
(City University, London, UK)
Cellular Protein Synthesis Control * **1**: 556–9
Lipoprotein Dynamics and Control * **4**: 2884–7
Protein Metabolism: Dynamics and Control * **6**: 3905–9

Crisp, V. H. C.
(Building Research Establishment, Watford, UK)
Light Switching **4**: 2764–9
Lighting Control **4**: 2770–5
Lighting Control: Applications **4**: 2775–7

Cronhjort, B.
(Royal Institute of Technology, Stockholm, Sweden)
Mineral Processing and Cement Industries: Process Control * **5**: 2989–94

Cronly-Dillon, J.
(UMIST, Manchester, UK)
Visual System: Psychophysics * **7**: 5070–8
Visual System: Structure and Function **7**: 5078–86

Crouch, P. E.
(University of Arizona, Tucson, Arizona, USA)
Realization of Finite Volterra Series **6**: 3956–62
Realization Theory, Nonlinear **6**: 3973–7

Cruz, J. B. Jr
(University of California, Irvine, California, USA)
Deterministic Incentives * **2**: 993–5
Sensitivity Analysis **6**: 4223–7

Cunningham, G. W.
(IBM Federal Systems Division, Bethesda, Maryland, USA)
Space-Shuttle Control Systems Reliability: Redundant Processing **7**: 4476–81

Cunningham, M. J.
(University of Manchester, Manchester, UK)
Speed and Acceleration Measurement **7**: 4509–13

Cury, J. E. R.
(LAAS–CNRS, Toulouse, France)
Decentralized Control: Geometric Approach **2**: 894–900

Cutler, M. M.
(IBM, Poughkeepsie, New York, USA)
Simulation Program Verification **6**: 4402–3

Czinczenheim, J.
(Avions Marcel Dassault–Breguet Aviation, Saint-Cloud, France)
Combat Aircraft: Long-Term Future * **1**: 618–23

Czogała, E.
(Silesian Technical University, Gliwice, Poland)
Measures of Fuzziness: A Survey * **5**: 2963–5

Daclin, E. R.
(Midirobots, Ramonville, France)
Aerospace Simulators **1**: 109–12

Dang Van Mien, H.
(Electricité de France, Clamart, France)
Power Plants: Nonlinear Modelling * **6**: 3772–7

D'Argenio, D. Z.
(University of Southern California, Los Angeles, California, USA)
Biomedical Systems Modelling: Experiment Design * **1**: 486–90

Dauphin-Tanguy, G.
(Institut Industriel du Nord, Villeneuve d'Ascq, France)
Multi-Timescale Systems: Dynamical Location * **5**: 3176–80

PID Controllers, Digital * **6**: 3701–3
Singular Perturbations: Boundary-Layer Problem * **7**: 4425–9
z-Transform Method **7**: 5189

David, Y.
(INRETS, Villeneuve D'Ascq, France)
Bus-Route Control **1**: 514–7
Road Information: Automatic Diversion Control **6**: 4079–81
Road Information: Communication **6**: 4081–2
Traffic Detectors and Recorders **7**: 4917–20

Davies, E. R.
(Royal Holloway and Bedford New College, Egham, UK)
Industrial Vision Systems: Segmentation of Gray-Scale Images **4**: 2479–84

Davies, W. L.
(Oxford Medical, Abingdon, UK)
Anesthesia: Feedback Control of Depth **1**: 277–81

Davis, M. H. A.
(Imperial College of Science and Technology, London, UK)
Optimal Stochastic Control: General Aspects **5**: 3517–23

Davison, E. J.
(University of Toronto, Toronto, Ontario, Canada)
Decentralized Control: Pole Assignment **2**: 900–2
Decentralized Fixed Modes **2**: 902–7
Robust Controller Design **6**: 4129–33
Robust Controller Design, Decentralized **6**: 4133–9

De Bra, D.
(Stanford University, Stanford, California, USA)
Drag-Free Satellites **2**: 1242–4

De Fornel, B.
(ENSEEIHT, Toulouse, France)
Variable Speed Induction Machines: Numerical Control **7**: 5019–27

de Glas, M.
(University of Paris VI, Paris, France)
Fuzzy σ-Fields in Fuzzy Measure Theory **3**: 1878–80
Fuzzy Systems, Dynamical **3**: 1895–8
Many-Valued Logics, Algebraic **5**: 2921–4

De Keyser, R. M. C.
(State University of Ghent, Ghent, Belgium)
Self-Tuners as Data Predictors **6**: 4168–72

de Luca, A.
(University of Rome, Rome, Italy)
Entropy Measures in Fuzzy Set Theory * **2**: 1467–73

de Paor, A. M.
(University College, Dublin, Republic of Ireland)
Cardiac Electrical Activity Modelling **1**: 532–6
Electrical Control in the Nineteenth Century *
2: 1398–402

Degani, R. T.
(LADSEB-CNR, Padua, Italy)
Computerized Electrocardiogram Diagnosis: Fuzzy Approach * **1**: 760

Delfour, M. C.
(University of Montreal, Montreal, Quebec, Canada)
Hereditary Systems **3**: 2147–51
Hereditary Systems Control: Introduction **3**: 2151–2
Optimal Control: Riccati Equations **5**: 3502–6

Denham, M. J.
(Kingston Polytechnic, Kingston-upon-Thames, UK)
Computer-Aided Design: Environments for Control Design **1**: 726–32

Deransart, P.
(INRIA, Le Chesnay, France)
Formal Languages **3**: 1712
Turing Machines **7**: 4958

Derigs, U.
(University of Bayreuth, Bayreuth, FRG)
Combinatorial Optimization: Independence Systems *
1: 624–6
Combinatorial Optimization Problems: Approximative Algorithms * **1**: 626–8
Combinatorial Optimization Problems: Branch-and-Bound Approach * **1**: 628–30
Discrete and Combinational Optimization: Introduction * **2**: 1080–3
Polyhedral Combinatorics and Cutting-Plane Methods * **6**: 3735–9

Descusse, J.
(CNRS-ENSM, Nantes, France)
Linear Quadruples (A, B; C, D): Geometric Approach
4: 2849–55

Desprès, R.
(RCE, Cergy-Pontoise, France)
Packet-Switching Networks **6**: 3581–90

Dexter, A. L.
(University of Oxford, Oxford, UK)
Heating Systems: Adaptive Control **3**: 2092–7

Di Nola, A.
(University of Naples, Naples, Italy)
Fuzzy Sets: Ordering by Fuzzy Entropy **3**: 1875–8

Dickson, J.
(University of Manchester, Manchester, UK)
Nuclear Energy **5**: 3413–6
Nuclear Reactors **5**: 3436–40

Dillon, T. S.
(La Trobe University, Bundoora, Victoria, Australia)
Power Systems: Maintenance Scheduling **6**: 3834–7

Dimitrov, V.
(Central Statistical Office, Sofia, Bulgaria)
Fuzzy Catastrophes **3**: 1786–8

Diosey, P. G.
(National Institute for Environmental Studies, Ibaraki, Japan)
Air Pollution: Wind Tunnel Simulation * **1**: 188–9
Bilinear Equations in Ecology **1**: 414–7

Dix, C. H.
(Harrogate, UK)
Electrical Noise: Principles and Measurement
2: 1403–5

Dodu, J. C.
(Electricité de France, Clamart, France)
Hydroelectricity: Production Planning * **3**: 2221–5
Power System Planning * **6**: 3784–90

Dopazo, J. F.
(American Electric Power, New York, USA)
Power Systems: Economic Dispatching * **6**: 3815–20
Power Systems: State Estimation * **6**: 3846–50

Dormeier, S.
(University of Paderborn, Paderborn, FRG)
Extruder Control **2**: 1583–6

Dost, M. H.
(IBM, San Jose, California, USA)
Simulation in Automatic Control **6**: 4310–1

Drazan, P. J.
(UWIST, Cardiff, UK)
End Effectors **2**: 1451
Industrial Robots **4**: 2392
Robot Arms **6**: 4082
Robot Operating System Design Using Syntax Graphs * **6**: 4089–93
Robotics **6**: 4093
Robots **6**: 4097
Robots: Teach Function **6**: 4123

Drouin, M.
(LSS-ESE, Gif sur Yvette, France)
Decomposition–Coordination: A New Method *
2: 952–6

Dubois, D.
(Paul Sabatier University, Toulouse, France)
Flexible Manufacturing Systems: Modelling and Simulation * **3**: 1642–9
Fuzzy Algebra * **3**: 1780–1

Fuzzy Events * **3**: 1800–2
Fuzzy Measures: Fuzzy Integral Approach * **3**: 1821–3
Fuzzy Numbers * **3**: 1829–31
Fuzzy Ranking * **3**: 1845–7

Duc, G.
(LSS-ESE, Gif Sur Yvette, France)

Aggregation and Suboptimal Control * **1**: 126–33

Düchting, W.
(University of Siegen, Siegen, FRG)

Tumor-Disease Treatment: Simulation **7**: 4956–8

Dufour, J.
(University of Savoie, Annecy, France)

Interaction Analysis: Informational Approach **4**: 2543–7

Dunnett, R. M.
(Central Electricity Research Laboratories, Leatherhead, UK)

Electric Generators: Unit Commitment **2**: 1383–8

Durst, F.
(University of Erlangen–Nuremberg, Erlangen, FRG)

Environmental Measurement: Optical Lasers for Long-Range Wind Velocity Determination **2**: 1528–32

Dutton, W. A.
(Noranda Research Centre, Pointe-Claire, Quebec, Canada)

Nonferrous Metallurgical Industry: Process Control * **5**: 3280–2

Dy Liacco, T. E.
(Dy Liacco Corporation, Cleveland, Ohio, USA)

Power Systems: On-Line Control **6**: 3842–6

Ebdon, L.
(Plymouth Polytechnic, Plymouth, UK)

Absorption Spectroscopy in Biology **1**: 8–13

Edmunds, J. M.
(UMIST, Manchester, UK)

Characteristic-Locus Method: Design Study **1**: 581–3

Egardt, B.
(ASEA Automation, Västerås, Sweden)

Self-Tuning and Model Reference Adaptive Systems: Interrelation **6**: 4184–7

Ekeland, I.
(University of Paris IX, Paris, France)

Variational Methods: Introduction **7**: 5041–7

El-Fattah, Y. M.
(Faculté des Sciences, Rabat, Morocco)

Learning Systems **4**: 2731–7

Ellison, D.
(Dundee College of Technology, Dundee, UK)

Simulation Modelling Formalism, Activity-Based **6**: 4333–4
Simulation of Activities-Based Models: Software **6**: 4392–3

Elloy, J. P.
(ENSM, Nantes, France)

Control Valve Sizing **2**: 828–32
Control Valve Technology **2**: 833–7

Elzas, M.
(Agricultural University, Wageningen, The Netherlands)

Combined Discrete and Continuous Models: Firmware **1**: 630–4
Simulation Methodology and Systems Control and Management **6**: 4313–9
Simulation Study Credibility **6**: 4406–11

Emre, E.
(Texas Tech University, Lubbock, Texas, USA)

Input–Output and Disturbance Decoupling **4**: 2522–3

Ephrath, A. R.
(Bell Communications Research, Piscataway, New Jersey, USA)

Human–Computer Interaction **3**: 2175–9

Erdem, U.
(Negretti Automation, Aylesbury, UK)

Force and Weight Measurement **3**: 1706–10

Escaffre, F.
(VERIDATAS, Toulouse, France)

Head-Up Displays and Trajectory Control **3**: 2080–9

España, M.
(Ciudad University, Mexico City, Mexico)

Distillation Columns: Bilinear Models **2**: 1116–21

Ewing, D. F.
(University of Hull, Kingston-upon-Hull, UK)

Gas Chromatography **3**: 1929–31
Liquid Chromatography **4**: 2887–9

Eykhoff, P.
(University of Technology, Eindhoven, The Netherlands)

Identification: History **4**: 2270–3

Fahmy, M. M.
(Queen's University, Kingston, Ontario, Canada)
Multidimensional Systems: Oscillations **5**: 3124–7

Fargier, P. H.
(INRETS, Arceuil, France)
Traffic Simulation: Macroscopic Models **7**: 4923–5

Fasol, K. H.
(University of Bochum, Bochum, FRG)
Dynamic-Systems Modelling: Basic Principles and Lumped-Parameter Systems * **2**: 1286–91
Dynamic-Systems Modelling: Distributed-Parameter Models and Discretization * **2**: 1292–7

Fatmi, M. B. E.
(Duke University, Durham, North Carolina, USA)
Communication Theory: Fuzzy Detection and Estimation * **1**: 651–65

Faysse, J.-M.
(INRETS, Arceuil, France)
Vessel Traffic Monitoring Systems **7**: 5051–5

Fell, R.
(University of Bradford, Bradford, UK)
Signal Analysis: Fundamentals **6**: 4277–81

Fenwick, P.
(St Thomas' Hospital, London, UK)
Electroencephalography **2**: 1410–3

Ferguson, J.
(University of Strathclyde, Glasgow, UK)
Viscosity Measurement **7**: 5056–60

Ferreira, P. M. G.
(Pontificia Catholic University, Rio de Janeiro, Brazil)
Internal Model Principle: Output Tracking and Disturbance Rejection **4**: 2567–70

Fieschi, M.
(Informatique Faculté de Médécin, Marseille, France)
Expert Systems in Medical Reasoning **2**: 1573–8

Fiksel, J.
(Teknowledge, Palo Alto, California, USA)
Systems Management **7**: 4785–9

Finkelstein, A. C. W.
(Royal College of Art, London, UK)
Instruments and Instrument Systems: Design Principles * **4**: 2533–6

Finkelstein, L.
(City University, London, UK)
Errors and Uncertainty in Measurements and Instruments **2**: 1553–9
Instruments and Instrument Systems: Concepts and Principles **4**: 2527–33
Instruments and Instrument Systems: Design Principles * **4**: 2533–6

Fisk, D. J.
(Department of the Environment, London, UK)
Anticipating Control **1**: 290
Building Management Systems **1**: 512–4
Critical Time Override **2**: 867
Deadband Control of Temperature **2**: 894
Degree Day **2**: 959
Demand Control **2**: 963
Duty Cycling **2**: 1262
Economizer Control **2**: 1342
Flame Safeguard Control **3**: 1632
Heat Pumps **3**: 2092
Humidistats **3**: 2206
Hydronic Systems **3**: 2234
Minimum On-Time **5**: 2998
Optimum Start Control of Space-Heating Systems **5**: 3554
Reset Control in Air Conditioning **6**: 4045
Reset Ratio in Air Conditioning **6**: 4045
Thermostats, Multistage **7**: 4881

Fleck, J.
(University of Edinburgh, Edinburgh, UK)
Robotics: Social Aspects **6**: 4093–7

Fletcher, K. S.
(The Foxboro Company, Foxboro, Massachusetts, USA)
pH Measurement **6**: 3683–8

Fliess, M.
(CNRS-ESE, Gif sur Yvette, France)
Functional Expansions **3**: 1769–79
Nonlinear Systems: Approximation by Simple State-Space Models * **5**: 3374–6
Nonlinear Systems: Local Realization Theory **5**: 3390–4

Florenzano, M.
(CEPREMAP, Paris, France)
Competitive Equilibrium **1**: 665–71

Foo, N. Y.
(University of Sydney, Sydney, New South Wales, Australia)
Homomorphism of Systems **3**: 2164–8

Forestier, J. P.
(CERT/DERA, Toulouse, France)
Industrial Robots: Manipulator Control * **4**: 2456–60

Forte, B.
(University of Waterloo, Waterloo, Ontario, Canada)
Information Theory * **4**: 2504–8

Fossard, A. J.
(ENSAE, Toulouse, France)
Multi-Timescale Systems **5**: 3171–6

Fraade, D. J.
(Burroughs Wellcome, Greenville, North Carolina, USA)
Pharmaceutical Industry: Process Control **6**: 3691–5

Francis, B. A.
(University of Toronto, Toronto, Ontario, Canada)
Internal Model Principle **4**: 2566–7

Franks, R. G. E.
(Du Pont, Wilmington, Delaware, USA)
Chemical Engineering: Simulation **1**: 583–5

Franta, W. R.
(University of Minnesota, Minneapolis, Minnesota, USA)
Process-Based Models: Simulation Software **6**: 3878–9
Simulation Modelling Formalism, Process-Based **6**: 4365–6

Fraser, N. M.
(University of Waterloo, Waterloo, Ontario, Canada)
Systems Management: Conflict Analysis * **7**: 4793–9

Frediani, S.
(CENS, Turin, Italy)
Kinetic Data Diagnosis in Biological Modelling * **4**: 2631–6

Fu, K. S.
(Late of Purdue University, West Lafayette, Indiana, USA)
Pattern Recognition: Decision-Theoretic Approaches **6**: 3634–45
Pattern Recognition: Syntactic Methods **6**: 3645–54
Structural Damage Assessment: Fuzzy Approach * **7**: 4667–9

Fuchs, J. J.
(IRISA, Rennes, France)
Explicit Self-Tuners: Convergence Analysis **2**: 1578–81
Non-Minimum-Phase Systems: Adaptive Control **5**: 3406–10

Fujimura, S,
(University of Tokyo, Tokyo, Japan)
Remote Sensing: Classification of Multispectral Images * **6**: 4029–33

Gabard, J. F.
(CERT/DERA, Toulouse, France)
Traffic Simulation: Microscopic Models **7**: 4925–7

Gabay, D.
(University of Paris VI and INRIA, Paris, France)
Constrained Optimization: Feasible-Direction Methods **2**: 786–9
Constrained Optimization: Lagrange-Multiplier Methods **2**: 789–93
Constrained Optimization: Lagrangian Methods Based on Sequential Quadratic Programming **2**: 793–7
Constrained Optimization: Penalty Methods **2**: 797–801
Linearly Constrained Optimization: Numerical Methods **4**: 2875–80
Optimal Control: Gradient-Related Numerical Solution of Optimal Control Problems **5**: 3481–7
Optimization of Functionals by Algorithms: Introduction **5**: 3546–7
Unconstrained Optimization: Basic Descent Method **7**: 4964–9
Unconstrained Optimization: Conjugate Gradient Methods **7**: 4969–71
Unconstrained Optimization: Newton and Quasi-Newton Methods **7**: 4971–5

Gaillet, A.
(CERT/DERA, Toulouse, France)
Warehouse Automation **7**: 5087–90

Gallopín, G. C.
(Fundación Bariloche, San Carlos de Bariloche, Rio Negro, Argentina)
Environment of a System **2**: 1486–9

Gansler, J. S.
(Analytic Sciences Corporation, Arlington, Virginia, USA)
System Acquisition and Procurement **7**: 4752–5

Garcia, C. B.
(University of Chicago, Chicago, Illinois, USA)
Path-Following Methods * **6**: 3631–4

Garcia, J. M.
(LAAS-CNRS, Toulouse, France)
Switching Networks: Real-Time Management * **7**: 4733–41

Gardner, R. H.
(Oak Ridge National Laboratory, Oak Ridge, Tennessee, USA)
Error Analysis and Sensitivity Analysis in Ecology **2**: 1549–52

Gascoigne, J. D.
(Loughborough University of Technology, Loughborough, UK)
Industrial Production Systems: Robot Integration * **4**: 2362–8

Gasparski, W.
(Polish Academy of Sciences, Warsaw, Poland)
Praxiology **6**: 3860–5

Gause, D. C.
(State University of New York, Binghamton, New York, USA)
Design of Systems * **2**: 987–92

Gauthier, P.
(CNET, Issy les Moulineaux, France)
Switching Networks: Traffic Routing and Real-Time Control * **7**: 4749–51

Gawthrop, P. J.
(University of Glasgow, Glasgow, UK)
Self-Tuning Control, Hybrid **6**: 4194–7
Self-Tuning Controllers: Input–Output Analysis **6**: 4205–9

Gaylor, J. D. S.
(University of Strathclyde, Glasgow, UK)
Artificial Kidneys **1**: 307–10
Artificial Lungs **1**: 312–5

Gentina, J. C.
(Institut Industriel du Nord, Villeneuve d'Ascq, France)
Petri Nets in Manufacturing * **6**: 3670–3

George, F. H.
(Brunel University, Uxbridge, UK)
Systems Concepts: History **7**: 4771–5

Gerard-Varet, L. A.
(EHESS, Marseilles, France)
Cooperative Game Theory **2**: 840–50

Geromel, J. C.
(University of Campinas, Campinas, Brazil)
Structure-Constrained Control: Parametric Optimization * **7**: 4678–85

Glaser, P. E.
(Arthur D. Little, Cambridge, Massachusetts, USA)
Solar Power Satellites * **7**: 4472–6

Glattfelder, A. H.
(Sulzer, Winterthur, Switzerland)
Hydroelectricity: Self-Selecting Control * **3**: 2225–8

Glavitsch, H.
(Swiss Federal Institute of Technology, Zürich, Switzerland)
Electric Generators: Load and Frequency Control **2**: 1375–80
Power System Operation **6**: 3780–4
Power Systems: Automatic Generation Control **6**: 3809–15

Glover, K.
(University of Cambridge, Cambridge, UK)
Identification: Frequency-Domain Methods **4**: 2264–70

Glowinski, R.
(University of Houston, Houston, Texas, USA)
Multigrid Methods **5**: 3135–40
Spectral Methods **7**: 4495–8

Godfrey, K. R.
(University of Warwick, Coventry, UK)
Identification: Correlation Methods **4**: 2245–53
Pharmacokinetics: Nonlinear Modelling **6**: 3695–9

Goeldel, C.
(CRAN, Nancy, France)
Stepping Motors * **7**: 4593–8

Gonzalez, R. C.
(University of Tennessee, Knoxville, Tennessee, USA)
Scene Analysis, Hierarchical * **6**: 4155–61

Goode, C. H.
(KDG Instruments, Crawley, UK)
Level Indication for Intermediate Storage **4**: 2737–43

Goodman, I. R.
(Naval Ocean Systems Center, San Diego, California, USA)
Identification of Fuzzy Sets with Random Sets **4**: 2293–301

Goodman, N.
(Sequoia Systems, Marlborough, Massachusetts, USA)
Distributed Database Systems: Concurrency Control * **2**: 1143–55

Goodwin, G. C.
(University of Newcastle, Newcastle, New South Wales, Australia)
Identification: Experiment Design **4**: 2257–64
Martingales and Semimartingales **5**: 2932–4
Stochastic Adaptive Systems Stability: Martingale Theory **7**: 4607–11

Gottlieb, G. L.
(University of Illinois, Chicago, Illinois, USA)
Biological Motor Systems: Nervous Control * **1**: 462–6

Gottwald, S.
(Karl Marx University, Leipzig, GDR)
Measures of Fuzziness: A Survey * **5**: 2963–5

Gould, F. J.
(University of Chicago, Chicago, Illinois, USA)
Path-Following Methods * **6**: 3631–4

Goursat, M.
(INRIA, Le Chesnay, France)
Optimal Stochastic Control: Numerical Methods * **5**: 3523–31

Graham, D.
(Custom Computer Consultants, Ypsilandi, Michigan, USA)
Automatic Flight Control Systems: History * **1**: 340–3

Grasse, K. A.
(University of Oklahoma, Norman, Oklahoma, USA)
Structural Stability of Control Systems **7**: 4674–7

Gray, J. O.
(University of Salford, Salford, UK)
Nonlinear Systems: Computer-Aided Design **5**: 3379–83

Gray, P.
(Southern Methodist University, Dallas, Texas, USA)
Office Automation **5**: 3443–8

Green, R. G.
(Bolton Institute of Higher Education, Bolton, UK)
Flow Measurement: Two-Phase Systems * **3**: 1685–94

Greenspan, D.
(University of Texas, Arlington, Texas, USA)
Simulation Modelling Formalism, Discrete Arithmetic-Based **6**: 4345–6

Greer, W.
(King Faisal Specialist Hospital and Research Centre, Riyadh, Saudi Arabia)
Lung Modelling **4**: 2898–901

Grimble, M. J.
(University of Strathclyde, Glasgow, UK)
Adaptive and Stochastic Control: Weighted Minimum Variance Method **1**: 38–45
Self-Tuners Design: Linear Quadratic Gaussian Theory **6**: 4172–6
Ship Positioning: Adaptive Control **6**: 4262–7

Gross, D.
(George Washington University, Washington, DC, USA)
Reliability of Systems: Statistical Design and Control * **6**: 4027–9

Grover, R.
(Agriculture Canada Research Station, Regina, Saskatchewan, Canada)
Herbicides in Agroecosystems **3**: 2143–7

Groves, M. R.
(Ortho Diagnostic Systems, Raritan, New Jersey, USA)
Coulter Counters **2**: 862–7

Grujić, Lj. T.
(University of Belgrade, Belgrade, Yugoslavia)
Interconnected Systems: Lyapunov Approach **4**: 2560–6

Guardabassi, G.
(Polytechnic of Milan and Center for Systems Theory, Milan, Italy)
Structure-Constrained Robust Controller Design * **7**: 4685–9

Guidorzi, R. P.
(University of Bologna, Bologna, Italy)
Model-Order Estimation in Ecological Modelling * **5**: 3040–4

Gupta, M. M.
(University of Saskatchewan, Saskatoon, Saskatchewan, Canada)
Adaptive Fuzzy Automation * **1**: 72
Adaptive Fuzzy Control System * **1**: 72–3
Adaptive Fuzzy Controller * **1**: 73
Approximate Reasoning (Fuzzy Reasoning) * **1**: 296
Composite Fuzzy Relational Equations: Solution Algorithms * **1**: 710
Composite Fuzzy Relations * **1**: 710
Control Rules: Competition * **2**: 812–3
Control Rules: Generation * **2**: 813
Crisp Controller * **2**: 867
Defuzzification Operators * **2**: 959
Direct Current Series Motors: Fuzzy Control * **2**: 1071–2
Direct Current Series Motors: Fuzzy Model * **2**: 1072–3
Eigen Fuzzy Sets * **2**: 1367
Expert Fuzzy Controller * **2**: 1568–9
Extension Principle * **2**: 1581
Fuzzification Operators * **3**: 1779
Fuzzy Algorithms * **3**: 1782
Fuzzy Automata * **3**: 1782
Fuzzy Automata: Identification * **3**: 1782
Fuzzy Cartesian Product * **3**: 1786
Fuzzy Conditional Inference * **3**: 1791–2
Fuzzy Conditional Inference: Compositional Rules * **3**: 1792
Fuzzy Connectives (Intersection and Union): General Class * **3**: 1792
Fuzzy Control Algorithm Completeness * **3**: 1792–3
Fuzzy Control Language * **3**: 1793
Fuzzy Control Processes: Convergence * **3**: 1793–4
Fuzzy Control Systems: Linguistic Analysis * **3**: 1794
Fuzzy Control Systems: Linguistic Synthesis * **3**: 1794–5
Fuzzy Controllers: Reproducibility Property * **3**: 1795
Fuzzy Discretization * **3**: 1800
Fuzzy Identification of Systems * **3**: 1805–6
Fuzzy Implication Operators * **3**: 1806
Fuzzy Integrals * **3**: 1810
Fuzzy Learning Systems * **3**: 1810
Fuzzy Linear Programming * **3**: 1810–1
Fuzzy Logic * **3**: 1811
Fuzzy Logic Controller * **3**: 1811–2
Fuzzy Logic Controller: Algebraic Analysis * **3**: 1812

Fuzzy Logic Controller: Good-Mapping Property * **3**: 1812
Fuzzy Logic Controller: Implementation of Parameters * **3**: 1812
Fuzzy Logic Controller: Interactivity * **3**: 1812–3
Fuzzy Logic Controller: Self-Improvement * **3**: 1813
Fuzzy Logic Controller: Self-Organization * **3**: 1813
Fuzzy Logic Controller: Sensitivity * **3**: 1813
Fuzzy Logic Controller: Stability * **3**: 1814
Fuzzy Logic Controller: Structure * **3**: 1814
Fuzzy Łukasiewicz Controller * **3**: 1815
Fuzzy Łukasiewicz Logic * **3**: 1815–6
Fuzzy Mapping on Fuzzy Sets * **3**: 1816
Fuzzy Mapping on Ordinary Sets * **3**: 1816
Fuzzy Models * **3**: 1825
Fuzzy Models: Accuracy * **3**: 1825
Fuzzy Models: Derivation * **3**: 1825–6
Fuzzy Models: Evaluation * **3**: 1826
Fuzzy Possibilistic Controller * **3**: 1838
Fuzzy Probabilistic Controller (Czogala's and Pedrycz's Controller) * **3**: 1839
Fuzzy Processes: Identification * **3**: 1843
Fuzzy Programming Languages * **3**: 1843
Fuzzy Relational Equations: Max–Min Fuzzy Relational Equations * **3**: 1852–3
Fuzzy Relations (Fuzzy Matrices) * **3**: 1853
Fuzzy Restrictions * **3**: 1853
Fuzzy Set Theory: Review **3**: 1867–72
Fuzzy Sets: α-Cut * **3**: 1872
Fuzzy Sets and Systems: Fundamentals **3**: 1872–4
Fuzzy Sets (Fuzzy Subsets) * **3**: 1874
Fuzzy Sets: Inclusion and Equality * **3**: 1875
Fuzzy Sets: Operations * **3**: 1875
Fuzzy Singletons * **3**: 1880
Fuzzy Systems: Control with Constraints * **3**: 1895
Fuzzy Systems (Discrete-Time Fuzzy Systems) * **3**: 1895
Fuzzy Systems: Simultaneous Identification and Control * **3**: 1902
Fuzzy Variables * **3**: 1905–6
Incompatibility Principle * **4**: 2343
Industrial Processes: Fuzzy Logic Controllers **4**: 2356–62
Industrial Production: Fuzzy Control Systems * **4**: 2362
Learning Fuzzy Controller * **4**: 2730–1
Linguistic Approximation * **4**: 2880
Linguistic Controller * **4**: 2880
Linguistic Hedges * **4**: 2880–1
Linguistic Models * **4**: 2881
Linguistic Variables * **4**: 2884
Look-Up Tables (Decision Tables) * **4**: 2898
Manual Process Control * **5**: 2918
Measures of Fuzziness: Conditions * **5**: 2965–6
Modus Ponens: Generalization * **5**: 3089
Modus Tollens: Generalization * **5**: 3089
Multidimensional Fuzzy Implication * **5**: 3124
Optimal Control of Fuzzy Systems * **5**: 3488–9
Optimal Fuzzy Target Control * **5**: 3515–6
Optimization in a Fuzzy Environment * **5**: 3540–1
Ordinary Mapping on Fuzzy Sets * **5**: 3563
Possibility Distribution Functions * **6**: 3769
Possibility Distributions * **6**: 3769
Prediction in Fuzzy Systems * **6**: 3866
Probabilistic Sets * **6**: 3878
Probabilistic Sets: Distribution Function * **6**: 3878
Relational Assignment Equations * **6**: 4000
Stabilization Control of Fuzzy Systems * **7**: 4529
Verbal Reports * **7**: 5047
Verbalization * **7**: 5047

Gurel, O.
(IBM, Cambridge, Massachusetts, USA)
Simulation and Graphics **6**: 4299–303

Gutenbaum, J.
(Polish Academy of Sciences, Warsaw, Poland)
Dynamic Multistage Processes: Multiobjective Control **2**: 1272–6

Guy, J. J.
(University of Salford, Salford, UK)
Gas Transmission Systems: Transient Analysis **3**: 1990–2

Haber, R.
(Technical University of Budapest, Budapest, Hungary)
Glass Industry: Adaptive Control * **3**: 2008–10

Haefner, J. W.
(Utah State University, Logan, Utah, USA)
Grammars in Ecology **3**: 2043–6

Haggard, M. P.
(University of Nottingham, Nottingham, UK)
Speech * **7**: 4504–9

Hagras, A. M.
(Cairo University, Giza, Egypt)
Complex Systems Control: Tracking Methods * **1**: 675–82

Haimes, Y. Y.
(Case Western Reserve University, Cleveland, Ohio, USA)
Multiple Objective Optimization: Pareto Optimality * **5**: 3156–65
Overlapping Coordination, Hierarchical * **5**: 3564–8
Surrogate Worth Trade-Off Method * **7**: 4703–9

Halfon, E.
(Canada Centre for Inland Waters, Burlington, Ontario, Canada)
Ecological Diversity and Information **2**: 1321–2
Ecological Modelling **2**: 1323–4
Ecological Modelling, Hierarchical **2**: 1329

Ecological Modelling: Law of the Minimum **2**: 1329–30
Ecological Modelling: Michaelis–Menten Models **2**: 1330
Ecological Modelling: Temperature Model Formulation **2**: 1333
Ecological Population Modelling: Leslie's Equations **2**: 1333
Ecological Population Modelling: Lotka–Volterra Equations **2**: 1334
Ecological Population Modelling: Volterra's Equations **2**: 1334–5
Ecosphere **2**: 1342–3
Ecosystems **2**: 1356
Food Chains **3**: 1700
Gnotobiotic Systems **3**: 2032
Maximum-Power Principle in Ecology **5**: 2956
Model-Order Estimation in Ecological Modelling * **5**: 3040–4
Nutrient Cycling **5**: 3440
Population Growth, Logistic **6**: 3755–6
Population in Ecology **6**: 3756
Primary Productivity **6**: 3877
Systems Ecology **7**: 4775

Halme, A.
(Helsinki University of Technology, Helsinki, Finland)
Fermenter Control **3**: 1615–9

Hämäläinen, R. P.
(Helsinki University of Technology, Helsinki, Finland)
Respiratory System: Optimal Control **6**: 4062–6

Hamy, M.
(ENSM, Nantes, France)
Servovalves, Electrohydraulic **6**: 4252–7

Hamza, M. H.
(University of Calgary, Calgary, Alberta, Canada)
Distributed Parameter Systems: Self-Tuning Control **2**: 1197–9

Hanafusa, H.
(Ritsumeikan University, Kyoto, Japan)
Articulated Robot Arms: Autonomous Contouring Control **1**: 296–301
Articulated Robot Arms: Electrohydraulic Servosystem Design **1**: 301–4

Handschin, E.
(University of Dortmund, Dortmund, FRG)
Power Systems: Load Forecasting **6**: 3830–4

Hang, C.-C.
(National University of Singapore, Kent Ridge, Singapore)
Model Reference Adaptive Systems Methods: Comparative Studies **5**: 3067–70

Haraguchi, H.
(University of Tokyo, Tokyo, Japan)
Inductively Coupled Plasma Emission Spectrometry **4**: 2343–7

Harashina, S.
(Tokyo Institute of Technology, Tokyo, Japan)
Environmental Systems Analysis for Policy Making **2**: 1539–42
Environmental Systems Analysis for Policy Making: A Support System **2**: 1542–5

Harmer, A. L.
(Battelle Research Center, Geneva, Switzerland)
Optical Fiber Sensors: General Aspects **5**: 3455–65
Optical Fiber Sensors: Interferometers **5**: 3465–71

Harris, C. J.
(University of Southampton, Southampton, UK)
Nonlinear Systems: Circle Criterion Extensions **5**: 3376–9

Harris, C. M.
(George Mason University, Fairfax, Virginia, USA)
Queuing Models **6**: 3922–5
Reliability of Systems: Statistical Design and Control * **6**: 4027–9

Haruyama, A.
(National Institute for Environmental Sciences, Ibaraki, Japan)
Environmental Databases: INFOTERRA **2**: 1503–6

Hasegawa, T.
(Kyoto University, Kyoto, Japan)
Environmental Data: Allocation Problems in Information Collecting Networks **2**: 1489

Hassan, M. F.
(Cairo University, Giza, Egypt)
Gaussian Random Variables **3**: 1998–9
Goal Coordination Method **3**: 2032–8
Large-Scale Systems: Multilevel Filter **4**: 2692–7
Mixed Method of Coordination **5**: 3012–9
Model Coordination Method **5**: 3035–40

Haurat, A.
(Institute de Productique, Besançon, France)
Flexible Manufacturing Systems: LMAC System for Control Language Generation **3**: 1637–42

Haurie, A.
(HEC, Montreal, Quebec, Canada)
Complex Systems Management: Game Theory **1**: 682–7

Hay, J.
(University of Salford, Salford, UK)
Simulators, Real-Time **6**: 4418

Heidepriem, J.
(University of Wuppertal, Wuppertal, FRG)
Rolling-Mill Control **6**: 4139–42

Hellman, W. D.
(Hillsboro, Oregon, USA)
Cybernetics: History **2**: 869–73

Henderson, I. S.
(Glasgow Royal Infirmary, Glasgow, UK)
Human Kidneys **3**: 2204–6

Henize, J.
(Henmark Managementsysteme, Bonn, FRG)
Simulation Modelling Formalism: Systems Dynamics **6**: 4366–8

Hennebert, C.
(RATP, Paris, France)
Automatic Train Control: Reliability and Safety Problems **1**: 360–2

Henry, R. M.
(University of Bradford, Bradford, UK)
Microprocessors in Instruments: Design Case Study **5**: 2980–4
Microprocessors in Instruments: General Aspects **5**: 2984–9

Hermes, H.
(University of Colorado, Boulder, Colorado, USA)
Bang–Bang Principle **1**: 410–3
Local Controllability **4**: 2893–8
Maximum Principle **5**: 2956–63
Optimal Control Theory: Introduction **5**: 3513
Singular Control: Higher-Order Conditions **7**: 4419–25
Transversality **7**: 4948–9

Hernández, D. B.
(Centre for Mathematical Research, Guanajuato, Mexico)
Model Structure Simplification **5**: 3079–82

Hesper, B.
(University of Utrecht, Utrecht, The Netherlands)
Simulation Modelling Formalism: Heterarchical Systems * **6**: 4350–3

Hetenyi, G. Jr
(University of Ottawa, Ottawa, Ontario, Canada)
Tracers and Metabolic Dynamics * **7**: 4901–6

Hetthéssy, J.
(Technical University of Budapest, Budapest, Hungary)
Cement Industry: Adaptive Control * **1**: 561–4

Hewit, J.
(Loughborough University of Technology, Loughborough, UK)
Robots: Coordination Control **6**: 4097–100

Heymann, M.
(Technion, Haifa, Israel)
Pole Assignment by State Feedback and Luenberger Observer **6**: 3717–20

Higashi, K.
(Government Industrial Research Institute, Osaka, Japan)
Water Pollution: Fingerprinting of Oil **7**: 5110–2

Hiiro, K.
(Government Industrial Research Institute, Osaka, Japan)
Ion-Selective Electrodes in Environmental Analysis **4**: 2604–8

Hill, R. R.
(University of Houston-Clear Lake, Houston, Texas, USA)
Simulation Modelling Formalism: Markov Chains **6**: 4355–6

Hipel, K. W.
(University of Waterloo, Waterloo, Ontario, Canada)
Fuzzy Multicriteria Modelling **3**: 1826–9
Systems Management: Conflict Analysis * **7**: 4793–9

Hirota, K.
(Hosei University, Tokyo, Japan)
Fuzzy Systems: Identification **3**: 1898–902
Subjective Entropy **7**: 4692–6

Hisdal, E.
(University of Oslo, Oslo, Norway)
Fuzzy Set Theory: Possibilities and Probabilities **3**: 1862–7

Ho, Y. C.
(Harvard University, Cambridge, Massachusetts, USA)
Stochastic Games and the Problem of Information **7**: 4621–2
Stochastic Incentive Problems **7**: 4627–31
Team Theory and Information Structure **7**: 4811–23

Hoffmann, U.
(Rhineland–Westphalia Technical University, Aachen, FRG)
Self-Tuning On–Off Control * **6**: 4209–11

Hofmeister, W.
(Siemens, Karlsruhe, FRG)
Pressure Control * **6**: 3872–5

Hogeweg, P.
(University of Utrecht, Utrecht, The Netherlands)
Simulation Modelling Formalism: Heterarchical Systems * **6**: 4350–3

Höhle, U.
(University of Wuppertal, Wuppertal, FRG)
Plausibility Measures **6**: 3707–10

Holford, N. H. G.
(University of Aukland, Aukland, New Zealand)
Drug Effect Prediction In Vivo **2**: 1244–9

Hollister, W. M.
(MIT, Cambridge, Massachusetts, USA)
Air Traffic Control Systems **1**: 196–202

Hoogendoorn, K.
(DSM Plastics, Sittard, The Netherlands)
Polymer Industry: Process Control **6**: 3739–45

Hope, G. S.
(University of Calgary, Calgary, Alberta, Canada)
Identification: Pseudo-Random Signal Method * **4**: 2307–10
Power System Component Identification: Pseudo-Random Signal Method * **6**: 3777–80

Horsfield, K.
(Cardiothoracic Institute, Midhurst, UK)
Lung Modelling: Measurement Aspects **4**: 2902–4

Huang, P. M.
(University of Saskatchewan, Saskatoon, Saskatchewan, Canada)
Abiotic Control Mechanisms in Terrestrial and Freshwater Environments **1**: 1–8
Agricultural Soils: Fate of Toxic Substances **1**: 143–8
Aluminum and the Fate of Nutrients and Toxic Substances in Terrestrial and Freshwater Environments **1**: 262–8
Limnology: Dispersion of Toxic Substances **4**: 2778–82

Hughes, F. M.
(UMIST, Manchester, UK)
Nuclear Power Station Boilers: Multivariable Control System Analysis and Design **5**: 3431–6

Hulbert, D. G.
(National Institute for Metallurgy, Randburg, South Africa)
Inverse Nyquist Array Design Method: Multivariable Control of a Milling Circuit **4**: 2601–4

Hultquist, P.
(University of Colorado, Denver, Colorado, USA)
Geophysical System Simulation **3**: 2005–6

Husson, R.
(CRAN, Nancy, France)
Motor Drives, Direct Current * **5**: 3092–7

Hutton, W. C.
(Polytechnic of Central London, London, UK)
Gait Analysis: Measurements **3**: 1910–3

Hyndman, B. W.
(Netherlands Institute for Preventive Medicine, Leiden, The Netherlands)
Cardiovascular System: Baroreflexes Model **1**: 536–42
Cardiovascular System: Hemodynamic Model **1**: 542–7

Iida, Y.
(Seikei University, Tokyo, Japan)
Gas Chromatography–Mass Spectrometry in Environmental Chemicals Analysis **3**: 1936–9

Ikeda, M.
(Kobe University, Kobe, Japan)
Overlapping Decentralized Control * **5**: 3568–72

Ikeda, S.
(University of Tokushima, Tokushima, Japan)
Environmental-Data Smoothing **2**: 1492–4

Imaeda, M.
(University of Tokushima, Tokushima, Japan)
Environmental-Data Compression **2**: 1489–92

Imbert, N.
(CERT/DERA, Toulouse, France)
Air Traffic Control Near Airports **1**: 189–96

Inamura, M.
(Saga University, Saga, Japan)
Remote Sensing: Classification of Multispectral Images * **6**: 4029–33

Ingram, D.
(Medical College of St Bartholomew's Hospital, London, UK)
Educational Computer Models in Medicine * **2**: 1361–6

Innis, G. S.
(Colorado State University, Fort Collins, Colorado, USA)
Ecological Modelling: New Perspectives **2**: 1330–3

Inoue, H.
(University of Tokyo, Tokyo, Japan)
Robots: Tactile and Force Sensing and Control **6**: 4113–7

Irisarri, G. D.
(American Electric Power, New York, USA)
Power Systems: Economic Dispatching * **6**: 3815–20

Irving, M. R.
(University of Durham, Durham, UK)
Power Systems: Integrated Software for Control * **6**: 3825–9

Irvoas, J.
(Aérospatiale, Toulouse, France)
Autopilots **1**: 380–8

Isermann, R.
(Darmstadt Technical University, Darmstadt, FRG)
Adaptive Controllers: Supervision Level * **1**: 64–72
Parameter-Adaptive Controllers: Applications **6**: 3604–11

Ishikawa, A.
(Aoyama Gakuin University, Tokyo, Japan)
Brains: Creative Functions * **1**: 508–11

Ishizuka, M.
(University of Tokyo, Tokyo, Japan)
Structural Damage Assessment: Fuzzy Approach * **7**: 4667–9

Isidori, A.
(University of Rome, Rome, Italy)
Nonlinear Decoupling **5**: 3282–8

Iwai, S.
(Kyoto University, Kyoto, Japan)
Analogical Inference * **1**: 272–6

Iwata, S.
(Kyoto University, Osaka, Japan)
Activation Analysis: Environmental Trace Metals **1**: 18–21

Izumi, T.
(National Industrial Research Institute, Kyushu, Japan)
Binary Random Signals and Random Signals **1**: 418–21

Jackson, K. M.
(Open Computer Services, Brighton, UK)
Gait Analysis and Locomotion Control **3**: 1907–10

Jacobs, O. L. R.
(University of Oxford, Oxford, UK)
Demand Analgesia: Modelling and Control of Postoperative Pain * **2**: 960–3
Optimal and Suboptimal Stochastic Control: Discrete-Time Systems **5**: 3471–5
pH Regulation **6**: 3688–91

Jaroń, J.
(Wroclaw Technical University, Wroclaw, Poland)
Adaptive Systems **1**: 91–3

Jarry-Lacombe, B.
(CNET, Issy les Moulineaux, France)
Switching Networks: Short-Term Planning * **7**: 4741–9

Jeffries, Clark
(La Ronge, Saskatchewan, Canada)
Directed Graphs: Stability Conditions **2**: 1078–80
Ecosystem Modelling: Qualitative Stability **2**: 1348–51

Johnston, J. S.
(Rosemount Ltd., Bognor Regis, UK)
Temperature Measurement: General Aspects **7**: 4848–51
Temperature Measurement: Types of Thermometer **7**: 4856–61
Temperature Measurement: Use and Installation of Equipment **7**: 4861–5

Jones, A. W.
(Florida Atlantic University, Boca Raton, Florida, USA)
Industrial Engineering and Operations Research: Simulation **4**: 2351–3

Jones, C. H.
(Royal Marsden Hospital, London, UK)
Thermal Imaging in Biology **7**: 4868–71

Jones, D. W.
(University of Bradford, Bradford, UK)
Nuclear Magnetic Resonance **5**: 3416–21

Jones, N. B.
(University of Leicester, Leicester, UK)
Muscles: Dynamic Models * **5**: 3243–8
Muscles: Nervous Control **5**: 3248–53

Jörgl, H. P.
(Technical University of Vienna, Vienna, Austria)
Dynamic-Systems Modelling: Basic Principles and Lumped-Parameter Systems * **2**: 1286–91
Dynamic-Systems Modelling: Distributed-Parameter Models and Discretization * **2**: 1292–7

Jumarie, G.
(University of Quebec, Montreal, Quebec, Canada)
Subjectivity: Human Communication **7**: 4696–8

Jurdjevic, V.
(University of Toronto, Toronto, Ontario, Canada)
Affine Controllability Systems **1**: 120–4

Jurvillier, J. C.
(INRETS, Arceuil, France)
Freeway Traffic Incidents: Automatic Detection **3**: 1734–6

Jury, E. I.
(University of Miami, Coral Gables, Florida, USA)
Sampled-Data Systems: History **6**: 4143–6

Jutila, E.
(Wärtsila, Helsinki, Finland)
Pulp and Paper Industry: Process Control **6**: 3909–17

Jutila, P.
(Helsinki University of Technology, Helsinki, Finland)
pH and Chemical Concentration Control **6**: 3680–3

Kacprzyk, J.
(Iona College, New Rochelle, New York, USA)
Stochastic Systems in Fuzzy Environments: Control **7**: 4657–61

Kailath, T.
(Stanford University, Stanford, California, USA)
Block Matrix Formulae **1**: 494–5
Constant-Parameter Systems: Fast Algorithms **2**: 781–5
Discrete-Time Processes: Innovations **2**: 1099–102
Kalman Filter Formulae: Variations **4**: 2615–8
Kalman Filtering: State-Space Models for Discrete-Time Processes **4**: 2622–8
Matrix Triangularization: Givens Transformations **5**: 2949–50
Matrix Triangularization: Householder Transformations **5**: 2950–1
Smoothing Algorithms **7**: 4463–6
Smoothing Algorithms, Two-Filter **7**: 4466–9
State-Space Modelling: Square Root Algorithms **7**: 4534–40

Kaleva, O.
(Tampere University of Technology, Tampere, Finland)
System Performance: Fuzzy Methods **7**: 4764–6

Kandel, A.
(Florida State University, Tallahassee, Florida, USA)
Fuzzy Switching Functions: Algebra **3**: 1883–8

Kania, A. A.
(Technical University of Kielce, Kielce, Poland)
Fuzzy Transformations **3**: 1902–5

Karcanias, N.
(City University, London, UK)
Linear Multivariable Systems: Zeros **4**: 2827–32
Linear Systems: Kronecker Canonical Forms and Invariants **4**: 2866–71

Kasahara, M.
(Kyoto University, Kyoto, Japan)
Air Pollution: Aerosol Particle Size **1**: 177–9

Kashiwagi, H.
(Kumamoto University, Kumamoto, Japan)
Maximum-Length Sequences **5**: 2951–6

Kaufmann, A.
(La Tronche, France)
Fuzzy Numbers: Applications **3**: 1831–5
Fuzzy Set Theory: An Introduction **3**: 1853–8

Kawata, Y.
(Kanazawa Institute of Technology, Ishikawa, Japan)
Earth–Atmosphere System: Identification of Optical Parameters * **2**: 1307–11
Ground Albedo Mapping: Invariant Embedding * **3**: 2046–53
Landsat Imaging: Removal of Atmospheric Effects * **4**: 2652–5

Kelton, W. D.
(University of Minnesota, Minneapolis, Minnesota, USA)
Stochastic Simulation: Initial Transient Techniques **7**: 4651–4

Kernévez, J. P.
(Compiègne University of Technology, Compiègne, France)
Optimal Control of Multistate Systems **5**: 3489–96

Kerre, E. E.
(State University of Ghent, Ghent, Belgium)
Electrocardiogram Diagnosis: Fuzzy Approach **2**: 1405–7

Kershenbaum, L. S.
(Imperial College of Science and Technology, London, UK)
Recursive Algorithms: Variable Forgetting Factors **6**: 3990–1

Keviczky, L.
(Hungarian Academy of Sciences, Budapest, Hungary)
Cement Industry: Adaptive Control * **1**: 561–4
Glass Industry: Adaptive Control * **3**: 2008–10
Load-Frequency Regulation: Adaptive Control * **4**: 2889–92
Self-Tuning and Extremal Control Systems **6**: 4180–4

Khalil, H.
(Michigan State University, East Lansing, Michigan, USA)
Singular Perturbations in Automatic Control **7**: 4435–9
State Feedback Control: Two-Stage Design by Slow–Fast Decomposition **7**: 4529–34

Kickert, R. N.
(Corvallis, Oregon, USA)
Agricultural Modelling: Field Applicability * **1**: 137–42
Chemical Stress in Agriculture: Modelling * **1**: 589–92

Kidd, P. T.
(UMIST, Manchester, UK)
Ship Propulsion Control **6**: 4267–71

Kijima, K.
(Tokyo Institute of Technology, Tokyo, Japan)
Goal-Oriented Systems **3**: 2038–43

Kikuchi, M.
(National Defense Medical College, Saitama, Japan)
Glucose Regulation: Adaptive Control * **3**: 2017–21

Kinerson, R.
(US Environmental Protection Agency, Washington, DC, USA)
Toxic Chemicals Assessment: Simulation Modelling * **7**: 4895–7

King, J.
(UWIST, Cardiff, UK)
Ship Control and Automation **6**: 4257–62

King-Smith, P. E.
(Ohio State University, Columbus, Ohio, USA)
Visual System: Psychophysics * **7**: 5070–8

Kirkbright, G. F.
(Late of UMIST, Manchester, UK)
Spectroscopy: Fundamentals and Applications * **7**: 4498–504

Kiszka, J. B.
(University of Saskatchewan, Saskatoon, Saskatchewan, Canada)
Adaptive Fuzzy Automation * **1**: 72
Adaptive Fuzzy Control System * **1**: 72–3
Adaptive Fuzzy Controller * **1**: 73
Approximate Reasoning (Fuzzy Reasoning) * **1**: 296
Composite Fuzzy Relational Equations: Solution Algorithms * **1**: 710
Composite Fuzzy Relations * **1**: 710
Control Rules: Competition * **2**: 812–3
Control Rules: Generation * **2**: 813
Crisp Controller * **2**: 867
Defuzzification Operators * **2**: 959
Direct Current Series Motors: Fuzzy Control * **2**: 1071–2
Direct Current Series Motors: Fuzzy Model * **2**: 1072–3
Eigen Fuzzy Sets * **2**: 1367
Expert Fuzzy Controller * **2**: 1568–9
Extension Principle * **2**: 1581
Fuzzification Operators * **3**: 1779
Fuzzy Algorithms * **3**: 1782
Fuzzy Automata * **3**: 1782
Fuzzy Automata: Identification * **3**: 1782
Fuzzy Cartesian Product * **3**: 1786
Fuzzy Conditional Inference * **3**: 1791–2
Fuzzy Conditional Inference: Compositional Rules * **3**: 1792
Fuzzy Connectives (Intersection and Union): General Class * **3**: 1792
Fuzzy Control Algorithm Completeness * **3**: 1792–3
Fuzzy Control Language * **3**: 1793
Fuzzy Control Processes: Convergence * **3**: 1793–4
Fuzzy Control Systems: Linguistic Analysis * **3**: 1794
Fuzzy Control Systems: Linguistic Synthesis * **3**: 1794–5
Fuzzy Controllers: Reproducibility Property * **3**: 1795
Fuzzy Discretization * **3**: 1800
Fuzzy Identification of Systems * **3**: 1805–6
Fuzzy Implication Operators * **3**: 1806
Fuzzy Integrals * **3**: 1810
Fuzzy Learning Systems * **3**: 1810
Fuzzy Linear Programming * **3**: 1810–1
Fuzzy Logic * **3**: 1811
Fuzzy Logic Controller * **3**: 1811–2
Fuzzy Logic Controller: Algebraic Analysis * **3**: 1812
Fuzzy Logic Controller: Good-Mapping Property * **3**: 1812
Fuzzy Logic Controller: Implementation of Parameters * **3**: 1812
Fuzzy Logic Controller: Interactivity * **3**: 1812–3
Fuzzy Logic Controller: Self-Improvement * **3**: 1813
Fuzzy Logic Controller: Self-Organization * **3**: 1813
Fuzzy Logic Controller: Sensitivity * **3**: 1813
Fuzzy Logic Controller: Stability * **3**: 1814
Fuzzy Logic Controller: Structure * **3**: 1814
Fuzzy Łukasiewicz Controller * **3**: 1815
Fuzzy Łukasiewicz Logic * **3**: 1815–6
Fuzzy Mapping on Fuzzy Sets * **3**: 1816
Fuzzy Mapping on Ordinary Sets * **3**: 1816
Fuzzy Models * **3**: 1825
Fuzzy Models: Accuracy * **3**: 1825
Fuzzy Models: Derivation * **3**: 1825–6
Fuzzy Models: Evaluation * **3**: 1826
Fuzzy Possibilistic Controller * **3**: 1838
Fuzzy Probabilistic Controller (Czogala's and Pedrycz's Controller) * **3**: 1839
Fuzzy Processes: Identification * **3**: 1843
Fuzzy Programming Languages * **3**: 1843
Fuzzy Relational Equations: Max–Min Fuzzy Relational Equations * **3**: 1852–3
Fuzzy Relations (Fuzzy Matrices) * **3**: 1853
Fuzzy Restrictions * **3**: 1853
Fuzzy Sets: α-Cut * **3**: 1872
Fuzzy Sets (Fuzzy Subsets) * **3**: 1874
Fuzzy Sets: Inclusion and Equality * **3**: 1875
Fuzzy Sets: Operations * **3**: 1875
Fuzzy Singletons * **3**: 1880
Fuzzy Systems: Control with Constraints * **3**: 1895
Fuzzy Systems (Discrete-Time Fuzzy Systems) * **3**: 1895
Fuzzy Systems: Simultaneous Identification and Control * **3**: 1902
Fuzzy Variables * **3**: 1905–6
Incompatibility Principle * **4**: 2343
Industrial Production: Fuzzy Control Systems * **4**: 2362
Learning Fuzzy Controller * **4**: 2730–1
Linguistic Approximation * **4**: 2880
Linguistic Controller * **4**: 2880
Linguistic Hedges * **4**: 2880–1
Linguistic Models * **4**: 2881
Linguistic Variables * **4**: 2884
Look-Up Tables (Decision Tables) * **4**: 2898

Manual Process Control * **5**: 2918
Measures of Fuzziness: Conditions * **5**: 2965–6
Modus Ponens: Generalization * **5**: 3089
Modus Tollens: Generalization * **5**: 3089
Multidimensional Fuzzy Implication * **5**: 3124
Optimal Control of Fuzzy Systems * **5**: 3488–9
Optimal Fuzzy Target Control * **5**: 3515–6
Optimization in a Fuzzy Environment * **5**: 3540–1
Ordinary Mapping on Fuzzy Sets * **5**: 3563
Possibility Distribution Functions * **6**: 3769
Possibility Distributions * **6**: 3769
Prediction in Fuzzy Systems * **6**: 3866
Probabilistic Sets * **6**: 3878
Probabilistic Sets: Distribution Function * **6**: 3878
Relational Assignment Equations * **6**: 4000
Stabilization Control of Fuzzy Systems * **7**: 4529
Verbal Reports * **7**: 5047
Verbalization * **7**: 5047

Kitamura, S.
(Kobe University, Kobe, Japan)
Laser Spectroscopy in Atmospheric Analysis * **4**: 2719–23
Remote Sensing: Laser Beam Analysis of Atmospheric Conditions * **6**: 4033–7

Kitney, R. I.
(Imperial College of Science and Technology, London, UK)
Thermoregulation: Modelling **7**: 4873–7
Thermoregulation: Physiological Control **7**: 4877–81

Klefenz, G.
(Hartmann and Braun, Minden, FRG)
Steam Generator Control **7**: 4577–80

Kleijnen, J. P. C.
(Brabant Catholic University, Tilburg, The Netherlands)
Model Behavior: Regression Metamodel Summarization **5**: 3024–30
Stochastic Simulation: Experiment Design **7**: 4648–51

Klein, L. R.
(University of Pennsylvania, Philadelphia, Pennsylvania, USA)
Economic Models **2**: 1337–42

Kleindorfer, P. R.
(University of Pennsylvania, Philadelphia, Pennsylvania, USA)
Discrete-Time Control **2**: 1093–9

Klement, E. P.
(Johannes Kepler University, Linz, Austria)
Fuzzy Measures: Classical Approach **3**: 1819–21

Klir, G. J.
(State University of New York, Binghamton, New York, USA)
Epistemological Hierarchy of Systems **2**: 1545–9
General Systems **3**: 1999–2000
General Systems Problem Solver **3**: 2000–3
International Federation for Systems Research: History **4**: 2570
Metamethodology of Systems **5**: 2971
Possibilistic Information **6**: 3766–8
Reconstructability Analysis **6**: 3983–90

Knight, R. B. D.
(National Physical Laboratory, Teddington, UK)
Harmonics **3**: 2076–80

Knowles, J. B.
(UK Atomic Energy Establishment, Winfrith, UK)
Direct Digital Control: System Design Procedures **2**: 1073–8

Kohout, L. J.
(Brunel University, Uxbridge, UK)
Computer Security Systems: Fuzzy Logics * **1**: 741–3
Fuzzy Implication Operators: A Survey * **3**: 1806–10
Relations, Mathematical * **6**: 4000–8

Koistinen, R.
(Helsinki University of Technology, Helsinki, Finland)
Evaporation Control **2**: 1566–8

Kojima, T.
(Kyoto University, Kyoto, Japan)
Gas Chromatography in Environmental Analysis **3**: 1931–3

Kokawa, M.
(Toyoda Gosei, Inazawa, Japan)
Many-Valued Logics: Applications **5**: 2924–7

Kolarz, G.
(Ludwig Boltzmann Institute for Rheumatology and Focal Diseases, Baden, Austria)
Computer-Assisted Medical Diagnostic Systems: Fuzzy Methods * **1**: 736–40

Koolman, M.
(Siemens, Karlsruhe, FRG)
Pressure Control * **6**: 3872–5

Korndörfer, V.
(Fraunhofer Institute, Stuttgart, FRG)
Personnel Aspects of Automation * **6**: 3659–64

Korte, B.
(University of Bonn, Bonn, FRG)
Combinatorial Optimization: Independence Systems * **1**: 624–6
Combinatorial Optimization Problems: Approximative Algorithms * **1**: 626–8
Combinatorial Optimization Problems: Branch-and-Bound Approach * **1**: 628–30

Discrete and Combinational Optimization: Introduction * **2**: 1080–3
Polyhedral Combinatorics and Cutting-Plane Methods * **6**: 3735–9

Kortela, U.
(Helsinki University of Technology, Helsinki, Finland)

Consistency, Density and Viscosity: Control * **2**: 777–81

Kouvaritakis, B.
(University of Oxford, Oxford, UK)

Characteristic-Locus and Multivariable-Root-Locus Design Methods **1**: 575–81

Krakowiak, S.
(IMAG, Saint-Martin-d'Hères, France)

Distributed Computer Systems: An Introduction **2**: 1126–32

Krzysztofowicz, R.
(University of Virginia, Charlottesville, Virginia, USA)

Risk Aversion **6**: 4070–5
Utility Measurements **7**: 4981–6
Utility Theory **7**: 4986–91
Water Resources: Systems Engineering **7**: 5135–40

Kuge, Y.
(Environmental Pollution Control Center, Osaka, Japan)

Water Pollution: Chemical Analysis **7**: 5109–10

Kusaka, T.
(Kanazawa Institute of Technology, Ishikawa, Japan)

Ground Albedo Mapping: Invariant Embedding * **3**: 2046–53
Landsat Imaging: Removal of Atmospheric Effects * **4**: 2652–5

Kushner, H. J.
(Brown University, Providence, Rhode Island, USA)

Recursive Stochastic Algorithms: Weak Convergence Theory **6**: 3994–7

Kuusisto, T.
(Valmet Process Automation, Tampere, Finland)

Fluid Flow Control **3**: 1695–700

Labarrère, M.
(ONERA-CERT, Toulouse, France)

Aircraft Sensor Failure: Detection by Analytic Redundancy **1**: 246–52
Wing Load Alleviation Modelling * **7**: 5164–70

Laborie, J. P.
(SNIAS, Toulouse, France)

Aircraft Instrumentation: Development of Cathode Ray Tube Technology * **1**: 238–46

Lachmann, K.-H.
(Diehl, Nuremberg, FRG)

Adaptive Controllers: Supervision Level * **1**: 64–72
Nonlinear Parameter-Adaptive Control Algorithms **5**: 3341–51

Ladet, P.
(LAG, Saint-Martin-d'Hères, France)

Distributed Data Processing Systems and Decentralized Control **2**: 1137–43

Lago, P. J.
(University of Porto, Porto, Portugal)

Muscles: Dynamic Models * **5**: 3243–8

Lagomasino, A.
(George Mason University, Fairfax, Virginia, USA)

Multiattribute Utility Theory in Decision Analysis * **5**: 3104–16

Lamba, S. S.
(Indian Institute of Technology, New Delhi, India)

Model Simplification: Frequency-Domain Approach **5**: 3071–7

Lamure, C.
(INRETS, Bron, France)

Fuel Conservation: Road Transport **3**: 1756–61

Landau, I. D.
(ENSIEG, Saint-Martin-d'Hères, France)

Adaptive Systems: Hyperstability and Positivity **1**: 93–7
Minimum-Phase Systems: Adaptive Controllers and Design **5**: 2998–3004
Model Reference Adaptive Control **5**: 3044–7
Parameter-Adaptive Algorithms: Feedback System Approach **6**: 3596–604

Lang, B.
(ENSMM, Besançon, France)

Optimal Control: Decomposition and Coordination using Asynchronous Iteration * **5**: 3475–81

Langlet, J.
(Thomson–CSF, Paris, France)

Aircraft Instrumentation: Development of Cathode Ray Tube Technology * **1**: 238–46

Lappus, G.
(Technical University of Munich, Munich, FRG)

Gas Transmission Systems: Supervision and Control * **3**: 1985–90

Larson, R. E.
(Systems Control Technology, Palo Alto, California, USA)
Dynamic Programming: Spatial Decomposition * **2**: 1279–82

Latour, P. R.
(Setpoint, Houston, Texas, USA)
Petrochemical Industry: Process Control **6**: 3673–80

Laub, A. J.
(University of California, Santa Barbara, California, USA)
Control System Design Computations: Numerical Aspects **2**: 813–28

Laughton, M. A.
(Queen Mary College, London, UK)
Energy Sources **2**: 1452–4
Energy Sources, Renewable **2**: 1454–8
Fossil-Fuel Energy **3**: 1714–9

Lausterer, G. K.
(Siemens, Karlsruhe, FRG)
Pressure Control * **6**: 3872–5

Le Gall, F.
(LAAS-CNRS, Toulouse, France)
Switching Networks: Modelling * **7**: 4726–32

Leach, K. G.
(Royal Military Hospital, Riyadh, Saudi Arabia)
Radioisotopic Imaging in Biology **6**: 3936–40

Lecomte, P. E.
(Aérospatiale, Paris, France)
Civil Aircraft: Long-Term Future **1**: 601–4

Lederer, P. J.
(University of Virginia, Charlottesville, Virginia, USA)
Management: Game Theory **5**: 2907–11

Lee, M. H.
(University College of Wales, Aberystwyth, UK)
Industrial Robots: Error Recovery **4**: 2444–7

Lefkowitz, I.
(Case Western Reserve University, Cleveland, Ohio, USA)
Multilayer Control: Vertical Decomposition **5**: 3140–5

Legros, J. F.
(CNET, Issy les Moulineaux, France)
Switching Networks: Short-Term Planning * **7**: 4741–9

Leigh, J. R.
(Polytechnic of Central London, London, UK)
Multivariable Systems: Functional Analysis **5**: 3201–6

Leininger, G. G.
(Standard Oil, Cleveland, Ohio, USA)
Inverse Nyquist Array Design Method: Dominance, Optimization and Sharing **4**: 2597–601

Lemaréchal, C.
(INRIA, Le Chesnay, France)
Optimization, Nondifferentiable **5**: 3541–6

Lesort, J. B.
(INRETS, Arcueil, France)
Traffic Simulation: Predictive Models **7**: 4927–30

Lettenmaier, D. P.
(University of Washington, Seattle, Washington, USA)
Hydrologic Forecasting **3**: 2228–34
Rainfall–Runoff Modelling **6**: 3940–4

Leung, Y.
(Chinese University of Hong Kong, Shatin, Hong Kong)
Conflict Resolution: Fuzzy-Ideal Concepts **2**: 773–6

Levien, R. E.
(Xerox Corporation, Stamford, Connecticut, USA)
Global and Regional Models **3**: 2015–6

Lévine, J.
(ENSMP, Fontainebleau, France)
Dynamic Duopoly Theory * **2**: 1262–9

Levine, S. H.
(Tufts University, Medford, Massachusetts, USA)
Markov Chains in Ecological Modelling **5**: 2927–30
Trophic Structures Modelling **7**: 4954–6

Libert, G. A.
(Polytechnical Faculty of Mons, Mons, Belgium)
Cluster Analysis, Nonmetric **1**: 606–8

Lin, C. S.
(Purdue University, West Lafayette, Indiana, USA)
Industrial Robots: An Optimal Control System * **4**: 2392–412

Lindenmayer, A.
(University of Utrecht, Utrecht, The Netherlands)
Lindenmayer Systems **4**: 2795–8

Linkens, D. A.
(University of Sheffield, Sheffield, UK)
Biological Measurement: Lung Function **1**: 457–9
Digestive Tract: Large Intestine Electrical Activity Modelling **2**: 1036–40
Mode Analysis: Coupled Nonlinear Oscillators **5**: 3019–24
Muscle Relaxation Control **5**: 3235–8

Lions, P. L.
(University of Paris IX, Paris, France)
Dynamic Programming: Introduction **2**: 1278–9
Hamilton–Jacobi–Bellman Equation **3**: 2063–8
Hamilton–Jacobi–Bellman Equation: Numerical Methods **3**: 2068–72
Optimality Principle **5**: 3535–8
Stochastic Hamilton–Jacobi–Bellman Equation **7**: 4622–7

Liu, D.
(Sichuan University, Chengdu, People's Republic of China)
Normal Fuzzy Measures and Fuzzy Random Measures * **5**: 3410–3

Liusosu D
(Sichuan University, Chengdu, People's Republic of China)
Fuzzy Probability * **3**: 1839–43

Ljung, L.
(Linköping University, Linköping, Sweden)
Adaptive Control: Positive-Real Condition **1**: 51–4
Adaptive Controllers: Convergence Analysis **1**: 54–60
Extended Kalman Filter **2**: 1581
Extended Least Squares Method **2**: 1581
Generalized Least Squares Method **3**: 2005
Identification: Asymptotic Theory **4**: 2235–9
Identification: Basic Problem **4**: 2239–45
Identification: Least Squares Method **4**: 2278–84
Identification: Maximum Likelihood Method **4**: 2284–7
Identification: Recursive Methods **4**: 2310–4
Identification: Time-Domain Methods **4**: 2314–20
Recursive Stochastic Algorithms: Ordinary Differential Equation Method **6**: 3991–4
Stochastic Adaptive Control Algorithms Convergence Analysis: Ordinary Differential Equation Approach **7**: 4602–7

Llibre, M.
(CERT, Toulouse, France)
Industrial Robots: Manipulator Control * **4**: 2456–60

Lobry, C.
(University of Nice, Nice, France)
Nonlinear System Theory: Geometric Methods **5**: 3362–6

Locatelli, A.
(Polytechnic of Milan and Centre for Systems Theory, Milan, Italy)
Structure-Constrained Robust Controller Design * **7**: 4685–9

Löfgren, L.
(University of Lund, Lund, Sweden)
Autology **1**: 326–33
Complexity of Systems **1**: 704–9
Computability **1**: 715–20
Decidability **2**: 932–8

Louis, J. P.
(ENS, Cachan, France)
Motor Drives, Direct Current * **5**: 3092–7

Luenberger, D. G.
(Stanford University, Stanford, California, USA)
Decomposition Methods **2**: 956–9
Linear Programming: An Introduction **4**: 2835–6
Linear Programming: Duality **4**: 2836–9
Simplex Method **6**: 4296–9

Luh, J. Y. S.
(Clemson University, Clemson, South Carolina, USA)
Industrial Robot Anatomy **4**: 2368–87
Industrial Robots: An Optimal Control System * **4**: 2392–412
Industrial Robots: Conventional Controller Design **4**: 2421–40

Lush, D. M.
(Ove Arup Partnership, London, UK)
Air Conditioning **1**: 148–50
Air Conditioning, Air Floor **1**: 150–2
Air Conditioning Control Systems **1**: 152–9
Air Conditioning, Dual Duct **1**: 159–61
Air Conditioning, Fan Coil **1**: 161–2
Air Conditioning, Induction Unit **1**: 162–5
Air Conditioning, Reversible Cycle Heat Pump **1**: 165–7
Air Conditioning Systems **1**: 167–71
Air Conditioning, Variable Air Volume **1**: 171–4
Humidity Control Systems **3**: 2206–9

Lynch, A. J.
(University of Queensland, Indooroopilly, Queensland, Australia)
Separation Process Control * **6**: 4248–52

Maccallum, N. R. L.
(University of Glasgow, Glasgow, UK)
Gas Turbines in Power Generation **3**: 1992–5

McCormack, W. M.
(Virginia Polytechnic Institute and State University, Blacksburg, Virginia, USA)
Simulation: Time-Advance Methods **6**: 4414–5

McEntire, P. L.
(Systems Control Technology, Palo Alto, California, USA)
Dynamic Programming: Spatial Decomposition * **2**: 1279–82

McIntyre, D. A.
(Electricity Council Research Centre, Capenhurst, UK)
Clothing Insulation **1**: 605–6
Comfort Equations **1**: 639–42
Comfort Indices **1**: 642–3
Comfort Indices: Standard Effective Temperature **1**: 644–5
Globe Thermometers **3**: 2016–7
Metabolic Rate **5**: 2970
Net Radiometers **5**: 3260–1
Thermal Comfort **7**: 4865–6
Thermal Environment **7**: 4866–8
Thermal Sensation **7**: 4871–2
Thermoregulation: General Aspects **7**: 4872–3

Mackay, D.
(University of Toronto, Toronto, Ontario, Canada)
Toxic Substances Modelling: Fugacity Concepts **7**: 4897–901

McKee, D. J.
(University of Queensland, Indooroopilly, Queensland, Australia)
Separation Process Control * **6**: 4248–52

McRuer, D.
(Systems Technology, Hawthorne, California, USA)
Automatic Flight Control Systems: History * **1**: 340–3

Mahmoud, M. S.
(University of Kuwait, Kuwait)
Costate Coordination Method **2**: 859–62
Singular Perturbations: Discrete Version **7**: 4429–34

Malanchuk, J. L.
(State University of New York, Plattsburgh, New York, USA)
Ecological Effects of Toxic Substances **2**: 1322–3
Nitrification Modelling in Lotic Systems **5**: 3264–6
Nitrogen Cycle Modelling: Nominal and Perturbed Dynamics in Watersheds **5**: 3266–8

Malik, O. P.
(University of Calgary, Calgary, Alberta, Canada)
Identification: Pseudo-Random Signal Method * **4**: 2307–10
Power System Component Identification: Pseudo-Random Signal Method * **6**: 3777–80

Malinowski, K.
(Warsaw University of Technology, Warsaw, Poland)
Steady-State Systems: On-Line Coordination **7**: 4566–71

Manczak, K.
(Polish Academy of Sciences, Warsaw, Poland)
Glass Industry: Process Control * **3**: 2011–4

Maral, G.
(ENST, Toulouse, France)
Digital Communications: Fundamentals **2**: 1043–51
Digital Communications: Satellite Systems * **2**: 1051–7

Marchant, G. R.
(Noranda Research Center, Pointe-Claire, Quebec, Canada)
Nonferrous Metallurgical Industry: Process Control * **5**: 3280–2

Marcus, S. I.
(University of Texas, Austin, Texas, USA)
Nonlinear Estimation **5**: 3293–303
Stochastic Integrals and Stochastic Calculus **7**: 4631–7

Marec, J.-P.
(ONERA, Châtillon, France)
Space-Vehicle Trajectories: Optimization **7**: 4481–90

Margolis, D. L.
(University of California, Davis, California, USA)
Simulation Modelling Formalism: Bond Graphs **6**: 4337–41

Mariano, M. J.
(State University of New York, Binghamton, New York, USA)
Computational Complexity **1**: 720–6

Mark, R. G.
(MIT, Cambridge, Massachusetts, USA)
Biological Measurement: Electrical Characteristics of the Heart **1**: 450–6

Marmarelis, V. Z.
(University of Southern California, Los Angeles, California, USA)
Biomedical Systems Modelling: Experiment Design * **1**: 486–90
Visual System: Nonlinear Modelling **7**: 5065–70

Marmorat, J. P.
(ENSMP, Valbonne, France)
Approximate Realization Theory: Comparison of Optimal Methods **1**: 294–6
Balanced Realization of a Linear Stationary Dynamical System **1**: 409–10
Realization Theory: Gauss–Markov Processes **6**: 3962–4

Marshall, J. E.
(University of Bath, Bath, UK)
Smith Predictors **7**: 4457–63

Mason, J.
(Pinderfields Hospital, Wakefield, UK)
Intracranial Pressure Regulation by Feedback **4**: 2580–3

Mates, R. E.
(State University of New York, Buffalo, New York, USA)
Coronary Blood Flow Modelling **2**: 850–3

Mathewson, S.
(Imperial College of Science and Technology, London, UK)
Interactive Simulation Model and Program Generation **4**: 2547–51

Matsuoka, Y.
(Kyoto University, Kyoto, Japan)
Environmental Measurement: Water Quality Monitoring **2**: 1532–6

Matthies, M.
(Association for Radiation and Environmental Research, Neuherberg, FRG)
Carbon-Cycle Modelling * **1**: 529–32
Radioecological Modelling * **6**: 3929–33
Radioecological Modelling in Agricultural Food Chains * **6**: 3933–6

Maxwell, M. S.
(Goddard Space Flight Center, Greenbelt, Maryland, USA)
Remote Sensing: Satellites * **6**: 4037–41

Mayne, D. Q.
(Imperial College of Science and Technology, London, UK)
Multivariable Systems: Design by Semi-Infinite Optimization * **5**: 3189–97

Medlock, R. S.
(Brown Boveri Kent, Luton, UK)
Flow Measurement: Flow and Volumetric Rate **3**: 1670–7
Flow Measurement: Mass Rate **3**: 1677–81
Flow Measurement, Quantity **3**: 1681–5

Mees, A. I.
(University of Western Australia, Nedlands, Western Australia, Australia)
Describing-Function Reliability **2**: 978–81

Meizel, D.
(Institut Industriel du Nord, Villeneuve d'Ascq, France)
PID Controllers * **6**: 3699–701

Merlin, A.
(Electricité de France, Clamart, France)
Hydroelectricity: Production Planning * **3**: 2221–5
Power System Planning * **6**: 3784–90

Meyer, C. R.
(La Chaux de Fonds, Switzerland)
Linear Continuous Model Manipulation Software **4**: 2798–800

Mezencev, R.
(University of Nantes, Nantes, France)
Hybrid Analog–Digital Computers **3**: 2214–7

Micháilesco, G.
(CNRS-ESE, Gif sur Yvette, France)
Aggregation and Suboptimal Control * **1**: 126–33

Michel, A. N.
(Iowa State University, Ames, Iowa, USA)
Interconnected Dynamical Systems: Stability **4**: 2551–6

Mieno, H.
(Rutgers University, Newark, New Jersey USA)
Brains: Creative Functions * **1**: 508–11

Mignot, F.
(University of Paris VI, Paris, France)
Eigenfunctions and Eigenvalues: Control **2**: 1367–72

Milanese, M.
(Turin Polytechnic, Turin, Italy)
Hepatobiliary Kinetics Modelling * **3**: 2139–43

Miller, D.
(Johnson Controls, Milwaukee, Wisconsin, USA)
Comfort Control: Modelling **1**: 638–9
Heating, Ventilation and Cooling Systems: Actuator Models **3**: 2097–9
Heating, Ventilation and Cooling Systems: Control Process Interactions **3**: 2105–8
Heating, Ventilation and Cooling Systems: Controller Tuning **3**: 2108–13
Pump Characteristics: Control Applications **6**: 3917–8

Millington, P. F.
(University of Strathclyde, Glasgow, UK)
Electron Microprobe Analysis * **2**: 1416–20
Scanning Electron Microscopy in Biology * **6**: 4149–52
Transmission Electron Microscopy in Biology * **7**: 4936–40

Million, G. S.
(Thames Water Authority, Reading, UK)
Water Supply Systems: Demand Forecasting **7**: 5154–5

Minch, E.
(State University of New York, Binghamton, New York, USA)
Design of Systems * **2**: 987–92

Mindt, W.
(Hoffmann–La Roche, Basel, Switzerland)
Biological Measurement: Blood Gas Analysis **1**: 437–40

Mitrani, I.
(AT & T Bell Laboratories, Murray Hill, New Jersey, USA)
Performance Modelling Methodology in Computing **6**: 3655–9

Mitson, R. B.
(Ministry of Agriculture, Fisheries and Food, Lowestoft, UK)
Telemetry, Underwater **7**: 4842–8

Mitter, S. K.
(MIT, Cambridge, Massachusetts, USA)
Nonlinear Filtering and Quantum Physics **5**: 3319–30

Miyamoto, S.
(University of Tsukuba, Ibaraki, Japan)
Spline Interpolation in Air Pollution Pattern Analysis **7**: 4517–8

Mizoguchi, T.
(National Institute for Environmental Studies, Ibaraki, Japan)
Air Pollution: Automatic Analyzers **1**: 181–3
Air Pollution: Monitoring Systems **1**: 184–5
Gas Chromatography–Mass Spectrometry in Environmental Analysis * **3**: 1933–6

Mizuike, A.
(Nagoya University, Nagoya, Japan)
Water Pollution: Trace Element Determination **7**: 5123–7

Mizumoto, M.
(Osaka Electro-Communication University, Osaka, Japan)
Fuzzy Reasoning Methods **3**: 1847–52

Molino, G.
(Turin University, Turin, Italy)
Hepatobiliary Kinetics Modelling * **3**: 2139–43

Monaco, S.
(University of Rome, Rome, Italy)
Input–Output Maps of Nonlinear Discrete-Time Systems * **4**: 2523–7

Moore, B. C. J.
(University of Cambridge, Cambridge, UK)
Auditory Perception **1**: 323–6

Moore, M. J.
(Central Electricity Research Laboratories, Leatherhead, UK)
Steam Thermal Units **7**: 4580–5

Moore, P. R.
(Loughborough University of Technology, Loughborough, UK)
Industrial Robots: Modular Systems * **4**: 2460–4

Moores, B. M.
(Christie Hospital, Manchester, UK)
X-Ray Imaging in Biology **7**: 5179–87

Morin, J. M.
(INRETS, Arceuil, France)
Traffic Corridor Control: General Principles **7**: 4910–3
Traffic Corridor Control: Ramp Assignment **7**: 4913–5
Traffic Corridor Control: Ramp Metering **7**: 4915–7

Morris, A. J.
(University of Newcastle-upon-Tyne, Newcastle-upon-Tyne, UK)
Distillation Columns: Adaptive Control * **2**: 1109–16
Multivariable and Multirate Self-Tuning Controllers * **5**: 3184–8

Mossino, J.
(University of Paris XI, Orsay, France)
Duality Methods **2**: 1257–62

Moulin, H.
(University of Paris IX, Paris, France)
Games, Noncooperative **3**: 1917–22
Pareto Optimum **6**: 3621–2
Social Choice Theory **7**: 4469–72

Moulin, T.
(ENSTA, Paris, France)
Simulation Modelling Formalism: Arithmetic Relators **6**: 4334–7

Moyes, R. B.
(Hull University, Kingston-upon-Hull, UK)
Chromatography **1**: 592–4

Moylan, P. J.
(University of Newcastle, Newcastle, New South Wales, Australia)
Stability: Dissipativeness Concept **7**: 4519–22

Mukaidono, M.
(Meiji University, Kawasaki-shi, Japan)
Fuzzy Switching Functions **3**: 1880–3

Munemori, M.
(University of Osaka Prefecture, Sakai, Japan)
Air Pollution: Chemiluminescence Monitoring Methods **1**: 183–4
Water Pollution: Organic Nitrogen Compound Determination **7**: 5122–3

Munro, N.
(UMIST, Manchester, UK)
Computer-Aided Design: Multivariable Systems **1**: 732–6
Inverse Nyquist Array Design Method: Applications **4**: 2588–97
Pole Assignment: A Review of Methods **6**: 3710–7

Murakami, Y.
(Osaka University, Osaka, Japan)
Laser Spectroscopy in Atmospheric Analysis * **4**: 2719–23
Remote Sensing: Laser Beam Analysis of Atmospheric Conditions * **6**: 4033–7

Murano, K.
(National Institute for Environmental Studies, Ibaraki, Japan)
Air Pollution: Aerosol Chemical Analysis **1**: 174–7

Murat, F.
(Pierre and Marie Curie University, Paris, France)
Control in Coefficients **2**: 808–12

Murata, T.
(University of Illinois, Chicago, Illinois, USA)
Petri Nets **6**: 3665–70

Murray-Smith, D. J.
(University of Glasgow, Glasgow, UK)
Muscle-Spindle Modelling and Identification * **5**: 3238–42

Mussi, P.
(INRIA, Sophia-Antipolis, France)
Homotopy **3**: 2168–9
Polyhedra **6**: 3734
Stationary Points **7**: 4555–6

Nagata, T.
(Kyushu University, Fukuoka, Japan)
Position Sensing in Three Dimensions: Laser Tracking **6**: 3762–6

Najim, K.
(ENSIGC, Toulouse, France)
Drying and Dewatering Control **2**: 1249–53

Nakamura, K.
(University of Virginia, Charlottesville, Virginia, USA)
Analogical Inference * **1**: 272–6

Nance, R. E.
(Virginia Polytechnic Institute and State University, Blacksburg, Virginia, USA)
Simulation Model Management Objectives and Requirements * **6**: 4328–33

Narendra, K. S.
(Yale University, New Haven, Connecticut, USA)
Adaptive Systems: Model Reference Approach **1**: 100–6
Large Stochastic Systems: Learning Automata **4**: 2714–9
Model Reference Adaptive Control Theory **5**: 3056–60
Model Reference Adaptive System Direct and Indirect Designs **5**: 3060–1
Model Reference Adaptive Systems: Augmented Error Signals **5**: 3061–5
Model Reference Adaptive Systems: Error Models **5**: 3065–7
Radio Telescopes: Adaptive Control **6**: 3927–9
Switching Networks: Learning Algorithms for Control Policy **7**: 4716–21

Naylor, A. W.
(University of Michigan, Ann Arbor, Michigan, USA)
Generalized Dependence **3**: 2003–5

Negoita, C. V.
(Hunter College, New York, USA)
Optimization: Fuzzy Set Theory **5**: 3538–40

Negre, Y.
(Aérospatiale, Toulouse, France)
Aircraft Automatic Landing in Bad Weather Conditions **1**: 207–13
Wing Load Alleviation Modelling * **7**: 5164–70

Neuhold, E. J.
(Technical University of Vienna, Vienna, Austria)
Database Management Systems: Introduction **2**: 884–93

Neuman, C. P.
(Carnegie Mellon University, Pittsburgh, Pennsylvania, USA)
Model Reference Adaptive Control: Microcomputer Implementation **5**: 3047–56

Newcomb, R. W.
(University of Maryland, College Park, Maryland, USA)
Linear Passive Network Synthesis **4**: 2832–5

Newton, O. L.
(Directorate of Design Assessment, Washington, DC, USA)
Simulation Models: Documentation **6**: 4372–4

Nguyen, H. T.
(New Mexico State University, Las Cruces, New Mexico, USA)
Linguistic Probabilities **4**: 2881–4
Possibility Measures **6**: 3769–72

Niemi, A. J.
(Helsinki University of Technology, Helsinki, Finland)
Mineral Processing and Cement Industries: Process Control * **5**: 2989–94

Nihtilä, R.
(Finnish Sugar, Espoo, Finland)
Sugar Industry: Process Control **7**: 4699–703

Nikiforuk, P. N.
(University of Saskatchewan, Saskatoon, Saskatchewan, Canada)

Normand-Cyrot, D.
(CNRS-ESE, Gif sur Yvette, France)

Nonlinear Systems: Approximation by Simple State-Space Models * **5**: 3374–6
Power Plants: Nonlinear Modelling * **6**: 3772–7

Nosadini, R.
(University of Padua, Padua, Italy)
Ketone Body Metabolism Modelling * **4**: 2628–31

Nour Eldin, H. A.
(University of Wuppertal, Wuppertal, FRG)
Linear Multivariable Systems Controllability and Observability: Numerical Aspects **4**: 2816–27

Nüesch, D. R.
(University of Zürich, Zürich, Switzerland)
Remote Sensing: Satellites * **6**: 4037–41

Ocone, D. L.
(Rutgers University, New Brunswick, New Jersey, USA)
Nonlinear Filtering: Iterated Ito Integral Expansions **5**: 3330–6

Oda, M.
(Chubu University, Kasugai, Japan)
Education Concepts: Fuzzy Aspects **2**: 1358–61

Odum, E. C.
(Santa Fe Community College, Gainesville, Florida, USA)
Energy Systems in Ecology * **2**: 1458–62

Odum, H. T.
(University of Florida, Gainesville, Florida, USA)
Energy Systems in Ecology * **2**: 1458–62

Ogawa, Y.
(National Institute for Environmental Studies, Ibaraki, Japan)
Air Pollution: Wind Tunnel Simulation * **1**: 188–9

Oguntade, O. O.
(University of Lagos, Abeokuta, Nigeria)
Pragmatic Fuzzy Systems **6**: 3855–7

Ohsato, A.
(Yokohama National University, Yokohama, Japan)
Composite Fuzzy Relations and Relational Equations **1**: 710–3

Oka, S.
(Shimadzu Corporation, Kyoto, Japan)
Air Pollution: Automatic Analysis of Hydrocarbons **1**: 179–81
Water Pollution Indices: Biochemical Oxygen Demand **7**: 5113–4
Water Pollution Indices: Chemical Oxygen Demand **7**: 5114
Water Pollution Indices: Dissolved Oxygen **7**: 5114–5
Water Pollution Indices: Total Organic Carbon **7**: 5115–6
Water Pollution Indices: Total Oxygen Demand **7**: 5116–7

Okada, N.
(Tottori University, Tottori, Japan)
Multivariate Analysis **5**: 3232–5

Okazaki, S.
(University of Tokushima, Tokushima, Japan)
Environmental Databases * **2**: 1500–3

Olkku, J.
(Technical Research Center of Finland, Espoo, Finland)
Food Industry: Process Control **3**: 1700–6

Omasa, K.
(National Institute for Environmental Studies, Ibaraki, Japan)
Environmental Measurement: Image Instrumentation for Evaluating Pollution Effects on Plants * **2**: 1516–22

Omatu, S.
(University of Tokushima, Tokushima, Japan)
Environmental Data Statistical Analysis: Physical Modelling Approach * **2**: 1497

O'Neill, R. V.
(Oak Ridge National Laboratory, Oak Ridge, Tennessee, USA)
Ecosystem Compartment Modelling **2**: 1343–5
Gaussian Random Variables * **3**: 1998–9
Stochastic Modelling in Ecology **7**: 4644–6

Opoitsev, V. I.
(Institute of Control Sciences, Moscow, USSR)
Structural Stability **7**: 4669–74

O'Reilly, J.
(University of Liverpool, Liverpool, UK)
State-Space Techniques: History **7**: 4551–4

Ören, T. I.
(University of Ottawa, Ottawa, Ontario, Canada)
Model Behavior: Type Taxonomy, Generation and Processing Techniques **5**: 3030–5
Simulation and Model Oriented Languages: Taxonomy **6**: 4303–6
Simulation Methodology: Top-Down Approach **6**: 4319–23
Simulation Models Symbolic Processing: Taxonomy **6**: 4377–81
Simulation Models: Taxonomy **6**: 4381–8
Simulation Taxonomy **6**: 4411–4

Otsuka, Y.
(Kyoto University of Education, Kyoto, Japan)
Environmental Measurement: Distribution of Chemical Elements in the Ocean * **2**: 1513–6

Otterman, J.
(Tel Aviv University, Ramat Aviv, Israel)
Remote Sensing: Satellites * **6**: 4037–41

Owens, D. H.
(University of Sheffield, Sheffield, UK)
Multivariable Systems: Dyadic Expansions and Multivariable Feedback Design **5**: 3197–201

Page, C. J.
(Coventry (Lanchester) Polytechnic, Coventry, UK)
Industrial Vision Systems, Three-Dimensional **4**: 2484–8

Papadimitriou, C. H.
(MIT, Cambridge, Massachusetts, USA)
Database Concurrency Control Theory **2**: 880–4

Papageorgiou, M.
(Technical University of Munich, Munich, FRG)
Freeway Traffic: Modelling, Surveillance and Multilayer Control * **3**: 1736–41

Pardoux, E.
(University of Provence, Marseille, France)
Stochastic Distributed Parameter Systems: Linear Estimation and Control **7**: 4617–21

Park, R. A.
(Rensselaer Polytechnic Institute, Troy, New York, USA)
Fate Modelling of Organic Chemicals: Case Study **3**: 1600–1

Parks, P. C.
(Royal Military College of Science, Shrivenham, UK)
Nonlinear Systems: Lyapunov Method **5**: 3394–8

Parnell, L.
(British Aerospace, Filton, UK)
Aircraft Engines: Survey of Systems **1**: 236–8

Patel, R. V.
(Concordia University, Montreal, Quebec, Canada)
Multivariable Systems: Minimal Design Problem **5**: 3215–20

Paterson, I. W. F.
(Whessoe Systems and Controls, Darlington, UK)
Level Measurement for Bulk Liquid Storage **4**: 2743–7

Patten, B. C.
(University of Georgia, Athens, Georgia, USA)
Environ Theory and Analysis **2**: 1479–86

Pedrycz, W.
(Silesian Technical University, Gliwice, Poland)
Fuzzy Systems: Analysis and Synthesis **3**: 1888–95

Pehrson, E.
(Swedish Institute of Computer Science, Stockholm, Sweden)
Concrete Mixing: Cautious Adaptive Control **2**: 771–3

Pélegrin, M.
(CERT, Toulouse, France)
Active Control Technology in Aircraft **1**: 21
Aerospace Control **1**: 106
Aerospace Systems and Control **1**: 112–20
Aircraft Artificial Control Load (Artificial Feel Spring) **1**: 207
Aircraft Automatic Landing Systems **1**: 213
Aircraft State Equations: Singular Perturbations **1**: 252
Angle of Attack **1**: 290
Attitude Control **1**: 319
Autopilot Modes **1**: 380
Control-Configured Vehicles **2**: 801
Data Links in Aerospace **2**: 879
Flexibility of Aerospace Vehicles **3**: 1632–3
Flight Envelope **3**: 1669
Flight Management Systems **3**: 1669
Fly-by-Wire Systems **3**: 1700
Four-Dimensional Navigation **3**: 1719
Full-Authority Control in Aerospace Systems **3**: 1769
Homing of Aerospace Vehicles **3**: 2164
Inertial Systems **4**: 2488
Simulation of Aerospace Systems **6**: 4393
Visual Flying Rules and Instrument Flying Rules **7**: 5060
Wing Load Alleviation **7**: 5164

Perkins, W. R.
(University of Illinois, Urbana, Illinois, USA)
Aggregation, Chained **1**: 133–7

Perrin, J. P.
(RATP, Paris, France)
Automatic Train Control: Line Supervision and Control **1**: 347–52
Automatic Train Control: Organization in Subways **1**: 352–60
Automatic Train Control: Train Driving **1**: 362–5
Automatic Train Control: Train Protection **1**: 365–7

Pethig, R.
(University College of North Wales, Bangor, UK)
Biological Measurement: Dielectric Properties of Soft Tissue **1**: 445–50

Petit, G.
(Aérospatiale, Marignane, France)
Helicopters: Long-Term Future * **3**: 2133–7

Petry, F. E.
(Tulane University, New Orleans, Louisiana, USA)
Fuzzy Databases * **3**: 1795–7

Peyrade, M.
(CNET, Issy les Moulineaux, France)
Switching Networks: General Structures **7**: 4710–6

Pham, D. T.
(University of Birmingham, Birmingham, UK)
Robots: Mechanical Structure **6**: 4100–8

Phillipson, P. H.
(University of Leicester, Leicester, UK)
Adaptive Systems: Lyapunov Synthesis **1**: 97–100

Pichler, F.
(State University of New York, Binghamton, New York, USA)
Dynamic Systems: A Survey **2**: 1282–6

Pinter, R. B.
(University of Washington, Seattle, Washington, USA)
Visual System Neural Networks: Feedback and Feedforward Lateral Inhibition **7**: 5060–5

Pironneau, O.
(INRIA, Le Chesnay, France)
Distributed Elliptic Systems: Optimal Design **2**: 1174–80

Pittarelli, M. A.
(State University of New York College of Technology, Utica, New York, USA)
Antisymmetric Relations **1**: 294
Behavior Systems **1**: 414
Data Systems **2**: 880
Directed Systems **2**: 1080
Emergent Properties **2**: 1450
Equivalence Relations **2**: 1549
Explanation **2**: 1578
Information Transmission **4**: 2509
Isomorphic Systems **4**: 2609
Law of Requisite Variety **4**: 2730
Linear Orderings **4**: 2832
Maximum-Entropy Principle **5**: 2951
Metasystems **5**: 2971
Monte Carlo Methods **5**: 3089
Neutral Systems **5**: 3264
Partial Orderings **6**: 3629
Partition of a Set **6**: 3629
Prediction **6**: 3865
Projections **6**: 3904
Quasiorderings **6**: 3922
Relational Join **6**: 4000
Retrodiction **6**: 4066
Semiotics **6**: 4223
Set-Covering Problems **6**: 4257
Set Covers **6**: 4257
Shannon Entropy **6**: 4257
Similarity **6**: 4294–6
Simplicial Complexes **6**: 4299
Source Systems **7**: 4476
State-Transition Systems **7**: 4554
Structure Systems **7**: 4692
Symmetric Relations **7**: 4751
Teleology **7**: 4848
Total Relations **7**: 4895
Transitivity of Relations **7**: 4935
U-Uncertainty **7**: 4959

Plateau, B.
(University of Paris XI, Orsay, France)
Distributed Database Systems: Performance Analysis of Consistency Preserving Algorithms * **2**: 1165–9

Poisson-Quinton, P.
(ONERA, Chatillon, France)
Active Control Technology in Aircraft: General Aspects **1**: 21–31

Polak, E.
(University of California, Berkeley, California, USA)
Multivariable Systems: Design by Semi-Infinite Optimization * **5**: 3189–97

Pollak, V.
(University of Saskatchewan, Saskatoon, Saskatchewan, Canada)
Biofeedback **1**: 435–7

Porat, B.
(Technion, Haifa, Israel)
Parameter Estimation: Basic Problem **6**: 3611–6
Stationary Time Series and their Spectra **7**: 4556–66

Porter, A. L.
(Georgia Institute of Technology, Atlanta, Georgia, USA)
Technological Forecasting * **7**: 4823–8

Postlethwaite, I.
(University of Oxford, Oxford, UK)
Multivariable Systems: Robustness in Design **5**: 3224–8

Powell, J. D.
(Stanford University, Stanford, California, USA)
Digital Control: Algorithm Design **2**: 1057–66
Digital Control: Practical Design Considerations **2**: 1066–71

Prade, H.
(Paul Sabatier University, Toulouse, France)
Fuzzy Algebra * **3**: 1780–1
Fuzzy Events * **3**: 1800–2
Fuzzy Measures: Fuzzy Integral Approach * **3**: 1821–3

Fuzzy Numbers * **3**: 1829–31
Fuzzy Ranking * **3**: 1845–7

Prevot, P.
(INSA, Villeurbanne, France)
Data Analysis and Interaction Evaluation **2**: 875–9

Prime, H. A.
(University of Birmingham, Birmingham, UK)
Frequency-Response Techniques: History **3**: 1745–9

Pritchard, R.
(University of Salford, Salford, UK)
Domestic Gas Equipment Control and Safety Systems **2**: 1212–4
Fuel-Gas Interchangeability **3**: 1766–9
Gas, Natural **3**: 1966–9
Industrial Gas Equipment Control and Safety Systems **4**: 2353–6

Probert, R.
(University of Ottawa, Ottawa, Ontario, Canada)
Large-Scale Software System Reliability **4**: 2678–81

Pronk, C.
(Lips, Drunen, The Netherlands)
Navigation Control **5**: 3256–60

Proth, J. M.
(INRIA, Vandoeuvre, France)
Flexible Manufacturing Systems **3**: 1633–7

Quadrat, J. P.
(INRIA, Le Chesnay, France)
Optimal Stochastic Control: Numerical Methods * **5**: 3523–31

Rada, R.
(National Library of Medicine, Bethseda, Maryland, USA)
Artificial Intelligence **1**: 304–7

Radu, L.
(American Electric Power, New York, USA)
Power Systems: State Estimation * **6**: 3846–50

Radziuk, J.
(Royal Victoria Hospital, Montreal, Quebec, Canada)
Hormonal Regulation **3**: 2169–75
Tracers and Metabolic Dynamics * **7**: 4901–6

Rake, H.
(Rhineland-Westphalia Technical University, Aachen, FRG)
Heat Exchanger Control **3**: 2089–92
Identification: Transient- and Frequency-Response Methods **4**: 2320–5
Self-Tuning On–Off Control * **6**: 4209–11

Ralescu, D.
(University of Cincinnati, Cincinnati, Ohio, USA)
Parameter Estimation: Fuzzy Logic **6**: 3616–9

Ramakrishna, A.
(Carnegie Mellon University, Pittsburgh, Pennsylvania, USA)
Large Linear Systems: Algebraic Approach to Decentralized Control **4**: 2666–71

Rambach, D.
(SNECMA, Moissy Cramayel, France)
Aircraft Engine Control: General Aspects **1**: 220–6

Randall, J. E.
(University of Mississippi, Jackson, Mississippi, USA)
Biological Systems Simulation * **1**: 466–8

Rapp, P. E.
(Medical College of Pennsylvania, Philadelphia, Pennsylvania, USA)
Biochemical Oscillators: Dynamic Behavior **1**: 424–9
Biochemical Systems Modelling: Switching Behavior **1**: 429–34
Chaotic Behavior in Physiological and Chemical Systems **1**: 567–70

Reckhow, K. H.
(Duke University, Durham, North Carolina, USA)
Error Analysis, First-Order **2**: 1552–3
Lake Ecosystem Modelling: Input–Output Models **4**: 2646–50
Trophic State Criteria **7**: 4952–3
Validation of Simulation Models: Philosophy and Statistical Methods of Confirmation **7**: 5011–5

Redon, S.
(Ecole Nationale de l'Aviation Civile, Toulouse, France)
Air Traffic Simulators: The CAUTRA Subsystem * **1**: 202–7

Reinisch, K.
(Ilmenau Technical University, Ilmenau, GDR)
Dynamic Systems: On-Line Coordination **2**: 1297–305

Rembold, U.
(University of Karlsruhe, Karlsruhe, FRG)
Industrial Assembly Robots: Programming Languages and Systems * **4**: 2347–51

Reutenauer, C.
(University of Paris VI, Paris, France)
Noncommuting Variables: Finite Automata **5**: 3268–72

Noncommuting Variables: Rational Generating Series **5**: 3272–6
Noncommuting Variables: Realization of Bilinear Systems **5**: 3276–9

Richard, J. P.
(Institut Industriel du Nord, Villeneuve d'Ascq, France)
State-Space Modelling: State-Space Transformation * **7**: 4540–7

Richetin, M.
(University of Clermont II, Aubière, France)
Structural Analysis: Graph Representation **7**: 4661–7

Rigas, A.
(University of Glasgow, Glasgow, UK)
Muscle-Spindle Modelling and Identification * **5**: 3238–42

Rine, D. C.
(Western Illinois University, Macomb, Illinois, USA)
Computer Security Systems: Possibility Theory **1**: 743–6

Rinks, D. B.
(Louisiana State University, Baton Rouge, Louisiana, USA)
Aggregate Planning Using Linguistic Variables **1**: 124–6

Robert, P. H.
(INRIA, Le Chesnay, France)
Closed Sets **1**: 605
Compact Sets **1**: 665
Semicontinuous Functions **6**: 4216

Roberts, P. D.
(City University, London, UK)
Combined System Optimization and Parameter Estimation **1**: 634–8

Rocha, A. F. da
(University of Campinas, Campinas, Brazil)
Brain Activity During Language Perception **1**: 502–5
Brains as Fuzzy Systems **1**: 505–8
Neural Fuzzy Learning Strategies **5**: 3261–4

Rocipon, F.
(Ecole Nationale de l'Aviation Civile, Toulouse, France)
Air Traffic Simulators: The CAUTRA Subsystem * **1**: 202–7

Rontani, B.
(Aérospatiale, Marignane, France)
Helicopters: Long-Term Future * **3**: 2133–7

Rose, K. A.
(University of Washington, Seattle, Washington, USA)
Sensitivity Analysis in Ecological Simulation Models **6**: 4230–5

Rosen, R.
(Dalhousie University, Halifax, Nova Scotia, Canada)
Anticipatory Systems **1**: 290–4

Rosenberg, J. R.
(University of Glasgow, Glasgow, UK)
Muscle-Spindle Modelling and Identification * **5**: 3238–42

Rosenbrock, H. H.
(UMIST, Manchester, UK)
Composite Systems **1**: 713–4
Inverse Nyquist Array Design Method **4**: 2584–8
Pole Assignment: Polynomial Matrices **6**: 3720–1
Polynomial Matrices in Systems Theory **6**: 3745–51

Rossini, F. A.
(Georgia Institute of Technology, Atlanta, Georgia, USA)
Technological Forecasting * **7**: 4823–8

Rouchaleau, Y.
(ENSMP, Valbonne, France)
Abstract Realization Theory **1**: 14–8
Partial Minimal Realization **6**: 3627–9
Realization Theory: Hankel Matrix Approach **6**: 3965–8
Realization Theory: Matrix Fraction Approach **6**: 3968–9

Ruddock, K. H.
(Imperial College of Science and Technology, London, UK)
Image Filtering by the Retinocortical Visual Pathways **4**: 2325–31

Ruff, E.
(British Aerospace, Filton, UK)
Aircraft Engine Control: System Design **1**: 231–6

Ruggiero, C.
(University of Genoa, Genoa, Italy)
Respiratory System: Modelling and Control of External Respiration **6**: 4051–5

Ruhm, K.
(Swiss Federal Institute of Technology, Zürich, Switzerland)
Oxygen Concentration Determination in Gases **5**: 3572–5
Oxygen Concentration Determination in Liquids **5**: 3575

Russell, D. L.
(University of Wisconsin, Madison, Wisconsin, USA)
Linear Hyperbolic Systems: Control Theory **4**: 2800–7
Linear Systems: Control Theory **4**: 2855–61

Ryan, E. P.
(University of Bath, Bath, UK)
Optimal Relay and Saturating Control Systems: History **5**: 3516–7

Rykiel, E. J. Jr
(Texas A & M University, College Station, Texas, USA)
Ecological Disturbance Theory **2**: 1311–8
Ecological Disturbances **2**: 1318–21
Ecosystems, Disturbed **2**: 1356–8

Safonov, M. G.
(University of Southern California, Los Angeles, California, USA)
Linear Systems: Robustness **4**: 2871–2

Sage, A. P.
(George Mason University, Fairfax, Virginia, USA)
Activity **1**: 38
Algorithms **1**: 262
Alternatives **1**: 262
Attribute Measures **1**: 323
Attributes **1**: 323
Axiology **1**: 406
Bayes Rule **1**: 414
Bounded Rationality **1**: 502
Cardinals **1**: 536
Certainty Equivalents **1**: 565
Closed Systems **1**: 605
Cognition **1**: 608
Collective Enquiry **1**: 608–18
Constraints **2**: 801
Contextual Relation **2**: 801
Correlation **2**: 853
Cost–Benefit and Cost–Effectiveness Analysis **2**: 854–9
Cost–Benefit Ratios **2**: 859
Coupling **2**: 867
Critical Activity **2**: 867
Critical Paths **2**: 867
Cybernetics **2**: 867
Data **2**: 875
Databases **2**: 893
Decision Analysis **2**: 938–43
Decision Support Systems **2**: 943–7
Decomposition **2**: 947
Deduction **2**: 959
Delphi Method **2**: 959
Descriptive Methods or Results **2**: 981
Design and Evaluation of Systems **2**: 982–7
Deterministic Systems **2**: 995
Diagnostic Inference **2**: 1003
Difference Equations, Ordinary **2**: 1003
Differential Equations **2**: 1003
Digraphs **2**: 1071
Directed Relations (Directed Graphs) **2**: 1080
Discrete Variables **2**: 1102
Dominance **2**: 1241
Dynamic Programming **2**: 1278
Dynamic Systems **2**: 1282
Economic Equilibrium **2**: 1337
Edges **2**: 1358
Effective Systems **2**: 1366
Efficient Frontiers **2**: 1367
Efficient Systems **2**: 1367
Empirical Results **2**: 1450
Endogenous Variables **2**: 1451
Estimation or Identification **2**: 1559
Evaluation **2**: 1559
Events **2**: 1568
Exogenous Variables **2**: 1568
Expert Systems **2**: 1569–73
Externality **2**: 1583
Extrapolation **2**: 1583
Fault Trees **3**: 1601
Feedback Loops **3**: 1607
Floats **3**: 1669
Forecasts **3**: 1711
Frames **3**: 1729
Fuzziness **3**: 1780
Game Theory **3**: 1913
Goal-Driven Approaches **3**: 2038
Graph Theory **3**: 2046
Group Decision Making and Voting **3**: 2053–62
Heuristics **3**: 2153
Hierarchy **3**: 2163
Human Information Processing **3**: 2186–92
Human Judgment and Decision Rules **3**: 2193–203
Hypotheses **3**: 2234
Impact Analysis and Hierarchical Inference **4**: 2331–9
Independence **4**: 2343
Independent Variables **4**: 2343
Inference **4**: 2492
Inference Engines **4**: 2492
Information Requirements Determination **4**: 2497–502
Input–Output Analysis **4**: 2517–22
Inputs **4**: 2527
Interacting Groups **4**: 2543
Interaction Matrices **4**: 2547
Interactive Approaches **4**: 2547
Iteration **4**: 2614
Knowledge Bases **4**: 2636
Knowledge Representation **4**: 2636–41
Large Stochastic Systems **4**: 2706–14
Latest Completion **4**: 2723
Latest Start **4**: 2723
Leontief Inverse **4**: 2737
Level of Significance **4**: 2754
Likelihood Ratio **4**: 2777
Management: Hypothesis Testing **5**: 2911–7

Market Clearing Equations **5**: 2927
Markov Processes **5**: 2932
Means–Ends Analysis **5**: 2963
Metarules **5**: 2971
Methodology **5**: 2973
Microeconomic Systems Analysis **5**: 2973–9
Models **5**: 3083
Multiattribute Utility Theory in Decision Analysis * **5**: 3104–16
Multidimensional Scaling **5**: 3124
Needs **5**: 3260
Net Present Value **5**: 3260
Network Methods **5**: 3261
Nodes **5**: 3268
Nondirected Relations **5**: 3279
Nonlinear Programming **5**: 3351
Normative Methods **5**: 3413
Null Hypothesis **5**: 3440
Objectives **5**: 3443
Objectives Trees **5**: 3443
Open Questions **5**: 3448
Operations Research **5**: 3449
Optima **5**: 3471
Options **5**: 3555
Ordinals **5**: 3555
Organization Charts **5**: 3564
Outputs **5**: 3564
Parameters **6**: 3621
Partition **6**: 3629
Pattern Recognition **6**: 3634
Perfectly Competitive Economy **6**: 3654–5
Performance Index **6**: 3655
Planning Horizons **6**: 3703
Preferential Independence **6**: 3871
Prescriptive Methods or Results **6**: 3872
Production Functions **6**: 3890
Production Rules **6**: 3890
Program Evaluation and Review Technique **6**: 3892–3
Public Goods **6**: 3909
Random Variables **6**: 3948
Rationality **6**: 3950
Regression **6**: 3998
Revealed Preference Theory **6**: 4066
Samples **6**: 4146
Satisficing **6**: 4149
Scenarios **6**: 4152–5
Scripts **6**: 4161
Sensitivity **6**: 4223
Server Occupancy **6**: 4252
Service Time **6**: 4252
Shadow Prices **6**: 4257
Slack **7**: 4454
Stability **7**: 4518
State Variables **7**: 4554
Stochastic Processes **7**: 4646
Strategies **7**: 4661
Structured Modelling **7**: 4692
Subjective Probability **7**: 4696
Suboptimal Solutions **7**: 4699
Substitution in Economics **7**: 4699
Surveys **7**: 4709
System Theory **7**: 4766
Systemic Processes **7**: 4766
Systems **7**: 4767
Systems Analysis **7**: 4767–71
Systems Engineering **7**: 4775
Systems Knowledge: Philosophical Perspectives **7**: 4779–85
Systems Management: Cognitive Styles **7**: 4790–3
Systems Methodology **7**: 4799–810
Systems Science **7**: 4810
Systems Synthesis **7**: 4810
Tatonnement **7**: 4811
Theory of the Firm **7**: 4865
Time Horizons **7**: 4892
Top-Down Methods **7**: 4894
Trade Off **7**: 4906
Transactions Tables **7**: 4935
Transients **7**: 4935
Transition Probability **7**: 4935
Transitivity **7**: 4935
Tree Structures **7**: 4949
Trees **7**: 4949–52
Utility **7**: 4981
Validation **7**: 4996
Value Systems **7**: 5019
Variables **7**: 5031
Verification **7**: 5047
Voting Paradox **7**: 5086
Welfare Functions **7**: 5162
Welfare or Normative Economics **7**: 5162–4

Saguez, C.
(INRIA, Le Chesnay, France)

Free Surfaces: Control **3**: 1729–34

Sallet, G.
(University of Metz, Metz, France)

Extension Techniques **2**: 1581–3
Lie Groups: Controllability **4**: 2756–9

Sampson, J. R.
(Late of University of Alberta, Edmonton, Alberta, Canada)

Cellular Space Models: Simulation Methods **1**: 559–61
Simulation Modelling Formalism: Cellular Automata **6**: 4341–5

Sandell, N. R. Jr
(Alphatech, Burlington, Massachusetts, USA)

Decentralized Stochastic Control **2**: 927–32
Linear Quadratic Gaussian Multivariable Control Designs: Robustness * **4**: 2842–6

Sanders, C. W.
(Middle South Services, New Orleans, Louisiana, USA)

Decentralized Observers: Design and Evaluation Methods **2**: 907–12

Sanderson, M. L.
(University of Salford, Salford, UK)

Amplifiers for Instrumentation **1**: 268–72
Electrical Measurement: Bridge Circuits **2**: 1402–3

Sandrin, P.
(Electricité de France, Clamart, France)

Hydroelectricity **3**: 2217–21

Sankar, S.
(Concordia University, Montreal, Quebec, Canada)

Mechanical System Simulation **5**: 2966–70

Sano, A.
(Keio University, Yokohama, Japan)

Glucose Regulation: Adaptive Control * **3**: 2017–21

Sargent, R. G.
(Syracuse University, Syracuse, New York, USA)

Validation of Simulation Models: General Approach **7**: 5008–11
Validation of Simulation Models: Statistical Approach **7**: 5015–9

Saridis, G. N.
(Rensselaer Polytechnic Institute, Troy, New York, USA)

Self-Organizing Control Systems **6**: 4161–5
Traffic Control: Intelligent Systems **7**: 4906–10

Sarna, S. K.
(VA Medical Center, Milwaukee, Wisconsin, USA)

Digestive Tract: Gastric Electrical Activity Modelling **2**: 1033–6
Digestive Tract: Small Intestine Electrical Activity Modelling **2**: 1040–3

Sasano, Y.
(National Institute for Environmental Studies, Ibaraki, Japan)

Meteorological Measurement Techniques **5**: 2971–3

Sasson, A. M.
(American Electric Power, New York, USA)

Power Systems: Economic Dispatching * **6**: 3815–20
Power Systems: State Estimation * **6**: 3846–50

Sastry, D.
(UMIST, Manchester, UK)

Large-Scale Systems: Parameter Estimation * **4**: 2702–6

Saxton, K. J. H.
(Brecon, UK)

Water Sources and Supply Systems **7**: 5140–3

Say, M. G.
(Herriot-Watt University, Edinburgh, UK)

Electric Generators, Rotating **2**: 1380–3

Scavia, D.
(Great Lakes Environmental Research Laboratory, Ann Arbor, Michigan, USA)

Dynamic Eutrophication Modelling: Variance Estimates **2**: 1269–72
Lake Ecosystem Modelling: Uncertainty **4**: 2650–2
Water Quality Modelling: Case Study **7**: 5129–35

Schaudel, D. E.
(Endress & Hauser, Maulburg, FRG)

Level Measurement: Liquids and Bulk Material **4**: 2747–54

Schaufelberger, W.
(Swiss Federal Institute of Technology, Zürich, Switzerland)

Hydroelectricity: Self-Selecting Control * **3**: 2225–8

Schiavoni, N.
(Polytechnic of Milan and Centre for Systems Theory, Milan, Italy)

Structure-Constrained Robust Controller Design * **7**: 4685–9

Schmeiser, B.
(Purdue University, West Lafayette, Indiana, USA)

Random Variate Generation **6**: 3948–50

Schmid, C.
(Ruhr University, Bochum, FRG)

Adaptive Controllers: Design Packages **1**: 60–3

Schmidt, G.
(Technical University of Munich, Munich, FRG)

Freeway Traffic: Modelling, Surveillance and Multilayer Control * **3**: 1736–41
Gas Transmission Systems: Supervision and Control * **3**: 1985–90

Schmidt, G. T.
(Charles Stark Draper Laboratory, Cambridge, Massachusetts, USA)

Inertial Systems: Selection and Integration **4**: 2488–92

Schraft, R. D.
(Fraunhofer Institute, Stuttgart, FRG)

Industrial Robot Arms * **4**: 2387–92
Industrial Robots: Applications * **4**: 2416–21
Robots: Tactile Sensors * **6**: 4118–23

Schreiber, F. A.
(University of Parma, Parma, Italy)

Distributed Database Systems: Efficiency **2**: 1155–60

Schriber, T. J.
(University of Michigan, Ann Arbor, Michigan, USA)
Simulation Output: Statistical Analysis * **6**: 4393–6

Sciacovelli, D.
(European Space Research and Technical Centre, Noordwijk, The Netherlands)
Flexible Spacecraft: Computer-Oriented Control Analysis **3**: 1649–65

Scrivens, J. H.
(ICI Petrochemicals and Plastics Division, Wilton, UK)
Mass Spectrometry **5**: 2934–44

Sekine, Y.
(University of Tokyo, Tokyo, Japan)
Power Systems: Microprocessor Monitoring and Control **6**: 3837–42

Selke, K.
(University of Hull, Kingston-upon-Hull, UK)
Robot Arms, Flexible * **6**: 4086–9

Sentis, R.
(INRIA, Le Chesnay, France)
Integrodifferential Systems **4**: 2536–41
Parabolic Systems **6**: 3591–6
Transport Equations **7**: 4940–4

Serazzi, G.
(IAN-CNR and University of Pavia, Pavia, Italy)
Computer Systems: Performance Measurement **1**: 746–51

Sharma, G. N.
(Anglian Water Authority, Huntingdon, UK)
Water Supply Systems: Suppression and Control of Transient Phenomena * **7**: 5155–6
Water Supply Systems: Transient Phenomena * **7**: 5156–61

Sheppard, L. C.
(University of Alabama, Birmingham, Alabama, USA)
Blood Pressure Hypertension Treatment: Feedback Control **1**: 495–8

Sheridan, T. B.
(MIT, Cambridge, Massachusetts, USA)
Automation: Social Effects **1**: 367–72
Human Factors Engineering **3**: 2179–86

Shibata, S.
(Government Industrial Research Institute, Nagoya, Japan)
Water Pollution: Inorganic Nitrogen Compound Determination **7**: 5117–22

Shieh, L. S.
(University of Houston, Houston, Texas, USA)
State-Space Self-Tuners * **7**: 4547–51

Shima, T.
(Keio University, Yokohama, Japan)
Overlapping Coordination, Hierarchical * **5**: 3564–8

Shimomura, S.
(University of Tokushima, Tokushima, Japan)
Environmental Databases * **2**: 1500–3

Shindo, J.
(National Institute for Environmental Studies, Ibaraki, Japan)
Gas Chromatography–Mass Spectrometry in Environmental Analysis * **3**: 1933–6

Shirai, Y.
(Electrotechnical Laboratory, Sakuramura, Japan)
Computer Vision by Three-Dimensional Rangefinding **1**: 755–60

Shono, T.
(Osaka University, Osaka, Japan)
Environmental Measurement: Odorous Compounds in Air **2**: 1527–8

Shrivastava, S. K.
(University of Newcastle-upon-Tyne, Newcastle-upon-Tyne, UK)
Distributed Computer Systems: Reliability **2**: 1132–7

Šiljak, D. D.
(University of Santa Clara, Santa Clara, California, USA)
Complex Ecosystems Stability **1**: 672–5
Interconnected Systems: Decentralized Control **4**: 2557–60
Overlapping Decentralized Control * **5**: 3568–72
Reliability of Control **6**: 4008–11

Silvert, W.
(Bedford Institute of Oceanography, Dartmouth, Nova Scotia, Canada)
Forecast Evaluation in Ecological Modelling **3**: 1710–1
Parameter Estimation in Ecological Modelling **6**: 3619–21
Perturbation Techniques in Ecosystem Modelling **6**: 3664–5
Size Dependence in Ecology **7**: 4452–4
Top-Down Modelling in Ecology **7**: 4894–5

Simpson, R. J.
(Preston Polytechnic, Preston, UK)
Electrical Control in the Nineteenth Century * **2**: 1398–402

Siret, J. M.
(CNRS-ESE, Gif sur Yvette, France)
Aggregation and Suboptimal Control * **1**: 126–33

Skala, H.
(University of Paderborn, Paderborn, FRG)
Many-Valued Logics **5**: 2918–21

Skoog, R. A.
(AT & T Bell Laboratories, Murray Hill, New Jersey, USA)
Switching Networks: Long-Term Planning **7**: 4721–6

Slate, J. B.
(Pacesetter Infusion, Sylmar, California, USA)
Blood Pressure Postoperative Treatment: Adaptive Control **1**: 498–502

Smets, P.
(Free University of Brussels, Brussels, Belgium)
Fuzzy Events: Probability **3**: 1802–5

Smith, J. R.
(University of Aberdeen, Aberdeen, UK)
Power System Stability **6**: 3795–802

Smith, R. H.
(Northumbrian Water, Thornaby, UK)
Water Supply Leakage Control **7**: 5143–7

Snodgrass, W. J.
(IEC Beak Consultants, Mississauga, Ontario, Canada)
Limnology: Modelling of Mixing and Reaction Characteristics **4**: 2782–9
Limnology: One- and Two-Box Models of Lake Response Time **4**: 2789–94
Water Bodies Time Scale Estimation: Mass Concepts **7**: 5090–4

Söderström, T.
(University of Uppsala, Uppsala, Sweden)
Identification: Instrumental Variable Techniques * **4**: 2273–8
Identification: Model Structure Determination **4**: 2287–93

Soeda, T.
(University of Tokushima, Tokushima, Japan)
Environmental Data Statistical Analysis: Nonphysical Modelling Approaches * **2**: 1494–7
Environmental Data Statistical Analysis: Physical Modelling Approach * **2**: 1497

Solheim, O. A.
(University of Trondheim, Trondheim, Norway)
Decentralized Stabilization: Block Gerschgorin Theorem **2**: 923–7

Soliman, H. M.
(Cairo University, Cairo, Egypt)
Decentralized Stabilization: A Review of Methods **2**: 916–23

Son, S. H.
(University of Maryland, College Park, Maryland, USA)
Distributed Database Systems: Failure Recovery Procedures * **2**: 1160–5

Song, J.
(Beijing Institute of Information and Control, Beijing, People's Republic of China)
Population Control **6**: 3751–5

Sontag, E. D.
(Rutgers University, New Brunswick, New Jersey, USA)
Nonlinear Discrete-Time Systems: Algebraic Theory **5**: 3288–93

Sorin, S.
(Louis Pasteur University, Strasbourg, France)
Games, Repeated **3**: 1922–6
Games with Infinitely Many Players **3**: 1926–9
Nash Equilibrium **5**: 3255–6

Soulas, C.
(INRETS, Arceuil, France)
Automated Guideway Transit Systems and Personal Rapid Transit Systems **1**: 334–40

Spillane, D. E. M.
(Polytechnic of North London, London, UK)
Spectroscopy: Fundamentals and Applications * **7**: 4498–504

Spiteri, P.
(Besançon University, Besançon, France)
Optimal Control: Decomposition and Coordination using Asynchronous Iteration * **5**: 3475–81

Spriet, J. A.
(University of Ghent, Ghent, Belgium)
Biochemical and Microbial Systems: Simulation **1**: 421–4
Structure/Parameter Identification by Simulation Experimentation * **7**: 4690–2

Stamm, A.
(Institute for Radiation and Environmental Research, Neuherberg, FRG)
Carbon-Cycle Modelling * **1**: 529–32

Stamm, K.
(Volkswagen, Wolfsburg, FRG)
Automotive Electronics * **1**: 372–80

Standridge, C. R.
(Pritsker & Associates, Lubbock, Texas, USA)
Databases for Simulation **2**: 893–4

Steer, M. W.
(University College, Dublin, Republic of Ireland)
Stereological Analysis of Heterogeneous Solids in Biology **7**: 4598–601

Stefanelli, M.
(University of Pavia, Pavia, Italy)
Ferrokinetic Modelling in the Body **3**: 1619–23

Stefani, G.
(University of Florence, Florence, Italy)
Multivalued Differential Equations in Control Systems * **5**: 3180–4

Stein, E. S.
(Federal Aviation Administration Technical Center, Atlantic City, New Jersey, USA)
Flight-Crew Workload **3**: 1665–9

Stein, W. E.
(Texas A & M University, College Station, Texas, USA)
Fuzzy Random Variables **3**: 1843–5

Sterling, M. J. H.
(University of Durham, Durham, UK)
Power Systems: Integrated Software for Control * **6**: 3825–9

Stimson, K. R.
(WRC Engineering, Swindon, UK)
Water Distribution Systems: Steady-State Analysis **7**: 5102–6

Stoica, P.
(Polytechnic Institute of Bucharest, Bucharest, Romania)
Identification: Instrumental Variable Techniques * **4**: 2273–8

Stott, B.
(Power Computer Applications, Mesa, Arizona, USA)
Power Systems: Steady-State Power Flow Modelling **6**: 3850–5

Sumpter, C.
(Loughborough University of Technology, Loughborough, UK)
Industrial Production Systems: Robot Integration * **4**: 2362–8

Sussmann, H. J.
(Rutgers University, New Brunswick, New Jersey, USA)
Lie Brackets **4**: 2754–6
Real Analyticity **6**: 3954–6
Universal Inputs **7**: 4975–6

Sutcliffe, D. S.
(National Physical Laboratory, Teddington, UK)
Frequency Measurement **3**: 1741–5

Suter, G. W.
(Oak Ridge National Laboratory, Oak Ridge, Tennessee, USA)
Gaussian Random Variables * **3**: 1998–9

Swanick, B. H.
(Leicester Polytechnic, Leicester, UK)
Fermentation Control * **3**: 1607–11
Fermentation Measurement * **3**: 1611–5

Swanson, G. D.
(University of Colorado Medical School, Denver, Colorado, USA)
Respiratory System: Control During Exercise **6**: 4045–51

Swierstra, S.
(European Organisation for the Safety of Air Navigation, Brussels, Belgium)
Fuel Conservation: Air Transport * **3**: 1749–56

Takahara, Y.
(Tokyo Institute of Technology, Tokyo, Japan)
Past Determinancy and Finite Memory **6**: 3629–31
Systems Interactions **7**: 4775–9

Takahashi, A.
(Denki Kagaku Keiki, Tokyo, Japan)
Environmental Measurement: Automatic and Continuous Analysis **2**: 1508–13

Takeda, E.
(Kobe University of Commerce, Kobe, Japan)
Multicriteria Decision Problem **5**: 3121–4

Takeuchi, N.
(National Institute for Environmental Studies, Ibaraki, Japan)
Laser Spectroscopy in Atmospheric Analysis * **4**: 2719–23
Remote Sensing: Laser Beam Analysis of Atmospheric Conditions * **6**: 4033–7

Talavage, J.
(Purdue University, West Lafayette, Indiana, USA)
Simulation Modelling Formalism: Hierarchical Decomposition **6**: 4353–5

Tamura, H.
(Osaka University, Osaka, Japan)
Multiattribute Utility Theory **5**: 3099–104

Multistage Decomposition Algorithms **5**: 3166–71

Tamura, M.
(National Institute for Environmental Studies, Ibaraki, Japan)
Environmental Measurement: Noise **2**: 1522–7

Tanaka, M.
(Osaka University, Osaka, Japan)
Water Pollution: Trace Organic Compound Determination **7**: 5127–9

Tanaka, S.
(University of Tokyo, Tokyo, Japan)
Infrared Spectrometry in Environmental Analysis * **4**: 2509–11

Tatnall, M. L.
(University of Salford, Salford, UK)
Anesthesia: Inhalational Agent Identification and Control **1**: 284–9

Tavares, H.
(University of Campinas, Campinas, Brazil)
Large-Scale Systems: Linear Programming **4**: 2686–92

Taylor, G. E.
(University of Hull, Kingston-upon-Hull, UK)
Robot Arms, Flexible * **6**: 4086–9

Taylor, P. M.
(University of Hull, Kingston-upon-Hull, UK)
Industrial Robots in Garment Manufacture **4**: 2447–9
Robot Arms, Flexible * **6**: 4086–9

Temam, R.
(University of Paris XI, Orsay, France)
Calculus of Variations **1**: 519–29

Tempo, R.
(Turin Polytechnic, Turin, Italy)
Identification Errors in Biomedicine * **4**: 2253–7

Teramae, N.
(University of Tokyo, Tokyo, Japan)
Infrared Spectrometry in Environmental Analysis * **4**: 2509–11

Termini, S.
(Institute of Cybernetics, Arco Felice, Italy)
Entropy Measures in Fuzzy Set Theory * **2**: 1467–73
Vagueness in Scientific Theories **7**: 4993–6

Terrisson, J. C.
(Centre National d'Etudes Spatiales, Toulouse, France)
Telemetry and Data Processing in Space Communication **7**: 4835–42

Tham, M. T.
(University of Newcastle-upon-Tyne, Newcastle-upon-Tyne, UK)
Distillation Columns: Adaptive Control * **2**: 1109–16
Multivariable and Multirate Self-Tuning Controllers * **5**: 3184–8

Thatcher, T. W.
(Loughborough University of Technology, Loughborough, UK)
Industrial Robots: Modular Systems * **4**: 2460–4

Thépot, J.
(EIASM, Brussels, Belgium)
Dynamic Duopoly Theory * **2**: 1262–9

Theuerkauf, H.
(Volkswagen, Wolfsburg, FRG)
Automotive Electronics * **1**: 372–80

Thoma, M.
(Technical University of Hanover, Hanover, FRG)
International Federation of Automatic Control: History **4**: 2571–5

Thomas, B.
(Gwent College of Higher Education, Newport, UK)
Industrial Robots: Laser-Beam Position Reference **4**: 2453–6

Thomason, M. G.
(University of Tennessee, Knoxville, Tennessee, USA)
Scene Analysis, Hierarchical * **6**: 4155–61

Thorley, A. R. D.
(City University, London, UK)
Water Supply Systems: Suppression and Control of Transient Phenomena * **7**: 5155–6
Water Supply Systems: Transient Phenomena * **7**: 5156–61

Thorn, R.
(South Australian Institute of Technology, Ingle Farm, South Australia, Australia)
Flow Measurement: Two-Phase Systems * **3**: 1685–94

Thornton, K. W.
(Ford, Thornton, Norton and Associates, Vicksburg, Mississippi, USA)
Sensitivity Analysis and Simulation Experimentation **6**: 4227–30

Tien, J. M.
(Rensselaer Polytechnic Institute, Troy, New York, USA)
Evaluation Design: Systems and Models Approach **2**: 1559–66

Timonen, E.
(Technical Research Center Finland, Oulu, Finland)
Consistency, Density and Viscosity: Control * **2**: 777–81

Tiran, J.
(The Hospital for Sick Children, Toronto, Ontario, Canada)
Glucose Regulation: Modelling and Control in Diabetics * **3**: 2025–32

Titli, A.
(LAAS-CNRS, Toulouse, France)
Complex Systems Methodology **1**: 687–704

Toates, F. M.
(Open University, Milton Keynes, UK)
Hunger and Thirst: Motivational Control Mechanisms **3**: 2209–14

Toffolo, G.
(University of Padua, Padua, Italy)
Ketone Body Metabolism Modelling * **4**: 2628–31

Togawa, T.
(Tokyo Medical and Dental University, Tokyo, Japan)
Temperature Measurement, Physiological **7**: 4851–6

Tondl, L.
(Design Institute PUDIS, Prague, Czechoslovakia)
Pragmatic Information **6**: 3858–60
Semantic Information **6**: 4214–6

Törn, A. A.
(Swedish University of Åbo, Åbo, Finland)
Simulation Modelling Formalism: Extended Petri-Net Graphs **6**: 4348–50

Toyota, H.
(University of Tokyo, Tokyo, Japan)
Remote Sensing: Classification of Multispectral Images * **6**: 4029–33

Trinder, J. R.
(Multitone Electronics, Basingstoke, UK)
Speech * **7**: 4504–9

Tripathi, S. K.
(University of Maryland, College Park, Maryland, USA)
Distributed Database Systems: Failure Recovery Procedures * **2**: 1160–5

Tropf, H.
(Vision Tools, St Leon, FRG)
Industrial Vision Systems **4**: 2469–73

Trowbridge, E. A.
(University of Sheffield, Sheffield, UK)
Biomechanics of Soft Tissue * **1**: 482–6

Tsay, Y. T.
(University of Houston, Houston, Texas, USA)
State-Space Self-Tuners * **7**: 4547–51

Tsypkin, Ya. Z.
(Institute of Control Sciences, Moscow, USSR)
Optimization Under Uncertainty **5**: 3551–4

Tzafestas, S. G.
(National Technical University, Athens, Greece)
Distributed Parameter Systems: Estimation and Control **2**: 1186–93
Partial Differential Equation Models: Numerical Solution **6**: 3622–6
Simulation Modelling Formalism: Partial Differential Equations **6**: 4360–5
Information Sources in Systems and Control **8**: 5575–86

Uchiyama, M.
(Tohoku University, Sendai, Japan)
Robot Arms: Dynamic Control **6**: 4082–6

Ueno, S.
(Kanazawa Institute of Technology, Ishikawa, Japan)
Earth–Atmosphere System: Identification of Optical Parameters * **2**: 1307–11
Ground Albedo Mapping: Invariant Embedding * **3**: 2046–53
Landsat Imaging: Removal of Atmospheric Effects * **4**: 2652–5

Ulanowicz, R. E.
(Chesapeake Biological Laboratory, Solomons, Maryland, USA)
Flow Network Ascendency and Self-Organization in Living Systems **3**: 1694–5

Umano, M.
(Okayama University of Science, Okayama, Japan)
Fuzzy Databases: Retrieval Processing **3**: 1798–800

Umetani, Y.
(Tokyo Institute of Technology, Tokyo, Japan)
Robots, Snakelike **6**: 4109–13

Unbehauen, H.
(Ruhr University, Bochum, FRG)
Discrete Model Reference Adaptive Systems: Design **2**: 1087–93

Ung, M.
(University of Southern California, Los Angeles, California, USA)
Aerospace Simulation: An Introduction **1**: 106–9

Unger, B. W.
(University of Calgary, Calgary, Alberta, Canada)
Computer Systems Simulation Models **1**: 751–5

Uronen, P.
(University of Oulu, Oulu, Finland)
Production Control in Process Industries **6**: 3884–90

Utsch, M. J.
(Institute of Medical Education, Bonn, FRG)
Educational Computer Models in Medicine * **2**: 1361–6

Vajk, I.
(Technical University of Budapest, Budapest, Hungary)
Load-Frequency Regulation: Adaptive Control * **4**: 2889–92

Vajta, M.
(Technical University of Budapest, Budapest, Hungary)
Load-Frequency Regulation: Adaptive Control * **4**: 2889–92

van Amerongen, J.
(Delft University of Technology, Delft, The Netherlands)
Ship Steering: Model Reference Adaptive Control **6**: 4275–7

van Cauwenberghe, A. R.
(University of Ghent, Ghent, Belgium)
Nuclear Power Plants: Self-Tuning Control **5**: 3425–30

van Duyl, W. A.
(Erasmus University, Rotterdam, The Netherlands)
Ureter Electrical Activity Modelling **7**: 4976–81

Vansteenkiste, G. C.
(University of Ghent, Ghent, Belgium)
Structure/Parameter Identification by Simulation Experimentation * **7**: 4690–2

Vardulakis, A. I. G.
(Aristotle University of Thessaloniki, Thessaloniki, Greece)
Smith–MacMillan Canonical Forms for Rational Matrices **7**: 4455–7

Verdegay, J.-L.
(University of Granada, Granada, Spain)
Fuzzy Mathematical Programming Problem: Resolution **3**: 1816–9

Verjus, J. P.
(University of Rennes I, Rennes, France)
Distributed Systems: Synchronization and Interprocess Communication **2**: 1202–9

Vidal, P.
(Aérospatiale, Marignane, France)
Helicopters: Flight Control Assistance **3**: 2124–33

Vidyasagar, M.
(University of Waterloo, Waterloo, Ontario, Canada)
Large-Scale Interconnected Systems: Input–Output Stability **4**: 2671–5
Large-Scale Interconnected Systems: Well Posedness **4**: 2675–8
Linear Systems: Stability **4**: 2872–5

Vincke, P.
(Free University of Brussels, Brussels, Belgium)
Multiobjective Decision * **5**: 3150–2

Viswanadham, N.
(Indian Institute of Science, Bangalore, India)
Large-Scale Systems: Observers **4**: 2697–702

Vriends, J. H. J. M.
(Royal Military Academy, Breda, The Netherlands)
Missile Guidance and Control **5**: 3006–12

Walker, R.
(Portsmouth Polytechnic, Portsmouth, UK)
Electric Current Measurement **2**: 1372–5
Electric Voltage Measurement: Basic Principles **2**: 1389–93
Electric Voltage Measurement: Commercial Systems **2**: 1393–8

Waller, K. V.
(Swedish University of Åbo, Åbo, Finland)
Distillation Column Control **2**: 1107–9

Walter, E.
(CNRS-ESE, Gif sur Yvette, France)
Biomedical Systems Modelling: Structural Identifiability **1**: 490–4

Walzer, P.
(Volkswagen, Wolfsburg, FRG)
Automotive Electronics * **1**: 372–80

Wang, P. P.
(Duke University, Durham, North Carolina, USA)
Communication Theory: Fuzzy Detection and Estimation * **1**: 651–65

Wang Pei-zhuang
(Beijing Normal University, Beijing, Peoples' Republic of China)
Random Sets in Fuzzy Set Theory **6**: 3945–7

Wanner, M. C.
(Fraunhofer Institute, Stuttgart, FRG)
Industrial Robot Arms * **4**: 2387–92

Ward, D. S.
(UCLA, Los Angeles, California, USA)
Respiratory System: Modelling and Identification * **6**: 4055–62

Warfield, J. N.
(University of Virginia, Charlottesville, Virginia, USA)
Interpretive Structure Modelling **4**: 2575–80

Warnecke, H. J.
(Fraunhofer Institute, Stuttgart, FRG)
Industrial Robot Arms * **4**: 2387–92
Industrial Robots: Applications * **4**: 2416–21
Robots: Tactile Sensors * **6**: 4118–23

Watson, V.
(University of Montana, Missoula, Montana, USA)
Environmental Problem Solving: Systems Concepts **2**: 1536–8
Lake Ecosystem Modelling **4**: 2642–6

Weiss, R. G.
(Ford Aerospace and Communications, College Park, Maryland, USA)
Large-Scale Systems: Development **4**: 2681–6

Wells, P. N. T.
(Bristol General Hospital, Bristol, UK)
Ultrasound Imaging in Biology **7**: 4959–64

Wellstead, P. E.
(UMIST, Manchester, UK)
Engine Speed: Self-Tuning Control **2**: 1462–7
Pole-Assignment Self-Tuning **6**: 3721–7

Wend, H. D.
(University of Duisburg, Duisburg, FRG)
Decomposition and Structural Properties **2**: 948–52

Weston, D.
(Sunderland and South Shields Water, Sunderland, UK)
Water Distribution Systems **7**: 5095–9

Weston, R. H.
(Loughborough University of Technology, Loughborough, UK)
Industrial Production Systems: Robot Integration * **4**: 2362–8
Industrial Robots: Modular Systems * **4**: 2460–4

Wever, P. A.
(Max Planck Institute for Psychiatry, Andechs, FRG)
Circadian Rhythms: Modelling **1**: 594–601

White, C. C. III
(University of Virginia, Charlottesville, Virginia, USA)
Linear Programming: Fundamentals **4**: 2839–42
Markov Decision Processes **5**: 2930–2
Stochastic Programming **7**: 4646–8

White, K. Preston Jr
(University of Virginia, Charlottesville, Virginia, USA)
Project Management: Network Models **6**: 3898–904
Spatial Allocation: Gravity Models **7**: 4491–5
System Dynamics Modelling **7**: 4755–62
Transportation Modelling **7**: 4944–8
Validation of Management Models **7**: 5001–5
Workshop Dynamic Modelling **7**: 5170–7

Wiberg, D.
(UCLA, Los Angeles, California, USA)
Respiratory System: Modelling and Identification * **6**: 4055–62

Wierzchoń, S. T.
(Polish Academy of Sciences, Warsaw, Poland)
Feature Selection: Fuzzy Classification **3**: 1601–4

Wilkinson, R.
(University of Strathclyde, Glasgow, UK)
Electron Microprobe Analysis * **2**: 1416–20
Scanning Electron Microscopy in Biology * **6**: 4149–52
Transmission Electron Microscopy in Biology * **7**: 4936–40

Willems, J. C.
(University of Groningen, Groningen, The Netherlands)
Pole Assignment: The General Problem **6**: 3727–34

Willems, J. L.
(University of Ghent, Ghent, Belgium)
Decentralized Stabilization **2**: 912–6

Williams, D.
(Liverpool Polytechnic, Liverpool, UK)
Fermentation Control * **3**: 1607–11
Fermentation Measurement * **3**: 1611–5

Williams, T. J.
(Purdue University, West Lafayette, Indiana, USA)
Steel Industry: Hierarchical Control **7**: 4585–93

Wilson, J. R.
(University of Texas, Austin, Texas, USA)
Stochastic Simulation: Variance-Reduction Techniques **7**: 4654–7

Wirth, E.
(Federal Health Office, Neuherberg, FRG)
Radioecological Modelling * **6**: 3929–33
Radioecological Modelling in Agricultural Food Chains * **6**: 3933–6

Wittenmark, B.
(Lund Institute of Technology, Lund, Sweden)
Self-Tuning Control: General Aspects **6**: 4190–3
Self-Tuning Proportional-Integral-Derivative Controllers: Pole Assignment Approach **6**: 4211–4

Wohl, J. G.
(Alphatech, Burlington, Massachusetts, USA)
Communication Command and Control Information Systems **1**: 645–51

Wonham, W. M.
(University of Toronto, Toronto, Ontario, Canada)
Linear Multivariable Control Systems Synthesis: Geometric Approach **4**: 2807–16

Wood, R. K.
(University of Alberta, Edmonton, Alberta, Canada)
Distillation Columns: Adaptive Control * **2**: 1109–16
Multivariable and Multirate Self-Tuning Controllers * **5**: 3184–8

Woollons, D. J.
(University of Exeter, Exeter, UK)
Electrocardiography, His Bundle **2**: 1407–10

Yacamini, R.
(University of Aberdeen, Aberdeen, UK)
Power Systems: High-Voltage DC Transmission **6**: 3820–5

Yagi, O.
(National Institute for Environmental Studies, Ibaraki, Japan)
Biodegradability Determination **1**: 434
Water Pollution Indices: Algal Growth Potential **7**: 5112–3

Yamamoto, T.
(Kyoto University of Education, Kyoto, Japan)
Environmental Measurement: Distribution of Chemical Elements in the Ocean * **2**: 1513–6

Yao, J. T. P.
(Purdue University, West Lafayette, Indiana, USA)
Structural Damage Assessment: Fuzzy Approach * **7**: 4667–9

Yasuhara, A.
(National Institute for Environmental Studies, Ibaraki, Japan)
Gas Chromatography–Mass Spectrometry in Environmental Analysis * **3**: 1933–6

Yasuoka, Y.
(National Institute for Environmental Studies, Ibaraki, Japan)
Remote Sensing: Water Quality **6**: 4041–5

Yokoyama, Y.
(Nagaoka Technological University, Nagaoka, Japan)
Photoacoustic Spectrometry in Environmental Analysis **6**: 3699

Yoshimura, T.
(University of Tokushima, Tokushima, Japan)
Environmental Data Statistical Analysis: Nonphysical Modelling Approaches * **2**: 1494–7

Young, S. J.
(UMIST, Manchester, UK)
Programming Languages, Real-Time **6**: 3893–8

Zakian, V.
(UMIST, Manchester, UK)
Multivariable Systems: Method of Inequalities **5**: 3206–15

Zecca, P.
(University of Florence, Florence, Italy)
Multivalued Differential Equations in Control Systems * **5**: 3180–4

Zeigler, B. P.
(University of Arizona, Tucson, Arizona, USA)
Automata Theory **1**: 333–4
Character Representation **1**: 575
Computer Architecture **1**: 736
Computer Memory Hierarchy **1**: 740
Data Structures **2**: 879–80
Discrete Mathematics and Continuous Mathematics **2**: 1087
Distributed Computer Systems **2**: 1125–6
Flowcharting in Computing **3**: 1695
Formalism **3**: 1712–3
Interactive Terminals **4**: 2551
Microprocessors **5**: 2980
Microprogramming **5**: 2989
Model Specification: Interfaces **5**: 3077–9
Models as System-Theoretic Specifications **5**: 3083–8
Multiprogramming **5**: 3165
Operating Systems (Computers) **5**: 3448–9
Parallelism and Concurrency in Computers **6**: 3596
Program Compilers **6**: 3892
Program Interpreters **6**: 3893
Programming Languages **6**: 3893
Programming Methodology **6**: 3898
Random Number Generation **6**: 3945

Semantics in Computing **6**: 4216
Simulation Methodology and Model Manipulation **6**: 4311–3
Simulation: Model Base Organization and Utilization **6**: 4323–6
Simulation Model Equivalences: System Specification Morphisms **6**: 4326–8
Simulation Modelling Support Environments **6**: 4370–2
Simulation Objectives: Experimental Frames and Validity **6**: 4388–92
Simulation Program Development: Structured Design Methodology **6**: 4396–402
Software Modules **7**: 4472
Synchronization in Computing **7**: 4751–2
Syntax in Computing **7**: 4752
System Model Specifications: Manipulation and Simplification **7**: 4762–4
Systems Software **7**: 4810
Utility Programs **7**: 4986

Zeleny, M.
(Fordham University at Lincoln Center, New York, USA)
Autopoiesis **1**: 393–400
Multicriteria Decision Making **5**: 3116–21
Simulation Models of Autopoiesis: Variable Structure **6**: 4374–7

Zimmermann, H.–J.
(Rhineland-Westphalia Technical University, Aachen, FRG)
Operations Research: Fuzzy Set Theory **5**: 3449–55

Zinober, A. S. I.
(University of Sheffield, Sheffield, UK)
Variable-Structure Systems: Adaptive Control **7**: 5027–31

Ziolkowski, A.
(Polish Academy of Sciences, Warsaw, Poland)
Glass Industry: Process Control * **3**: 2011–4

Zühlke, D.
(Rhineland–Westphalia Technical University, Aachen, FRG)
Industrial Robots: Programming Languages **4**: 2464–9

AUTHOR CITATION INDEX

The Author Index has been compiled so that the reader can proceed either directly to the page where the author's work is cited, or to the reference itself in the bibliography of the relevant article. For each name, the page numbers for the bibliographic citation are given first, followed by the page number(s) in parentheses where that reference is cited in text. Where a name is referred to only in text, and not in a bibliography, the page number appears only in parentheses.

The accuracy of the spelling of authors' names has been affected by the use of different initials by some authors, or a different spelling of their name in different papers or review articles (sometimes this may arise from a transliteration process), and by those journals which give only one initial to each author.

Aarna O, **4**: 2516 (2516)
Aarts N J M, **7**: 4871 (4868)
Abbiss J B, **2**: 1532 (1530)
Abbott M M, **7**: 4901 (4898)
Abdel-Fattah Y M, **2**: 1253 (1251)
Abe H, **3**: 2021 (2017)
Abele E, **6**: 4123 (4120)
Abele J, **2**: 1532 (1528)
Abelson A, **2**: 1160 (1160)
Abelson R P, **4**: 2640 (2638), 2641 (2637)
Abernathy W J, **7**: 4832 (4830, 4832), 4833 (4828, 4832)
Abeysekera S A, **3**: 1693 (1688)
Abignoli M, **7**: 4598 (4598)
Abo F, **2**: 1522 (1517, 1518)
Abouelwafa M S A, **3**: 1693 (1691); **5**: 3421 (3419)
Abou-Kandil H, **2**: 956 (955)
Aboulnasr T T, **5**: 3127
Abragam A, **5**: 3425 (3422, 3424)
Abraham R, **2**: 1583 (1582); **4**: 2759 (2757)
Abrahamson S, **2**: 1366 (1364, 1366)
Abramsky Z, **1**: 142 (141)
Abramson N, **1**: 91 (87)
Accad Y, **7**: 4888 (4887)
Ackerman E, **3**: 2025 (2022)
Ackermann J, **6**: 3720 (3718), 3734 (3733)
Ackermann R, **5**: 2921 (2918)
Ackley S I, **5**: 3272 (3268)
Ackoff R L, **4**: 2352 (2351), 2353 (2352); **5**: 3034 (3033)
Acock B, **1**: 142 (140, 141)
Aczél J, **4**: 2508 (2504, 2505, 2507)
Adachi N, **1**: 675 (673)
Adam G, **1**: 433 (432)
Adam N R, **4**: 2353 (2351)
Adamjan V M, **1**: 296 (294, 295)
Adamo J M, **3**: 1845 (1845); **5**: 3454
Adamovsky G, **5**: 3470 (3469)
Adams A P, **1**: 440
Adams D, **2**: 965
Adams G, **3**: 1822 (1821), 1880 (1878)
Adams M S, **7**: 4897 (4896)
Adams S S, **1**: 142 (139, 140)
Adey A J, **6**: 3870 (3867)
Adiba M, **2**: 1160 (1159)
Adizes I, **7**: 4832 (4830)
Adlard E R, **7**: 5112
Adlassnig K-P, **1**: 740 (736)
Adler R A, **7**: 5051 (5048)
Adriano D C, **2**: 1355 (1352); **5**: 2929 (2928); **6**: 4237 (4235); **7**: 4956 (4956)
Aertsen A, **6**: 4167 (4166)
Aeyels D, **5**: 3341 (3339, 3340, 3341)
Afgan N H, **2**: 1416 (1413)
Agar W E, **1**: 294 (293)
Agarwal G C, **1**: 466 (466); **5**: 3242 (3241)
Agarwal R C, **2**: 1078 (1077)
Agathoklis P, **5**: 3127 (3125, 3127)
Agee W S, **7**: 4539 (4536)
Aggarwal J, **7**: 4528
Aggarwal S, **3**: 2168 (2167); **6**: 4403 (4403), 4411 (4406, 4407)
Agin G J, **4**: 2385 (2382)
Agin G T, **6**: 3766 (3762)
Agmon S, **6**: 4223 (4218)
Agrawal G P, **5**: 3471 (3466)
Agrawala A K, **2**: 1165 (1164); **6**: 4159; **7**: 5018 (5017)
Aguilar-Martin J, **3**: 1604 (1603)
Ahava O, **6**: 4204 (4200)
Ahlberg H H, **4**: 2412 (2402)
Ahlberg J H, **7**: 4518 (4518)
Ahlers G, **1**: 569 (568)
Ahlstrom R C, **5**: 2942 (2934, 2935)
Ahmed K, **2**: 1366 (1362, 1365)
Ahmed S W, **2**: 1105 (1105)
Aho A V, **1**: 725 (721, 722, 723); **2**: 1083 (1082), 1153 (1148); **5**: 3034 (3030); **6**: 3652 (3645, 3648, 3649), 3892, 3893, 4159 (4158)
Ahola H, **6**: 3916 (3913)
Aiba S, **3**: 1611 (1609), 1614 (1614), 1619 (1616)
Aida K, **2**: 1366 (1363, 1366)
Aida S, **1**: 372 (371)
Aidarous E, **2**: 807 (802); **6**: 4244 (4238)
Aidinis S J, **1**: 502 (499)
Aiello N, **4**: 2338 (2338)
Aiga I, **2**: 1522 (1517, 1518, 1519, 1520, 1521)

Aiken R, **5**: 3559 (3558)
Aikman A R, **5**: 3421 (3419)
Aiko T, **6**: 4033 (4031)
Ainsworth J D, **6**: 3825 (3823)
Aird T J, **2**: 825 (824)
Airy G B, **5**: (3383); **7**: (4515, 4517)
Aizawa K, **4**: 2798 (2797)
Aizawa T, **5**: 2994 (2993)
Aizerman M A, **2**: 1478 (1476); **3**: (1749); **5**: 3319 (3319), 3373 (3372), 3388 (3387); **7**: 4529 (4527)
Akagi S, **3**: 1766 (1761, 1762, 1763, 1764, 1765)
Akaike H, **2**: 1106, 1497 (1494); **4**: 2239 (2237), 2287 (2287), 2292 (2291, 2292); **5**: 3232 (3230); **6**: 4172 (4170); **7**: 4566 (4563, 4566), 5001 (5000)
Akashi H, **1**: 682 (679); **2**: 879 (878, 879), 1311 (1307, 1309), 1542 (1540); **3**: 1700 (1699), 1736, 2151; **4**: 2264 (2263), 2560 (2559), 2827 (2821, 2826); **5**: 2994 (2992), 3351 (3343, 3344), 3436, 3506 (3505); **6**: 3610 (3604), 3611 (3610), 4011 (4009), 4211 (4210); **7**: 4481, 4593 (4588), 4689 (4688, 4689), 4731, 4740 (4738), 4751 (4751)
Akima H, **7**: 4518 (4518)
Akimoto H, **1**: 187 (187)
Akitt J W, **5**: 3421 (3418)
Akizuki J, **5**: 3388 (3388)
Akramov T, **6**: 3744 (3740)
Al Chalabi L A M, **4**: 2253 (2249); **5**: 2984 (2980)
Al-Haboubi A, **1**: 478 (477)
Al-Janabi T H, **5**: 3215 (3206)
Al-Naib U, **5**: 3197 (3189), 3215 (3206, 3207, 3213), 3383 (3381)
al-Shaikh A, **6**: 3916 (3913)
Al-Swaha A, **5**: 3214 (3206)
Alb P F, **6**: 3968
Albaiges J, **2**: 1516 (1515)
Albanese J R, **3**: 1601 (1600); **7**: 4897 (4896)
Albers J, **7**: 4749 (4745)
Albert E, **7**: 4871 (4868)
Albert K, **5**: 3421 (3419)
Albert R E, **7**: 4876 (4875)
Alberti K G M M, **4**: 2631 (2629, 2631)
Albertson V D, **6**: 3820 (3818)
Albion N, **3**: 2137
Albisser A M, **3**: 2021 (2017), 2032 (2026, 2027, 2030, 2032), 2174 (2170), 2175 (2172)
Albouy M, **1**: 687 (686)
Albouy Y, **6**: 3789
Albrecht P F, **6**: 3825 (3821)
Albright C M, **6**: 3682 (3681)
Albrycht J, **6**: 4346 (4346)
Albus J S, **6**: 4100, 4123
Albuyeh F, **2**: 1388 (1387)
Alder H, **5**: 2942 (2935)
Aldermeshian M, **4**: 2522 (2520)
Aldrich J F, **6**: 3820 (3817)
Aleksandrov P S, **6**: 3734
Alengrin G, **6**: 4176 (4176)
Alevisakis G, **2**: 1568 (1568)
Alexander A B, **1**: 437
Alexander B, **1**: 201
Alexander C, **2**: 992 (988)
Alexander D K, **4**: 2769 (2769)
Alexander H, **1**: 482 (480)
Alexander J, **5**: 3464 (3460)
Alexander R K, **5**: 3134 (3131)
Alexander R M, **6**: 4066 (4062)
Alexander S, **3**: 2164
Alexandris N, **2**: 1431 (1427), 1440 (1440), 1445 (1441)
Alfrey A C, **1**: 268 (265)
Ali F, **6**: 3643, 3652, 4159 (4158), 4160 (4158)
Alian S E-D, **5**: 3024 (3023, 3024)
Alioth A, **6**: 3663 (3661)
Alj A, **1**: 687 (686)
Alkins J, **2**: 1573
Alkon D L, **6**: 4167 (4167)
Allain C, **6**: 4257
Allaire P E, **1**: 482 (480)
Allais M, **3**: 2202 (2194, 2197), 22022203 (2194, 2195); **5**: 3104 (3103)
Allal L K, **2**: 1577 (1574)
Allan D W, **3**: 1745
Allan J F, **6**: 4274 (4274)
Allan R, **4**: 2473 (2472)
Allan R N, **3**: 1954 (1949); **6**: 3795, 4015, 4020, 4026
Allan W B, **5**: 3464 (3458)
Allbritton E J, **5**: 2994 (2992)
Allemong J J, **5**: 3180 (3176); **6**: 3850 (3849)
Allen G I, **1**: 466 (465)
Allen G L, **2**: 1248 (1248)
Allen J C, **1**: 674 (674)
Allen K R, **7**: 4645 (4645)
Allen M, **7**: 5114
Allen R, **4**: 2582 (2582)
Allen R C, **3**: 2174 (2169)
Allen T J, **7**: 4832 (4831)
Alles W, **1**: 393 (391)
Allidina A Y, **5**: 3430 (3426, 3430); **6**: 3727 (3725), 4176 (4174), 4189 (4187, 4188, 4189)
Allinger D F, **5**: 3302 (3298)
Allison G, **7**: 4784 (4783, 4784)
Allott P R, **1**: 284 (281)
Alluisi E A, **3**: 1668 (1665, 1667)
Allwright D J, **1**: 428 (427), 433 (430)
Allwright J C, **3**: 2163 (2158)
Almgren Jr F J, **1**: 529 (526)
Alper P, **5**: 3351 (3342)
Alpeter L L, **3**: 1958
Alsaç O, **6**: 3829 (3828), 3846 (3845), 3850 (3849), 3855 (3854)
Alsberg P A, **2**: 1153 (1151)
Alsholm H, **2**: 1568 (1568)
Alsina C, **2**: 1473 (1468); **3**: 1781 (1780)
Alspach D L, **4**: 2714 (2709)
Alt F L, **6**: 4159
Alter S L, **2**: 947
Altieri M A, **2**: 1317 (1317)
Altman P L, **1**: 450 (448, 449)
Altura B M, **1**: 477 (475, 477)

Alvarez W C, **1**: (460)
Alves R B, **5**: 3465 (3460)
Ambarzumian V A, **3**: 2052 (2048)
Ambler A P, **4**: 2386 (2374), 2488
Ambrosetti A, **1**: 529 (527, 528)
Ameling W, **1**: 633 (632)
Amend W J C, **2**: 1248 (1248)
Ames W, **2**: 1192 (1186)
Ames W F, **6**: 3626 (3623, 3625), 4302 (4300, 4301), 4303, 4364 (4363)
Ames W G, **7**: 5161 (5158)
Amiot L W, **6**: 3762 (3762)
Amlaner C J, **7**: 4848 (4847)
Amouroux M, **2**: 808 (802, 803, 804, 805), 1192; **6**: 4244 (4238, 4242), 4245 (4238, 4241)
Ampère A M, **2**: (867)
Amrehn H, **6**: 3744 (3741, 3743)
Amsterdamski S, **3**: 1862
Amundson R G, **2**: 1521 (1521)
Anagnostopoulos G, **1**: 340
Anbib M, **4**: 2517 (2514)
Anbumani K, **6**: 4183 (4182)
Ancker C J Jr, **7**: 4654 (4652, 4653)
Andeiev N, **6**: 4213 (4212)
Anděl J, **4**: 2292 (2287)
Anders O U, **7**: 5094 (5093)
Andersen A, **7**: 5155 (5154)
Andersen A C, **3**: 2032 (2027)
Andersen B W, **3**: 2099 (2099)
Andersen J A, **2**: 1416 (1413, 1414, 1415, 1416)
Anderson B D O, **1**: 105 (102), 703 (694, 695); **2**: 825 (814), 902 (901), 907 (903, 906), 916 (914, 915, 916), 923 (918); **4**: 2618 (2616), 2627, 2671, 2835, 2865 (2865), 2875; **5**: 3060 (3057), 3220 (3220), 3508; **7**: 4466 (4466), 4551 (4548), 4554 (4552, 4554), 4684 (4678)
Anderson D, **4**: 2347 (2345)
Anderson D H, **1**: 493 (493)
Anderson H A, **7**: 5018 (5015)
Anderson J, **6**: 4007 (4007)
Anderson J R, **3**: 2191 (2186)
Anderson L, **5**: 3176 (3174)
Anderson N C, **7**: 4980 (4977, 4979)
Anderson R F, **4**: 2342
Anderson R J, **2**: 972 (970)
Anderson R L, **5**: 3180 (3176); **7**: 4920
Anderson R W, **5**: 2993 (2991); **6**: 4252 (4250)
Anderson T, **2**: 1136 (1136); **4**: 2447 (2445)
Anderson T W, **2**: 1105 (1103); **5**: 3235; **7**: 4566 (4556, 4557, 4558, 4559, 4560, 4561, 4566)
Andersson G, **7**: 4996 (4995)
Ando S, **1**: 301 (298)
Ando T, **1**: 495 (494); **3**: 1614 (1614)
André F, **2**: 1209 (1208, 1209)
Andre G, **6**: 4123
Andre P, **6**: 4108 (4107)
Andres R, **3**: 2025 (2022), 2174 (2171)
Andrew A M, **1**: 399; **6**: 4167 (4166)
Andrew E R, **1**: 450 (449, 450)
Andrew W G, **7**: 5060 (5059)
Andrews B J, **4**: 2901 (2899)
Andrews H C, **6**: 3643 (3636)
Andrews R W, **6**: 4396 (4395)
Andrietti F, **1**: 674 (674)
Andronov A A, **1**: 570 (568); **5**: 3388 (3384)
Anelli R, **2**: 1388 (1386)
Anger H O, **6**: (3938)
Angers D, **5**: 3242 (3241)
Angrand F, **2**: 1179 (1174, 1177, 1178)
Ankumah-Saikoom E, **5**: 3232 (3230, 3232)
Anlauf J, **1**: 380 (376)
Anliker M, **1**: 445 (441, 442, 443, 444)
Annaswamy A M, **5**: 3067 (3066, 3067)
Annet F A, **7**: 5051 (5048)
Anschütz-Kaempfe H F J, **6**: 4275 (4271)
Anscombe P J, **7**: 5015 (5011)
Anthony J M, **3**: 1781 (1781)
Anthony M, **5**: 3238 (3235, 3236)
Anthony R N, **2**: 947 (943)
Antonini A, **5**: 2942 (2935)
Antonioni T N, **5**: 3282 (3282)
Antsaklis P, **5**: 3220 (3220)
Aoki K, **1**: 184 (184); **2**: 1513
Aoki M, **1**: 132 (127), 137 (133), 703 (699); **2**: 916 (912), 1098 (1095, 1097); **4**: 2559 (2559); **5**: 3475 (3471), 3572 (3569); **7**: 4684 (4683)
Aoki T, **7**: 5123 (5122)
Aoki Y, **2**: 1507; **7**: 5162
Apostel L, **5**: 3454 (3450)
Appel B R, **1**: 176
Appel U, **1**: 85 (84)
Appleton E V, **5**: (3384)
Aqleh K A, **6**: 4061 (4057, 4060, 4061), 4062 (4061)
Arafeh S A, **4**: 2706 (2702, 2703); **5**: 3144
Aragona J, **3**: 1611 (1609)
Arakawa J, **6**: 3889 (3889)
Arakawa M, **5**: 3425 (3425)
Araki M, **1**: 674 (674); **4**: 2556 (2554), 2678 (2677); **5**: 3572 (3569)
Arant J B, **2**: 839
Arbel A F, **1**: 272 (268, 272)
Arbib M A, **1**: 18, 294; **2**: 1286; **3**: 2168 (2165, 2167); **4**: 2866 (2865); **5**: 3082 (3080), 3272 (3268), 3292 (3289), 3394 (3390); **6**: 3734 (3733), 3751 (3749), 3776, 3968, 3977 (3974), 4323 (4319), 4345 (4342, 4343, 4344); **7**: 4554 (4552, 4553)
Arcelli C, **7**: 4883
Archer M, **2**: 1402 (1400)
Archetti F, **6**: 4204
Archibald G, **7**: 5155 (5155)
Archibald R D, **6**: 3904
Ardell G G, **1**: 588 (588)
Ardema M D, **7**: 4429, 4439
Arden B, **5**: 3035 (3030)
Arditty H J, **5**: 3464, 3470 (3465, 3466, 3467)
Ardran G M, **7**: 5186 (5180)
Arduini C, **3**: 1664 (1657, 1663), 1665 (1651)
Argyris C, **7**: 4784 (4783)
Arigoni A O, **3**: 1791 (1788, 1789); **4**: 2659 (2657, 2658)

Arimoto S, **4**: 2386 (2376), 2412 (2393); **6**: 4128 (4124, 4125, 4128)
Aris R, **1**: 424 (422)
Arkadev A G, **6**: 3643
Arkin G F, **1**: 142 (141)
Armentano V A, **1**: 703 (695); **2**: 923 (917); **7**: 4689 (4688, 4689)
Armiger W B, **3**: 1619 (1618)
Armijo L, **5**: 3196 (3194); **7**: 4969 (4965)
Armitage D W, **1**: 450 (448)
Armstrong E H, **3**: (1745)
Armstrong J S, **3**: 2202 (2196); **7**: 4827 (4824)
Armstrong R F, **2**: 1366 (1364)
Armstrong W W, **4**: 2385 (2379)
Arnold G P, **7**: 4848 (4847)
Arnold G W, **2**: 1537 (1537)
Arnold K E, **3**: 1958
Arnold L, **7**: 4637 (4631, 4633, 4634, 4635)
Arnold M J, **5**: 3282 (3280)
Arnold V, **1**: 323 (321)
Arnold V I, **3**: 1778 (1773)
Arnold W F, **2**: 826; **4**: 2622 (2621)
Aron M, **7**: 4927 (4927)
Aronowsky J S, **4**: 2353 (2351, 2352)
Aronsson G, **7**: 4677 (4675, 4676)
Arpaci V S, **3**: 2104 (2099)
Arrow K J, **1**: 671 (666, 668, 669); **2**: 1261 (1257, 1258, 1260); **3**: 2060 (2054, 2055); **5**: 3115 (3105), 3163 (3159); **6**: 4075; **7**: 4472 (4471), 4822 (4818)
Arseven E, **2**: 1106
Artle R P, **7**: 4495 (4495)
Artola M, **2**: 1210 (1210), 1211 (1210)
Artstein Z, **6**: 3619 (3618)
Arunachalam V, **2**: 1057 (1055)
Arzbaecher R C, **1**: 769 (760, 761)
Asada H, **4**: 2440 (2431, 2432, 2439)
Asai K, **3**: 1603 (1603), 1802 (1801), 1819 (1816), 1838
Asakawa S, **2**: 1507
Asawa C K, **5**: 3464 (3459)
Asbjørnsen O A, **1**: 588 (586, 587, 588), 589 (587); **2**: 1416 (1413, 1414, 1415)
Asbury A J, **5**: 3238 (3235, 3236, 3237)
Asbury D L, **7**: 5186 (5186)
Ascher W, **7**: 4827 (4824, 4825)
Aschoff J, **7**: 4880 (4877)
Ash G R, **1**: 735 (734); **7**: 4726 (4723), 4740 (4735), 4751 (4751)
Ash R H, **1**: 735 (734); **3**: 2113 (2111)
Ashby W R, **1**: 93 (93), 675 (672); **2**: 869 (868), 992 (990), 1355 (1353); **4**: 2497 (2492, 2495), 2543 (2542); **6**: 3768 (3766), 3989 (3984), 4007 (4007), 4353 (4350); **7**: 4645 (4644)
Ashfield A W, **3**: 2014
Ashkenas I, **1**: 120, 343
Ashley H, **7**: 5170 (5165)
Ashley J R, **6**: 3626 (3623)
Ashman R B, **1**: 474 (471)
Asholm O, **6**: 3916 (3913)
Asquith P D, **3**: 2003 (2000)
Asscher N, **2**: 849 (847)
Assilian S, **1**: 126 (124, 125), 664 (664), 713 (712); **4**: 2361 (2359, 2360), 2362 (2360); **5**: 3454 (3453)
Astarita G, **7**: 5060 (5059)
Astrahan M M, **2**: 1174 (1170)
Åström K J, **1**: 45 (39), 47, 51 (48), 54 (54), 493 (491), 588 (586); **2**: 773 (772), 1116 (1110), 1199 (1197), 1257 (1257); **3**: 2010 (2010), 2021 (2018); **4**: 2239 (2238), 2245 (2242), 2273 (2272), 2283 (2279, 2282, 2283), 2287 (2287), 2292 (2287, 2291), 2319 (2317, 2318), 2627 (2627), 2892; **5**: 3006 (3004, 3006), 3188 (3184), 3232 (3229), 3351 (3341), 3410 (3407, 3409), 3430 (3426), 3523 (3519); **6**: 3591 (3590), 3610 (3604), 3727 (3721, 3722, 3723, 3724, 3725), 3871 (3870), 3991 (3990, 3991), 3994 (3994), 4164 (4164), 4172 (4169, 4171), 4176 (4172, 4174, 4175), 4180 (4176, 4179), 4183 (4181, 4182), 4186 (4185, 4186), 4193 (4190, 4191, 4193), 4197 (4194), 4199 (4198), 4204 (4200, 4203), 4209 (4207), 4211 (4209), 4213 (4211, 4212), 4214 (4212); **7**: 4607 (4603, 4606), 4611 (4607), 4894 (4893)
Atal B S, **1**: 80 (78); **7**: 4509 (4506, 4507)
Athans M, **1**: 51 (51); **2**: 912 (908), 916, 927 (924), (930, 931), 932 (931), 956 (952), 1099 (1095, 1097), 1192 (1191); **4**: 2560 (2559), 2713 (2713), 2846 (2842, 2843, 2844); **5**: 3228 (3225, 3226), (3517), 3572 (3568); **6**: 4165 (4162); **7**: 4554 (4551, 4552), 4684 (4678, 4683), 4685 (4678, 4680, 4684), 4689 (4685), 4822 (4821)
Atherton D P, **5**: 3024 (3024), 3319 (3319), 3373 (3369, 3370, 3371, 3372), 3383 (3380, 3381), 3388 (3385, 3387, 3388), 3389 (3388), 3390 (3385, 3388); **6**: 4138 (4133); **7**: 4452 (4444, 4445, 4448, 4450, 4451, 4452)
Atkin R H, **6**: 3922 (3920)
Atkins G L, **3**: 2174 (2171); **7**: 4905 (4903)
Atkins M S, **6**: 4369 (4368)
Atkinson B, **1**: 424 (423)
Atkinson J, **7**: 5077 (5075)
Atkinson P, **7**: 4964
Atkinson R, **1**: 187 (187)
Atkinson R C, **1**: 276 (272); **7**: 4721 (4717)
Atlas D, **6**: 4041 (4037)
Attar R, **2**: 1154 (1151), 1165 (1164)
Attouch H, **2**: 812 (810), 1032
Atwater M, **6**: 3663 (3662)
Atwood M E, **3**: 2179
Aubin J P, **1**: 671; **2**: 849 (843, 845), 1033 (1023); **3**: 1917 (1913, 1914, 1917); **4**: 2342; **5**: 2997 (2996, 2997), 3184 (3181), 3530 (3525); **6**: 3622; **7**: 5046 (5043, 5045)
Auble G T, **1**: 675 (674); **2**: 1348 (1345, 1346), 1486 (1480)
Aubry J F, **5**: 3097
Audoin C, **3**: 1745
Audus L J, **3**: 2146
Auer M T, **1**: 261 (258, 260), 262 (258, 259, 260, 261); **7**: 5007 (5007)
Aulich H, **5**: 3464 (3461)
Aumann R J, **1**: 413 (410), 687 (683); **2**: 849 (840, 841, 842, 843, 844, 845, 846, 847, 848); **3**: 1917 (1913), 1926 (1923); **5**: 3256 (3255); **6**: 3619 (3617), 3772 (3769)
Auret F D, **6**: 4346 (4346)
Auslander D M, **5**: 3055 (3049); **6**: 3703 (3702)

Auslender A, **2**: 956 (954)
Auslender L, **3**: 1723; **7**: 4498 (4496)
Austin R E, **3**: 1623 (1619)
Austine M J, **3**: 2014
Avasthey P, **1**: 546 (544)
Averbukh A, **6**: 4411 (4406)
Avez A, **5**: 3329 (3325)
Avezaat C J J, **4**: 2582 (2580)
Avogaro A, **3**: 2025 (2025); **4**: 2631 (2629, 2631)
Avramovic B, **5**: 3176; **7**: 4439 (4439)
Avron M, **2**: 1522 (1517, 1521)
Avruch L I, **3**: 2032 (2027), 2175 (2172)
Avrunin G S, **3**: 2202 (2196)
Avula X J R, **4**: 2301 (2293, 2301)
Axelby G, **4**: 2575
Axelrod R M, **4**: 2640 (2637, 2638); **7**: 4784 (4783)
Ayala F J, **1**: 556 (554)
Aylmer S F, **3**: 1954 (1949, 1951)
Ayoub A K, **2**: 1388 (1387)
Ayres R U, **6**: 4097; **7**: 4827 (4826)
Ayscough P B, **7**: 4504 (4499)
Aytaç K Z, **6**: 4322
Azarbaijani S, **2**: 1329 (1327, 1328)
Azencott R, **1**: 91 (89, 90, 91)
Aziz A K, **3**: 1632
Azorín F, **2**: 1106

B

Baaklini N, **4**: 2362 (2359)
Baas J M, **5**: 3156 (3156)
Baas S M, **3**: 1847 (1846)
Baba M L, **7**: 4832 (4830), 4833 (4830)
Baba N, **4**: 2719 (2718)
Babary J P, **2**: 808 (801, 803, 805), 1192 (1186, 1187, 1189), 1193 (1186, 1187, 1188), 1199 (1197, 1199); **5**: 3496 (3494); **6**: 4244 (4242), 4245 (4241)
Babb A L, **1**: 310 (309)
Babbage C, **3**: (2175, 2178)
Babich A F, **6**: 4415 (4414)
Babloyantz A, **1**: 433 (432), 570 (569)
Babuŝka I, **3**: 1632
Bacastow R, **1**: 531 (529, 531)
Baccelli F, **2**: 1169 (1167, 1168, 1169)
Bacciotti A, **5**: 3184 (3181)
Bachelart S, **1**: 570 (569)
Bachmann W, **2**: 952 (951)
Bachofer B T, **6**: 4041 (4038)
Bach-y-Rita P, **2**: 1596 (1593)
Backer E, **2**: 1473 (1473)
Backus J W, **3**: 1712 (1712)
Backus M M, **5**: 3130 (3130)
Bacon F, **1**: (552)
Bacon J F, **3**: 1998
Badal D Z, **2**: 1154
Badawi F A, **7**: 4468 (4467)
Badr O, **1**: 681 (676)
Badran W A, **6**: 3690 (3690)
Badre A, **3**: 2178
Baer J L, **1**: 754 (753)
Baertschi A J, **2**: 1590 (1587, 1589)
Baes C F, **2**: 1344 (1343); **4**: 2782 (2779); **6**: 3933 (3931), 3936
Bagnall J, **2**: 992 (991)
Bahadur R R, **1**: 91 (91); **2**: 1106 (1103, 1104)
Baheti R S, **4**: 2253 (2252)
Bahlman S H, **1**: 289 (285)
Bahr G F, **3**: 1825 (1824)
Bahrami K, **1**: 417 (414, 415); **2**: 1337 (1336)
Bailar J C, **7**: 5186 (5185)
Bailey F N, **2**: 1305 (1299, 1302, 1304); **4**: 2556 (2555); **5**: 3144; **7**: 4571 (4568, 4569, 4571)
Bailey J B W, **5**: 3282 (3282)
Bailey J E, **1**: 424 (422); **3**: 1619 (1615)
Bailey N T J, **6**: 3762 (3756, 3757); **7**: 4645 (4644, 4645)
Bailey R W, **3**: 2186 (2185)
Bailey S J, **2**: 832, 837; **4**: 2464 (2462)
Baillieul J, **1**: 567
Baily C S, **7**: 4856 (4853)
Bainbridge L, **4**: 2361
Baïocchi C, **2**: 1211 (1210); **4**: 2342; **7**: 5036 (5033)
Baird J C, **3**: 1668 (1667)
Bairstow L, **1**: (340)
Bajaj D, **7**: 4732
Bajcsy R, **4**: 2487
Bajd T, **3**: 1910 (1910)
Baker D L, **7**: 4920
Baker D N, **1**: 142 (138, 139, 140, 141)
Baker G L, **5**: 3077 (3074)
Baker H D, **7**: 4865
Baker K R, **2**: 1099 (1098)
Baker L E, **1**: 450 (445)
Baker N H, **7**: 4865
Baker R C, **3**: 1693 (1685)
Bakhvalov N S, **5**: 3139 (3135)
Balachandran W, **3**: 1693 (1691)
Balakrishna S, **4**: 2292 (2287)
Balakrishnan A V, **2**: 1186 (1185); **5**: 3205 (3205), 3523 (3521, 3523); **6**: 4211 (4210), 4223 (4219, 4222)
Balas M J, **3**: 1665 (1649, 1650)
Balasse E O, **4**: 2631 (2628)
Balasubramanian R, **5**: 3024 (3024), 3373 (3371), 3388 (3387)
Balbo G, **6**: 3659
Balchen J G, **6**: 4266 (4264)
Balci O, **4**: 2550 (2550); **6**: 4332 (4329, 4330, 4332), 4333; **7**: 5010, 5018 (5017, 5018)
Balck K, **6**: 4160 (4158)
Baldissera C, **2**: 1160
Balduino M A, **1**: 508; **5**: 3264 (3263)
Baldwin J F, **1**: 746 (743); **3**: 1847 (1846), 1852 (1847, 1848, 1850), 1905 (1905); **5**: 2926 (2925, 2926), 3540 (3539)
Balescu R, **1**: 429 (424)
Balestrini S J, **5**: 2942
Baley K W, **3**: 1992 (1991)
Balinski M L, **5**: 3545 (3544)
Ball G H, **3**: 1902 (1901)
Ball G W, **5**: 2942 (2935, 2936)
Ballard D H, **4**: 2473 (2471), 2484 (2483), 2487
Ballard E O, **5**: 3465 (3458)
Ballester A, **6**: 4007 (4005)
Ballik E, **4**: 2723 (2721, 2722)
Ballinger D G, **7**: 5114
Balslev I, **5**: 3579 (3578)
Balzer R, **2**: 1573
Bamberger W, **1**: 638 (635); **4**: 2253 (2253); **5**: 3351 (3342); **6**: 4183 (4180)
Bammi D, **7**: 4495 (4495)
Banathy B, **6**: 3989 (3985)
Banâtre J P, **2**: 1132 (1126)
Banâtre M, **2**: 1132 (1126)
Bandle C, **2**: 1372 (1372); **4**: 2613 (2609, 2611, 2612, 2613)
Bandler W, **1**: 743 (741, 742), 746 (743); **2**: 1473 (1473); **3**: 1809 (1807, 1808, 1809), 1810 (1807), 1852 (1848), 1867 (1865, 1866), 1872 (1868); **5**: 3264; **6**: 4007 (4002, 4004, 4005, 4007)
Bandyopadhyay R, **4**: 2501 (2499)
Banerjee S, **3**: 1693 (1688)
Banichuk N, **2**: 1179 (1174, 1177)

Banin A, **1**: 148 (143)
Banks H T, **1**: 428 (425, 427); **2**: 1186 (1183); **3**: 2150, 2152 (2152); **5**: 3506
Banks J, **6**: 4395 (4395)
Banks S P, **6**: 4245
Banon G, **3**: 1822 (1822)
Banowetz, **3**: 1761 (1760)
Bansal U S, **3**: 1779 (1776)
Banse K, **7**: 4454 (4453)
Banwell C M, **1**: 13 (8)
Bányász Cs, **6**: 4184 (4180), 4213 (4212)
Baptistella L F B, **1**: 703 (699, 703); **5**: 3156 (3153, 3154, 3155)
Bar-Hillel Y, **6**: 4216 (4214, 4215)
Bar-Shalom Y, **2**: 1257 (1254, 1255, 1256, 1257); **5**: 3035 (3033); **6**: 4011 (4011), 4165 (4162, 4163), 4193 (4190)
Bar-Yosef B, **1**: 142 (138, 139)
Baradello C S, **5**: 3055 (3048, 3049, 3050, 3051)
Baram Y, **7**: 5001 (4998)
Barankin E D, **5**: 3163 (3159)
Baras J S, **5**: 3303 (3301), 3330 (3325); **7**: 4929 (4929)
Barash M M, **3**: 1648 (1644)
Baratchart L, **1**: 17, 296 (294, 295)
Barbara V P, **7**: 4785 (4780)
Barbenel J C, **1**: 445 (441, 442, 444), 478 (474), 482 (479, 480, 482)
Barber B, **2**: 1577 (1575)
Barber C R, **7**: 4851 (4851)
Barber J, **2**: 1521 (1518)
Barber M C, **5**: 2929 (2927); **6**: 4237 (4235); **7**: 4956 (4956)
Barbier L, **6**: 3801 (3795)
Barbot A, **1**: 226
Barbu V, **3**: 1734; **7**: 5041 (5038, 5039, 5040, 5041)
Barbur J L, **4**: 2330 (2326, 2327, 2328)
Barclay S, **4**: 2338 (2335, 2337)
Bard Y, **1**: 489 (487, 488); **5**: 3030 (3027)
Bardakjian B L, **2**: 1040 (1036, 1037, 1038, 1039)
Bardos C, **7**: 4944 (4941, 4943)
Barel M, **1**: 633 (632); **2**: 1087 (1085, 1086), 1201 (1199)
Barelli A E, **5**: 3464 (3458)
Bariff M L, **7**: 4793 (4791, 4792)
Barker D C, **3**: 2117 (2117)
Barker H A, **2**: 1311 (1307); **4**: 2253 (2252); **5**: 2956 (2955), 3188 (3188), 3242 (3241); **6**: 4172 (4172)
Barker I, **5**: 2993 (2991)
Barker K, **1**: 755 (753)
Barker R H, **6**: 4146 (4145)
Barkhausen H, **3**: (1745)
Barkmeyer E, **6**: 4123
Barles G, **3**: 2068 (2066)
Barlow A J, **5**: 3464 (3461)
Barlow H B, **7**: 5070 (5070), 5086
Barlow R C, **1**: 623
Barlow R E, **6**: 4011 (4008, 4009), 4029 (4027, 4029); **7**: 4766 (4764, 4765)
Barman J F, **1**: 580 (577)
Barmish B R, **2**: 1023 (1022)
Barnes B V, **2**: 1321 (1318)
Barnes C C, **6**: 3809
Barnes D P, **4**: 2447 (2447)
Barnes G R, **2**: 1590 (1590), 1595 (1591, 1593, 1594, 1595)
Barnes J A, **3**: 1745
Barnes J G P, **6**: 3897 (3895)
Barnes J L, **3**: (1748)
Barnes R M, **4**: 2347 (2343)
Barnett J A, **3**: 1694 (1692)
Barnett S, **5**: 3214 (3207, 3208, 3213), 3398, 3401 (3400), 3402 (3399, 3400, 3401), 3441 (3441); **7**: 4551 (4549, 4550)
Barney G C, **2**: 1424 (1423), 1427 (1425), 1431 (1427, 1430), 1434, 1440 (1436, 1440), 1445 (1441, 1442); **7**: 5051 (5048)
Barnthouse L W, **2**: 1539 (1538, 1539), 1552
Baron D P, **3**: 2202 (2196)
Baron S, **2**: 1020 (1017); **7**: 4784 (4784)
Barosi G, **3**: 1623 (1621, 1622)
Barr A, **1**: 307 (307); **2**: 1573; **4**: 2640 (2636); **7**: 4784 (4784)
Barr C C, **2**: 1595 (1594)
Barraud A Y, **6**: 3983 (3982)
Barrero A, **6**: 3652 (3646)
Barrett A H, **7**: 4856 (4855), 4871 (4870)
Barrett E, **6**: 3653 (3652)
Barrett E C, **6**: 4045
Barrett J F, **3**: 1778 (1769, 1777); **5**: 3351 (3342), 3376 (3374)
Barrett M F, **5**: 3228
Barrett P, **6**: 4159 (4156)
Barringer A R, **2**: 1521 (1518)
Barron F H, **3**: 2191 (2186)
Barrow H G, **6**: 4160 (4157, 4159)
Barry W, **2**: 1595 (1593, 1594)
Bars R, **3**: 2010 (2008)
Bart A J, **4**: 2901 (2899)
Bart J A, **6**: 4061 (4055, 4056)
Barta S M, **2**: 932 (931)
Bartels P H, **3**: 1825 (1824)
Bartels R H, **2**: 826 (823); **6**: 4299
Bartels R M, **4**: 2905
Bartelstone H J, **1**: 542 (540, 541)
Barth T F W, **2**: 1516 (1513)
Bartlett F, **5**: 3264 (3263)
Bartlett K G, **2**: 1532 (1528)
Bartlett M R, **7**: 5078
Bartlett M S, **2**: 1106; **4**: 2270 (2267); **6**: 3621 (3619), 4159; **7**: 4645 (4645)
Bartolin R, **1**: 769 (762, 765); **3**: 1604 (1601, 1603)
Barton D, **5**: 3560 (3560)
Barton R N, **4**: 2631 (2628, 2631)
Barton T H, **5**: 3097
Barwell F T, **1**: 352, 360, 362, 365, 367
Barwise J, **1**: 333 (326, 331, 332), 720 (715)
Basak P K, **6**: 3806 (3804)
Başar T, **1**: 687 (682, 684, 685, 687); **2**: 995 (993, 994, 995), 1268 (1263, 1264); **5**: 3149 (3146, 3147, 3148, 3149); **7**: 4622 (4622), 4630 (4629), 4631, 4822
Bascur O, **5**: 29932994 (2991)
Bashkerov O A, **2**: 1478 (1476)

Basile G, **1**: 137 (134); **4**: 2816 (2807)
Basile L J, **4**: 2511 (2509)
Basile P S, **3**: 1718 (1715)
Basili V R, **4**: 2681 (2679)
Basill G, **4**: 2702 (2699)
Basin S L, **2**: 1551 (1550)
Baskett F, **2**: 1169
Basmajian J V, **1**: 437; **3**: 1910 (1908)
Basor E, **2**: 1413 (1412)
Basore B L, **3**: 1695 (1694)
Bass L, **3**: 2143 (2141)
Bass R W, **5**: 3388 (3386)
Bassan Y, **3**: 2076
Bassani J W M, **1**: 508 (507)
Basserman R W, **7**: 4956 (4954, 4956)
Bassett M, **1**: 179 (179)
Basseville M, **1**: 85 (82), 86 (82, 83, 84, 85)
Bassu S, **5**: (3132)
Batchelor B G, **4**: 2479; **6**: 3643
Batchelor D W, **3**: 1694 (1692)
Batelle M F, **2**: 1348 (1346)
Bateman L F, **1**: 245
Bates J F, **2**: 965, 969, 978
Bates R G, **6**: 3687 (3685)
Bateson G, **2**: (870, 872)
Bathe K J, **3**: 1632
Batley G E, **7**: 5127
Bator J M, **2**: 866 (865)
Batson V M, **1**: 246
Battaglia F C, **7**: 4905 (4905)
Batteau D W, **1**: 326 (325)
Battisson M J, **6**: 3806 (3804)
Battocletti J H, **5**: 3421 (3419)
Battye C K, **3**: 1910 (1908)
Baudet G M, **2**: 1125 (1123); **3**: 2163 (2154)
Baudier E, **2**: 849 (843)
Baudois D, **1**: 682 (679)
Baudouin A P, **2**: 1181 (1181)
Bauer H, **3**: 1802 (1801), 1821 (1819, 1820)
Baughman G L, **7**: 4901 (4899)
Bauman E J, **2**: 956 (952)
Bauman R P, **7**: 4504
Bauman W C, **1**: 177 (175)
Baumbick R J, **5**: 3464 (3460)
Baumert L, **1**: 725 (725)
Baur U, **4**: 2253 (2253)
Bauzil G, **6**: 4123
Bax N J, **6**: 4234 (4233)
Baxley R A, **6**: 3679 (3675)
Bay T C W, **1**: 564 (561)
Bayer E, **5**: 3421 (3419)
Bayer R, **2**: 1132 (1129), 1136 (1134, 1135), 1154 (1153), 1174 (1169, 1170, 1171)
Bayes R, **2**: 1478 (1474)
Bayoumi M M, **1**: 45 (42); **5**: 3232 (3229, 3230, 3232)
Bayraktar B A, **2**: 1454 (1453); **3**: 1718 (1715)
Bazaraa M S, **4**: 2692, 2841
Bazukis E V, **2**: 1356 (1356)
Beach B H, **3**: 2202 (2195)
Bean A E, **7**: 4751 (4751)
Bean A N, **7**: 4741 (4739)
Bear F E, **1**: 7 (1, 2)
Beard C S, **3**: 2099 (2099)
Bech M I, **6**: 4261 (4259)
Beck M B, **7**: 5008 (5006, 5007)
Beck M S, **3**: 1693 (1688, 1689, 1690, 1691, 1692, 1693)
Beck W, **5**: 3464 (3461)
Becker D P, **4**: 2582 (2580, 2581)
Becker F F, **7**: 4958 (4957)
Becker G, **5**: 2944 (2935)
Becker J C, **2**: 1043 (1041)
Becker M, **3**: 1693 (1688)
Becker N G, **1**: 417 (414, 415)
Becker P W, **6**: 3643, 4015 (4012, 4014)
Becker R G, **5**: 3196 (3189), 3215 (3206), 3383 (3382)
Becker W H, **7**: 4856 (4852)
Beckett J T, **4**: 2385 (2382)
Beckman R, **2**: 1551; **6**: 4235 (4231, 4233)
Beckwith T G, **3**: 1710
Becquerel A H, **6**: (3936)
Bedford R E, **7**: 4851
Bedford R J, **2**: 1434 (1434), 1445 (1441)
Bedocs L, **4**: 2769 (2769)
Beebe J R, **2**: 1427 (1425)
Beedon D, **3**: 1998
Beek J, **5**: 3268 (3266)
Beer S, **2**: 869 (869); **4**: 2352 (2351); **5**: 3035 (3034); **7**: (4771)
Beghelli S, **5**: 3044 (3040, 3043, 3044); **6**: 3776 (3772)
Begovich C L, **1**: 142 (140, 141), 592 (590, 591)
Behiri J, **1**: 473 (468, 473)
Behrens W W, **2**: 1538 (1536); **3**: 2016 (2015); **7**: 4762 (4756)
Behringer R P, **1**: 569 (568)
Beilner H, **6**: 3670 (3668)
Bejczy A K, **4**: 2440 (2427, 2432, 2434, 2437, 2438), 2460 (2458)
Bekey G A, **1**: 493 (492), 502 (499); **3**: 2228 (2225); **5**: 3238 (3237), 3248 (3245); **6**: 3776 (3772), 3991 (3991), 4172 (4170, 4172), 4176 (4172), 4184 (4182), 4213 (4212, 4213), 4310 (4308), 4411 (4406); **7**: 4551 (4548), 4692
Bel G, **3**: 1648 (1645, 1647, 1648)
Belady L A, **4**: 2681 (2679)
Bélanger P R, **1**: 51 (49, 50); **6**: 3994 (3994), 4176 (4172); **7**: 4894 (4892)
Belevitch V, **4**: 2835; **6**: 3751 (3748)
Belford G C, **2**: 1154 (1153)
Belforte G, **3**: 2143 (2140, 2141, 2142, 2143); **4**: 2257 (2256), 2635 (2634)
Bell D C, **3**: 2202, 22022203 (2194, 2195)
Bell D E, **5**: 3104 (3102), 3124 (3122), 3152; **7**: 4985
Bell D J, **4**: 2872 (2872); **5**: 3228 (3227), 3508 (3508); **6**: 4133 (4129); **7**: 4425 (4425)
Bell G, **1**: 433 (432)
Bell G I, **7**: 4944
Bell J L, **3**: 1791 (1789)
Belleni-Morante A, **5**: 3082 (3081)

Belletrutti J J, **1**: 580 (575, 577); **5**: 3383 (3380, 3381), 3436 (3432); **6**: 3717 (3710)
Bellhouse B J, **1**: 315 (314)
Bellhouse F H, **1**: 315 (314)
Bellini A, **5**: 3279 (3276)
Belliston L, **2**: 992 (988, 989)
Bellman R, **1**: 409 (407), 493 (491), 664 (651), 725 (724); **2**: 1099 (1095, 1097), 1211 (1209), 1257 (1255), 1278, 1279 (1278), 1282 (1279), 1311 (1307, 1308), 1337 (1336); **3**: 1714, 1781 (1780), 1819 (1816), 1825, 2052 (2048), 2068, 2072 (2069), 2150, 2152 (2152); **4**: 2556 (2555), 2565 (2560), 2664, 2719 (2715), 2865 (2865); **5**: 2921 (2921), 3215 (3206), 3454 (3451), 3475 (3471), 3506, 3512 (3511), (3516), 3538 (3535); **6**: 4165 (4162); **7**: 4529 (4527), (4551, 4552), 4626 (4623), 4661 (4657, 4658), 5153 (5153)
Belloni F L, **2**: 853 (852)
Bellos I, **4**: 2386 (2374)
Bellville J W, **6**: 4061 (4057, 4060, 4061), 4062 (4059, 4061)
Belogus D, **6**: 4411 (4407)
Ben-David M, **3**: 1887
Ben Israel A, **5**: 3362
Ben-Tal A, **5**: 3164 (3161), 3362
Benayoun R, **5**: 3152 (3151), 3164 (3157)
Benbasat I, **7**: 4793 (4792)
Benci V, **1**: 529 (528)
Bendat J S, **4**: 2253 (2249), 2310 (2307, 2309); **6**: 4281 (4279)
Bender E A, **5**: 3082 (3079)
Bender F E, **3**: 1705 (1701)
Bendixson I, **5**: (3384)
Bendow B, **5**: 3464 (3459)
Beneken J E W, **1**: 546 (542, 545, 546)
Beneš V E, **5**: 3302 (3300, 3301), 3329 (3322, 3324), 3523 (3523)
Benet L Z, **2**: 1248 (1247, 1248)
Benett S, **4**: 2473 (2472)
Bengtsson G, **4**: 2567 (2566), 2570 (2569)
Benham C J, **1**: 433 (432)
Benilan Ph, **6**: 4223 (4222)
Benît A, **3**: 1756 (1753)
Benjamin J R, **2**: 1553 (1552); **7**: 5015 (5014)
Benjamin R W, **1**: 142 (138, 139, 140, 141)
Bennet J R, **3**: 1958
Bennet W, **2**: 1181 (1181)
Bennett D, **2**: 1537 (1537)
Bennett F M, **6**: 4051 (4048, 4049)
Bennett J, **5**: 3079 (3078)
Bennett J L, **2**: 947
Bennett J R, **7**: 5134 (5133), 5135 (5130, 5133, 5134)
Bennett L L, **3**: 2174 (2173)
Bennett P G, **7**: 4798 (4798)
Bennett S, **2**: 1402 (1402); **3**: 1607 (1604, 1607), 1749 (1745); **6**: 4274 (4272); **7**: 4517 (4513, 4514, 4515, 4516, 4517), 4577 (4573)
Benoît A, **1**: 196 (193); **3**: 1756 (1750, 1751, 1752, 1753, 1754)
Benoit J, **2**: 1573
Benrejeb M, **7**: 4547
Benson A J, **2**: 1590 (1590), 1595 (1593, 1594, 1595)
Benson H P, **5**: 3164 (3159, 3160, 3163)
Benson R S, **2**: 1466 (1463)
Bensoussan A, **1**: 409; **2**: 812 (809), 1020 (1017, 1018), 1121 (1116), 1179, 1180, 1192 (1191, 1192), 1211, 1268 (1264); **3**: 1648 (1644), 1714, 1921, 2151, 2152 (2152); **4**: 2342 (2342), 2559 (2559), 2583, 2664 (2663); **5**: 3205, 3279 (3279), 3310 (3309), 3316 (3313, 3314), 3376 (3374, 3376), 3502, 3506, 3523 (3522), 3530 (3525, 3530), 3535; **6**: 3596 (3593, 3594, 3595), 3706 (3706), 3755, (3890), 4070, 4244 (4238, 4239, 4240), 4248 (4248); **7**: 4468 (4467), 4612, 4617 (4615, 4617), 4621 (4617, 4618, 4620), 4644 (4643), 4944 (4943), 5036 (5032)
Bentley F F, **7**: 4504 (4499)
Benton R, **3**: 2117 (2114, 2117); **4**: 2901 (2899)
Bentz A P, **7**: 5112
Benveniste A, **1**: 80 (77, 78, 79, 80), 86 (82, 83, 84, 85)
Benz H, **1**: 633 (632), 634 (632)
Benzecri J P, **2**: 879 (877)
Benzinger T H, **7**: 4856 (4853), 4876 (4873), 4880 (4879)
Berbari E J, **2**: 1410 (1409)
Berbee H, **3**: 1929 (1929)
Berecz I, **5**: 2942 (2935)
Berens H, **6**: 4223
Berestycki H, **1**: 529 (528, 529)
Berg C R, **6**: 4302 (4300)
Berg E, **4**: 2386 (2374)
Berg H, **1**: 634 (632)
Berge C, **5**: 3550
Bergen A R, **2**: 981 (981); **5**: 3319 (3319), 3388 (3387), 3389 (3386); **7**: 4452 (4450)
Berger A R, **7**: 4856 (4852)
Berger B, **6**: 4142
Berger J O, **6**: 3616 (3613)
Berger T, **2**: 1492
Bergh R A, **5**: 3470 (3466, 3467)
Bergin T G, **1**: 399 (394)
Bergland G D, **1**: 754 (754); **3**: 1723; **5**: 3558 (3556)
Bergman D, **2**: 812 (809, 810, 812)
Bergman R N, **3**: 2025 (2022, 2023), 2032 (2026), 2174 (2171), 2175 (2174)
Bergmann S, **5**: 3351 (3341)
Bergner P-E E, **3**: 1623 (1620); **7**: 4905 (4903, 4904)
Bergström J, **1**: 312 (310)
Bergstrom T C, **1**: 671 (668)
Bergter F, **1**: 424 (423)
Berhaegen M, **4**: 2622 (2621)
Berheide W, **6**: 4306 (4303)
Berkenbosch A, **6**: 4061 (4057, 4061), 4062 (4059)
Berkey G A, **5**: 3410 (3408)
Berkowitz L D, **2**: 1016 (1012); **5**: 3512 (3512)
Berkowitz W D, **2**: 1410 (1408)
Berkson J, **2**: 1106
Berkus M, **6**: 4345 (4343)
Berman M, **3**: 2174 (2171, 2172); **4**: 2331 (2325), 2887; **7**: 4905 (4902, 4905), 4906 (4905)
Bernadou M, **3**: 1632
Bernard C, **3**: 2174 (2169); **7**: (4877)
Bernardelli H, **6**: (3752)

Bernas R, **5**: 2942
Berne R M, **2**: 853 (851)
Berner R A, **7**: 5094 (5090)
Bernhard P, **2**: 1008 (1005, 1006, 1008), 1016 (1012, 1013), 1020 (1018), 1023 (1022), 1268 (1266); **3**: 2163 (2153, 2161); **4**: 2849; **6**: 4070
Bernhard R, **6**: 4011 (4008)
Bernorio M, **4**: 2385 (2374)
Bernotat R K, **3**: 2186 (2181)
Bernoulli D, **7**: (4986)
Bernstein D A, **2**: 1209 (1208)
Bernstein D S, **1**: 132 (132); **6**: 4418 (4418)
Bernstein P A, **2**: 883 (881, 883), 884 (881), 1136 (1133, 1134), 1154 (1143, 1145, 1146, 1147, 1148, 1151, 1152, 1153), 1155 (1145), 1160 (1159), 1165 (1164), 1169 (1166, 1168)
Bernussou J, **1**: 703 (695); **2**: 923 (922); **7**: 4684 (4679), 4685 (4678, 4680, 4682, 4683, 4684), 4731, 4740 (4738), 4741
Berrada M, **3**: 1648 (1645)
Berridge M J, **1**: 428 (425)
Berrie T W, **3**: 1718
Berson A S, **1**: 769 (760, 761)
Berstad H, **5**: 2994 (2991)
Bertelé U, **2**: 1282 (1279, 1280)
Berthelot P E M, **7**: (5117)
Berthier P, **2**: 879 (878, 879)
Berthod M, **1**: 575 (573, 574); **6**: 3645
Berthouex P M, **2**: 1553 (1552); **6**: 4229 (4228)
Bertram J E, **6**: 4146 (4146); **7**: 4529 (4528), 4685 (4684)
Bertrand P, **1**: 132 (128, 129, 130, 131), 493 (492), 703 (698, 699); **2**: 956 (952, 954, 955)
Bertsekas D P, **1**: 405 (400, 403, 404, 405); **2**: 793 (791, 793), 797 (795), 801 (801), 1099 (1093, 1095, 1096, 1098), 1125 (1122, 1123, 1125), 1388 (1388); **3**: 1648 (1644); **5**: 2932, 3475 (3471), 3487 (3487), 3510 (3509), 3523 (3518, 3519); **6**: 4310 (4309); **7**: 4612, 4741 (4736)
Berzuini C, **3**: 1623 (1619, 1620)
Bessemer H, **6**: 4275 (4272)
Bessis M, **1**: 445 (444)
Bessman J D, **2**: 866 (865)
Best E, **2**: 1136 (1133)
Beth E W, **1**: 333 (326)
Bethe A, **6**: 4066 (4063)
Bettayeb M M, **1**: 296 (294, 295)
Bettencourt V M Jr, **5**: 3030 (3029)
Bettini J P, **5**: 3470 (3467)
Bettman J R, **3**: 2191 (2186)
Bevegard B S, **1**: 542 (540)
Bevegård S, **1**: 546 (544)
Bevilacqua F, **4**: 2385 (2379)
Bewley L V, **6**: 3806 (3803)
Bewley T, **3**: 1926 (1924)
Beynon J H, **5**: 2942 (2937)
Bezdek J C, **1**: 608 (607); **2**: 1106 (1105); **3**: 1603 (1601, 1602, 1603), 1831 (1830), 1838, 1874; **6**: 3643 (3641)
Beziac G, **3**: 2137
Bhaktavatsalam R, **4**: 2719 (2718)
Bhanu B, **4**: 2473 (2472)
Bhargava B, **2**: 1165
Bhargava B K, **6**: 3653 (3648), 4160 (4158)
Bhargava V K, **2**: 1057 (1055, 1056)
Bharucha-Reid A T, **6**: 4011 (4011), 4356; **7**: 4621 (4617)
Bhatia N P, **3**: 1898 (1896)
Bhattacharya S P, **4**: 2702 (2699)
Bhattacharyya G K, **1**: 86 (82)
Bhuta P G, **5**: 3470 (3468)
Bianchini R M, **5**: 3184 (3182)
Bickford J P, **6**: 3806 (3802, 3804, 3805)
Bickford R G, **1**: 281 (279)
Bickley W G, **4**: 2905
Bidulock D S, **1**: 755 (752, 753, 754)
Biemann K, **3**: 1936 (1933), (1938)
Bierman G J, **1**: 51 (50); **2**: 826 (819); **4**: 2283 (2280, 2281), 2622 (2620, 2621), 2892 (2891); **5**: 2950 (2950); **6**: 4204 (4202); **7**: 4466 (4464), 4539 (4535, 4536, 4537, 4539), 4540 (4535)
Biermann A W, **6**: 4159 (4159)
Bigelow J, **2**: 873 (871, 872); **4**: (2271)
Bigger J, **2**: 1458
Biggers K B, **6**: 4108 (4104)
Biggs M C, **2**: 797 (795)
Bihain A, **5**: 3546
Biller H, **2**: 893 (891)
Billera L J, **2**: 849 (843, 845, 846)
Billings S A, **2**: 1093 (1087, 1092), 1116 (1110); **3**: 2113 (2113); **5**: 3475 (3473); **6**: 3776 (3772), 4180 (4179), 4199 (4198), 4204 (4199), 4266 (4264); **7**: 4551 (4548), 5031 (5029, 5030)
Billings W D, **7**: 4897 (4896)
Billingsley J, **6**: 3870 (3867, 3868)
Billingsley P, **6**: 3997 (3994)
Billinton R, **6**: 3795, 3834 (3831, 3832), 4015, 4020, 4026
Billman K W, **7**: 4476
Bilsen F A, **1**: 326 (325)
Binder Z, **1**: 681 (676), 682 (678, 679, 681)
Binford T O, **1**: 347 (343), 759 (759); **4**: 2473; **6**: 3766 (3762)
Bintz, **3**: 1761 (1760)
Birat J P, **3**: 1734 (1734)
Birch D, **7**: 4833 (4830)
Birch R, **7**: 5186 (5180)
Birdsall T G, **1**: 664 (654, 659)
Birk J R, **4**: 2385 (2382, 2383); **6**: 4123
Birkhead W G, **2**: 839
Birkhoff G, **2**: 1286 (1283); **4**: 2342, 2871 (2866, 2867); **5**: 2921 (2918); **6**: 3616 (3616)
Birnbaum L, **2**: 1573
Biroli M, **2**: 1211 (1210)
Birrell A D, **2**: 1132 (1126)
Birta L G, **6**: 4310 (4309); **7**: 4692 (4691)
Birtles A B, **3**: 2096 (2096)
Birtwistle G, **1**: 755 (753); **4**: 2550 (2547, 2550); **6**: 4350 (4349)
Bischoff K B, **2**: 1291, 1297; **4**: 2517 (2512)
Biscoe P V, **1**: 142
Bishop Y M M, **2**: 1106; **5**: 2917
Bismut J, **7**: 4822 (4817)
Bismut J M, **5**: 3523 (3522); **7**: 4612, 4644 (4643)

Bisogni J J Jr, **4**: 2782 (2781)
Bisplinghoff R L, **7**: 5170 (5165)
Bisschop J, **6**: 4332 (4329)
Biswas A K, **2**: 1538 (1537); **7**: 5008 (5006, 5007), 5140, 5177 (5171)
Bixler H J, **5**: 2942 (2936)
Björck Å, **2**: 826 (816); **5**: 3558 (3556); **6**: 4359 (4358)
Björkström A, **1**: 531 (529)
Björn L O, **2**: 1521 (1521)
Bjorner D, **3**: 1712 (1712)
Blaauw A K, **7**: 4715 (4714)
Blachere H T, **3**: 1611 (1609)
Black D, **3**: 2060 (2057)
Black E, **6**: 3909 (3908)
Black F O, **2**: 1596 (1594)
Black H S, **3**: 1749 (1746, 1747); **4**: (2271)
Black M, **5**: 2923 (2921)
Black M M, **1**: 478 (477), 486 (483, 484, 486)
Black W R, **1**: 268 (266)
Blackburn J D, **6**: 3706
Blackburn J F, **6**: 4257
Blackburn S, **7**: 5123 (5122)
Blackith R E, **2**: 1106
Blackman R B, **3**: (1748); **4**: 2270 (2267)
Blackwell B, **1**: 437
Blackwell D, **2**: 1099 (1095); **3**: 1926 (1923); **5**: 3163 (3159)
Blagodatskih V I, **2**: 1033; **5**: 3184 (3181, 3182)
Blahut R E, **1**: 91 (91)
Blais G, **7**: 5011 (5008), 5019
Blake E A, **3**: 1700
Blake G, **7**: 5008 (5006)
Blakemore C, **4**: 2330 (2326); **6**: 4167 (4167)
Blakeslee T R, **3**: 2202 (2201)
Blalock A B, **1**: 556 (554)
Blalock H M Jr, **1**: 556 (554); **5**: 3235 (3234)
Blanch H W, **3**: 1611 (1609)
Blanchard B S, **2**: 1003 (996); **4**: 2764; **6**: 3904; **7**: 4771
Blanchard M, **2**: 1143
Blanchard N, **3**: 1785 (1783, 1785)
Bland D R, **6**: 4142 (4140, 4142)
Blankenship G L, **2**: 1192 (1192); **5**: 3310 (3304, 3306, 3307, 3308); **6**: 3997 (3996); **7**: 4434 (4432, 4433), 4944 (4943)
Blanning R W, **6**: 4229 (4228), 4332 (4329)
Blaquière A, **1**: 687 (685); **2**: 1010 (1010), 1269 (1264); **5**: 3388 (3384); **7**: 4980 (4978)
Blaser A, **6**: 3653
Blasgen M W, **2**: 1154 (1144), 1174 (1170)
Blass E M, **3**: 2213 (2213)
Blattner M, **4**: 2798 (2797)
Blatz P J, **1**: 477 (475, 477)
Blauert J, **1**: 326 (325)
Bledsoe L J, **2**: 1538 (1537)
Bleich H L, **2**: 1366 (1363)
Bligh A S, **6**: 3940 (3936)
Bligh J, **7**: 4873, 4876 (4876)
Blin J M, **3**: 1847 (1847); **6**: 3643
Bliss D, **1**: 428 (426, 427, 428)
Bliss G A, **5**: 3512 (3512)
Bloch F, **5**: (3417)
Blockley D I, **5**: 3454 (3451, 3454)
Bloom S, **4**: 2775 (2773)
Bloore G C, **4**: 2362
Bloss S, **4**: 2646 (2646)
Blum H, **7**: 4883 (4881)
Blum J, **2**: 1179 (1174); **6**: 3994 (3992)
Blum M, **1**: 709 (708)
Blume C, **4**: 2351 (2347), 2469 (2468)
Blume V C, **4**: 2385 (2374)
Blumenthal R, **1**: 433 (432)
Blumentritt W, **7**: 4917 (4916)
Blunden W R, **7**: 4947
Bobellier P A, **7**: 4935 (4933)
Bobillot G, **6**: 3643
Bobrovsky B Z, **5**: 3310 (3305, 3306)
Bobrow D G, **4**: 2338 (2338), 2473
Bochenski I M, **4**: 2579 (2575, 2577)
Bochman G V, **2**: 1209 (1209)
Böchner S, **3**: 1728
Bock O, **2**: 1590 (1587, 1590)
Bode H W, **1**: (732, 734); **3**: 1749 (1746, 1747); **4**: (2271); **5**: 3215 (3206), 3228 (3224)
Bodenstein G, **1**: 86 (84)
Bodley C S, **3**: 1665 (1651, 1653)
Bodvarsson G S, **3**: 2006 (2006)
Boehne E W, **6**: 3806 (3805)
Bogard J S, **1**: 176 (175)
Bogdanoff J L, **2**: 1479 (1474)
Bogdanov A A, **1**: 399 (394)
Bogner I, **5**: (3516)
Bogoliubov N, **5**: 3024 (3020), 3389 (3384, 3385)
Bogumil R J, **3**: 2174 (2174)
Bohatka S, **5**: 2942 (2935)
Bohlin T, **2**: 773 (772); **4**: 2287 (2287), 2307 (2302), 2319 (2317, 2318); **7**: 5001 (4999)
Bohm D, **1**: 333 (332)
Bohn E V, **5**: (3387)
Bohr N, **1**: 333 (330)
Böhret H, **1**: 393 (393)
Boidin M, **7**: 5027
Boinodiris S, **6**: 4302 (4301)
Boisserie J M, **3**: 1632
Boissonnat J D, **1**: 347 (345, 347)
Boiteux A, **1**: 433 (431)
Bojanowski R, **2**: 1516 (1516)
Boland J P, **7**: 4856 (4853)
Bolc L, **4**: 2473
Boldyreff W E, **2**: 839; **3**: 1958 (1957)
Bolie V W, **3**: 2025 (2022)
Bolin B, **1**: 531 (529)
Boll F, **7**: 5086 (5080)
Bolles R C, **3**: 2213 (2212)
Bollinger K E, **5**: 3215 (3206)
Bolt G H, **1**: 148 (143, 144)
Bolton A G, **6**: 4289 (4288)
Bolton E, **6**: 3806 (3804)

Bolton W, **3**: 1710
Boltyanski V G, **2**: 1099, 1278; **5**: 2963 (2956, 2959), 3513 (3511), 3515 (3514, 3515), 3517 (3516), (3517); **6**: 3956 (3956), 4000 (3999)
Boltz D F, **7**: 5122
Bolzoni M L, **1**: 755 (752)
Bompiani E, **7**: (4517)
Bona B, **3**: 2143 (2140, 2141, 2143); **4**: 2635 (2634)
Bonadonna G, **2**: 1479 (1478)
Bonatti I S, **7**: 4731, 4740 (4738)
Bonczek R H, **2**: 947
Bondavera O N, **2**: 849 (844)
Bondesson U, **2**: 963 (962)
Bonfield W, **1**: 473 (468, 469, 470, 471, 472, 473), 474 (468, 469, 470)
Bongard M, **6**: 3643
Bongiorno J J, **4**: 2570 (2568)
Bongulielmi A P, **6**: 4304 (4303)
Bonin J C, **3**: 1642 (1640)
Böning B, **1**: 380 (376)
Bonissone P P, **3**: 1847 (1846)
Bonivento C, **6**: 3983 (3983)
Bonka H, **6**: 3936
Bonnans J F, **5**: 3496 (3494)
Bonnard B, **1**: 124 (123, 124), 323 (320, 321, 323); **2**: 1583; **4**: 2759 (2758); **5**: 2948 (2948)
Bonne U, **2**: 1226 (1214, 1215, 1216, 1217, 1223, 1225), 1233 (1228, 1232, 1233), 1239 (1236), 1241
Bonnemain, **3**: (1606)
Bonner S, **4**: 2351 (2348)
Bonnier G, **7**: 4851
Book W J, **6**: 4088 (4086)
Boole G, **4**: (2577)
Booler R V, **7**: 5186 (5186)
Boone W W, **2**: 937 (933)
Booth D A, **3**: 2213 (2210, 2211, 2212)
Booth R S, **7**: 4897 (4896)
Booth T L, **3**: 2045 (2045); **6**: 3653 (3646)
Boothby W A, **5**: 3288 (3284)
Boothby W M, **4**: 2756 (2754); **5**: 2948 (2944, 2948), 3082 (3081); **6**: 3977 (3973)
Booton R C Jr, **5**: 3388 (3385, 3388)
Borbé T, **6**: 4337 (4336)
Borcherding K, **3**: 2192 (2186)
Bordry F, **7**: 5027
Borel E, **1**: 687 (682)
Borelli G A, **2**: (870)
Borish M, **7**: 5077 (5077)
Borisovic J G, **3**: 2150 (2148), 2152
Borisson U, **1**: 51 (49, 50), 564 (562); **2**: 1116 (1110), 1199 (1197), 1257 (1257); **4**: 2892; **5**: 2993 (2990), 3060 (3059), 3188 (3184), 3232 (3229), 3430 (3426); **6**: 3591 (3590), 3727 (3723), 3994 (3994), 4186 (4185), 4193 (4193), 4197 (4194), 4199 (4198); **7**: 4607 (4603)
Borkowski L, **5**: 2923 (2921)
Borman S A, **5**: 2942
Bormann F H, **2**: 1318, 1321 (1318, 1320), 1358; **5**: 3268 (3266)
Born M, **4**: 2723 (2720)
Bornard G, **7**: 4677 (4677)
Borne P, **5**: 3406 (3402, 3405); **7**: 4429 (4428), 4547
Borodin A, **1**: 725 (722)
Boroush M A, **7**: 5177 (5171)
Borrill P A, **3**: 1961
Borrisson U, **7**: 4611 (4607)
Bortins R, **5**: 3508 (3507, 3508)
Bortolan G, **1**: 769; **3**: 1847
Bortolucci A, **2**: 1479 (1478)
Bortolussi M, **3**: 1668 (1666)
Borun R, **1**: 769 (761)
Borůvka O, **6**: 4007 (4003, 4006)
Bose A, **2**: 1388 (1387)
Bose B K, **5**: 3092
Bose N K, **4**: 2875; **5**: 3130 (3128, 3129, 3130), 3131 (3129), 3134 (3131, 3132), 3135 (3133); **6**: 3972 (3970, 3971, 3972)
Bose R C, **6**: 4294 (4291)
Bosley M J, **4**: 2800
Bossel H, **1**: 740 (738)
Bosserman R W, **2**: 1346 (1345), 1348 (1345, 1346), 1355 (1352, 1353, 1355), 1486 (1480); **3**: 1601 (1600), 1862 (1859); **5**: 2930 (2929); **6**: 4237 (4235), 4238 (4235)
Boston J F, **1**: 585 (583)
Bosveld M, **4**: 2347 (2345)
Boswell G G J, **6**: 3940 (3937)
Both W, **1**: 588 (588)
Botkin D B, **2**: 1317 (1315, 1317), 1321 (1318); **7**: 4645 (4645)
Bott R, **7**: 5046 (5043)
Botta O, **3**: 1785 (1785)
Bottomley P A, **1**: 450 (449, 450); **2**: 1521 (1518)
Botz C K, **3**: 2032 (2030), 2174 (2170)
Bouch W E, **3**: 1975
Boucher T O, **6**: 3904
Boucher W I, **2**: 1542 (1540); **6**: 4155
Bouchon B, **3**: 1838 (1836, 1837)
Boudon R, **1**: 556 (554)
Bouhou A, **6**: 4336
Boullé D, **6**: 3954 (3951, 3953)
Boulton M, **7**: (4513)
Boumans P W J M, **4**: 2347 (2345)
Bourbaki N, **3**: 1778 (1777); **6**: 4007 (4001)
Bourbin Y, **5**: 3470 (3466)
Bourdeau L, **6**: (3861)
Bourdon P, **2**: 1121 (1116); **5**: 3292 (3289), 3376 (3374, 3375); **6**: 3776 (3772, 3773)
Bourke P J, **2**: 1532 (1528)
Bourne A J, **4**: 2447 (2445); **6**: 4015
Bourne D E, **3**: 2014
Bourne J R, **6**: 4160 (4158)
Bouroche J M, **2**: 879 (878, 879)
Bousquet M, **2**: 1057 (1052)
Boustang N M, **5**: 3508 (3507, 3508)
Bove A A, **2**: 853
Bovet D P, **1**: 754 (753)
Bovey F A, **7**: 4504
Bowden C R, **3**: 2025 (2023), 2174 (2171)
Bowen H C, **4**: 2353 (2352)

ESC8-D*

Bowen R D, **5**: 2943 (2942)
Bowen U C, **3**: 2191 (2186)
Bower G H, **7**: 4721 (4717)
Bowes C, **4**: 2904 (2902)
Bowland B J, **1**: 735 (735); **4**: 2596 (2589); **5**: 3383 (3380)
Bowler B C, **7**: 4439 (4439)
Bowran P, **6**: 4284 (4283)
Bowyer A, **1**: 347 (346)
Box E P, **7**: 4566 (4562, 4565, 4566)
Box G E P, **1**: 489, 588 (588); **2**: 1497 (1494); **4**: 2307 (2304); **6**: 3949 (3949), 4229 (4228), 4235 (4232); **7**: 5001 (4998), 5015 (5013, 5014), 5155 (5154)
Boxenbaum H G, **6**: 3698 (3697)
Boyarsky S, **7**: 4980 (4976)
Boyce P R, **4**: 2769 (2767), 2777 (2776)
Boycott B B, **7**: 5070 (5066)
Boye-Møller K D, **3**: 1985
Boylan M G, **7**: 4833 (4828, 4829, 4830, 4831)
Boyle B J, **2**: 1577 (1574)
Boyle J M, **2**: 826, 827 (815, 818, 824)
Boyle T J, **3**: 1611 (1609)
Boylen C W, **7**: 5008 (5007)
Bozler E, **7**: 4980 (4976)
Bozler G, **1**: 490 (488)
Braae M, **3**: 1905 (1903); **5**: 2993 (2990)
Braakman R, **4**: 2582 (2581, 2582)
Bracchi G, **2**: 1155 (1150), 1160 (1159)
Braccio M, **3**: 2080
Brackmann R T, **5**: 2943 (2934)
Braddick O, **7**: 5077 (5075)
Bradley D, **6**: 4252 (4251)
Bradley S E, **1**: 547 (544)
Bradley S P, **4**: 2841
Bradlow H S, **5**: 3238 (3237)
Bradshaw J D, **4**: 2723 (2721)
Brady M M, **1**: 450 (449)
Brady N C, **1**: 7 (2, 5)
Braeckman M, **7**: 4703 (4701)
Braffirt P, **3**: 1712 (1712)
Brain A E, **1**: 759 (755); **4**: 2487; **6**: 4123
Brain T J C, **3**: 1693 (1692)
Braitenberg V, **1**: (464)
Braithwaite R B, **7**: 4774 (4771)
Brakkee A J M, **1**: 482 (480)
Brameller A, **2**: 1388; **3**: 1954 (1949), 1958 (1957), 1990 (1989); **7**: 5099, 5106 (5103, 5104, 5105)
Brams G W, **2**: 1143
Brams S J, **2**: 850 (849)
Bramwell D G, **3**: 1958
Brandeau M L, **2**: 1565 (1559)
Brandon D M, **4**: 2800 (2799)
Brandt A, **1**: 85 (84); **5**: 3139 (3135, 3136, 3138, 3139)
Brandt G B, **5**: 3464 (3460)
Brandwood L, **2**: 1106
Brans J M, **7**: 4981 (4976, 4979)
Bransby M L, **7**: 4577 (4573, 4577)
Brantzaw M, **5**: 3425 (3425)
Brasch F M, **2**: 902 (900); **6**: 3717 (3713), 3734 (3733)
Brash M, **6**: 3720 (3720)
Bratley P, **6**: 3950 (3948); **7**: 4656 (4655, 4656)
Brauer W, **6**: 3670 (3665, 3667)
Braun A C, **7**: 4958 (4957)
Braun L Jr, **6**: 4165 (4162)
Brauner N, **2**: 1586 (1585, 1586)
Braunstein M L, **3**: 2203 (2202)
Braunwald E, **1**: 542 (537, 541)
Braverman E M, **2**: 1478 (1476); **6**: 3643
Brayer J M, **6**: 3653 (3645, 3646, 3652)
Brayer K, **2**: 1087 (1084)
Brayton D B, **2**: 1532 (1528)
Brdyś M, **2**: 1305 (1299, 1302, 1304); **7**: 4571 (4568, 4569, 4571)
Breakwell J V, **2**: 1008 (1006, 1007), 1017 (1013, 1014); **6**: 4070; **7**: 4490
Brebbia C A, **2**: 1179 (1176)
Brebner D F, **7**: 4880 (4879)
Brebner G G, **5**: 3012
Brechot A, **7**: 4749 (4745)
Breddermann R, **6**: 4211 (4210)
Bree D W, **7**: 5109 (5108)
Breed H, **5**: 3465 (3462)
Breguet A L, **6**: (4143)
Brehmer B, **3**: 2191 (2190)
Breidenbach G, **5**: 2943 (2935, 2942)
Breitenecker F, **6**: 4311
B'Rells W F, **6**: 3815 (3812)
Brémaud P, **5**: 3302 (3302), 3303 (3294); **7**: 4637 (4637)
Bremer H, **6**: 4088 (4086)
Bremermann H J, **3**: 1825 (1824); **4**: 2497 (2497)
Brenci M, **5**: 3465 (3461)
Brender R F, **1**: 561 (561)
Brener J, **1**: 437
Brennan R D, **6**: 4305 (4303)
Brenner B M, **3**: 2206 (2206)
Brent R P, **2**: 1102 (1102)
Bres E, **6**: 3663 (3662)
Bressan A, **5**: 3184 (3182)
Bressan M, **2**: 1407 (1405)
Breton A, **1**: 687 (685); **4**: 2342; **6**: 3789 (3788)
Bretschger H J, **6**: 4066
Bretschi J, **4**: 2479
Brewer H B, **4**: 2887
Brewer J W, **1**: 417 (414, 415, 416); **2**: 1329 (1327, 1328), 1337 (1335, 1336, 1337); **6**: 3762 (3759, 3761), 3972 (3972)
Brézis H, **2**: 1033, 1179 (1177), 1261 (1259); **5**: 2997 (2997); **6**: 4223 (4218, 4220, 4221, 4222); **7**: 5036 (5033, 5034, 5036)
Briander P A, **6**: 3940 (3939)
Bribiesca E, **6**: 4159 (4157)
Brice C R, **1**: 759 (757); **6**: 4159
Bridgland T F, **2**: 1486 (1482)
Brierley C W, **3**: 1998
Briggs P A N, **4**: 2253 (2252); **5**: 2956 (2955)
Briggs W, **3**: 1975
Brigham E O, **5**: 3558 (3555)
Bright J R, **6**: 4155; **7**: 4827 (4824)
Brihaye C, **6**: 3940 (3940)

Brill N, **2**: 969
Briller S A, **1**: 769 (760, 761)
Brillhart J, **5**: 2956 (2952)
Brilliant M, **3**: 1778
Brillouin L, **2**: 1478 (1475); **6**: 3768 (3766), 3860 (3859)
Brinch P, **3**: 1642 (1638)
Brinch-Hansen P, **2**: 1143
Brinkman M O, **5**: 2942 (2935)
Brinkman U A T, **7**: 5129 (5129)
Brioschi F, **2**: 1282 (1279, 1280)
Briot M, **6**: 4123
Brisbin I L, **2**: 1355 (1352); **5**: 2929 (2928); **6**: 4237 (4235); **7**: 4956 (4956)
Brissebras J M, **7**: 4749 (4748)
Brissey G M, **4**: 2511 (2510)
Bristol E H, **3**: 2108 (2108); **6**: 3703 (3701), 3884
Britt H I, **1**: 585 (583)
Brittain E F H, **7**: 4504
Britten A, **1**: 445 (444)
Britten R J, **1**: 559
Britton J P, **6**: 3855 (3854)
Broadbent D T, **4**: 2575 (2573)
Broadhurst A, **3**: 2192
Broadhurst S W, **2**: 1087 (1084)
Brock M, **4**: 2582 (2581)
Brock P, **5**: 3559 (3555)
Brocker Th, **7**: 4674 (4672)
Brockett R W, **1**: 417 (414), 567; **2**: 826 (814); **3**: 1778; **4**: 2527 (2523), 2756 (2756), 2759 (2756, 2757), 2849, 2865 (2862, 2865), 2898 (2895); **5**: 2948 (2948), 2949 (2948), 3272 (3268), 3279 (3276), 3329 (3323), 3376 (3374), 3388 (3387), 3398 (3396, 3397), 3514 (3514); **6**: 3734 (3734), 3961 (3957, 3958, 3959), 4000 (3999, 4000), 4070; **7**: 5192
Brockhaus R, **1**: 220 (220), 393 (392)
Brockmeyer E, **7**: 4731 (4730)
Brode H, **5**: 3575
Brodie P F, **7**: 4454 (4453)
Brodie S E, **7**: 5065 (5062)
Brodsky N, **7**: 4833 (4829, 4830)
Brodsky R F, **3**: 1665 (1649)
Brody D A, **1**: 769 (760, 761)
Broecker W S, **2**: 1516
Broekhuyse H, **3**: 2032 (2026, 2027, 2030)
Broekstra G, **4**: 2497 (2495, 2496); **6**: 3768 (3766, 3768); **7**: 4779 (4779)
Brogan W L, **6**: 4284 (4283)
Brok S W, **1**: 633 (632)
Bromm B, **2**: 1590
Bron B, **1**: 633 (632), 634 (632)
Bronner F, **3**: 2174 (2170)
Brook R, **5**: 3579 (3578)
Brooke O G, **7**: 4856 (4853)
Brooks D E, **1**: 445 (444)
Brooks F P, **2**: 992 (992); **4**: 2685 (2684)
Brooks J, **7**: 4716 (4710)
Brooks N H, **4**: 2789 (2788)
Brooks R A, **1**: 759 (759); **4**: 2447 (2447)
Brooks R E, **5**: 3465 (3462)
Brooks V B, **5**: 3253
Brosio E, **5**: 3421 (3418)
Brown A J, **3**: 2234 (2233)
Brown B B, **1**: 437
Brown B H, **1**: 462 (460, 461); **2**: 1043 (1042); **5**: 3238 (3235, 3236)
Brown B M, **5**: 2934 (2933)
Brown C G, **2**: 1532 (1528)
Brown C M, **4**: 2473 (2471), 2487, 2488
Brown D F, **1**: 177 (174)
Brown D M, **1**: 437
Brown E W, **7**: (4889)
Brown G, **5**: 3440
Brown G S, **3**: (1748)
Brown H L, **5**: 2943 (2935)
Brown I A, **1**: 482 (479)
Brown J B, **6**: 4235 (4230)
Brown J H U, **3**: 2174 (2170)
Brown J L, **7**: 5078
Brown J R, **6**: 3904 (3904)
Brown K, **3**: 2213 (2210)
Brown L S, **4**: 2596 (2589)
Brown M P, **2**: 1057 (1056)
Brown P K H, **3**: 1979
Brown R, **2**: 942 (938, 939); **3**: 2203 (2195); **4**: 2338 (2338)
Brown R F, **4**: 2253 (2251); **6**: 3698 (3696); **7**: 4905 (4904)
Brown R H, **4**: 2641 (2638)
Brown R J, **4**: 2713 (2710)
Brown S, **6**: 4155
Brown T G, **3**: 1910 (1909)
Brown W A, **5**: 3238 (3235, 3236)
Brown W C, **7**: 4476 (4473)
Brown W S, **1**: 504 (502)
Browne J, **3**: 1648 (1642)
Browne J J, **2**: 1431 (1428)
Browning S R, **2**: 1388
Brownlee M, **4**. 2631 (2628)
Brownwell G L, **7**: 4905 (4902)
Broyden C G, **6**: 4310 (4308); **7**: 4974 (4972, 4973)
Bruck Y M, **5**: 3130 (3128)
Bruckmann G, **3**: 2016 (2016)
Bruell S C, **6**: 3659
Bruggenwert M G M, **1**: 148 (143, 144)
Brümmer F, **4**: 2473 (2470)
Brundiers H-J, **1**: 633 (632), 634 (632)
Bruni C, **1**: 417 (414, 415, 416); **5**: 3279 (3276)
Brunner T O, **1**: 570 (568)
Brunovský P, **2**: 1583 (1582); **4**: 2608 (2608), 2609 (2608), 2756 (2756), 2871 (2870); **5**: 3184 (3183), 3488 (3487), 3514 (3514); **6**: 3751 (3749), 3956 (3956), 4000 (3999); **7**: 4677 (4676, 4677)
Brush C F, **2**: 1402 (1402)
Bryan G H, **1**: (340)
Bryant D J, **6**: 4159
Bryant G F, **5**: 3223 (3221, 3222); **6**: 4142 (4140, 4141)
Bryant R E, **2**: 1201 (1200)
Bryant R M, **6**: 4405 (4405)

Bryce A L G, **3**: 1637
Bryce D B, **1**: 252 (248)
Brylinsky M, **2**: 1551 (1549)
Bryson A E, **2**: 1020 (1017), 1099 (1093); **4**: 2618 (2618); **5**: 3180; **6**: 4070; **7**: 4466 (4464, 4465), 4540 (4535, 4536, 4539), (4553), 5170
Bubeck K B, **4**: 2452 (2451)
Bubenko J, **3**: 2225
Bucaro J A, **5**: 3464, 3470 (3467, 3468, 3469)
Buchanan B G, **2**: 1573; **4**: 2640 (2637)
Bucharles A, **1**: 251 (251)
Buchin V A, **1**: 588 (586)
Buchthal F, **3**: 1910 (1908, 1909, 1910)
Buck R C, **2**: 1087; **5**: 3082 (3082)
Buckes B P, **3**: 1797
Buckles B P, **3**: 1797
Buckley G W, **6**: 3802 (3797)
Buckley P S, **2**: 1109 (1108); **3**: 2108 (2108)
Bucolo R J, **3**: 2032 (2026)
Bucy R S, **4**: 2283 (2281); **5**: 3303 (3293, 3294, 3300), 3310 (3305), 3330 (3320, 3322, 3323); **6**: 3717 (3713), 3734 (3733)
Buddrus J, **5**: 3421 (3419)
Budiansky B, **5**: 3470 (3467)
Budzikiewicz H, **5**: 2943 (2937)
Budzynski T, **1**: 437
Buehring W A, **3**: 2202 (2196)
Buell J, **4**: 2901 (2899); **6**: 4061 (4055, 4056)
Buhler H, **3**: 1614 (1612); **5**: 3097
Bührer R E, **1**: 633 (632), 634 (632)
Buhrmann H, **4**: 2754
Bula R J, **1**: 142 (140)
Bulbring E, **1**: 462 (459)
Bulford D E, **1**: 201 (201)
Bulirsch B, **4**: 2827 (2817)
Bulirsch R, **2**: 827 (816)
Bulkin B J, **7**: 4504
Bullinger H-J, **6**: 3663 (3660)
Bullingham R E S, **2**: 963 (961, 962, 963)
Bullock T E, **6**: 3626 (3623)
Bunce J W, **6**: 3972 (3972)
Bunch J R, **2**: 826 (817, 818, 820, 824); **6**: 4299
Bunge M, **1**: 556 (553, 554, 555); **2**: 1332 (1331), 1346 (1346); **3**: 2164; **6**: 3865 (3862), 3998
Bunow B, **1**: 433 (431)
Bunyard G B, **5**: 2943 (2935)
Buras E M, **1**: 482 (480)
Burch R H, **7**: 5149
Burcharet C W, **4**: 2455 (2453)
Burchi R S, **6**: 4302 (4300)
Burdess J S, **6**: 4086 (4085), 4100
Burdette E C, **1**: 450 (448, 449)
Burg J P, **5**: 3130 (3130); **7**: 4566 (4564)
Burger H C, **1**: 535 (532)
Burger J, **2**: 808; **6**: 4311 (4311)
Burgers J M, **1**: 294 (293)
Burges S J, **4**: 2652 (2651)
Burgess C, **7**: 4504
Burgess D, **7**: 4871 (4869)
Burghes D H, **7**: 4771
Burghes D N, **1**: 552 (550)
Burish F J, **2**: 839
Burkholder D, **5**: 2943 (2935)
Burkov V N, **1**: 37
Burks A W, **1**: 709 (704, 707), 720 (715); **6**: 4345 (4342, 4343), 4353 (4350), 4377 (4375)
Burmeister E, **2**: 1342
Burnett T C, **5**: 3282 (3282)
Burns A, **4**: 2550 (2550)
Burns J A, **3**: 2150 (2149), 2152; **5**: 3506
Burns J C, **2**: 1559
Burns L A, **2**: 1323 (1323); **7**: 4897 (4896)
Burns M, **4**: 2338 (2338)
Burns W K, **5**: 3470 (3466)
Burris H R, **2**: 1451 (1450, 1451)
Burroff P G, **2**: 1311 (1307, 1309, 1310, 1311)
Burrows W D, **1**: 268 (265)
Burrus C S, **2**: 1078 (1077)
Burstein A H, **1**: 474 (468, 469, 471)
Burt G, **5**: 2943 (2935)
Burt S K, **6**: 4302 (4302)
Burton A C, **7**: 4876 (4875), 4880 (4878)
Burton D, **7**: 4509 (4508)
Burton G J, **4**: 2330, 2331 (2327, 2328)
Burton R W, **2**: 1551 (1550)
Burzynski S, **5**: 2943 (2934)
Busacker R G, **2**: 1355 (1352)
Busbridge I W, **3**: 2052 (2046)
Busby D E, **2**: 1595 (1594, 1595)
Bush B, **7**: 5112
Bush R R, **1**: 276 (272); **7**: 4721 (4717)
Bush V, **2**: (870); **6**: (4311)
Bushaw D, **3**: 1898 (1896)
Bushaw D W, **5**: 3512 (3511), (3516)
Butcher H R, **7**: 4980 (4978)
Butkovskii A G, **2**: 808 (802), 1186 (1182); **4**: 2865 (2865); **5**: 3205 (3205), (3517)
Butler J F, **4**: 2511 (2509)
Butler J N, **6**: 3687 (3687)
Butler N C, **3**: 1958
Butter C D, **5**: 3470 (3469)
Butterfield F E, **4**: 2491
Buttleman H W, **6**: 4160 (4158)
Butzer Ph, **6**: 4223
Buxbaum A, **5**: 3097
Buxbaum J P, **1**: 251 (248)
Buxton J N, **1**: 755 (753); **6**: 4305
Buzacott J A, **3**: 1648 (1645), 1649 (1646)
Buzo A, **7**: 4509 (4507)
Byatt W J, **1**: 664; **3**: 1825 (1824), 1845 (1845), 1888
Bydder G M, **5**: 3425 (3425)
Byer R L, **4**: 2723 (2720); **6**: 4036 (4036)
Byrne G D, **2**: 1181 (1181)
Byrnes C I, **2**: 827 (823); **4**: 2865 (2865), 2866 (2865); **5**: 3303 (3301); **6**: 3734 (3734), 3972 (3972)

C

Caballero P A, **7**: 4749 (4745)
Cabanac M, **7**: 4866 (4865), 4872 (4871)
Cacak R K, **7**: 4856 (4856)
Cacoullos T, **2**: 1106 (1104)
Cacuci D G, **2**: 1551 (1549)
Cadzow J A, **7**: 4566 (4565)
Caesar J, **7**: 4905 (4902)
Caffarelli L A, **4**: 2343
Cage J M, **2**: 1403 (1402)
Cahill G F Jr, **3**: 2032 (2030)
Cahn M F, **2**: 1565 (1559, 1564)
Cain F L, **1**: 450 (448, 449)
Caines P E, **1**: 60 (58, 59, 60); **2**: 1581 (1579, 1580); **4**: 2239 (2236); **5**: 3006 (3005), 3060 (3056, 3059), 3061 (3061), 3065 (3062); **6**: 3929 (3929), 4193 (4192), 4209 (4206, 4207); **7**: 4607 (4603, 4606), 4611 (4607)
Cairns J Jr, **2**: 1318 (1314), 1321; **4**: 2646; **7**: 4901 (4899)
Cale W G Jr, **2**: 1326 (1324, 1325, 1326), 1348 (1345, 1346), 1355 (1352, 1353), 1486 (1480), 1551; **6**: 4238 (4235); **7**: 4454 (4453), 5008 (5005)
Calhoun D W, **3**: 1825 (1824)
Cali J P, **1**: 21 (20)
Callahan A, **6**: 4168 (4166)
Callender A, **6**: 3884 (3883)
Callender E, **4**: 2782 (2779)
Callier F M, **4**: 2570 (2570)
Calmet J, **5**: 3560 (3559)
Calovic M S, **5**: 3572 (3569)
Calvert T W, **6**: 3645
Calvet J L, **1**: 703 (692); **3**: 2163 (2155, 2157, 2158, 2161, 2162)
Calvin M, **1**: 7 (4)
Calzarossa M C, **1**: 755 (752)
Cameron K, **7**: 4785 (4783)
Cameron S, **4**: 2497 (2497); **6**: 4168 (4167)
Cameron T J, **6**: 4311
Cammen L M, **7**: 4454 (4453)
Campanella S J, **2**: 1057
Campani P, **4**: 2511 (2510)
Campbell D T, **2**: 1565 (1561)
Campbell F W, **4**: 2330 (2326); **7**: 5077 (5075)
Campbell I D, **1**: 13 (13)
Campbell J E, **2**: 1551 (1551)
Campbell N T, **5**: 3464 (3459)
Campion P J, **2**: 1559
Can A, **7**: 4871 (4868)
Canale R P, **1**: 261 (258, 260), 262 (258, 259, 260, 261); **2**: 1272 (1269), 1552 (1549); **4**: 2652 (2650, 2651), 2794 (2789); **7**: 5007 (5007)
Cannon C R, **5**: 3425 (3425)
Cannon J R, **6**: 4244 (4238)
Cannon R, **6**: 4089 (4086)
Cannon R H, **2**: 1291, 1297
Cannon R L, **4**: 2551 (2548)
Cannon W B, **3**: 2174 (2169)
Cannonito F, **2**: 937 (933)
Canon M D, **2**: 1099 (1093, 1094)
Cansever D H, **7**: 4630 (4629)
Cantrell C A, **1**: 184 (184)
Canuto C, **3**: 1723 (1722)
Capasso A, **6**: 3954 (3953)
Capelo A, **4**: 2342
Caplan S R, **1**: 428
Carathéodory C, **5**: (3366); **6**: 4070 (4067); **7**: 5046 (5044)
Caravani P, **6**: 4244 (4238)
Carberry J C, **5**: 3082 (3081)
Carbone R, **7**: 5155 (5154)
Carbonell J G, **2**: 1573; **4**: 2640 (2637, 2639); **7**: 4784 (4783)
Card S K, **2**: 1573; **3**: 2179
Cardús D, **1**: 542 (537); **6**: 4007 (4005)
Cardwell R H, **7**: 4726 (4723)
Carew T E, **1**: 486 (484)
Carey G F, **3**: 1632
Carey T T, **4**: 2685 (2684)
Carleson L, **3**: 1723 (1720)
Carlitz L, **5**: 3130 (3128)
Carlson C M, **6**: 4302 (4300)
Carlson E D, **2**: 947 (945, 946); **4**: 2353 (2351), 2501 (2498); **5**: 3079 (3078); **6**: 4372 (4371)
Carlson N A, **7**: 4540 (4536)
Carlson R E, **7**: 4953 (4953)
Carnap R, **2**: 1478 (1475, 1476); **3**: 1805 (1802); **6**: 4216 (4214, 4215)
Carney J H, **2**: 1272 (1272), 1551 (1549, 1550), 1552 (1551); **6**: 4235 (4230)
Caro C G, **1**: 433 (432), 445 (441)
Caroll D, **6**: 4369 (4368)
Carome E F, **5**: 3464 (3459), 3470 (3468)
Caron J Y, **7**: 5031 (5029)
Carotenuto L, **6**: 3626 (3625)
Carpenter G A, **1**: 570 (569)
Carpenter S R, **2**: 1566 (1560); **7**: 5177 (5171)
Carpentier F, **6**: 3801 (3795)
Carpentier J, **6**: 3846, 3954 (3951, 3952, 3953)
Carpentor S R, **7**: 4827 (4826)
Carr E H, **1**: 556 (554)
Carr J, **6**: 3780 (3777)
Carr R, **6**: 4088 (4086)
Carranza F A, **2**: 973 (972)
Carr-Brion K G, **5**: 2943 (2934)
Carrier W H, **1**: 150; **3**: 2209
Carrit D E, **7**: 5115
Carroll J M, **1**: 372 (369)
Carroll J S, **3**: 2203 (2202)
Carroll R, **1**: 100 (99)
Carroll R L, **1**: 105 (102)
Carslaw H S, **5**: 3099; **6**: 4364 (4364)

Carson E R, **3**: 2025 (2022, 2023, 2024), 2174 (2173); **4**: 2631 (2629, 2630), 2887 (2886); **6**: 3909 (3905, 3907, 3908)
Carson J S II, **6**: 4369 (4368), 4395 (4395)
Cartan E, **5**: 3394 (3391, 3393)
Carter A P, **3**: 2016 (2016)
Carter G A, **4**: 2361 (2359), 2362 (2359, 2360)
Carter M M, **7**: 4948
Carter T M, **6**: 3916 (3914)
Carter V, **3**: 1862
Carter W P L, **1**: 187 (187)
Cartledge C J, **1**: 633 (633)
Cartwright D E, **2**: 1542 (1541); **4**: 2579 (2577), 2640 (2639); **7**: 4892 (4889, 4891), 4956 (4955, 4956)
Cartwright M L, **1**: 570 (568)
Carvallo M, **6**: 4337 (4335)
Carver M B, **1**: 682 (681); **2**: 1181 (1181); **6**: 4310 (4309)
Casanova M A, **2**: 883 (881), 1154 (1145, 1149, 1150)
Case J H, **2**: 1268 (1264)
Case K, **4**: 2385 (2383)
Case K M, **7**: 4944
Case T J, **1**: 674 (673)
Cass N M, **5**: 3238 (3235, 3236)
Cassandras C, **3**: 1648 (1647)
Castaing C, **1**: 413 (410); **2**: 1033
Castañon D A, **2**: 1192 (1191); **3**: 2163 (2153); **7**: 4466 (4465)
Castel C, **7**: 4751
Castellan N J, **3**: 2192 (2186)
Casten R G, **1**: 674 (673)
Casti J L, **1**: 409, 674; **2**: 1286; **3**: 1714; **4**: 2866 (2862); **6**: 3922 (3920)
Caswell H, **2**: 1332 (1331); **7**: 5008 (5005)
Catheron A R, **3**: 1700 (1696)
Cats G J, **2**: 1497 (1497)
Cauchy A L, **3**: (1745)
Caulfield J T, **7**: 4481
Cautreels W, **1**: 176 (176)
Cavaillé J B, **3**: 1648 (1645, 1646)
Cavalli F, **2**: 1479 (1478)
Cavallo R E, **3**: 2003 (2000); **6**: 3768 (3766), 3989 (3984, 3985, 3986, 3987, 3988), 4372 (4371)
Cavill I, **3**: 1623 (1619)
Cawthorne T, **2**: 1590 (1587)
Cazzola M, **3**: 1623 (1621, 1622)
Céa J, **2**: 812 (810, 812), 956 (954), 1179 (1174, 1175), 1261 (1258)
Cegrell T, **1**: 51 (49, 50); **6**: 3591 (3590)
Cekirge H M, **5**: 3099
Cellier F E, **2**: 1192 (1191); **5**: 3035 (3034), 3559 (3558); **6**: 3626 (3622, 3623), 4304 (4303), 4305 (4303), 4306 (4303), 4313 (4312), 4322 (4321), 4359 (4357, 4358), 4360 (4359), 4372 (4370, 4371), 4388; **7**: 5011 (5008)
Cellina A, **2**: 1033 (1023)
Cena K, **1**: 606 (606), 645 (644); **7**: 4872 (4871)
Cerasi E, **3**: 2174 (2173)
Ceri S, **2**: 1160 (1159)
Cerretelli P, **6**: 4051 (4048)
Cerruti U, **3**: 1781 (1781)
Cesari L, **2**: 981 (980); **5**: 2963 (2958), (3181), 3512 (3511); **7**: 4529
Cessenat M, **7**: 4944 (4943)
Chabali Z, **2**: 1590 (1587)
Chackett K F, **6**: 3940 (3937)
Chacon E A, **4**: 2547 (2547)
Chadwick G, **7**: 4495
Chadwick J, **5**: (3414)
Chagny J, **5**: 2943 (2935)
Chaikin S E, **1**: 570 (568)
Chaitin G J, **1**: 725 (720)
Chakrabarti A K, **7**: 4833 (4831)
Chalabi Z, **2**: 1590 (1587, 1588)
Chalage G, **6**: 3889 (3888), 3916 (3912)
Chalk C R, **1**: 220
Chamberlain D A, **2**: 1410
Chamberlain R M, **2**: 797 (796)
Chamberlin D D, **2**: 1174 (1170)
Chamia M, **6**: 3806 (3802)
Chan A, **2**: 1165
Chan D S K, **5**: 3127 (3125)
Chan K, **4**: 2723 (2722)
Chan L S, **2**: 1106
Chan R Y, **7**: 5070 (5066)
Chan S P, **5**: 3055 (3052, 3053, 3054)
Chan T F, **5**: 3139 (3139)
Chanas S, **5**: 3454 (3454)
Chance B, **1**: 535 (535)
Chand S, **6**: 3706
Chandler J A, **2**: 1420 (1419)
Chandler R E, **7**: 4927
Chandrasekeran B, **2**: 1573
Chandrasekhar S, **2**: 785 (782); **3**: 2052 (2048, 2050); **7**: 4944
Chandrashekar M, **7**: 5106 (5103)
Chandy K M, **2**: 1087 (1084), 1169, 1201 (1200)
Chang C C, **2**: 937 (934); **5**: (2918), 2921 (2920, 2921)
Chang Chin-Liang, **3**: 1888
Chang K W, **7**: 4534 (4530)
Chang M F, **3**: 1761 (1760)
Chang N S, **6**: 3653 (3650, 3652)
Chang P R, **4**: 2412 (2401)
Chang P Y-T, **4**: 2659 (2656)
Chang R C, **7**: 5129 (5128)
Chang S C, **7**: 4631 (4629, 4630)
Chang S K, **2**: 1361; **3**: 1781 (1780, 1781), 1802 (1801), 1809 (1808, 1809); **5**: 3455; **6**: 3653 (3652), 4007; **7**: 4766
Chang S S L, **3**: 1898 (1895)
Chang T L, **5**: 3127 (3126)
Chang T M S, **1**: 312 (310, 311)
Chang T N, **6**: 4139 (4135); **7**: 4689 (4687)
Chang T S, **7**: 4630, 4631 (4629, 4630), 4822 (4815)
Changeux J P, **1**: 429 (424), 433 (432)
Chankong V, **5**: 3115 (3115), 3124 (3123), 3164 (3157, 3158, 3160, 3161, 3163); **7**: 4709 (4704, 4705, 4706, 4707, 4708, 4709)
Chao Y W, **7**: 5018 (5017)
Chapanis A, **1**: 372 (369)
Chapin F S Jr, **7**: 4495

Chapin N, **3**: 1695
Chapman J R, **3**: 1938
Chapman M, **7**: 4871 (4868)
Chapman R C, **7**: 4645 (4645)
Chapra S C, **4**: 2649 (2648, 2649), 2650 (2649), 2794; **7**: 4953 (4953)
Char B W, **5**: 3560 (3559)
Charalambous C, **5**: 3545 (3544)
Charlesworth B, **6**: 3762 (3756, 3757, 3760, 3761)
Charlson M, **2**: 1458
Charlson R J, **1**: 177 (176)
Charm S E, **1**: 445 (443)
Charnes A, **5**: 3164
Charnes O, **2**: 850 (845)
Charnetski J R, **5**: 3115 (3114)
Charrier P, **2**: 1211 (1210)
Chase J V, **7**: 4848 (4846)
Chase W P, **2**: 1003
Chastang J, **6**: 4336 (4335), 4337
Chatterji B N, **1**: 575
Chaudhry M H, **7**: 5161 (5157)
Chaudhuri B B, **3**: 1825; **7**: 4766
Chaudhuri D K R, **6**: 4294 (4291)
Chauprade R, **5**: 3097
Chavent G, **2**: 812 (809, 812), 1197 (1197)
Chazan D, **2**: 1125 (1123); **3**: 2163 (2154)
Cheeseman C L, **7**: 4848 (4847)
Cheik Obeid N E, **5**: 3480
Cheliutskin A, **5**: 3144; **6**: 3889 (3889)
Chemouil P, **4**: 2706; **7**: 4443 (4441, 4442), 4751
Chen C C, **1**: 585 (583)
Chen C H, **6**: 3643 (3636, 3641), 3653 (3645, 3646, 3648, 3652)
Chen C L, **5**: 3470 (3466)
Chen C S, **3**: 1888
Chen C T, **6**: 3717 (3713, 3714)
Chen C W, **4**: 2646 (2644, 2645)
Chen D I, **5**: 3149 (3147, 3148)
Chen F Y, **6**: 4108 (4103)
Chen G T, **1**: 589 (587)
Chen J C P, **7**: 4703 (4699)
Chen K, **4**: 2522; **7**: 5177 (5171)
Chen K T, **5**: 3279 (3277), 3394 (3390, 3391); **7**: 4425 (4421)
Chen M, **6**: 3850 (3850)
Chen M J, **5**: 3196
Chen N, **4**: 2385 (2382)
Chen P N T, **3**: 2117
Chen P P, **2**: 893 (891)
Chen R M M, **4**: 2706 (2702, 2703, 2705)
Chen T C, **5**: 3077 (3076)
Chen W H, **6**: 4244 (4238, 4240)
Chen W L, **6**: 4417
Chen Y, **6**: 3909 (3905)
Chen Zhen-Yu, **1**: 63 (60)
Chenais D, **2**: 1179 (1177)
Cheney E W, **5**: 3545 (3545)
Cheng D C, **7**: 5060 (5059)
Cheng G, **5**: 3514 (3514)
Cheng L, **4**: 2570 (2569)
Cheng S, **5**: 3413 (3410, 3411)
Cheng W K, **2**: 1154 (1153)
Cheremisinoff N P, **4**: 2754; **6**: 3875
Chernavskii D S, **1**: 433 (432)
Cherne R E, **1**: 150; **3**: 2209
Cherniack N S, **6**: 4066 (4062)
Cherniavsky E A, **2**: 1454 (1453); **3**: 1718 (1715)
Chernoff H, **1**: 91; **2**: 1106 (1103); **5**: 3535 (3533)
Cherrington A D, **3**: 2175 (2171); **4**: 2631 (2628)
Cherry C, **6**: 3768 (3766), 3860 (3859)
Cherry R D, **2**: 1516 (1516)
Chervany J J, **7**: 4793 (4793)
Chervany N L, **7**: 4793 (4793)
Chesters M S, **7**: 5186 (5182, 5184)
Chestnut H, **2**: 1003; **4**: 2575 (2571, 2575); **6**: 3870 (3866, 3867, 3868)
Chetayev N G, **5**: 3398
Cheung C H, **2**: 1388
Chew G, **1**: 333 (332)
Chiandussi L, **7**: 4905 (4902)
Chiba S, **7**: 4509 (4508)
Chien R T, **6**: 4161 (4157)
Chien S, **1**: 445 (442, 443)
Chien Y T, **6**: 3643 (3636)
Chikan A, **6**: 4388
Chilcoat R T, **1**: 281 (277, 279, 280), 284 (281, 283), 289 (284)
Child G I, **2**: 1344 (1343, 1344)
Child T F, **5**: 3421 (3419)
Childers D G, **4**: 2270 (2265, 2269); **7**: 4566 (4564)
Childress M E, **3**: 1668 (1666)
Chiles W D, **3**: 1668 (1665, 1667)
Chimura H, **2**: 1361
Chin L, **2**: 912 (910)
Chin S S, **5**: 3012
Chintapalli P, **6**: 3916 (3913)
Chiou C T, **4**: 2782 (2779)
Chippendale A M, **5**: 3421 (3418)
Chisholm R A, **7**: 5147 (5146, 5147)
Chislett M A, **6**: 4262 (4259)
Chiu C-L, **7**: 5140
Chlamtac I, **1**: 755 (752)
Cho C H, **3**: 2099 (2099)
Cho Y S, **5**: 3379 (3377)
Choksy N H, **3**: 2150, 2152 (2152); **5**: 3506
Chomsky N, **3**: 1712 (1712), 2045 (2045)
Chong C Y, **2**: 932 (931); **4**: 2713 (2713); **7**: 4684 (4678), 4822
Chorba R W, **3**: 2191 (2186)
Chorley R A, **1**: 245
Chou C, **2**: 1165 (1165)
Chou S M, **6**: 3653 (3646, 3652)
Choudbury S K, **7**: 4452 (4448)
Chow E Y, **1**: 86 (84); **3**: 1736, 1741 (1738, 1739)
Chow G A, **2**: 1342
Chow G C, **5**: 3170 (3170)
Chow H Z, **5**: 3077 (3073)
Chow J H, **1**: 132; **5**: 3180 (3176), 3502; **7**: 4439 (4438, 4439), 4443 (4439, 4440, 4442), 4534 (4530, 4531, 4532, 4533)

Chow V T, **3**: 2234 (2229); **7**: 5153 (5153)
Chow W L, **5**: 3366 (3366)
Chretien J, **3**: 1665 (1660)
Chrisman D A, **1**: 482 (480)
Christakis A N, **7**: 5177 (5171)
Christensen E E, **6**: 3940 (3940)
Christensen G S, **3**: 2225 (2222)
Christensen J, **1**: 462; **2**: 1043 (1040)
Christensen R A, **2**: 1478 (1473, 1474, 1475, 1476, 1477, 1478), 1479 (1478); **6**: 3989
Christopher T G, **1**: 310 (309)
Christov N D, **2**: 827 (821); **4**: 2827 (2826)
Chu B M, **1**: 477 (475, 477)
Chu K C, **2**: 932 (930); **5**: 3156 (3152); **7**: 4622 (4622), 4822 (4815, 4820)
Chu W W, **2**: 1282 (1281)
Chua L O, **2**: 981 (981); **3**: 1778 (1776), 1779 (1776); **7**: 4522 (4519)
Chujo Y, **2**: 1497 (1494, 1497)
Chukwuenye C O, **4**: 2608 (2607)
Chung E K, **1**: 456 (455)
Chung K L, **7**: 4637 (4631)
Church A, **1**: 333 (326, 330), 709 (704), 720 (715); **2**: 937 (933, 934); **7**: 4958 (4958)
Church R, **5**: 3164 (3160)
Churchill D A, **4**: 2356 (2356)
Churchman C W, **7**: 4784 (4779), 4789 (4786)
Chvátal V, **6**: 3739 (3737)
Chyung D H, **2**: 1211 (1209)
Ciarlet P, **3**: 1632; **5**: 3487 (3484), 3530 (3525); **6**: 3626 (3626)
Ciborra C, **6**: 4097
Cichowlas W M, **3**: 1694 (1691)
Ciechanowicz W, **2**: 1192 (1191)
Cielo P G, **5**: 3470 (3467)
Cioranescu D, **2**: 1179 (1177)
Clague A D H, **5**: 3421 (3418)
Claisse F, **7**: 4504 (4500)
Clancy S J, **6**: 3962 (3961)
Clanton H, **2**: 1577 (1574)
Clark A S, **2**: 839
Clark C, **7**: 4577 (4574), 5046 (5045)
Clark C M, **3**: 1887, 1888
Clark D K, **6**: 4045 (4045)
Clark D L, **1**: 281 (280)
Clark D S, **1**: 80 (77); **5**: 3531 (3529); **6**: 3994 (3992), 3997 (3995)
Clark E A, **1**: 473 (469)
Clark F E, **1**: 142 (141)
Clark J A, **1**: 606 (606), 645 (644); **7**: 4872 (4871)
Clark J E, **3**: 1862
Clark J M C, **5**: 3303 (3295), 3329 (3320, 3323), 3523 (3522)
Clark J R, **3**: 1862
Clark K B, **7**: 4832 (4832)
Clark K L, **6**: 4411 (4410)
Clark R, **7**: 4577 (4575)
Clark R E, **7**: 4906 (4903)
Clark R N, **1**: 251 (248)
Clark R P, **7**: 4871 (4869, 4870)
Clark W E, **1**: 177 (175)
Clarke A A H, **3**: 1981 (1980)
Clarke D, **6**: 4261 (4258, 4259)
Clarke D J, **3**: 1998
Clarke D W, **1**: 51 (49), 63 (63); **2**: 1116 (1110, 1111); **3**: 2005 (2005), 2113 (2112); **4**: 2283 (2283), 2319 (2319); **5**: 2956 (2955), 3004 (3002), 3006 (3005, 3006), 3188 (3184, 3188), 3232 (3229), 3410 (3407), 3430 (3426, 3428); **6**: 3727 (3721, 3723, 3724, 3725), 3871 (3870, 3871), 3898 (3894), 4172 (4171, 4172), 4176 (4174, 4175), 4180 (4176, 4180), 4189 (4187), 4193 (4190, 4193), 4197 (4194), 4199 (4198), 4204 (4200), 4209 (4205, 4207), 4211 (4209)
Clarke F H, **2**: 1033; **5**: 3184 (3182, 3183), 3362 (3359, 3360), 3545 (3541); **7**: 5046 (5045, 5046)
Clarke W L, **3**: 2021 (2017)
Clarkson D T, **1**: 268 (264)
Clarkson S F, **2**: 1345 (1343)
Clarkson S P, **5**: 3440 (3440)
Claude D, **5**: 3288 (3286, 3287)
Claxton G F, **3**: 1998
Clayton G B, **1**: 272 (270); **6**: 4284 (4283), 4289 (4286)
Clayton N M, **6**: 3806 (3806)
Cleeve P, **3**: 1975
Clemens A H, **3**: 2021 (2017), 2175 (2170)
Clement K, **4**: 2263 (2263)
Clement R E, **7**: 5129 (5127)
Clement W E, **3**: 1668 (1667)
Clements D J, **1**: 703 (694); **2**: 907 (903), 916 (914, 915, 916); **5**: 3508; **6**: 4271 (4269)
Clements D P, **1**: 743, 746 (744)
Clements F E, **2**: 1321 (1318, 1320)
Clements K A, **6**: 3850 (3848)
Clements P J, **2**: 1248 (1247)
Clements R G, **2**: 1344 (1344)
Clementson A T, **4**: 2550 (2549); **6**: 4393 (4392)
Clerget M, **1**: 296 (294, 295); **7**: 4468 (4467)
Clerici A, **6**: 3806 (3806)
Clesceri L S, **3**: 1601 (1600); **7**: 4897 (4896)
Clesceri N L, **4**: 2646 (2645)
Cleveland A, **3**: 1979
Cleveland C J, **2**: 1462
Clifford D A, **7**: 5117
Clifford W J, **6**: 4417
Clifton J, **1**: 769 (761)
Clifton J A, **2**: 1043 (1040)
Clinch D A L, **7**: 4577 (4574)
Clinch J M, **2**: 1233 (1228), 1239 (1236)
Cline A K, **2**: 826 (817, 820)
Cline D M, **2**: 1323 (1323); **7**: 4897 (4896)
Clocksin W F, **4**: 2487
Close C M, **2**: 1291, 1297
Closs G D, **2**: 1434 (1431, 1434)
Clymer A B, **4**: (2272); **6**: 4417
Coale A J, **6**: 3755
Coales J F, **3**: 1607 (1606); **4**: 2560 (2559), 2575 (2571, 2575); **5**: (3516); **6**: 3870 (3866, 3867, 3868)
Coates P B, **7**: 4861 (4858)

Coates P C, **2**: 1248 (1247)
Coats K H, **3**: 2006 (2006)
Coats R, **3**: 1995
Cobb L, **6**: 3772 (3769, 3771)
Cobb P W, **3**: 2186 (2180)
Cobb W, **2**: 1590 (1587)
Cobby J M, **1**: 142 (139, 140)
Cobelli C, **1**: 493 (492); **3**: 2021 (2018), 2025 (2022, 2023, 2024, 2025), 2174 (2171, 2173); **4**: 2631 (2629, 2630, 2631), 2887 (2886); **6**: 3909 (3905, 3907); **7**: 4905 (4904)
Cobourn W G, **1**: 176 (175)
Cochet L, **5**: 2943 (2935)
Cochin I, **2**: 1291, 1297; **5**: 2970
Cochran W G, **2**: 1106; **7**: 5015 (5014)
Cochran W T, **3**: 1723
Cochrane J L, **5**: 3121 (3121), 3164 (3161), 3165 (3157), 3568
Cocivera M, **5**: 3421 (3419)
Cocker V, **7**: 5155 (5155)
Cockroft J P, **3**: 2117 (2114, 2117)
Codd C F, **6**: 3653
Codd E F, **3**: 1797, 1800 (1798); **4**: 2659 (2656, 2658); **6**: 4007, 4345 (4343)
Coddington E A, **5**: 3563
Code C F, **2**: 1043 (1040)
Codrington R S, **5**: 3421 (3419)
Cody W J, **2**: 826 (814)
Coelho C A D, **5**: 3215 (3206)
Coffman E G, **2**: 1154, 1169 (1167); **5**: 3449; **6**: 3596; **7**: 4752
Coffman W P, **7**: 4956 (4955)
Cogneau M, **6**: 3940 (3940)
Cogswell F N, **7**: 5060 (5059)
Cohen A I, **2**: 1388
Cohen D, **7**: 4646 (4645)
Cohen E G D, **1**: 567; **7**: 4529 (4528)
Cohen G, **1**: 405 (400, 402, 403, 404, 405); **2**: 956 (952); **3**: 1648 (1646), 1838 (1836, 1837), 2163 (2153, 2155); **4**: 2849; **7**: 4684 (4678)
Cohen H, **3**: 1995
Cohen J, **1**: 720 (715)
Cohen J K, **5**: 3310 (3305)
Cohen J W, **7**: 4741 (4735)
Cohen L J, **3**: 2191 (2190)
Cohen M, **1**: 482 (480); **4**: 2501 (2498)
Cohen P, **7**: 4585 (4581)
Cohen P J, **5**: (2918)
Cohen P R, **1**: 307 (307); **2**: 1573; **4**: 2640 (2636); **7**: 4784 (4784)
Cohen R L, **7**: 4504 (4500)
Cohen S I, **7**: 4832 (4831)
Cohn N, **6**: 3815 (3811)
Cohon J L, **5**: 3115 (3115), 3121 (3121), 3164 (3160, 3161); **7**: 4709 (4708), 5140
Cojocariu R, **5**: 2993 (2991)
Cokelet G R, **1**: 445 (441, 442, 443, 444)
Coker J E, **5**: 3464 (3461)
Colantuoni G, **6**: 4245
Colburn W A, **6**: 3698 (3697)
Colby J A, **2**: 1329 (1329)
Colby K M, **4**: 2640 (2638), 2641
Cole C V, **1**: 142 (141)
Cole H S D, **7**: 4762 (4756)
Cole J H, **5**: 3470 (3467, 3468, 3469)
Cole W G, **5**: 2930 (2929); **7**: 4956 (4954, 4956)
Coleman D C, **1**: 142 (138, 139, 140, 141)
Coleman R, **5**: 3215 (3213)
Coleman T G, **1**: 467 (467), 468 (466); **2**: 1366 (1361)
Coles J R, **5**: 3238 (3235, 3236)
Colijn H, **3**: 1710
Collett P, **1**: 567
Collier D, **7**: 4577 (4574)
Collier K I, **1**: 623
Collier P, **2**: 1248 (1247)
Collier P I, **3**: 2076
Collier R S, **3**: 2006 (2006)
Colli Franzone P, **3**: 1623 (1619, 1620)
Collins A, **4**: 2338 (2338)
Collins C C, **2**: 1596 (1593)
Collins C D, **7**: 5008 (5006, 5007)
Collins D C, **5**: 3165 (3159)
Collins G B, **5**: 3436 (3434)
Collins J C, **7**: 4856 (4853)
Collins J F, **3**: 1736
Collins P, **2**: 1552 (1549)
Collis R T H, **6**: 4036 (4033, 4034, 4036)
Collmeyer A J, **2**: 1154 (1147)
Colomb A L, **2**: 1268 (1266)
Colongo C, **3**: 2089
Colonius F, **3**: 2150, 2152 (2152); **5**: 3506
Colton C K, **1**: 315 (314), 433 (431)
Colton K W, **2**: 1565 (1559)
Combs J F, **5**: 2943 (2935)
Côme G M, **5**: 3082 (3081)
Comes M, **1**: 196 (194)
Comfort A, **1**: 294 (294)
Comhaire F, **7**: 4871 (4870)
Comincioli V, **2**: 1211 (1210); **7**: 5036 (5033)
Commault C, **1**: 132 (127); **4**: 2855 (2849)
Commoner F, **6**: 3670 (3668)
Compere E L, **7**: 5094 (5093)
Comroe J H, **6**: 4051 (4045)
Comstock C, **7**: 4434 (4430)
Comstock F L, **6**: 3989 (3984)
Comstock J P, **6**: 4261 (4258)
Comte P, **5**: 3480 (3478)
Conant R C, **4**: 2497 (2494, 2495, 2496), 2547 (2545); **6**: 3768 (3766)
Concepcion A I, **2**: 1201 (1200)
Concus P, **7**: 4971 (4969)
Condon R E, **2**: 1036 (1033), 1043 (1040)
Condorcet M J A N, **3**: (2151)
Conforti G, **5**: 3465 (3461)
Conley R T, **7**: 4504
Conn A R, **5**: 3545 (3544)
Connell J H, **2**: 1317 (1313, 1315, 1316)
Connelly M E, **1**: 120, 201 (201)
Connolly C I, **3**: 1601 (1600); **7**: 4897 (4896)

Connolly J P, **4**: 2646 (2644, 2645)
Connolly T, **7**: 4785 (4783)
Connor N E, **3**: 1965; **4**: 2356 (2353)
Conolly J E, **6**: 4261 (4260)
Conover W J, **2**: 1551; **6**: 4235 (4231, 4233)
Conroy T F, **1**: 755 (752)
Considine D M, **3**: 1710; **4**: 2516 (2511); **6**: 3682 (3681)
Constantine L, **2**: 992 (987); **7**: 4472
Constantinides A, **3**: 1611 (1609)
Constantinides G M, **2**: 1099 (1096)
Constantiou C E, **7**: 4980 (4976)
Constanza V, **5**: 3292 (3289)
Constatinou C E, **7**: 4980 (4976)
Conte G, **6**: 3972 (3972)
Contensou P, **7**: 4490
Conti F, **5**: 3421 (3418)
Conti R, **7**: 5041 (5038)
Conway J, **1**: (559, 560); **6**: (4342)
Conway J H, **6**: (4376)
Conway R W, **6**: 4415 (4414); **7**: 4654 (4652)
Cook P A, **4**: 2872 (2872); **5**: 3215 (3206), 3228 (3227), 3319 (3319), 3373, 3379 (3377), (3387); **7**: 4452 (4444), 4463 (4462), 4547 (4541)
Cook S A, **2**: 1083 (1082)
Cook T H, **1**: 482 (480)
Cooke E D, **7**: 4871 (4870)
Cooke K L, **2**: 1211 (1209); **3**: 2150, 2152 (2152); **5**: 3506
Cooke M, **4**: 2511 (2510)
Cooksey J D, **1**: 769 (761)
Cooley B, **6**: 4374
Cooley J W, **3**: 1723; **4**: 2270 (2265); **7**: 4498 (4496)
Cooley W W, **2**: 879 (876); **3**: 1723 (1722)
Cooling J E, **6**: 4271 (4268)
Coolsaet B L R A, **7**: 4981 (4976, 4979)
Coombs C H, **3**: 2060 (2058), 2202 (2196)
Coombs M J, **3**: 2179
Cooney C L, **3**: 1611 (1609), 1615 (1613, 1614), 1619 (1618); **5**: 2943 (2935, 2942), 2944 (2935, 2942)
Cooney D O, **1**: 310 (309)
Cooper G E, **3**: 1668 (1666, 1667)
Cooper H R, **6**: 4252 (4250)
Cooper J W, **5**: 3421 (3419)
Cooper K E, **7**: 4856 (4853), 4880 (4879)
Cooper L, **4**: 2412 (2397)
Cooper S J, **3**: 2213 (2210)
Cooper T G, **2**: 1192 (1189)
Cooper W, **5**: 3164
Cooper W D, **3**: 2080
Cooperman J A, **1**: 751 (751)
Cope S N, **6**: 4204 (4200)
Copeland B, **6**: 4133 (4132)
Coppel W A, **7**: 4439 (4437, 4438)
Corbeel D, **6**: 3673 (3670, 3671, 3672)
Cordero A O, **6**: 3991 (3991)
Corduneanu C, **7**: 4529 (4528)
Corfmat J P, **2**: 899 (895), 902 (900), 907 (902, 903), 916 (915), 923 (917); **4**: 2559 (2559), 2671 (2666, 2668)
Corijnen G E, **6**: 4411 (4407)
Corley N W, **5**: 3164 (3161)
Cornacchio J, **1**: 709 (709)
Cornell C A, **2**: 1553 (1552); **6**: 4229 (4228); **7**: 5015 (5014)
Cornell D, **3**: 1975
Cornell E, **7**: 5027
Cornet B, **2**: 1033; **5**: 3550
Cornew R W, **7**: 4876 (4873), 4877, 4880 (4880)
Cornfield J, **2**: 1106
Cornsweet T, **7**: 5077
Corrado G, **3**: 1665 (1653, 1660)
Correa G O, **4**: 2292 (2291)
Corrigan J R, **3**: 2191 (2190)
Corsi P, **2**: 1342
Corson R C, **5**: 3282 (3281)
Cortey-Dumont P, **5**: 3530 (3525)
Corwin W M, **5**: 2980
Côté G, **6**: 3789 (3788)
Cote Y, **7**: 4726 (4723)
Cottinet M, **1**: 517
Cottle R W, **1**: 495 (494)
Cottle W, **4**: 2343
Cotton F A, **1**: 7 (4), 268 (262); **7**: 5094 (5090, 5092)
Cotton I W, **4**: 2367 (2365)
Cottrell M, **1**: 91 (91)
Coughanowr D R, **6**: 3680
Coulam C M, **1**: 546 (544)
Coulam W J, **5**: 3282 (3282)
Coulbeck B, **7**: 5106 (5104), 5109 (5107, 5108), 5149, 5153 (5151, 5153)
Coulter W H, **2**: 866 (862, 866)
Coulter W M, **2**: 867 (866)
Coupé G M, **1**: 323 (320); **6**: 4148 (4147)
Courant R, **2**: 801 (798), 1261 (1257); **4**: 2807 (2807); **6**: 4364 (4360, 4361)
Courconbetis C A, **2**: 1169 (1169)
Courcoulas J H, **6**: 4142 (4141)
Courdesses M, **2**: 808 (802); **6**: 4245 (4238, 4241)
Courtenay J M, **1**: 312 (311)
Courtiol B, **1**: 51 (50)
Courtney J M, **1**: 312 (310, 312)
Courtney R G, **2**: 1445 (1441)
Courtois P J, **6**: 3659
Courvoisier M, **2**: 1143
Cousins T, **3**: 1693 (1687)
Coussedierre M, **1**: 246
Cowan J D, **1**: 433 (432), 434 (432)
Cowardin L M, **3**: 1862
Cowell D F, **6**: 4411 (4410)
Cowell W, **2**: 826 (815, 824)
Cowin S C, **1**: 474 (471)
Cowper C J, **1**: 594
Cox D R, **2**: 1106
Cox M, **3**: 1756 (1751)
Cox R J, **1**: 437
Coxeter H, **6**: 4336 (4334)
Coyle R G, **7**: 4762 (4755, 4761)
Craig A B, **6**: 4051 (4046)

Craig J J, **4**: 2385 (2380), 2386 (2380); **6**: 4100, 4108 (4103, 4104, 4105)
Craig N L, **1**: 176 (175)
Craig P P, **2**: 1329 (1327)
Craik F I M, **7**: 4793 (4791)
Cramér H, **1**: 91; **4**: 2273 (2271), 2287 (2284, 2285)
Cramp D G, **3**: 1623 (1621, 1622), 2025 (2022), 2143 (2140, 2142), 2174 (2170, 2171, 2173), 2175 (2170, 2171, 2172); **4**: 2887 (2886); **6**: 3909 (3905, 3907); **7**: 4906 (4905)
Crandall M G, **2**: 1017 (1014, 1015, 1016), 1279 (1278); **3**: 2068 (2065, 2066), 2072 (2069, 2072); **6**: 4223 (4220, 4222); **7**: 4626 (4624)
Crane C, **6**: 4108 (4105)
Crane C L, **6**: 4261 (4258)
Crane M A, **6**: 4395 (4395); **7**: 4654 (4653)
Cranefield P F, **1**: 456 (451, 452)
Crank J, **5**: 3099
Cranston W I, **7**: 4856 (4852, 4853)
Crawford G F, **4**: 2488
Creagan C C, **3**: 1611 (1609)
Creel C L, **1**: 433 (431)
Crellin G, **2**: 1479 (1475)
Cremer M, **3**: 1736, 1741 (1737, 1738)
Cremers A B, **7**: 4822
Cremers P, **5**: 3030 (3027)
Cremieu M, **6**: 4275 (4273)
Cremoux J P, **7**: 4749 (4748)
Crenson M A, **7**: 4762 (4756), 5005 (5001, 5002)
Crevasse L, **1**: 769 (761)
Crim K O, **4**: 2579 (2576, 2579)
Crisp V H C, **4**: 2769 (2765, 2767, 2768, 2769), 2775 (2772, 2773, 2774), 2777 (2775, 2776)
Crissey B L, **7**: 4762 (4756), 5005 (5001, 5002)
Cristerna R, **6**: 3789
Cristescu N, **5**: 3099
Crittenden A L, **1**: 177 (176)
Crochiere R E, **1**: 80 (78)
Croft D J, **2**: 1407 (1405)
Crofts C E, **1**: 486
Crompton H, **3**: 1958 (1957)
Crompton R E B, **2**: 1402 (1399)
Cromwell T H, **1**: 289 (285)
Cronan C S, **1**: 268 (264)
Cronhjort B T, **5**: 2993 (2993)
Cronin L E, **4**: 2782 (2780)
Cronn D, **1**: 177 (176)
Cronshaw D, **1**: 751 (750)
Crooks L E, **5**: 3421 (3419, 3420), 3425 (3425)
Crosbie R E, **1**: 633 (633); **5**: 3035 (3033)
Crosbie R J, **7**: 4876 (4873, 4875)
Crosby D G, **1**: 148 (143)
Cross H, **3**: 1954 (1948)
Crossley D A, **2**: 1344 (1344)
Crossley F R E, **6**: 4108 (4104)
Crossley T R, **6**: 3717 (3711)
Crothers E J, **7**: 4721 (4717)
Crouch P E, **3**: 1778 (1770); **4**: 2527 (2526); **6**: 3961 (3958, 3959, 3960, 3961), 3977 (3973, 3976)
Crouhy M, **1**: 409; **3**: 1714; **4**: 2583; **6**: 3706 (3706), (3890)
Crouzeix M, **3**: 1627
Crowder H, **2**: 826 (825)
Crowl D A, **6**: 3916 (3912)
Crum R, **3**: 2203 (2202); **6**: 4075
Cruon R, **6**: 4029 (4029); **7**: 4766 (4765)
Cruz J B Jr, **1**: 137, 405 (400, 403, 404, 405), 687 (684, 687), 703 (701); **2**: 995 (994, 995), 1192 (1191); **5**: 3149 (3147, 3148, 3149), 3156 (3152), 3228 (3225), 3530 (3530); **6**: 4227 (4224, 4225, 4226, 4227); **7**: 4443 (4443), 4631, 4822
Cruz J C, **6**: 4139 (4133)
Cruz-Orive L M, **7**: 4601 (4599)
Csáki F, **4**: 2239 (2237), 2292 (2292)
Császár Z, **3**: 2010 (2008, 2010)
Csiszár I, **2**: 1473 (1473)
Csuros L, **6**: 3806 (3805)
Cuenod M A, **7**: 4689 (4687, 4689)
Cugiani M, **6**: 4204
Cugy A, **7**: 5090
Cukier R I, **6**: 4235 (4230)
Culbert S A, **7**: 4833 (4830)
Culik K II, **4**: 2798 (2795, 2796, 2798)
Cullen B F, **1**: 289 (285)
Cullen D J, **1**: 289 (285)
Cullum C D Jr, **2**: 1099 (1093, 1094)
Cullum I D, **6**: 3940 (3940)
Culp R D, **7**: 4490
Culshaw B, **5**: 3464, 3470 (3469)
Cumming G, **4**: 2904 (2902, 2903)
Cummins K W, **7**: 4956 (4955)
Cuninghame-Green R, **3**: 1648 (1646)
Cunningham D J C, **6**: 4061 (4055, 4057)
Cunningham J R, **7**: 5186 (5179, 5180, 5182, 5185)
Cunningham P, **3**: 1756 (1750)
Cunningham P T, **1**: 176 (175)
Cunningham R, **4**: 2488
Cunningham T B, **1**: 251 (250)
Curl C M, **1**: 315 (314)
Curran A H, **2**: 1214 (1213)
Currey J D, **1**: 474 (468, 469)
Curry D L, **3**: 2174 (2173)
Curry T S, **6**: 3940 (3940)
Curtain R F, **2**: 1186 (1185); **3**: 1665 (1650); **4**: 2523 (2522, 2523), 2861; **6**: 4245 (4238, 4239, 4240, 4242); **7**: 4621 (4617)
Curtis L F, **6**: 4045
Cury J E R, **2**: 900 (895, 896, 897)
Cutler C R, **1**: 588 (588); **6**: 3679 (3675)
Cutler M M, **6**: 4403 (4402, 4403), 4411 (4406, 4407)
Cygler M, **7**: 4571 (4568, 4571)
Cynader M, **4**: 2331 (2325)
Cyre W R, **1**: 633 (632)
Cyrous J, **4**: 2608 (2607)
Czeiner N, **3**: 2010 (2008, 2010)
Czeisler C A, **1**: 601 (601)
Czinczel A, **1**: 380 (376)
Czogała E, **1**: 713 (710, 713); **2**: 1473 (1469); **3**: 1894 (1891, 1892); **5**: 2965 (2965), 3540 (3538)

D

Da Prato, **6**: 4223
Da Roche A F G, **6**: 3940 (3936)
Da Silva K M C, **6**: 4055 (4052)
Dabes J N, **1**: 424 (423)
Dacorogna B, **1**: 529 (527)
DaCunha N D, **5**: 3164 (3159)
Dadam P, **2**: 1165
Dahhou B, **5**: 3004 (3004)
Dahl O J, **1**: 755 (753); **6**: 3898
Dahlin E B, **3**: 2113 (2112)
Dahlman R C, **2**: 1344 (1343, 1344)
Dahlqvist G, **2**: 826 (816); **5**: 3134 (3134), 3558 (3556); **6**: 4359 (4358)
Dahlqvist S A, **2**: 1109 (1108), 1116 (1110)
Dahlstrom B, **2**: 963 (962)
Dainty J C, **7**: 5186 (5181, 5183)
Dairaku K, **3**: 1619 (1618)
Dakin J P, **5**: 3464 (3457, 3459)
Dale M B, **3**: 2045 (2045)
Dale R S, **6**: 4271 (4269)
D'Alessandro P, **2**: 1121 (1116, 1119); **5**: 3279 (3276); **6**: 3961 (3959), 3977 (3973)
Dali S, **1**: (705)
Dalkey N C, **1**: 618 (614)
Dalla Libera F, **2**: 1160 (1160)
Dalla Volta S, **2**: 1407 (1405)
Dallery Y, **3**: 1648 (1644)
Dalton G E, **2**: 1537 (1537)
Daly C H, **1**: 482 (480)
Damato A N, **2**: 1410 (1408)
d'Ambrosio U, **3**: 1898 (1896)
Damen A A H, **6**: 3983 (3979)
Damji S W H, **5**: 3421 (3419)
Damlamian A, **2**: 1032
Damokosh-Giordano A, **6**: 4066 (4062)
Damrauer L, **1**: 184 (184)
Dand I W, **6**: 4261 (4259)
Dandekar A J, **1**: 246
Dandridge A, **5**: 3464, 3470 (3466, 3467, 3468, 3469), 3471 (3469)
Danforth H, **6**: 3916 (3913)
Dang Van Mien H, **2**: 1121 (1116); **5**: 3292 (3289), 3293 (3289), 3376 (3374); **6**: 3776 (3772, 3773)
Daniel C, **7**: 4651 (4648)
Daniel C L, **1**: 478 (477), 486 (483, 484, 486)
Daniel D J, **2**: 1521 (1521)
Daniel E E, **2**: 1036 (1033, 1034, 1035, 1036), 1040 (1036), 1043 (1040, 1041)
Daniel J N, **7**: 4969
Daniel J W, **2**: 1125 (1124); **5**: 3487 (3482, 3486); **7**: 4971 (4969)
Daniel R W, **5**: 3228
Daniell P J, **4**: 2270 (2267); **5**: 3388 (3385)
Daniels J, **1**: 674 (672)
Danielsen S, **5**: 2994 (2991)
Danielsson L, **2**: 1532 (1528)
Danilin Y M, **5**: 3546 (3544)
Danly B G, **1**: 570 (568)
Danskin J M, **2**: 1023 (1021)
Danter E, **1**: 152
Dantzig G B, **1**: 405 (400); **2**: 956 (952), 958, 959, 1388 (1387); **4**: 2836, 2839, 2841 (2841); **5**: 3170 (3167); **6**: 3739 (3736), 4299; **7**: 4666 (4661, 4664)
Dao T K, **6**: 4165 (4162, 4163)
D'Appolito J, **4**: 2713 (2710)
Darby M L, **6**: 3680 (3675, 3676)
Dardy H D, **5**: 3470 (3468)
D'Argenio D Z, **1**: 489 (487)
Darmon C A, **3**: 1665 (1656)
Darnall K R, **1**: 187 (187)
Daróczy Z, **4**: 2508 (2504, 2505); **6**: 3768
Dart A R, **3**: 1998
Darwin G H, **7**: (4890)
Darwish M G, **2**: 923 (919, 920, 922); **4**: 2697 (2692, 2695); **5**: 3019 (3017, 3018, 3019)
Das P, **7**: 4709 (4708)
Das Gupta S, **2**: 1105, 1106
DaSilveira M A, **2**: 1192 (1190, 1191)
D'Aspremont C, **2**: 850 (847)
Datardina S P, **2**: 1040 (1037); **5**: 3024 (3024)
Date C J, **2**: 893 (888), 947 (946), 1165 (1161); **4**: 2659 (2656)
Datko R, **3**: 2150 (2149); **5**: 3506
Datta K B, **4**: 2523 (2522)
Datta P K, **1**: 473 (469, 470, 471, 472)
Daubenspeck J A, **6**: 4066 (4065)
Dauer J, **5**: 3184 (3183, 3184)
Dauphin-Tanguy G, **5**: 3180; **7**: 4429 (4428), 4547
Dautov S A, **5**: 3134 (3131)
Dave J V, **3**: 2052 (2050)
Davenport H W, **6**: 3690 (3688)
Davenport J M, **2**: 1551 (1550)
Davenport R, **3**: 1943 (1942), 1958 (1957)
David A J, **3**: 2097 (2094)
David F N, **2**: 1106
David R H, **3**: 2076
David Y, **7**: 4927 (4926, 4927)
Davidon W C, **7**: 4974 (4972)
Davidovac Z, **3**: 2174 (2170)
Davidson D, **1**: 556 (554)
Davidson E H, **1**: 559
Davidson G A, **4**: 2386 (2382)
Davidson P J, **2**: 1445 (1441)
Davies A C, **1**: 421
Davies A M, **7**: 4888 (4884, 4886, 4888)
Davies B E, **1**: 268 (267)
Davies C T, **2**: 1136 (1136)
Davies D E N, **5**: 3470 (3469)
Davies E R, **4**: 2484 (2480, 2482, 2484); **6**: 3940 (3936)
Davies P, **7**: 4920

Davies P L, **5**: 3425 (3425)
Davies R B, **1**: 86 (83)
Davies R M, **3**: 1969 (1969)
Davies W D T, **1**: 498 (496); **4**: 2325 (2323)
Davies W L D, **2**: 963 (961, 962)
Davio M, **3**: 1883 (1881)
Davis A L, **7**: 5106 (5103)
Davis C J, **1**: 633 (632)
Davis C M, **5**: 3470 (3469), 3471 (3469)
Davis D D, **4**: 2723 (2721)
Davis D R, **7**: 5140 (5136)
Davis E W, **6**: 3904
Davis G B, **3**: 2192 (2186); **4**: 2501 (2499)
Davis J A, **6**: 4051 (4049)
Davis J C, **7**: 5015 (5014)
Davis J D, **3**: 2213 (2212)
Davis J E, **7**: 5060 (5059)
Davis J L, **5**: 3470 (3466)
Davis L S, **6**: 4159 (4157)
Davis M, **1**: 333, 720 (717, 718, 720), 725 (721); **2**: 850 (846), 937 (933, 936); **7**: 4612 (4612)
Davis M H A, **5**: 3303 (3301), 3329 (3320), 3336 (3336), 3523 (3519, 3521, 3522)
Davis P W, **6**: 3850 (3848)
Davis R, **2**: 1573; **4**: 2640 (2637)
Davis R B, **4**: 2782 (2781)
Davis R L, **3**: 2060 (2058)
Davis S M, **6**: 4033
Davison B, **7**: 4944
Davison E J, **1**: 132 (127), 580 (578), 704 (694), 735 (734); **2**: 902 (900, 901, 902), 907 (902, 903, 904, 906, 907), 916 (913, 914), 923 (917), 1043 (1041); **4**: 2559 (2559), 2560 (2559), 2570 (2570), 2671 (2666, 2668), 2702 (2699), 2827 (2825); **5**: 3196 (3189, 3191), 3220 (3215), 3228 (3227, 3228); **6**: 3717 (3713), 4132 (4129, 4130), 4133 (4129, 4130, 4131, 4132), 4138 (4133, 4134, 4135, 4137, 4138), 4139 (4133, 4135, 4136, 4137, 4138); **7**: 4684 (4683), 4689 (4686, 4687, 4688)
Davison S H, **6**: 3734 (3734)
Davison T W, **7**: 4871 (4868)
Davisson L D, **2**: 1492
Davson H, **7**: 5070 (5066, 5070), 5086
Davy H, **2**: 1402 (1399)
Davy R W, **4**: 2253 (2252)
Dawes C, **5**: 2993 (2991)
Dawes R M, **3**: 2191 (2190), 2202 (2195, 2196); **5**: 3115 (3113)
Dawson J K, **2**: 1458
Day J D, **2**: 1153 (1151)
Day J W Jr, **2**: 1538 (1537); **7**: 4454 (4453), 4897 (4897), 5008 (5005)
Day R H, **3**: 2202 (2195)
Dayal V, **2**: 1165
D'Azzo J J, **1**: 735 (732, 734)
De M, **3**: 1829 (1826)
De Angelis D L, **2**: 1539 (1538); **7**: 4645 (4645)
de Boor C, **5**: 3134 (3134); **7**: 4518 (4518)
De Carlo R, **2**: 916, 923 (922); **4**: 2559 (2559)
de Claris N, **4**: 2866 (2864); **6**: 3629 (3627)
de Cordova S, **2**: 1244 (1243)
de Does M, **4**: 2798 (2797)
De Figueiredo R J P, **3**: 1778 (1776)
De Finetti B, **3**: 1791 (1790)
De Fitts K E, **4**: 2338 (2337)
De Forest L, **3**: (1745)
De Fornel B, **7**: 5027
De Fronzo R A, **3**: 2025 (2022)
de Glas M, **3**: 1880 (1878, 1879), 1898 (1896, 1897); **5**: 2923 (2923), 2924 (2923)
de Goede J, **6**: 4061 (4057, 4061), 4062 (4059)
De Jager P A, **5**: 3421 (3419)
de Jong L S, **6**: 3983 (3979)
de Jong R H, **1**: 289 (285)
De Keyser R M C, **3**: 2096 (2094); **5**: 3430 (3426, 3428); **6**: 4172 (4170, 4171, 4172), (4172)
de Koning H W, **1**: 185
de la Brassinne M, **1**: 482 (480)
de Lange H, **4**: 2331 (2326)
de Larminat P, **1**: 60 (59); **2**: 1581 (1580); **5**: 3410 (3407, 3408); **6**: 3701
De Los Reyes G, **4**: 2385 (2380)
De Los Rios F J, **7**: 4749 (4745)
de Luca A, **2**: 1473 (1467, 1468, 1469, 1472, 1473), 1479 (1477); **3**: 1785 (1783), 1858 (1855), 1878 (1876); **5**: 2965 (2963, 2964, 2965)
De Meo E, **2**: 1458
de Monasterio F M, **4**: 2331 (2326)
de Montgolfier J, **5**: 3152 (3151), 3164 (3157)
De Mori R, **1**: 769 (764); **6**: 4159 (4158)
de Paor A M, **7**: 4462 (4460)
de Santis R, **6**: 4262 (4260)
de Saussure G, **2**: 1552 (1549)
de Valois K K, **4**: 2331 (2327)
De Vries G, **6**: 3626 (3623, 3625)
de Vries H, **2**: 967
De Wispelaere R, **3**: 1756
de Wit C T, **6**: 4411 (4410)
Deal K R, **2**: 1269 (1267)
Deam R J, **3**: 1718 (1715, 1716)
Dean F, **3**: 1975 (1972)
Deards S R, **6**: 3751 (3748)
Debaille S, **7**: 4927 (4927)
Debreu G, **1**: 671 (666, 668, 669); **3**: 1917 (1913); **5**: 3115 (3105), 3256 (3255); **6**: 3619 (3617)
Debruyn H P, **2**: 1116 (1112)
Dechamps P Y, **3**: 2052 (2050)
Dechene R L, **3**: 1694 (1691)
Deckert J C, **1**: 86 (84), 251 (251)
Declairis R N, **3**: 1862 (1859)
Deconinck H, **5**: 3139 (3139)
Decorpo J J, **5**: 2943 (2935)
DeCosse J, **2**: 1036 (1033), 1043 (1040)
Deemer W L Jr, **4**: 2353 (2352)
Deepak A L, **2**: 1311 (1307)
Defares J G, **4**: 2901 (2899)
DeFazio T L, **4**: 2385 (2380)
Defosse C, **1**: 7 (3, 4)
Degani R, **1**: 769 (764); **2**: 1407 (1405, 1406, 1407); **3**: 1847

Degen R J, **6**: 4369 (4369)
Degens E T, **1**: 531 (529); **7**: 5123 (5122)
Degn H, **1**: 429 (424), 433 (431), 570 (569); **5**: 3579 (3578)
DeGroot M H, **7**: 4622 (4621), 4990 (4988)
Dehl G, **7**: 4958 (4957)
DeHoff R L, **2**: 1244 (1243)
Deirmendjian D, **4**: 2655 (2655)
Dekker L, **1**: 424 (422), 633 (632), 634 (632); **2**: 1201 (1201); **3**: 2217; **6**: 4319 (4316), 4323 (4321), 4333 (4329), 4411 (4406)
Dekkiche E A, **2**: 879 (879)
DeKlerk L F W, **3**: 2192 (2187)
Del Castillo D, **1**: 142 (140, 141)
Deland E C, **2**: 1366 (1364)
Delaney H D, **3**: 2191 (2186)
Delbecq A L, **1**: 618; **3**: 2061 (2060); **4**: 2579 (2576)
Delbourg P, **3**: 1769 (1766)
Delbrouck L E N, **7**: 4731, 4741 (4734)
Delebecque F, **3**: 2225
Deleris M, **4**: 2547 (2547)
Delestre B, **7**: 4749 (4745)
Delforge J, **1**: 493 (491)
Delfosse C M, **2**: 894 (894); **6**: 4333 (4329, 4330, 4331), 4369 (4368), 4370 (4369); **7**: 4654 (4652, 4653)
Delfour M C, **2**: 1211 (1209); **3**: 2151 (2148, 2149, 2150), 2152 (2152); **5**: 3506 (3505)
DeLisi C, **1**: 494 (493)
Delisle G Y, **5**: 3242 (3241)
Dell R B, **2**: 1366 (1364)
DellAcqua R, **7**: 5112
Delobel C, **2**: 1160 (1157)
Delorme M, **3**: 1785 (1785)
Delosme J, **2**: 1102 (1102)
Delpuech J-J, **5**: 3421 (3418)
Delsarte P, **5**: 3134 (3132, 3134), 3135 (3133, 3134)
Delves L M, **2**: 826 (824, 825)
Demange G, **7**: 4472 (4472)
DeMarre D A, **1**: 13 (9)
Dembo R S, **2**: 826 (825)
deMello F P, **6**: 3815 (3812)
Demengel F, **1**: 529 (526)
Demerjian K L, **1**: 187 (185)
Demers K P, **4**: 2452 (2451, 2452)
Demers R, **2**: 1590 (1587, 1589)
Deminet J, **6**: 4405 (4404)
Demjanov V F, **5**: 3545 (3543)
DeMontigny C, **1**: 570 (569)
DeMorgan A, **4**: (2577)
Dempsey J R, **3**: 2006 (2006)
Dempster A P, **3**: 1785 (1785), 1822 (1821); **4**: 2883 (2881); **6**: 3772 (3769, 3770); **7**: 4669 (4667)
Dempster M A H, **2**: 959; **5**: 3303 (3301), 3336 (3336); **6**: 4148 (4147)
Dempster W T, **1**: 474 (469)
Den Hartog D, **5**: 3388 (3384)
Denardo E, **2**: 1099 (1098)
Denavit J, **6**: 4088 (4087)
Denbigh K, **1**: 433 (431), 588 (586)
Denes P, **1**: 456 (455)
Denham M J, **4**: 2855 (2855), 2871 (2871)
Denney R C, **1**: 594
Denning D E, **1**: 743 (741)
Denning P J, **1**: 743 (741), 755 (753); **5**: 3449; **6**: 3596; **7**: 4752
Dennis C, **1**: 142
Dennis J E Jr, **6**: 4310 (4308); **7**: 4975 (4973, 4974)
Dennis L P, **6**: 3780 (3777)
Dennis T D, **5**: 2980
Densham A B, **3**: 1969 (1968, 1969)
Denslow J S, **2**: 1317
Denson J S, **2**: 1366 (1364, 1366)
Deo N, **4**: 2556 (2552)
Deo U, **6**: 4310 (4309)
DePalma G F, **6**: 3653 (3648)
Derby S L, **6**: 4075
Derevitskii D P, **1**: 80 (77)
Derksen H E, **4**: 2901 (2899)
Derman C, **5**: 2932
DeRoos F, **4**: 2511 (2510)
DeRose A J, **1**: 594
DeRusso P M, **6**: 3870 (3868)
Desai M N, **1**: 86 (84), 251 (251)
Desai U B, **7**: 4466 (4463, 4465), 4468 (4467), 4469 (4466, 4468)
Desalu A A, **2**: 1500
Descartes R, **2**: (870, 872)
Deschamps P J, **7**: 4731
Descheneau C, **1**: 561 (560, 561); **6**: 4345 (4344)
Descusse J, **4**: 2855 (2852, 2853, 2854, 2855); **7**: 4547
Desforges D T, **6**: 4086 (4084), 4100
Deshayes J, **1**: 86 (84)
Deshpande J G, **6**: 4165 (4162, 4163)
Deshpaude P B, **3**: 2113 (2111)
Desoer C A, **1**: 580 (577); **2**: 1286, 1348 (1346), 1486 (1480); **3**: 2228 (2226, 2227); **4**: 2570 (2569, 2570), 2671 (2670), 2675 (2672, 2673), 2678 (2676), 2832, 2875; **5**: 3196 (3192), 3215 (3208, 3209), 3228 (3225), 3319 (3317), 3379 (3377), 3441 (3441); **6**: 4209 (4206), 4227 (4225), 4355 (4354)
Dessimoz J, **6**: 4123
Dettman E H, **4**: 2646 (2645)
Deutsch R, **4**: 2697 (2693)
Devaud F M, **7**: 5031 (5029)
Devijver P, **4**: 2635 (2635); **6**: 3643 (3637)
Devine J G, **7**: 5106 (5103)
Devroye L P, **4**: 2719 (2719); **6**: 3950 (3948)
Devry H, **3**: 1756 (1750)
Dewar J, **3**: 1958
DeWerth D W, **2**: 1233 (1233)
Dewilde P, **2**: 785 (784)
DeWispelare A, **5**: 3115
Dexter A L, **3**: 1992 (1991), 2096 (2094, 2095, 2096), 2097 (2096)
Deyst J J, **1**: 86 (84), 251 (251); **2**: 1257 (1256)
D'Hooge H D, **5**: 2980
Di Nola A, **3**: 1781 (1781), 1878 (1876, 1877); **5**: 3421 (3418)

Di Pillo G, **1**: 417 (414, 415, 416); **5**: 3279 (3276); **6**: 4244 (4238)
Di Stefano N, **2**: 1211
Diamant N E, **2**: 1043 (1041)
Diaz C M, **5**: 3282 (3282)
Diaz J I, **4**: 2613 (2610)
Diaz R J, **4**: 2464 (2462)
Dickinson B W, **5**: 3292 (3289)
Dickinson C J, **1**: 468 (466, 467); **2**: 1366 (1362, 1364, 1365), 1410; **4**: 2901 (2899)
Dickinson K W, **7**: 4920
Dickinson R M, **7**: 4476 (4473)
Dicks J L, **2**: 1057 (1056)
Dickson F N, **7**: 4504 (4499)
Dickson G W, **7**: 4793 (4793)
Dickson K L, **4**: 2646; **7**: 4901 (4899)
Diday E, **1**: 608 (608); **6**: 3643 (3641)
Diebold J, **7**: 4775 (4774)
Diemer F P, **2**: 1311 (1307, 1309); **3**: 2053 (2048, 2049, 2052)
Diesendorf W, **6**: 3806 (3803)
Diesing P, **7**: 4785 (4781)
Dietrich C F, **2**: 1559
Dietrich D, **7**: 5027
Dietz H, **4**: 2582 (2581)
Dieudonné J, **5**: 3376 (3374)
Diffenbach J, **7**: 4789 (4788)
Diffie W, **2**: 1132 (1130)
DiGiano F A, **4**: 2646 (2646)
Dijkstra E W, **2**: (883); **3**: 1642 (1640), 1712 (1712); **6**: 3898; **7**: 4472
Dijstelbergen H H, **3**: 1693 (1687)
Dillard J K, **6**: 3806 (3806)
Dillmann R, **4**: 2351 (2347), 2385 (2374)
Dillon J A Jr, **2**: 1346 (1345), 1355
Dillon P J, **4**: 2646 (2643), 2650 (2648), 2794 (2789)
Dillon T S, **2**: 1388 (1386, 1387); **3**: 2225; **6**: 3834 (3832), 3837 (3836)
Dillon W R, **2**: 1106 (1103, 1104, 1105)
Dils R, **5**: 3464 (3460)
Dimirovski G M, **5**: 3383 (3382)
Dinh T V, **5**: 3464 (3463)
Dinsel D L, **1**: 181
Dintenfass L, **1**: 445 (441)
Director S W, **6**: 4322 (4319)
Disanto A R, **6**: 3698 (3697)
Dishkant H, **3**: 1825 (1824)
Dishon M, **7**: 4646 (4644)
DiStefano J J III, **1**: 493 (493); **3**: 2143 (2140); **4**: 2257 (2254), 2263 (2263); **7**: 4905 (4904)
DiToro D M, **4**: 2646 (2644, 2645)
Dittmar E A, **7**: 4905 (4905)
Dittmer D S, **1**: 450 (448, 449)
Dixon J B, **1**: 7 (1, 2), 8 (1, 2, 3, 5)
Dixon K R, **1**: 142 (140, 141), 592 (590, 591)
Dixon L C W, **7**: 4975 (4973)
Dixon R C, **2**: 1057 (1056); **5**: 2956 (2955)
Djerassi C, **5**: 2943 (2937)
Djibo H, **6**: 3954 (3953)
Djorovic M V, **5**: 3572 (3569)
Dlaboha I, **4**: 2449 (2448, 2449)
Dobois D, **3**: 1785 (1783)
Dobson C M, **1**: 13 (13)
Dobzhansky T, **1**: 556 (554)
Dockner E, **2**: 1269 (1263, 1266)
Dodds S J, **6**: 3870
Dodu J C, **5**: 3530 (3529); **6**: 3789 (3788, 3789)
Doebelin E O, **2**: 1291, 1297; **4**: 2533 (2528); **5**: 2970, 2973 (2972)
Doedel E J, **5**: 3496 (3490, 3492, 3496)
Doemens G, **4**: 2473
Doepel P S, **6**: 3806 (3805)
Doering B, **6**: 4306 (4303)
Doetsch G, **4**: 2664 (2659)
Doetsch K H, **6**: 4088 (4086)
Dogramaci A, **4**: 2353 (2351)
Doherty M E, **7**: 4785 (4783)
Doktor R H, **7**: 4793 (4791)
Dolan W M, **1**: 289 (285)
Dolecki S, **4**: 2861
Doll J R, **7**: 4490
Dollezhal' N A, **5**: 3440
Dombi J, **3**: 1781 (1781)
Domican E H, **3**: 1981
Dommel H W, **6**: 3829 (3829)
Donachie R P, **7**: 5106 (5104)
Donald D E, **1**: 542 (540)
Donald M W, **1**: 437
Donath M, **6**: 4108 (4104)
Done J N, **7**: 5112
Dong E Jr, **4**: 2314 (2311)
Dongarra J, **2**: 826 (817, 818, 820, 824), 827 (815, 818, 824)
Donnell M L, **3**: 1845 (1845), 1847 (1846)
Donoghue E A, **2**: 1449 (1449)
Donoghue J F, **5**: 3144
Donovan J J, **4**: 2504 (2503); **5**: 3165, 3449; **7**: 4752, 4810
Donsker M D, **1**: 91 (89)
Doob J L, **5**: 2934 (2932, 2933); **7**: 4644 (4638), 4741 (4736)
Doodson A T, **7**: 4892 (4889)
Dooner M, **4**: 2385 (2383)
Dopazo J F, **6**: 3820 (3818), 3850 (3846, 3848, 3849)
Doraiswamy P C, **1**: 592 (591)
Dorato P, **2**: 1099 (1095)
Doremus R H, **6**: 3687 (3683)
Dorf R C, **2**: 1291, 1297
Dorling C, **7**: 5031 (5028)
Dormeier S, **2**: 1586 (1584, 1585)
Dorn H C, **5**: 3421 (3419)
Dorny C V, **3**: 1894 (1892)
Dorrah H T, **5**: 3388 (3387)
Dorrenbacker J, **2**: 826
Dorsey A J, **7**: 4929 (4929)
Doscher T M, **3**: 2006 (2006)
Doss H, **7**: 4637 (4635)
Doss S K, **2**: 1181 (1181)

Dos Santos S M, **2**: 1424 (1423), 1431 (1427, 1428, 1430), 1434, 1440 (1436, 1440), 1445 (1441, 1442); **7**: 5051 (5048)
Dost M H, **6**: 4311
Dote Y, **7**: 5027
Dotson W L, **6**: 3933
Doty K W, **2**: 1282 (1279)
Douce J L, **5**: 3373 (3370), 3390 (3385); **7**: 4452 (4451)
Doucet P G, **4**: 2798 (2796)
Douglis A, **6**: 4223 (4218)
Doukidis G I, **4**: 2550 (2550)
Douklias N, **5**: 3464 (3461)
Doulliez P, **6**: 3789
Dovas T, **7**: 5186 (5186)
Doverspike R D, **7**: 4749 (4748)
Dowd D L, **3**: 2006 (2005)
Dowdle J R, **4**: 2846 (2845)
Dowling J E, **7**: 5070 (5066)
Downie I C, **5**: 3242 (3241)
Downing D J, **2**: (1550)
Downing D M, **4**: 2517 (2516)
Downs R M, **4**: 2640 (2638)
Downs T, **7**: 4732
Doyle G J, **1**: 176 (175)
Doyle J C, **1**: 736 (732, 735); **2**: 826 (822); **4**: 2846 (2845), 2872 (2872); **5**: 3196, 3228 (3225, 3226, 3227, 3228); **6**: 4227 (4224)
Doyle L B, **1**: 276 (275)
Doyle T W, **2**: 1317 (1315, 1317)
Dozono S, **6**: 4155 (4154)
Dracup J A, **7**: 5140
Draffan I W, **2**: 1160
Drake B, **3**: 1705 (1701)
Draper C S, **4**: 2492 (2488)
Draper N R, **5**: 3030 (3027, 3028); **6**: 3621 (3620)
Drazan P J, **4**: 2464
Draženović B, **7**: 5031 (5028)
Drebbel C, **3**: (1604, 1606)
Dreben B, **2**: 937 (933)
Dredge J, **2**: 1402 (1399, 1400, 1401)
Dregn H E, **2**: 1358
Drenick R, **7**: 4822 (4822)
Dresher M, **2**: 849 (845, 846), 850 (847); **3**: 1926 (1926)
Dressler J, **7**: 4716 (4714)
Dressler M, **7**: 5129 (5128)
Drew D R, **7**: 4933
Dreyfoos W D, **4**: 2452 (2451)
Dreyfus H L, **4**: 2501 (2501)
Dreyfus J L, **3**: 2191 (2188)
Dreyfus S E, **1**: 409, 725 (724); **3**: 1714; **4**: 2501 (2501); **7**: 4793 (4792), 5153 (5153)
Dreyfuss R G, **3**: 2014
Dreze J H, **2**: 849 (841, 844, 847, 848)
Driskell L R, **3**: 1700
Dritz K, **2**: 826
Driver M J, **7**: 4793 (4791, 4792)
Droop M R, **4**: 2646 (2644)
Dror Y, **2**: 1542 (1540)
Drouilhet P R, **1**: 201 (200), 202
Drouin M, **2**: 956 (952, 954, 955)
Drucker D C, **5**: 3470 (3467)
Drueke T, **1**: 310 (308)
Druffel L E, **1**: 755 (753)
Drukker W, **1**: 312 (310)
Drusiani R, **3**: 1975
Drysdale D B, **6**: 4061 (4057)
D'Silva A P, **1**: 177 (174)
D'Souza C, **4**: 2469 (2468)
Du Charme W M, **4**: 2338 (2337)
Dubé Y, **5**: 2993 (2991)
Dubeau F, **5**: 3550
Dubin M W, **7**: 5070 (5066)
Dubke M, **5**: 3139 (3136)
Dubois D, **1**: 740 (738), 769; **2**: 1407 (1405, 1407), 1549 (1547); **3**: 1603 (1602, 1603), 1648 (1642, 1646, 1647), 1781 (1780, 1781), 1785 (1785), 1802 (1802), 1822 (1822), 1831 (1829, 1830), 1835, 1847 (1846, 1847), 1858 (1856), 1872 (1868), 1874, 1883 (1880), 1894 (1891); **5**: 2921 (2919), 3156 (3154, 3155), 3454 (3451); **6**: 3989 (3984); **7**: 4766 (4765)
DuBois D F, **4**: 2550 (2548)
Dubois D M, **7**: 4895 (4895)
Dubois L, **1**: 177 (176)
Dubois R J, **1**: 177 (175)
Dubosq L J, **2**: 1402 (1400)
DuBourdieu D J, **2**: 1154 (1153)
Dubowsky S, **6**: 4086 (4084), 4100
Dubreuil M, **1**: 561 (561)
Dubs A, **7**: 5065 (5063, 5064)
Dubuisson B, **6**: 3644
Duc G, **1**: 132 (131)
Düchting W, **7**: 4958 (4957, 4958)
Duckstein L, **2**: 1553 (1552); **3**: 1829 (1826); **5**: 3115 (3112), 3164 (3157, 3163), 3165 (3157); **7**: 5140 (5135)
Ducornet D, **7**: 4913 (4913)
Ducrest C, **5**: 2942 (2935)
Duda R O, **1**: 575 (574), 664 (654); **2**: 1106 (1105), 1332 (1332), 1573; **3**: 1603 (1602, 1603); **4**: 2385 (2382), 2473, 2484 (2481, 2483), 2487; **6**: 3643 (3637), 3653 (3650, 3651), 4123, 4159 (4156)
Dudani S A, **4**: 2385 (2382)
Duderstadt J, **7**: 4944
Duerr B, **6**: 4159 (4158)
Duever M J, **2**: 1344 (1344)
Duff I S, **4**: 2827 (2825)
Duff J, **1**: 63 (60); **6**: 4204 (4200)
Duff M J B, **4**: 2473 (2472)
Duffing G, **5**: 3388 (3383, 3384, 3385)
Duffy G G, **2**: 781 (779)
Duffy I, **4**: 2391
Duffy J, **6**: 4108 (4105)
Duffy J J, **2**: 1497 (1494)
Dufort E C, **6**: 3626 (3625)
Dufour J, **1**: 703 (689); **2**: 879 (878, 879); **4**: 2547 (2544, 2546, 2547)
Dufournaud C M, **7**: 4799 (4794)
Dugard L, **5**: 3004 (3003); **6**: 4193
Duhon R E, **2**: 781 (779)

Dukhovny M, **3**: 1648 (1646)
Dulac P, **5**: (3384)
Dumont G A, **1**: 51 (49, 50); **6**: 3994 (3994), 4213 (4213); **7**: 4894 (4892)
Dumont J E, **2**: 1249 (1245)
Dumontel M, **3**: 1665 (1660)
Duncan T E, **5**: 3303 (3293, 3294, 3299)
Dunford N, **4**: 2343 (2342); **5**: 3205 (3202); **6**: 4223
Dunklin L J, **7**: 4856 (4853)
Dunn B E, **5**: 2943 (2935)
Dunn D J, **2**: 1106
Dunn I J, **3**: 1614 (1612); **5**: 2943 (2935)
Dunn J C, **1**: 608 (607, 608); **3**: 1603 (1602), 1825 (1824), 1874
Dunn J M, **5**: 2921 (2921)
Dunn K L, **1**: 474 (471)
Dunn M, **1**: 769 (761)
Dunn O J, **2**: 1106
Dunn R, **1**: 246
Dunne M J, **1**: 301 (297)
Dunnet A, **7**: 4913 (4912, 4913)
Dunnett R M, **2**: 1388; **7**: 4577 (4575)
Dunnette M D, **3**: 2192 (2186)
Dunning J M, **2**: 972 (971, 972)
Duong N, **3**: 2097 (2094)
Durand J, **6**: 4155
Durbeck R C, **1**: 638 (634)
Durbin J, **7**: 4566 (4564, 4566)
Durnin J V G A, **5**: 2970
Durrandeau M, **1**: 120
Durst F, **2**: 1532; **3**: 1693 (1686)
Durst R A, **6**: 3688 (3685)
Durup J, **1**: 428 (424)
Dussaix A M, **1**: 687 (686)
Duthie H L, **1**: 462 (460, 461); **2**: 1040 (1036), 1043 (1042)
Dutilh J R, **3**: (1749); **5**: 3389 (3385); **7**: 4452 (4443)
Dutta Majumder D, **3**: 1825
Dutton B C, **3**: 1769 (1766, 1768, 1769)
Dutton J, **7**: 4833 (4831)
Dutton J M, **6**: 4411 (4406)
Dutton W A, **5**: 3282 (3282)
Duvaut G, **2**: 1261; **4**: 2664; **5**: 3535 (3534); **7**: 5036 (5031, 5034, 5035, 5036)
Duyff J W, **4**: 2901 (2899)
Dwek M G, **6**: 3806 (3804)
Dwolatzky B, **1**: 63 (60); **6**: 4204 (4200)
Dwyer T A W, **3**: 1778 (1776)
Dyckhoff H, **3**: 1781 (1781)
Dyer D S, **5**: 3115 (3106, 3111, 3113)
Dyer J A, **6**: 4302 (4301)
Dyer J S, **5**: 3124 (3123), 3164 (3157), 3568 (3567); **6**: 4075; **7**: 4990
Dyer P, **7**: 4540 (4535, 4536, 4538)
DyLiacco T E, **5**: 3144; **6**: 3846 (3842, 3846)
Dym H, **2**: 785 (784)
Dynkin E B, **2**: 1099 (1095); **5**: 3523 (3519), 3535 (3533); **7**: 4612
Dzubay T G, **1**: 177 (174, 175)
Dzúrik R, **1**: 312 (310)

E

Eager D, **2**: 1165 (1161, 1165)
Eames C, **2**: (988)
Earl C F, **6**: 4108 (4106)
Eaton R P, **3**: 2174 (2169)
Eatough D J, **1**: 177 (175)
Eaves B C, **5**: 3196 (3195); **6**: 3634
Ebanks B R, **2**: 1473 (1470)
Ebdon L, **1**: 13 (12)
Eberhard P, **1**: 440
Ebert K H, **6**: 3745 (3740, 3742)
Eccles E S, **1**: 226
Eccles J C, **1**: 466 (465); **5**: 3253 (3249, 3250)
Echlin P, **2**: 1420; **6**: 4152
Echman D P, **2**: 832
Eckberg A E, **4**: 2871 (2870)
Ecker J G, **5**: 3164 (3161)
Eckhaus W, **7**: 4429 (4427)
Eckman J, **6**: 4355 (4354)
Eckmann J-P, **1**: 567, 570 (568)
Eckstein E C, **2**: 866 (865)
Eda H, **6**: 4261 (4258)
Edden A C, **7**: 4892 (4889)
Edelbaum T N, **2**: 1337 (1335); **7**: 4490
Edelman G M, **1**: 326 (324)
Edelman S K, **6**: 4051 (4049)
Edelstein B B, **1**: 433 (431)
Edelstein M, **2**: 812 (810)
Eden C, **4**: 2640 (2637, 2638, 2639); **7**: 4785 (4783)
Eden M, **6**: 3644
Edgar A T, **3**: 1979
Edgar T F, **2**: 1109 (1108, 1109); **6**: 4193 (4190), 4214 (4212, 4213)
Edgerton M T, **1**: 482 (480)
Edgington D N, **4**: 2782 (2779)
Edholm O G, **7**: 4871 (4869), 4880 (4878)
Edie L, **4**: 2353 (2352)
Edie L C, **7**: 4922
Edington J W, **7**: 4940
Edison T A, **2**: 1402 (1399, 1400)
Edlich R F, **1**: 482 (480)
Edlund S G, **6**: 3889 (3888, 3889)
Edmonds J, **1**: 626 (625); **6**: 3739 (3737, 3738)
Edmunds J M, **1**: 580 (575, 578, 579), 736 (732); **2**: 1466 (1464); **5**: 3228 (3225, 3226), 3383 (3381); **6**: 3727 (3722, 3724, 3725), 4193 (4190, 4193), 4197 (4194)
Edmunds L, **1**: 429 (424, 426)
Edrich J, **7**: 4856 (4856), 4871 (4870)
Edwards D, **2**: 992 (988, 989)
Edwards D H, **7**: 5065 (5062, 5065)
Edwards E, **3**: 2186 (2183)
Edwards J A, **2**: 1282 (1281)
Edwards J B, **5**: 3130 (3129), 3135 (3134)
Edwards J J, **6**: 4160 (4159)
Edwards J K, **6**: 4142 (4141)
Edwards M J, **5**: 3215 (3213)
Edwards N T, **2**: 1345 (1343)
Edwards R, **2**: 1078 (1074, 1077, 1078)
Edwards S J, **7**: 5114
Edwards W, **2**: 942 (941); **3**: 2202 (2196)
Edwin K W, **2**: 1388 (1386, 1387, 1388); **6**: 3834 (3833)
Efroymson M A, **2**: 1416 (1415)
Efstathiou J, **1**: 769 (766); **3**: 1829 (1826)
Egan G T, **6**: 3837 (3836)
Egan J A, **7**: 5112
Egan J T, **6**: 4302 (4302)
Egardt B, **1**: 51 (48), 54 (54), 60 (56, 59); **5**: 3004 (3003), 3061 (3061), 3065 (3062); **6**: 4186 (4186), 4209 (4206, 4207); **7**: 4607 (4603), 4611 (4607)
Egeland O, **6**: 3991 (3991)
Eger E I II, **1**: 281 (277), 289 (285)
Eggemeir F T, **3**: 1668 (1666), 1669 (1666)
Egger M D, **6**: 4168 (4167)
Egloff D A, **2**: 1355 (1353); **4**: 2646 (2645); **7**: 5008 (5005, 5007)
Eguchi H, **2**: 1521 (1519)
Ehlers W, **1**: 142 (138)
Ehrmann S T, **6**: 3850 (3847, 3848)
Eichhorst P, **4**: 2798 (2797)
Eichna L W, **7**: 4856 (4852), 4876 (4876)
Eickhoff W, **5**: 3470 (3469)
Eigen M, **6**: 4167 (4167)
Eikemeier H, **1**: 433 (432)
Eilbert R F, **2**: 1478 (1478), 1479 (1478)
Eilenberg S, **1**: 18; **5**: 3272 (3268, 3272), 3276 (3273), 3394 (3393)
Einhorn H J, **3**: 2202 (2199, 2200); **5**: 3115
Einthoven W, **1**: (532)
Eisenbeis S A, **6**: 4160 (4157)
Eisenberg S, **4**: 2887
Eisenfeld J, **1**: 493 (493), 494 (493)
Eisenman G, **6**: 3687 (3683, 3686)
Eising R, **5**: 3135 (3131); **6**: 3734 (3733), 3972 (3970)
Eisner A, **2**: 1244 (1244)
Ekeland I, **1**: 405 (400), 529 (524, 525, 526, 527); **2**: 1033, 1261 (1257, 1261); **5**: 3184 (3182), 3530 (3526); **7**: 5046 (5043, 5044, 5045), 5047 (5044, 5045, 5046)
Ekiel J, **5**: 3248 (3243, 3244, 3247)
Ekström M P, **5**: 3130 (3130)
El-Abiad A H, **2**: 1388 (1387)
El-Agizi N G, **5**: 3127 (3125, 3126, 3127)
El Attar R A, **1**: 132 (127)
El Awtani A W, **5**: 3480
El-Bagoury M A, **5**: 3232 (3229, 3230, 3231, 3232)
El-Fattah Y M, **4**: 2737
El Geldawi F, **5**: 3389 (3385)
El Ghamrawy E, **2**: 969
El-Ghezawi O M E, **7**: 5031 (5029, 5030)
El-Hawary M E, **3**: 2225 (2222)
El-Hefnawy A, **5**: 3097
El Jai A, **2**: 808 (802, 807); **6**: 4245 (4244)

El Karoui N, **5**: 3523 (3521); **7**: 4612 (4612)
El Moudni A, **7**: 4547
El-Sherbing M, **4**: 2723 (2721, 2722)
Elam J J, **6**: 4332 (4329, 4331, 4332)
Elbrazaz Z, **1**: 132
Elcock E, **4**: 2640 (2637)
Eldor M, **7**: 4929 (4929)
Eldridge F L, **6**: 4061 (4056)
Elenbaas J R, **3**: 1975
Elftman H, **3**: 1910 (1908)
Elgerd O, **6**: 3815 (3811, 3812), 3855 (3851, 3852)
Elhardt E, **2**: 1154 (1153)
Elias P E, **2**: 1473 (1473); **4**: 2497 (2494)
Eliott R J, **2**: 1008 (1006), 1017 (1016)
Elis Norden K, **3**: 1710
Eljkman E G G J, **3**: 2192 (2187)
Ell P J, **6**: 3940 (3940)
Ellenson J L, **2**: 1521 (1521)
Elliot H, **5**: 3220 (3220); **6**: 4197 (4194)
Elliott C T, **7**: 4871 (4869)
Elliott D L, **5**: 3341 (3340)
Elliott H, **5**: 3060 (3059)
Elliott R J, **5**: 3523 (3520, 3521, 3523); **7**: 4637 (4631, 4632)
Ellis C A, **2**: 1154, 1169 (1166, 1167)
Ellis F, **6**: 4142 (4140, 4142)
Ellis F P, **1**: 643 (642)
Ellis G A, **3**: 1668 (1667)
Ellis J E, **1**: 638 (638)
Elloy J P, **3**: 1642 (1640)
Elmaghraby S E, **6**: 3904
Elmquist H, **1**: 732 (727)
Elsasser W M, **2**: 1479 (1474)
Elsayed E A, **6**: 3904
El'sgol'c L E, **3**: 2151, 2152 (2152); **5**: 3506
Elsilä M, **6**: 3916 (3914)
Elstein A S, **2**: 1577 (1574)
Elterman L, **4**: 2655 (2655)
Elton C S, **1**: 674 (672); **7**: 4454 (4453), (4954)
Elveback L R, **2**: 1043 (1040)
Elving P J, **7**: 4504
Elzas M S, **1**: 561 (561), 634 (632); **2**: 1552 (1550); **5**: 3035 (3030, 3032, 3033), 3088 (3083); **6**: 4229 (4228), 4313, 4322 (4321), 4323 (4321), 4326, 4353 (4350, 4351, 4352, 4353), 4372 (4371), 4381, 4388 (4386), 4392 (4390), 4411 (4406, 4410), 4414; **7**: 4692 (4691), 4764 (4762), 5008 (5007), 5011 (5008)
Emami-Naeini A, **2**: 826 (822)
Emanuel W R, **6**: 4229 (4228)
Embree P, **5**: 3130 (3130)
Emel'yanov S V, **7**: 5031 (5027)
Emery F, **7**: 4833 (4828)
Emre E, **4**: 2523 (2523), 2702 (2699); **6**: 3969
Emshoff J R, **6**: 4415 (4414); **7**: 4785 (4780)
Emslie Smith D, **2**: 1410 (1408)
Endo T, **5**: 3024 (3020, 3023)
Endrenyi L, **1**: 489 (487)
Enever K J, **7**: 5156, 5161 (5157)
Eng R S, **4**: 2511 (2509)
Engel B T, **1**: 437
Engelberg G P, **6**: 4418 (4418)
Engelberger J, **1**: (371)
Engelbrecht-Wiggans R, **5**: 2911 (2909)
Engelken E J, **2**: 1596 (1594, 1595)
Engeman R M, **6**: 4051 (4048)
England G, **3**: 1718 (1718)
England J W, **4**: 2508 (2506, 2507)
English J W, **6**: 4266 (4262)
English M J, **2**: 1410
En-Naamani A, **2**: 879 (879)
Enomoto S, **5**: 2943
Enright W, **2**: 826 (821)
Enroth-Cugell C, **4**: 2331 (2325)
Enslow P H, **2**: 1126
Entance P, **3**: 1694 (1692)
Enzer S, **4**: 2338 (2334)
Eppen G, **6**: 3706 (3705)
Epstein A N, **3**: 2213 (2210), 2214 (2211)
Epstein G, **5**: 2921 (2921)
Epstein R L, **2**: 937 (937)
Erdem U, **3**: 1710 (1707, 1709)
Erickson K M, **3**: 2174 (2169)
Erickson L E, **3**: 1611 (1609)
Erickson M D, **7**: 5129 (5128)
Ericson R, **1**: 333 (332)
Ericsson K A, **3**: 2191
Erikson J, **5**: 3351 (3341)
Eriksson L, **6**: 3889 (3889)
Erisman A M, **2**: 826 (821); **6**: 3846 (3845)
Erlang A K, **6**: (3924)
Ermentrout G B, **1**: 433 (432)
Ernst G, **3**: 1700
Ernst R R, **5**: 3425 (3423)
Ershler J, **3**: 1648 (1644)
Ershov A A, **5**: 3554 (3552)
Erzberger H, **7**: 5031 (5029)
Esat I I, **6**: 4108 (4107)
Escaffre F, **3**: 2089
Escanye J M, **7**: 4856 (4856)
Escudero L F, **2**: 1106 (1104, 1105)
España M D, **2**: 1121 (1117, 1118, 1119); **5**: 3376 (3374); **6**: 3776 (3772)
Espiau B, **1**: 86 (82); **6**: 4123
Espinas A V, **6**: (3861)
Estes W K, **1**: 276 (272); **3**: 2191
Estrin G E, **1**: 754 (753)
Eswaran K P, **2**: 883 (881, 882), 1154 (1145, 1147), 1174 (1171)
Etkin B, **1**: 108, 220 (220)
Ettlie J E, **7**: 4833 (4831)
Etzioni A, **3**: 2202 (2202)
Etzwiler D D, **3**: 2032 (2030)
Euler L, **4**: (2577, 2659); **6**: (3751)
Eustache H, **5**: 2942 (2935)
Evans A J, **1**: 201 (200)
Evans F C, **2**: 1356 (1356)
Evans F G, **1**: 474 (471)
Evans F J, **2**: 952 (950), 1329 (1327)
Evans J H, **1**: 478 (474), 482 (479, 480)

Evans J P, **5**: 3164 (3161)
Evans L, **3**: 1761 (1760)
Evans L B, **1**: 585 (583); **7**: 4703 (4701)
Evans L C, **2**: 1279 (1278); **3**: 2068 (2065); **7**: 4626 (4624), 4627 (4626)
Evans R, **2**: 1479 (1475)
Evans W E, **6**: 4037 (4034)
Evans W R, **1**: (732, 734); **3**: 1749 (1748)
Evarts E V, **1**: 466 (463)
Eveland J D, **7**: 4833 (4828, 4829, 4830, 4831)
Even S, **6**: 3670 (3668)
Everall P F, **7**: 4915 (4915), 4917 (4916)
Everett D, **4**: 2253 (2248)
Eversheim W, **4**: 2469 (2465)
Ewart T G, **3**: 2174 (2170)
Eweda E, **1**: 80 (77, 78, 79)
Ewing D J, **6**: 4245
Ewing K L, **7**: 4871 (4868)
Ewing L, **3**: 1943
Extance P, **5**: 3464 (3463)
Eykhoff P, **3**: 1894 (1888, 1891), 2143 (2140); **4**: 2245 (2242), 2257 (2255), 2273 (2272), 2283 (2279, 2282, 2283), 2307 (2302, 2306), 2319 (2319), 2325 (2325), 2575 (2573), 2635 (2633), 2697 (2693); **5**: 3248 (3246, 3247), 3475 (3473); **6**: 4180 (4179), 4199 (4198); **7**: 4692 (4690, 4691)
Ezekial S, **5**: 3464, 3470 (3465, 3466, 3467)
Ezri M D, **1**: 456 (455)

F

Fabian M, **4**: 2331 (2325)
Fabrycky W J, **2**: 1003 (996); **4**: 2764; **6**: 3904; **7**: 4771
Faddeev D K, **1**: 735 (734); **6**: 4237 (4235); **7**: 4695 (4694)
Faddeeva V N, **1**: 735 (734); **6**: 4237 (4235)
Fadden E J, **1**: 634 (632)
Fagan J, **6**: 3850 (3850)
Fagel M M, **5**: 3164 (3157)
Fagen R E, **2**: 1489 (1488)
Fagerlund C, **2**: 963 (962)
Fagerström T, **4**: 2782 (2780)
Fahmy M M, **5**: 3127 (3125, 3126, 3127)
Fairchild B T, **6**: 4417
Faires R A, **6**: 3940 (3937)
Falb P F, **3**: 2108 (2108)
Falb P L, **1**: 18, 294, 735 (734); **2**: 1286; **4**: 2517 (2514), 2523 (2522), 2827 (2826), 2865 (2865), 2866 (2865); **5**: 3272 (3268), 3389 (3387), 3394 (3390), (3517); **6**: 3734 (3733), 3751 (3749), 3776, 3977 (3974); **7**: 4554 (4551, 4552, 4553)
Falciari R, **5**: 3465 (3461)
Falconer D D, **1**: 80 (75, 76, 78); **6**: 4204 (4203)
Falek D, **4**: 2385 (2374)
Falgarone F, **3**: 2225; **6**: 3789 (3788)
Falk G, **2**: 1125 (1122, 1123); **4**: 2385 (2382)
Falkenborg D H, **2**: 1538 (1537); **4**: 2646 (2645)
Falkenburg D R, **6**: 3755, 4088 (4087)
Fallert D L, **5**: 2943 (2935)
Fallows J, **7**: 4755
Fallside F A, **3**: 1779 (1777); **7**: 5098, 5150
Falster P, **2**: 952 (950)
Fambro D B, **3**: 1736
Fan L T, **3**: 1611 (1609)
Fan T I, **6**: 3653
Fandel G, **5**: 3164 (3161)
Fanel G, **2**: 1276 (1273)
Fang C S, **4**: 2511 (2510)
Fang L, **7**: 4799 (4798)
Fang Li-Zhang, **1**: 63 (60)
Fanger P O, **1**: 159, 639, 642 (639); **7**: 4866 (4865), 4868
Fanneløp T K, **3**: 1992 (1992)
Fannin D R, **6**: 4281 (4278)
Fant C G M, **7**: 4509 (4504)
Fantin J, **2**: 923 (919, 920, 922)
Farcot J J L, **3**: (1745)
Fargier P H, **7**: 4913 (4913)
Farhi L E, **2**: 1366 (1362)
Farines J M, **7**: 5027
Farison J B, **6**: 4165 (4162)
Farkas D F, **3**: 1705 (1701)
Farley J, **6**: 4167 (4167)
Farmer A, **3**: 1862
Farmer E D, **2**: 1388 (1386); **6**: 3829 (3826)
Farmer W M, **2**: 1532 (1528)
Farnsworth R L, **6**: 4128 (4125)
Farquhar J W, **3**: 2025 (2022)
Farquhar P H, **3**: 2202 (2195); **5**: 3104 (3103); **7**: 4990 (4990)
Farquharson R, **3**: 1921 (1918)
Farrell E J, **4**: 2901 (2899)
Farrell J H, **2**: 978 (978)
Farrington D P, **2**: 1565 (1563)
Farris D R, **2**: 942 (941); **3**: 2060 (2053), 2096 (2093, 2094)
Fasol K H, **2**: 1291, 1297; **3**: 2228 (2226); **5**: 3389 (3387)
Fassel V A, **1**: 177 (174); **4**: 2347 (2343, 2345)
Fatt I, **5**: 3579 (3578)
Fattorini H O, **2**: 1186 (1185); **5**: (3517)
Faucheux C, **1**: 399
Faude D, **5**: 3416
Faulconer A, **1**: 281 (279)
Faulkner F D, **4**: 2846 (2843)
Faurre P, **6**: 3964
Favier G, **6**: 4176 (4176)
Fawn D L, **3**: 1723
Fayolle G, **1**: 91 (87)
Fazzolare R A, **3**: 1761 (1758)
Feather N, **3**: 2192 (2186), 2202 (2196)
Fechner G T, **3**: (1667)
Federer H, **5**: 3538 (3537)
Federov V V, **4**: 2263 (2263)
Federspil G, **3**: 2021 (2018), 2025 (2022, 2024)
Fedorenko R P, **5**: 3139 (3135)
Fedra K, **7**: 5008 (5005)
Feely R A, **2**: 1516 (1516)
Feeney M K, **1**: 456 (451)
Feferman S, **2**: 937 (937)
Fehér A, **3**: 2010 (2008, 2010)
Feher K, **2**: 1057 (1055)
Fehlberg E, **5**: 3559 (3557)
Feichtinger G, **2**: 1269 (1263, 1264, 1266, 1267)
Feick G, **4**: 2782 (2778)
Feigenbaum E A, **1**: 307 (307); **2**: 1573; **4**: 2543 (2543), 2640 (2636); **6**: 4168 (4166); **7**: 4784 (4784)
Feigenbaum M J, **1**: 567, 570 (568)
Feigl E O, **2**: 853 (851, 852)
Feinberg A, **5**: 3124 (3123), 3164 (3157), 3568 (3567)
Feinberg J, **3**: 1958
Feinberg S E, **2**: 1106
Feingold D G, **2**: 927 (924)
Feinmesser D, **1**: 132 (127)
Feinstein A, **4**: 2497 (2494), 2508
Feitknecht W, **4**: 2782 (2778, 2779)
Feld'baum A A, **2**: 1257 (1255); **4**: 2273 (2272); **5**: 3475 (3471), (3511), (3516)
Feldman J, **4**: 2543 (2543); **6**: 4168 (4166)
Feldman J A, **6**: 4159 (4159)
Feldman M P, **3**: 2192
Felix M, **2**: 781 (779)

Fell R, **4**: 2253 (2249)
Feller I, **7**: 4833 (4831)
Feller W, **1**: 512; **6**: 4411 (4410); **7**: 4645 (4644), 4835
Feltham H G, **3**: 1943
Femenia J, **3**: 1766 (1766)
Fenchel T, **7**: 4454 (4453)
Fenchel W, **1**: 529 (524); **2**: 1262 (1257, 1258)
Fenelon J F, **2**: 879 (877)
Feng D, **2**: 1372 (1368)
Fenn W O, **6**: 4051 (4046), 4066 (4063)
Fennema C L, **1**: 759 (757); **6**: 4159
Fenselau A, **4**: 2631 (2628)
Fenton T P, **2**: 1116 (1110, 1111); **5**: 3188 (3184); **6**: 3991 (3990)
Ferbeck D, **1**: 360, 362, 365, 367
Ferfera A, **3**: 1779 (1772)
Fergason J L, **7**: 4871 (4868)
Fergeau R, **5**: 2944
Ferguson A M, **6**: 4261 (4259)
Ferguson I J, **5**: 3196 (3189, 3191), 3228 (3228); **6**: 4133 (4131, 4132), 4139 (4135); **7**: 4689 (4687, 4688)
Ferguson J, **7**: 5060 (5059)
Ferguson T S, **6**: 3616 (3612)
Ferguson-Pell M W, **4**: 2901 (2899)
Ferin M, **3**: 2174 (2174)
Fernandes R A, **6**: 3954 (3951, 3952)
Fernandez C, **2**: 1595 (1592, 1594)
Fernandez-Ballesteros M L, **3**: 1910 (1908, 1909, 1910)
Féron R, **6**: 3619 (3617)
Ferrano G, **5**: 3130
Ferrari D, **1**: 751 (748, 749)
Ferraro J R, **4**: 2511 (2509)
Ferrarotto D, **7**: 4905 (4905)
Ferraté G, **2**: 1586 (1584); **6**: 3610 (3610)
Ferré M, **6**: 4336 (4335), 4337
Ferreira P G, **4**: 2570 (2568, 2569, 2570)
Ferreira V A, **1**: 142 (138)
Ferrell W R, **3**: 2186 (2179)
Ferrier D, **1**: (463)
Fessas P, **4**: 2671 (2666, 2669)
Fessenden E, **7**: 4876 (4873, 4875)
Festathiou J, **3**: 1852 (1849)
Fettweis A, **5**: (3132), 31343135 (3132), 3135 (3132); **6**: 3972 (3972)
Feuer A, **7**: 4611 (4607)
Feynman R P, **1**: 294 (292); **5**: (3320), 3329 (3322)
Fiacco A V, **1**: 529 (526, 527); **2**: 801 (798, 799); **5**: 3362, 3455 (3452), 3550; **6**: 4310 (4309)
Fick G, **3**: 2174 (2173)
Fidelus K, **5**: 3248 (3243, 3244, 3247)
Fiechter A, **6**: 3834 (3832)
Fiedler M, **4**: 2565 (2561, 2562)
Fieguth W, **5**: 3389 (3388)
Field J B, **3**: 2174 (2172)
Fielden C, **1**: 514
Fielder M, **4**: 2556 (2553, 2555)
Fields J N, **5**: 3464 (3459)
Fieli W K, **3**: 2202 (2196)
Fienberg S E, **5**: 2917
Fieschi D, **2**: 1577 (1575)
Fieschi M, **2**: 1577 (1575)
Fife P C, **2**: 1329 (1327); **6**: 3762 (3756, 3758)
Figalli G, **5**: 3279 (3276)
Fiksel J, **7**: 4789 (4786, 4788, 4789)
Fildes R, **7**: 5155 (5154)
Filipiak J, **7**: 4751
Filippov A F, **1**: 413 (413); **2**: 1033; **4**: 2609 (2608); **5**: 2963 (2958), 3184 (3180, 3181, 3183), 3512 (3511)
Filipski A, **6**: 3653 (3648)
Filleau J B, **1**: 196 (193)
Finch A, **7**: 4504 (4499)
Finch J A, **6**: 4252 (4251)
Fincham A E, **3**: 1990 (1986)
Fincham W, **4**: 2901 (2899)
Findeisen W, **2**: 956 (952), 1305 (1299, 1302, 1304); **3**: 1741 (1739); **5**: 3144; **7**: 4571 (4567, 4568, 4569, 4571)
Findlay B L, **4**: 2331 (2327)
Findlay I A, **2**: 978 (978)
Findlay M G, **3**: 2202 (2196, 2200)
Findlay N C, **5**: 3116 (3114)
Findler N V, **4**: 2640 (2637, 2638)
Fine T L, **1**: 725 (720); **3**: 1805 (1802, 1803)
Finegood D T, **3**: 2025 (2023)
Finess H, **1**: 633 (632)
Finger J M, **4**: 2353 (2352); **7**: 5010 (5009)
Finger T H, **3**: 2014
Fink L H, **6**: 3820 (3816), 3834 (3832)
Finkel R A, **4**: 2412 (2393, 2402), 2440 (2439)
Finkelstein A C W, **4**: 2536 (2533)
Finkelstein L, **3**: 2025 (2022, 2023, 2024); **4**: 2533 (2528, 2529), 2536 (2533), 2631 (2629, 2630); **6**: 3909 (3905)
Finlay J B, **1**: 482 (479, 480)
Finlayson-Pitts B J, **1**: 187 (185); **2**: 1513
Finn J T, **2**: 1321 (1318, 1320), 1348 (1345, 1346), 1355 (1352, 1353), 1486 (1480); **5**: 2929 (2928), 2930 (2929); **6**: 4237 (4235, 4236), 4238 (4235); **7**: 4956 (4954, 4956)
Finn R K, **1**: 424 (423)
Fiocco G, **6**: 4036 (4033)
Fiori C, **2**: 1420; **6**: 4152
Firla A, **3**: 1990 (1989)
Firoozan F, **4**: 2284 (2283)
Fisch M H, **1**: 399 (394)
Fischer E A, **5**: 3416
Fischer G, **3**: 2202 (2196)
Fischer H B, **4**: 2789 (2788)
Fischer M, **6**: 3727 (3724)
Fischer M J, **2**: 1165 (1164)
Fischhoff B, **2**: 987; **3**: 2202 (2196), 2203 (2196); **5**: 3116 (3113); **6**: 4075
Fishburn P C, **2**: 942 (941); **3**: 2060 (2060), 2202 (2196, 2197, 2198, 2200, 2202), 2203 (2196); **5**: 3104 (3102, 3103), 3115 (3105, 3107, 3111, 3113, 3114), 3152; **6**: 4075; **7**: 4985, 4991 (4987, 4988)
Fisher D G, **2**: 1568 (1567); **6**: 3916 (3913)
Fisher F H, **7**: 4848 (4844)
Fisher M, **2**: 937 (937)
Fisher M L, **1**: 630 (629)

Fisher R A, **2**: 1106 (1103); **4**: 2273 (2271), 2287 (2284); **6**: 3762 (3759)
Fisher W D, **4**: 2386 (2380), 2440 (2426)
Fishman G S, **3**: 2076; **6**: 4359 (4359), 4395 (4395); **7**: 4654 (4652, 4653), 4656 (4656), 5005 (5001), 5018 (5017, 5018)
Fisk D J, **1**: 159, 514 (514); **3**: 2209
Fite W L, **5**: 2943 (2934)
Fittes B A, **6**: 4160
Fitts P M, **3**: 2186 (2184)
Fitz R, **4**: 2580 (2576, 2579)
Fitzgerald R V, **5**: 3464 (3463)
Fitz-Gibbon C T, **5**: 2917
Fitzgibbons P J, **1**: 326 (325)
FitzHugh R, **7**: 4980 (4978)
Fitzsimons W A, **4**: 2356 (2353)
Fix G J, **3**: 1632
Fjeld M, **1**: 588 (586); **2**: 781 (779); **6**: 3591 (3591), 4204 (4200)
Flake R H, **5**: 3351 (3342)
Flanagan J L, **1**: 80 (78); **7**: 4509 (4506)
Flatley J, **5**: 3471 (3467)
Fleck J, **6**: 4097
Fleck L, **3**: 1862
Fleishman E A, **3**: 2192 (2186)
Fleming M, **3**: 2060 (2057)
Fleming P D, **3**: 2006 (2006)
Fleming P T, **2**: 1431 (1428)
Fleming R J, **2**: 923 (919)
Fleming R T, **6**: 4302 (4301)
Fleming W H, **2**: 1008 (1005, 1006), 1017 (1014, 1016), 1279; **3**: 2068, 2072 (2069); **5**: 3316 (3312), 3329 (3320, 3322), 3523 (3518, 3520, 3523), 3530 (3530), 3538; **7**: 4612, 4627 (4623, 4624), 4644 (4643)
Fletcher H, **2**: 1527 (1523)
Fletcher J F, **6**: 3933
Fletcher K S, **6**: 3688 (3686)
Fletcher L R, **6**: 3717 (3710)
Fletcher P T, **2**: 1445 (1441)
Fletcher R, **2**: 793 (790), 801 (801); **3**: 1623 (1620); **4**: 2600 (2598), 2879 (2877, 2879); **5**: 3487 (3482, 3485, 3486), 3546 (3544); **6**: 4310 (4308); **7**: 4969, 4971 (4969), 4975 (4972, 4973)
Fletcher W I, **1**: 575
Fliess M, **1**: 493 (492); **2**: 1121 (1116); **3**: 1779 (1769, 1770, 1771, 1772, 1774, 1775, 1777, 1778); **4**: 2527 (2523, 2526, 2527), 2609 (2608); **5**: 3272 (3268), 3276 (3274, 3275), 3279 (3276, 3279), 3288 (3287), 3292 (3289, 3290, 3291), 3293 (3291, 3292), 3376 (3374, 3375, 3376), 3394 (3390, 3391, 3392, 3393); **6**: 3776 (3772, 3773), 3961 (3957, 3958, 3961), 3972 (3970), 3977 (3974)
Flippin R S, **4**: 2782 (2782)
Flondor P, **3**: 1822 (1821)
Florenzano M, **1**: 671
Flowers B, **5**: 3416
Flowers N C, **1**: 769 (760, 761); **2**: 1410 (1409)
Floyd M, **4**: 2782 (2781)
Floyd M A, **1**: 177 (174)
Floyd R, **3**: 1712 (1712)
Flügge-Lotz I, **5**: 3389 (3385, 3386), (3516, 3517); **6**: 4274
Flurscheim C H, **6**: 3806 (3803)
Flynn M J, **2**: 1451 (1450)
Fochtman E G, **1**: 179 (177)
Foex P, **5**: 2943 (2935)
Fogarty L E, **6**: 4417
Fogel L J, **6**: 4168 (4166)
Foias C, **6**: 4223 (4220)
Foisseau J, **1**: 120
Foith J P, **4**: 2473 (2470, 2471)
Foley J D, **3**: 2179; **6**: 4302 (4301)
Folger T A, **6**: 3768 (3768), 3989 (3984)
Folkman J, **7**: 4958 (4957)
Fomin S, **5**: 3019 (3017)
Fontaine H, **1**: 682 (679)
Fontaine T D, **5**: 2956 (2956)
Fontijn A, **1**: 184 (183)
Foo N Y, **3**: 2168 (2167)
Foo S H, **3**: 1693 (1691)
Foord A G, **1**: 638 (634)
Foord T R, **2**: 1403 (1402, 1403)
Foote R S, **7**: 4922
Forbat L N, **2**: 1595 (1594, 1595)
Forbes C D, **1**: 445 (441, 442, 444)
Ford D E, **6**: 4229 (4228), 4230 (4228, 4229)
Ford H, **6**: 4142 (4140, 4142)
Fordyce W E, **6**: 4051 (4048, 4049)
Foreman K F, **6**: 3806 (3805)
Forestier J P, **3**: 1648 (1645); **7**: 4751 (4751)
Forker E L, **3**: 2143 (2141)
Forman L, **5**: 2942
Forman R T T, **2**: 1317
Fornasini E, **5**: 3293 (3289); **6**: 3972 (3971)
Forney G D Jr, **4**: 2871 (2867); **5**: 3220 (3215, 3217); **6**: 3751 (3749)
Forrest J J M, **6**: 4299
Forrester J N, **7**: 4495 (4495)
Forrester J W, **2**: 1332 (1330), 1355 (1352), 1538 (1536); **3**: 2016 (2015); **4**: 2353 (2351); **6**: 4359 (4359), 4368 (4368); **7**: 4762 (4755, 4756), 4775 (4774), 4956 (4954)
Forsberg A S, **2**: 1521 (1521)
Forsén S, **1**: 13 (13)
Forssberg K, **5**: 2993 (2991)
Forster R, **4**: 2775 (2772)
Forstner H, **5**: 3579 (3575, 3576, 3577, 3578)
Förstner U, **4**: 2782 (2778, 2780)
Forsythe G E, **2**: 826 (815, 817); **3**: 1627; **7**: 5161 (5158)
Fort J-C, **1**: 91 (91)
Forte B, **4**: 2508 (2507); **6**: 3768 (3766)
Fortescue T R, **2**: 1466 (1464); **5**: 3232 (3231); **6**: 3991 (3990, 3991)
Fortmann T E, **4**: 2866 (2862); **5**: 3220 (3215); **6**: 3720 (3720)
Forwood B, **2**: 1431 (1427)
Fosco G, **3**: 1664 (1657)
Fosdick L D, **2**: 826 (824, 825)
Fosha C E, **6**: 3815 (3811, 3812)
Foss B, **6**: 3991 (3991)

ESC8-E

Fossard A J, **1**: 132 (127), 196 (193, 194), 388, 703 (697); **3**: 2108 (2108); **5**: 3176; **7**: 4443 (4441)
Foster A, **4**: 2347 (2345)
Foster A M, **2**: 808 (803)
Foster D H, **4**: 2331 (2327, 2328, 2329)
Foster D M, **7**: 4905 (4905)
Foster D W, **4**: 2631 (2628)
Foster K E, **1**: 177 (175)
Foster K R, **1**: 450 (448, 449)
Foster L V, **5**: 3220 (3219)
Foster R O, **3**: 2174 (2173)
Foster T H, **2**: 1521 (1518)
Fotakis J, **1**: 45 (38)
Foucault J B L, **2**: 1402 (1399, 1400)
Fouin P, **7**: 4749 (4745)
Foulard C, **2**: 781 (779); **4**: 2737; **6**: 3889 (3888), 3916 (3912)
Fouletier J, **5**: 3575 (3573)
Foulk C R, **4**: 2502 (2498); **7**: 4785
Fourade H E, **1**: 289 (285)
Fourier J B J, **3**: (1745); **4**: (2270)
Fowler S W, **2**: 1516 (1516)
Fox B L, **6**: 3950 (3948); **7**: 4656 (4655, 4656)
Fox D, **6**: 4302 (4301)
Fox J, **3**: 1712 (1712), 1862 (1859)
Fox J R, **7**: 4755
Fox K A, **2**: 1342 (1338)
Fox L, **6**: 3626 (3623, 3624, 3625)
Fox M A, **1**: 177 (176)
Fox R H, **7**: 4856 (4853, 4854), 4873
Fox S, **2**: 1165
Foxworth T G, **3**: 2089
Foz W C, **1**: 664 (654, 659)
Fraade D J, **6**: 3695
Fradkov A L, **1**: 80 (77)
Fraeijs de Veubeke B, **7**: 4490
Frahm H, **6**: 4274 (4272, 4273)
Franchi L, **6**: 3954 (3952)
Franchimout C, **1**: 482 (480)
Francioni J M, **3**: 1888
Francis B A, **4**: 2567 (2566), 2570 (2569, 2570); **5**: 3197 (3190), 3228 (3228); **7**: 4689 (4686)
Francis J G F, **2**: 826 (818)
Francis R F, **3**: 1979 (1978)
Franck G, **3**: 2092 (2091)
Franco A, **1**: 682 (679)
Franco A S, **7**: 4892
Francoise G, **5**: 2943 (2935)
Françozo E, **1**: 505 (502)
Frank A A, **1**: 633 (632)
Frank E, **1**: 535 (534)
Frank K H, **1**: 360, 362, 367
Frank L K, **2**: 873
Frank U, **7**: 5112
Franke R, **4**: 2846 (2843)
Frankel S P, **6**: 3626 (3625)
Frankel V H, **1**: 474 (468)
Franklin D E, **4**: 2385 (2383)
Franklin G F, **2**: 1066 (1057, 1060), 1071 (1068, 1070), 1078 (1074)
Franklin H D, **1**: 588 (588)
Franklin M A, **2**: 1497 (1494)
Franklin M W, **6**: 4252 (4250)
Frankowsa H, **5**: 3184 (3181)
Franks R G E, **1**: 585 (585); **5**: 3082 (3079)
Franksen O I, **2**: 952 (950)
Frant M S, **4**: 2511 (2509)
Franta W R, **1**: 755 (752); **2**: 1087 (1085); **6**: 3879; **7**: 4935 (4934)
Franz E H, **2**: 1321 (1318, 1320)
Franzozo E, **1**: 507; **5**: 3264 (3263)
Fraser D C, **7**: 4468 (4466)
Fraser N M, **7**: 4799 (4793, 4794, 4795, 4796, 4797, 4798)
Fraser R S, **3**: 2052 (2050); **6**: 4041
Frayman M, **1**: 417 (414, 415)
Frazer A S, **4**: 2794 (2789)
Frazer F, **3**: 1969 (1967)
Frazer W F, **2**: 1538 (1537)
Frazier M, **7**: 4466 (4464, 4465)
Frazzon S, **2**: 1160 (1160)
Frederick D K, **2**: 1291, 1297
Frederico P J, **6**: 3734
Fredrickson A G, **1**: 424 (422)
Freed V H, **3**: 2147; **4**: 2782 (2779)
Freedman H T, **5**: 3124 (3123), 3164 (3159), 3568; **7**: 4709 (4704, 4708, 4709), 5140
Freedman M I, **5**: 3389 (3387); **7**: 4439 (4437), 4637 (4635)
Freedman P L, **4**: 2794 (2789)
Freehafer J, **1**: 352, 360, 362, 365, 367
Freeling A N S, **3**: 1845 (1845); **4**: 2501 (2498, 2499); **5**: 3115
Freeman C, **7**: 4762 (4756)
Freeman C F, **7**: 4833 (4830)
Freeman H, **6**: 4159 (4157)
Freeman H E, **2**: 1566 (1560)
Freeman P, **4**: 2685 (2681, 2685); **6**: 3898
Freeman S K, **7**: 4504 (4499)
Frehse J, **7**: 4617 (4615)
Frei R W, **7**: 5129 (5129)
Frein Y, **3**: 1648 (1644)
Freinkel N, **3**: 2032 (2030)
French C A, **1**: 437
Frerking M A, **4**: 2723 (2722)
Freudenberg J S, **5**: 3228 (3225); **6**: 4227 (4224, 4225)
Freudenstein F, **4**: 2387 (2373)
Freudenthal H, **5**: 3454 (3450)
Freund J E, **7**: 5019 (5016)
Frey B M, **2**: 1248 (1248)
Frey F J, **2**: 1248 (1248)
Frey W, **3**: (1748)
Friberg L, **1**: 148 (144)
Fricke H, **1**: 450 (447)
Fried W R, **1**: 201
Friedberg R, **2**: 937 (937)
Friedland B, **1**: 132; **4**: 2622 (2620); **7**: 4429 (4428)

Friedlander B, **2**: 785 (782, 783, 784), 1102 (1102); **4**: 2277 (2273, 2275), 2278 (2273, 2275), 2284 (2282), 2618 (2617); **7**: 4466 (4465), 4469 (4466, 4468)
Friedlander S K, **1**: 179 (178, 179)
Friedlin M I, **1**: 91 (89, 91)
Friedman A, **1**: 512, 687 (685); **2**: 1008 (1005, 1006), 1186 (1185); **4**: 2342, 2343; **5**: 3535; **6**: 3596 (3592, 3595), 4364 (4360); **7**: 4612, 4617 (4617), 4637 (4631), 4644 (4643), 5041 (5040)
Friedman E A, **1**: 312 (310)
Friedman J W, **1**: 687 (682, 684)
Friedman M, **5**: 3121 (3116)
Friedman M I, **3**: 2213 (2210)
Frieland C K, **1**: 547 (544)
Friis M, **4**: 2646 (2645)
Fris I, **4**: 2798 (2796)
Frisby J P, **7**: 5078
Frissel M J, **5**: 3268 (3266); **6**: 3933 (3932)
Fritsch G T, **1**: (463)
Fritsch K, **2**: 1241; **5**: 3470 (3469)
Fritz J, **2**: 1099 (1093)
Fritz J S, **7**: 5110, 5129 (5127, 5128)
Frohberg K, **6**: 4359 (4358)
Froment G F, **6**: 3745 (3740)
Fromovitz S, **2**: 1099 (1093)
Frost P J, **6**: 3898 (3894)
Frost T H, **1**: 310 (308), 312 (310)
Froude W, **6**: 4275 (4272)
Fry A J, **7**: 4856 (4853, 4854)
Fryback D, **4**: 2338 (2337)
Fryer J S, **5**: 3030 (3029)
Fu K S, **1**: 372 (360), 665 (664), 746 (743); **3**: 1781 (1780), 1802 (1801), 1822 (1821), 1825 (1824), 1829 (1827), 1852 (1847, 1851), 1862 (1859), 1872 (1868), 1874, 1905 (1904), 2045 (2045); **4**: 2339 (2338), 2635 (2634), 2883 (2881); **6**: 3643 (3636, 3637, 3641, 3642), 3644 (3635, 3636, 3637, 3641, 3643), 3645 (3636, 3643), 3653 (3645, 3646, 3647, 3648, 3649, 3650, 3651, 3652), 3654 (3645, 3646, 3648, 3650, 3652), 3772 (3770), 4159 (4158), 4160 (4157, 4158), 4161 (4158), 4165 (4161, 4163); **7**: 4661 (4659), 4669
Fuchs H, **1**: 347 (345)
Fuchs J J, **2**: 1581 (1579, 1580); **5**: 3065 (3062), 3410 (3406, 3407, 3408, 3409)
Fuchs M, **2**: 1521 (1517, 1518)
Fugelso M A, **4**: 2453 (2450)
Fugier-Garrel C, **6**: 3643
Fuhrmann P A, **5**: 3205 (3205); **6**: 3751 (3746, 3747), 3969, 3972 (3971)
Fujihara Y, **6**: 4051 (4048)
Fujii S, **6**: 4108 (4104)
Fujimasa I, **2**: 1366 (1363, 1366)
Fujimura S, **6**: 4033 (4031)
Fujisaki M, **5**: 3303 (3293, 3297), 3329 (3319)
Fujishige S, **1**: 626 (624)
Fujishima C, **7**: 4593 (4588)
Fujiwara M, **7**: 4518 (4518)
Fujiwara R, **6**: 3654 (3652)
Fukaya H, **1**: 477 (477)
Fukuda A, **5**: 3130 (3128)
Fukunaga K, **4**: 2635 (2632, 2635); **6**: 3644 (3636), 4033
Fukushima M, **2**: 1282; **3**: 1829 (1826), 1847 (1847)
Fukushima Y, **1**: 592 (590, 591)
Fulkerson D R, **6**: 3734 (3734), 3739 (3736)
Fuller A T, **1**: 421 (419); **3**: 1749 (1745); **5**: 3389 (3383, 3384, 3388), 3517 (3516, 3517); **6**: 3870 (3867, 3868); **7**: 4462 (4458, 4461, 4462), 4517 (4514, 4515, 4517)
Fullerton G D, **7**: 4964
Funada S, **2**: 1521 (1517)
Funck-Brentano J-L, **1**: 310 (308), 312 (310)
Fung A K W, **1**: 589 (587)
Fung L W, **3**: 1781 (1780), 1825 (1824), 1829 (1827); **6**: 3653 (3648), 4160 (4158); **7**: 4661 (4659)
Fung P T K, **1**: 45 (42); **6**: 4266 (4264)
Fung Y C, **1**: 445 (441, 442, 443, 444), 477 (475, 477), 486 (484); **2**: 853
Funk J P, **5**: 3261 (3261)
Fuortes M G, **7**: 5070 (5066, 5070)
Furhmann P, **4**: 2866 (2865)
Furman A, **2**: 1416 (1413)
Furman G G, **7**: 5065 (5062, 5065)
Furukawa K, **5**: 2943 (2935)
Furuya T, **4**: 2723 (2722)
Fusfeld H I, **7**: 4833 (4830)
Fuwa K, **3**: 1936 (1934)
Fyfe C A, **5**: 3421 (3419)
Fymat A L, **2**: 1311 (1307)

G

Gabard J F, **7**: 4913 (4912), 4927 (4926, 4927)
Gabasov R, **5**: 3206
Gabay D, **2**: 788 (786, 788), 789 (788), 797 (795, 796); **3**: 1921
Gabe D R, **4**: 2607 (2607)
Gabor D, **1**: 535 (534)
Gabriel R, **7**: 5027
Gächter R, **4**: 2646 (2643)
Gaden E L, **3**: 1611 (1609)
Gafarian A V, **7**: 4654 (4652, 4653)
Gagge A P, **1**: 642 (639), 643, 645 (644); **7**: 4876 (4876), 4877 (4873, 4874), 4881 (4880)
Gaines B R, **1**: 126 (125, 126), 575, 608 (607), 664 (664), 713 (710, 712, 713), 743 (741), 746 (743); **2**: 1333 (1331), 1407 (1405), 1549 (1545); **3**: 1809 (1808), 1821 (1819, 1821), 1822 (1821), 1867 (1863), 1872 (1868), 1874; **4**: 2301 (2299), 2361; **5**: 2926 (2926), 3454 (3451, 3453, 3454); **6**: 3772 (3770), 3865 (3863), 3947, 4007 (4007), 4167 (4166); **7**: 4661 (4659), 4996 (4995)
Gairola A, **6**: 4123 (4120)
Gaitonde N Y, **1**: 588 (588)
Gakuhari Y, **5**: 2993 (2993)
Gal T, **2**: 1276 (1273); **5**: 3164 (3160, 3161)
Galanter E, **1**: 276 (272)
Gale D, **1**: 671 (667, 668)
Galiana F D, **6**: 3815 (3811, 3812), 3834 (3832)
Galileo Galilei, **1**: (552)
Galing B, **5**: 3116; **7**: 4785 (4783)
Gall W E, **1**: 326 (324)
Gallager R G, **2**: 1051 (1046); **4**: 2508 (2506, 2507)
Gallagher J S, **7**: 4585 (4581)
Gallagher R H, **1**: 477 (477); **2**: 1179 (1174); **6**: 3626 (3625)
Galletti P M, **1**: 315 (315)
Gallianpur G, **5**: 3329 (3319)
Gallier P W, **1**: 585 (583)
Gallopín G C, **1**: 674; **2**: 1355 (1352, 1353, 1355), 1486 (1481), 1489 (1488); **7**: 4956 (4954)
Gallouet T, **4**: 2613 (2610)
Galonsky R S, **1**: 312 (310)
Galvani L, **2**: 1402 (1398)
Gambrell R P, **4**: 2782 (2780)
Gamkrelidze R V, **2**: 1099, 1278; **5**: 2963 (2956, 2959), 3512 (3511), 3513 (3511), 3515 (3514, 3515), 3517 (3516)
Gammage R B, **5**: 3464 (3463)
Gander R E, **3**: 2032 (2030)
Gandy R, **1**: 720 (715)
Gann D S, **3**: 2174 (2170)
Gannon D, **2**: 826 (818)
Ganoe W H, **3**: 1910 (1910)
Ganslaw M J, **7**: 4929 (4928)
Gansler J S, **7**: 4755
Gantmacher F R, **3**: 1779 (1770); **4**: 2827 (2817), 2832 (2828), 2871 (2866); **5**: 3024 (3021), 3276 (3274), 3279 (3276), 3388 (3387); **6**: 3751 (3745, 3749); **7**: 4457 (4455, 4456), 4529 (4527), 4547 (4541)
Gantz D, **7**: 5011 (5008), 5019
Ganzerli-Valentini M T, **4**: 2607 (2607)
Garabedian P R, **4**: 2613 (2610), 2807 (2807); **6**: 3626 (3623, 3625), 4364 (4360, 4361)
Garabedian V, **3**: 2225
Garay R, **1**: 433 (432)
Garbow B S, **2**: 826 (818, 820, 824), 827 (815, 818, 824); **5**: 3559 (3555)
Garbutt A B, **3**: 1975
Garcia A, **6**: 3850 (3850)
Garcia C B, **6**: 3634
Garcia C E, **1**: 588 (588)
Garcia J F, **3**: 1623 (1619)
Garcia J M, **7**: 4731, 4740 (4738), 4741 (4735, 4736, 4737, 4738), 4751
Garcia-Molina H, **2**: 883 (881), 1154 (1153)
Gardenier J S, **6**: 4417
Gardner D, **7**: 5127
Gardner G E, **6**: 3806 (3804)
Gardner M, **1**: 561 (559); **6**: 4345 (4342), 4377 (4376)
Gardner M F, **3**: (1748)
Gardner M R, **1**: 675 (672); **2**: 1355 (1353); **7**: 4645 (4644)
Gardner R H, **2**: 1272 (1272), 1326 (1324, 1325, 1326), 1333 (1332), 1344 (1343, 1344), 1539 (1538, 1539), 1551 (1549, 1550, 1551), 1552 (1550, 1551); **6**: 4229 (4228), 4235 (4230, 4231, 4234)
Garey M R, **1**: 725 (723, 724); **2**: 884 (882), 1083 (1082, 1083)
Garfinkel R S, **1**: 630 (628)
Garlet M, **6**: 3789
Garnell P, **5**: 3012
Garner K, **5**: 2994 (2991)
Garnir J, **5**: 3082 (3081)
Garrett C J R, **7**: 4889 (4888)
Garrettson L K, **6**: 3698 (3698)
Garrity S D, **4**: 2685 (2684)
Garside B K, **4**: 2511 (2510), 2723 (2721, 2722)
Garten C T Jr, **2**: 1344 (1343, 1344), 1551 (1551)
Gartner K P, **3**: 2186 (2181)
Gartner N H, **7**: 4925
Garven H C, **5**: 3282 (3282)
Garver L L, **6**: 3837 (3835)
Garvin P, **1**: 399 (394)
Garzillo A, **2**: 1388 (1386); **6**: 3954 (3952)
Gascoigne J D, **4**: 2368 (2367), 2464 (2461, 2463, 2464)
Gascoyne P R C, **1**: 450 (450)
Gaskill J D, **5**: 3130 (3128)
Gasnier J, **1**: 86 (82)
Gasparski W, **6**: 3865 (3861)
Gass S, **5**: 3164 (3159, 3160)
Gass S I, **6**: 4326 (4323), 4374; **7**: 4771, 5005 (5002, 5003), 5010 (5008)

Gates D M, **2**: 1462, 1521 (1517)
Gates P N, **7**: 4504 (4499)
Gates S B, **1**: (340)
Gatewood L C, **3**: 2025 (2022)
Gaudiano F F, **5**: 3077 (3072, 3073)
Gaudin J L, **6**: 3954 (3953)
Gauer O H, **1**: 547 (544)
Gause D C, **2**: 992 (991)
Gauss K F, **4**: (2271), 2283 (2278)
Gautherie M, **7**: 4856 (4856), 4871 (4868)
Gauthier J P, **7**: 4677 (4677)
Gauthier P, **7**: 4751
Gautier J P, **2**: 781 (779)
Gauvin J, **5**: 3550
Gauvrit M, **1**: 120; **7**: 5027
Gavalas G R, **5**: 3082 (3081)
Gavelin G, **2**: 781 (779)
Gaver D P, **2**: 1440 (1436)
Gawthrop P J, **1**: 51 (49), 63 (63); **2**: 1116 (1110, 1111); **5**: 3004 (3002), 3006 (3006), 3188 (3184), 3232 (3229), 3410 (3407), 3430 (3426, 3428); **6**: 3691, 3727 (3721, 3723, 3725), 3871 (3870, 3871), 4172 (4171), 4176 (4174, 4175), 4180 (4176, 4178, 4180), 4189 (4187), 4193 (4190, 4193), 4197 (4194, 4195, 4196, 4197), 4204 (4200), 4209 (4205, 4206, 4207, 4208), 4211 (4209); **7**: 4462 (4460)
Gay B W, **4**: 2511 (2510)
Gaylor J D S, **1**: 312 (310, 311, 312)
Gaze R M, **6**: 4167 (4167)
Gazis D C, **7**: 4925, 4927 (4925, 4926), 4933
Gazley C, **6**: 4303 (4302)
Gbuert A M, **2**: 1420 (1419)
Gear C W, **2**: 1181 (1181); **5**: 3559 (3556, 3558); **6**: 4359 (4358)
Gebben V D, **6**: 4341
Geddes L A, **1**: 450 (445)
Gedling P, **6**: 4261 (4258, 4259)
Gee G, **1**: 486 (484)
Geering H P, **2**: 1276 (1274); **5**: 3171 (3170); **6**: 4139 (4133)
Gehrig E H, **6**: 3789
Gehrs C W, **2**: 1539 (1539), 1552
Geiger A, **5**: 3421 (3418)
Geiger I, **1**: 380
Geiselman P J, **3**: 2213 (2212)
Geisselmann H, **4**: 2473 (2470)
Geisser S, **2**: 1106 (1105)
Gelb A, **4**: 2622, 2652 (2650); **5**: 3373 (3369), 3389 (3385); **7**: 4452 (4451, 4452)
Gelbard F, **1**: 179 (179)
Gelenbe E, **1**: 91 (87); **2**: 1154 (1153), 1169 (1168); **6**: 3659, 3670 (3668)
Gelfand I M, **2**: 1186 (1184)
Gelfand R, **6**: 4061 (4056)
Gelinas R J, **2**: 1181 (1181)
Gelpke G J, **2**: 1107 (1104, 1105)
Gelsema E S, **6**: 3644
Geman S, **5**: 3310 (3309)
Gembiki F, **5**: 3164 (3157)
Gemperline D, **7**: 4709 (4709)
Gendreau R M, **4**: 2511 (2510)
Genesio R, **1**: 132 (127); **4**: 2257 (2254)
Genin Y, **5**: 3134 (3132, 3134), 3135 (3133, 3134)
Genna S, **6**: 3940 (3939)
Genrich H J, **6**: 3670 (3669), 4350 (4349)
Genthe W K, **3**: 1693 (1688)
Gentil S, **7**: 5008 (5006)
Gentilhomme Y, **4**: 2301 (2299)
Gentina J C, **5**: 3406 (3402, 3405); **6**: 3673 (3670, 3671, 3672); **7**: 4547
Gentle E J, **1**: 109
Gentle J E, **6**: 3950 (3948)
Gentleman M, **2**: 826 (818)
Gentleman W M, **3**: 1723; **5**: 2950 (2949, 2950); **7**: 4540 (4536)
Gentner D, **4**: 2338 (2338)
Gentry J B, **2**: 1323 (1323), 1344 (1344), 1345 (1344)
Geoffrion A M, **2**: 956 (952), 1276 (1273); **5**: 3124 (3123), 3152, 3164 (3157, 3159, 3160, 3163), 3568 (3567)
George D A, **3**: 1779 (1769, 1776, 1777); **6**: 3961 (3957)
George D B, **5**: 3282 (3282)
George E P, **7**: 4980 (4977)
George J H, **7**: 4452 (4448)
George K R, **5**: 3389 (3388)
George W O, **7**: 4504
Gerard F, **1**: 687 (685); **2**: 1010 (1010)
Gerardin L, **6**: 4155
Gerardy R, **2**: 1549
Gerber H J, **4**: 2449 (2448)
Gerber N, **6**: 3698 (3698)
Gerbrandy J, **7**: 4856 (4852)
Gerland H, **1**: 340
Gerlough D L, **7**: 4927 (4925)
Germain F, **7**: 4468 (4467)
Gero J S, **2**: 1431 (1427)
Geromel J C, **2**: 923 (922); **5**: 3156 (3153, 3154); **7**: 4684 (4679), 4685 (4678, 4679, 4680, 4682, 4683, 4684)
Gershon M E, **3**: 1829 (1826)
Gershwin S B, **3**: 1648 (1644), 1736, 1741 (1738, 1739)
Gertier J, **4**: 2609 (2608)
Gertler J, **2**: 1143; **5**: 3188 (3188); **6**: 3690 (3690), 4176 (4173); **7**: 4540 (4537)
Gertner I, **5**: 3303 (3294)
Gerwin D, **3**: 2192 (2186)
Geschka H, **1**: 618
Geselowitz D B, **1**: 456 (453), 536 (532, 534, 535), 769 (760, 761)
Gesing W, **2**: 902 (901, 902), 907 (907); **6**: 4133 (4129, 4130), 4139 (4135, 4136, 4138)
Gessaman M P, **2**: 1106
Gessaman P H, **2**: 1106
Gessner U, **3**: 2175 (2173)
Gettys C, **4**: 2338 (2335)
Gevarter W B, **2**: 1573
Gevers M, **4**: 2292 (2288, 2291), 2293 (2288), 2618 (2615)
Gevers M R, **6**: 4244 (4238)
Ghizzetti A, **7**: 5036 (5033)
Ghose T K, **1**: 424 (423)

Ghosh A K, **1**: 535 (535)
Ghosh D, **3**: 1741 (1738)
Ghosh S P, **3**: 2217
Giacoletto L J, **1**: 272 (272)
Giallorenzi T G, **5**: 3464, 3470 (3466, 3467, 3468, 3469), 3471 (3469)
Giannessi F, **4**: 2343
Giauque R D, **1**: 177 (174)
Gibaldi M, **6**: 3698 (3695, 3697)
Gibbard A, **7**: 4472 (4472)
Gibbon R B, **3**: 1979
Gibbons J D, **7**: 5018 (5017)
Gibbons R M, **3**: 1969 (1968, 1969)
Gibbs C H, **2**: 978
Gibbs J, **3**: 2214 (2211)
Gibson F, **1**: 482 (479)
Gibson G H, **3**: 1975, 1981
Gibson J, **3**: 1961
Gibson J E, **5**: 3389 (3385, 3387, 3388)
Gibson T, **1**: 478 (474), 482 (479, 480)
Gick M L, **3**: 2202 (2201)
Gien D, **5**: 3480
Giertz M, **3**: 1781 (1780)
Giese D A, **6**: 4160 (4158)
Giesecke C, **3**: 1718 (1716)
Giesy J P Jr, **2**: 1323 (1323)
Gifford D K, **2**: 1154 (1151)
Gifkins M R, **5**: 2980
Gift D A, **2**: 1479 (1478)
Giglio R J, **6**: 3790
Giguere M T, **7**: 5129 (5128)
Gikhman I I, **7**: 4613, 4637 (4631, 4636), 4644 (4643)
Gilb T, **3**: 2179
Gilbart J W, **5**: 3070 (3068)
Gilbert B, **6**: 4284 (4284)
Gilbert E G, **1**: 735 (734); **3**: 1779 (1776)
Gilbert E O, **1**: 634 (632); **3**: 2217
Gilbert E S, **2**: 1106
Gilbert G A, **1**: 201
Gilbert G G, **4**: 2523 (2522)
Gilbert M G, **3**: 1769 (1767)
Gilbert M T, **4**: 2889
Gilby J, **4**: 2456 (2454)
Gilchrist T, **1**: 312 (310, 311, 312)
Giles R, **4**: 2301 (2293)
Gilinsky A S, **4**: 2331 (2326)
Gill A, **1**: 575; **5**: 3035 (3031); **6**: 3631 (3630)
Gill A M, **2**: 1318
Gill P E, **2**: 797 (797); **4**: 2879 (2877), 2880 (2879); **6**: 4310 (4309); **7**: 4975 (4973)
Gilles G, **4**: 2547 (2547)
Gilley J R, **7**: 4897 (4896)
Gilliand J L, **1**: 564 (561)
Gillies D B, **2**: 850 (844, 847)
Gilliland E R, **1**: 445 (443, 444)
Gilmore J P, **3**: 2206
Gilmore R, **1**: 550
Gilmore R S, **1**: 474 (471)
Giloi W K, **1**: 634 (632)
Gilpin M E, **1**: 675 (672, 674)
Gimonet B, **1**: 251 (251); **7**: 5170 (5168)
Gini G, **4**: 2385 (2374), 2447 (2447)
Gini M, **4**: 2385 (2374), 2447 (2447)
Ginocchio R, **3**: 2221
Gioan A, **2**: 1179 (1174)
Giordano C, **1**: 312 (311, 312)
Giorgiadis P I, **1**: 634 (632)
Giorgis E A, **3**: 1969 (1966)
Gips J, **6**: 3653
Girard A, **7**: 4726 (4723), 4741 (4736, 4737), 4749 (4746)
Girard R, **6**: 3954 (3951, 3953)
Girault C, **6**: 3670 (3665)
Girsanov I V, **3**: 2008 (2007); **5**: 3303 (3294)
Giss G N, **4**: 2511 (2510)
Gist C S, **2**: 1344 (1344)
Gitlin R, **1**: 80 (78)
Giush E, **1**: 529 (526)
Givellari F, **2**: 1160 (1160)
Givenbaum D, **6**: 3734 (3734)
Gizycki R von, **6**: 4097
Glansdorff P, **3**: 2174 (2174)
Glasberg B R, **1**: 326 (324)
Glaser P E, **7**: 4476 (4473)
Glass L, **1**: 570 (568, 569)
Glass R L, **4**: 2681 (2679)
Glass T E, **5**: 3421 (3419)
Glassey C R, **5**: 3170 (3166)
Glasstone S, **5**: 3416; **7**: 4944
Glattfelder A H, **3**: 2228 (2225); **6**: 4204
Glatzel J, **2**: 1226 (1216, 1217, 1225)
Glavitsch H, **6**: 3815 (3811, 3812, 3813), 3846
Gleason G J, **4**: 2385 (2382)
Gleason R E, **3**: 2174 (2173)
Glegg G L, **2**: 992 (991)
Glemme B, **4**: 2444 (2441); **6**: 4108 (4102)
Glick D, **1**: 13 (13)
Glick G, **1**: 542 (537)
Glicksberg I L, **5**: 3256 (3255), 3512 (3511), (3516)
Glicksman A S, **6**: 4302 (4302)
Glickstein M, **7**: 5070 (5066)
Gligor V D, **2**: 1154 (1150)
Glimm J, **5**: 3329 (3320)
Gloterman H L, **4**: 2646 (2646)
Glover J R Jr, **4**: 2314 (2311)
Glover K, **1**: 132 (132), 493 (491); **2**: 1099 (1096); **4**: 2270 (2270), 2292 (2291)
Glover R D, **1**: 252 (251)
Glowinski A, **7**: 4716 (4715)
Glowinski R, **2**: 956 (954), 1262 (1260); **3**: 1632; **5**: 3496 (3495, 3496); **7**: 4498 (4497), 5036 (5033)
Glumineau A, **3**: 2217
Glushkov V M, **6**: 4167 (4166)
Gnaiger E, **5**: 3579 (3575, 3576, 3577, 3578)
Gnauck A H, **2**: 1351 (1348)
Gnavi G, **6**: 3972 (3971)
Gobetz F W, **7**: 4490
Godal R C, **4**: 2301 (2300)
Godbee H W, **7**: 5094 (5093)

Gödel K, **1**: 333 (326, 328, 329), 720 (715, 717); **2**: 937 (933, 934, 935); **3**: (1790); **7**: (4993)
Godfrey K R, **1**: 493 (493); **4**: 2245 (2241), 2253 (2250, 2251, 2252); **6**: 3698 (3696)
Godin G, **7**: 4892
Godron M, **2**: 1317, 1318, 1358
Godwin A M, **2**: 1427 (1425)
Godwin M, **2**: 1449 (1447)
Goedecke W, **5**: 3351 (3342, 3344); **6**: 3610 (3608)
Goeke W C, **5**: 3131 (3129)
Goeldel C, **7**: 4598 (4596, 4598)
Goeller H F, **2**: 1344 (1343)
Goertz R C, **4**: 2385 (2379)
Goff M R, **7**: 4871 (4870)
Goffin J L, **5**: 3546
Goguen J A, **1**: 18 (17), 399 (394), 746 (743); **2**: 1473; **3**: 1712 (1712), 1905 (1903, 1905); **5**: 2921 (2919), 2924 (2921); **7**: 4996 (4995)
Goh B S, **1**: 675 (673)
Gohberg I, **2**: 826
Göhring B, **4**: 2293 (2287); **7**: 5001 (4996)
Goicoechea A, **3**: 1829 (1826); **5**: 3164 (3157, 3163); **7**: 5140 (5135)
Goitein B, **3**: 2202 (2196)
Gold B, **6**: 4289 (4286)
Gold H J, **1**: 468 (466)
Gold R, **5**: 2956 (2953)
Goldberg B B, **7**: 4964
Goldberg E D, **2**: 1516
Goldberg J M, **2**: 1595 (1592, 1594)
Goldberg M E, **4**: 2331 (2325)
Goldbeter A, **1**: 428, 429 (426, 428)
Goldblatt R, **7**: 4929, 4930 (4928)
Golden J P, **5**: 3560 (3559)
Goldenberg A, **6**: 4133 (4129, 4130, 4131), 4139 (4136, 4137)
Goldfarb D, **4**: 2880 (2877); **7**: 4975 (4972)
Goldfarb L C, **5**: 3389 (3385); **7**: 4452 (4443)
Goldfarb W, **2**: 937 (933)
Goldmann R, **4**: 2351 (2350)
Goldsmith C H, **2**: 1366 (1362)
Goldsmith J P, **1**: 319
Goldsmith R, **7**: 4856 (4853)
Goldstein A A, **5**: 3545 (3545); **7**: 4969 (4966)
Goldstein A L, **7**: 5117
Goldstein G D, **2**: 869 (868); **6**: 4165 (4161)
Goldstein H, **4**: 2412 (2394), 2440 (2432)
Goldstein J I, **2**: 1420; **6**: 4152
Goldstein M C, **2**: 1106 (1103, 1104, 1105)
Goldstein R A, **2**: 1344 (1343); **7**: 4897 (4896)
Goldstern S R, **5**: 3464 (3463)
Goldwater M H, **3**: 1979, 1990 (1986)
Golet F C, **3**: 1862
Golley F B, **2**: 1317 (1314), 1344 (1344); **7**: 4897 (4896)
Gollub J P, **1**: 570 (568)
Golomb S W, **1**: 725 (725); **5**: 2956
Golstine H H, **2**: (872)
Golterman H L, **4**: 2646 (2645), 2782 (2780)
Golub G H, **2**: 826 (818, 819, 820, 821, 823); **5**: 3134 (3134); **6**: 4299; **7**: 4540 (4535), 4971 (4969)
Golubitsky M, **7**: 4674 (4672)
Gomes J A C, **7**: 4716 (4714)
Gomez P, **6**: 3990
Gomory R E, **6**: 3739 (3736)
Gondran M, **2**: 1083 (1080)
Gonzaga C, **5**: 3196 (3193, 3195)
Gonzalez J, **5**: 3012
Gonzalez R, **3**: 2068 (2064), 2072 (2071); **5**: 3480 (3478)
Gonzalez R C, **4**: 2484 (2480); **6**: 3644, 3645, 3652 (3646), 3654 (3648), 4033, 4160 (4157, 4158, 4159), 4161
Gonzalez R F, **6**: 3904
Good I J, **2**: 1479 (1476); **3**: 1874
Goodin W R, **1**: 187 (187); **7**: 4518 (4517)
Goodlin R C, **4**: 2314 (2311)
Goodman D, **5**: 3135 (3132)
Goodman E D, **1**: 592 (590)
Goodman F R, **2**: 1458
Goodman I R, **3**: 1838; **4**: 2301 (2293, 2294, 2295, 2300, 2301), 2883 (2881); **6**: 3619 (3617), 3772 (3769, 3770), 3947 (3945)
Goodman J, **3**: 1979
Goodman L A, **3**: 2060 (2056, 2057)
Goodman M R, **7**: 4762
Goodman N, **2**: 883 (883), 1136 (1133, 1134), 1154 (1143, 1147, 1148, 1151, 1152, 1153), 1160 (1159), 1165 (1164), 1169 (1166, 1168), 1209 (1208)
Goodman N J, **1**: 437
Goodman T J, **4**: 2301 (2300)
Goodrich G J, **6**: 4261 (4260)
Goodson R E, **2**: 1197; **3**: 2014; **6**: 4245 (4241)
Goodwin B C, **1**: 429 (427)
Goodwin D N, **3**: 1736
Goodwin G C, **1**: 60 (58, 59, 60), 489 (487); **2**: 1581 (1579, 1580); **4**: 2239 (2237), 2245 (2242), 2264 (2263), 2283 (2283), 2307 (2303), 2319; **5**: 3006 (3006), 3060 (3056, 3059), 3061 (3061), 3065 (3062), 3410 (3407, 3408), 3475 (3473); **6**: 3929 (3929), 4061 (4057), 4193 (4190, 4192, 4193), 4209 (4206, 4207); **7**: 4607 (4603, 4606), 4611 (4607, 4608, 4610, 4611), 5001 (4998, 4999)
Goodwin R H, **2**: 1318 (1317)
Goodwin S, **7**: 5147 (5143, 5145)
Gopinath B, **4**: 2702 (2698); **7**: 4741
Gordon G, **6**: 4302 (4300), 4369 (4368); **7**: 4935 (4933)
Gordon H R, **6**: 4045 (4045)
Gordon J A, **7**: 5008 (5006)
Gordon K J, **4**: 2550 (2550)
Gordon L, **2**: 1106
Gordon R F, **4**: 2550 (2550)
Gordon T J, **4**: 2338 (2334)
Goren B, **3**: 1910 (1910)
Gori-Giorgi C, **4**: 2756 (2756), 2816 (2807); **5**: 3288 (3285, 3286, 3287)
Goring C A I, **3**: 2146
Goris M L, **6**: 3940 (3939)
Goriya Y, **3**: 2021 (2017)
Gorman R R, **1**: 301 (297)
Gorostiza, **7**: 4627 (4624)
Gorry G A, **2**: 1577 (1574)

Gorsline G W, **1**: 736, 740
Gosiewski A, **3**: 1990 (1989)
Goss W C, **5**: 3470 (3466)
Gossian D M, **6**: 4088 (4086)
Gossman G, **5**: 2993 (2990)
Gostony G, **7**: 4733
Gottlieb D, **7**: 4498 (4496)
Gottlieb G L, **1**: 466 (466); **5**: 3242 (3241)
Gottlieb M, **5**: 3464 (3460); **6**: 3916 (3913), 4346 (4346)
Gotto A M Jr, **4**: 2887
Gottschalk C W, **7**: 4980 (4976)
Gottschalk W H, **2**: 1286 (1283)
Gottwald S, **2**: 1473 (1469); **3**: 1785 (1783, 1785), 1894 (1891); **5**: (2918), 2965 (2965)
Gould F, **6**: 3706 (3705)
Gould F J, **6**: 3634
Gould L A, **5**: 3336 (3330)
Gould P, **6**: 3922
Gould R F, **3**: 2146
Goulding F S, **1**: 177 (174)
Gourdon J, **6**: 3644
Goursat M, **1**: 80 (77); **5**: 3530 (3525, 3529)
Gouvernet J, **1**: 769 (762, 765); **3**: 1604 (1601, 1603)
Gouy G, **6**: 4146 (4143)
Govindjee, **2**: 1522 (1521)
Gower J C, **7**: 4645 (4644)
Graber T M, **2**: 965
Graebner K E, **4**: 2453 (2452)
Graedel T E, **2**: 1528 (1527)
Graeme J G, **1**: 272 (269, 270)
Graf F L, **3**: 1958
Graham C H, **7**: 5078
Graham D, **1**: 120, 343
Graham E B, **3**: 1975
Graham J M, **1**: 262 (259, 260)
Graham K D, **4**: (2805)
Graham L T, **7**: 5070 (5066, 5070)
Graham N, **7**: 5065 (5060, 5064)
Graham R A, **1**: 176 (175)
Graham R E, **6**: 4165 (4162)
Graham R M, **2**: 1132 (1129), 1136 (1134, 1135), 1154, 1174 (1169, 1170, 1171)
Graham W J, **3**: 2096 (2096), 2097 (2096)
Graham-Bryce I J, **3**: 2147
Grahame R, **1**: 482 (480)
Graindorge P, **5**: 3470 (3467)
Grainger R G, **7**: 5186 (5182)
Gramme Z T, **2**: 1402 (1400)
Grams R R, **4**: 2681 (2680)
Grandpierre M, **7**: 5027
Granelli J G, **7**: 4504
Granero-Porati M I, **1**: 675 (672)
Granger H J, **1**: 468 (466)
Granit R, **1**: 466 (466); **5**: 3253 (3250, 3251, 3252)
Granovsky B L, **7**: 4656 (4655)
Grant E L, **7**: 4726 (4724)
Grant W A, **1**: 150; **3**: 2209
Grashuis J L, **7**: 4981 (4976, 4979)
Grason J, **6**: 4415 (4414)
Grasse K, **5**: 3184 (3181, 3182)
Grasse K A, **7**: 4677 (4675, 4676, 4677)
Grasselli O M, **5**: 3341 (3339)
Grasshoff K, **7**: 5127
Grateloup G, **2**: 1192 (1189); **7**: 4666 (4665)
Gratzer G, **3**: 2168 (2168)
Grauer M, **2**: 1276 (1274)
Graunt J, **6**: (3751)
Gravel R L, **5**: 3464 (3464), 3465 (3459)
Gravenstein J S, **1**: 440
Graves C, **1**: 570 (568)
Graves R L, **4**: 2880 (2877); **6**: 3739 (3736)
Gray A H Jr, **1**: 86 (85); **7**: 4509 (4507)
Gray D O, **7**: 4833 (4832)
Gray J M, **3**: (1745)
Gray J N, **2**: 883 (881, 882), 1136 (1134, 1135), 1154 (1144, 1145, 1147), 1165 (1161, 1164), 1174 (1170, 1171)
Gray J O, **5**: 3215 (3206), 3383 (3380, 3381, 3382), 3389 (3386)
Gray J S, **4**: 2901 (2898); **6**: 4061 (4055)
Gray R M, **2**: 1492
Grayson L, **7**: 5031 (5029)
Graziani F, **3**: 1664 (1657, 1663)
Greco C, **6**: 4172 (4172)
Greco G, **5**: 3264
Green A, **1**: 437
Green A E, **4**: 2447 (2445); **6**: 4015
Green E, **1**: 437
Green J H, **1**: 535 (532, 534)
Green J R, **5**: 2911 (2908); **7**: 4630, 4631
Green P B, **4**: 2798 (2795)
Green R G, **3**: 1693 (1691, 1693)
Green T B, **2**: 987; **7**: 4809
Green T R G, **3**: 2179
Green W L, **6**: 4311 (4311)
Greenaway D S, **6**: 4302 (4301)
Greenberg A E, **7**: 5122
Greenberg H, **7**: 4709 (4708), 4922
Greenberg H J, **6**: 4313, 4333 (4329), 4380
Greenberg M, **3**: 2169
Greenberg R R, **1**: 177 (174)
Greenberg S G, **2**: 808 (802)
Greenberger M, **7**: 4762 (4756), 5005 (5001, 5002)
Greene C S, **3**: 1736, 1741 (1739)
Greene J C, **2**: 973 (969, 970, 972)
Greenfield J, **1**: 769 (761)
Greenfield S, **4**: 2347 (2343)
Greenspan D, **6**: 4346 (4346)
Greenspan E, **2**: 1552 (1549)
Greenstadt J L, **7**: 4975 (4974)
Greenwood A, **6**: 3806 (3802)
Greenwood M, **7**: 4645 (4645)
Greer W, **4**: 2901 (2899)
Greeson P E, **3**: 1862
Gregg E C, **7**: 5186 (5182)
Gregor I, **5**: 3135 (3132)
Gregory B A, **3**: 2080
Gregory D H, **3**: 1961
Gregory G A, **1**: 289 (285)

Gregory S A, **2**: 992 (991)
Gregson M J, **5**: 3319 (3319)
Gregson R A M, **1**: 276 (272)
Greig J M, **3**: 1943
Greiner R, **1**: 759 (759)
Grenander U, **6**: 3644
Grene M, **6**: 4168 (4167)
Grenoble D E, **1**: 474 (471)
Grewal M S, **1**: 493 (491)
Grey J, **7**: 4476
Griegelionis B I, **7**: 4617
Grieve D W, **3**: 1910 (1910)
Griffeth N D, **2**: 1165 (1164)
Griffin A W J, **1**: 736 (734)
Griffin L, **3**: 1761 (1758)
Griffith I, **1**: 437
Griffith R E, **4**: 2412 (2393, 2397)
Griffiths D F, **3**: 1627
Griffiths I D, **2**: 1527 (1524)
Griffiths P R, **4**: 2511 (2510); **7**: 4504 (4499)
Grigorov L N, **1**: 433 (432)
Grillner S, **3**: 1910 (1909)
Grimble M J, **1**: 45 (38, 39, 42, 43); **5**: 3410 (3407); **6**: 4176 (4173, 4174), 4261 (4259), 4266 (4264), 4267 (4262, 4264); **7**: 4462 (4460)
Grime J P, **2**: 1317, 1321 (1318, 1320)
Grimes R W, **6**: 4294
Grippo L, **6**: 4244 (4238)
Grisse J, **3**: 2092 (2091)
Grisvard P, **6**: 4223
Grob R I, **3**: 1931
Grob R L, **1**: 594
Gröbner W, **3**: 1779 (1773, 1774, 1777); **5**: 3394 (3392)
Grochla E, **6**: 4333 (4329, 4332)
Grodins F S, **4**: 2901 (2898, 2899); **6**: 4051 (4048, 4049), 4061 (4055, 4056, 4060), 4066 (4062, 4065)
Grodsky G M, **3**: 2174 (2173)
Grodzovskii G L, **7**: 4490
Groen J J, **2**: 1595 (1592), 1596 (1592)
Grom E, **5**: 3421 (3419)
Grombach P, **7**: 5149
Gromley K, **6**: 3745 (3741)
Gronli B, **5**: 2993 (2991); **6**: 4252 (4250)
Groover M P, **3**: 1648 (1642, 1644)
Gros C, **7**: 4856 (4856)
Gros H, **7**: 4703 (4701)
Grosjean D, **1**: 177 (176)
Gross D, **6**: 3925 (3923)
Gross J F, **1**: 477 (477)
Gross M G, **2**: 1516
Gross O A, **5**: 3512 (3511), (3516)
Grossberg S, **1**: 570 (569)
Grossin J, **1**: 388
Grosso C A, **6**: (3861)
Grote J D, **2**: 1017 (1013)
Grötschel M, **6**: 3739 (3738)
Grotstollen H, **5**: 3097
Grouchko D, **6**: 4029 (4029); **7**: 4766 (4765)
Grove E L, **7**: 4504
Grover R, **3**: 2146
Groves F R Jr, **1**: 589 (585)
Groves M R, **2**: 866 (865), 867 (866)
Groves R H, **2**: 1318
Groves T, **2**: 995 (995); **7**: 4822 (4819)
Gruca A, **1**: 132 (128)
Grujić L T, **4**: 2565 (2564); **5**: 3406 (3405); **7**: 4534 (4530), 4547
Grumondz V T, **5**: 3389 (3386)
Grundler H C, **1**: 262 (259)
Grundy S M, **1**: 493 (493); **4**: 2887; **7**: 4906 (4905)
Grùsser O J, **7**: 5078 (5075)
Grynpas M D, **1**: 473 (469, 473)
Gualtierotti T, **2**: 1596 (1592)
Guanella G, **6**: 4146 (4143)
Guardabassi G, **2**: 1093 (1090); **7**: 4689 (4687, 4688, 4689)
Guardo R, **1**: 536 (534)
Guckenheimer J, **5**: 3035 (3032); **7**: 4529 (4526)
Gudan F, **6**: 4370 (4369)
Gudefin E, **7**: 4598 (4596)
Gudmundson C, **2**: 1568 (1568)
Guegue C, **3**: 2108 (2108)
Guentert T W, **2**: 1248 (1247)
Guenzi W D, **1**: 148 (143, 144); **3**: 2147
Guerchet P, **2**: 900 (895, 896)
Guérin J P, **1**: 132 (127)
Guest M J, **4**: 2631 (2628)
Guevara L, **7**: 4905 (4902)
Guevara M R, **1**:. 570 (568, 569)
Gugelmann A, **1**: 531 (529)
Guidorzi R P, **4**: 2292 (2288, 2291); **5**: 3044 (3040, 3042, 3043, 3044); **6**: 3776 (3772), 3983 (3983); **7**: 5170 (5166)
Guigou J-L, **5**: 3152
Guild N C F, **3**: 1847 (1846)
Guile A E, **6**: 3784
Guillaume M, **6**: 3940 (3940)
Guillemin V, **5**: 3394 (3390, 3393); **7**: 4674 (4672)
Guillon M, **3**: 2133; **6**: 4257
Guinzy N J, **4**: 2706 (2703), 2713 (2711)
Guiver J P, **5**: 3130 (3129), 3135 (3133); **6**: 3972 (3971)
Gukhman A A, **6**: 4296
Gul'ko F B, **6**: 3870 (3867)
Gully S W, **4**: 2846 (2843, 2844)
Gulrajani R M, **1**: 536 (534)
Gumowski B, **1**: 588 (588)
Gundersen H J G, **7**: 4601 (4599)
Gunning A J, **1**: 315 (314)
Gunther B, **7**: 4454 (4453)
Gupal A M, **5**: 3546 (3541)
Gupta A K, **2**: 1106 (1105)
Gupta M M, **1**: 126 (125, 126), 276 (272, 276), 575, 608 (607), 664 (664), 713 (710, 712, 713), 740 (736); **2**: 776 (775), 1106, 1473 (1472, 1473), 1549 (1547), 1578 (1575); **3**: 1785 (1785), 1797, 1800 (1798, 1800), 1809 (1809), 1810 (1807), 1819, 1821 (1819, 1821), 1822 (1821), 1829 (1826), 1831 (1830, 1831), 1835, 1838, 1845 (1845), 1847 (1846), 1852 (1847, 1848, 1850), 1858, 1867 (1864), 1872 (1868), 1874, 1878 (1876), 1888, 1902 (1900); **4**: 2301 (2295, 2299, 2300), 2361, 2501 (2499),

2884 (2883); **5**: 2926 (2926), 2956, 2965 (2963, 2964, 2965), 3264 (3262), 3454 (3451, 3454), 3540 (3539, 3540); **6**: 3768 (3766), 3772 (3770), 3947 (3945), 3990 (3984), 4008 (4005); **7**: 4661 (4659), 4695 (4693), 4696 (4695), 4789 (4788), 4996 (4994, 4995)
Gupta N K, **4**: 2245 (2242), 2319 (2318)
Gurpide E, **7**: 4905 (4904)
Gurtin M, **1**: 529 (522)
Gurwitz R F, **6**: 4302 (4301)
Gustafson C L, **5**: 3196 (3192)
Gustafson D E, **3**: 1838
Gustafson D H, **1**: 618; **4**: 2579 (2576)
Gustavson F G, **2**: 1102 (1102)
Gustavsson I, **4**: 2307 (2302, 2306), 2314 (2312, 2313)
Gutenbaum J, **2**: 1276 (1272, 1274, 1276); **5**: 3454 (3454)
Gutmanis I, **4**: 2522
Guttinger W, **1**: 433 (432)
Gutz S, **1**: 755 (754)
Guy C R, **6**: 3870 (3868)
Guy J J, **3**: 1965, 1992 (1990, 1991); **4**: 2356 (2353)
Guyenne T D, **3**: 2052 (2050)
Guyton A C, **1**: 468 (466), 542 (540); **2**: 853 (851), 1366 (1361); **3**: 2175 (2171), 2214 (2213); **4**: 2901 (2899)
Guyton J R, **3**: 2174 (2173)
Guzman A, **6**: 4159 (4157)
Gylys V B, **2**: 1282 (1281)

H

Haanstra J W, **4**: 2685 (2684)
Haapoja V, **6**: 3916 (3913)
Haar L, **7**: 4585 (4581)
Haas L, **2**: 1174 (1171)
Haase F, **1**: 63 (63)
Haba Y, **3**: 2053 (2048, 2049); **4**: 2655 (2652)
Habbema J D F, **2**: 1107 (1104, 1105)
Haber R, **3**: 2010 (2008, 2010); **4**: 2892; **5**: 3351 (3342); **6**: 4184 (4180, 4181)
Haberer J, **7**: 4828 (4824)
Haberman R, **7**: 4923
Habermayer M, **3**: 2010 (2008)
Habets P, **1**: 675 (673); **5**: 3402 (3399, 3400); **7**: 4529 (4525)
Haccoun D, **2**: 1057 (1055, 1056)
Hack M, **6**: 3673 (3671)
Hackbush W, **5**: 3139 (3136, 3139)
Hacker W, **6**: 3663 (3660)
Hackett C, **1**: 142 (139)
Hacking I, **3**: 1805 (1805)
Hadamard J S, **2**: 1179 (1174), 1372 (1371); **4**: (2609); **6**: 4364 (4360)
Haddad E K, **1**: 433 (430)
Haddad G, **2**: 1033
Haddad R A, **1**: 251 (248)
Haddock G, **1**: 494 (491)
Hadler M I, **1**: 508; **5**: 3264 (3263)
Hadzilacos V, **2**: 884 (883)
Haefner J W, **3**: 2045 (2044, 2045)
Haelg W, **1**: 633 (632)
Haelsman L P, **6**: 4284 (4283)
Haenschke D G, **7**: 4726 (4723)
Haettich W, **6**: 4159 (4158)
Häfele W, **5**: 3416
Haftari A, **5**: 3568 (3568)
Hagander P, **6**: 4267 (4265)
Hagedoi P, **2**: 1016 (1012)
Hagedorn P, **2**: 1008 (1006), 1017 (1013)
Hagen O, **3**: 2202 (2194, 2197), 22022203 (2194, 2195); **5**: 3104 (3103)
Hägglund T, **6**: 4213 (4212)
Hagiwara S, **6**: 4117
Hagras A N, **1**: 682 (679)
Hague B, **2**: 1403 (1402, 1403)
Hague M J, **4**: 2361 (2359)
Hague R, **3**: 2147
Hahn C E W, **1**: 440; **5**: 2943 (2935)
Hahn F H, **1**: 671 (669)
Hahn G M, **1**: 417 (414, 415), 450 (449, 450); **2**: 1337 (1336)
Hahn V, **1**: 63 (62); **5**: 3398
Hahn W, **2**: 1212; **4**: 2556 (2551, 2555); **5**: 3398, 3402 (3399, 3400, 3401), 3406 (3405); **7**: 4529 (4525)
Haik M, **1**: 176
Haimes Y Y, **1**: 638 (634, 636); **3**: 2053 (2048, 2049), 2192 (2186); **5**: 3115 (3115), 3124 (3123), 3144, 3164 (3157, 3158, 3159, 3160, 3161, 3163), 3568 (3564, 3567, 3568); **7**: 4709 (4704, 4705, 4706, 4707, 4708, 4709), 4771, 4810, 5140
Haines R W, **3**: 2108 (2106)
Hainsworth B D, **3**: 1700 (1696)
Haith D A, **2**: 1538 (1537); **7**: 5140
Hajdasinski A K, **6**: 3983 (3979)
Hajek B, **5**: 3303 (3293, 3294); **7**: 4637 (4631)
Hajek O, **5**: 3184 (3183)
Hájek P, **5**: 2921 (2918, 2919)
Hakathorn R, **2**: 947 (947)
Halanay A, **2**: 1211 (1209); **3**: 2151, 2152 (2152); **5**: 3506
Halback R E, **5**: 3421 (3419)
Hale H W, **4**: 2556 (2552, 2555)
Hale J K, **2**: 1211 (1209); **3**: 2151 (2150), 2152 (2152); **4**: 2609 (2608); **5**: 3506; **7**: 4529 (4528)
Hales D B, **5**: 3282 (3282)
Haley K B, **4**: 2353 (2352)
Halfman R L, **7**: 5170 (5165)
Halfon E, **1**: 417 (414, 415), 675 (673, 674); **2**: 1326 (1324), 1329 (1327), 1337 (1336), 1351 (1348, 1350), 1355 (1352); **3**: 2003; **4**: 2646 (2645); **5**: 3044 (3040, 3043, 3044); **6**: 3989 (3984); **7**: 4454 (4453), 5008 (5007)
Halford G J, **2**: 1405
Halg W, **1**: 634 (632)
Halin H J, **1**: 633 (632), 634 (632); **5**: 3559 (3557)
Halkin H, **1**: 413 (411); **2**: 1099 (1093)
Hall A C, **3**: (1748)
Hall A D, **2**: 1489 (1488); **4**: 2685 (2684)
Hall A S, **5**: 3425 (3425)
Hall C A S, **2**: 1462, 1538 (1537); **7**: 4454 (4453), 4897 (4897), 5008 (5005)
Hall D J, **3**: 1902 (1901)
Hall E L, **4**: 2484 (2482, 2483); **6**: 4160 (4157)
Hall H H, **2**: 1403 (1402)
Hall J L, **7**: 4509
Hall P, **4**: 2659 (2657); **5**: 2934 (2932)
Hall W A, **5**: 3124 (3123), 3164 (3157, 3158, 3159), 3568; **7**: 4709 (4704, 4708, 4709), 5140
Hall W G, **3**: 2213 (2213)
Haller M A, **7**: 4598 (4596, 4598)
Halliday T R, **3**: 2213 (2211, 2212)
Hallum A, **1**: 437
Halme A, **3**: 1619 (1618), 1779 (1776); **6**: 4204 (4200)
Halsall J R, **6**: 4108 (4107)
Halttunen J, **3**: 1700
Ham J, **2**: 1249 (1247); **5**: 3238 (3236); **6**: 3699 (3698)
Ham J M, **6**: 4142 (4141)
Hamacher H, **3**: 1781 (1781), 1819
Hamaker J W, **3**: 2146
Hamakoga M, **2**: 1521 (1519)
Hämäläinen J J, **6**: 4066 (4062)

Hämäläinen R P, 6: 4066 (4062, 4064, 4065)
Hamam Y M, 3: 1954 (1949)
Hamanaka K, 5: 3464 (3459)
Hamasaki Y, 5: 3464 (3461)
Hamblin P F, 7: 4889 (4887)
Hamel B, 5: 3373 (3370), 3389 (3386)
Hamer J W, 6: 3744 (3740)
Hamid A, 3: 1694 (1687)
Hamidi M, 2: 808
Hamielec A E, 6: 3745 (3740)
Hamilton B, 3: 2014
Hamilton S, 3: 2192 (2186)
Hamilton W F, 7: 4793 (4791)
Hamilton W R, 2: 1033 (1031)
Hamlyn D W, 3: 1985
Hammarling S J, 2: 826 (823)
Hammel H T, 7: 4876 (4873, 4875), 4880 (4878, 4879)
Hammer H, 1: 564 (561)
Hammer M, 2: 1154 (1144), 1165 (1164)
Hammer U T, 4: 2782 (2778, 2780)
Hamming R W, 6: 4289
Hammond J S III, 3: 2192
Hammond K R, 1: 372 (369); 3: 2192, 2202 (2196); 5: 3115 (3113)
Hammond P H, 4: 2287 (2287), 2319 (2317, 2318); 5: 3390 (3386), (3516)
Hammons A S, 2: 1323 (1322)
Hammoude A, 5: 2993 (2992)
Hamza M, 5: 3464 (3463)
Hamza M H, 1: 682 (681); 2: 1199 (1197); 5: 3055 (3048, 3050)
Han J-Y, 6: 3626 (3623)
Han K W, 5: 3077 (3076)
Han S P, 2: 797 (795)
Hanafusa H, 1: 301 (298)
Hanani M Z, 6: 4302 (4301)
Hance R J, 3: 2147
Hancock R A, 4: 2356 (2353, 2354)
Hancox R, 5: 3416
Hand D J, 2: 1106 (1104, 1105)
Hand R C, 2: 1410 (1409)
Handa J, 3: 2202 (2197)
Handschin E J, 4: 2892; 6: 3829 (3826), 3834 (3832), 3846 (3845), 3850 (3847)
Hang C-C, 5: 3070 (3068, 3069, 3070); 7: 4462 (4460), 5031 (5029)
Hanken A F G, 5: 3035 (3033)
Hanks H, 2: 992 (988, 989)
Hanna G R, 2: 1590 (1587, 1589)
Hannaford J F, 3: 2234 (2233)
Hannah E H, 3: 1992 (1991)
Hannan E J, 4: 2292 (2292), 2293 (2288); 7: 4566 (4566)
Hannapel J, 7: 4980 (4980), 4981
Hannigan F J, 3: 2104 (2104), 2117 (2117)
Hannon B, 2: 1486 (1479); 5: 2929 (2929); 6: 4237 (4235); 7: 4956 (4956)
Hanrahan D J, 4: 2301 (2300)
Hansen A R, 6: 4160
Hansen D R, 3: 1829 (1826); 5: 3164 (3157, 3163); 7: 5140 (5135)
Hansen E H, 7: 5122
Hansen L D, 1: 177 (175)
Hansen P B, 5: 3165, 3449; 6: 3596
Hansen T, 6: 3634
Hansen W S, 1: 639 (639)
Hanson J A, 7: 5161
Hanson J D, 1: 142 (139, 140)
Hanson R J, 2: 827 (817, 818); 4: 2283 (2280); 6: 3983 (3978, 3979); 7: 4540 (4535)
Hanson W G, 7: 4495
Hansotia B J, 7: 4648 (4646)
Hanst P L, 4: 2511 (2510)
Hanuzet B, 4: 2343
Hao Bai-Lin, 1: 567
Hapiot J C, 7: 5027
Happ H H, 2: 1388 (1387); 6: 3820 (3817), 3954 (3951, 3952)
Happel J, 1: 494 (493)
Haque S I, 6: 3929 (3927)
Haralick R M, 4: 2484 (2481); 6: 3644
Harary F, 2: 1346 (1345), 1348 (1345, 1346), 1355 (1352); 4: 2579 (2577), 2640 (2639); 7: 4666 (4662), 4956 (4955, 4956)
Harasawa H, 2: 1544 (1542, 1543)
Harashina S, 2: 1542 (1540), 1544 (1542, 1543)
Harbert J, 6: 3940 (3936)
Harden Jones F R, 7: 4848 (4844, 4847)
Hardie D W F, 6: 3884
Harding G C H, 7: 4454 (4453)
Hardt R, 6: 3956 (3956)
Hardy G F, 2: 1479 (1475, 1476)
Hardy G H, 4: 2614 (2611)
Hardy J D, 7: 4856 (4853, 4854), 4876 (4873, 4874, 4875, 4876), 4877 (4873, 4874, 4875, 4876), 4880 (4879, 4880), 4881 (4880)
Hardy N W, 4: 2447 (2447)
Hargreaves P, 3: 1910 (1910)
Haring G, 1: 755 (752)
Harker A B, 1: 176 (175), 177 (175)
Harkness J, 1: 445 (441, 442)
Harleman D R F, 4: 2646 (2646)
Harlow C A, 6: 4160 (4157)
Harlow K C, 7: 5098
Harman H H, 5: 3235
Harman R T C, 3: 1995
Harmer A L, 5: 3464 (3462)
Harmer M, 2: 963 (962)
Harmon L D, 1: 575; 6: 3644 (3636); 7: 5078 (5075)
Harmon P, 2: 1573
Harms H, 5: 3464 (3461)
Harmsen R, 2: 1552
Harmston A T, 2: 1087 (1084)
Haroules G G, 1: 201
Harper B, 7: 4871 (4869)
Harper M D, 3: 1979
Harper R P, 3: 1668 (1666, 1667)
Harrington J W, 2: 1590 (1587)

Harriott P, 2: 1109 (1109); 3: 2092 (2090)
Harris C, 2: 1545 (1544)
Harris C J, 2: 1093 (1087, 1092), 1116 (1110), 1431 (1427); 3: 2113 (2113); 5: 3303 (3301), 3336 (3336), 3379 (3377, 3378), 3475 (3473); 6: 4180 (4179), 4199 (4198), 4204 (4199), 4266 (4264); 7: 4551 (4548)
Harris C M, 6: 3925 (3923)
Harris D B, 3: 1693 (1692)
Harris G P, 7: 5094 (5090, 5092)
Harris H, 3: (1748)
Harris J A, 3: 1769 (1766, 1767, 1768, 1769)
Harris J D, 2: 1106 (1105); 3: 1838
Harris M, 1: 556 (554); 6: 4108
Harris R F, 4: 2646 (2645)
Harris R K, 7: 4504 (4499)
Harris T R, 7: 4935 (4934)
Harris W C, 1: 181
Harris W F, 2: 1344 (1343)
Harrison A, 2: 978
Harrison D C, 1: 456 (455)
Harrison G H, 4: 2347 (2344)
Harrison G W, 1: 675 (674); 2: 1326 (1324)
Harrison H L, 1: 639 (639)
Harrison J S, 3: 1615 (1612)
Harrison J T, 3: 1943
Harrison L I, 6: 3698 (3697)
Harsanyi J C, 2: 850 (843, 849), 1269 (1263); 3: 1926, 2061 (2057, 2060); 5: (2908), 2911 (2910)
Hart A, 4: 2769 (2769)
Hart C E, 6: 3855 (3853)
Hart H L A, 1: 556 (553)
Hart J, 6: 4302 (4302)
Hart P E, 1: 575 (574), 664 (654); 2: 1106 (1105); 3: 1603 (1602, 1603); 4: 2385 (2382), 2473, 2484 (2481, 2483); 6: 3643 (3637), 3653 (3650, 3651)
Hart S, 3: 1929 (1928)
Hart S C, 3: 1668 (1666)
Hartenberg R J, 6: 4088 (4087)
Hartikainen O, 2: 1568 (1567); 6: 3916 (3913)
Hartl W, 7: 5186 (5180)
Hartley G S, 3: 2147
Hartley R V L, 6: 3768 (3766)
Hartmanis J, 4: 2497
Hartmann G L, 2: 827 (822); 4: 2846; 5: 3197, 3228 (3225); 6: 4176 (4173)
Hartnell J P, 2: 1253 (1249)
Hartnell W, 7: (4516)
Hartoch G, 4: 2386 (2379)
Hartree D R, 6: 3884 (3883)
Hartshorn C, 6: 4007 (4001)
Hartvig P, 2: 963 (962)
Harvey R J, 3: 2206
Hasegawa K, 6: 4108 (4107)
Hasegawa T, 2: 1489 (1489)
Haselden G G, 3: 1975
Hashida O, 7: 4751 (4750)
Hashimoto Y, 2: 1521 (1517, 1518), 1522 (1517, 1518, 1519, 1520, 1521)
Haskell E F, 2: 1486 (1480, 1481)
Has'minskii R Z, 1: 86 (82, 84)
Hassan A B, 7: 4948
Hassan G A, 2: 1586 (1585)
Hassan M F, 1: 703 (696); 2: 862 (859, 861, 862), 923 (922); 4: 2697 (2692, 2695), 2706 (2702, 2704), 2713 (2712); 5: 3019 (3016, 3017, 3018, 3019)
Hastings A, 2: 1317 (1315)
Hastings S, 1: 429 (428)
Hastings-James R, 1: 45 (39); 5: 3006 (3005)
Hatfield J J, 1: 246
Hattori M, 2: 1154 (1153), 1445 (1441)
Haug E V, 2: 1179 (1174)
Haurat A, 3: 1642 (1638, 1640)
Haurie A, 1: 687 (684, 685, 686, 687); 5: 3149 (3149)
Hausler D W, 5: 3421 (3419)
Häusler G, 5: 3130 (3129)
Hausmann D, 1: 626 (625), 628 (627); 2: 1083 (1080)
Hausner A, 3: 2217
Haussmann U, 5: 3523 (3522, 3523); 7: 4621 (4621), 4644 (4643)
Haut R C, 1: 486 (483)
Hautus M L J, 4: 2523 (2522, 2523), 2849 (2847); 6: 3734 (3733), 3969, 4070 (4067)
Havata O, 2: 1497 (1494)
Havelock R G, 2: 1348 (1345, 1346); 7: 4833 (4830)
Havránek T, 5: 2921 (2918, 2919)
Haw J F, 5: 3421 (3419)
Hawkins A D, 7: 4848 (4847)
Hawkins W, 2: 1099 (1097)
Hax A C, 4: 2841
Hay J L, 1: 633 (633); 5: 3035 (3033)
Hay W, 1: 755
Hay W W Jr, 7: 4905 (4905)
Hayashi C, 5: 3235 (3233), 3389 (3384)
Hayashi K, 2: 1489 (1489)
Hayat M A, 2: 1420; 6: 4152
Haydock K P, 6: 4235 (4230, 4231)
Hayek F A von, 1: 399 (394)
Hayes B, 1: 319
Hayes J E, 2: 1573; 3: 1786 (1783), 1821 (1820)
Hayes J K, 1: 289 (284)
Hayes J P, 1: 740; 5: 2989
Hayes M H B, 1: 7 (3)
Hayes P J, 4: 2640 (2640)
Hayes R G, 3: 2096 (2094, 2095)
Hayes R H, 7: 4833 (4828, 4832)
Hayes W C, 1: 474 (469, 473)
Hayes-Roth F, 2: 1573; 4: 2640 (2640)
Haynes G, 7: 4425 (4424)
Hays G R, 5: 3421 (3418)
Hays W L, 5: 2917
Hayton G E, 4: 2832 (2828, 2829), 2871 (2869); 6: 3721 (3720, 3721), 3734 (3734), 3751 (3750, 3751)
Hayward H, 4: 2338 (2334)
Haywood R W, 7: 4585 (4580)
Hazelton T W, 1: 592 (589, 590)
Hazen H L, 3: (1748); 5: 3389 (3385, 3386), (3516); 6: 4146 (4145)

Hazewinkel M, **5**: 3293 (3291, 3292), 3303 (3299), 3310 (3304, 3305, 3306), 3329 (3327), 3330 (3325, 3327)
Hazony D, **4**: 2835
Healey F B, **7**: 4856 (4853)
Heaman S, **2**: 963 (962)
Hearfield L, **3**: 1975
Hearn R H, **4**: 2314 (2311)
Hearon J Z, **7**: 4905 (4904)
Heath M, **3**: 1975
Heaviside O, **2**: (870); **3**: (1745); **4**: 2664 (2659)
Hebb D O, **6**: (4344)
Hecht H G, **7**: 4504 (4499)
Heck W W, **1**: 592 (590)
Heddle J F, **7**: 5114
Heddon K, **3**: 1969
Hedlund G A, **2**: 1286 (1283)
Hedqvist T, **1**: 51 (49, 50); **6**: 3591 (3590)
Hedström B, **2**: 1568 (1568)
Heftmann E, **1**: 594
Heginbotham W B, **4**: 2385 (2383); **6**: 4108 (4102)
Hegner N S, **5**: 3164 (3161)
Hegner S J, **6**: 4411 (4407)
Hehra R K, **4**: 2292 (2287, 2291)
Heicklen J, **1**: 187 (185)
Heidari M, **7**: 5153 (5153)
Heidelberger P, **6**: 4302 (4300), 4395 (4395)
Heike T, **3**: 1969
Heikkilä S, **6**: 3889 (3885, 3888)
Heimann W, **7**: 4851
Heims S, **2**: 873 (872)
Heinonen P, **5**: 2994 (2991)
Heinzle E, **5**: 2943 (2935)
Heise D, **5**: 2944 (2935)
Heister M, **4**: 2827 (2821, 2823, 2824, 2825)
Heitz G, **5**: 2943 (2935)
Heitzman G W, **3**: 1601 (1600); **7**: 4897 (4896)
Held R, **7**: 5077 (5075), 5078
Helfant R H, **2**: 1410 (1408)
Helleman R H G, **1**: 567; **7**: 4529 (4528)
Heller A, **3**: (1938)
Heller D E, **2**: 826 (818)
Heller H, **2**: 1154 (1153)
Heller S R, **3**: 1936 (1933); **5**: 2943
Hellerman H, **1**: 755 (752)
Hellman M E, **1**: 726 (720); **2**: 1132 (1130)
Hellman W D, **2**: 873 (872)
Helly W, **7**: 4495, 4927 (4926), 4948
Helmer O, **2**: 1545
Helmholtz H L F von, **5**: (3383)
Helms D, **4**: 2455 (2453)
Helms H D, **3**: 1723
Helson H, **5**: 3130 (3129)
Helton I W, **1**: 132 (132)
Helton J C, **2**: 1551 (1551); **6**: 4235 (4230)
Hemami H, **6**: 4128 (4125)
Hemker P W, **5**: 3082 (3081); **7**: 4434 (4430)
Hemming M A, **5**: 3421 (3419)
Hemmingsen A M, **7**: 4454 (4453)
Hemp J, **3**: 1693 (1685)
Hendee W R, **7**: 4856 (4856)
Hendershott M C, **7**: 4889 (4887)
Henderson F M, **3**: 2234 (2230)
Henderson H V, **1**: 495 (494)
Henderson J C, **6**: 4332 (4329, 4331, 4332)
Henderson J H, **3**: 2006 (2006)
Henderson J M, **7**: 5164
Hendriks D W, **6**: 4252 (4250)
Hengstenberg J, **5**: 3575
Henham A, **3**: 1761
Heninger K L, **4**: 2501 (2500)
Henkel R E, **5**: 2917
Henkin L, **3**: 1791 (1790)
Hennel M A, **2**: 826 (824, 825)
Hennequin P L, **1**: 91 (89)
Hennessy T G, **3**: 1623 (1619)
Hennet J C, **7**: 4741 (4736, 4737, 4738)
Henning K, **3**: 2092 (2091)
Hennion B, **7**: 4732 (4727), 4751 (4749, 4750, 4751)
Henrici P, **2**: 826 (816)
Henriksen J O, **6**: 4370 (4369); **7**: 4935 (4934)
Henry C L, **2**: 1033
Henry D, **5**: 3496 (3496)
Henry J, **3**: 1734 (1734)
Henry J J, **7**: 4913 (4912), 4927 (4926, 4927)
Henry R M, **4**: 2253 (2249); **5**: 2984 (2980)
Henry W, **3**: (1606)
Hensel H, **7**: 4872 (4871), 4873
Henshaw E T, **7**: 5186 (5181)
Hepburn J S A, **4**: 2567 (2566)
Herbandson H H, **7**: 4897 (4896)
Herbillon A, **1**: 7 (3, 4)
Herbrandson H H, **3**: 1601 (1600)
Herbst J F, **5**: 2993 (2990, 2991), 3430 (3426); **6**: 3776 (3772), 3954 (3952, 3953)
Herceg E E, **3**: 1710
Hercules D M, **7**: 4504 (4499)
Herdman T L, **3**: 2150 (2149), 2152; **5**: 3506
Herget C J, **2**: 826 (825)
Herman D, **2**: 1209 (1205, 1208, 1209)
Herman F, **3**: 1761 (1760)
Herman G T, **4**: 2798 (2795, 2796); **5**: 3425 (3423)
Herman M, **3**: 2052 (2050)
Herman P G, **7**: 5186 (5182)
Herman R C, **7**: 4922, 4927 (4925, 4926)
Hermann R, **3**: 1779 (1770); **4**: 2866 (2865); **5**: 3288 (3285), 3341 (3337, 3338), 3366 (3366), 3394 (3390, 3393); **6**: 3734 (3733), 3977 (3975, 3976); **7**: 4425 (4420)
Hermes H, **1**: 323 (321, 323), 413 (411, 412), 720 (715); **4**: 2609 (2608), 2756 (2756), 2898 (2893, 2894, 2895, 2896); **5**: 2963 (2958, 2963), 3206, (3517); **7**: 4425 (4420, 4421, 4422, 4424), 4677 (4675)
Hermite C, **3**: (1745); **7**: (4517)
Hernández D B, **5**: 3082 (3081)
Heron of Alexandria, **3**: (1604, 1605)
Herrara A O, **3**: 2016 (2016)
Herrero J L, **6**: 4284 (4282)
Herrmann J A, **1**: 437
Herschel C, **6**: (3882)

Hershey E A, **4**: 2502 (2500)
Hershey J C, **3**: 2202 (2197, 2199)
Hertel R H, **3**: 1936 (1933)
Hertz A, **5**: 3530 (3529); **6**: 3789 (3788)
Hertz H S, **3**: 1936 (1933)
Hertzberg T, **1**: 589 (587)
Hertzman A B, **7**: 4876 (4875), 4880 (4880)
Herzberg G, **7**: 4504
Herzog H, **5**: 3421 (3419)
Heslot A, **5**: 3329 (3325)
Hesper B, **6**: 4353 (4350, 4351, 4352, 4353)
Hess B, **1**: 429 (424, 426), 433 (431), 535 (535)
Hesse M, **1**: 720 (715)
Hesseborn B, **2**: 781 (779)
Hessel S J, **7**: 5186 (5182)
Hesselink W H, **6**: 3717 (3710, 3717), 3734 (3733)
Hessert R L, **1**: 312 (310)
Hestenes M R, **2**: 793 (790); **5**: 3487 (3486); **7**: 4971 (4969)
Hetenyi G Jr, **3**: 2025 (2022), 2175 (2170, 2171, 2172); **7**: 4905 (4905), 4906 (4905)
Hetrick D L, **2**: 1186 (1184)
Hetsroni G, **3**: 1693 (1685)
Hett J M, **2**: 1344 (1343)
Hetthéssy J, **1**: 564 (561, 562); **3**: 2010 (2008, 2010); **4**: 2892 (2891); **5**: 3006 (3004), 3188 (3184), 3232 (3229); **6**: 4184 (4181, 4182)
Hetzner W A, **7**: 4833 (4828, 4829, 4830, 4831, 4832)
Heun K D, **6**: 3631 (3631)
Heunis A J, **5**: 3196 (3189), 3215 (3206), 3224 (3223), 3383 (3382)
Hevner A R, **2**: 1160 (1159)
Hewit J R, **4**: 2455 (2453); **6**: 4086 (4085), 4100
Hewitt G F, **3**: 1693 (1685, 1690)
Hewkin P F, **6**: 3691 (3690)
Hewlett J D, **5**: 3268 (3266)
Hey E N, **3**: 2017 (2016)
Heyde C C, **5**: 2934 (2932, 2933)
Heymann M, **6**: 3720 (3718), 3734 (3733), 3969
Heymans N, **1**: 542 (536)
Hibbard T N, **7**: 4822
Hibbs A, **5**: 3329 (3322)
Hibon M, **7**: 5155 (5154)
Hick B, **7**: (4516)
Hicken P, **4**: 2904 (2903)
Hickin J, **1**: 132 (128); **5**: 3077 (3077)
Hickman T R, **5**: 3470 (3467)
Hicks C R, **6**: 4235 (4232)
Hicks J R, **2**: 1342 (1341)
Hicks M, **1**: 433 (431)
Hicks N J, **2**: 1351 (1349)
Hicks T, **3**: 1668 (1666)
Hida E, **7**: 5070 (5066)
Hidy M, **1**: 179 (179)
Higashi M, **1**: 713 (713); **5**: 2929 (2928); **6**: 3768 (3767, 3768), 3989 (3984, 3985, 3987, 3988, 3989)
Higgins J, **1**: 429 (426)
Higgins T J, **2**: 808 (802); **6**: 4245
Higgo J J W, **2**: 1516 (1516)
Higgs P, **2**: 1402 (1401)
Highland H J, **6**: 4370 (4368); **7**: 5018 (5017)
Higuchi S, **1**: 713 (712); **3**: 1604 (1602); **5**: 2927 (2924)
Hijab O, **5**: (3305), 3310 (3305)
Hikata T, **4**: 2560 (2559)
Hilbert D, **2**: 1261 (1257); **4**: 2807 (2807); **6**: 4364 (4360, 4361)
Hilburn J L, **6**: 4284
Hildebrand S G, **2**: 1539 (1539), 1552
Hildebrandt J, **1**: 433 (432), 477 (477); **6**: 4051 (4048)
Hildenbrand W, **1**: 671
Hildreth E, **4**: 2331 (2329), 2484 (2481)
Hilger M, **1**: 564 (561); **4**: 2892 (2891); **5**: 3232 (3229)
Hill A V, **2**: 1248 (1246); **5**: 3248 (3243)
Hill B C, **5**: 3464 (3458)
Hill D A, **4**: 2479
Hill D J, **1**: 60 (60); **4**: (2672); **7**: 4522 (4520)
Hill D W, **5**: 3579 (3575, 3578)
Hill I D, **6**: 3898 (3894)
Hill J, **2**: 1355
Hill J D, **4**: 2580 (2579)
Hill R C, **5**: 3421 (3418)
Hill R R, **6**: 4356 (4356)
Hille E, **3**: 1779 (1771); **5**: 3206 (3205)
Hillerström K, **2**: 967
Hillier F S, **4**: 2842; **6**: 3904
Hillman R E, **1**: 246
Hills M, **2**: 1106
Hills P R, **6**: (4393)
Hillstrom K E, **2**: 827
Hilpinen J, **6**: 3860 (3859)
Hilton P, **1**: 434 (430)
Hiltz S R, **3**: 2179
Himmelberg C J, **1**: 413 (410); **2**: 1099 (1096)
Himmelblau D M, **2**: 1192, 1291, 1297; **3**: 2038 (2033); **4**: 2412 (2402, 2406), 2517 (2512); **5**: 3019 (3012); **7**: 4666
Hinderer K, **2**: 1099 (1093, 1095)
Hindmarsh A C, **2**: 1181 (1181); **5**: 3559 (3558)
Hinds C J, **2**: 1366 (1364)
Hine G, **6**: 4261 (4258, 4259)
Hines H B, **3**: 1939
Hines W G S, **1**: 86 (82)
Hinkley D V, **1**: 86 (81); **2**: 1106
Hinkley E D, **4**: 2723 (2722); **6**: 4036 (4034, 4036), 4037 (4034, 4035)
Hinrichsen D, **2**: 827 (823); **4**: 2827 (2820); **5**: 3288 (3286), 3366 (3366); **7**: 4547 (4541)
Hintikka J, **6**: 3860 (3858, 3859), 4216 (4215, 4216)
Hipel K W, **3**: 1829 (1826, 1827, 1829), 1862 (1859); **7**: 4799 (4793, 4794, 4795, 4796, 4797, 4798)
Hipeli R, **6**: 3916 (3913)
Hirai K, **4**: 2560 (2559)
Hirama M, **5**: 3464 (3463)
Hirano C, **1**: 592 (590, 591)
Hirano S, **6**: 4261 (4258)
Hirata H, **5**: 3044
Hironaka H, **6**: 3956 (3956)
Hirose S, **4**: 2391; **6**: 4108 (4107, 4108), 4113 (4109, 4112), 4128 (4124)

Hirota K, **3**: 1902 (1900); **7**: 4695 (4693), 4696 (4695)
Hirsch M, **2**: 1351 (1350)
Hirschberg D, **3**: 1712 (1712)
Hirschorn R M, **4**: 2756 (2756), 2759 (2759); **5**: 3288 (3285, 3287)
Hirsh C, **5**: 3139 (3139)
Hisdal E, **3**: 1867 (1862, 1863, 1866); **6**: 3768 (3768)
Hishiguch H, **6**: 4261 (4260)
Hitch H P Y, **1**: 31 (21)
Hitchman M L, **5**: 3579 (3575, 3576, 3578)
Hites R A, **3**: 1936 (1933)
Hitz C, **4**: 2866 (2862)
Hitz K L, **3**: 1648 (1644)
Hitzig E, **1**: (463)
Hixson W, **2**: 1595 (1594, 1595)
Hladick J, **4**: 2664
Hmamed A, **2**: 923 (923)
Ho B L, **6**: 3983 (3978)
Ho J K, **2**: 959; **5**: 3164 (3157)
Ho K, **1**: 512 (512)
Ho Y C, **1**: 124 (122), 495 (494), 687 (685); **2**: 907 (903), 932 (930), 995 (994, 995), 1020 (1017), 1099 (1093), 1192 (1191), 1269 (1263); **3**: 1648 (1647); **4**: 2618 (2615); **5**: 3156 (3152), 3164 (3158, 3161), 3180; **6**: 4070; **7**: 4622 (4622), 4630, 4631 (4628, 4629, 4630), 4822 (4815, 4818, 4820, 4821, 4822), 5170
Hoare C A R, **1**: 755 (754); **3**: 1712 (1712); **6**: 3893, 3898
Hoather R C, **7**: 5099
Hobbie J E, **6**: 4235 (4234); **7**: 4646 (4644)
Hobson E, **6**: 3790 (3789)
Hochhausen W, **2**: 1402 (1401)
Hocken R D, **7**: 4462 (4462)
Hocker G B, **5**: 3470 (3467, 3468, 3469)
Hodges H F, **1**: 142 (140, 141)
Hodges J L Jr, **2**: 1106 (1103)
Hodges L F, **1**: 755 (752)
Hodges T, **1**: 592 (591)
Hodgkin A A, **1**: 508 (506)
Hodgkin A L, **7**: 4980 (4977)
Hodgson D C, **4**: 2479
Hodouin D, **5**: 2993 (2991)
Hodrien R C, **3**: 1961
Hoel P G, **2**: 1106
Hoenich N A, **1**: 310 (308)
Hoenig H, **1**: 564 (561)
Hoenninger J, **5**: 3425 (3425)
Hoff A C, **5**: 3248 (3244)
Hoffenberg R, **6**: 3909 (3908)
Hoffer E P, **2**: 1366 (1362, 1363, 1365, 1366)
Hoffman B F, **1**: 456 (451)
Hoffman D B, **6**: 4302 (4300)
Hoffman F O, **2**: 1551 (1551); **6**: 3933 (3931), 3936
Hoffman J P, **1**: 262 (259, 260)
Hoffman J R, **2**: 894
Hoffman K, **1**: 433 (433)
Hoffman L J, **1**: 746 (744)
Hoffmann A, **6**: 3739 (3736)
Hoffmann U, **6**: 4211 (4210, 4211)
Hoffmann W, **5**: 3012
Hofman D, **2**: 1559
Hofmann A F, **3**: 2143 (2142, 2143)
Hofmann D, **5**: 2943 (2935)
Hofstadter D, **1**: 333 (332)
Hofstadter R, **7**: 4692 (4691)
Hoft R G, **7**: 5027
Hoftijzer P J, **6**: 3744 (3742)
Hogarth R M, **3**: 2061 (2060); **4**: 2338 (2337); **5**: 3115
Hogeweg P, **1**: 561 (560, 561); **6**: 4345 (4344), 4353 (4350, 4351, 4352, 4353); **7**: 4764 (4763), 4958 (4957)
Hogg J S, **7**: 4703 (4703)
Hogg W R, **2**: 866 (866)
Hogo H, **1**: 188 (187)
Höhle U, **3**: 1781 (1781), 1785 (1783), 1802 (1802), 1809 (1809), 1831 (1831); **4**: 2301 (2296, 2297, 2299, 2300); **6**: 3710 (3708)
Hohn F E, **6**: 3706 (3706)
Hoing S Y, **4**: 2452 (2451, 2452)
Holdeman J T, **1**: 142 (138)
Holden A V, **1**: 570 (568)
Holford N H G, **2**: 1248 (1246, 1247, 1248), 1249 (1247)
Holland P W, **2**: 1106; **5**: 2917
Hollerbach J M, **4**: 2385 (2378, 2379), 2412 (2392), 2440 (2438); **6**: 4086 (4083)
Hollerith H, **3**: (2178)
Holley E R, **4**: 2789 (2785)
Holliday D V, **3**: 1992 (1991)
Holliday I E, **4**: 2331 (2326)
Holling C S, **2**: 1545 (1543)
Hollingham J, **4**: 2452 (2451)
Holloman G H, **4**: 2901 (2899)
Holloway C A, **2**: 942 (938, 939)
Holloway G A, **5**: 3465
Holm R A, **6**: 3916 (3913)
Holmes F H, **2**: 1402
Holmes J N, **7**: 4509 (4505, 4507)
Holmes P, **7**: 4529 (4526)
Holmes V, **2**: 1087
Holmgren A, **1**: 542 (540), 546 (544)
Holsapple C W, **2**: 947
Holst J, **1**: 54 (54); **7**: 4607 (4603)
Holst P A, **6**: 4417
Holt A W, **6**: 3670 (3668)
Holt C C, **4**: 2353 (2352)
Holt D A, **1**: 142 (140)
Holt J P, **1**: 547 (544)
Holt R C, **2**: 1154 (1147)
Holtby L B, **7**: 4454 (4453)
Holter R M, **6**: 4041 (4038, 4040)
Holton J R, **2**: 1272 (1272)
Holtslag A A M, **2**: 1497 (1497)
Holtzman J M, **2**: 1099 (1093); **7**: 4741 (4734)
Holyoak K J, **3**: 2202 (2201)
Holz M, **5**: 3421 (3418)
Honda K, **6**: 4261 (4258)
Honderd G, **2**: 1388
Honey R C, **6**: 4037 (4034)
Honoré A M, **1**: 556 (553)

Hood D W, **1**: 7 (2)
Hood J D, **2**: 1590 (1590), 1595 (1594), 1596 (1594)
Hood W C, **5**: 3235 (3234)
Hoogendoorn K, **6**: 3744
Hooke R, **6**: 4310 (4308)
Hooker W W, **3**: 1665 (1653)
Hoos I R, **7**: 4789 (4786)
Hopcroft E, **2**: 1153 (1148)
Hopcroft J E, **1**: 334, 725 (721, 722, 723); **2**: 1083 (1082); **3**: 2045 (2044); **4**: 2798 (2795, 2796); **5**: 3034 (3030)
Hope G S, **4**: 2310 (2309); **6**: 3780 (3777)
Hopf W, **6**: 3663 (3660)
Hopfe M W, **6**: 4333 (4329)
Hopfgarten S, **2**: 1305 (1302, 1303)
Hopkin A M, **5**: 3389 (3385), (3516)
Hopkins C E, **2**: 1106
Hopkins C M, **3**: 1736
Hopkinson J, **7**: 4577 (4577)
Hoppe H A, **2**: 1516 (1514, 1516)
Hoppeler H, **7**: 4601 (4599)
Hoppensteadt F C, **6**: 3762 (3756, 3757); **7**: 4434 (4430, 4431, 4432), 4439
Hopper M J, **6**: 3983 (3982)
Horan L G, **2**: 1410 (1409)
Hordijk A, **2**: 1099 (1095)
Horgan J D, **4**: 2901 (2899)
Horler D N H, **2**: 1521 (1518)
Hörmander L, **3**: 1728 (1724)
Horn A R, **2**: 1043 (1042)
Horn B K P, **4**: 2386 (2379, 2382), 2487; **6**: 4100
Hornbach D, **4**: 2580 (2576, 2579)
Hornbein T F, **6**: 4051 (4046, 4049)
Hornberger G, **2**: 1333 (1331)
Horne R A, **1**: 7 (1); **4**: 2782 (2778)
Horner M G, **3**: 2113 (2112)
Hornung C, **6**: 4302 (4301)
Horovitz R, **5**: 3004 (3004)
Horowitz E, **1**: 725 (721, 722, 724, 725)
Horowitz I M, **4**: 2872; **5**: 3196 (3189), 3228 (3227), 3383 (3382), 3475 (3473)
Horsfield K, **4**: 2904 (2902, 2903)
Horsley D F A, **7**: 4703 (4700, 4703)
Horsthemke W, **1**: 433 (431, 432)
Hort W, **6**: 4275 (4273)
Horton P W, **6**: 3940 (3936)
Horvath S M, **7**: 4876 (4876)
Hosaka M, **1**: 759 (759)
Hoshi H, **3**: 1611 (1609)
Hosoe S, **4**: 2827 (2821)
Hosono E, **5**: 2943 (2935)
Houk J C, **7**: 4877, 4880 (4880)
Houkowsky A, **5**: (3383)
Houpis C H, **1**: 735 (732, 734)
Houpt P K, **3**: 1736, 1741 (1738, 1739)
Householder A S, **2**: 826 (817); **7**: 4906 (4903)
Houston E J, **2**: 1402 (1402)
Hovis W A, **6**: 4045 (4045)
Howard B V, **4**: 2887; **7**: 4906 (4905)
Howard L S, **7**: 5027
Howard N, **7**: 4799 (4794, 4795)
Howard R, **2**: 1099 (1095, 1097); **5**: 2932
Howard R A, **1**: 409; **2**: 942 (938); **3**: 1714, 2061 (2053), 2202 (2195); **5**: 3115 (3113)
Howard R W, **1**: 343
Howarth F, **7**: 5147 (5144)
Howden W E, **4**: 2681 (2680)
Howe C W, **7**: 5155 (5154)
Howe I, **5**: 2943 (2942)
Howe M, **3**: 1943, 1958 (1955)
Howe R, **5**: 3559 (3558); **6**: 4359 (4358)
Howe R M, **1**: 634 (632); **3**: 2217
Howell F G, **2**: 1323 (1323), 1344 (1344), 1345 (1344)
Howell J A, **7**: 5122
Howell S C, **3**: 2192 (2186)
Howland H R, **6**: 4245
Hoyle G M, **1**: 167
Hrubesh L W, **7**: 4504 (4499)
Hsia T C, **4**: 2283 (2283)
Hsia Y, **7**: 5078
Hsiao G C, **7**: 4434 (4430)
Hsing C C, **5**: 3060 (3059)
Hsu C H, **6**: 3717 (3713, 3714)
Hsu D A, **7**: 5018
Hsu P H, **1**: 268 (262, 263, 266, 267)
Hu S S, **2**: 1181
Hu T C, **2**: 850 (845)
Huang H, **1**: 80 (77, 79); **6**: 3997 (3996)
Huang L K, **4**: 2511 (2510)
Huang P M, **1**: 7 (1, 2, 3, 4, 5, 6, 7), 148 (143), 268 (263, 264, 265, 266, 267); **4**: 2782 (2778, 2780)
Huang T S, **6**: 4160 (4158)
Hubbard M, **2**: 1329 (1327, 1328)
Hubbert M K, **2**: 1454 (1452); **3**: 1718 (1714)
Hubel D H, **4**: 2331 (2325, 2327)
Huber G, **5**: 2944
Huber G P, **3**: 2061 (2060), 2192 (2186), 2202 (2195); **4**: 2501 (2501); **5**: 3115 (3112); **7**: 4785, 4793 (4792)
Huber M J, **7**: 4927 (4925)
Huber O, **3**: 2192
Huber P J, **5**: 3554 (3552)
Hubka D E, **5**: 2980
Huch A, **1**: 440
Huch R, **1**: 440
Hudson N, **7**: 5060 (5059)
Huebler J E, **3**: 1958
Hueckel M H, **4**: 2386 (2382)
Huelsman L P, **1**: 272 (269, 270)
Huet J D, **6**: 4336 (4335)
Huff D D, **2**: 1272 (1272); **4**: 2646 (2645)
Huff R L, **3**: 1623 (1619)
Hughes F M, **5**: 3430 (3426, 3430), 3436 (3434); **6**: 3727 (3725), 4176 (4174), 4189 (4187, 4188, 4189)
Hughes H W D, **3**: 1998
Hughes J, **3**: 1913, 1985
Hughes J R, **2**: 1413 (1411)
Hughes J S, **3**: 1887
Hughes P H, **1**: 751 (750)
Hughes R, **1**: 450 (446, 447); **5**: 3470 (3467, 3468)

Hughes T P, **1**: 343; **6**: 4274 (4272)
Hugill A L, **2**: 1403 (1403)
Huglen R, **2**: 1416 (1414, 1416)
Huguenin F, **6**: 4204
Hulbert D G, **4**: 2604 (2601); **5**: 2993 (2990)
Hulburt H M, **5**: 3082 (3079)
Hulko G, **2**: 1192 (1191)
Hull C J, **2**: 963 (962)
Hull L G, **6**: 4370 (4368)
Hullett W, **6**: 4364 (4363)
Hultquist P F, **3**: 2006 (2006)
Humby E, **4**: 2517 (2516)
Hume D, **1**: (552); **3**: (1607); **6**: (4166)
Hume S H, **5**: 2944 (2934)
Hummel R, **4**: 2387 (2382), 2484 (2482); **6**: 4160 (4157)
Humphrey A E, **3**: 1614 (1614), 1619 (1616)
Humphrey T L, **7**: 4930 (4928)
Humphreys M A, **1**: 152, 606 (605); **3**: 2017 (2017); **7**: 4866 (4866)
Humphreys W F, **7**: 4454 (4453)
Hunder G G, **2**: 1248 (1248)
Hung Y S, **5**: 3228 (3227)
Hunk J C, **7**: 4876 (4873)
Hunt D R G, **4**: 2769 (2765, 2766, 2767)
Hunt H A, **6**: 4097
Hunt H W, **1**: 142 (141); **6**: 4235 (4230, 4231)
Hunt J J, **3**: 2052 (2050)
Hunt K H, **6**: 4108 (4101, 4105, 4106)
Hunt L R, **4**: 2898 (2897)
Hunt S, **2**: 1424 (1423), 1440 (1440)
Hunt T L, **6**: 4097
Hunt V R, **6**: 3917 (3911, 3915)
Hunter J S, **6**: 4235 (4232); **7**: 5015 (5013, 5014), 5018
Hunter P, **3**: 1958 (1957)
Hunter W G, **6**: 4229 (4228), 4235 (4232); **7**: 5015 (5013, 5014)
Hunter W J, **6**: 4229 (4228)
Huntingdon R, **7**: 5099, 5106 (5102)
Huovinen N, **6**: 3916 (3914)
Hurd R G, **1**: 142 (140)
Hurewicz W, **6**: 4146 (4145)
Hurley E L, **2**: 866 (865)
Hurley J, **2**: 867 (863)
Hurrion R D, **6**: 4302 (4300)
Hurteau R, **2**: 923 (922)
Hurty W C, **5**: 3390 (3386)
Hurwich L M, **7**: 5078
Hurwicz L, **2**: 1261 (1257, 1258, 1260); **5**: 2911 (2908, 2909); **7**: 4822 (4818)
Hurwitz A, **3**: (1745); **7**: (4517)
Husar J D, **1**: 176 (175)
Husar R B, **1**: 176 (175)
Husband R K, **5**: 3379 (3377, 3378)
Hüseyin Ö, **2**: 923 (921); **4**: 2559 (2559), 2560 (2559); **7**: 4685 (4683)
Husic F, **7**: 4833 (4828)
Husson R, **5**: 3097
Hutchings H J, **5**: 3470 (3469)
Hutchins L A, **5**: 2980
Hutchinson D, **2**: 1132 (1128)
Hutchinson E A, **3**: 1975
Hutchinson G B, **7**: 4948
Hutchinson G E, **2**: 873; **7**: 4956 (4954)
Hutchinson J, **3**: 1700
Hutchison C, **4**: 2713 (2710)
Hutchisson J T, **1**: 312 (310)
Hutt S H, **4**: 2356 (2355)
Hutton M F, **1**: 132; **5**: 3077 (3076); **7**: 4429 (4428)
Hutzinger O, **1**: 7 (2, 5, 6, 7), 148 (143, 144); **3**: 1939
Hutzler V P, **5**: 3550
Huxley A F, **1**: 508 (506); **7**: 4980 (4977)
Huxley M W, **1**: 142 (141)
Huysman J H, **7**: 4793 (4791)
Hwang C L, **3**: 2143 (2143); **5**: 3156 (3152); **7**: 4877, 4880 (4880)
Hwang J, **1**: 651 (650)
Hyafil L, **1**: 725 (722)
Hyde J E, **2**: 1596 (1593)
Hyland D C, **1**: 132 (132)
Hylland A, **3**: 2061 (2060)
Hyman J M, **2**: 1181 (1181)
Hymans S H, **2**: 1342
Hyndman B W, **1**: 542 (539)

I

Iannino A, **4**: 2681 (2679)
Ibaraki T, **2**: 1489 (1489)
Ibarra O H, **1**: 628 (627)
Iberall A S, **3**: 2175 (2171)
Ibragimov I A, **1**: 86 (82, 84)
Ibrahim T A S, **1**: 735 (734); **4**: 2597 (2595)
Ichbiah J B, **3**: 1642 (1638); **6**: 3898 (3896)
Ichbiah J D, **2**: 1143
Ichikawa A, **3**: 2151, 2152; **5**: 3506 (3502, 3505); **6**: 4245 (4238, 4239, 4240, 4242, 4243)
Ichikawa K, **1**: 511
Ider Y Z, **3**: 2025 (2023), 2174 (2171)
Ieki M, **5**: 2943 (2935)
Iggo A, **1**: 466 (463)
Iglehart D L, **6**: 4395 (4395); **7**: 4654 (4653)
Ignall E, **4**: 2353 (2351)
Ignizio J P, **5**: 3121 (3121), 3124 (3122), 3164 (3157)
Ihalainen H, **3**: 1700
Iivarinen T, **5**: 2994 (2991)
Ike E E, **4**: 2331 (2326)
Ikebe Y, **2**: 827 (815, 818, 824); **7**: 4518 (4518)
Ikeda E, **2**: 1434 (1431)
Ikeda M, **1**: 675 (673, 674); **2**: 923 (922); **4**: 2559 (2558, 2559), 2702 (2702); **5**: 3572 (3569, 3570, 3571); **6**: 4011 (4008, 4010), 4161 (4157)
Ikeda N, **7**: 4627 (4623), 4637 (4631)
Ikeland I, **6**: 4070
Ikeuchi K, **1**: 759 (759); **4**: 2487
Ilin V A, **7**: 4490
Imai H, **5**: 3130 (3128)
Iman R L, **2**: 1551 (1550, 1551); **6**: 4235 (4230)
Imberger J, **4**: 2789 (2788)
Imbert J F, **3**: 1665
Imbert N, **1**: 196 (193, 194); **3**: 1756 (1753)
Imboden D M, **4**: 2646 (2643), 2794 (2789)
Inaba H, **4**: 2723 (2722); **5**: 3464 (3463); **6**: 4036 (4036)
Inagaki T, **3**: 1611 (1609)
Inamura M, **6**: 4033
Inasaka F, **1**: 713 (710, 712, 713); **3**: 1902 (1902), 1905 (1903); **5**: 2927 (2925)
Indyke B H, **3**: 1601 (1600); **7**: 4897 (4896)
Ingham A E, **4**: (2804)
Inglis Y, **1**: 437
Ingold W, **3**: 1614 (1612)
Ingram D, **2**: 1366 (1362, 1364, 1365)
Inkielman M, **2**: 1276 (1272, 1276)
Innis G S, **1**: 142 (141); **2**: 1326 (1324), 1332 (1330, 1332), 1348 (1345, 1346); **3**: 2045 (2045); **6**: 4235 (4230, 4231); **7**: 5008 (5005)
Innorta M, **2**: 1388 (1386); **6**: 3954 (3952)
Ino T, **2**: 1521 (1518)
Inooka H, **5**: 3215 (3208, 3213)
Inoue H, **4**: 2386 (2380); **6**: 4117
Inoue Y, **7**: 4593 (4588)
Insel P A, **3**: 2174 (2171)
Installe M J, **6**: 4244 (4238)
Intriligator M D, **1**: 671 (668, 669)
Invzuca I, **2**: 1434 (1431)
Ioannou P, **1**: 105 (105)
Ioffe A D, **1**: 529 (527); **2**: 1033
Ioh H, **4**: 2723 (2722)
Ion D C, **2**: 1454 (1452), 1458; **3**: 1718
Ionescu M I, **1**: 315 (313)
Ionescu T, **2**: 1093 (1089); **5**: 3055 (3048, 3051)
Iosifescu M, **7**: 4721 (4717)
Ipsen I C F, **2**: 826 (818)
Ireson W G, **7**: 4726 (4724)
Iri M, **1**: 626 (624)
Irvine T F, **2**: 1253 (1249)
Irving D C, **2**: 1551 (1550)
Irving E, **6**: 3983 (3978)
Irving M, **4**: 2827 (2823)
Irving M R, **6**: 3829 (3826, 3827)
Isaacs R, **1**: 687 (682); **2**: 1010 (1009), 1017 (1013); **7**: 4856 (4853, 4854)
Isaacson D L, **6**: 4356 (4356)
Isacson A, **1**: 633 (632), 634 (632)
Isaksen L, **5**: 3572 (3568)
Isard J O, **6**: 3687 (3686)
Isard W, **2**: 776 (775)
Isermann M, **5**: 3164 (3160, 3161)
Isermann R, **1**: 47, 63 (60), 72 (64, 68), 638 (635, 638); **2**: 879 (879), 1116 (1110, 1111); **3**: 1990 (1987, 1989), 2097 (2094, 2095); **4**: 2253 (2253), 2257 (2254, 2255), 2263 (2263), 2325 (2324); **5**: 3188 (3188), 3351 (3341, 3343, 3344, 3345, 3346), 3475 (3473); **6**: 3610 (3604, 3605, 3607, 3608, 3610), 3611 (3610), 3703 (3702), 4183 (4180), 4184 (4181, 4182), 4204; **7**: 4692 (4690)
Ishida T, **6**: 4108 (4104)
Ishigai S, **3**: 1766 (1761, 1762, 1763)
Ishihara H, **2**: 1497 (1494, 1497)
Ishii M, **6**: 3766 (3762)
Ishikawa A, **1**: 511 (508, 509, 510)
Ishikawa K, **7**: 4881 (4880)
Ishikawa T, **3**: 1614 (1614)
Ishikawa Y, **3**: 1614 (1614)
Ishizuka M, **3**: 1802 (1801); **4**: 2883 (2881); **6**: 3772 (3770); **7**: 4669
Isidori A, **2**: 827 (823), 1121 (1116, 1119); **4**: 2527 (2525, 2526, 2527), 2756 (2756), 2816 (2807); **5**: 3279 (3276), 3288 (3285, 3286, 3287), 3293 (3289), 3341 (3338, 3339), 3366 (3366); **6**: 3961 (3959), 3977 (3973)
Isner A, **2**: 1568 (1568)
Isobe T, **7**: 4593 (4588)
Iswandhi G I, **6**: 4128 (4124)
Itkis U, **7**: 5031 (5027)
Ito H, **3**: 1936 (1934)
Ito K, **3**: 1766 (1765); **5**: 3336 (3330, 3332), 3413 (3412)
Ito M, **7**: 4463 (4460)
Itoh K, **2**: 1154

Itty C, **7**: 4856 (4856)
Iung C, **5**: 3097
Ivakhnenko A G, **2**: 1497 (1494)
Ivanoff A, **6**: 3884 (3883)
Ivanov H, **7**: 4490
Ivarson W R, **4**: 2646 (2645)
Ivashkin V V, **7**: 4490
Ives B, **3**: 2192 (2186)
Ives D J G, **6**: 3688 (3683)
Iveson R H, **6**: 3954 (3952)
Iwai S, **1**: 276 (272, 276); **4**: 2640 (2639)
Iwaki H, **1**: 142 (140)
Iyengar R D, **5**: 2944 (2935)
Izatt S, **1**: 177 (175)
Izawa Y, **6**: 4036 (4035)
Izumi T, **1**: 421

J

Jablonowski J, **6**: 4123 (4120)
Jabr H A, **4**: 2570 (2568)
Jackson C, **2**: 1431 (1428)
Jackson D A, **5**: 3470 (3469)
Jackson D C, **7**: 4876 (4875), 4880 (4879)
Jackson D F, **7**: 4871 (4869)
Jackson K, **6**: 3898 (3895)
Jackson K M, **3**: 1910 (1907, 1908)
Jackson L, **1**: 769 (761)
Jackson M L, **1**: 7 (1, 2), 268 (262)
Jackson R L, **4**: 2887; **6**: 3806 (3804)
Jackson T J, **3**: 2234 (2233)
Jacob F, **1**: 429 (425), 433 (432), 559 (558)
Jacob G, **3**: 1779 (1772); **5**: 3272 (3268)
Jacobi G T, **2**: 869 (868); **6**: 4165 (4161)
Jacobi K G J, **2**: 1033 (1031); **7**: (5045)
Jacobs A, **3**: 1623 (1619)
Jacobs D A H, **4**: 2880 (2879); **7**: 4975 (4973)
Jacobs I M, **2**: 1051 (1048, 1051)
Jacobs M, **7**: 4833 (4830, 4831)
Jacobs O L R, **2**: 963 (961, 962, 963); **5**: 3303 (3301), 3329 (3323), 3330, 3336 (3336), 3475 (3475); **6**: 3690 (3690), 3691 (3690); **7**: 4822 (4817)
Jacobsen S C, **1**: 312 (311); **6**: 4108 (4104)
Jacobson D H, **2**: 1257 (1256); **5**: 3508 (3508); **7**: 4425 (4425), 4661
Jacobson M, **3**: 1761
Jacobson R D, **2**: 1226 (1225)
Jacoby S L S, **4**: 2412 (2393, 2397, 2399); **5**: 3082 (3079)
Jacod J, **7**: 4637 (4631, 4632)
Jacowitz L A, **7**: 4481
Jacquard A, **1**: 675 (674)
Jacquart R, **1**: 120
Jacquez J A, **1**: 417 (414), 493; **2**: 1329 (1327), 1344 (1343); **3**: 2143 (2140); **4**: 2635 (2632)
Jacquot P, **5**: 2942 (2935)
Jaeger J C, **5**: 3099; **6**: 4364 (4364)
Jaffe A, **5**: 3329 (3320)
Jaffe S, **4**: 2832 (2829, 2831)
Jagannathan E V, **7**: 5140 (5138)
Jähn B, **2**: 1586 (1584)
Jahn J, **5**: 3164 (3161)
Jahn K U, **5**: (2918)
Jahoda M, **7**: 4762 (4756)
Jain R, **1**: 769 (766); **2**: 1407 (1405, 1407)
Jakeman A, **4**: 2277 (2276), 2293 (2290)
Jaklevic J M, **1**: 177 (174)
Jakob W, **4**: 2351
Jakobsen R J, **4**: 2511 (2510)
Jakubczyk B, **4**: 2756 (2756); **5**: 3293 (3291), 3394 (3390); **6**: 3977 (3973, 3974)
Jakubik A T, **6**: 3933 (3932)
Jakubovič V A, **2**: 1020 (1020)
Jakubowski R, **6**: 3644, 3653
James A T, **5**: 2943 (2934)
James D J G, **5**: 3070 (3068, 3069)
James H M, **3**: 1749 (1748, 1749); **5**: (3388); **6**: 4146 (4145)
James K A, **5**: 3464 (3460, 3461)
Jameson A, **5**: 3139 (3139)
Jameson D A, **3**: 2045 (2044)
Jämsä S-L, **2**: 781 (781); **5**: 2994 (2991)
Jamshidi M, **4**: 2559 (2559)
Janak J F, **7**: 4645 (4645)
Janex A, **1**: 682 (678)
Janin R, **4**: 2609 (2608)
Janini G M, **1**: 177 (176)
Janis I L, **2**: 987
Janssen J E, **2**: 1226 (1214, 1217, 1223, 1225)
Jansson B, **7**: 4958 (4957)
Jantsch E, **1**: 333 (331), 399, 400, 709 (705, 706, 709); **2**: 1348; **6**: 4167 (4167)
Janz G J, **6**: 3688 (3683)
Jaquette D L, **2**: 1337 (1335, 1336)
Jaquette S C, **2**: 1551 (1550)
Jarny Y, **6**: 4311 (4311)
Jaroń J, **1**: 93 (92)
Jarrit P H, **6**: 3940 (3940)
Jarvis J F, **6**: 3644, 3653
Jarvis J J, **4**: 2692, 2841
Jarvis R A, **1**: 347 (343); **4**: 2487
Jarzynski J, **5**: 3470 (3467, 3468, 3469)
Jasper G W, **2**: 1214 (1213); **3**: 1965
Jategoankar R V, **4**: 2292 (2287)
Javdan M R, **2**: 956 (952)
Jávor A, **6**: 4322
Jawari S, **1**: 682 (681)
Jayant N S, **1**: 80 (78); **2**: 1051 (1046)
Jayawant B V, **7**: 4452 (4451)
Jaynes E T, **2**: 1479 (1474); **4**: 2270 (2269), 2508 (2507); **6**: 3989 (3988)
Jazwinski A H, **4**: 2622 (2619, 2620), 2627 (2623); **5**: 3475 (3472)
Jeanneau J L, **1**: 60 (59)
Jedyak L, **1**: 633 (632)
Jeeves T A, **6**: 4310 (4308)
Jeffers J N R, **1**: 552 (550); **6**: 3933 (3933)
Jeffries C, **2**: 1080 (1079, 1080), 1351 (1348, 1350)
Jelalian A V, **6**: 4036 (4034)
Jenaleev A K, **1**: 37
Jenkins G M, **1**: 588 (588); **2**: 1492, 1494, 1497 (1494); **4**: 2245 (2242), 2270 (2266, 2268), 2307 (2304), 2310 (2308, 2309); **6**: 3949 (3949); **7**: 4566 (4562, 4565, 4566), 5001 (4998), 5155 (5154)
Jenkins G N, **2**: 965
Jenkins J D, **7**: 5150
Jenkins J M, **1**: 769 (760)
Jenkins K T, **2**: 1445 (1445)
Jenkyns H A, **1**: 626 (625)
Jenner R, **2**: 1410

Jennergren L P, **2**: 995 (995), 1332 (1330); **5**: 2911 (2908)
Jennett B, **4**: 2582 (2581)
Jensen F, **1**: 333 (326); **6**: 4015 (4012)
Jensen J I, **6**: 4061 (4057)
Jensen K, **6**: 3670 (3669), 4289 (4288), 4350 (4349)
Jensen S, **4**: 2782 (2781)
Jenssen N A, **6**: 4266 (4264)
Jeppson R W, **7**: 5106 (5103), 5153 (5151)
Jerlov N G, **6**: 4045
Jernelöv A, **4**: 2782 (2780, 2781)
Jervis M W, **7**: 4577 (4574)
Jessen P F, **3**: 1969 (1968)
Jette C L, **6**: 4302 (4301)
Jewkes J, **7**: 4833 (4830)
Jex H R, **3**: 1668 (1666, 1667)
Jilani M A, **4**: 2386 (2380)
Jobe W E, **7**: 4856 (4856)
Johannsen G, **3**: 1669 (1665, 1666, 1667), 2186 (2180)
Johansson B, **1**: 542 (540)
Johansson T B, **1**: 177 (174)
John E R, **1**: 505 (502); **5**: 3264 (3263)
John S C, **2**: 1106
Johns H E, **7**: 5186 (5179, 5180, 5182, 5185)
Johnson A E, **2**: 1226 (1223, 1225)
Johnson C C, **4**: 2511 (2510); **5**: 3464 (3463)
Johnson C R Jr, **1**: 51 (48), 81 (78, 79); **5**: 3055 (3048)
Johnson D E, **6**: 4284
Johnson D S, **1**: 725 (723, 724); **2**: 884 (882), 1083 (1082, 1083)
Johnson E C, **5**: 3389 (3386); **7**: 4452 (4450), 4833 (4828, 4829, 4830, 4831)
Johnson E F, **5**: 3292 (3289)
Johnson E M, **3**: 2202 (2195); **5**: 3115 (3112)
Johnson G R, **6**: 4108 (4106)
Johnson J, **6**: 3904 (3904)
Johnson J H, **2**: 1402 (1399)
Johnson M A, **1**: 45 (38)
Johnson M E, **6**: 3950
Johnson P C, **1**: 477 (477)
Johnson R A, **1**: 86 (82); **3**: 1669 (1666)
Johnson R C, **2**: 1388 (1387); **6**: 3806 (3805)
Johnson R F, **7**: 5008 (5005, 5006)
Johnson R L, **5**: 3470 (3468)
Johnson R M, **6**: 4415 (4414)
Johnson R N, **2**: 1590 (1587, 1589)
Johnson R W, **6**: 3990 (3988)
Johnson S A, **1**: 176 (175)
Johnson S M, **6**: 3739 (3736)
Johnson T L, **7**: 4685 (4678), 4689 (4685)
Johnson W C, **7**: 4645 (4645)
Johnson W L, **5**: 3272 (3268)
Johnston F D, **1**: 535 (532), 536 (532)
Johnston R L, **2**: 826 (815)
Joly G, **5**: 3496 (3495, 3496); **6**: 3789
Joly J, **4**: 2343
Jonckheere E A, **1**: 132 (132)
Jones A J, **1**: 687 (682, 683)
Jones A K, **2**: 1132 (1129)
Jones B, **2**: 1441 (1436); **6**: 3989 (3984, 3985)
Jones B E, **2**: 1403 (1403); **4**: 2533 (2528, 2529); **7**: 4513
Jones B L, **3**: 1943
Jones C E, **2**: 1366 (1361)
Jones C H, **7**: 4871 (4869, 4870)
Jones D, **5**: 2943 (2935)
Jones D D, **5**: 3035 (3032)
Jones D W, **5**: 3421 (3417, 3418, 3419)
Jones E A, **6**: 3909 (3907)
Jones G A, **3**: 1979
Jones G E, **4**: 2356 (2354)
Jones H, **7**: 4827
Jones H L, **1**: 86 (82, 84), 251 (248), 252 (249)
Jones J, **2**: 1272 (1272)
Jones J C, **2**: 992 (988, 991)
Jones N B, **1**: 482 (480); **5**: 3238 (3235), 3248
Jones N D, **4**: 2798 (2797)
Jones R E, **3**: 1694 (1692)
Jones R P, **4**: 2253 (2252); **6**: 3698 (3696)
Jones R W, **4**: 2901 (2898)
Jones S, **4**: 2640 (2637, 2638); **7**: 4785 (4783)
Jones S K, **4**: 2800
Jones W E, **5**: 2943 (2935)
Jones W P, **1**: 150, 171, 174; **3**: 2209
Jongkees L B W, **2**: 1596 (1592)
Jongkind W, **6**: 4148 (4147)
Jonkman J H G, **2**: 1248 (1247)
Jonsson B, **1**: 542 (540), 546 (544)
Joos D, **5**: 3012
Joos G, **5**: 3097
Jordan C F, **2**: 1344 (1343)
Jordan J R, **3**: 1693 (1690)
Jordan K E, **7**: 4434 (4430)
Jordan M M, **4**: 2901 (2899)
Jordan R L, **6**: 4041 (4039)
Jordan W S, **1**: 289 (284)
Jorgansen S B, **4**: 2263 (2263)
Jorgens K, **7**: 4944 (4942)
Jorgensen S, **2**: 1269 (1263, 1266)
Jørgensen S E, **4**: 2646 (2645); **7**: 5008 (5006, 5007)
Jörgl H P, **2**: 1291, 1297
Jorre D, **5**: 2943 (2935)
Joseph J, **3**: 1910 (1908)
Joseph P D, **6**: 3717 (3713)
Joseph P E, **2**: 894 (894)
Joshi A K, **6**: 3653 (3648)
Joss J, **5**: 3559 (3557)
Joubert M, **2**: 1577 (1575)
Jouron C, **2**: 1372 (1371)
Jouve M, **2**: 1165
Jover J M, **7**: 4540 (4537)
Joy D C, **2**: 1420; **6**: 4152
Joy D S, **7**: 5094 (5093)
Joy M, **7**: 4901 (4901)
Joyce J W, **2**: 1248 (1248)
Joyner R W, **7**: 4980 (4977, 4979)
Joynson R, **6**: 3644, 3653
Judd R P, **6**: 4088 (4087)
Judge A W, **2**: 1466 (1463)
Jufer M, **7**: 4598 (4595, 4598)

Juillerat R, **2**: 1244 (1243)
Julesz B, **7**: 5078 (5075)
Julien J M, **2**: 837
Jumarie G, **7**: 4698 (4698)
Jung A, **3**: 2174 (2170, 2171)
Jung C G, **7**: (4790)
Jung L, **4**: 2292 (2287, 2291)
Junge C E, **1**: 179 (179)
Jungers P, **1**: 310 (308)
Jurdjevic V, **1**: 124 (122, 124); **2**: 1583 (1581); **4**: 2756 (2755), 2759 (2756, 2757, 2758); **5**: 2948 (2948), 2949 (2948), 3288 (3285), 3366 (3365, 3366); **6**: 3956 (3955); **7**: 4425 (4420)
Jurevich E I, **6**: 4123
Jürgensen H, **4**: 2798 (2797)
Jurvillier J C, **3**: 1736
Jury E I, **2**: 1078 (1074); **5**: 3055 (3051), 3127 (3125, 3127), 3135 (3132, 3133), 3319 (3319), 3559 (3558); **6**: 4146 (4145), 4289, 4359 (4358); **7**: 5192 (5191)
Jusko W J, **6**: 3698 (3698)
Justice J H, **4**: 2508 (2507)
Jutard A, **6**: 4257
Jutila E A A, **5**: 2994 (2992); **6**: 3889 (3885, 3888), 3916 (3911, 3912, 3913, 3914)
Jutila J, **5**: 2943 (2934)
Jutila P, **6**: 3682 (3681)
Jutze G A, **1**: 177 (175)

K

Kacewicz B Z, **4**: 2257 (2256)
Kachel V, **2**: 867 (864, 865)
Kacprzyk J, **5**: 3454 (3454), 3540 (3539); **7**: 4661 (4658, 4659, 4660)
Kaczmarek L K, **1**: 570 (569)
Kaczorek T, **6**: 3972 (3972)
Kadish A H, **3**: 2025 (2022)
Kadoi H, **3**: 1766 (1763)
Kaesser R, **6**: 3682 (3681)
Kafkafi U, **1**: 148 (143)
Kafker A H, **7**: 4726 (4723), 4740 (4735), 4751 (4751)
Kagami I, **1**: 424 (423)
Kaggerud I, **5**: 2993 (2991); **6**: 4252 (4250)
Kagiwada H H, **2**: 1311 (1307, 1309); **3**: 2052; **7**: 4895 (4895)
Kågström B, **2**: 826 (821)
Kahan B C, **7**: 4935 (4933)
Kahan G, **3**: 1705 (1701)
Kahan W, **2**: 826 (819, 820)
Kahn C D, **1**: 588 (588)
Kahn C R, **3**: 2174 (2172, 2173)
Kahn D A, **5**: 3389 (3385), 3464 (3459)
Kahn H, **6**: 4155; **7**: 4657 (4655)
Kahn K C, **5**: 2980
Kahn M E, **4**: 2386 (2376), 2412 (2393)
Kahneman D, **3**: 2192 (2189), 2202 (2196), 2203 (2197, 2198, 2199); **4**: 2338 (2337); **5**: 3104 (3103, 3104), 3115 (3113)
Kahr A, **2**: 942 (938, 939)
Kaihara S, **2**: 1577 (1574)
Kaikin C E, **5**: (3384)
Kailath T, **1**: 495 (494), 714 (714); **2**: 785 (782, 783, 784), 1102 (1102); **4**: 2618 (2615, 2617), 2628 (2627), 2671 (2669), 2832 (2830), 2866 (2865), 2871 (2869, 2871); **5**: 3220 (3215, 3216, 3217, 3218), 3303 (3293, 3301), (3319); **6**: 3751 (3751); **7**: 4457 (4456), 4466 (4463, 4465), 4469 (4466, 4467, 4468), 4540 (4537), 4551 (4548), 4554 (4552, 4553)
Kaiser K W, **6**: 4245 (4242)
Kak A C, **4**: 2473 (2472); **6**: 3654, 4160 (4157)
Kakutani S, **1**: 671 (667)
Kalaba R, **2**: 1311 (1307, 1308, 1309); **3**: 1825, 2052 (2048); **4**: 2664; **5**: 3215 (3206); **7**: 4529 (4527)
Kalaba R E, **4**: 2264 (2263)
Kalai E, **5**: (2910)
Kalenich W A, **6**: 4322 (4319)
Kaleva O, **7**: 4766 (4765)
Kaley G, **1**: 477 (475, 477)
Kallala M, **3**: 1809 (1809); **6**: 4007 (4007)
Kallianpur G, **5**: 3303 (3293, 3294, 3297, 3298, 3299), 3336 (3332, 3333, 3334, 3335); **7**: 4637 (4631)
Kallio M J, **2**: 959
Kallman R F, **1**: 417 (414, 415); **2**: 1337 (1336)
Kallström C G, **5**: 3351 (3341)
Kalman R E, **1**: 18, 124 (122), 294; **2**: 912 (908), 1099 (1097), 1286, 1332 (1332); **3**: 1862 (1859), 2113 (2112); **4**: 2273, 2283 (2281), 2301 (2293, 2301), 2517 (2514), 2559 (2557), 2565 (2563), 2628 (2624), 2827 (2816, 2826), 2866 (2862, 2863, 2864, 2865), 2871 (2870, 2871); **5**: (2948, 3229), 3272 (3268), 3293 (3289), 3303 (3293, 3300), 3319 (3319), (3323, 3374), 3389 (3385), 3394 (3390), 3475 (3473); **6**: 3629 (3627), 3720 (3718), 3734 (3733), 3751 (3748, 3749), 3776, 3968, 3972 (3970), 3977 (3974), 3983 (3978), 4070, 4146 (4146), 4186 (4185), 4193 (4190), 4204; **7**: 4529 (4528), 4554 (4552, 4553), 4685 (4684)
Kalmanson D, **2**: 1407 (1405)
Kalocsai R, **1**: 687 (685)
Kaltenbach J C, **6**: 3789
Kaltenecker H, **1**: 638 (638); **2**: 879 (879), 1116 (1110, 1111); **3**: 1990 (1987, 1989), 2097 (2094, 2095); **5**: 3188 (3188)
Kalton N J, **2**: 1008 (1006), 1017 (1016)
Kam W Y, **5**: 3188
Kamburowsky J, **5**: 3454 (3454)
Kameda T, **3**: 1887
Kamen E W, **5**: 3130 (3129), 3135 (3133)
Kaminski P G, **4**: 2618 (2618); **7**: 4540 (4535, 4536, 4539)
Kamiyama N, **3**: 1614 (1614)
Kamke E, **2**: 1351 (1350); **4**: 2565 (2560); **5**: 3560 (3559)
Kamp Y, **5**: 3134 (3132, 3134), 3135 (3133, 3134)
Kamp-Neilson L, **4**: 2646 (2645)
Kampé de Fériet J, **6**: 3772 (3770)
Kampschulte B, **7**: 5027
Kanade T, **4**: 2440 (2431, 2432, 2439)
Kanal L, **4**: 2635 (2635)
Kanal L N, **6**: 3644 (3635, 3637, 3641, 3643), 4160
Kandel A, **1**: 664; **2**: 1473 (1473); **3**: 1809 (1809), 1810 (1807), 1822 (1821), 1825 (1824), 1838, 1845 (1845), 1852 (1848), 1872 (1868), 1883 (1880), 1887, 1888; **5**: (2918), 3264; **6**: 4007 (4007); **7**: 4766
Kandel E R, **7**: 5065 (5061, 5065)
Kane J, **7**: 5177 (5171, 5172)
Kaneko A, **2**: 1154 (1153)
Kanellakis P C, **2**: 884 (881, 883), 1154, 1155
Kang G, **4**: 2351 (2348)
Kania A A, **3**: 1905 (1903, 1904, 1905)
Kanjilal P P, **6**: 4172 (4171)
Kannan R, **2**: 981 (980)
Kant I, **2**: 1486 (1481); **6**: (4166)
Kantrow A M, **7**: 4832 (4832), 4833 (4832)
Kao F F, **6**: 4061 (4056)
Kao I, **1**: 424 (423)
Kaplan D, **3**: 2017
Kaplan J L, **7**: 4439 (4437)
Kaplin M F, **3**: 2203
Kapur K C, **6**: 4015
Karaa F, **7**: 5140 (5139)
Karasek F W, **3**: 1939; **7**: 5129 (5127)

Karatzas I, 5: 3523 (3523)
Karcanias N, 1: 580 (578); 4: 2832 (2828, 2829, 2830, 2831), 2871 (2867, 2868, 2869, 2870); 6: 3751 (3748); 7: 4457 (4455, 4456)
Karhunen K, 2: 1479 (1477)
Karjalainen T, 6: 4204 (4200)
Karlin S, 2: 1262 (1259); 5: 3115 (3105)
Karmarkar J S, 5: 3196 (3189)
Karmarkar N, 4: 2692 (2686); 6: 3739 (3736)
Karmarkar U S, 3: 2203 (2197, 2198); 5: 3104 (3104)
Karnopp D C, 2: 1329 (1327, 1328); 6: 4341 (4337, 4341)
Karotkin E H, 1: 319
Karp R M, 1: 628 (628), 725 (725); 2: 884 (882)
Karp S, 1: 201
Karplus W J, 1: 424 (422), 634 (633); 6: 3626 (3622), 4306 (4303), 4319 (4316), 4411 (4406); 7: 4692 (4691)
Karrakchou J, 3: 2151 (2149); 5: 3506
Kartaschoff P, 3: 1745
Karttunen P, 2: 781 (779)
Kasemo B, 5: 2943 (2935)
Kashani H R, 7: 4910 (4908, 4909, 4910)
Kashef B, 2: 1211
Kashiwagi H, 5: 2956 (2953, 2954)
Kashyap R L, 4: 2307 (2302, 2303, 2307); 6: 4160 (4158); 7: 4692 (4691), 5001 (4998, 4999)
Kasprzak A, 6: 3644, 3653
Kassin S M, 3: 2203 (2196)
Kassirer J P, 2: 1577 (1574)
Kassovski N N, 2: 1023 (1022)
Kastner M P, 2: 932 (930); 7: 4822 (4818, 4820)
Kastning C, 2: 1083 (1080)
Kasuda T, 3: 2105 (2101)
Katebi M R, 4: 2706
Katebi S D, 5: 3383 (3382)
Kato I, 6: 4108 (4104)
Kato Y, 6: 4223 (4217, 4219, 4220, 4222)
Katoh M, 5: 3464 (3461)
Katoh N, 2: 1489 (1489)
Katz A M, 1: 456 (452, 453)
Katz B, 1: 456 (451)
Katz D F, 2: 1179 (1177)
Katz D L, 3: 1975
Katz J L, 1: 474 (471); 3: 1669 (1666, 1668)
Katz M, 7: 4996 (4995)
Katz S S, 7: 4732 (4731)
Katz W E, 7: 5117
Katzenbeisser R, 6: 4211 (4211)
Katzenelson J, 1: 580 (577); 5: 3336 (3330)
Katzman L, 2: 1226 (1217, 1225)
Katzner D W, 2: 1342 (1338)
Katzur R, 5: 3310 (3305, 3306)
Kauffmann J M, 6: 4108 (4107)
Kaufman A, 3: 1835; 6: 4302 (4301)
Kaufman B, 7: 5186 (5182)
Kaufman D D, 3: 2147
Kaufman H, 6: 3870 (3868); 7: 4833 (4829, 4830)
Kaufman L, 5: 3421 (3420)
Kaufman R, 2: 1462
Kaufman R D, 6: 4061 (4057, 4060), 4062 (4061)
Kaufman Y J, 3: 2052 (2050)
Kaufmann A, 1: 276 (275), 664 (651); 3: 1831 (1831), 1835, 1847 (1846), 1858, 1862 (1859), 1872 (1867), 1874; 5: 2965 (2964, 2965); 6: 3857 (3856), 3865 (3861), 4029 (4029); 7: 4766 (4764, 4765)
Kaufmann L, 5: 3425 (3425)
Kaunitz J, 4: 2314 (2311)
Kauppinen S, 6: 3916 (3912, 3913)
Kavanagh R, 3: 1719 (1715)
Kavanagh R J, 7: 4452 (4451)
Kavrakoglu I, 3: 1718 (1716)
Kawahara M, 1: 301 (297)
Kawamori R, 3: 2021 (2017)
Kawamura H, 2: 1522 (1518)
Kawashima Y, 7: 4881 (4880)
Kawata Y, 2: 1311 (1307); 3: 2052 (2050), 2053 (2048, 2049); 4: 2655 (2652)
Kawazu S, 2: 1154
Kay D, 7: 4940
Kay P, 4: 2685 (2683, 2685)
Kay R A S, 4: 2775 (2773)
Kay R N, 4: 2685 (2685)
Kaye S V, 6: 3933 (3931), 3936
Kayton M, 1: 201
Kazakov I Y, 5: 3389 (3385)
Kazda L F, 5: (3516)
Kazmierczak H, 4: 2473
Keane J M, 2: 781 (780)
Keane M, 4: 2508 (2506)
Kearney P C, 3: 2147
Keatinge W R, 7: 4856 (4853)
Kechris A S, 3: 1926 (1926)
Kedem B, 1: 86 (82)
Kedem Z, 1: 347 (345); 2: 884 (882), 1155
Keegan H J, 2: 1521 (1517)
Keeling C D, 1: 531 (529, 531)
Keen P G W, 2: 947 (944, 947), 987; 4: 2681 (2680); 7: 4785 (4783), 4793 (4792)
Keenan C, 5: 2942 (2935)
Keenan R, 4: 2582 (2580)
Keene D E, 3: 1958 (1957)
Keener J P, 1: 570 (569)
Keeney R L, 1: 372 (360); 2: 942 (941), 1565 (1563); 3: 2061 (2056, 2060), 2202 (2196), 2203 (2194, 2196); 5: 3104 (3099, 3100, 3101), 3115 (3104, 3105, 3107, 3108, 3112), 3121 (3121), 3124 (3122, 3123), 3152, 3164 (3157); 6: 4075; 7: 4985, 4986, 4991
Keeri-Szanto M, 2: 963 (962)
Keey R B, 2: 1253 (1249)
Kefalas A G, 2: 1003
Keidlig S, 3: 2143 (2141)
Keinath T M, 2: 1553 (1552); 6: 4229 (4228)
Keisler H J, 1: 333 (326, 331, 332), 720 (715); 2: 937 (934); 5: 2921 (2920, 2921)
Keith G A, 2: 832
Kell G S, 7: 4585 (4581)
Keller E L, 2: 1595 (1594, 1595)
Keller H B, 5: 3139 (3139), 3496 (3496)
Keller H E M, 7: 4927 (4926)

Keller J, **7**: 4944
Keller J B, **2**: 1372 (1369, 1371); **4**: 2614 (2613)
Keller T, **5**: 3421 (3419)
Keller U, **4**: 2631 (2628)
Kelley H J, **7**: 4425 (4424), 4439 (4439)
Kelley J E, **5**: 3546 (3545)
Kelley P E, **1**: 177 (176)
Kelley P L, **6**: 4036 (4036)
Kelley R, **4**: 2385 (2382); **6**: 4123
Kelley R A, **7**: 4897 (4897)
Kelly C W, **4**: 2338 (2335, 2337)
Kelly E A, **6**: 3898 (3894)
Kelly F P, **6**: 3659; **7**: 4741 (4737)
Kelly G E, **2**: 1226 (1217, 1225)
Kelly J J, **2**: 1431 (1428)
Kelly M, **6**: 4160 (4157)
Kelly P, **7**: 4833 (4828, 4830, 4831)
Kelly R G, **3**: 1693 (1690)
Kelman A W, **2**: 1248 (1247); **6**: 3699 (3698)
Kelton W D, **4**: 2353 (2352); **6**: 4396 (4395); **7**: 4654 (4651, 4652, 4653), 5005 (5002, 5003), 5008 (5006), 5018 (5017, 5018)
Kelvin (Lord), **7**: (4849)
Kember N F, **6**: 4302 (4301)
Kemeny J G, **5**: 2929 (2927, 2928); **6**: 4356
Keméy T, **4**: 2253 (2249)
Kemp D R, **4**: 2449 (2448, 2449)
Kemp J F, **6**: 4261 (4261)
Kemp R A, **6**: 4061 (4056)
Kemp W M, **7**: 4956 (4955)
Kempe S, **1**: 531 (529)
Kempf K G, **4**: 2447 (2447)
Kempthorne O, **2**: 1106
Kemter H, **7**: 4703 (4703)
Kendall D G, **7**: 4645 (4644)
Kendall D N, **7**: 4504 (4499)
Kendall E J M, **3**: 1693 (1691); **5**: 3421 (3419)
Kendall M G, **2**: 1106; **4**: 2307 (2303, 2305); **6**: 3621 (3619); **7**: 5001 (4998, 4999)
Kendall Pye E, **1**: 535 (535)
Kendler M B, **6**: 4204
Kenedi R M, **1**: 312 (310, 311, 312), 478 (474, 477), 482 (479, 480); **3**: 1913
Keng J, **6**: 3653 (3646, 3652)
Kennedy W J Jr, **6**: 3950 (3948)
Kenyon N D, **4**: 2511 (2510)
Keravnou E, **3**: 1809 (1808); **6**: 4007 (4007)
Kercher J R, **1**: 592 (590); **2**: 1345 (1343); **7**: 4956 (4955)
Kerckhoffs E J H, **1**: 633 (632), 634 (632); **2**: 1201 (1201); **3**: 2217; **6**: 4323
Kerekes J J, **4**: 2646 (2644)
Kerfoot W C, **7**: 4454 (4453, 4454)
Kernahan P, **1**: 450 (449, 450)
Kerneis J, **7**: 4749
Kernévez J P, **1**: 433 (432); **2**: 1186 (1183); **5**: 3496 (3491, 3492, 3495, 3496)
Kernighan B W, **1**: 732 (730); **7**: 4666 (4665)
Kerr A, **3**: 1947
Kerr D N S, **1**: 310 (308)
Kerr D S, **4**: 2504 (2502)
Kerr S R, **7**: 4454 (4453)
Kerr T H, **2**: 912 (910)
Kerre E E, **2**: 1407 (1406, 1407)
Kershaw D S, **5**: 3487 (3486); **7**: 4971 (4969)
Kershenbaum L S, **2**: 1466 (1464); **5**: 3232 (3231); **6**: 3991 (3990, 3991)
Kerslake D McK, **7**: 4868 (4867), 4880 (4879)
Kersten L, **6**: 4108 (4107)
Kerwin W J, **6**: 4284 (4283)
Kessel W C, **3**: 1838
Kessis J J, **4**: 2392
Kessler F M, **5**: 3070 (3070)
Kessler K, **2**: 808 (802); **6**: 4245 (4240)
Kessler R C, **5**: 2943 (2935)
Ketler W R, **2**: 1551 (1550)
Ketner P, **1**: 531 (529)
Kettler D A, **7**: 4726 (4723)
Keulen H Van, **1**: 142 (138, 139, 140, 141)
Keviczky L, **1**: 564 (561, 562); **3**: 2010 (2008, 2010); **4**: 2609 (2608), 2892 (2891); **5**: 3006 (3004), 3188 (3184, 3188), 3232 (3229), 3351 (3342); **6**: 3690 (3690), 4176 (4173), 4184 (4180, 4181, 4182), 4213 (4212); **7**: 4540 (4537)
Keyfitz N, **6**: 3755, 3762 (3756, 3757, 3758, 3759, 3760, 3762)
Keyzer F, **5**: 3030 (3029)
Kezer A, **5**: 3070 (3068)
Khachian L G, **6**: 3739 (3736)
Khaikin S E, **5**: 3388 (3384)
Khali H K, **7**: 4439
Khalid B M N, **2**: 1388 (1387)
Khalid R A, **4**: 2782 (2780)
Khalifa I H, **5**: 3067 (3066, 3067)
Khalil H K, **2**: 923 (923); **4**: 2560 (2559); **5**: 3176 (3176); **7**: 4439 (4436), 4534 (4529, 4532, 4533, 4534)
Khalili S, **3**: 1802 (1801), 1872 (1871), 1880 (1878)
Khan O, **6**: 3940 (3940)
Kharatishvili G L, **3**: 2152 (2152); **5**: 3506
Khargonekar P P, **6**: 3972 (3971, 3972)
Khasminskii R E Z, **5**: 3554 (3551); **6**: 3994 (3992); **7**: 4944 (4943)
Khinchine A I, **4**: (2272)
Khodaverdian E, **2**: 1388
Khoshnevis B, **7**: 4935 (4934)
Khosla P K, **5**: 3055
Khotanzad A, **7**: 4910 (4908, 4909, 4910)
Kibbey A H, **7**: 5094 (5093)
Kickert R N, **1**: 142 (137)
Kickert W J M, **1**: 126 (125); **3**: 1874; **4**: 2361 (2359, 2360); **5**: 3454, 3540 (3539)
Kidal H, **4**: 2723 (2720)
Kidd P T, **6**: 4271 (4270)
Kidd R A, **5**: 2994 (2993)
Kieser M S, **7**: 5007 (5007)
Kijima K, **3**: 2043 (2041, 2043)
Kikuchi H, **6**: 4128 (4124)
Kikuchi M, **3**: 2021 (2017, 2018)
Kilar L A, **6**: 3806 (3806)
Kilbridge M D, **7**: 4495

Kildal H, **6**: 4036 (4036)
Kilgour D M, **7**: 4799 (4794, 4798)
Killick R D, **6**: 4142 (4141)
Killus J P, **1**: 188 (187)
Kilmann R H, **7**: 4793 (4791)
Kilmer V J, **1**: 7 (2, 5)
Kiloh L G, **2**: 1413 (1411)
Kim Ch E, **1**: 628 (627)
Kim H H, **6**: 4045
Kim M, **1**: 417 (414, 415); **2**: 1337 (1336)
Kim S, **6**: 3820 (3817, 3819)
Kimbark E W, **6**: 3825 (3821)
Kimberly J, **7**: 4833 (4830)
Kimemia J, **3**: 1648 (1644)
Kimmel L V, **4**: 2794 (2789)
Kimura H, **6**: 3734 (3734)
Kimura M, **1**: 759 (759)
Kinderlehrer D, **1**: 529 (524); **4**: 2343; **7**: 5036 (5033)
Kindig N B, **6**: 4051 (4048, 4049)
Kindler J, **3**: 2053 (2048, 2049); **7**: 5140
King D, **2**: 1573
King G S, **5**: 3470 (3467)
King J J, **4**: 2640 (2637)
King J R, **6**: 3889 (3888)
King P F, **2**: 1154 (1147)
King P J, **3**: 1898 (1895), 1943 (1940), 1958 (1955, 1957); **4**: 2362 (2359, 2360)
King R, **5**: 2993 (2991), 2994 (2991)
King R E, **3**: 1611 (1609)
King R P, **6**: 3683 (3682)
King W F, **2**: 1174 (1170)
King-Hele D G, **3**: 2006 (2005)
Kingma Y J, **2**: 1036 (1033, 1034, 1035, 1036), 1043 (1040, 1041); **7**: 4981 (4976, 4979)
Kingsley S A, **5**: 3470 (3469)
Kinkade R G, **3**: 2186 (2181)
Kinlen P J, **5**: 2943 (2935)
Kinncunan P, **6**: 4123 (4120)
Kinoshita H, **7**: 4980 (4977)
Kiparissides C, **6**: 3745 (3740)
Kipins D H, **3**: 2021 (2017)
Kipphan H, **3**: 1694 (1689)
Kirby B J, **7**: 5041 (5038)
Kirby O B, **2**: 1186 (1183)
Kirchmayer L K, **6**: 3820 (3817, 3818), 3846 (3845)
Kirchner F F, **5**: 3421 (3419)
Kirchner W B, **4**: 2650 (2648), 2794 (2789)
Kiritani K, **1**: 142 (141), 592 (590, 591)
Kirkbright G F, **7**: 4504
Kirkland J J, **4**: 2889
Kirklin J W, **1**: 498 (495)
Kirkpatrick C A, **6**: 3904
Kirkpatrick M, **7**: 4856 (4852)
Kirkwood C W, **3**: 2061 (2056, 2060)
Kirrilova F M, **5**: 3206
Kirshbaum P J, **3**: 2179
Kisiel C C, **2**: 1553 (1552); **5**: 3165 (3157)
Kiszka J B, **2**: 1473 (1473); **3**: 1809 (1809), 1810 (1807), 1852 (1848), 1872 (1868)
Kitabata T, **3**: 1611 (1609)
Kitamura S, **2**: 1197; **6**: 4036 (4035), 4245 (4238, 4241)
Kitanidis P K, **3**: 2234 (2231)
Kitano Y, **1**: 7 (2)
Kitchen R L, **2**: 1462
Kitney R I, **1**: 542 (539); **7**: 4877 (4873)
Kitsunai T, **3**: 1614 (1614)
Kittler J, **4**: 2635 (2635); **6**: 3644 (3641, 3643), 4160 (4159)
Kittrick J A, **1**: 7 (2)
Kitzinger C, **7**: 4876 (4873)
Kiviat P J, **6**: 4348 (4348), 4402, 4405 (4404), 4415 (4414); **7**: 5005 (5001)
Kiwiel K C, **5**: 3546 (3544)
Kiyozumi R, **2**: 1497 (1494, 1497)
Kizka J B, **5**: 3264
Klaczko S, **1**: 740 (738)
Klapisch R, **5**: 2943, 2944
Klarhoefer C, **1**: 380
Klaua D, **5**: (2918)
Klee V, **1**: 347; **2**: 1080 (1079, 1080)
Kleene S C, **1**: 333 (326), 720 (718, 719); **2**: 937 (933, 935); **5**: (2919); **7**: 4958 (4958)
Klefenz G, **7**: 4580
Kleijnen J P C, **5**: 3030 (3027, 3028, 3029); **6**: 4395 (4395); **7**: 4651 (4648, 4650, 4651), 4657 (4655)
Klein D A, **1**: 142 (141)
Klein F, **5**: (2946)
Klein G A, **3**: 2203 (2201); **4**: 2501 (2501)
Klein L, **1**: 201
Klein L R, **2**: 1342 (1338)
Klein M, **6**: 3706
Klein R E, **6**: 4244 (4238), 4245
Kleindorfer G B, **2**: 1099 (1093, 1094); **6**: (3892)
Kleindorfer P R, **2**: 1099 (1093, 1094, 1095, 1096, 1097, 1098); **6**: (3892)
Kleinman D, **5**: 3264 (3263)
Kleinman D L, **3**: 2163 (2162); **7**: 4462 (4460), 4685 (4678)
Kleinman R, **1**: 437
Kleinmuntz B, **3**: 2192 (2190)
Kleinmuntz D N, **3**: 2192 (2190)
Kleinrock L, **3**: 1648 (1645); **6**: 3659, 3925; **7**: 4716, 4741 (4735), 4835
Kleir G A, **4**: 2501 (2501)
Klema V C, **2**: 826 (819, 820, 821, 822), 827 (815, 818, 824), 912 (910)
Klement E P, **1**: 664; **3**: 1781 (1780, 1781), 1786 (1783), 1802 (1801), 1809 (1809), 1810 (1809), 1821 (1820, 1821), 1843 (1842), 1880 (1878); **4**: 2301 (2294); **5**: 3413 (3410); **6**: 3710 (3707, 3709, 3710)
Klennerman L, **3**: 1913
Klett J D, **6**: 4036 (4035)
Klijerhout J J, **7**: 4913
Klimisch H J, **1**: 177 (176)
Kline J R, **2**: 1344 (1343)
Klinger A, **6**: 3653 (3648, 3649, 3650)
Klinke R, **7**: 5078 (5075)
Klinkmann H, **1**: 312 (310)

Klir G J, **1**: 93 (93), 399 (394), 561 (561), 709 (707, 709), 713 (713), 720 (718, 719); **2**: 992 (987), 1333 (1331), 1486 (1480), 1489 (1486, 1487, 1488), 1549 (1546, 1547, 1548); **3**: 1781 (1781), 2003 (2000, 2001), 2164; **4**: 2497; **5**: 2971 (2971), 3044 (3042), 3088 (3083); **6**: 3768 (3766, 3767, 3768), 3865 (3863), 3989 (3984, 3985, 3986, 3987, 3988, 3989), 3990, 4007 (4007), 4066 (4062), 4167 (4166), 4229 (4228), 4313, 4319, 4322 (4319), 4323 (4321), 4326, 4336, 4353 (4350, 4351, 4352, 4353), 4372 (4371), 4377 (4377), 4388, 4392, 4411 (4406, 4407, 4409), 4414; **7**: 4764, 5008 (5007)
Klitin O A, **6**: 3820 (3818), 3850 (3846, 3848, 3849)
Kliv G J, **2**: 1552 (1550)
Klok F, **7**: 5047 (5044)
Klopatek J M, **2**: 1346 (1345), 1355 (1353, 1355); **3**: 1601 (1600); **6**: 4237 (4235)
Klopfstein G, **3**: 2089
Klotz P, **1**: 177 (175)
Kluckner P D, **1**: 177 (174)
Klug A, **2**: 893 (887)
Klugt N van der, **6**: 3933 (3932)
Kluver H, **2**: (872)
Knapp A W, **6**: 4356
Knapp C H, **3**: 1741 (1738)
Knapp H, **3**: 1779 (1773, 1774, 1777)
Knapp K T, **4**: 2511 (2510)
Knapp R, **2**: 1358
Kneppo P, **4**: 2253 (2253)
Knets I V, **1**: 474 (471)
Knight B W, **7**: 5065 (5060, 5062, 5064)
Knight J, **7**: 5086
Knight J A G, **3**: 1637 (1635)
Knighten G V, **4**: 2723 (2721, 2722)
Knipling E B, **2**: 1521 (1517)
Knisel W G, **1**: 142 (138); **7**: 4897 (4896)
Kniseley R N, **4**: 2347 (2343)
Knobloch H W, **2**: 1008 (1006), 1016 (1012), 1017 (1013); **7**: 4425 (4420)
Knopfmacher J, **2**: 1473 (1468); **3**: 1878 (1876)
Knowles A, **7**: 4504
Knowles A E, **3**: 1985 (1982, 1983)
Knowles G, **2**: 1366 (1364); **4**: 2861
Knowles J B, **2**: 1078 (1073, 1074, 1075, 1077, 1078)
Knox J M, **5**: 3464 (3460)
Knudsen J, **6**: 3825 (3821)
Knuth D E, **2**: 884 (881); **3**: 1712 (1712); **5**: 3272 (3268); **6**: 3893, 3945
Knutti D F, **6**: 4108 (4104)
Kobayashi H, **1**: 755 (752); **4**: 2386 (2382); **7**: 4823 (4821)
Kobayashi J, **2**: 1516 (1516)
Kobayashi R, **3**: 1975
Kobayashi T, **2**: 1197
Kobayasi T, **5**: 3464 (3463)
Kobe K A, **4**: 2517 (2511)
Koberg D, **2**: 992 (991)
Kobrinsky A A, **6**: 4108 (4105)
Kobrinsky A E, **6**: 4108 (4105)
Koch B, **2**: 1532 (1528)
Koch G, **1**: 417 (414, 415, 416); **5**: 3279 (3276)
Koch W, **3**: 2017
Kochanska M, **3**: 1872
Kochar A K, **2**: 1586 (1585)
Kochenberger G A, **6**: 4075
Kochenburger R J, **5**: 3389 (3385), (3516); **7**: 4452 (4443)
Kocher D C, **2**: 1551
Kochs H D, **2**: 1388 (1386, 1387)
Koda M, **2**: 1500
Kodama H, **1**: 8 (1, 3)
Kodama S, **4**: 2559 (2559)
Koditschek D E, **4**: 2719 (2718)
Koegler W, **5**: 2943 (2934, 2935)
Koehret B, **2**: 1253 (1249)
Koenig M, **5**: 3480 (3477)
Koenigsberg E, **3**: 1648 (1646)
Köerner J, **6**: 3780 (3779)
Kofakis P, **1**: 347 (347)
Koga R, **4**: 2723 (2721)
Koga S, **1**: 424 (423)
Kogan B Ya, **6**: 3870 (3867)
Kogure J, **2**: 1522 (1518)
Koh R C Y, **4**: 2789 (2788)
Kohavi Z, **5**: 3272 (3268)
Kohberger R C, **6**: 4235 (4231)
Kohl W, **6**: 3663 (3660, 3662), 3664 (3662)
Kohlberg E, **2**: 850 (848); **3**: 1926 (1924); **5**: 3256
Köhler A, **1**: 185
Kohler C F, **4**: 2511 (2510)
Kohler W, **2**: (872)
Kohler W H, **2**: 1126, 1136 (1133), 1165 (1161), 1174 (1171), 1209 (1208, 1209)
Kohli S, **1**: 176 (175)
Kohn A, **2**: 1179 (1174)
Kohn R, **1**: 529
Kohn R V, **2**: 812 (809)
Köhne M, **3**: 1990 (1987)
Kohnert R L, **4**: 2782 (2779)
Kohonen T, **1**: 511 (508); **6**: 4167 (4166)
Kohout L J, **1**: 743 (741, 742), 746 (743); **3**: 1809 (1807, 1808, 1809), 1867 (1865, 1866); **5**: 2926 (2926); **6**: 4007 (4002, 4004, 4005, 4007)
Koide S, **6**: 4245 (4238)
Koistinen R, **2**: 1568 (1567); **6**: 3916 (3913)
Koivo A, **2**: 1192 (1191)
Koivo H N, **5**: 2993 (2991), 3188 (3184), 3232 (3229, 3230); **6**: 3727 (3723), 4184 (4182)
Kokawa M, **5**: 2926 (2926)
Kokotovic P V, **1**: 105 (105), 132; **5**: 3082 (3081), 3176 (3174, 3176), 3180 (3176), 3502; **6**: 4227; **7**: 4439 (4435, 4438, 4439), 4443 (4439, 4440, 4441, 4442, 4443), 4534 (4529, 4530, 4531, 4532, 4533, 4534), 5031 (5028), 5153 (5153)
Kolarz G, **1**: 740 (736)
Kolence K W, **1**: 751 (749)
Kolers P A, **6**: 3644
Kolesar P, **4**: 2353 (2351)
Kolff W J, **1**: 312 (311)

Kolmogorov A N, **1**: 709 (708); **2**: 1549 (1547); **4**: (2271); **5**: 3019 (3017), (3387); **6**: 3990
Kolostori J, **1**: 564 (561); **4**: 2892 (2891); **5**: 3232 (3229)
Kolthoff I M, **7**: 4504
Komorník J, **5**: 3514 (3514)
Komura Y, **6**: 4223 (4220)
Koncz L, **3**: 2174 (2173)
Kondo B, **4**: 2556 (2554)
Kondo J, **2**: 1522 (1521)
Kondo K, **3**: 2021 (2017, 2018)
Kondo T, **2**: 1497 (1494, 1497)
Kondratiev V V, **1**: 37
Kønig M, **2**: 1532 (1528)
Konishi Y, **6**: 4223 (4222)
Konstantinov M M, **2**: 827 (821); **4**: 2827 (2826)
Konstantinov M S, **6**: 4108 (4103, 4105)
Konz S A, **7**: 4877, 4880 (4880)
Koo K P, **5**: 3464 (3459)
Koonce J F, **4**: 2646 (2645)
Koopman B O, **7**: 4789 (4786)
Koopmans T C, **4**: 2273 (2272); **5**: 3235 (3234); **6**: 4299
Kopacek P, **6**: 4311
Kopociński B, **2**: 1106
Kopp R E, **7**: 4425 (4424)
Koppel L B, **6**: 3680, 4245 (4242)
Kopstein S, **4**: 2473 (2472)
Korbinski N E, **1**: 334
Korendyasev A I, **6**: 4108 (4107)
Korfhage R, **2**: 1451 (1451)
Korn G A, **1**: 421; **5**: 3351 (3344); **6**: 4359 (4358)
Korn T M, **6**: 4359 (4358)
Korndörfer V, **6**: 3663 (3663)
Kõrner S, **5**: 2921 (2919)
Kornhuber H H, **1**: 466 (465)
Korostelev A P, **1**: 80 (80)
Korpela K, **6**: 4184 (4182)
Korsak A J, **6**: 3820 (3819)
Kortanek K O, **1**: 529 (526, 527); **2**: 850 (845); **5**: 3455 (3452)
Korte B, **1**: 626 (625), 628 (627)
Kortela U K J, **2**: 781 (779); **6**: 3683 (3682)
Kos C M, **2**: 1596 (1594, 1595)
Kosaka M, **4**: 2723 (2721)
Koschitzky H, **2**: 1590 (1587, 1590)
Kossen N W F, **1**: 424
Kosten L, **7**: 4733
Koster B, **3**: 1665 (1660)
Kosut R L, **2**: 932 (931); **7**: 4685 (4678)
Kosy D, **1**: 755 (752)
Kotarbinski T, **6**: 3865 (3861, 3862, 3865)
Kou S R, **5**: 3341 (3340)
Kouada I A, **5**: 3164 (3161)
Kouchoukos N T, **1**: 498 (495)
Kouda A, **5**: 3024
Koudstaal J, **5**: 2993 (2990)
Koutsou A, **6**: 4093 (4089)
Kouvaritakis B, **1**: 45 (38), 580 (575, 577, 578, 579, 580), 583 (581, 582), 735 (734); **4**: 2832 (2828, 2829, 2831), 2871 (2869, 2870); **5**: 3228 (3226), 3379 (3378)
Kouzmak G E, **7**: 4490
Kovács K, **4**: 2892
Kovalevsky V A, **1**: 575; **6**: 3644
Kovarik Z V, **2**: 927 (924)
Kovasin K K, **5**: 2943 (2935)
Kowalik J S, **2**: 827 (818); **4**: 2412 (2393, 2397, 2399); **5**: 3082 (3079)
Kowalski N, **2**: 1595 (1593, 1594)
Kozak J J, **1**: 433 (432)
Kozak L M, **1**: 268 (263, 266)
Kozietulski M, **2**: 1305 (1298, 1302)
Kozin F, **2**: 1479 (1474)
Kozlowski D R, **1**: 623
Kozyakin V S, **7**: 4674 (4672)
Krainak J C, **2**: 932 (931); **7**: 4822 (4815)
Kraiss K-F, **6**: 4306 (4303)
Krakauer L, **4**: (2382)
Kramer A, **3**: 1705 (1701)
Kramer J, **2**: 1143
Kramer K J, **3**: 2174 (2171)
Kramer P J, **2**: 1521 (1518), 1522 (1521)
Kranc G M, **2**: 1078 (1078)
Krantz D H, **5**: 3115 (3105, 3107, 3109); **7**: 4991 (4987)
Krantzberg M, **7**: 4833 (4828, 4830, 4831)
Krasnoleskii M A, **7**: 4674 (4672), 5047 (5042)
Krasnow H S, **4**: 2353 (2352)
Krasovskii A A, **6**: 3870
Krasovskii N N, **2**: 1008 (1005, 1006); **3**: 2151, 2152 (2152); **4**: 2609 (2608); **5**: 3206, (3387), 3506, 3512 (3511); **6**: 3994 (3993), 4011 (4010)
Kraus G, **4**: 2511 (2511)
Krause G M, **3**: 1736, 1741 (1739)
Krausz A S, **5**: 3099
Kravaris C, **6**: 4364 (4363)
Krebs J E, **2**: 1345 (1343); **5**: 3440 (3440)
Krée P, **2**: 812
Kreer J B, **7**: 4929 (4928)
Krein S G, **7**: 4529 (4528)
Kreindler E, **6**: 4227 (4225)
Kreiss H O, **3**: 1723 (1722); **7**: 4498 (4496)
Kreisselmeier G, **1**: 105 (102), 106 (105); **2**: 1581 (1580); **5**: 3060 (3059), 3061 (3061), 3410 (3408)
Krelle W, **2**: 1262 (1258)
Kremer J N, **7**: 4897 (4896)
Krener A J, **1**: 323 (321), 413 (413); **3**: 1779 (1770, 1777); **4**: 2756 (2756), 2816 (2807); **5**: 3288 (3285, 3286, 3287), 3341 (3337, 3338), 3394 (3390, 3393), (3516); **6**: 3961 (3957, 3958), 3977 (3975, 3976); **7**: 4425 (4420, 4422, 4423, 4424), 4677 (4674)
Kreps D M, **7**: 4622 (4622), 4661 (4660)
Kreuzer L B, **4**: 2511 (2510), 2723 (2721, 2722)
Kreysa G, **5**: 2943 (2935, 2942)
Kreyszig E, **5**: 3206 (3202)
Kriebel C H, **2**: 1099 (1093); **6**: (3892)
Krief J P, **7**: 5170 (5168)
Krieger A J, **2**: 1479 (1476, 1478)
Kripke S, **1**: 333 (329)
Krishnaiah P R, **2**: 1106; **7**: 4566 (4563)
Krishnan K R, **7**: 4726 (4723), 4740 (4735), 4751 (4751)

Krishnaprasad P A, **5**: 3329 (3327)
Kristt D A, **1**: 437
Kron-Morelli R, **1**: 675 (672)
Kronauer R E, **1**: 601 (601)
Kronecker L, **6**: 3629
Kronmal R A, **6**: 3950 (3949)
Kroone H B, **2**: 969
Krtolica R, **4**: 2559 (2557, 2558, 2559)
Krueger P, **1**: 639 (639)
Krüger G J, **5**: 3421 (3419)
Krumpholz G R, **6**: 3850 (3848)
Krupa S V, **1**: 142 (139, 140), 592 (590)
Kruskal J B, **5**: 3235; **6**: 3739 (3736)
Kruskal W H, **2**: 1106
Kruzic P G, **7**: 5177 (5171)
Krycuk G, **5**: 3465 (3462)
Krylov N V, **5**: 3024 (3020), 3523 (3520); **7**: 4613, 4627 (4623, 4626)
Krylov U, **5**: 3389 (3384, 3385)
Kryter K D, **2**: 1527 (1525)
Krzanowski W J, **2**: 1106
Krzysztofowicz R, **5**: 3115 (3112); **6**: 4075; **7**: 5140 (5136, 5138, 5139)
Kshirsagar A M, **2**: 1106 (1104)
Ktesibios of Alexandria, **3**: (1604)
Ku R T, **4**: 2723 (2722); **6**: 4036 (4036)
Ku Y H, **5**: 3389 (3384, 3385)
Kubicek M, **5**: 3496 (3495)
Kubota H, **7**: 4779 (4775, 4777)
Kubrusly C S, **2**: 808, 1197, 1199 (1199); **6**: 4245
Kucera J, **5**: 3366 (3366)
Kučera V, **1**: 45 (39, 42); **4**: 2849; **5**: 3006 (3004), 3514 (3514); **6**: 3727 (3725), 3871 (3870), 4176 (4173)
Kuck D J, **2**: 827 (818)
Kuczura A, **7**: 4732 (4731), 4741 (4734)
Kudrewicz J, **5**: 3389 (3386)
Kudva P, **1**: 106 (102); **4**: 2702 (2698, 2699); **5**: 3067 (3065, 3066)
Kue Z-Y, **6**: (3861)
Kuehn T J, **7**: 4828 (4824)
Kuenzel N T, **2**: 1345 (1343)
Kuffler J S, **3**: (1606)
Kuhlberg J, **1**: 226
Kuhlman H W, **3**: 1943
Kuhn H W, **2**: 850 (848), 1020 (1017, 1018), 1262 (1257, 1258); **3**: 1917 (1913, 1917), 1926 (1922); **4**: 2559 (2559); **5**: 3163 (3159), 3164 (3159, 3161, 3163), 3256 (3255); **6**: 3739 (3736), 3820 (3818)
Kuhn J R D, **7**: 4799 (4794, 4798)
Kuhn T S, **2**: 1538 (1536); **3**: 1862; **7**: (4953)
Kühne R D, **3**: 1736
Kuhura Y, **5**: 3464 (3461)
Kuiper J P, **1**: 482 (480)
Kujawski E, **2**: 1552 (1549)
Kuki K, **3**: 1619 (1618)
Kukulka C G, **1**: 437
Kulakowski B T, **3**: 2097 (2094)
Kulhanek L, **3**: 2214 (2213)
Kulhavy R, **6**: 3991 (3991)
Kulikowski C A, **2**: 1573, 1578 (1573)
Kulkarni A V, **6**: 3644 (3643)
Kullback J E, **2**: 1479 (1476)
Kullback S, **4**: 2508
Kulpa Z, **4**: 2473
Kumar A, **5**: 3425 (3423)
Kumar D, **2**: 1272 (1272)
Kumar P R, **6**: 4176 (4172)
Kumar R, **1**: 176 (175)
Kumar S, **6**: 4245 (4238)
Kumar S P, **2**: 827 (818)
Kun S C, **5**: 3487 (3485)
Kundel H L, **7**: 5186 (5182, 5183)
Kundu S S, **1**: 142 (140, 141)
Kunen K, **1**: 720 (715)
Kung H T, **1**: 725 (722); **2**: 826 (818), 884 (881, 882), 1154 (1149); **7**: 4540 (4537)
Kung S-Y, **1**: 296 (294); **2**: 785 (783), 1102 (1102); **5**: 3077 (3077), 3220 (3215, 3217, 3218)
Kunii T L, **3**: 1800 (1798); **6**: 3653 (3648, 3649, 3650, 3652)
Kunisch K, **2**: 1211 (1209)
Kunita H, **2**: 1583 (1581, 1583); **5**: 3303 (3293, 3297), 3329 (3319), 3330 (3326); **7**: 4644 (4642)
Kunnen M, **7**: 4871 (4870)
Kunreuther H C, **2**: 1099 (1098); **6**: 3706, (3890)
Kunzi H P, **2**: 1262 (1258)
Kuo B C, **3**: 2113 (2112); **5**: 3097; **7**: 4598 (4595, 4598)
Kuo J C W, **5**: 3082 (3079)
Küpfmüller K, **4**: (2321)
Kupka I, **1**: 124 (122, 124); **2**: 1583 (1581); **4**: 2527 (2527), 2759 (2758); **5**: 2948 (2948), 2949 (2948), 3366 (3365)
Kurator W G, **6**: 4333 (4329)
Kuratowski C, **6**: 3734
Kurcyusz S, **1**: 406 (403)
Kurkjian A L, **3**: 1736, 1741 (1738, 1739)
Kurland G S, **1**: 445 (443)
Kurono S, **6**: 4108 (4104)
Kurose J F, **4**: 2550 (2550), 2551 (2548)
Kurosu K, **1**: 713 (710, 712, 713); **3**: 1902 (1902), 1905 (1903); **5**: 2927 (2925)
Kurpinski K, **2**: 1305 (1298, 1299, 1301, 1302)
Kurtaran B, **7**: 4822 (4821)
Kurtz T G, **6**: 3997 (3994)
Kurtze T, **4**: 2582 (2581)
Kurz H, **1**: 63 (60); **3**: 2097 (2094); **5**: 3351 (3341); **6**: 3610 (3608), 3683 (3681), 4204
Kusaka T, **3**: 2052 (2050), 2053 (2048, 2049); **4**: 2655 (2652)
Kushner H J, **1**: 80 (77, 79), 91 (86, 91); **2**: 1099 (1096); **5**: 3303 (3293), 3523 (3519, 3522), 3531 (3529); **6**: 3994 (3992), 3997 (3994, 3995, 3996), 4209 (4207); **7**: 4611 (4607), 4613, 4644 (4643)
Kuss H, **2**: 1165
Kutz M P, **7**: 4810
Kuwat Y, **6**: 3631 (3631)
Kuzdrall P J, **6**: 4370 (4369)
Kuz'min V B, **2**: 1407 (1406)
Kuznicki J T, **3**: 1669 (1666)

Kwak N K, **6**: 4370 (4369)
Kwakernaak H, **2**: 826 (814); **3**: 1831 (1830), 1845 (1845), 1847 (1846), 1872 (1871); **4**: 2245 (2240), 2846 (2845); **5**: 3156 (3156), 3228 (3225, 3227, 3228), 3316 (3316), 3475 (3472); **6**: 3619 (3617), 3755; **7**: 4554 (4552), 5170 (5169)
Kwatny H G, **6**: 3820 (3816)
Kwong Ng Kee K F, **1**: 268 (263, 264, 266)
Kwong R, **3**: 2151, 2152; **5**: 3506 (3502, 3505)
Ky Fan, **2**: 850 (844), 1262 (1259); **5**: 2998 (2997), (3255)
Kyburg H E Jr, **7**: 5015 (5012)
Kydes A S, **1**: 493
Kyuma K, **5**: 3464 (3459)

L

La Budde R A, **6**: 4346 (4346)
Labarrère M, **1**: 251 (251); **7**: 5170 (5168)
Labetoulle J, **1**: 91 (87)
Labiale G, **3**: 1761
Labrujere T E, **2**: 1179 (1176)
Labutti G, **4**: 2511 (2509)
Lacey J A, **3**: 1961
Lachenbruch P A, **2**: 1105 (1105), 1106 (1103, 1104, 1105)
Lachman L, **6**: 3695
Lachmann K H, **1**: 72 (64, 67, 68); **5**: 3351 (3341, 3342, 3343, 3344); **6**: 3610 (3610), 3611 (3610), 4184 (4182)
Lacroix J M, **1**: 437
Ladde G S, **1**: 675 (674); **6**: 4011 (4009, 4010)
Ladyzenskaya O A, **6**: 3596 (3592)
Laface P, **6**: 4159 (4158)
Lafferty E, **5**: 3560 (3559)
Laffont J J, **5**: 2911 (2908); **7**: 4630, 4631 (4630)
Lafleur V, **2**: 1087 (1084)
Lafrenière L, **6**: 4336, 4337 (4335)
Lagakos N, **5**: 3464 (3459), 3470 (3467, 3468, 3469)
Lagasse J, **5**: 3097
Lago P J, **5**: 3238 (3235), 3248
Lagomasino A, **5**: 3116; **7**: 4785 (4783)
Laha A K, **6**: 4302 (4301)
Lahey R T, **3**: 1693 (1688); **5**: 3465 (3462)
Lai M Y, **2**: 1154 (1153)
Lai T L, **4**: 2283 (2279)
Lainiotis D G, **2**: 1197; **3**: 1990 (1987); **4**: 2277 (2276), 2292 (2287, 2291), 2307 (2302), 2713 (2710); **5**: 3487 (3485); **6**: 4088 (4086), 4165 (4162, 4163), 4187 (4186); **7**: 4469 (4466), 4621 (4617, 4618, 4620), 4692 (4690), 5001 (5000)
Laird J B, **3**: 1975, 1979
Laks M, **1**: 769 (761)
Lakshmikantham V, **5**: 3376 (3374)
Lakshminarayan S, **6**: 4061 (4060)
Lakshmivarahan S, **4**: 2719 (2718), 2737 (2731); **7**: 4721 (4717, 4718)
Lallement G, **5**: 3394 (3393)
Laloy M, **1**: 675 (673); **5**: 3402 (3399, 3400); **7**: 4529 (4525)
Lalui A, **1**: 638 (638)
Lam D C L, **4**: 2794 (2789)
Lam K P, **6**: 3645, 4176 (4172), 4199 (4199)
Lam T Y, **6**: 3972 (3971)
LaMarra N, **6**: 4051 (4049)
Lamarre Y, **1**: 570 (569)
Lamb H, **5**: 3260 (3257)
Lamba S S, **1**: 132 (127); **5**: 3077 (3075, 3077); **7**: 4452 (4451)
Lamberson L R, **6**: 4015
Lambert G M S, **3**: 1969 (1968)
Lambert J D, **5**: 3559 (3556, 3557); **6**: 4359 (4358, 4359)
Lambert J P F, **1**: 177 (174)
Lambert J R, **1**: 142 (138, 139)
Lambert M, **1**: 623
Lambert R S, **1**: 177 (175)
Lambertsen C J, **6**: 4061 (4056)
Lambeth D M, **4**: 2452 (2452)
Lammer P, **2**: 963 (962, 963)
Lamnabhi M, **3**: 1779 (1770, 1771)
Lamnabhi-Lagarrigue F, **3**: 1779 (1770); **4**: 2609 (2608)
Lamont J W, **6**: 3954 (3952)
Lampard D G, **5**: 3238 (3235, 3236)
Lamport L, **2**: 884 (881), 1132 (1131), 1154 (1145), 1174 (1173)
Lamprecht I, **7**: 4698
Lampson B W, **2**: 1132 (1126, 1132), 1136 (1134, 1135, 1136), 1154 (1144), 1174 (1170, 1171, 1172, 1173)
Lamure C, **3**: 1761 (1758)
Lan H S, **5**: 3238 (3235)
Lanam D H, **5**: 3560 (3560)
Lancaster F H, **5**: 2993 (2991, 2993), 2994 (2991)
Lancaster P, **2**: 826
Lance J C, **1**: 148 (147)
Lanchester F W, **1**: (340)
Landau I D, **1**: 51 (48, 50), 60 (56, 58), 97 (96), 100 (99, 100), 106 (102); **2**: 1093 (1090), 1121 (1117, 1118, 1119); **3**: 1779; **4**: 2314 (2312); **5**: 3004 (3002, 3003, 3004), 3047, 3055 (3048, 3051), 3061 (3060, 3061), 3070 (3068, 3070), 3071 (3070), 3272 (3268), 3376 (3374); **6**: 3604 (3599, 3600, 3601, 3602, 3603, 3604), 3776 (3772), 4165 (4163), 4186 (4184, 4186), 4277 (4276); **7**: 4547, 5031 (5029)
Landauer J P, **1**: 634 (632)
Landaw E M, **1**: 489 (487, 488)
Lander L, **7**: 4674 (4672)
Landsberg J J, **1**: 142 (141)
Landwehr C E, **1**: 743 (741); **6**: 4355 (4353)
Lane L J, **1**: 142 (138)
Lane P, **2**: 1355 (1352)
Lang B, **5**: 3480
Lang S B, **1**: 474 (471)
Lang W W, **3**: 1723
Langdon F J, **2**: 1527 (1524)
Lange B O, **2**: 1244 (1242, 1243)
Lange D L, **4**: 2901 (2899)
Lange F H, **6**: 4281 (4281)
Lange O, **1**: 634 (632); **6**: 3865 (3861)
Lange O L, **7**: 4897 (4896)
Langenheim J H, **2**: 1486 (1480, 1481), 1489 (1487)
Langenhop C E, **6**: 3734 (3733)
Langer G, **5**: 2942 (2935)
Lanir Y, **1**: 478 (477), 486 (484)
Lansing J C Jr, **6**: 4041 (4038)
Lao R C, **1**: 177 (176)
Lapalme G, **2**: 1132 (1126)
Lapidus L, **2**: 1181 (1181); **3**: 1611 (1609)
Lapiere C M, **1**: 482 (480)
Laplace P S, **2**: 1479 (1475); **3**: (1745); **4**: 2664 (2659)
Lapp S A, **4**: 2517 (2516)

Lappus G, **3**: 1990 (1987, 1988, 1989)
Larcher W, **2**: 1522 (1521)
Larcombe M H E, **6**: 4123 (4120)
Larichev O I, **5**: 3152 (3151), 3164 (3157)
Larimore M G, **1**: 81 (78, 79)
Larke E C, **6**: 4142
LaRoe E T, **3**: 1862
Laroque G, **5**: 3550
Larrecq M, **3**: 1734 (1734)
Larsen A, **2**: 1416 (1413, 1415)
Larsen D P, **4**: 2650 (2649), 2794 (2789)
Larsen E W, **7**: 4944
Larsen I M, **4**: (2359)
Larsen M, **6**: 3904 (3904)
Larsen R I, **1**: 592 (590)
Larsen W A, **7**: 5015 (5014)
Larson R C, **2**: 1565 (1564)
Larson R E, **1**: 409; **2**: 1282 (1279); **3**: 1714, 1990 (1989); **6**: 3820 (3819, 3820); **7**: 5153 (5152)
Larson R R, **1**: 252 (251)
Larson S, **3**: 22022203 (2194, 2195)
LaSalle J P, **1**: 413 (410, 411, 412); **3**: 1898 (1897); **4**: 2609 (2608), 2898 (2893); **5**: 2963 (2963), 3206, 3389 (3384, 3387), 3398, 3512 (3511), (3516, 3517); **6**: (3866); **7**: 4425 (4420), 4529 (4525, 4528), 4677 (4675)
Lasdon L, **3**: 2163 (2159); **5**: 3156 (3154), 3164 (3158); **7**: 4709 (4708)
Lasdon L S, **1**: 405 (400); **4**: 2692; **7**: 4709 (4705)
Lasker G E, **1**: 511 (509, 510); **3**: 1604 (1601, 1602, 1603), 1810 (1807, 1809), 1831 (1831), 1862 (1859); **4**: 2301 (2296, 2300); **5**: 3455 (3454); **6**: 3989 (3984, 3985, 3986, 3987, 3988), 4337 (4336); **7**: 4832 (4830)
Laskowski J, **5**: 2993 (2991); **6**: 4252 (4250)
Lasry J M, **7**: 5047 (5046)
Lassahn G D, **3**: 1693 (1690)
Lassen N A, **7**: 4906 (4902)
Lassiter R R, **2**: 1323 (1323); **7**: 4897 (4896), 4901 (4899)
Laszlo C A, **2**: 1590 (1587, 1589)
Latour P R, **6**: 3679 (3674, 3679), 3680 (3675, 3676)
Latsis S J, **3**: 2192
Latzel J, **5**: 2943 (2935)
Lau D, **7**: 4906 (4904)
Lau S H, **2**: 1410 (1408)
Laub A J, **1**: 735 (735); **2**: 826 (819, 820, 821, 822, 823, 825), 827 (819, 822, 823, 825), 912 (910); **3**: 2163 (2153, 2155); **4**: 2622 (2621), 2846 (2843); **5**: 3196, 3197, 3228 (3225); **6**: 4176 (4173)
Laue H J, **5**: 3416
Lauer G S, **2**: 1388 (1388)
Lauer H, **5**: 3389 (3385)
Laughhunn D J, **3**: 2203 (2202); **6**: 4075
Laughlin S B, **7**: 5065 (5063, 5064)
Laughton M A, **2**: 1454 (1453); **3**: 1718 (1715, 1716); **6**: 3789 (3788), 3790 (3789)
Laugt H, **7**: 4749
Laundy M, **6**: 3789
Laurance N, **2**: 1466
Laurence J H, **3**: 1623 (1619)
Laurent F, **5**: 3406 (3402, 3405)
Lauritsen T N, **6**: 4066 (4065)
Lausen G, **2**: 884 (883)
Lautenbach K, **6**: 3670 (3669), 4350 (4349)
Lavenberg S S, **7**: 4654 (4652, 4653), 4657 (4656)
Laverhne J, **3**: 2089
Lavi A, **4**: 2601 (2598); **5**: 3092
Lavi N, **6**: 3940 (3940)
Lavie D, **6**: 4108 (4107)
Lavie R, **2**: 1586 (1585, 1586)
Lavigne J R, **2**: 781 (778, 779)
Lavison P, **6**: 3644
Lavoie D, **7**: 4833 (4830)
Lavoine J, **4**: 2664
Law A M, **4**: 2353 (2352); **6**: 4396 (4395); **7**: 4654 (4651, 4653), 5005 (5002, 5003), 5008 (5006), 5018 (5017, 5018)
Law F M, **7**: 5099
Law J M, **3**: 2014
Lawden D F, **6**: 4146 (4145); **7**: 4490
Lawler E L, **1**: 725 (725); **2**: 1083 (1080), 1125 (1123), 1388 (1387); **5**: 3224 (3222)
Lawless J F, **6**: 4029 (4029)
Lawlor L R, **1**: 675 (672)
Lawrence A E, **4**: 2782 (2781)
Lawrence J, **4**: 2353 (2351)
Lawrence P, **1**: 769 (761)
Lawson C L, **2**: 827 (817, 818); **4**: 2283 (2280); **6**: 3983 (3978, 3979); **7**: 4540 (4535)
Lawson W D, **7**: 4871 (4869)
Lawvere W, **1**: 333 (332)
Lax P D, **6**: 4364 (4360, 4361)
Layard P R G, **5**: 2979; **6**: 3655; **7**: 5164
Layberry R A, **7**: 4905 (4905)
Lazowska E D, **1**: 755
Lazzara R, **2**: 1410 (1409)
Le Chevalier F, **6**: 3643
Le Corbeiller P, **5**: (3384)
Le Foll J, **2**: 1179 (1174)
Le Gall F, **7**: 4731, 4732 (4729, 4731), 4740 (4738), 4741 (4738), 4751 (4750)
Le Magnen J, **3**: 2213 (2210, 2212)
Le Quément J, **6**: 4097
Le Roy A, **6**: 3954 (3953)
Lea W, **7**: 4509 (4507)
Leach K G, **6**: 3940 (3936)
Leach L B, **2**: 972 (971, 972)
Leaf C F, **3**: 2234 (2233)
Leahy D, **1**: 177 (175)
Leak W B, **7**: 4645 (4645)
Leal A, **4**: 2501 (2499)
Lean D R S, **5**: 3044 (3043)
Leather J, **3**: 1718 (1715)
Léauté H, **5**: (3383, 3385, 3386, 3516)
Leavenworth R S, **7**: 4726 (4724)
Lebacque P, **7**: 4929 (4929)
Leban R, **2**: 1269
Lebart L, **1**: 608 (607, 608); **2**: 879 (877)
Leberling H, **3**: 1819 (1818)
Leblanc H, **5**: 2921 (2920)
Leblanc P, **2**: 884 (881)

LeBlanc P J, **7**: 5114
Lebrun M, **6**: 4257
Lechermeier G, **7**: 4751
Lecomte P, **1**: 220
Lecourtier Y, **1**: 493 (492), 494 (492, 493)
Lederberg J, **2**: 1573
Lederer P, **3**: 2225
Ledley R S, **6**: 4160 (4158)
Leduc S, **1**: 399 (394)
Lee B W, **5**: 3319 (3319)
Lee C S G, **4**: 2440 (2438); **7**: 4910 (4906)
Lee D N, **5**: 3165 (3161)
Lee D T, **4**: 2284 (2282)
Lee E B, **2**: 1211 (1209); **3**: (1606, 1747); **5**: 2963 (2962, 2963), 3497, 3514 (3514), 3515 (3514, 3515), (3517)
Lee E S, **5**: 2943
Lee H, **2**: 1154 (1153)
Lee H C, **6**: 3653 (3646)
Lee H H, **6**: 4245 (4242)
Lee K H, **2**: 827 (823)
Lee K Y, **2**: 1192 (1189)
Lee M H, **4**: 2447 (2447)
Lee P A, **2**: 1136 (1133, 1134, 1136), 1174 (1169, 1170); **4**: 2447 (2445)
Lee R C K, **4**: 2273 (2272), 2618 (2615); **6**: 4165 (4162, 4163)
Lee R C T, **1**: 746 (743); **3**: 1888
Lee S C, **1**: 664 (664); **3**: 1883 (1880), 1888; **5**: 3055 (3050); **7**: 4766
Lee S H, **6**: 3795
Lee S M, **2**: 987; **5**: 3121 (3121), 3124 (3122), 3164 (3157)
Lee T S, **1**: 445 (443)
Lee W H, **4**: 2846 (2843, 2844)
Lee Y H, **5**: 3579 (3578)
Lee Y W, **1**: 489 (488)
Lees F P, **3**: 2186 (2183); **4**: 2800
Lefever R, **1**: 429 (426), 433 (431, 432)
Lefevre H C, **5**: 3470 (3466, 3467)
Lefferts E J, **4**: 2622 (2622)
Leffler J W, **2**: 1323 (1323)
Leffler N, **6**: 3683 (3682)
Lefkowitz I, **2**: 956 (952); **3**: 1741 (1739); **5**: 3144 (3140, 3142); **6**: 3889 (3889); **7**: 4571 (4567)
Lefohn A S, **4**: 2511 (2510)
Lefschetz S, **3**: 1898 (1897); **5**: 2963 (2963), 3389 (3384, 3387), 3398, 3512 (3511); **7**: 4425 (4420), 4529 (4525)
Legall P, **7**: 4733
Legendre A M, **4**: (2271)
Legg N J, **5**: 3425 (3425)
Leggett R W, **7**: 5008 (5005, 5006), 5015 (5014)
Legros J F, **7**: 4749 (4745)
Leguay C, **4**: 2342
Leguyade H, **6**: 4336 (4336), 4337 (4335, 4336)
Lehman M M, **4**: 2681 (2679)
Lehnigk S H, **5**: 3398
Lehtinen F G, **4**: 2601 (2598)
Lehto O, **5**: 3488 (3487); **6**: 4000 (3999, 4000)
Lehtomäki K J, **6**: 3683 (3682)
Lehtomaki N A, **4**: 2846 (2843, 2844); **5**: 3228 (3225, 3226)
Leibel B S, **3**: 2032 (2030), 2174 (2170)
Leiber H, **1**: 380 (376)
Leibler R A, **2**: 1479 (1476)
Leibniz G W, **4**: (2577)
Leibowitz H W, **3**: 1669 (1666); **7**: 5077 (5075), 5078
Leigh J R, **5**: 3206 (3202)
Leininger G G, **1**: 735 (735); **4**: 2601 (2598, 2600); **6**: 4139 (4133)
Leipnik R, **4**: 2331 (2329)
Leiponen M, **5**: 3282 (3282)
Leitmann G, **1**: 687 (685, 686); **2**: 1010 (1010), 1269 (1264); **5**: (3517); **7**: 4425 (4424)
Leiviskä K, **5**: 3170 (3170); **6**: 3889 (3885, 3887, 3888), 3916 (3911, 3914)
LeLann G, **2**: 1154 (1153), 1169
LeLetty L, **2**: 1192 (1186, 1187, 1189), 1193 (1186, 1187, 1188), 1199 (1197, 1199); **6**: 4245 (4241)
Lemaire B, **2**: 1262 (1259)
Lemaitre M, **1**: 120
Lemaréchal C, **1**: 405 (405); **2**: 797 (796); **5**: 3531 (3529), 3546 (3544, 3545)
Lemieux P F, **7**: 5106 (5103)
Lemoine A J, **6**: 4395 (4395); **7**: 4654 (4653)
Lemoine M, **1**: 120
Lenat D B, **2**: 1573
Lendaris G G, **7**: 5177 (5172)
Lenox T G, **1**: 226
Lentes H P, **6**: 3663
Lenton R L, **7**: 5140
Leon A, **4**: 2601 (2598)
Leon G J, **7**: 4732
Leonardo da Vinci, **3**: (1675)
Leondes C T, **2**: 785 (782, 784), 1186 (1182), 1257 (1257); **4**: 2277 (2273, 2275), 2491, 2492 (2491), 2570 (2569, 2570), 2618 (2616), 2671 (2670); **5**: 3012, 3144; **6**: 4139 (4133), 4193 (4190); **7**: 4439 (4439), 4540 (4535)
Leonhard W, **5**: 3097; **7**: 5027
Leontief W W, **2**: 1342, 1486 (1479); **3**: 2016 (2016); **4**: 2522 (2517); **5**: 2929 (2929); **6**: 4237 (4235)
Lepage W R, **4**: 2664
Lepeschkin E, **1**: 769 (760, 761)
Leppard G G, **1**: 7 (5)
Lepschy A, **1**: 493 (492)
Leray J, **1**: 529 (520, 521)
Lerman A, **4**: 2646 (2646); **7**: 5094 (5093)
Lerner A Ya, **5**: (3511); **6**: 3870 (3867)
Lerner D, **1**: 556
Lerski R A, **7**: 4964
Lesea A, **6**: 4294
Lesiak C, **3**: 1779 (1770, 1777); **6**: 3961 (3957, 3958)
Leslie C M, **2**: 1233 (1233)
Leslie D H, **5**: 3470 (3469)
Leslie E G, **6**: (3752)
Leslie P H, **2**: 1333 (1333); **6**: (3751, 3760); **7**: 4645 (4644)
Lesmo L, **1**: 769 (763, 764)
Lesort J B, **7**: 4929 (4929)
Lesourne J, **2**: 1269
Less S M, **7**: 4809

Lessem A S, **6**: 4230 (4228)
Letov A M, **5**: 3389 (3387)
Lettenmaier D P, **4**: 2652 (2651)
Letty L, **5**: 3496 (3494)
Leuer J F, **6**: 3820 (3817)
Leunberger D G, **2**: 1099 (1093)
Leung Y, **2**: 776 (773, 774, 775)
Leusen I, **1**: 542 (540)
Lev-Ari H, **2**: 785 (783, 784), 1102 (1102)
Leveau B, **2**: 781 (779)
LeVeen H H, **1**: 450 (448)
Levenshtein V I, **6**: 3653 (3649)
Levenspiel O, **1**: 589 (586); **4**: 2789 (2785)
Leventhal L A, **5**: 2980, 2989
Levèque D, **3**: 1648 (1644)
Levick W R, **7**: 5070 (5070)
Levin H R W, **2**: 1388
Levin M, **3**: 1736, 1741 (1739)
Levin M J, **4**: 2273 (2272)
Levin R, **2**: 1132 (1126)
Levin R I, **6**: 3904
Levin S A, **1**: 675 (674); **2**: 1317 (1316), 1318, 1321, 1538 (1537); **7**: 4646 (4645)
Levin S S, **2**: 1355 (1352)
Levine H B, **6**: 4235 (4230)
Lévine J, **2**: 1269 (1263, 1264, 1265, 1266, 1267)
Levine M D, **4**: 2488
Levine R D, **4**: 2508 (2507)
Levine R L, **6**: 3989 (3988)
Levine S H, **5**: 2930 (2928); **7**: 4956 (4955, 4956)
Levine W S, **5**: 3572 (3568); **7**: 4685 (4678, 4680, 4684), 4689 (4685), 4929 (4929)
Levins R, **2**: 1355 (1352); **7**: 4956 (4954)
Levinson E, **5**: 3389 (3385); **7**: 4452 (4451)
Levinson N, **1**: 413 (411); **2**: (871); **5**: 3563; **7**: 4566 (4563)
Levinson S E, **7**: 4509 (4508)
Levis A H, **2**: 1099 (1095); **5**: 3572 (3568)
Levy A H, **6**: 4007 (4007)
Levy B, **7**: 4466 (4465)
Lévy B C, **1**: 714 (714); **6**: 3751 (3751)
Levy D, **1**: 512 (511)
Levy G, **2**: 1248 (1247)
Levy M N, **2**: 853 (851)
Lévy P, **3**: 1779 (1776, 1777); **5**: 2934 (2932), 3376 (3374); **7**: (4638)
Levy W J, **1**: 281 (279)
Lewandowski A, **2**: 1276 (1274); **3**: 1990 (1989)
Lewandowski R, **7**: 5155 (5154)
Lewins J, **3**: 1693 (1688)
Lewis B L, **3**: 2192 (2186)
Lewis B R, **6**: 4302 (4301)
Lewis C K, **5**: 3282 (3282)
Lewis E, **6**: 4257
Lewis E V, **3**: 1766 (1766)
Lewis F L, **4**: 2835 (2835); **5**: 3475 (3472)
Lewis G N, **2**: 1479 (1474); **7**: 4901 (4898)
Lewis J B, **1**: 181
Lewis J M, **2**: 1155 (1150)
Lewis P A W, **3**: 1723; **6**: 4302 (4300)
Lewis P M, **2**: 884 (881, 882)
Lewis R A, **4**: 2440 (2432)
Lewis R M, **5**: 3310 (3305)
Lewis R M II, **2**: 1155 (1145)
Lewis R W, **3**: 2117 (2117)
Lewis T, **2**: 1410
Lewis T G, **5**: 2956 (2954)
Lewis Buck N L, **3**: 1710
Lewontin R C, **7**: 4646 (4645)
Lhote F, **3**: 2163 (2153); **5**: 3480, 3481; **6**: 4108 (4107)
Li C H, **1**: 473 (470, 471), 474 (469, 470)
Li R Y, **6**: 3653
Li T-Y, **1**: 567 (565, 566)
Lian D, **6**: 4108 (4104)
Liano F, **1**: 310 (308)
Liapis A I, **4**: 2301 (2293, 2301)
Liaw W K, **4**: 2782 (2778, 2780)
Libby R, **3**: 2192 (2186), 2203 (2196)
Libchaber A, **1**: 570 (568)
Liberman S, **6**: 3806 (3802)
Libert G, **1**: 608 (607)
Liblong B, **4**: 2550 (2547)
Licence J V, **3**: 1998
Licht S, **1**: 450 (447, 449)
Lichtenstein S, **3**: 2203 (2196); **5**: 3116 (3113); **6**: 4075
Liddicoat R T, **1**: 474 (469)
Liddicoat T J, **5**: 3464 (3457)
Lidgate D, **2**: 1388 (1387)
Lidskii E A, **6**: 4011 (4010)
Lidstone G J, **2**: 1479 (1476)
Lieber Z, **2**: 1099 (1094, 1095, 1098)
Lieberman C, **4**: 2447 (2447)
Lieberman E B, **7**: 4929, 4930 (4928)
Lieberman G J, **4**: 2842; **6**: 3904
Lieberman H A, **6**: 3695
Lieberman L I, **4**: 2386 (2374), 2469 (2466), 2487
Liebhafsky H A, **7**: 4504
Liégeois A, **6**: 4086
Liénard A, **5**: (3383, 3384)
Lieuson H Y, **2**: 1116 (1110, 1111); **5**: 3188 (3188)
Lifshin E, **2**: 1420; **6**: 4152
Ligda M G H, **6**: 4036 (4033)
Light T S, **6**: 3688 (3685, 3686)
Lightfoot E N, **1**: 315 (314)
Lighthill M J, **7**: 4922 (4921)
Lightstone A H, **3**: 1791
Liguori C, **5**: 3470 (3469)
Likens G E, **2**: 1321 (1318, 1320), 1358; **5**: 3268 (3266)
Liljenquist J E, **3**: 2174 (2171); **4**: 2631 (2628)
Lillebuen B, **2**: 1416 (1414, 1416)
Lillien L, **2**: 1165
Lilly J C, **1**: 437
Lilly J K, **7**: 4856 (4853)
Lim E, **2**: 1479 (1478)
Lim H C, **3**: 1611 (1609); **6**: 4245 (4242)
Lim K W, **6**: 4197 (4194), 4209 (4205, 4208)

Limebeer D J N, **1**: 296 (294); **4**: 2588 (2585), 2832 (2830), 2871 (2869); **5**: 3224 (3223); **7**: 4457 (4455, 4456)
Lin C S, **4**: 2386 (2369, 2377, 2378, 2379), 2412 (2401), 2440 (2434, 2438, 2439)
Lin C T, **2**: 952 (949)
Lin D W, **5**: 3077 (3077)
Lin I J, **6**: 4252 (4251)
Lin J G, **5**: 3164 (3158, 3161), 3165 (3158)
Lin K C, **5**: 2943 (2935)
Lin P M, **7**: 4732
Lin S, **7**: 4666 (4665)
Lin S C, **5**: 3470 (3466)
Lin T H, **1**: 51 (48)
Lin T L, **7**: 4929 (4929)
Lin W T K, **2**: 1154 (1153)
Lin Y H, **1**: 60 (56, 58, 59), 106 (101, 104); **5**: 3055 (3048, 3050), 3060 (3056, 3057, 3059), 3061 (3061), 3065 (3062, 3063, 3064, 3065), 3067 (3066), 3070 (3070); **6**: 3929 (3929), 4209 (4206); **7**: 4611 (4607)
Linaweaver F P, **7**: 5155 (5154)
Lind J F, **2**: 1040 (1036)
Lind N C, **7**: 4799 (4794)
Lindberg D A B, **2**: 1577 (1574)
Lindblom B E, **7**: 4509 (4505, 4507)
Lindemann R L, **2**: 1344 (1343); **7**: 4956 (4954)
Linden K J, **4**: 2511 (2509)
Linden R J, **7**: 4873
Linden T A, **1**: 755 (753)
Lindenberg S, **1**: 333 (330)
Lindenhoker D, **1**: 437
Lindenmayer A, **3**: 2046 (2045); **4**: 2798 (2795, 2797, 2798); **6**: 4353 (4351); **7**: 4958 (4957)
Lindenstrauss J, **1**: 413 (411)
Linderstrøm-Lang K, **6**: 3691 (3688)
Lindgren A S, **5**: 3389 (3386)
Lindgren O H, **2**: 1479 (1478)
Lindley D, **4**: 2338 (2338)
Lindley D V, **3**: 2203 (2195)
Lindman B, **1**: 13 (13)
Lindner D K, **1**: 137 (134, 135, 136, 137)
Lindorff D P, **1**: 100 (99), 105 (102)
Lindquist A L, **2**: 827 (823); **7**: 4468 (4467), 4469 (4467)
Lindroth P, **7**: 5123 (5122)
Lindsay B G, **2**: 1154 (1144), 1174 (1170, 1171)
Lindsay R K, **2**: 1573
Lindsay S J, **3**: 1669 (1666)
Lindsay W L, **1**: 7 (3, 4), 148 (143)
Lindsey W C, **1**: 80 (80), 91 (86, 87)
Lindstedt A, **5**: 3389 (3383, 3384)
Lindstrom K, **4**: 2747
Lindström L H, **5**: 3464 (3458)
Ling C H, **3**: 1781 (1780)
Ling K, **5**: 2943 (2935)
Ling R Y-S, **2**: 912
Liniger W, **2**: 1305 (1298, 1302)
Linkens D A, **1**: 429 (427), 459 (457, 459), 535 (535); **2**: 1040 (1036, 1037), 1043 (1042); **5**: 3024 (3024), 3238 (3235, 3236, 3237), 3253 (3249, 3252); **7**: 4577 (4573)
Linkins P W, **3**: 1665 (1652)
Linnemann A, **4**: 2827 (2820)
Linnemann H, **2**: 923 (921); **3**: 2016 (2016)
Linnett D T, **3**: 1975
Linstone H A, **1**: 372 (360), 618; **2**: 1542 (1541), 1545 (1543); **7**: 4785 (4782), 4827 (4825)
Linton T D, **2**: 912 (910)
Linvill W K, **6**: 4146 (4145)
Linzey P T, **2**: 1431 (1427)
Lion B, **7**: 4749 (4746)
Lions J L, **1**: 529 (519, 520, 521); **2**: 808 (802, 803, 804), 812 (809), 1121 (1116), 1179 (1174, 1177), 1180, 1186 (1182, 1185), 1192 (1191, 1192), 1211, 1261, 1262 (1260); **3**: 1632, 1648 (1644), 1728 (1727), 1734 (1730), 2151, 2152 (2152); **4**: 2342 (2342), 2343, 2559 (2559), 2664 (2661, 2662, 2663), 2807 (2806, 2807), 2861, 2866 (2865); **5**: 3279 (3279), 3310 (3309), 3376 (3374, 3376), 3487 (3483, 3484), 3496 (3489, 3492, 3493, 3495, 3496), 3506, (3517), 3523 (3522), 3530 (3525), 3535 (3534); **6**: 3596 (3592, 3593, 3594, 3595), 3755, 4070 (4067), 4223 (4216, 4219, 4222), 4248 (4248); **7**: 4468 (4467), 4498 (4497), 4612, 4617, 4644 (4643), 4944 (4943), 5036 (5031, 5032, 5033, 5034, 5035, 5036), 5041 (5037, 5038)
Lions P L, **1**: 529 (529); **2**: 1017 (1014, 1015, 1016), 1279 (1278); **3**: 2068 (2064, 2065, 2066, 2067), 2072 (2069, 2070, 2071, 2072); **5**: 3523 (3520), 3531 (3526), 3538; **7**: 4613, 4626 (4624), 4627 (4623, 4624, 4625, 4626), 5047 (5045)
Lipinski H, **7**: 5177 (5172)
Lipo T, **7**: 5027
Lippman S A, **2**: 1099 (1094); **6**: (3892)
Lippmann G, **1**: (532)
Lipski W, **2**: 884 (883)
Lipson J D, **3**: 1723 (1722)
Lipster R S, **5**: 3336 (3333, 3334, 3335)
Liptak B G, **7**: 4865
Liptser R S, **5**: 3303 (3293, 3294, 3299); **7**: 4637 (4631, 4633)
Lishka R J, **7**: 5114
Liskov B, **2**: 1136 (1134); **3**: 1642 (1638); **4**: 2681 (2680)
List E J, **4**: 2789 (2788)
Litovitz T, **5**: 3464 (3459)
Little R C, **1**: 456 (451)
Little R W, **1**: 486 (483)
Littlewood A B, **3**: 1931
Littlewood B, **4**: 2681 (2679)
Littlewood J E, **1**: 570 (568); **4**: 2614 (2611)
Litwin W, **2**: 1160 (1157)
Litz L, **1**: 132 (128)
Liu C-H, **5**: 3303 (3301), 3310 (3304, 3306, 3307, 3308)
Liu D, **5**: 3413 (3410, 3411)
Liu L K, **5**: 3188 (3188)
Liu L Y, **3**: 2217
Liu M A, **2**: 1165 (1165)
Liu X, **6**: 3947
Livesley R K, **5**: 3373 (3370), 3390 (3385)
Livingston G K, **5**: 3464 (3460)
Livingston J S, **4**: 2685 (2681)
Livingston W K, **2**: 873
Ljung B, **1**: 542 (540)

Ljung L, **1**: 80 (75, 76, 77); **2**: 785 (782, 783, 784), 1102 (1102), 1199 (1197), 1257 (1257); **4**: 2239 (2236, 2239), 2245 (2242), 2270 (2270), 2277 (2276, 2277), 2284, 2292 (2288, 2291), 2293 (2288, 2291), 2307 (2306), 2314 (2310, 2312, 2313), 2319 (2317, 2319), 2892; **5**: 3004 (3003), 3061 (3061), 3430 (3426); **6**: 3994 (3991, 3992, 3993, 3994), 4184 (4183), 4186 (4185), 4187 (4186), 4197 (4194), 4204 (4203), 4209 (4207); **7**: 4466 (4463), 4469 (4466, 4467, 4468), 4551 (4548), 4607 (4603, 4604, 4606), 4611 (4607, 4608), 4692 (4690), 4894 (4892, 4893)
Llewellyn G J, **3**: 1693 (1692)
Llinas R, **1**: 466 (464)
Lloyd A C, **1**: 187 (187)
Lloyd B B, **6**: 4061 (4055, 4056)
Lloyd R, **5**: 3012
Lloyd Morgan C, **3**: 2164
Lo J T, **5**: 3336 (3335)
Lo K L, **3**: 1990 (1989)
Lobbia R N, **6**: 4165 (4162)
Lobry C, **2**: 1583 (1582); **3**: 1779 (1770); **4**: 2759 (2757); **5**: 3184 (3182), 3366 (3366); **7**: 4677 (4676, 4677)
Locatelli A, **2**: 1093 (1090); **7**: 4689 (4687, 4688, 4689)
Loche J R, **3**: 1601 (1600)
Locher P R, **5**: 3425 (3423)
Locke A S, **5**: 3012
Locke J, **6**: (4166)
Locker A, **1**: 333
Lockett J, **4**: 2664
Lodge J H, **5**: 3127 (3125, 3126)
Lodzinski A, **2**: 1276 (1274)
Löe H, **2**: 973 (971)
Loeb J M, **5**: 3389 (3385)
Loeb J N, **2**: 1249 (1245)
Loeb M, **2**: 995 (995); **7**: 4822 (4819)
Loehe J R, **7**: 4897 (4896)
Loève M, **2**: 1478 (1474), 1479 (1477); **5**: 3303 (3295)
Loeve W, **2**: 1179 (1176)
Löfgren L, **1**: 333 (326, 329, 331, 332), 709 (704, 705, 706, 707, 708, 709), 720 (718, 719); **2**: 937 (937); **3**: 2003 (2000)
Logan B F Jr, **4**: 2331 (2330)
Logofet D O, **1**: 675 (674); **2**: 1080 (1080), 1351 (1348)
Lohman G, **2**: 1165
Lohnes P R, **2**: 879 (876)
Loible J M, **3**: 2014
Łojasiewicz S Jr, **4**: 2609; **5**: 3184 (3182); **6**: 3956 (3956)
Lolwen P, **5**: 3184
Lombari T, **5**: 3279 (3276)
Lomer W M, **5**: 3416
Lomet D B, **2**: 1155 (1150), 1174 (1170)
Lomow G, **1**: 755 (753)
Long R S, **5**: 3060 (3059); **6**: 4193 (4193)
Long S J, **1**: 177 (174)
Longhurst A R, **6**: 3621 (3621); **7**: 4895 (4895)
Longmuir R J, **3**: 1958
Longobardo G S, **6**: 4066 (4062)
Loo S G, **2**: 1473 (1469)
Loomis R G, **2**: 1479 (1475)
Loomis R S, **1**: 142 (139, 140)
Loos R, **5**: 3560 (3559)
Lootsma F A, **7**: 4975 (4973)
Looze D P, **5**: 3228 (3225); **6**: 4227 (4224, 4225, 4226)
Loparo K A, **5**: 3303 (3301); **7**: 4709 (4709)
Lopez A M, **3**: 2113 (2111, 2112)
Lorange P, **6**: 4075
Lord R E, **2**: 827 (818)
Lorente N R de, **2**: (872)
Lorenz E N, **1**: (566)
Lorenzen M W, **4**: 2794 (2789)
Lorie R A, **2**: 883 (881, 882), 1154 (1144, 1145, 1147), 1174 (1170, 1171)
Loś J, **6**: 3710 (3707)
Losito M P, **4**: 2292 (2291); **5**: 3044 (3042)
Lotka A J, **5**: 2956 (2956); **6**: (3751)
Lottin J, **7**: 4741 (4740), 4751 (4751)
Loucks D P, **2**: 1538 (1537); **7**: 5140
Loucks O L, **4**: 2646 (2645)
Loughhead J, **4**: 2904 (2902)
Louis J P, **5**: 3097
Louis-Guerin C, **3**: 1917 (1914)
Lourtie P M G, **1**: 583 (581); **4**: 2597 (2588)
Lousard P D, **7**: 4749 (4746)
Loute E, **2**: 959
Lovasz L, **6**: 3739 (3738)
Love L J C, **1**: 177 (174)
Love P E, **2**: 1454 (1452); **3**: 1719 (1715)
Lovelace D E, **3**: 1769 (1767, 1768)
Low S S, **6**: 3806 (3805)
Lowdenslager D, **5**: 3130 (3129)
Lowe G D O, **1**: 445 (441, 442, 444)
Lowe K G, **2**: 1410 (1408)
Lowen R, **3**: 1781 (1781), 1802 (1801); **6**: 3710 (3707, 3709, 3710)
Lowenstein O, **2**: 1595 (1592)
Lowman P D, **6**: 4041
Lowry I S, **7**: 4495 (4493, 4494)
Lowther J A, **3**: 1998
Lozada F, **2**: 1248 (1248)
Lozano R, **5**: 3004 (3002, 3003); **6**: 3604 (3602)
Lozano-Pérez T, **3**: 1642 (1638); **4**: 2386 (2373), 2447 (2447), 2460 (2459)
Lozier J C, **5**: 3389 (3386)
Lu K M, **4**: 2892
Lu L, **3**: 1838
Lu R, **5**: 2989 (2987)
Lu S Y, **6**: 3653 (3648, 3649, 3650, 3651, 3652), 4160 (4158)
Lu Y C, **1**: 550
Lubbock J K, **3**: 1779 (1776)
Lucas H C Jr, **3**: 2192 (2186)
Lucas H L, **1**: 489
Lucas P G, **2**: 1388
Lucas W F, **2**: 850 (847)
Luce R D, **1**: 276 (272); **2**: 850 (840, 843, 846, 847), 942; **3**: 1921 (1918), 2061 (2056); **5**: 2911 (2907, 2910), 3115 (3105, 3107, 3109), 3116 (3109); **7**: 4991 (4987)
Luce R L, **3**: 1926 (1923)
Luckham D C, **2**: 884 (881)
Lucky R W, **1**: 80 (76)

Lüders G, **1**: 106 (102); **4**: 2314 (2310)
Ludwig N, **1**: 340
Luenberger D G, **2**: 789 (786, 788), 801 (799), 932 (930, 931), 959 (958), 1125 (1124); **3**: 1990 (1987), 2163 (2155, 2156); **4**: 2692, 2697 (2694), 2702 (2697), 2836 (2836), 2839 (2837, 2839), 2871 (2871); **5**: 3156 (3153); **6**: 3717 (3713), 3720 (3719, 3720), 3734 (3733), 4299; **7**: (4553), 4709 (4705), 4969, 4971 (4969), 4975 (4974)
Lugliani R, **6**: 4051 (4050)
Luh J Y S, **4**: 2386 (2369, 2377, 2378, 2379, 2380, 2382), 2412 (2392, 2401, 2402), 2440 (2421, 2426, 2434, 2438, 2439); **6**: 4086 (4083), 4100
Luh P, **7**: 4822 (4821)
Luh P B, **1**: 687 (685); **2**: 995 (994, 995); **7**: 4631 (4628, 4629, 4630)
Lui C L, **1**: 725 (722)
Luk J Y K, **7**: 4925
Łukasiewicz J, **3**: (1785); **5**: 2921 (2918), (2924)
Luker P A, **4**: 2550 (2550)
Lukes D, **2**: 1269 (1264)
Lull H W, **5**: 3268 (3266)
Lumer G, **6**: 4223 (4218)
Luminet J P, **6**: 4336 (4335)
Lund T, **5**: 3464 (3463)
Lundan A, **1**: 564 (561)
Lundberg K, **2**: 781 (777); **5**: 3464 (3458)
Lundeen H C, **2**: 978
Lundgren J R, **6**: 4333 (4329)
Lundin R A, **6**: 3706
Lundquist F, **6**: 3699 (3697)
Lundstrom B, **4**: 2444 (2441)
Lundstrom G, **6**: 4108 (4102)
Lung W S, **4**: 2794 (2789)
Lunn J N, **1**: 284 (281)
Lunnon P W, **3**: 1958
Luno J, **1**: 310 (308)
Lur'e A I, **5**: 3389 (3387)
Lush D M, **1**: 159, 174
Lusk E J, **7**: 4793 (4791, 4792)
Luss D, **1**: 589 (587)
Lutfi R A, **1**: 326 (324)
Lütke-Daldrup B, **6**: 3834 (3832)
Lutte N P, **6**: 4204
Luttinger J M, **4**: 2614 (2611)
Lutzeyer W, **7**: 4980 (4980), 4981
Luukkanen J J, **6**: 3683 (3682)
Luxmoore R J, **1**: 142 (138, 140, 141), 592 (590, 591)
Luxon B, **3**: 2143 (2141)
Luyben W L, **1**: 589 (586); **2**: 1109 (1108)
Luzi A D, **7**: 4785
Lyapunov A M, **3**: (1895); **4**: 2565 (2560); **5**: 3389 (3384, 3387), 3398 (3395); **7**: (4517, 4551)
Lyle J H, **5**: 3470 (3469)
Lyles M A, **7**: 4785
Lynch A J, **2**: 781 (780); **6**: 4252 (4251)
Lynch H W, **1**: 751 (751)
Lynch N A, **2**: 1165 (1164)
Lynch P, **6**: 3850 (3847)
Lyndon R, **2**: 937 (933)
Lyndon-Bell R M, **7**: 4504 (4499)
Lynes J A, **4**: 2769 (2769)
Lyness F K, **3**: 1981 (1980), 1998
Lynn P A, **6**: 4281 (4278), 4294
Lyon D, **3**: 2203 (2201)
Lyons M G, **3**: 1665 (1656)
Lyubchik L M, **4**: 2737 (2731)

M

Ma P R, **2**: 1282 (1281)
Maarleveld A, **2**: 1109 (1108, 1109)
Maas A I R, **4**: 2582 (2582)
Maas H, **7**: 4851
Maas J, **6**: 3954 (3952, 3953)
Maas S J, **1**: 142 (141)
McAdams W H, **3**: 2105 (2101)
Macagno M, **2**: 1532
McAllister D F, **2**: 1087
MacArthur J W, **3**: 2117 (2114, 2117)
MacArthur R H, **1**: 675 (672, 674)
McArthur W, **4**: 2775 (2773)
McBean E A, **7**: 5140 (5138)
MacBean P, **4**: 2871 (2867, 2868, 2869)
McBride L E, **7**: 5170 (5166)
McBrooks C, **6**: 4061 (4056)
Maccacaro G A, **2**: 1479 (1475)
McCaffrey R J, **4**: 2782 (2780)
McCallion H, **6**: 4108 (4106)
McCandless W, **7**: 4749 (4745)
McCarthy J, **2**: 869 (868); **4**: 2640 (2640)
McCartney B S, **7**: 4848 (4844)
McCarty P L, **5**: 3266 (3265); **7**: 5114
Macchi O, **1**: 80 (77, 78, 79)
McClellan J H, **5**: 3130 (3129)
McClelland G H, **1**: 372 (369); **3**: 2192, 2202 (2196); **5**: 3115 (3113)
McClelland J L, **6**: 4167 (4167)
McClintock R B, **7**: 4585 (4581)
McCloy D, **6**: 4108, 4257
McCluney W R, **6**: 4045 (4045)
McClure W, **7**: 4877 (4875)
MacColl L A, **5**: 3389 (3385, 3386), (3516); **6**: 4146 (4145)
McCollister G M, **2**: 1497 (1494)
McComas A J, **2**: 1413 (1411)
McConnell R K, **2**: 1479 (1478)
McCook R D, **7**: 4877, 4880 (4879, 4880)
McCool J M, **1**: 81 (78, 79); **4**: 2314 (2311)
McCorduck P, **1**: 307
McCormack W M, **6**: 4415 (4415)
McCormick E J, **3**: 2186 (2180, 2185)
McCormick G P, **2**: 801 (798, 799); **4**: 2880 (2879); **5**: 3362; **6**: 4310 (4309)
McCracken D, **6**: 3898 (3895)
McCracken M, **7**: 4529 (4526)
MacCrimmon K R, **3**: 2192 (2186), 22022203 (2194, 2195)
McCulloch A, **4**: 2631 (2629, 2631)
McCulloch W S, **2**: 869 (868), 873 (870, 871, 872); **6**: 4167 (4165)
McCullough R L, **7**: 5060 (5059)
McDanell J P, **5**: 3508 (3508)
McDonald D, **5**: 3389 (3386), 3513 (3510)
MacDonald D C, **5**: (3516)
MacDonald D W, **7**: 4848 (4847)
Macdonald G A, **3**: 1681 (1680)
MacDonald I C, **7**: 4856 (4853, 4854)
McDonald J P, **6**: 3820 (3816)
MacDonald N, **1**: 429 (428)
McDonnell J D, **3**: 1668 (1666, 1667)
McDonnell J P, **7**: 4425 (4425)
McDowell-Boyer L M, **6**: 3933 (3931), 3936
MacDuffee C C, **7**: 4457 (4455)
Macedo P, **5**: 3464 (3459)
McElhaney J H, **1**: 474 (469)
MacElroy R D, **6**: 4302 (4302)
McEntire P L, **2**: 1282 (1279)
MacEwan S R, **2**: 1181 (1181)
McEwen J, **6**: 3983 (3978)
McFadden D, **4**: 2559 (2559)
McFadden W H, **3**: 1939; **5**: 2943 (2936)
McFarland D J, **3**: 2213 (2213)
McFarland F J, **3**: 2213 (2213)
McFarland R A, **1**: 437
MacFarlane A G J, **1**: 580 (575, 577, 578, 580), 583 (581, 582), 735 (732, 734, 735), 736 (732); **3**: 1749 (1745, 1748); **4**: 2832 (2828, 2830, 2831); **5**: 3201 (3197, 3199), 3228 (3224, 3225, 3226, 3227), 3383 (3380, 3381), (3387), 3436 (3431, 3432), 3441 (3441); **6**: 3717 (3711), 3751 (3748); **7**: 4554 (4551)
Macfarlane D E, **2**: 1248 (1245)
Macfarlane P W, **2**: 1407 (1405)
McFee R, **1**: 535 (532), 536 (532, 534)
McGarry J D, **4**: 2631 (2628)
McGee V E, **7**: 4827 (4826)
McGeer E G, **1**: 466 (465)
McGeer P L, **1**: 466 (465)
McGhee R B, **4**: 2386 (2379, 2382); **6**: 4128 (4124)
McGhee T, **7**: 5099
McGill R, **7**: 5015 (5014)
McGill W B, **1**: 142 (141)
McGill W J, **4**: 2497 (2492), 2547 (2545)
McGinnis J T, **2**: 1344 (1344)
McGrath E J, **2**: 1551 (1550)
MacGregor J F, **6**: 3745 (3740)
McGuire C B, **7**: 4822
McGuire E A, **3**: 2174 (2172)
Machado J, **3**: 2225 (2223)
McHale J B, **6**: 4271 (4269)
McHarness D D, **5**: 3266 (3265)
Machate R D, **2**: 1388 (1387, 1388)
Machell F W, **7**: 4822 (4815)
Macho D, **4**: 2713 (2710)
McIntire C D, **2**: 1329 (1329)
McIntosh J E A, **3**: 2174 (2174)
McIntosh R P, **2**: 1318 (1314), 1321, 1322 (1321); **3**: 2174 (2174)
McIntyre D A, **1**: 642 (639); **7**: 4866 (4865), 4868 (4868)
McIntyre H C, **5**: 3440
McIvor M C, **5**: 3421 (3419)
McJones P, **2**: 1154 (1144)
McKague R L, **5**: 3282 (3281)

Mackay A L, **1**: 674 (672)
Mackay D, **7**: 4901 (4898, 4899, 4901)
McKay E M, **2**: 1427 (1425), 1445 (1441, 1443, 1444, 1445)
McKay M D, **2**: 1551; **6**: 4235 (4231, 4233)
MacKean H, **5**: 3535 (3533)
MacKean H P, **1**: 512
Mackechnie Jarvis C, **2**: 1402 (1399, 1400)
McKee D, **5**: 2993 (2991)
McKeeney J L, **7**: 4793 (4792)
MacKellar A C, **1**: 201 (200)
McKennan G T, **4**: 2769 (2769)
McKenny J L, **4**: 2353 (2352)
Mackenzie K D, **7**: 4785
McKenzie L W, **1**: 671 (668, 669)
McKeoun D Jr, **6**: 4160 (4159)
McKeown M, **5**: 2943 (2934)
McKersie R B, **5**: 2911 (2911)
Mackey M C, **1**: 570 (568, 569); **2**: 1186 (1184)
Mackey P J, **5**: 3282 (3282)
McKim R H, **2**: 992 (991)
McKinion J M, **1**: 142 (140, 141)
McKinney G T, **1**: 142 (140)
Mackintosh N J, **5**: 3264 (3263)
Macko D, **1**: 405 (400, 404); **2**: 862 (861), 956 (952); **3**: 2043 (2039), 2163 (2153, 2155, 2159); **5**: 3568 (3564, 3568); **6**: 4322 (4319); **7**: 4571 (4570), 4666 (4661)
McLachlan N W, **4**: 2331 (2329)
McLafferty F W, **3**: 1936 (1933), (1938); **5**: 2943
Maclaine C G, **5**: 3242 (3241)
MacLane G, **4**: 2871 (2866, 2867)
MacLane S, **6**: 3616 (3616)
Maclaren M D, **6**: 3945
McLaren R W, **6**: 4165 (4163)
McLaughlin C D, **7**: 4749 (4745)
McLean E R, **6**: 4332 (4329, 4331, 4332)
McLean T P, **7**: 4871 (4869)
MacLennan D N, **7**: 4848 (4847)
McLeod J, **2**: 1431 (1427)
McLeod J M, **2**: 1318
McLeod R S, **1**: 735 (734)
MacLinden L, **2**: 1262 (1259)
McLoughlin J, **6**: 3691 (3688)
McLoughlin J B, **7**: 4495 (4495)
McMahon D H, **5**: 3464 (3457, 3464), 3465 (3459)
McMahon G W, **5**: 3470 (3467)
McMillan B, **7**: 4457 (4456)
Macmillan T I, **1**: 315 (314)
McMillian C, **6**: 3904
MacMurray S B, **1**: 315 (314)
McMurtrie R, **4**: 2293 (2290)
McMurtry G J, **6**: 4165 (4163)
MacNair E A, **4**: 2550 (2550), 2551 (2548, 2549)
McNamee J, **4**: 2905
MacNeal R H, **5**: (3516)
McNeill G V, **4**: 2769 (2769)
McNiven M, **3**: 1947
McPherson M, **6**: 4341 (4337, 4341)
McPherson S P, **1**: 319
McQuaker N R, **1**: 177 (174)
McQuay H J, **2**: 963 (962, 963)
McQuillan J, **2**: 1125 (1122, 1123)
McQuiston F C, **6**: 3918 (3917)
MacRae D Jr, **2**: 1566 (1563)
McRae G J, **1**: 187 (187); **7**: 4518 (4517)
McReynolds S R, **7**: 4540 (4535, 4536, 4538)
McRuer D T, **1**: 120, 220 (220), 343, 393
McShan D L, **6**: 4302 (4302)
MacShane E, **1**: 529 (526, 527)
McShane E J, **5**: 3082 (3082), 3513 (3512); **7**: 4637 (4635)
McShane W R, **7**: 4929, 4930 (4928)
McWilliam P N, **5**: 3242 (3241)
MacWilliams F J, **5**: 2956 (2955)
Maday Y, **7**: 4498 (4496)
Maddison R N, **4**: 2659 (2656)
Madison B L, **5**: 3421 (3418)
Madnick S E, **4**: 2504 (2503); **5**: 3165, 3449; **7**: 4752
Madrigal O S, **7**: 5018 (5017)
Madsen N K, **2**: 1181 (1181)
Madsen R W, **6**: 4356 (4356)
Maeda T, **2**: 969
Maeda Y, **1**: 184 (184); **2**: 1513
Mäenpää J, **4**: 2798 (2795)
Maes D, **7**: 4703 (4700)
Maffezzoni C, **7**: 4689 (4687, 4689)
Magalhaes M F, **1**: 682 (679)
Magee J, **2**: 1143
Mageirou E F, **2**: 1020 (1019, 1020)
Magenes E, **2**: 1186 (1185); **3**: 1728 (1727), 1734 (1730); **4**: 2664 (2662); **6**: 3596 (3592), 4223; **7**: 5036 (5033)
Magill D T, **4**: 2713 (2710)
Magnanti T, **4**: 2841
Magni J F, **1**: 703 (697); **5**: 3176; **7**: 4443 (4441)
Magora A, **3**: 1910 (1910)
Magyar J J, **5**: 2943 (2935)
Mah R S H, **4**: 2517 (2516)
Mahaffy J M, **1**: 428 (425, 427)
Mahalanabis A K, **3**: 1802, 1805 (1803, 1804); **4**: 2702 (2702); **6**: 4138 (4133)
Mahalanobis P C, **2**: 1106 (1103)
Mahar T J, **6**: 4346 (4346)
Mahdmina A, **6**: 3717 (3713)
Maher J F, **1**: 312 (310)
Mahesh J K, **3**: 2117 (2114, 2117)
Mahmoud M G, **1**: 703 (697)
Mahmoud M S, **2**: 862 (859, 861); **4**: 2697 (2692, 2695), 2713 (2709); **5**: 3019 (3017, 3018, 3019), 3040 (3039), 3077 (3074), 3144; **7**: 4434 (4430, 4433, 4434)
Mahmoud S, **3**: 2175 (2171)
Mahmoud S A, **2**: 1160
Mahoney E E, **7**: 4930 (4928)
Maier J, **2**: 1402 (1399)
Maier M, **4**: 2511 (2511)
Maijanen J, **5**: 2994 (2991)
Mailloux G E, **1**: 536 (534)
Maine L, **3**: 1943
Maitra A, **3**: 1926
MaizzaNeto O, **6**: 4088 (4086)
Majkowski J, **6**: 4235 (4230)

Majone G, **2**: 987; **7**: 4809
Major D C, **7**: 5140
Majumder D D, **7**: 4766
Makhonine V A, **6**: 4159 (4158)
Makhoul J, **7**: 4509 (4505)
Maki A W, **4**: 2646; **7**: 4901 (4899)
Makila P M, **1**: 589 (586, 587)
Makino H, **4**: 2392; **6**: 4108 (4105)
Mako D, **3**: 2038; **5**: 3019 (3015), 3040 (3035, 3036)
Makridakis S, **1**: 399; **3**: 2192; **4**: 2338 (2337); **7**: 4827 (4826), 5155 (5154)
Maksa G, **4**: 2508 (2505)
Malabre M, **4**: 2816 (2807), 2855 (2852, 2854, 2855)
Malakooti B, **5**: 3165 (3157)
Malandrakis C G, **6**: 4244 (4242), 4245 (4242)
Malanowski K, **2**: 812 (810, 812)
Malcev A, **6**: 3962 (3960)
Malcev A I, **3**: 1791 (1790)
Malchesky P S, **1**: 312 (312)
Malcolm D G, **4**: 2353 (2351)
Malcolm M A, **2**: 826 (815)
Maldonado T, **6**: (3862)
Malebranche H, **2**: 808, 1199 (1199); **6**: 4245
Malenge J P, **5**: 3480 (3477)
Malet M A, **2**: 1445 (1441)
Malgouyres G, **1**: 91 (91)
Maliakal J C, **5**: 2943 (2934)
Maling G C Jr, **3**: 1723
Malinowski K, **1**: 703 (693); **2**: 1305 (1299, 1302, 1304); **7**: 4571 (4568, 4569, 4571)
Malloum A, **4**: 2547 (2545, 2547)
Malmeisters A, **1**: 474 (471)
Maloney T E, **2**: 1538 (1537); **4**: 2646 (2645)
Malozemov V N, **5**: 3545 (3543)
Malsburg C von der, **6**: 4167 (4166), 4168 (4167)
Malthus T R, **6**: 3755 (3751), 3762 (3756, 3761)
Malueg K W, **4**: 2794 (2789)
Maly K, **2**: 1087 (1085)
Mamdani E H, **1**: 126 (124, 125), 664 (664), 713 (712); **3**: 1898 (1895); **4**: 2361 (2359, 2360), 2362 (2359, 2360); **5**: 3156 (3155), 3454 (3453)
Mamode A, **3**: 1665
Mamoud M S, **2**: 956 (952)
Man N K, **1**: 310 (308)
Mancini J L, **4**: 2646 (2645)
Mancy K H, **7**: 5115
Mandel P, **6**: 4261 (4258)
Mandel W, **1**: 769 (761)
Manes E G, **5**: 3082 (3080); **6**: 4328 (4328)
Manfield D R, **7**: 4732
Mangasarian O L, **1**: 628 (627); **2**: 797 (795), 1099 (1093); **5**: 3165 (3161), 3546
Manitius A, **2**: 1211 (1209); **3**: 2150, 2151 (2149), 2152 (2152); **5**: 3506 (3502)
Mankin J B, **2**: 1272 (1272), 1333 (1332), 1344 (1343), 1551 (1549, 1550); **6**: 4229 (4228), 4235 (4230, 4231, 4234); **7**: 4897 (4896), 5008 (5005)
Manlapig E, **5**: 2993 (2990)
Mann L, **2**: 987
Mann N R, **6**: 4029 (4027)
Manne A S, **6**: 3706; **7**: 4726 (4725)
Manning E, **2**: 1201 (1200)
Manning E G, **2**: 1087 (1084)
Manolitsakis I, **5**: 3130 (3128)
Manor R, **4**: 2501 (2500); **5**: 3116
Mansfield E, **5**: 2979; **6**: 3655; **7**: 4833 (4828), 5164
Mansfield P, **5**: 3425 (3422, 3424, 3425)
Mansfield P H, **7**: 4513
Mansfield T A, **2**: 1521 (1521)
Manson-Hing L R, **2**: 976
Mansour M, **2**: 1276 (1274); **5**: 3127 (3125, 3127), 3135 (3133), 3171 (3170); **6**: 4139 (4133)
Mantel R J, **7**: 4716 (4714)
Manzoni G, **6**: 3790
Mapelli P, **1**: 755 (752)
Mapleson W W, **1**: 284 (281, 283), 319
Marable J H, **2**: 1552 (1549)
Maral G, **2**: 1057 (1052)
Marannino P, **6**: 3954 (3952, 3953)
March J G, **7**: 4785 (4783, 4784)
Marchal C, **2**: 1017 (1014); **5**: 3488 (3487); **7**: 4490
Marchant C, **3**: 1975
Marchant G R, **5**: 3282 (3282)
Marchesini G, **2**: 1121 (1116); **5**: 3279 (3276), 3293 (3289), 3376 (3374, 3375); **6**: 3776 (3772, 3773), 3777 (3772)
Marciniak A, **6**: 4346 (4346)
Marcus M, **4**: 2613 (2612), 2614 (2611)
Marcus M L, **2**: 853 (851, 853)
Marcus S, **7**: 4822 (4815)
Marcus S I, **2**: 932 (931); **5**: 3293 (3289), 3303 (3299, 3301, 3302), 3310 (3304, 3306, 3307, 3308), 3329 (3327), 3330 (3325, 3327), 3336 (3336); **7**: 4637 (4635)
Marcy H T, **3**: (1748); **5**: 3389 (3385)
Marec J-P, **7**: 4490
Margalef R, **5**: 2956 (2956)
Margetts A R, **7**: 4848 (4847)
Margolis I B, **3**: 1888
Margulies G, **3**: 1665 (1653)
Margulis A R, **5**: 3421 (3420), 3425 (3425)
Mari A, **3**: 2025 (2022, 2024)
Maria G A, **5**: 3127 (3126)
Mariani E, **6**: 3954 (3952, 3953)
Mariani G, **4**: 2887 (2886); **6**: 3909 (3907)
Marinho J L, **6**: 3846 (3845)
Marinos P N, **1**: 664 (664); **3**: 1888
Maritsas D G, **1**: 634 (632)
Markel J D, **1**: 86 (85); **7**: 4509 (4507)
Markiewicz B R, **4**: 2412 (2394)
Markley F L, **4**: 2622 (2622)
Markov M D, **6**: 4108 (4105)
Markowitz H, **3**: 2060 (2056, 2057)
Markowitz H M, **2**: 894 (894); **6**: 4348 (4348), 4402, 4405 (4404)
Marks D H, **5**: 3164 (3161)
Marks J, **2**: 1410
Marks P, **4**: 2362
Marks R M, **1**: 482 (480)

Marks S, **2**: 1106
Markus L, **5**: 2963 (2962, 2963), 3497, 3514 (3514), 3515 (3514, 3515), (3517)
Marliss E B, **3**: 2032 (2030)
Marlow K C, **7**: 5150
Marmarelis P Z, **1**: 490 (488, 489); **7**: 5070 (5066)
Marmarelis V Z, **1**: 490 (488, 489); **7**: 5070 (5066)
Marmorat J P, **6**: 3964
Marotto F R, **1**: 567 (566)
Marr A G, **1**: 428 (426, 427, 428)
Marr D C, **1**: 466 (464); **4**: 2331 (2329, 2330), 2484 (2481); **7**: 5065 (5060, 5062, 5064)
Marris J H S, **3**: 1943, 1958 (1955)
Marro G, **1**: 137 (134); **4**: 2702 (2699), 2816 (2807); **6**: 3983 (3983)
Marrocco A, **2**: 1179 (1174, 1177, 1178)
Mars P, **7**: 4751 (4751)
Marsaglia G, **2**: 1087 (1086); **6**: 3945
Marschak J, **4**: 2560 (2557); **5**: 2911 (2908); **6**: 3860 (3859); **7**: 4991
Marsden J, **7**: 4529 (4526)
Marsden J E, **2**: 1583 (1582); **4**: 2759 (2757)
Marsh D J, **3**: 2032 (2026)
Marsh J T, **1**: 504 (502)
Marsh M L, **1**: 502 (499)
Marsh O, **3**: 2175 (2174)
Marshak J, **2**: 932 (931)
Marshall A W, **7**: 4657 (4655)
Marshall J, **5**: 3425 (3425)
Marshall J E, **5**: 3238 (3237); **7**: 4462 (4462), 4463 (4460, 4462)
Marshall L F, **1**: 502 (499)
Marshall M, **7**: 5186 (5180)
Marshall P M, **5**: 3464 (3460)
Marshall R J, **1**: 542 (540), 547 (544)
Marshall W, **5**: 3416
Martell A E, **1**: 7 (4), 8 (4)
Martell P E, **4**: 2782 (2779)
Martella G, **2**: 1160 (1157)
Martin A J P, **5**: 2943 (2934)
Martin B V, **7**: 4933
Martin C, **4**: 2866 (2865)
Martin C F, **5**: 3303 (3301); **6**: 3734 (3733)
Martin C J, **1**: 477 (477)
Martin D A, **3**: 1926 (1926)
Martin D W, **7**: 5106 (5103)
Martin G D, **6**: 3680 (3675, 3676)
Martin G J, **5**: 3421 (3418)
Martin H R, **6**: 4257
Martin J, **2**: 947; **3**: 2179; **4**: 2504 (2504)
Martin J A, **3**: 1736
Martin J C, **2**: 808 (802)
Martin M L, **5**: 3421 (3418)
Martin M R, **7**: 4944
Martin N F G, **4**: 2508 (2506, 2507)
Martin P, **6**: 3789
Martin R J, **3**: 1998
Martin R S, **2**: 827 (818)
Martin W E, **5**: 3464 (3463)
Martin-Sanchez J M, **5**: 3055 (3048)
Martinelli M, **5**: 3470 (3469)
Martinez A, **1**: 450 (449, 450)
Martino J, **1**: 618 (610, 614, 615)
Martino J P, **7**: 4827 (4825, 4826, 4827)
Martins H, **6**: 4123
Marto P J, **7**: 4585 (4583)
Marton L L, **7**: (4936)
Marzetta T L, **5**: 3130 (3129)
Masaki I, **1**: 301 (297)
Maschler M, **2**: 849 (845, 846), 850 (846, 848)
Mas-Colell A, **1**: 671 (666, 667, 668)
Masher D P, **7**: 4917 (4916)
Masinter L, **1**: 732 (726)
Maskin E, **5**: 3116; **7**: 4631 (4630)
Mason A D, **3**: 1998
Mason C E, **6**: (3882), 3884 (3883)
Mason H L, **2**: 1486 (1480, 1481), 1489 (1487)
Mason L G, **7**: 4721 (4716), 4741 (4736, 4737)
Mason M T, **4**: 2386 (2380); **6**: 4100
Mason R, **1**: 333 (332)
Mason R E A, **4**: 2685 (2684)
Mason R O, **4**: 2640 (2637); **7**: 4785 (4780), 4793 (4790, 4792)
Mason R Q, **3**: 2203
Masri S F, **6**: 4310 (4308)
Massé P, **3**: 2225 (2224)
Massey J B, **7**: 5186 (5179, 5180, 5182, 5184, 5185)
Massey J L, **5**: 3220 (3215)
Massie E, **1**: 456 (454, 455), 769 (761)
Massler M, **2**: 973 (971)
Massoth G J, **2**: 1516 (1516)
Mastroberti R, **7**: 4651 (4651)
Masubuchi M, **4**: 2575 (2573)
Masud A S M, **3**: 2143 (2143); **5**: 3156 (3152)
Matas R, **1**: (315)
Mates R E, **2**: 853 (852)
Matheron G, **6**: 3619 (3617)
Matheson J E, **3**: 2202 (2195)
Mathewson S C, **4**: 2550 (2549, 2550); **6**: 4333 (4329), 4393, (4393)
Mathiesen R, **6**: 3916 (3914)
Mathieu O, **7**: 4601 (4599)
Matijasevic J, **2**: 937 (933)
Matis J H, **1**: 417 (414); **2**: 1329 (1327), 1344 (1343), 1486 (1482, 1483)
Matrosov V M, **4**: 2556 (2555), 2565 (2560), 2566 (2560)
Matsuda F, **6**: 4123
Matsuda H, **5**: 2943
Matsugamo S, **6**: 4123
Matsui T, **2**: 1521 (1519)
Matsuoka Y, **1**: 262 (259)
Matsuzawa T, **7**: 4751 (4750)
Mattessich R, **6**: 3865 (3865)
Matthews D E, **4**: 2798 (2797)
Matthews P B C, **5**: 3242 (3238, 3240, 3242)
Matthews P J, **7**: 5112
Matthews P R, **6**: 4294
Mattila O, **1**: 564 (561); **5**: 2993 (2991)

Mattson J S, **7**: 5112
Maturana H R, **1**: 399 (394); **7**: 4996 (4995)
Matusita K, **2**: 1106
Matyas R, **2**: 1057 (1055, 1056)
Maudarbocus A Y, **4**: 2331 (2328)
Maurer H A, **5**: 3012
Maurer J, **1**: 570 (568)
Maurras J F, **3**: 2225 (2223)
Maxfield M, **6**: 4168 (4166)
Maxim H, **1**: (341); **2**: 1402 (1401)
Maxwell J C, **3**: (1745); **5**: (3383); **6**: (3881); **7**: (4517)
Maxwell M S, **6**: 4041 (4037)
Maxwell M W, **1**: 352, 360, 362, 365, 367
Maxwell W L, **6**: 4415 (4414)
May A D, **7**: 4913 (4912), 4927 (4926)
May R M, **1**: 567, 570 (568), 675 (672, 674); **2**: 1321, 1322 (1321), 1332 (1331), 1334 (1334), 1355 (1353); **6**: 3756; **7**: 4646 (4645)
Maybeck P S, **1**: 45 (38); **4**: 2622 (2620)
Maybee J, **2**: 1080 (1079)
Maybee J S, **6**: 4313, 4333 (4329), 4380
Mayeda H, **2**: 952 (949, 950)
Mayer G E, **4**: 2386 (2371, 2372), 2412 (2402), 2440 (2423, 2438); **6**: 4100
Mayer R, **4**: 2551 (2550)
Mayhew Y R, **7**: 4585 (4580)
Maynard H B, **4**: 2386 (2383)
Maynard P, **5**: 2943 (2935)
Maynard Smith J, **1**: 559 (558); **2**: 1324 (1323); **6**: 3756 (3756)
Mayne A J, **4**: 2367 (2365)
Mayne D, **4**: 2898 (2895)
Mayne D Q, **1**: 417 (414), 580 (575), 736 (735); **2**: 797 (796); **4**: 2277 (2273), 2284 (2283), 2759 (2756, 2757); **5**: 3196 (3189), 3197 (3189, 3193, 3196), 3215 (3206, 3213), 3224 (3223), 3383 (3381, 3382); **6**: 3991 (3991); **7**: 4469 (4466, 4468)
Mayoh B, **1**: 333 (326)
Mayr O, **2**: 1402; **3**: 1607 (1604, 1605, 1606, 1607); **5**: 3215 (3206); **6**: 3884 (3883), 4146 (4143); **7**: 4517 (4513, 4516, 4517)
Mazza G P, **6**: 3806 (3806)
Mazziotto G, **5**: 3303 (3294)
Mead M, **2**: 869 (868), 873 (872)
Mead R, **6**: 4310 (4308)
Meade G P, **7**: 4703 (4699)
Meadow C T, **3**: 2179
Meadows D H, **2**: 1538 (1536); **3**: 2016 (2015, 2016); **7**: 4762 (4756)
Meadows D L, **2**: 1538 (1536); **3**: 2016 (2015); **7**: 4762 (4756)
Medanic J V, **1**: 133, 137 (134, 135, 137); **2**: 1020 (1020); **4**: 2849
Medeiros D J, **4**: 2453 (2451)
Meditch J S, **2**: 1497 (1494); **3**: 1999; **4**: 2697 (2692, 2693); **7**: 4466 (4463)
Meek B L, **6**: 3898 (3894)
Meeraus A, **6**: 4332 (4329), 4333 (4329)
Mees A I, **1**: 429 (427, 428), 570 (568, 569); **2**: 981 (980, 981); **3**: 1779 (1776); **5**: 3319 (3319), 3373 (3369), 3383 (3380), 3389 (3386); **7**: 4452 (4450)
Meetham G W, **3**: 1995
Megson N C, **6**: 3870 (3868)
Mehl J W, **2**: 1174 (1170)
Mehlmann A, **2**: 1269 (1264)
Mehra R K, **1**: 251 (249), 490 (487); **4**: 2245 (2242), 2264 (2263), 2277 (2276), 2307 (2302), 2319 (2318), 2713 (2710); **6**: 4061 (4060), 4187 (4186), 4245; **7**: 4469 (4466), 4692 (4690, 4691), 5001 (5000)
Mehta D P, **1**: 639 (639)
Meidner H, **2**: 1521 (1521)
Meier L, **2**: 1257; **6**: 4165 (4163)
Meier R, **3**: 1990 (1989)
Meillou J C, **5**: 3480
Meirelles Ribeiro M, **2**: 808 (801)
Meirovitch L, **3**: 1665 (1650, 1651)
Meiry J, **2**: 1595 (1594)
Meis G J, **7**: 4913 (4912, 4913)
Meisel W, **6**: 3644
Meisel W S, **2**: 1479
Meiselman H J, **1**: 445 (444)
Meissinger H F, **5**: 3248 (3245)
Meissner J, **7**: 5060 (5059)
Meister R, **5**: 3464 (3459)
Meixel G O, **3**: 2117 (2114)
Mejer H, **4**: 2646 (2645)
Mel H C, **2**: 867 (865)
Melama H, **2**: 781 (781); **5**: 2994 (2991)
Melamed B, **3**: 2168 (2167)
Melamed M R, **2**: 867 (864, 865)
Melbourne W G, **7**: 4490
Melcher J R, **6**: 4364 (4363)
Mellen R H, **7**: 4848 (4844)
Meller F H, **7**: 5117
Mellichamp D A, **6**: 4418 (4418)
Melling A, **3**: 1693 (1686)
Melnikov N N, **3**: 2147
Melsa J L, **3**: 1999, 2096 (2093, 2094); **4**: 2517 (2515), 2628, 2697 (2692), 2713 (2706, 2708, 2709, 2710, 2711), 2714 (2709, 2711), 2800; **5**: 2917, 3224 (3220, 3223); **6**: 4133 (4129)
Meltzer B, **4**: 2640 (2640); **6**: 4160 (4158)
Melvill Jones G, **2**: 1595 (1593, 1594, 1595)
Memmi G, **6**: 3670 (3667)
Menad M, **5**: 3238 (3237)
Menaker M, **1**: 433 (433)
Menaldi J L, **3**: 2068 (2064); **4**: 2343; **6**: 4248 (4248)
Menasce D A, **2**: 1155 (1150, 1153), 1169 (1166)
Mendel J M, **4**: 2284 (2283); **6**: 3644, 4165 (4163)
Mendelsohn M L, **2**: 867 (864, 865); **6**: 3644
Mendelson E, **1**: 333 (327), 720 (715)
Mendu S R, **5**: 3568 (3568)
Menga G, **6**: 4172 (4172)
Menger C, **1**: 399 (394)
Menges G, **2**: 1586 (1586)
Mensonen K, **6**: 3916 (3912, 3913)
Menzel R, **7**: 5065 (5062)

Mepuis J M, **7**: 4716 (4714)
Merchant M E, **3**: 1637 (1634, 1635)
Mercier B, **3**: 2072 (2069, 2070); **5**: 3531 (3526)
Mercier C A, **6**: (3861)
Mercier H T, **4**: 2650 (2649)
Meredith B, **7**: 4749 (4745)
Meredith D D, **7**: 5153 (5153)
Meredith F W, **1**: (341)
Meredith W J, **7**: 5186 (5179, 5180, 5182, 5184, 5185)
Meriam J L, **1**: 109
Merikallio R A, **4**: 2353 (2352)
Merilo M, **3**: 1694 (1691)
Meriluoto T, **5**: 3282 (3280)
Merlin A, **2**: 1388; **6**: 3789 (3788, 3789)
Mermelstein M D, **5**: 3464 (3461)
Merrill E W, **1**: 445 (443, 444)
Merrill H W, **1**: 751 (751)
Merritt H E, **6**: 4257
Mertens J F, **3**: 1926 (1924, 1925, 1926), 1929 (1928); **5**: 3256
Merton P A, **1**: (465)
Merz A W, **2**: 1008 (1006), 1017 (1013, 1014)
Mesarovic M D, **1**: 93 (92), 294 (291), 405 (400, 404); **2**: 862 (861), 956 (952), 1286 (1283), 1348 (1346), 1486 (1480); **3**: 2016 (2016), 2038, 2043 (2039, 2042), 2163 (2153, 2155, 2159); **4**: 2713 (2710); **5**: 3019 (3015), 3035 (3033), 3040 (3035, 3036), 3088 (3083), 3144; **6**: 3631 (3629), 4165 (4161), 4322 (4319), 4355 (4354, 4355); **7**: 4571 (4570), 4666 (4661), 4779 (4776)
Mesch F, **3**: 1694 (1689)
Meschia G, **7**: 4905 (4905)
Meslier F, **3**: 2225
Mesmer R E, **4**: 2782 (2779)
Message G M, **3**: 1939
Messham S E, **3**: 1961
Messina P, **2**: 827
Messing R A, **1**: 424 (422)
Mester U, **7**: 4851
Metcalfe J, **6**: 3909 (3908)
Métivier M, **1**: 80 (75, 76, 77); **7**: 4637 (4631)
Metra K, **3**: 1614 (1612)
Metro S J, **1**: 181
Mettin F, **6**: 4123 (4121)
Metwally M M, **5**: 3572 (3569)
Metzler C M, **6**: 3699 (3698)
Metzler J A, **3**: 1975
Meunier J F, **7**: 5090
Meurman O, **6**: 3916 (3913)
Meyer C A, **7**: 4585 (4581)
Meyer C R, **4**: 2800
Meyer G, **1**: 323 (320, 321)
Meyer J, **4**: 2386 (2374)
Meyer P A, **7**: 4637 (4631, 4632)
Meyer R R, **1**: 628 (627); **2**: 797 (795); **5**: 3546
Meyer W S, **6**: 3820 (3818), 3855 (3854)
Mezaache A L, **6**: 4333 (4329, 4330, 4331)
Mezencev R, **3**: 2217
Mezzalama M, **6**: 4159 (4158)
Miaram G A, **3**: 2076
Michael E, **1**: 671 (667)
Michaels D, **1**: 13 (9)
Michaelson R, **2**: 1568 (1568)
Michailesco G, **1**: 132 (128, 129, 130, 131), 703 (698, 699)
Michalak P, **7**: 4571 (4568)
Michalski R S, **2**: 1573; **4**: 2640 (2637, 2639); **7**: 4784 (4783)
Michel A N, **2**: 981 (981); **4**: 2556 (2551, 2552, 2554, 2555, 2556), 2675 (2671, 2672, 2675); **5**: 3056
Michel C, **4**: 2338 (2335)
Michel J, **2**: 1179 (1174)
Michel J G L, **5**: 3389 (3385)
Michel J M, **4**: 2800
Michelman E H, **1**: 746 (744)
Michie D, **2**: 1573; **3**: 1786 (1783), 1821 (1820); **4**: 2640 (2637, 2640); **6**: 4160 (4158)
Michon J A, **3**: 2192 (2187)
Mickel M, **2**: 1451 (1450, 1451)
Mickell J, **4**: 2582 (2580)
Mickle M, **4**: 2522 (2520)
Middlebrooks E J, **2**: 1538 (1537); **4**: 2646 (2645)
Middleditch B S, **3**: 1939
Middleton D, **1**: 490 (486, 488)
Midgley D, **7**: 5122
Miele A, **2**: 1179 (1174)
Miellou J C, **1**: 405 (402); **2**: 956 (954), 1125 (1123); **3**: 2163 (2153); **5**: 3480 (3478), 3481 (3478)
Mieno H, **1**: 511 (508, 509, 510)
Miernyk W, **4**: 2522
Miesel W S, **5**: 3165 (3159)
Mifflin R, **1**: 405 (405); **5**: 3196 (3193), 3531 (3529), 3546 (3541, 3544, 3545)
Mignat F, **4**: 2343 (2342)
Mignot A, **3**: 1627
Mignot F, **2**: 1372 (1368); **6**: 3596 (3595); **7**: 5041 (5039)
Mihram G A, **7**: 5008 (5005)
Mikana K, **6**: 4261 (4260)
Mikelis N E, **6**: 4261 (4259)
Mikhailov A V, **3**: (1748)
Mikhailov N N, **6**: 3870 (3867)
Mikhlin S G, **3**: 1627
Mikulich L I, **3**: 1786 (1783), 1821 (1820)
Milanese M, **1**: 132 (127); **3**: 2143 (2140, 2141, 2142, 2143); **4**: 2257 (2254, 2256), 2635 (2634)
Milborrow D J, **2**: 1458
Miles C D, **2**: 1521 (1521)
Miles G E, **1**: 142 (140)
Miles J W, **7**: 4889 (4884, 4886)
Miles R, **7**: 4833 (4830)
Milgram D L, **6**: 4160 (4158)
Milgram M, **1**: 703 (690); **7**: 4666 (4665)
Milgrom P R, **5**: 2911 (2909)
Milhorn H T, **4**: 2901 (2899)
Miller A J, **2**: 1427 (1427), 1434 (1431)
Miller C, **3**: 2203 (2196)
Miller C W, **2**: 1551 (1551)
Miller D E, **3**: 2117 (2117)
Miller D M, **7**: 4996 (4995)

Miller D R, **6**: 4229 (4228), 4235 (4230, 4231); **7**: 5008 (5006)
Miller E, **4**: 2681 (2680)
Miller G, **1**: 651 (646)
Miller G A, **2**: 992 (990); **3**: 2186 (2181)
Miller G M, **6**: 4294
Miller H H, **4**: 2782 (2782)
Miller I, **7**: 5019 (5016)
Miller J A, **2**: 1445 (1445)
Miller J C, **4**: 2501 (2500)
Miller J D, **4**: 2582 (2580)
Miller J G, **2**: 1348 (1346); **4**: 2497 (2497)
Miller J J H, **5**: 3082 (3081); **7**: 4434 (4430)
Miller J R III, **2**: 942 (941)
Miller K, **2**: 1181 (1181)
Miller K B, **2**: 1410 (1409)
Miller L W, **2**: 1099; **6**: 4332 (4329, 4331, 4332)
Miller M L, **4**: 2338 (2338)
Miller R A, **2**: 1578 (1573)
Miller R D, **2**: 1249 (1247); **5**: 3238 (3236); **6**: 3699 (3698)
Miller R E, **3**: 2202 (2195)
Miller R G Jr, **7**: 5018 (5017)
Miller R J, **1**: 226
Miller R K, **2**: 981 (981); **4**: 2556 (2551, 2552, 2554, 2555, 2556), 2675 (2671, 2672, 2675)
Miller R R, **6**: 4302 (4300)
Miller S G, **7**: 4481
Miller S M, **6**: 4097
Miller W, **2**: 827 (814, 815); **3**: 1756 (1754)
Miller W E, **5**: 2993 (2993)
Millington P F, **1**: 482 (480)
Millis N F, **3**: 1614 (1614), 1619 (1616)
Millman R S, **4**: 2756 (2756); **5**: 2949 (2948); **6**: 3734 (3734), 4000 (3999, 4000)
Mills H, **6**: 4333 (4330)
Mills H D, **4**: 2501 (2500)
Mills R J, **6**: 3815 (3812)
Mills W H, **5**: 3514 (3514)
Millstein R, **2**: 1165 (1165)
Millstein R E, **2**: 1155 (1153)
Milne G W A, **5**: 2943
Milner S D, **2**: 1366 (1365)
Milner-Brown H S, **5**: 3248 (3245)
Milnor J, **7**: 5047 (5043)
Milnor J W, **2**: 850 (843)
Milnor W R, **2**: 853 (851)
Milroy R, **1**: 326 (324)
Milsum J H, **7**: 4877, 4880 (4880)
Milton G W, **2**: 1043 (1041)
Milward R C, **5**: 3421 (3418)
Miminis G S, **2**: 827 (821); **6**: 3734 (3733)
Minami S, **2**: 1154
Minamikawa T, **2**: 1366 (1363, 1366)
Minderhoud J M, **4**: 2582 (2581, 2582)
Mindt W, **1**: 440
Mine H, **2**: 1282 (1279); **3**: 1829 (1826), 1847 (1847)
Miner R J, **2**: 894 (894)
Minorsky N, **1**: 433 (431); **3**: (1749); **5**: 3389 (3384); **6**: 4275 (4273)
Minoura T, **2**: 1155
Minoux M, **2**: 1083 (1080); **7**: 4716 (4714), 4726 (4726), 4749 (4745, 4748)
Minsky M, **4**: 2640 (2638, 2639), 2641 (2637); **6**: 3644, 4160 (4159), 4167 (4166)
Minsky M L, **6**: 4322 (4319)
Mintzberg H, **2**: 947 (945); **7**: 4789 (4787)
Miranker W, **2**: 1125 (1123); **3**: 2163 (2154)
Miranker W L, **7**: 4434 (4430, 4431, 4432)
Miriani J, **2**: 1132 (1128)
Mirica S, **4**: 2609 (2608); **5**: 3184 (3183)
Mischan E J, **7**: 5164
Mischel W, **7**: 4793 (4792)
Mischenko E F, **2**: 1099, 1278; **5**: 3513 (3511), 3515 (3514, 3515), 3517 (3516)
Mischke T M, **1**: 176 (175)
Mises L von, **6**: 3865 (3861)
Mishkin E, **6**: 4165 (4162)
Mishra R N, **1**: 132
Misra J, **2**: 1087 (1084), 1201 (1200)
Misra K B, **7**: 4766 (4765)
Missenard A, **1**: 643 (643)
Missler S R, **3**: 1939
Mitchell A I, **7**: 4925
Mitchell A R, **3**: 1627, 1632
Mitchell D, **7**: 4856 (4855)
Mitchell E E, **4**: 2800 (2798)
Mitchell T M, **2**: 1573; **4**: 2640 (2637); **7**: 4784 (4783)
Mitereff S D, **6**: 3884 (3883)
Mitra S K, **5**: 3130 (3130)
Mitrani I, **6**: 3659
Mitroff I I, **3**: 2203; **4**: 2640 (2637); **7**: 4785 (4780)
Mitroff J I, **7**: 4793 (4790, 4792)
Mitsch W J, **2**: 1355 (1353, 1355); **3**: 1601 (1600); **6**: 4237 (4235)
Mitsch W M, **2**: 1346 (1345)
Mitson R B, **7**: 4848 (4844, 4845, 4847)
Mitten L, **1**: 725 (725)
Mitter S, **2**: 907 (903), 1192 (1191); **7**: 4822 (4822)
Mitter S K, **2**: 1121 (1116); **3**: 2151 (2150), 2152; **5**: 3164 (3158, 3161), 3279 (3276), 3302 (3298), 3303 (3301), 3310 (3305), 3329 (3320, 3322), 3330 (3320, 3321, 3322, 3323), 3336 (3336), 3376 (3374, 3375), 3487 (3483), 3506 (3505); **6**: 3776 (3772, 3773), 3777 (3772)
Mitzel G E, **3**: 1779 (1778); **6**: 3962 (3957, 3961)
Miura M, **7**: 4443 (4439)
Miwa K, **3**: 1614 (1614)
Miwa T, **2**: 1366 (1363, 1366)
Miyahara V S, **7**: 5113
Miyai S, **1**: 142 (141)
Miyakawa H, **5**: 3130 (3128)
Miyake N, **1**: 301 (298)
Miyamoto H, **7**: 4598 (4598)
Miyamoto S, **7**: 4518 (4518)
Miyamoto T, **5**: 3464 (3461)
Miyazaki F, **6**: 4128 (4124, 4125, 4128)
Miyazaki S, **5**: 3248 (3245)

Miyazaki Y Y, **6**: 3889 (3889)
Miyazawa N, **7**: 4593 (4588)
Mizano T, **4**: 2775 (2772)
Mizoguchi T, **3**: 1936 (1934)
Mizuike A, **7**: 5127
Mizumoto M, **3**: 1831 (1829, 1830), 1852 (1848, 1849, 1851)
Mizumoto N, **3**: 1835, 1858
Mizutani T, **6**: 4108 (4107)
M'Kendrick A G, **6**: 3762 (3756, 3758)
Moayer B, **6**: 4160 (4158)
Mobley F F, **2**: 1244 (1243)
Mock T J, **7**: 4793 (4791, 4792)
Modell H I, **2**: 1366 (1362)
Moden P E, **5**: 3006 (3006)
Moder J J, **6**: 3904
Modigliani F, **6**: 3706 (3706)
Moeller R P, **5**: 3470 (3466)
Mogee M, **7**: 4833 (4830)
Mohan C, **2**: 1174 (1171)
Mohler R D, **3**: 1611 (1609)
Mohler R R, **1**: 417 (414, 416); **4**: 2253 (2252); **5**: 3279 (3276); **6**: 3961 (3959)
Mohondas N, **1**: 445 (444)
Mohr R, **5**: 3464 (3459)
Mohtadi C, **5**: 3188 (3188); **6**: 4172 (4172)
Moir T J, **1**: 45 (42, 43); **6**: (4174), 4267
Moisil C G, **5**: (2918, 2919), 2924 (2922)
Molaug O, **6**: 4108 (4107)
Mole R H, **7**: 5177 (5171, 5173)
Moler C B, **2**: 826 (815, 817, 818, 820, 824), 827 (814, 815, 818, 821, 822, 824)
Molet E, **3**: 1712 (1712)
Molinari B P, **1**: 137 (134); **4**: 2849
Molinaro P, **3**: 1642 (1640)
Molinie L, **7**: (4514)
Molino G, **3**: 2143 (2140, 2142, 2143)
Moller K, **1**: 333 (326)
Mollon J, **7**: 5086
Molnar G D, **3**: 2025 (2022)
Moltz D M, **5**: 2943
Monaco S, **4**: 2527 (2524, 2526, 2527), 2756 (2756), 2816 (2807); **5**: 3288 (3285, 3286, 3287)
Monarchi D E, **5**: 3165 (3157)
Monash E A, **3**: 2006 (2006)
Monga T N, **1**: 437
Monge G, **1**: 626 (625)
Moniz W B, **5**: 3421 (3418)
Monnier B, **1**: 682 (678, 681)
Monod J, **1**: 424 (423), 429 (424, 425), 433 (432), 556 (554), 559 (558); **4**: 2789 (2783)
Monod J L, **3**: (1616)
Monopoli R V, **1**: 60 (57, 58, 59), 106 (101, 104); **2**: 1093 (1089, 1092); **5**: 3055 (3048, 3049, 3050, 3051), 3060 (3059), 3065 (3062, 3063, 3064), 3067 (3066), 3070 (3068, 3069, 3070), 3475 (3473); **6**: 3929 (3927, 3929), 4180 (4179), 4187 (4185, 4186), 4193 (4190), 4204 (4203)
Monro S, **4**: 2314 (2310)
Montague G A, **5**: 3188 (3188)
Montague J T, **5**: 2980
Monteyne R, **7**: 4871 (4870)
Montgomery D C, **5**: 3030 (3029)
Montgomery F J, **1**: 289 (285, 286)
Montgomery H, **3**: 2192 (2186)
Montgomery H A C, **7**: 5116
Montgomery R C, **1**: 252 (248)
Montgomery W A, **2**: 1155, 1165
Monticelli A, **6**: 3850 (3850)
Montroll E W, **7**: 4927 (4925, 4926)
Moody J L, **2**: 1272 (1269), 1552 (1549); **4**: 2652 (2650, 2651)
Moog C H, **2**: 900 (895, 896, 897)
Mook D T, **5**: 3390 (3385)
Mookerjee P, **2**: 1257 (1257)
Mooney H A, **2**: 1318, 1358
Moore B C, **1**: 410; **2**: 827 (819, 821, 822); **4**: 2827 (2826); **5**: 3220 (3215)
Moore B C J, **1**: 326 (323, 324, 325)
Moore D H, **2**: 1106
Moore D J, **4**: 2253 (2251)
Moore E, **6**: (4375)
Moore E F, **6**: 4011 (4008)
Moore E L, **4**: 2253 (2252)
Moore G L, **1**: 246
Moore J, **4**: 2865 (2865)
Moore J B, **1**: 703 (695); **2**: 825 (814), 902 (901), 916 (915), 923 (918); **4**: 2618 (2616), 2627; **7**: 4466 (4466), 4551 (4548), 4554 (4554), 4684 (4678)
Moore J G, **7**: 5094 (5093)
Moore L H, **3**: 1648 (1647)
Moore M J, **7**: 4585 (4582)
Moore P R, **4**: 2464 (2461, 2463, 2464)
Moore R A, **2**: 963 (963)
Moore R C, **7**: 4920
Moore R E, **3**: 1831 (1829); **7**: 4873, 4876 (4876)
Moore R L, **3**: 1700; **6**: 3691 (3690)
Moore-Ede M C, **1**: 601 (601)
Moores B M, **7**: 5186 (5181, 5182, 5184, 5186)
Moo-Young M, **1**: 424 (423)
Mopper K, **7**: 5123 (5122)
Moppett D J, **4**: 2356 (2355)
Mora P M, **6**: 3762 (3762)
Moraal J, **6**: 4306 (4303)
Morais B G, **3**: 1665 (1649)
Moran E J, **6**: 4302 (4301)
Moran T P, **2**: 1573; **3**: 2179
Morari M, **1**: 588 (588), 589 (587); **4**: 2517 (2515)
Moray N, **3**: 1668 (1667), 1669 (1665, 1666, 1667)
Mordoff K F, **4**: 2453 (2452)
Moré J J, **2**: 827; **6**: 4310 (4308); **7**: 4975 (4973)
More J W A, **7**: 4980 (4977, 4979)
Morecki A, **5**: 3248 (3243, 3244, 3247)
Moreggia V, **1**: 301 (297)
Moreigne O, **7**: 4429
Morel A, **6**: 4045
Moreno C, **3**: 1734 (1731); **7**: 5041 (5040)
Morf F, **6**: 4204 (4203)
Morf M, **2**: 785 (782, 783, 784), 1102 (1102); **4**: 2284 (2282)
Morgan A P, **1**: 106 (102); **5**: 3060 (3057)

Morgan B S, **4**: 2523 (2522)
Morgan C A, **4**: 2511 (2510)
Morgan D, **2**: 785 (783)
Morgan D L, **5**: 3238 (3235)
Morgan J J, **1**: 8 (1, 2, 3, 4, 5, 7); **4**: 2782 (2779); **5**: 3266 (3264)
Morgan K, **3**: 2117 (2117)
Morgan P A, **5**: 2943 (2935)
Morgan S K, **7**: 4842
Morganthaler G W, **4**: 2353
Morgenstern O, **1**: 687 (682); **2**: 850 (840, 843, 847), 1269 (1262); **3**: 2061 (2053), 2203 (2193, 2196); **5**: 2911 (2907), 2998 (2995), 3104 (3099, 3103), 3116 (3105, 3112); **6**: 4075; **7**: 4799 (4794), 4991
Mori S, **1**: 575 (573, 574); **5**: 3024 (3020, 3023); **6**: 3645
Mori Y, **5**: 3104 (3104)
Morice P, **2**: 1179 (1177)
Morimoto T, **2**: 1521 (1517)
Morin J M, **7**: 4913 (4911, 4912, 4913), 4915 (4914)
Morin T L, **2**: 1388 (1387); **5**: 3164 (3159, 3160, 3163)
Morineau A, **1**: 608 (607, 608)
Morisaku F, **7**: 4654 (4652, 4653)
Morishima T, **3**: 2021 (2017), 2175 (2169)
Morison D F, **5**: 3235
Moriuchi T, **5**: 2956 (2953)
Morley P, **7**: 4964
Moro A, **5**: 3303 (3296), 3330
Morowitz H J, **2**: 1106, 1107
Morrey Jr C B, **1**: 529 (520, 524)
Morris A J, **2**: 1116 (1110, 1111, 1115); **5**: 3188 (3184, 3188); **6**: 3991 (3990)
Morris C, **6**: 3860 (3858)
Morris H, **7**: 4513
Morris J H, **5**: 3272 (3268)
Morris J R W, **3**: 1910 (1910)
Morris L L, **5**: 2917
Morris P, **1**: 289 (285, 286, 289)
Morris P A, **2**: 943 (941); **3**: 2203 (2195)
Morris P J, **5**: 3425 (3422, 3425)
Morris R L, **5**: 3055 (3048, 3049, 3050, 3051, 3052, 3053, 3054)
Morris W T, **4**: 2353 (2351)
Morrisett J D, **4**: 2887
Morrison D B, **2**: 1311 (1307, 1309, 1310, 1311)
Morrison E L, **2**: 1445 (1441)
Morrison R, **6**: 4284
Morrison R J, **5**: 3464 (3459)
Morse A C, **6**: 4257
Morse A S, **2**: 899 (895), 900, 902 (900), 907 (902, 903), 916 (915), 923 (917); **4**: 2523 (2522), 2559 (2559), 2671 (2666, 2668), 2816 (2807), 2855 (2849, 2851, 2853, 2854), 2871 (2868, 2869); **5**: (3059), 3060 (3056, 3059), 3061 (3061), 3065 (3062); **7**: 4611 (4607)
Morse M, **7**: 5047 (5042)
Morse R I Jr, **4**: 2550 (2550)
Morsztyn K, **3**: 2225; **6**: 3834 (3832), 3837 (3836)
Mörtel K, **4**: 2351 (2347)
Mortenson R E, **2**: 894 (894); **5**: 3303 (3293, 3294, 3299)
Mortimer J A, **6**: 4345 (4344)
Morton K W, **3**: 1627
Morton M S S, **2**: 987; **7**: 4785 (4783)
Morton T E, **6**: 3706 (3706), (3890)
Mosca E, **6**: 4172 (4172)
Mosco U, **4**: 2343
Mosel P, **7**: 4703 (4700)
Moses R A, **7**: 5086
Mosher S, **7**: 4454 (4453)
Moshman J, **6**: 3904 (3904)
Moskowitz H, **4**: 2338 (2337)
Moskowitz M, **3**: 2192 (2186)
Moss F K, **3**: 2186 (2180)
Mossino J, **2**: 1262 (1260, 1261)
Mossino J A, **4**: 2614 (2613)
Mosteller F, **2**: 879 (878); **7**: 4721 (4717), 5015 (5015)
Mostellor W C, **3**: 1943 (1942)
Motard R L, **4**: 2517 (2516)
Motell E, **5**: 3421 (3419)
Motora S, **6**: 4275 (4273)
Moulin H, **3**: 1921 (1919); **7**: 4472 (4472)
Moulin P, **7**: 4842
Moulin T, **1**: 333 (332); **6**: 4336, 4337 (4335, 4336)
Moura J, **5**: 3330 (3320, 3322, 3323)
Moura J M F, **5**: 3310 (3305)
Movich R C, **4**: 2453 (2451, 2452)
Moyer H G, **7**: 4425 (4424)
Moylan P J, **4**: 2675 (2673), 2678 (2678); **7**: 4522 (4519, 4520), 4554 (4552)
M'Pherson P K, **6**: 3865 (3862)
M'Saad M, **5**: 3004 (3004)
Muchnik A, **2**: 937 (937)
Mudge T N, **4**: 2440 (2438)
Mudgett D R, **5**: (3059)
Muehlner D J, **4**: 2723 (2722)
Mueller C G, **7**: 5078
Mueller J A, **4**: 2794 (2789)
Mueller J L, **6**: 4045 (4045)
Mueller K H, **1**: 80 (78)
Mühlemann H R, **2**: 969
Mui J K, **6**: 4160 (4157)
Mujtaba S, **4**: 2351 (2350)
Mukai S, **3**: 2053 (2050, 2051)
Mukaidono M, **3**: 1883 (1881, 1882), 1888
Mulady B, **5**: 2943 (2934)
Mulder T, **1**: 437
Mulik J D, **1**: 177 (175); **4**: 2511 (2510); **7**: 5122
Mullaney P F, **2**: 867 (864, 865)
Mullenders E, **5**: 3030 (3029)
Muller E, **7**: 4472 (4471)
Müller G, **4**: 2575 (2571)
Müller J, **1**: (532)
Müller N, **1**: 740 (738)
Mullholland R J, **3**: 1695 (1694)
Mullineux N, **6**: 3806 (3802)
Mullins C, **5**: 3448
Multala R, **5**: 2944 (2935)
Mulvey J M, **2**: 826 (825)
Mumpower J L, **1**: 372 (369); **3**: 2192, 2202 (2196); **5**: 3115 (3113)
Munack A, **2**: 1192 (1189)

Mundy J L, **6**: 3644, 3653
Munemori M, **1**: 184 (184); **2**: 1513; **7**: 5123 (5122)
Munk W H, **7**: 4892 (4889, 4891)
Munro G, **7**: 5046 (5045)
Munro I, **1**: 725 (722)
Munro M C, **4**: 2501 (2499)
Munro N, **1**: 583 (581), 735 (734, 735), 736 (732, 734, 735); **4**: 2588 (2587), 2596 (2589, 2590, 2593), 2597 (2588, 2595), 2872 (2872); **5**: 3201 (3197, 3200), 3220 (3215), 3228 (3227), 3383 (3380); **6**: 3717 (3711, 3712, 3713, 3715), 4133 (4129), 4271 (4270); **7**: 4554 (4554)
Munro P W, **6**: 4167 (4167)
Munson D M, **1**: 606 (606)
Munson J H, **1**: 575 (574)
Munson W A, **2**: 1527 (1523)
Munstedt H, **7**: 5060 (5059)
Muntz R R, **2**: 1155 (1150), 1169
Murakami Y, **6**: 4036 (4035)
Muralidharan R, **2**: 995 (995)
Murat F, **2**: 812 (809, 810, 812), 1179 (1175, 1177), 1372 (1371)
Murat H, **2**: 1179 (1177)
Murata T, **3**: 2021 (2017); **6**: 3670 (3667, 3668)
Murata Y, **1**: 142 (140)
Muratet G, **2**: 1253 (1250)
Muratori T, **4**: 2292 (2291); **5**: 3044 (3042)
Murayama Y, **1**: 713 (710, 712, 713); **3**: 1902 (1902), 1905 (1903); **5**: 2927 (2925)
Murdoch D M, **7**: 5117
Murdoch P, **6**: 3717 (3713), 3720 (3720)
Murli A, **2**: 827
Muro S, **2**: 1489 (1489)
Murphy A, **5**: 2970
Murphy A H, **3**: 1711 (1711)
Murphy J M D, **5**: 3092; **7**: 5027
Murphy K L, **4**: 2789 (2787, 2788)
Murray F J, **5**: 3559 (3555)
Murray G D, **2**: 1106
Murray G P, **3**: 1910 (1908)
Murray J, **4**: 2570 (2570)
Murray J J, **5**: 3055
Murray J M, **4**: 2449 (2448)
Murray L S, **2**: 1107 (1104, 1105)
Murray R P, **7**: 4726 (4723)
Murray R T, **5**: 3464 (3460)
Murray W, **1**: 580 (580); **2**: 797 (797); **4**: 2879 (2877), 2880 (2879); **6**: 4310 (4309); **7**: 4975 (4973)
Murray-Smith D J, **1**: 633 (633); **4**: 2901 (2899); **5**: 3242 (3240, 3241, 3242)
Murrill P W, **3**: 2113 (2110, 2111)
Murtagh B A, **2**: 959; **4**: 2880 (2879)
Murty K, **4**: 2842
Murty K L, **1**: 474 (471)
Musa J D, **4**: 2681 (2679)
Muschik G M, **1**: 177 (176)
Mushin W W, **1**: 319
Musselman K, **5**: 3165 (3157)
Mutter I F, **3**: 1943
Muybridge E, **3**: 1910 (1907)
Myers G J, **4**: 2681 (2680), 2685 (2685)
Myers J D, **2**: 1578 (1573)
Myers P C, **7**: 4856 (4855), 4871 (4870)
Myers R H, **7**: 5005 (5003)
Myerson R B, **5**: 2911 (2908, 2909, 2910)
Myhill J, **6**: 4377 (4375)
Myhrhaug B, **1**: 755 (753)
Mynatt C R, **7**: 4785 (4783)
Myshkis A D, **3**: 2151, 2152 (2151, 2152); **5**: 3506
Myskis A, **2**: 1211 (1209)

N

Nababan S, **2**: 1211 (1210)
Nada M D, **1**: 459 (457)
Nader A, **2**: 781 (779)
Nadkarni D, **6**: 3909 (3908)
Nadler D A, **3**: 2192 (2186)
Nagai S, **3**: 1611 (1609)
Nagai T, **7**: 4980 (4976, 4977)
Nagakiri S, **2**: 1197
Nagano T, **6**: 3977 (3975); **7**: 4425 (4420)
Nagao Y, **2**: 776 (775)
Nagata T, **6**: 3766 (3762)
Nagel E, **5**: 3455 (3450); **6**: 3998
Naghshineh S, **4**: 2330, 2331 (2327)
Nagl M, **4**: 2798 (2798)
Nahi N E, **3**: 1741 (1738)
Nahmias S, **3**: 1831 (1829, 1830), 1835, 1845 (1845), 1858 (1856)
Naidu D S, **7**: 4434 (4430)
Nair K, **3**: 2203 (2196)
Naito M, **2**: 1542 (1540); **7**: 4518 (4518)
Najim K, **2**: 1253 (1249, 1250, 1251); **5**: 3004 (3004)
Najim M, **2**: 1253 (1249, 1251)
Naka K I, **7**: 5070 (5066)
Nakada M, **3**: 2076
Nakagiri S, **6**: 4245 (4238, 4241)
Nakahara T, **5**: 3464 (3461)
Nakai T, **1**: 713 (710, 713)
Nakajima S, **7**: 4751 (4750)
Nakamizo T, **4**: 2325 (2325)
Nakamura A, **4**: 2798 (2797)
Nakamura G, **5**: 2994 (2993)
Nakamura H, **2**: 1522 (1518)
Nakamura K, **1**: 276 (272, 276); **4**: 2640 (2639)
Nakamura Y, **5**: 3104 (3099, 3101, 3102, 3103, 3104), 3116 (3107); **7**: 5162
Nakanishi T, **2**: 1155 (1153), 1169 (1166)
Nakano B, **3**: 2043 (2041, 2043); **6**: 3631 (3631); **7**: 4779 (4775, 4777)
Nakano D, **2**: 1549 (1547)
Nakase K, **5**: 2994 (2993)
Nakhla N B, **5**: 3383 (3381)
Nałécz M, **4**: 2575 (2573)
Nalimov V V, **4**: 2301 (2300); **6**: 4322 (4319)
Nance R E, **4**: 2353 (2352), 2550 (2550); **6**: 4306 (4303), 4332 (4329, 4330), 4333 (4329, 4330, 4331), 4374, 4402 (4396), 4415 (4414)
Nanda S K, **7**: 4709 (4709)
Naparstek A, **1**: 433 (432)
Narasimham G V L, **2**: 1342 (1338)
Narendra K S, **1**: 51 (48), 60 (56, 58, 59), 100, 106 (101, 102, 104, 105), 124 (122); **4**: 2314 (2310), 2719 (2716, 2717, 2718), 2737 (2731); **5**: 3055 (3048, 3049, 3050, 3054), 3060 (3056, 3057, 3059), 3061 (3061), 3065 (3062, 3063, 3064, 3065), 3067 (3065, 3066, 3067), 3070 (3070), 3319 (3318), 3379 (3377), 3390 (3387), 3475 (3473); **6**: 3929 (3927, 3929), 4165 (4163), 4180 (4179), 4193 (4190), 4204 (4203), 4209 (4206); **7**: 4452 (4450), 4611 (4607), 4721 (4716, 4717, 4719, 4720), 4741 (4739), 4751 (4751)
Narotam M D, **1**: 633 (633)
Narula O S, **2**: 1410
Nash J, **1**: 687 (684); **3**: 1921 (1920); **5**: (2910)
Nash J F, **3**: 2061 (2057); **5**: 3256 (3255)
Nash S, **2**: 826 (823)
Nasi M, **6**: 3680 (3675, 3676)
Naslin P, **1**: 388 (380, 383)
Nathwani J S, **7**: 5094 (5093)
Nauchno-Issledovatelsky V, **3**: 1681 (1681)
Naughton K V H, **1**: 502 (499)
Naumann E, **7**: 4953 (4952)
Naunin D, **7**: 5027
Naur P, **3**: 1712 (1712)
Navarrete N R Jr, **3**: 1829 (1826), 1847 (1847)
Nay J N, **4**: 2685 (2683, 2685)
Nayfeh A H, **5**: 3390 (3385)
Naylor A W, **2**: 1521 (1518); **3**: 2005 (2003, 2005)
Naylor T H, **4**: 2353 (2351, 2352); **7**: 5010 (5009)
Nazer Y, **2**: 1116 (1110, 1111); **5**: 3188 (3184, 3188); **6**: 3991 (3990)
Nead M W, **4**: 2622 (2621)
Neat R C, **3**: 1694 (1692)
Nederbragt G W, **1**: 184 (184)
Nedzelnitsky O V, **4**: 2719 (2718); **7**: 4721 (4716, 4719, 4720)
Needham D, **3**: 1943
Needham R M, **2**: 1132 (1126)
Neely W B, **7**: 4901 (4899)
Neergaard K, **1**: 459 (457)
Neff T P, **3**: 1888
Negoiţă C V, **1**: 713 (711); **2**: 1479 (1477); **3**: 1604 (1602, 1603), 1781 (1781), 1819 (1817), 1822 (1821), 1874, 1878 (1877), 1888, 1894 (1889, 1893), 1902 (1898), 1905 (1905); **4**: 2301 (2299); **5**: 3264 (3262), 3540 (3538, 3539, 3540); **6**: 3619 (3616, 3618)
Negrao A I, **4**: 2292 (2287)
Nehari Z, **4**: 2614 (2611)
Nehrir M H, **6**: 3870
Neil E, **1**: 542 (536); **6**: 4051 (4050)
Neilson R M, **7**: 5094 (5093)
Neiswander R S, **5**: (3516)
Nelder J A, **6**: 4310 (4308)
Nelems J M, **1**: 315 (314)
Nelson A R, **5**: 3464 (3457, 3464)
Nelson C V, **1**: 456 (453), 535 (534), 536 (532, 534, 535)
Nelson D E, **3**: 1723
Nelson F A, **5**: 3421 (3419)
Nelson J W, **1**: 177 (174)
Nelson L W, **2**: 1226 (1214), 1239 (1234), 1241; **3**: 2105 (2101, 2103)
Nelson N, **7**: 4876 (4876)
Nelson T S, **2**: 1043 (1041)

Nelson W, **1**: 769 (761)
Nemhauser G L, **1**: 630 (628); **2**: 1282 (1279), 1388 (1387)
Nemirovskii A S, **5**: 3546 (3544)
Nemitz W C, **3**: 1838
Nemitzky V V, **2**: 1286 (1283); **5**: (3384)
Nemoto T, **7**: 4856 (4854, 4855)
Nenchev D N, **6**: 4108 (4105)
Nerem R M, **2**: 853 (852)
Nerode A, **1**: 294 (291)
Nesline F W, **5**: 3012
Nesselroade J R, **7**: 4833 (4830)
Nesset J E, **6**: 4252 (4251)
Nessler J, **7**: (5118)
Netravali A N, **1**: 80 (78)
Nettamo K, **6**: 3916 (3914)
Neubert H K P, **2**: 1403 (1403); **3**: 1710; **7**: 4513
Neuhauser G, **1**: 725 (724)
Neuhold E J, **2**: 893 (891, 892), 1160 (1159)
Neuman C P, **5**: 3055 (3048, 3049, 3050, 3052, 3053, 3054), 3056
Neumann E, **1**: 433 (431)
Neumark S, **1**: (340)
Neuschütz E, **6**: 4142
Neustadt L W, **2**: 1337 (1335); **5**: 3513 (3511)
Neuts M F, **2**: 1337 (1335)
Nevelson M B, **6**: 3994 (3992)
Nevelson M V, **5**: 3554 (3551)
Neveu J, **2**: 1581 (1579); **5**: 2934 (2932), 3316 (3311)
Nevins J L, **4**: 2386, 2460 (2459); **6**: 4123 (4121)
New J L, **3**: 2191 (2186)
Newbower R S, **1**: 440
Newbury D E, **2**: 1420; **6**: 4152
Newcomb R W, **4**: 2835 (2835); **6**: 4284 (4283)
Newell A, **2**: 1573; **3**: 2179; **4**: 2641 (2636, 2637)
Newell R B, **2**: 1568 (1567)
Newhouse S, **1**: (567)
Newman E J, **1**: 142 (141)
Newman L, **1**: 177 (175); **7**: 4913 (4912, 4913)
Newsom W B, **2**: 987; **7**: 4809
Newsom Davis J, **6**: 4055 (4052)
Newton J, **7**: 5155 (5154)
Ney P, **5**: 3336 (3332)
Neyman A, **3**: 1926 (1924), 1929 (1927)
Neyman J, **6**: 3820 (3818); **7**: 4646 (4644)
Ng E, **5**: 3560 (3559)
Ng K C, **5**: 3238 (3236)
Ng S K, **5**: 3336 (3335)
Ng T S, **4**: 2264 (2263)
Ng Y K, **7**: 5164
Nguyen D H, **6**: 3626 (3625)
Nguyen H T, **3**: 1831 (1830), 1838; **4**: 2301 (2293, 2294, 2300, 2301), 2883, 2884 (2883); **6**: 3768 (3768), 3772 (3770), 3947 (3945); **7**: 4789
Nguyen N G, **6**: 4088 (4086)
Nguyen V T, **6**: 3954 (3951, 3953)
Nguyen V V, **4**: 2239 (2236)
Nibler J W, **4**: 2723 (2721, 2722)
Nichols N B, **1**: (732, 734); **3**: 1749 (1748, 1749), 2113 (2111); **6**: 3701 (3700), 3884 (3883), 4146 (4145)
Nichols S T, **4**: 2310 (2309); **6**: 3780 (3777)
Nicholson H, **5**: 3383 (3381); **7**: 4930 (4929)
Nicodemo L, **7**: 5060 (5059)
Nicolaisen P, **6**: 4123 (4120)
Nicolis G, **1**: 429 (426, 428), 433 (432); **2**: 1333 (1330)
Nicolo V, **1**: 301 (297)
Niczgodka M, **5**: 3454 (3454)
Nieder M, **5**: 3421 (3419)
Niehaus R J, **6**: 3663 (3662)
Nielsen M, **4**: 2798 (2797)
Nielson N R, **3**: 2192 (2186); **6**: 4370 (4368)
Niemann H, **6**: 3644
Niemi A J, **2**: 1568 (1567); **3**: 1700 (1699), 1902 (1902), 1905 (1903); **4**: 2756 (2756); **5**: 2927 (2925), 2956, 2993 (2993), 2994 (2991), 3061 (3061); **6**: 3683 (3681), 3691 (3689), 3889 (3888, 3889), 3890 (3889), 3916 (3913)
Nienu A, **7**: 4823 (4821)
Niepold R, **4**: 2473 (2470)
Niering W A, **2**: 1318 (1317)
Nieto Vesperinas M, **5**: 3130 (3128)
Niezgodka M, **2**: 1276 (1272, 1276)
Nightingale J M, **5**: 3253 (3249, 3252)
Nihtilä M, **5**: 2994 (2991)
Niiniluoto I, **1**: 333 (331)
Nijkamp P, **2**: 1542; **5**: 3124 (3122), 3568 (3567)
Nijmeijer H, **4**: 2816 (2807); **5**: 3288 (3285, 3286)
Nijssen G M, **2**: 893 (891), 1155 (1150), 1160 (1159)
Nikiforov I V, **1**: 86 (82, 83)
Nikiforuk P N, **3**: 1847 (1846); **5**: 2956
Nikoliç Z J, **6**: 4165 (4163)
Nilsen R N, **6**: 4306 (4303)
Nilson E N, **7**: 4518 (4518)
Nilsson G E, **5**: 3464 (3459)
Nilsson M, **4**: 2473 (2472)
Nilsson N J, **1**: 307 (306); **2**: 1573; **6**: 3644 (3637, 3639)
Nimmo-Smith I, **1**: 326 (324)
Niordson F I, **2**: 1372 (1369, 1371)
Nirenberg L, **1**: 529 (527); **2**: 1261 (1259); **5**: 2997 (2997); **6**: 4223 (4218)
Nisenfeld A E, **2**: 1109 (1108)
Nishi I, **5**: 2943 (2935)
Nishi Y, **1**: 642 (639), 643, 645 (644)
Nishida T, **3**: 1819
Nishihara Y, **2**: 1154 (1153)
Nishino K, **2**: 1497 (1494, 1497)
Nishioka S, **2**: 1544 (1542, 1543)
Nishizawa Y, **3**: 1611 (1609)
Nisio M, **2**: 1017 (1015); **3**: 2072 (2071); **5**: 3316 (3314), 3523 (3522, 3523); **7**: 4613, 4627 (4623, 4626)
Nitecki Z, **7**: 4529
Nitzan A, **1**: 433 (431)
Nitzan D, **1**: 759 (755); **4**: 2487; **6**: 4123, 4159 (4156)
Niwa N, **2**: 1521 (1517)
Nix H A, **1**: 142 (139)
Nixon S W, **7**: 4897 (4896)
Noble B, **1**: 45 (41); **6**: 4237 (4235)
Noble D, **1**: 456 (451), 536 (535)

Noble I R, **2**: 1318 (1314, 1315, 1316), 1321; **3**: 2046 (2045)
Noda H, **2**: 1507
Nof S Y, **3**: 1648 (1644); **4**: 2386 (2379, 2383, 2384, 2385)
Nohel J A, **3**: 2151, 2152 (2152); **5**: 3506
Nokes M A, **5**: 3464 (3458)
Nolan R L, **7**: 4651 (4651)
Nolte J, **2**: 1154 (1153)
Noma E, **3**: 1668 (1667)
Nomoto K, **6**: 4261 (4258)
Nomura T, **5**: 3130 (3128)
Noonan F, **6**: 3790
Noordergraaf A, **1**: 468 (467)
Nordberg G F, **1**: 148 (144)
Nordby C, **7**: 5027
Norden C P, **6**: 4275 (4273)
Nordon H V, **5**: 2943 (2935)
Noriki S, **2**: 1516 (1516)
Norins A I, **4**: 2901 (2898)
Norkin V I, **5**: 3546 (3541)
Norman D A, **1**: 732 (730, 732); **4**: 2641 (2637)
Norman E, **1**: 417 (416)
Norman G E, **5**: 3282 (3281)
Norman K, **3**: 2006 (2006)
Norman M, **2**: 963 (962)
Norman M F, **7**: 4721 (4717)
Norman R, **4**: 2640 (2639); **7**: 4956 (4955, 4956)
Norman R F, **4**: 2579 (2577)
Norman V D, **6**: 4075
Normand-Cyrot D, **2**: 1121 (1116); **3**: 1779 (1773); **4**: 2527 (2524, 2526, 2527); **5**: 3279 (3279), 3292 (3289, 3291), 3293 (3289), 3376 (3374, 3376); **6**: 3776 (3772)
Norrbin N H, **6**: 4261 (4258, 4259)
Norrie D H, **6**: 3626 (3623, 3625)
Norris P E, **5**: 2943 (2934)
North D W, **2**: 943 (938, 939); **3**: 2061 (2053)
North J F, **1**: 482 (479)
Northend C A, **6**: 4037 (4034)
Northrop F S C, **2**: (872)
Nortoft-Thomson J E, **6**: 4261 (4259)
Norton H N, **3**: 1710
Norton J L, **6**: 4230 (4229)
Norton J P, **1**: 494 (491)
Norwich K H, **7**: 4906 (4904)
Norwood V T, **6**: 4041 (4038)
Nosadini R, **3**: 2025 (2025); **4**: 2631 (2629, 2631)
Nose M, **6**: 4128 (4124)
Nosé Y, **1**: 312 (312)
Notley R G, **7**: 4980 (4976)
Noton A R M, **4**: 2713 (2712); **5**: (3516); **6**: 3870 (3866, 3867)
Nottale L, **6**: 4337 (4335, 4336)
Nour Eldin H A, **4**: 2827 (2817, 2818, 2819, 2820, 2821, 2824, 2825)
Novakov T, **1**: 176 (175)
Novin D, **3**: 2213 (2210, 2212)
Novin-Hirbod S, **1**: 736 (734); **6**: 3717 (3713, 3715, 3717)
Novosel'tseva Zh A, **6**: 3870 (3867)
Novotnak J F, **6**: 4370 (4369)
Nowakowska M, **3**: 1825; **6**: (3862), 3865
Nudelmann S, **7**: 5186 (5186)
Nüesch D R, **6**: 4041
Nuhoshita M, **5**: 3464 (3459)
Nunn R H, **7**: 4585 (4583)
Nunnally J, **6**: 3940 (3940)
Nunney L, **1**: 675 (674)
Nurmi H, **3**: 1829 (1828); **7**: 4703 (4701)
Nuspl P, **2**: 1057 (1055, 1056)
Nutt G J, **1**: 751
Nye W T, **1**: 732 (730)
Nygaard K, **1**: 755 (753)
Nyman K E, **1**: 564 (561); **5**: 2994 (2993)
Nyne W T, **1**: 736 (735)
Nyquist G, **2**: 967
Nyquist H, **1**: (732, 734); **3**: 1749 (1746, 1747); **4**: (2271); **5**: 3228 (3224), (3387), 3441 (3440)
Nystrom P C, **7**: 4785 (4781)

O

Oates C, **2**: 1466 (1462)
Oatley K, **3**: 2214 (2213)
Obeid A A, **2**: 969
Oberer E, **7**: 4726 (4723)
Öberg R A, **5**: 3464 (3459)
Obidegwu S N, **5**: 2956 (2955)
O'Block R P, **7**: 4495
Ochi T, **2**: 1431 (1428)
Ocone D L, **5**: 3303 (3297, 3300, 3301), 3310 (3307), 3330 (3325, 3326), 3336 (3332, 3334, 3335, 3336)
O'Connor D J, **4**: 2646 (2645), 2794 (2789)
O'Connor P, **1**: 474 (468, 469, 470)
O'Connor P D T, **6**: 4015
Oda M, **2**: 1361
Oda S, **4**: 2391; **6**: 4113 (4112)
Odell A P, **3**: 2052 (2049); **4**: 2655 (2652)
Odell P L, **2**: 1326 (1324); **7**: 4454 (4453)
Oden J T, **3**: 1632
Odum E C, **2**: 1462
Odum E P, **1**: 675 (672); **2**: 1321 (1318, 1320), 1348, 1355 (1352); **6**: (3877), 4237; **7**: 4956 (4954, 4955)
Odum H T, **2**: 1333 (1330), 1344 (1344), 1462; **5**: 2956 (2956); **6**: 4359 (4358); **7**: 4956 (4954)
Oehme F, **5**: 3579 (3575, 3578)
Oeschger H, **1**: 531 (529)
Ogata K, **1**: 284 (281); **2**: 1291, 1297
Oguntade O O, **6**: 3857 (3855, 3856, 3857)
Oguztoreli M N, **2**: 1211 (1209); **3**: 2151, 2152 (2152); **5**: 3506
O'Hara C L, **1**: 181
Ohashi M, **3**: 1614 (1614)
O'Haver T C, **7**: 4504
Ohe S, **2**: 1497 (1494, 1497)
Ohlander R B, **6**: 4160 (4157)
Ohlin L E, **2**: 1565 (1563)
Ohman S, **7**: 4509 (4505, 4507)
Ohno K, **2**: 1282 (1279)
Ohno Y, **5**: 3465 (3461)
Ohsato A, **1**: 713 (710, 712, 713); **3**: 1902 (1901, 1902)
Ohshima A, **2**: 1361
Ohteru S, **4**: 2386 (2382)
Oi K, **7**: 4518 (4518)
Oja H, **1**: 177 (176)
Okada K, **7**: 4751 (4750)
Okada T, **6**: 4108 (4104)
Okamoto K, **2**: 1516 (1514, 1515, 1516)
Okazaki H, **4**: 2655 (2652)
Okazaki M, **2**: 1516 (1514, 1515, 1516)
O'Keefe R, **4**: 2550 (2550)
Okono J, **6**: 3889 (3889)
Okoshi T, **6**: 4160
Okuda M, **1**: 187 (187)
Okuda T, **3**: 1603 (1603), 1802 (1801), 1819 (1816), 1838
Okumoto K, **4**: 2681 (2679)
Okun E A, **7**: 5115
Olberg R M, **7**: 5065 (5062)
Oldberg S T, **2**: 1479 (1478)
Oldenbourg R C, **6**: 4146 (4145)
Oldenburger R, **4**: 2575 (2571); **5**: 3389 (3385), 3390 (3385, 3388), 3513 (3511), 3517 (3516, 3517)
O'Leary D P, **2**: 1596 (1594); **7**: 4971 (4969)
Olech C, **2**: 1033; **5**: 3184 (3181)
Olenik S C, **7**: 4709 (4709)
Olesky D D, **1**: 675 (674); **2**: 927 (924)
Olievier C N, **6**: 4061 (4057, 4061)
Oliger J, **3**: 1723 (1722); **7**: 4498 (4496)
Oliver B M, **2**: 1403 (1402)
Oliver G C, **1**: 769 (760, 761)
Oliver R M, **7**: 4948
Oliver W K, **2**: 1568 (1567)
Olkku J, **3**: 1706 (1701)
Olle T W, **2**: 893 (891)
Ollero A, **5**: 3156 (3155)
Ollila A, **6**: 3916 (3913)
Ollila K, **6**: 3916 (3912, 3913)
Ollis P F, **1**: 424 (422); **3**: 1619 (1615)
Olmsted J M H, **6**: 3710 (3708)
Olner P, **7**: 5147 (5144)
Olsder G H, **2**: 1008 (1006), 1016 (1012), 1017 (1013)
Olsder G J, **1**: 687 (685); **2**: 995 (993, 994, 995); **5**: 3149 (3146, 3147, 3148, 3149); **7**: 4622 (4622), 4631 (4628, 4629), 4822 (4821)
Olsder G T, **1**: 687 (682, 684)
Olsen C F, **4**: 2800 (2798)
Olsen J, **7**: 4785 (4783, 4784)
Olsen L F, **1**: 429 (424), 433 (431), 570 (569)
Olsen S A, **4**: 2580 (2575)
Olsen T O, **5**: 2993 (2991), 2994 (2991); **6**: 4252 (4250)
Olshen R A, **2**: 1106
Olson J S, **2**: 1344 (1343, 1344); **7**: 4897 (4896)
Olson W P, **3**: 2006 (2005)
Olsson G, **4**: 2307 (2302)
Olszowka A J, **2**: 1366 (1362)
O'Malley R E Jr, **1**: 132 (129); **5**: 3082 (3081), 3176, 3180 (3176), 3502; **7**: 4434 (4430), 4439 (4435, 4437, 4438), 4443 (4439, 4441), 4534
Oman C M, **2**: 1591 (1590), 1596 (1592, 1594)
O'Mara J, **6**: 4302 (4301)
Omasa K, **2**: 1522 (1517, 1518, 1519, 1520, 1521)
Omatu S, **2**: 1500 (1499); **6**: 4245 (4238)
O'Melia C R, **4**: 2646 (2643), 2794 (2789)
Omens G, **7**: 4833 (4828)
Omerod W G, **7**: 4577 (4577)
O'Moore R R, **2**: 1577 (1575)
Omori M, **2**: 1507
Omori T, **6**: 4033
O'Neill J, **3**: 2164
O'Neill J T, **6**: 4302 (4302)
O'Neill R P, **6**: 4333 (4329)

O'Neill R V, **2**: 1272 (1272), 1326 (1324, 1325, 1326), 1333 (1332), 1344 (1343, 1344), 1462, 1539 (1538, 1539), 1551 (1549, 1550), 1552 (1550, 1551); **6**: 4229 (4228), 4235 (4230, 4231, 4234); **7**: 4454 (4453), 4646 (4644), 4897 (4896), 5008 (5005)
Onoe M, **2**: 1522 (1518, 1521)
Onyiriuka E C, **4**: 2511 (2510)
Ookhtens M, **3**: 2175 (2174)
Oonishi Y, **7**: 4518 (4518)
Oosterkamp W J, **7**: 5186 (5180)
Oppelt W, **3**: (1749); **5**: 3390 (3385); **7**: 4452 (4443)
Oppenheim A V, **5**: 3131 (3130); **6**: 4281 (4278, 4280), 4289 (4285); **7**: 4566 (4560)
Oppenheim P, **1**: 333 (330)
Oppenheimer K R, **5**: 3165 (3157)
Orander P C, **2**: 1410 (1409)
Orava P J, **6**: 3691 (3689)
Orchard-Hays W, **6**: 4299
Orden A, **6**: 4299
O'Regan R G, **6**: 4051 (4050)
O'Reilly J, **7**: 4439, 4443 (4443), 4534 (4532, 4534), 4554 (4553, 4554)
O'Reilly J G, **2**: 1282 (1279)
Oren S S, **7**: 4975 (4974)
Ören T I, **1**: 561 (561); **2**: 894 (894), 1552 (1550); **4**: 2353 (2351, 2352), 2681 (2681); **5**: 3035 (3030, 3032, 3033), 3088 (3083); **6**: 4229 (4228), 4302 (4301), 4303 (4300), 4306 (4303), 4313 (4312), 4322 (4319, 4320, 4321), 4323 (4320, 4321), 4326, 4333 (4329, 4330, 4331), 4353 (4350, 4351, 4352, 4353), 4369 (4368), 4370 (4369), 4372 (4370, 4371), 4381, 4388 (4385, 4386), 4392, 4402 (4399), 4403 (4402, 4403), 4406 (4404), 4411 (4406), 4414; **7**: 4654 (4652, 4653), 4692 (4691), 4764 (4762), 5008 (5007), 5010 (5008), 5011 (5008)
Orie N G M, **2**: 1248 (1247)
Orilovsky G N, **3**: 1910 (1909)
Orin D E, **4**: 2386 (2379); **6**: 4086
Orito H, **6**: 4261 (4260)
Orlando V A, **1**: 202 (201)
Orlob G T, **4**: 2646 (2645)
Orloci L, **3**: 2045 (2045)
Orlov A I, **4**: 2301 (2300)
Orlovsky S A, **3**: 1819, 1847 (1846)
Ormerod P, **2**: 1342
Ornstein D S, **4**: 2508 (2506)
Orr C H, **7**: 5109 (5108), 5153 (5153)
Orsini M, **6**: 4176 (4176)
Orszag S A, **7**: 4498 (4496)
Ort M G, **7**: 5186 (5182)
Ortega J M, **1**: 405 (402); **2**: 956 (954), 1125 (1122, 1124); **3**: 2163 (2155); **5**: 3487 (3482); **7**: 4969
Ortendah D A, **5**: 3425 (3425)
Orthlieb M P, **7**: 4913 (4912)
Ortmans J, **7**: 5047 (5044)
Orton B M, **2**: 1545 (1544)
Osborne J C, **4**: 2887
Osburn P V, **5**: 3070 (3068)
Oscarson D W, **1**: 7 (3, 4); **4**: 2782 (2780)
Oscarsson O, **1**: 466 (463)
O'Shaughnessy J, **5**: 3035 (3033)
O'Shea J, **5**: 2993 (2990, 2992), 29932994 (2991), 2994 (2991)
O'Shea M, **7**: 5065 (5062)
Oshima M, **1**: 760 (757, 758)
Oshima Y, **6**: 4117
Ossendoth U, **6**: 4211 (4211)
Osserman R, **4**: 2614 (2609)
Oster G, **2**: 1318 (1316)
Oster G F, **1**: 567
Osterberg E L, **3**: 1958
Østerby R, **7**: 4601 (4599)
Østergaard J J, **1**: 126 (125, 126); **4**: 2362 (2359, 2360)
Osterweil L J, **2**: 826
Ostrowski A M, **2**: 827 (820); **4**: 2601 (2600)
Ostrowski J J, **6**: 3865 (3861)
O'Sullivan G, **2**: 963 (962, 963)
Othmer H G, **1**: 429 (427), 433 (429, 430, 431, 432)
Otis A B, **6**: 4066 (4063)
Otis E, **7**: (5047)
Otsuka Y, **2**: 1516 (1514, 1515, 1516)
Ott W E, **2**: 1078 (1076)
Ott W R, **2**: 1538 (1537)
Otterman J, **6**: 4041 (4037, 4038)
Ottestad P, **2**: 1106
Ottley T W, **5**: 2943 (2935, 2936)
Ouazzani T, **2**: 1253 (1249)
Ouellette D V, **1**: 495 (494)
Ouimet Y, **7**: 4726 (4723)
Oumamar A, **5**: 3097
Ousset A, **7**: 4749 (4745)
Outerbridge J, **2**: 1590 (1587, 1589)
Outrata J, **5**: 3546 (3542)
Ovchinnikov S V, **3**: 1838 (1836)
Overstreet C M, **4**: 2550 (2550); **6**: 4306 (4303), 4333 (4329, 4330, 4331), 4402 (4398)
Overstreet R E, **6**: 4411 (4406)
Overton W S, **2**: 1329 (1329)
Owen G, **2**: 1269
Owen R C, **6**: 3829 (3826)
Owens D H, **4**: 2832 (2831); **5**: 3130 (3129), 3135 (3134), 3201 (3197, 3198, 3199, 3200, 3201), 3215 (3206, 3207)
Owens G, **7**: 4622 (4622)
Owlett J, **4**: 2659 (2657)
Özgüner Ü, **2**: 902 (900, 901, 902), 907 (903, 904, 906); **4**: 2560 (2559), 2671 (2666); **6**: 4139 (4133, 4136, 4137)
Özmizrak M, **6**: 4388 (4385, 4386)

P

Paakkinen U, **5**: 2994 (2991); **6**: 4252 (4250)
Paasila M, **6**: 3916 (3912)
Pachuta D G, **1**: 177 (174)
Pacini G, **1**: 769 (764); **2**: 1407 (1405, 1406, 1407); **3**: 2021 (2018), 2025 (2022, 2024), 2174 (2173)
Paddock D, **2**: 826
Padmanabhan L, **6**: 4245
Padulo L, **6**: 4323 (4319)
Page C, **6**: 3653 (3648)
Page D M, **3**: 1995
Page E S, **1**: 86 (81)
Page F M, **1**: 433 (431)
Page R K, **4**: 2769 (2769)
Pagurek B, **3**: 2175 (2171); **4**: 2284; **7**: 4905 (4905)
Paice D A, **5**: 3092
Paidy S R, **5**: 3156 (3152)
Paige C C, **2**: 827 (821, 824); **4**: 2827 (2825, 2826); **6**: 3734 (3733)
Paine R, **7**: 4646 (4645)
Paine R T, **2**: 1317 (1316), 1321
Painter P R, **1**: 428 (426, 427, 428), 429 (424, 426)
Pal J, **1**: 132
Palacis F G, **2**: 1169
Palais R S, **1**: 529 (527); **7**: 5047 (5043)
Palaniswami M, **1**: 60 (60)
Palazzo T, **3**: 1637
Paley R E, **3**: 1728 (1725); **4**: 2664
Palka J, **7**: 5065 (5062, 5063)
Pallet E H J, **3**: 2133
Pallis C A, **5**: 3425 (3425)
Pallu de la Barrière R, **4**: 2664
Palm G, **6**: 4167 (4166)
Palm R D, **5**: 3464 (3463)
Palmay F V, **6**: 4139 (4135)
Palmes E D, **7**: 4876 (4875, 4876)
Pandya R N, **4**: 2284
Pang C K, **2**: 1388 (1387)
Pang S C, **6**: 3940 (3939)
Pantle A J, **4**: 2331 (2329)
Panuska V, **4**: 2284 (2283)
Panzieri F, **2**: 1136 (1135, 1136)
Papa L J, **1**: 181
Papadimitriou C H, **1**: 628, 630; **2**: 884 (881, 882, 883), 1083 (1080), 1125 (1123), 1154, 1155 (1145, 1146)
Papadoupolous N, **7**: 4822
Papageorgiou G, **2**: 1522 (1521)
Papageorgiou M, **2**: 862 (862); **3**: 1741 (1737, 1738, 1739)
Papaleski R, **5**: (3384)
Papanicolaou G C, **2**: 812 (809), 1179; **4**: 2664 (2663); **6**: 3997 (3996); **7**: 4944 (4943)
Papavassilopoulos G P, **2**: 995 (994), 1268 (1266); **7**: 4631, 4822
Papazoglu M P, **1**: 634 (632)
Papert S, **6**: 3644, 4167 (4166)
Papon J, **6**: 3680 (3675)
Papoulis A, **1**: 421; **6**: 4281 (4278)
Papp A, **5**: 3464 (3461)
Pappas T, **2**: 827 (823); **4**: 2622 (2621)
Pappis C P, **1**: 713 (712); **4**: 2362 (2359)
Papuchon M, **5**: 3470 (3466)
Paranjape M A, **7**: 4454 (4453)
Pardo L T, **6**: 3893
Pardoux E, **5**: 3303 (3296, 3299), 3310, 3316 (3312, 3313), 3329 (3320), 3330 (3322), 3523 (3523); **7**: 4621 (4618, 4619), 4637 (4636)
Parent M, **4**: 2385 (2374)
Parente R, **3**: 1779 (1777)
Pareto V, **5**: (3157), 3165 (3157)
Parfitt R L, **1**: 7, 268 (266, 267)
Pargen E, **7**: 5155 (5154)
Parikh R, **7**: 4996 (4995)
Paris L, **6**: 3790
Park A C, **3**: 1665 (1651)
Park C, **2**: 1479 (1478); **3**: 2097 (2094)
Park C R, **7**: 4876 (4876)
Park C W, **3**: 2203 (2195)
Park D M, **2**: 884 (881)
Park R A, **3**: 1601 (1600); **4**: 2646 (2645); **5**: 3044 (3040); **7**: 4897 (4896), 5008 (5006)
Park T, **7**: 4646 (4644)
Parke M E, **7**: 4889 (4887)
Parker A M, **6**: 3802 (3796)
Parker C A, **7**: 4504
Parker G A, **4**: 2253 (2252), 2456 (2454)
Parker J, **1**: 755 (752, 753, 754)
Parker J D, **6**: 3918 (3917)
Parker R B, **5**: 2942 (2935, 2936)
Parker R P, **7**: 5186 (5181, 5182, 5184)
Parkes A S, **3**: 1958
Parks D L, **3**: 1669 (1667)
Parks P C, **1**: 54 (51), 60 (58), 100; **5**: 3070 (3068, 3069), 3303 (3301), 3336 (3336), 3398; **6**: 4187 (4184, 4185); **7**: 5031 (5029)
Parlett B N, **2**: 827 (818)
Parlow H, **2**: 1431 (1428)
Parmater J Q, **6**: 4302 (4301)
Parnaby J, **2**: 1586 (1585)
Parnas D L, **4**: 2681 (2678, 2680); **6**: 4333 (4332), 4402 (4396), 4415 (4414)
Parnas R S, **1**: 588 (588)
Parr J M, **6**: 3703 (3702)
Parrott P G, **6**: 3806 (3804)
Parry C M, **4**: 2769 (2769)
Parry R D, **3**: 1958
Parslow R D, **3**: 1809 (1808, 1809)
Parson M L, **4**: 2347 (2345)
Parsons F M, **1**: 312 (310)
Parthasarathy T, **2**: 1099 (1096); **3**: 1926
Parton W J, **1**: 142 (138, 139, 140)
Parungao A, **1**: 536 (534)

Parzen E, **5**: 3232 (3230); **7**: 4566 (4563)
Pascal B, **3**: (2175)
Pasekov V P, **1**: 675 (674)
Pashigian B P, **6**: 3706 (3705)
Pask G, **1**: 333 (332); **6**: 4167 (4166)
Pasquill F, **6**: 3933 (3930)
Passmore R, **5**: 2970
Pasternak H J, **2**: 1445 (1445)
Pastrick H L, **5**: 3012
Patani A, **2**: 1226 (1215, 1216, 1217, 1225), 1233 (1228, 1232, 1233), 1239 (1236), 1241
Patankar S V, **3**: 2117 (2114)
Pate W, **5**: 2993 (2991)
Patel C G, **5**: 3077 (3073)
Patel C K N, **4**: 2511 (2510)
Patel D J, **1**: 478 (477), 486 (484)
Patel N R, **3**: 2014
Patel R V, **1**: 583 (581), 736 (732); **2**: 827 (821); **4**: 2827 (2826); **5**: 3201 (3197, 3200), 3220 (3215); **6**: 3717 (3712)
Paterson I W F, **4**: 2747
Paterson M S, **2**: 884 (881)
Paterson S, **7**: 4901 (4899, 4901)
Paterson W, **6**: 3784
Patnaik L M, **6**: 4183 (4182)
Paton W D M, **5**: 3238 (3236)
Patrick E A, **4**: 2636 (2632); **6**: 3644
Patrick W H Jr, **4**: 2782 (2780)
Pattee H, **1**: 333 (329), 709 (709)
Patten B C, **1**: 417 (414), 674, 675 (674); **2**: 1323 (1323), 1326 (1324), 1329 (1327, 1329), 1333 (1330), 1344 (1343), 1345 (1343, 1344), 1348 (1345, 1346), 1355 (1352, 1353, 1355), 1356 (1353), 1486 (1479, 1480, 1481, 1482, 1483, 1485), 1489, 1538 (1537), 1551 (1549); **4**: 2646 (2645); **5**: 2930 (2929); **6**: 4237 (4235), 4238 (4235); **7**: 4646 (4645), 4956 (4954, 4956), 5008 (5005, 5007)
Patterson R D, **1**: 326 (324)
Patterson T A, **5**: 2943 (2934)
Patton A D, **2**: 1388 (1387)
Patton H D, **1**: 433 (432)
Patton M D, **5**: 3253 (3248, 3250, 3252)
Patton R J, **6**: 4267 (4262)
Pau L F, **6**: 3644 (3643)
Pauker S G, **2**: 1577 (1577)
Paul H, **4**: 2392
Paul J P, **3**: 1913; **4**: 2901 (2899)
Paul M, **2**: 1116 (1115), 1132 (1126, 1132), 1136 (1136), 1174 (1170, 1171, 1172, 1173); **5**: 3188 (3188)
Paul R, **4**: 2386 (2374)
Paul R J, **4**: 2550 (2550)
Paul R P C, **4**: 2386 (2369, 2371, 2372, 2373, 2374, 2375, 2378, 2379, 2380, 2383, 2384, 2385), 2387 (2380), 2412 (2392, 2393, 2394, 2395, 2401, 2402), 2440 (2421, 2423, 2424, 2426, 2427, 2430, 2431, 2432, 2434, 2437, 2438, 2439), 2460 (2456), 2469 (2466); **6**: 4086 (4083), 4088 (4086, 4087, 4088), 4100
Paulsen K A, **2**: 1416 (1414, 1415, 1416)
Pauly P, **2**: 1342
Pavel M, **6**: 3644
Pavese F, **7**: 4851
Pavitt K L R, **7**: 4762 (4756)
Pavlidis T, **1**: 433 (431, 433); **6**: 3643, 3644, 3652, 3653 (3648), 4159 (4158), 4160 (4157, 4158); **7**: 4883
Pavlov A A, **5**: (3517)
Pavon M, **7**: 4468 (4467)
Paxton B R, **5**: 2994 (2993)
Payne D N, **5**: 3464 (3461)
Payne H J, **1**: 137 (133); **3**: 1736, 1741 (1737, 1739); **4**: 2523 (2522); **5**: 3220 (3215), 3572 (3568); **7**: 4913 (4912)
Payne J P, **5**: 3579 (3575, 3578)
Payne J W, **3**: 2192 (2186), 2203 (2200, 2202); **6**: 4075; **7**: 4793 (4791)
Payne L E, **4**: 2614 (2609, 2611, 2612)
Payne M, **5**: 3165 (3159)
Payne P A, **1**: 482 (480, 482)
Payne R L, **1**: 489 (487); **4**: 2239 (2237), 2245 (2242), 2264 (2263), 2283 (2283), 2307 (2303), 2319; **6**: 4061 (4057); **7**: 5001 (4998, 4999)
Payne W H, **5**: 2956 (2954)
Paynter H M, **6**: 3870 (3866), (4337)
Paz A, **4**: 2798 (2797)
Pazy A, **6**: 4223 (4220)
Peaceman D W, **3**: 2006 (2006)
Peacock J K, **2**: 1087 (1084), 1201 (1200)
Pearce J, **2**: 963 (962)
Pearl J, **4**: 2338 (2336, 2338), 2501 (2499)
Pearlman J G, **5**: 3293 (3289)
Pearson J B, **2**: 902 (900); **4**: 2570 (2569); **6**: 3717 (3713), 3720 (3720), 3734 (3733)
Pearson J D, **2**: 956 (952); **3**: 2038; **5**: 3019 (3016, 3017), 3040 (3039)
Pearson K G, **5**: 3253 (3250, 3251, 3252)
Pearson K S, **2**: 1527 (1525)
Peart R M, **1**: 142 (140)
Pease D L, **7**: 4940
Pecan E V, **1**: 532 (529)
Pechkowski J C, **5**: 3224 (3220, 3223)
Peck C C, **1**: 490 (488)
Peczkowski J L, **6**: 4133 (4129)
Pederson H C, **2**: 797 (796)
Pedgen C D, **3**: 1648 (1647, 1648)
Pedley T J, **1**: 433 (432), 445 (441)
Pedotti A, **3**: 1910 (1908)
Pedretti A, **1**: 333 (332)
Pedrycz W, **1**: 713 (710, 712, 713); **2**: 1473 (1469); **3**: 1781 (1781), 1894 (1889, 1890, 1891, 1892); **5**: 2965 (2965), 3540 (3538); **6**: 4007 (4007)
Peele R W, **2**: 1552 (1549)
Pegden C D, **3**: 1648 (1648); **7**: 4935 (4934)
Pegden D, **4**: 2353 (2352)
Pegler S M, **4**: 2356 (2354)
Pehl R H, **1**: 177 (174)
Pehrson B, **2**: 773 (773)
Peirce C S, **4**: (2577); **6**: 4007 (4001)
Peiss C N, **7**: 4877, 4880 (4879, 4880)
Peixoto M M, **1**: 552; **3**: 1898 (1895); **7**: 4674 (4670)
Pekeris C L, **7**: 4888 (4887)
Pelagatti G, **2**: 1160 (1159)

Peleg B, **2**: 849 (842, 843, 847), 850 (846, 847, 848); **7**: 4472 (4470)
Pélegrin M, **1**: 196 (193); **5**: (3388); **7**: 4476
Pellionisz A, **1**: 466 (464)
Peltola H, **6**: 3916 (3914)
Pemberton T J, **2**: 837
Pennders R, **6**: 3933 (3932)
Penné J, **4**: 2392
Pennefather P, **2**: 1248 (1245)
Pennington K S, **6**: 3766 (3762)
Penrose L S, **2**: 1106
Penrose R, **1**: 333 (332)
Penttilä H, **5**: 2993 (2991)
Penttinen J, **2**: 781 (781)
Peperstraete J A, **2**: 1116 (1112)
Peppler H J, **3**: 1611 (1608), 1615 (1612)
Peray K E, **4**: 2362
Perdon P M, **6**: 3972 (3972)
Perdue E M, **4**: 2782 (2778)
Pereira B de B, **2**: 1106
Peres P L D, **7**: 4685 (4683)
Perez M G, **4**: 2292 (2287)
Peringer P, **3**: 1611 (1609)
Perkins S W, **1**: 490 (488)
Perkins W A, **4**: 2473 (2469, 2471)
Perkins W R, **1**: 133, 137 (134, 135, 137); **6**: 4227 (4224, 4226)
Perks R, **5**: 3238 (3235, 3236)
Perl W, **7**: 4906 (4902)
Perley C R, **5**: 2943 (2935, 2942), 2944 (2935, 2942)
Perlis H J, **4**: 2706 (2702, 2703, 2705)
Perlman D, **1**: 424 (423)
Pernebo L, **1**: 410; **5**: 3197 (3190)
Perone N, **1**: 445 (441, 442, 443, 444)
Perram J W, **1**: 433 (431), 570 (569)
Perret R, **1**: 681 (676), 682 (679)
Perrier D, **6**: 3698 (3695)
Perrin J P, **1**: 352, 360, 362, 365, 367
Perron M, **6**: 3916 (3912, 3913)
Perry A M, **5**: 3416
Perry J A, **1**: 594
Perry J F, **6**: 3916 (3913)
Person H B, **3**: 2191 (2186)
Persoon E, **6**: 3644, 3654 (3648, 3650)
Persoz H, **6**: 3789
Pervozvanskii A A, **5**: 3390 (3388)
Peschon J, **1**: 251 (249); **6**: 3789
Pesin Ya B, **1**: 570 (568)
Pestel E, **3**: 2016 (2016)
Peter D, **7**: 5186 (5180)
Peterka V, **4**: 2277 (2277); **5**: (3229); **6**: 3727 (3721), 4193 (4190), 4199, 4204 (4202); **7**: 5001 (4998)
Peters G, **5**: 3559 (3555); **7**: 5106 (5103)
Peters L J, **4**: 2501 (2500)
Peters L S, **7**: 4833 (4830)
Peters T J, **7**: 4833 (4831)
Petersen R A, **6**: 4346 (4346)
Peterson A V Jr, **6**: 3950 (3949)
Peterson B B, **1**: 100, 106 (105); **5**: 3060 (3059), 3067 (3066, 3067)
Peterson C, **2**: 942 (938, 939); **4**: 2338 (2335)
Peterson C A, **6**: 3834 (3831, 3832)
Peterson D W, **2**: 1099 (1098)
Peterson E, **6**: 3890 (3889)
Peterson J I, **5**: 3464 (3463)
Peterson J L, **2**: 1143; **6**: 3670 (3665), 4350 (4348)
Peterson J N, **1**: 585 (583)
Peterson J T, **1**: 187 (185)
Peterson N M, **6**: 3855 (3854)
Peterson S, **6**: 4108 (4104)
Peterson T G, **1**: 751; **6**: 3652 (3649)
Peterson V J, **4**: 2347 (2345)
Peterson W W, **1**: 421 (418), 664 (654, 659); **4**: 2253 (2248); **5**: 2956 (2952); **6**: 4294
Pethig R, **1**: 450 (448, 450)
Petit G, **3**: 2123
Petkov P H, **2**: 827 (821); **4**: 2827 (2826)
Petri C, **6**: (4348)
Petri C A, **6**: 3670 (3665)
Petri P A, **3**: 2016 (2016)
Petrie C J S, **7**: 5060 (5058)
Petrillo G A, **1**: 570 (568)
Petro P P, **3**: 1943
Petroff P D, **5**: 3421 (3418)
Petrov B N, **4**: 2239 (2237), 2292 (2292)
Petrovic M, **6**: (3862)
Petrovskiĭ I G, **6**: 4364 (4360)
Petry F E, **3**: 1797
Petry J, **4**: 2827 (2826)
Pettersson A, **4**: 2310 (2309)
Pettifor M J, **5**: 2944 (2935)
Pettigrew J D, **7**: 5086 (5083)
Petzold L R, **5**: 3559 (3558); **6**: 4359 (4358)
Pew R, **3**: 1669 (1666, 1667)
Peyret R, **7**: 4498 (4496)
Pfaltz J L, **6**: 3654 (3646)
Pfeifer H J, **2**: 1532 (1528)
Pfeiffer E F, **3**: 2175 (2170)
Pfeiffer H G, **7**: 4504
Pfeiffer P E, **4**: 2338 (2335, 2337)
Pfitscher G H, **5**: 3097
Pfouts R W, **7**: 4822 (4818)
Pfurtscheller G, **2**: 1413 (1411)
Phalen R F, **4**: 2904 (2903, 2904)
Pham D T, **6**: 4108 (4102, 4106)
Pham P T N, **6**: 4108 (4106)
Phatak A V, **3**: 1668 (1666, 1667)
Phelps E S, **1**: 687 (686)
Philbrick G, **6**: (3883)
Philip J, **5**: 3165 (3161)
Philipps L S, **3**: 2025 (2023)
Philippson M, **1**: 450 (447)
Philips C R, **7**: 5094 (5093)
Phillips C G, **1**: 466 (463)
Phillips C L, **6**: 3703 (3702)
Phillips C R, **6**: 3904
Phillips D T, **6**: 3904

Phillips J R, **2**: 894
Phillips K A, **4**: 2607 (2606)
Phillips R E, **1**: 456 (451)
Phillips R G, **7**: 4434 (4430, 4433, 4434)
Phillips R S, **2**: 1424 (1423), 1441 (1440); **3**: 1749 (1748, 1749), 1779 (1771); **5**: 3206 (3205); **6**: 4146 (4145), 4223 (4218); **7**: 5051 (5048)
Phillips R W, **2**: 969
Philon of Byzantium, **3**: (1604, 1605)
Phinney D E, **1**: 592 (591)
Phua K, **6**: 3834 (3832)
Pianigiani G, **2**: 1033
Picard C F, **4**: 2508 (2508)
Picard D, **1**: 86 (84)
Picard J, **5**: 3310 (3310)
Picci G, **4**: 2292 (2287); **7**: 4469 (4467)
Pichai V, **2**: 916
Pichler F, **2**: 1286 (1283); **5**: 3088 (3083); **6**: 4167 (4166), 4381; **7**: 4764 (4762)
Pick R, **2**: 1466 (1463)
Pickel P F, **6**: 3972 (3971)
Pickering E W, **3**: 1958
Pickering T R, **7**: (4516)
Pickett S T A, **2**: 1318 (1313), 1321 (1318), 1358
Picoy R, **7**: 4749 (4748)
Pidd M, **6**: 4333 (4330)
Pielou E C, **2**: 1322 (1321); **6**: 3755
Pierard G E, **1**: 482 (480)
Pierce J G, **2**: 1337 (1335, 1336)
Pierre N A, **1**: 400
Piersol A G, **4**: 2253 (2249), 2310 (2307, 2309); **6**: 4281 (4279)
Pierson D J, **6**: 4061 (4060)
Pietarinen J, **6**: 3860 (3859)
Pietkiewicz-Saldan H, **2**: 1276 (1272, 1276)
Pietrzak-David M, **7**: 5027
Pietrzykowski T, **6**: 4310 (4309)
Pigeon R F, **2**: 1344 (1344)
Piggott J L, **3**: 1979 (1977), 1981
Pike K A, **5**: 2994 (2993)
Pilato S F, **1**: 601 (601)
Pilcher F R, **5**: 3044 (3042)
Pillo J, **1**: 618
Pilsworth B W, **1**: 746 (743); **5**: 2926 (2925, 2926), 3540 (3539)
Pilwat G, **2**: 867 (865, 866)
Pimbley P, **1**: 433 (432)
Pimenides T G, **6**: 3972 (3970)
Pincock D G, **7**: 4848 (4848)
Pinfold W J, **2**: 1431 (1431)
Pingle K K, **4**: 2386 (2382)
Pinglot D, **5**: 2943 (2935)
Pinkava V, **3**: 1809 (1809)
Pinney E, **3**: 2151, 2152 (2152); **5**: 3506
Pinter R B, **7**: 5065 (5062, 5063, 5065)
Pipberger H V, **1**: 769 (760, 761)
Pipes L A, **7**: 4927 (4925)
Pircher M, **1**: 251 (251)
Pironneau O, **2**: 1179 (1174, 1176, 1177, 1178)
Pirt S J, **3**: 1619 (1616)
Pisoni D B, **3**: 2192 (2186)
Pita E G, **6**: 3918 (3917)
Pitt C W, **5**: 3470 (3469)
Pitt G D, **3**: 1694 (1692); **5**: 3464 (3463)
Pittarelli M, **6**: 3989 (3984, 3985, 3987, 3988)
Pitts D E, **6**: 4041 (4037)
Pitts J N Jr, **1**: 176 (175), 177 (176), 187 (185, 187); **2**: 1513
Pitts R F, **1**: 312 (310); **3**: 2206
Pitts W, **2**: 869 (868), 873 (871, 872); **6**: 4167 (4165)
Pizzo J T, **4**: 2412 (2393, 2397, 2399)
Plackett G F, **6**: 4180 (4179)
Plagenhoef S, **3**: 1910 (1907, 1910)
Plambeck J A, **5**: 3575 (3573)
Plant R E, **1**: 508 (507)
Plaskowski A, **3**: 1693 (1690, 1693)
Plateau B, **2**: 1154, 1169 (1167, 1168)
Platt J, **2**: 1348 (1346)
Platt T, **7**: 4454 (4453, 4454)
Platzman G W, **7**: 4889 (4887)
Plauger P J, **1**: 732 (730)
Plesser T, **1**: 429 (424, 426)
Plott C R, **3**: 2061 (2060)
Ployette F, **2**: 1132 (1126)
Plummer A P N, **4**: 2484 (2482)
Plumptre P H, **2**: 1388
Pnuli A, **6**: 3670 (3668)
Poelaert D H, **3**: 1665 (1656)
Poelstra P, **6**: 3933 (3932)
Poettmann F H, **3**: 1975
Pohjanpalo H, **1**: 494 (492)
Poincaré H, **3**: 1898 (1895, 1896); **5**: 3390 (3383, 3384); **7**: (4551)
Pointon A J, **7**: 4513
Poisson-Quinton P, **1**: 31 (21); **7**: 5170
Poisson S D, **6**: (4146)
Polak E, **1**: 736 (735); **2**: 789 (786), 797 (796), 1099 (1093, 1094, 1095), 1125 (1124); **4**: 2278 (2273); **5**: 3164 (3159), 3165 (3159), 3196 (3193, 3195), 3197 (3189, 3191, 3193, 3196), 3215 (3213), 3224 (3223), 3383 (3381), 3487 (3482); **6**: 4310 (4308); **7**: 4969, 4971 (4969)
Polge R J, **5**: 3092
Polis M, **5**: 2993 (2990, 2992), 2994 (2991)
Polis M P, **2**: 1197; **6**: 4245 (4238, 4241)
Pollack F J, **5**: 2980
Pollack R A, **1**: 687 (686)
Pollard J H, **6**: 3755
Polonsky K, **3**: 2175 (2169)
Pólya G, **4**: 2614 (2609, 2610, 2611)
Polyak B T, **1**: 80 (80), 405 (405); **2**: 794 (790), 801 (798), 1125 (1124); **4**: 2737 (2734); **5**: 3531 (3529), 3546 (3544), 3554 (3551, 3552, 3553)
Polyak S, **7**: 5086 (5082)
Polyakova M S, **1**: 433 (432)
Pomraning G C, **7**: 4944
Poncelet J V, **7**: (4517)
Ponssard J P, **3**: 1926 (1923)

Pontryagin L S, 2: 1009 (1004, 1007), 1023 (1022), 1099, 1278; 5: 2963 (2956, 2959), 3513 (3511), 3515 (3514, 3515), 3517 (3516); 6: 3870 (3867); 7: (4551, 4552)
Poole F, 2: 1160
Poole J H, 3: 1958
Poole T G, 4: 2551; 6: 4393
Pooley A, 1: 142 (140)
Poon C S, 6: 4051 (4046)
Poon P T, 3: 2052 (2052)
Pope G G, 6: 3626 (3625)
Popek G J, 2: 1155
Pople H E Jr, 2: 1578 (1573)
Popov V M, 1: 97 (95); 2: 1093 (1089); 3: (1749); 4: 2560 (2559), 2566 (2563), 2827 (2817, 2825), 2871 (2871); 5: 3319, 3373 (3372), 3390 (3387); 6: 3734 (3733); 7: (4552)
Popov Y P, 2: (1210); 7: (4527)
Popovich R P, 1: 310 (309)
Popp R, 3: 2203 (2201)
Popper K, 1: 709 (704); 3: 1805 (1802, 1803); 6: (4215); 7: 4996 (4995), 5015 (5012)
Popplestone R J, 4: 2386 (2374), 2488
Poranski C F Jr, 5: 3421 (3418)
Porati A, 1: 675 (672)
Porter A, 5: 3389 (3385); 6: 3884 (3883)
Porter A L, 2: 1566 (1560); 7: 4827 (4826), 5177 (5171)
Porter B, 1: 100 (99); 6: 3717 (3711)
Porter C T, 7: 4517 (4515)
Porter D W, 4: 2556 (2554)
Porter G B, 6: 3654 (3646)
Porter J H, 5: 3272 (3268)
Porter R, 1: 466 (463)
Porter W A, 2: 808 (804)
Porteus E L, 2: 1099 (1095)
Posa L, 1: 725 (725)
Posbergh T A, 2: 1388 (1388)
Posch B, 3: 1741 (1738)
Posner M I, 3: 2192 (2186)
Post E, 1: 720 (715); 2: 937 (933, 934, 936); 5: 2924 (2921), (2924); 7: 4958 (4958)
Post R B, 3: 1669 (1666)
Postlethwaite I, 1: 580 (575, 577), 736 (732); 4: 2832 (2831), 2872 (2872); 5: 3228 (3224, 3225, 3226)
Poston T, 1: 550 (548); 3: 1788; 7: 4674 (4671, 4672)
Potapov V P, 2: 785 (784)
Potchen E J, 2: 1479 (1478)
Potier D, 2: 884 (881); 3: 1648 (1648)
Potkonjak V, 6: 4086
Potter J E, 4: 2655 (2652); 7: 4468 (4466), 4540 (4534, 4536)
Potts G, 3: 2192 (2186)
Potts R B, 7: 4927 (4925, 4926), 4948
Potts R E E, 3: 1998
Potts R O, 1: 482 (480)
Pouget J, 3: 2221
Pouliquen R, 2: 1121 (1116, 1119)
Pounds D, 1: 450 (449, 450)
Poupard R E, 7: 4481
Powell B A, 2: 1440 (1436)
Powell D, 6: 3916 (3913)
Powell J D, 2: 1066 (1057, 1060), 1071 (1068, 1070), 1244 (1244)
Powell K R, 5: 2943 (2934)
Powell M J D, 2: 794 (790), 797 (795, 796); 4: 2601 (2598); 6: 4310 (4308); 7: 4975 (4972, 4973, 4974)
Powell R A, 1: 675 (674)
Powell T, 3: 2175 (2170, 2173)
Power H M, 2: 1402
Powers C D, 2: 1539 (1538)
Powers D L, 2: 927 (924)
Powers G J, 4: 2517 (2516)
Powers W F, 2: 1272 (1269), 1552 (1549); 4: 2652 (2650, 2651); 5: 3508 (3507, 3508); 7: 4425 (4425)
Poyneer R D, 1: 251 (250)
Poznyak A S, 4: 2737 (2731); 5: 3554 (3554)
Pozzi G A, 7: 5036 (5033)
Prade H, 1: 740 (738), 769; 2: 1407 (1405, 1407), 1549 (1547); 3: 1603 (1602, 1603), 1781 (1780, 1781), 1785 (1785), 1797, 1802 (1802), 1822 (1822), 1831 (1829, 1830), 1835, 1847 (1846, 1847), 1858 (1856), 1872 (1868), 1874, 1883 (1880), 1894 (1891); 5: 2921 (2919), 3156 (3154, 3155), 3454 (3451, 3452, 3454); 6: 3989 (3984); 7: 4766 (4765)
Pradin B, 2: 1192 (1190, 1191)
Pradisthayon T, 5: 2956 (2955)
Praetorius H M, 1: 86 (84)
Prager D, 5: 3410 (3407); 6: 4193 (4190, 4193); 7: 4833 (4828)
Prager D L, 2: 1466 (1463, 1464); 5: 3188 (3184), 3232 (3229); 6: 3727 (3721, 3722, 3723, 3724, 3725)
Prager J M, 6: 4160 (4157)
Prager P, 6: 4176 (4175)
Prajoux R, 5: 3097
Prakash A, 7: 4454 (4453)
Pratt A W, 7: 4876 (4873)
Pratt C W, 7: 4716 (4714), 4726 (4722), 4741 (4734), 4749 (4746)
Pratt J D, 6: 3884
Pratt J W, 5: 3116; 6: 4075; 7: 4991
Pratt L H, 2: 1522 (1517)
Pratt R H, 3: 1694 (1692)
Pratt T W, 1: 725 (722); 6: 3893, 4216; 7: 4752
Pratt V R, 5: 3272 (3268)
Pratt W K, 4: 2484 (2482); 5: 3131 (3130)
Prausnitz J M, 7: 4901 (4898)
Prengle H W, 4: 2511 (2510)
Prenter P M, 5: 3082 (3081)
Preparata F P, 3: 1883 (1881), 1888
Preston S, 4: 2904 (2902)
Prestrud M C, 3: 2052 (2048)
Preti G P, 5: 3465 (3461)
Prevot P, 2: 879 (878, 879)
Prewitt J M S, 6: 3644
Price B J, 2: 866 (865)
Price C F, 5: 3070 (3068)
Price D J, 4: 2582 (2581)
Price K, 6: 4160
Price T, 2: 1154 (1144)

Price T G, **2**: 1174 (1170)
Price W C, **6**: 4261 (4259)
Price W L, **6**: 4310 (4308)
Priest R, **5**: 3470 (3469)
Priestley M B, **4**: 2270 (2265, 2266, 2267)
Priestly M, **2**: 1413
Prigg J A, **3**: 1769 (1767)
Prigogine I, **1**: 429 (424), 709 (706); **2**: 1333 (1330); **3**: 2174 (2174); **5**: 2956 (2956); **6**: 4168 (4167)
Prime H A, **3**: 1749 (1745)
Pringle R M, **6**: 3717 (3714)
Prionas S, **1**: 450 (449, 450)
Prior D A, **3**: 1979
Priouret P, **1**: 80 (75, 76, 77)
Pritchard A J, **2**: 1186 (1185); **3**: 1665 (1650); **4**: 2861; **6**: 4245 (4242)
Pritchard R, **3**: 1965; **4**: 2356 (2353)
Pritsker A A B, **2**: 894 (894); **3**: 1648 (1647, 1648); **5**: 3035 (3034); **6**: 4313 (4312), 4372 (4370, 4371), 4406 (4404), 4415 (4414); **7**: 4654 (4652, 4653), 4935 (4934)
Pro R, **2**: 1106
Proakis J G, **1**: 80 (75)
Probert R L, **4**: 2681 (2680)
Probst A R, **7**: 4935 (4933)
Probst J, **3**: 2092 (2091)
Procyk T J, **4**: 2362
Proell R R, **5**: (3383)
Profos P, **3**: (1748); **7**: 4580
Pronk C, **5**: 3260 (3260)
Proschan F, **6**: 4011 (4008, 4009), 4029 (4027, 4029); **7**: 4766 (4764, 4765)
Prosser C L, **7**: 4980 (4976, 4977)
Proth J M, **1**: 409; **3**: 1714; **4**: 2583; **6**: 3706 (3706), (3890)
Protinski H, **3**: 2203 (2201)
Proudfoot C G, **6**: 3690 (3690), 3691
Pruess K, **3**: 2006 (2006)
Pruitt D G, **3**: 2202 (2196)
Pryde A, **4**: 2889
Przelecki M, **6**: 3865 (3865)
Pshenichny B N, **2**: 1023 (1022); **5**: 3546 (3544)
Pszczolowski T, **6**: 3865 (3861)
Ptak V, **4**: 2556 (2553, 2555), 2565 (2561, 2562)
Puckett R, **1**: 177 (175)
Puech C, **5**: 3470 (3466)
Puech P, **2**: 1410
Puel J P, **4**: 2343 (2342); **6**: 3596 (3595); **7**: 5041 (5039)
Puente E A, **2**: 1586 (1584); **6**: 3610 (3610)
Pugachev V S, **5**: (3387)
Pugh A, **4**: 2449 (2448), 2479
Pugh A C, **1**: 714 (713); **6**: 3751 (3747)
Pugh A L, **6**: 4368 (4368)
Pugh D O, **3**: 1998 (1997)
Pugh W J, **6**: 3962 (3961)
Pulkkinen T, **6**: 3916 (3914)
Pullan B R, **7**: 5186 (5181, 5182, 5184)
Pungor E Jr, **5**: 2943 (2935, 2942)
Purbrick J A, **6**: 4123
Purcell E M, **5**: (3417)
Purdom P B, **3**: 1637
Puri M L, **3**: 1822 (1822), 1845 (1845); **6**: 3619 (3617, 3618), 3990
Purnell J H, **5**: 2943 (2935)
Purviance J E, **1**: 703 (695)
Putley E H, **7**: 4856 (4854)
Putnam F W, **6**: 3909
Putz U, **5**: 3092
Putzoly F, **2**: 1174 (1170)
Putzulo F, **2**: 1154 (1144)
Putzulo G R, **2**: 1154
Pykett I L, **5**: 3425 (3424)
Pyster A, **6**: 4160 (4158)

Q

Quade E S, **2**: 987, 1542 (1540); **4**: 2353 (2351); **6**: 4155; **7**: 4809
Quadrat J P, **3**: 1648 (1646), 2072 (2069), 2225; **5**: 3530 (3525, 3529), 3531 (3525, 3526, 3527, 3528); **6**: 3789 (3788); **7**: 4613
Quandt R E, **7**: 5164
Quarteroni A, **3**: 1723 (1722)
Quastel D M J, **2**: 1248 (1245)
Quastler H, **2**: (872)
Quazza G, **6**: 3846
Quick W H, **5**: 3464 (3460, 3461)
Quillen D, **6**: 3972 (3971)
Quillian M R, **4**: 2641 (2637)
Quin J P, **4**: (2804)
Quine W, **2**: 937 (933)
Quinn B G, **4**: 2292 (2292)
Quinn R E, **7**: 4785 (4783)
Quinton H, **7**: 4785 (4780)
Quiram E R, **1**: 181
Quirk J, **2**: 1081 (1079); **5**: 2979; **6**: 3655; **7**: 5164
Quirk J P, **2**: 777
Quisell B J, **7**: 4905 (4905)
Qureshi Z H, **4**: 2264 (2263)
Qvist J, **1**: 315 (314)

R

Raabe O G, **4**: 2904 (2903, 2904)
Rabin M, **2**: 937 (933, 934, 937)
Rabiner L R, **6**: 4289 (4286); **7**: 4509 (4508)
Rabinowitz P H, **1**: 529 (527, 528); **2**: 981 (981); **5**: 3496 (3496); **7**: 5047 (5042)
Rabone N S, **3**: 1965, 1985
Rabotnov T A, **2**: 1358
Rachal C M, **7**: 5106 (5104)
Radcliff R A, **3**: 2192 (2187)
Radcliffe K, **7**: 4504 (4499)
Radecki T, **4**: 2301 (2300)
Rademacher H, **5**: 3362 (3359)
Rademaker O, **2**: 1109 (1108, 1109)
Rader B, **7**: 4856 (4852)
Rader C M, **3**: 1723; **6**: 4289
Radhe D, **5**: 2943 (2935)
Radke F, **1**: 63 (63); **3**: 2097; **6**: 3611 (3607, 3610)
Radner R, **2**: 932 (931); **4**: 2560 (2557); **5**: 2911 (2908); **7**: 4822 (4814, 4818)
Radnitzky G, **6**: (3862); **7**: 4996 (4995)
Radnor M, **7**: 4833 (4831)
Radouane L, **2**: 923 (923)
Radu L, **6**: 3850 (3849)
Radziuk J, **3**: 2025 (2022), 2175 (2169, 2170, 2171, 2172); **7**: 4906 (4904, 4905)
Raffort L, **5**: 3480
Ragade R K, **2**: 1346 (1345), 1355, 1473 (1472); **3**: 1819, 1831 (1830), 1835, 1838, 1845 (1845), 1852 (1847, 1848, 1850), 1858, 1862 (1859), 1867 (1864), 1872 (1868), 1874, 1878 (1876), 1888, 1902 (1900); **4**: 2301 (2300), 2361, 2884 (2883); **5**: 2965 (2963, 2964, 2965), 3540 (3539); **6**: 4008 (4005); **7**: 4695 (4693), 4789 (4788)
Ragazzini J B, **2**: 1078 (1074)
Ragazzini J R, **6**: 4146 (4145)
Rahn H, **6**: 4066 (4063)
Rai V R, **3**: 1611 (1609)
Raibert M H, **4**: 2385 (2380), 2386 (2379, 2380), 2460 (2458); **6**: 4100, 4128 (4124)
Raiconi G, **6**: 3626 (3625)
Raidt H, **2**: 1532 (1528)
Raiffa H, **1**: 372 (360); **2**: 850 (840, 843), 942 (941), 943 (938, 939, 941), 1257 (1255), 1565 (1563); **3**: 1921 (1918), 2061 (2053, 2056), 2202, 22022203 (2194, 2195), 2203 (2194, 2196); **5**: 2911 (2907, 2910), 3104 (3099, 3100, 3101), 3115 (3104, 3105, 3107, 3108, 3112), 3121 (3121), 3124 (3122, 3123), 3152, 3164 (3157); **7**: 4985, 4986, 4991
Rajagopalan P K, **7**: 4434 (4430)
Rajahalli P, **5**: 2944 (2935)
Rajala D W, **4**: 2501 (2499)
Rajala M, **2**: 781 (779)
Rajamani K, **5**: 2993 (2990, 2991)
Rajkovic V, **1**: 769 (766); **3**: 1829 (1826)
Rake H, **3**: 2092 (2091); **4**: 2245 (2241), 2253 (2251, 2252)
Raksanyi A, **1**: 494 (493)
Ralescu D A, **1**: 713 (711); **2**: 1479 (1477); **3**: 1604 (1602, 1603), 1781 (1781), 1819 (1817), 1822 (1821, 1822), 1845 (1845), 1874, 1880 (1878), 1888, 1894 (1889, 1893), 1902 (1898); **4**: 2301 (2299, 2300); **5**: 3264 (3262), 3540 (3538, 3539); **6**: 3619 (3616, 3617, 3618), 3990
Ralston A, **2**: 1416 (1415)
Ramachandra Rao A, **4**: 2307 (2302, 2303, 2307); **7**: 4692 (4691), 5001 (4998)
Ramadan S M, **1**: 570 (568)
Ramadge P J, **1**: 60 (58, 59, 60); **2**: 1581 (1579, 1580); **5**: 3060 (3056, 3059), 3061 (3061), 3065 (3062); **6**: 3929 (3929), 4193 (4192), 4209 (4206, 4207); **7**: 4607 (4603, 4606), 4611 (4607)
Ramaker B L, **1**: 588 (588); **6**: 3679 (3675)
Ramakrishna A, **4**: 2560 (2559), 2671 (2666, 2669, 2670), 2702 (2698, 2699, 2702)
Ramakrishnan R, **2**: 1366 (1364)
Ramaley L, **5**: 2943 (2935)
Ramamoorthy C V, **4**: 2681 (2679); **6**: 3893
Ramani N, **5**: 3383 (3381), 3390 (3386)
Ramaz A, **2**: 781 (779)
Rambaut J P, **4**: 2392
Rametti L B, **5**: 3238 (3237)
Ramfjord S P, **2**: 973 (971)
Ramirez W F, **6**: 4364 (4363)
Ramkrishna D, **1**: 424 (423)
Ramm F, **2**: 1586 (1586)
Ramon F, **7**: 4980 (4977, 4979)
Ramon Y Cajal S, **7**: 5070 (5066)
Ramsey A R J, **3**: 1607 (1606)
Ramsey H R, **3**: 2179
Ramsey N F, **3**: 1745
Ramskow-Hansen J J, **5**: 3464 (3461)
Ramslo K, **5**: 2993 (2991)
Rand M C, **7**: 5122
Randall J E, **1**: 468 (466, 467); **2**: 1366 (1362)
Randall W C, **7**: 4877 (4875), 4880 (4879, 4880)
Randell B, **2**: 1136 (1133, 1134, 1136), 1174 (1169, 1170)
Randell W L, **2**: 1402 (1399)
Randers J, **2**: 1538 (1536); **3**: 2016 (2015); **7**: 4762 (4756)
Randow R von, **2**: 1083 (1080)
Rane D S, **4**: 2800 (2799)
Rankama K, **1**: 7 (1), 148 (144)
Ranke L, **1**: (554)
Rans L L, **2**: 1479 (1478)
Ransom R, **7**: 4958 (4957)
Ransom R J, **1**: 561 (559); **6**: 4345 (4341, 4343, 4344, 4345)
Ranyard R H, **3**: 2203 (2199, 2201)
Rao A S, **1**: 132 (127); **5**: 3077 (3075)
Rao C R, **2**: 1107 (1103); **4**: 2273 (2271)
Rao D B, **7**: 4889 (4885, 4887)
Rao H S, **7**: 5109 (5108)
Rao K, **6**: 4160 (4158)
Rao R, **3**: 2010, 2045 (2045)
Rao S V, **1**: 132 (127); **5**: 3077 (3075, 3077)

Rao T C, **6**: 4252 (4251)
Raol J R, **4**: 2292 (2287)
Rapaport J, **7**: 4833 (4828)
Rapp P E, **1**: 428 (425), 429 (427, 428), 433 (429), 570 (569); **2**: 981 (980)
Rapp Y, **7**: 4716 (4714)
Rashleigh S C, **5**: 3470 (3469, 3470), 3471 (3470)
Rasiowa H, **5**: 2921 (2918, 2919)
Rasmussen J, **3**: 1669 (1666, 1667), 2203 (2193); **4**: 2501
Rastegar J, **4**: 2456 (2453); **6**: 4108 (4105)
Rastogi L K, **2**: 1568 (1567)
Rathmill K, **3**: 1648 (1642)
Ratliff F, **7**: 5065 (5060, 5062, 5064), 5070 (5068)
Ratliff R, **7**: 5186 (5182)
Rau N S, **6**: 4139 (4135)
Raub W F, **4**: 2901 (2899)
Rauch H E, **7**: 4466 (4465)
Rault A, **2**: 1121 (1116, 1119); **3**: 1669 (1667); **6**: 3680 (3675)
Raviart P A, **3**: 1632
Ravidran R, **6**: 4088 (4086)
Ravindran A, **6**: 3904
Rawls J, **3**: 2061 (2060)
Rawson H, **3**: 2014
Rawson R O, **7**: 4877, 4880 (4879, 4880)
Ray W H, **2**: 808, 1197; **3**: 1990 (1987); **5**: 3487 (3485); **6**: 3744 (3740), 3745 (3740, 3742), 4088 (4086), 4245; **7**: 4621 (4617, 4618, 4620)
Raybould R H, **3**: 1947 (1946), 1975
Rayleigh (Lord), **1**: 536 (532); **3**: (1745); **5**: (3383); **7**: 4554 (4551)
Raymond F H, **6**: 4146 (4145)
Rayner A A, **6**: 3717 (3714)
Rea D P, **7**: 4452 (4451)
Ream A K, **1**: 440
Rearick L W, **6**: 4370 (4369)
Reasbeck M P, **2**: 963 (961, 962, 963)
Réaumur R A F de, **3**: (1606)
Reaven G M, **3**: 2025 (2022)
Reay D A, **1**: 167
Rebhuhn D, **7**: 4677 (4674)
Reboulet C, **3**: 1665 (1660); **7**: 5027
Reckhow K H, **4**: 2650 (2649); **7**: 4953 (4953)
Rector F J, **3**: 2206 (2206)
Reddy D R, **6**: 3644, 4160
Reddy J N, **3**: 1632
Reddy V R, **1**: 142 (140, 141)
Redford J B, **5**: 3253 (3250, 3251, 3252)
Redheffer R, **2**: 1351 (1350)
Redmond M J, **1**: 633 (632)
Redpath W, **5**: 3416
Ree R, **3**: 1779 (1772)
Reed D P, **2**: 1155 (1152), 1174 (1173)
Reed G, **3**: 1611 (1608), 1615 (1612)
Reed J R, **6**: 3806 (3802)
Reed S, **3**: 2192; **7**: 4793 (4792)
Rees D, **7**: 4463 (4462)
Reese S E, **3**: 1958
Reethof G, **6**: 4257
Reeve C, **2**: 1160 (1159)
Reeve E B, **3**: 2214 (2213); **6**: 3909 (3905)
Reeves C, **3**: 1623 (1620)
Reeves C M, **6**: 4310 (4308)
Reeves R G, **6**: 4045
Reeves W R, **6**: 4346 (4346)
Regan F J, **1**: 257
Reggiani M G, **2**: 1326 (1324)
Rehmann J T, **3**: 1669 (1666, 1668)
Rehmus P H, **1**: 429 (428)
Reiber K, **7**: 5186 (5180)
Reichardt W, **4**: 2331 (2327)
Reichenbach H, **1**: 333 (330); **5**: (2918)
Reichert K, **6**: 3829 (3829)
Reichert T A, **2**: 1479 (1476, 1478)
Reichertz P L, **2**: 1577 (1575)
Reichle D E, **2**: 1344 (1344), 1345 (1343)
Reid G, **3**: 1669 (1666)
Reid J, **4**: 2511 (2510), 2723 (2721, 2722)
Reid J K, **2**: 827
Reid K J, **5**: 2994 (2991)
Reid R W, **5**: 3165 (3160)
Reid W K, **7**: 5112
Reiersøl O, **4**: 2277 (2273)
Reiffen B, **2**: 1051 (1051)
Reilley C N, **7**: 5115
Reilly C A, **5**: 3421 (3419)
Reilly D T, **1**: 474 (468, 469, 471)
Reiman M I, **2**: 1169 (1169)
Reinganum J F, **2**: 1269 (1268)
Reinhart K G, **5**: 3268 (3266)
Reinhold A G, **4**: 2386 (2382)
Reinhold B, **5**: 2944 (2935)
Reinisch K, **2**: 1305 (1298, 1299, 1300, 1301, 1302, 1303); **5**: 3171 (3170)
Reinsch C, **2**: 826 (820), 828 (819)
Reischl P, **6**: 4051 (4048), 4061 (4060)
Reiser A, **2**: 1154 (1153)
Reiser M, **3**: 1648 (1646)
Reisig W, **6**: 3670 (3665), 4350 (4348)
Reiste T, **1**: 570 (568)
Reithmaier L W, **1**: 109
Reitman J, **6**: 4302 (4301)
Rejańnek M, **1**: 675 (672)
Rekasius Z V, **4**: 2523 (2522)
Rellstab L, **6**: 4275 (4273)
Rembold U, **4**: 2351 (2347)
Remde J R, **7**: 4509 (4506, 4507)
Renda A, **7**: 4789 (4788)
Rendell-Baker L, **1**: 319
Renteux J L, **3**: 1756 (1750)
Rényi A, **6**: 3768 (3766)
Repperger D, **2**: 1192 (1191)
Reps T, **5**: 3079 (3078)
Rescher N, **2**: 1545; **4**: 2501 (2500); **5**: 2921 (2918), 2924 (2921, 2922), 3116
Respondyk W, **4**: 2756 (2756)
Restle F, **3**: 2191 (2190), 2202 (2196)
Retallack D G, **6**: 3717 (3711)

Reuck A V, **7**: 5086
Reul H, **1**: 312 (311)
Reuschel R, **7**: 4927 (4925)
Reuss J O, **1**: 142 (141)
Reuter J H, **4**: 2782 (2778)
Reuver H A, **5**: 3035 (3033)
Reverdy J, **2**: 1211 (1210)
Revesz G, **7**: 5186 (5182)
Revington A M, **7**: 4577 (4576)
Rexstad E, **2**: 1332 (1332)
Rey D, **1**: 682 (678, 679)
Reyment R A, **2**: 1106
Reynolds W J, **4**: 2901 (2899)
Rhee S W, **3**: 2006 (2006)
Rheinboldt W C, **1**: 405 (402); **2**: 956 (954), 1125 (1122, 1124); **3**: 2163 (2155); **5**: 3487 (3482); **7**: 4969
Rheinland S, **1**: 634 (632)
Rhodes I B, **2**: 932 (930, 931)
Ribbends D, **3**: 1712 (1712)
Ribbens-Pavella M, **6**: 3846 (3845)
Ribière G, **6**: 4310 (4308)
Ricciardi G, **2**: 1333 (1331)
Ricciardi L, **1**: 399 (394); **3**: 1781 (1781); **6**: 4066 (4062), 4167 (4166), 4377 (4377)
Rice C B F, **1**: 594
Rice J G, **4**: 2551 (2550)
Rice J R, **2**: 827 (815); **5**: 3470 (3467)
Rich C I, **1**: 7 (2, 5), 268 (263, 266)
Rich E, **1**: 307 (306); **2**: 1573
Rich L G, **4**: 2789
Richalet J, **2**: 1121 (1116, 1119); **6**: 3680 (3675)
Richard A G, **3**: 1958
Richard J P, **7**: 4547 (4544, 4545)
Richard L A, **6**: 3680 (3676)
Richard S F, **6**: 4075
Richards L W, **1**: 177 (175)
Richards P I, **7**: 4922
Richardson H, **5**: 2970
Richardson T H, **2**: 1355 (1353); **4**: 2646 (2645); **7**: 5008 (5005, 5007)
Richer I, **2**: 1125 (1122, 1123)
Richetin M, **1**: 703 (689, 690); **4**: 2547 (2545, 2546, 2547); **7**: 4666 (4664, 4665)
Richey G K, **1**: 623
Richmond I J, **5**: 3421 (3418)
Richter G, **2**: 1532 (1530)
Richter P H, **1**: 429 (428)
Richter S, **2**: 923 (922)
Richtmyer R D, **3**: 1627
Rickel D, **1**: 177 (175)
Ricker W E, **7**: 4646 (4645)
Ricketts C, **3**: 1623 (1619)
Ricklefs R E, **7**: 4956 (4955)
Rickman S M, **3**: 1888
Riddle W, **4**: 2681 (2680)
Rideout V C, **1**: 498 (496), 542 (537), 633 (632)
Ridge M D, **1**: 482 (480)
Ridgeway J M, **6**: 4235 (4230)
Riegelman S, **6**: 3698 (3697)
Riemann F, **2**: 867 (865, 866)
Riera T, **2**: 1473 (1470, 1471); **3**: 1878 (1876)
Ries D, **2**: 1155 (1153), 1165
Riffenburgh R H, **7**: 4646 (4645)
Rigas A, **5**: 3242 (3240, 3241, 3242)
Rigerl K H, **6**: 3889 (3888, 3889)
Riggs L, **7**: 5078
Rigler F H, **4**: 2650 (2648), 2794 (2789)
Riguet J, **6**: 4008 (4000, 4002)
Rijnsdorp J E, **2**: 1109 (1108, 1109)
Rimmer S J, **1**: 459 (459); **5**: 3238 (3237)
Rinaldi S, **2**: 1093 (1090); **5**: 3171 (3170)
Rine D C, **1**: 743, 746 (743, 744, 746)
Rines G A, **5**: 3464 (3459)
Ring E F J, **7**: 4871 (4868)
Ringle M, **1**: 307
Rinks D B, **1**: 126 (126)
Riordon J S, **2**: 1160; **3**: 2175 (2171); **4**: 2737 (2731, 2736); **6**: 4165 (4163)
Rippin D W T, **6**: 3745 (3740)
Riseman E M, **6**: 4160
Rishel R W, **2**: 1279; **3**: 2068, 2072 (2069); **5**: 3523 (3518, 3520, 3523), 3538; **7**: 4627 (4623, 4624), 4644 (4643)
Rissanen J, **4**: 2292 (2288, 2291, 2292), 2560 (2559); **6**: 3629 (3629), 3734 (3733), 3983 (3979); **7**: 4566 (4563), 5001 (5000)
Risson J, **7**: 4749 (4745)
Ritch G P, **3**: 1736
Ritchie B, **1**: 755
Ritson P J, **3**: 1958
Rittenhouse J W, **6**: 3806 (3804, 3805)
Ritz T J, **5**: (3385)
Rivett P, **2**: 1291, 1297; **4**: 2353
Rivlin R S, **1**: 486 (483)
Rix C J, **4**: 2607 (2606)
Rizvi M H, **6**: 3997 (3996)
Robb C, **7**: 4848 (4847)
Robbins H, **4**: 2314 (2310), 2737
Robbins J, **1**: 80 (78)
Roberge F A, **1**: 536 (534)
Robert F, **5**: 3406 (3402)
Robert J, **7**: 4856 (4856)
Robert P, **2**: 1169 (1169)
Roberts A, **1**: 675 (672)
Roberts B M, **3**: 1623 (1619)
Roberts D V, **1**: 424 (422)
Roberts F, **7**: 4523 (4523)
Roberts F S, **2**: 1355 (1352); **4**: 2641 (2637, 2638); **5**: 2930 (2927), 3116 (3105); **7**: 4956 (4955), 4986
Roberts G F I, **3**: 1969
Roberts J D, **7**: 4551 (4550)
Roberts K, **7**: 4472 (4469)
Roberts L G, **4**: 2440 (2422)
Roberts P A, **3**: 1637
Roberts P D, **1**: 638 (634, 636, 638); **7**: 4571 (4571)
Roberts R A, **1**: 664 (659)
Roberts S D, **2**: 894 (894)
Roberts T D M, **5**: 3253 (3249, 3250, 3251, 3252)

Roberts V, **2**: 1458
Roberts W H, **1**: 150; **3**: 2209
Robertson A, **7**: 5008 (5007)
Robertson D I, **1**: 517 (516); **7**: 4925
Robertson J A L, **5**: 3440
Robertson J B, **1**: 246
Robertson J S, **7**: 4905 (4902)
Robertson K, **5**: 2993 (2990)
Robertson N R E, **2**: 967 (967)
Robertson R E, **3**: 1665 (1653)
Robes P, **6**: 4123
Robey D, **4**: 2501 (2501)
Robin G C, **3**: 1910 (1910)
Robin M, **4**: 2343; **5**: 3316 (3314)
Robinson A, **3**: 1791 (1789); **5**: 2921 (2921)
Robinson A M, **4**: 2631 (2628)
Robinson C F, **5**: 2944 (2934)
Robinson D A, **2**: 1595 (1594), 1596 (1595)
Robinson D M, **3**: 1958
Robinson D W, **2**: 1527 (1525)
Robinson E A, **2**: 785 (783), 1494; **4**: 2270 (2264)
Robinson G S, **6**: 4160
Robinson J T, **2**: 1154 (1149)
Robinson R, **7**: 4880 (4880)
Robinson S, **5**: 3551 (3548); **7**: 4877 (4876)
Robinson S M, **1**: 628 (627); **2**: 797 (795), 850 (845); **5**: 3546
Robles de Medina E O, **1**: 769 (761)
Robson J G, **4**: 2331 (2325)
Rocha A F, **1**: 505 (502), 508 (507); **5**: 3264 (3262, 3263)
Rockafellar R T, **1**: 406 (400, 405), 529 (524); **2**: 793 (793), 1262 (1257, 1258); **3**: 1917 (1913, 1916, 1917); **5**: 3362; **6**: 3622; **7**: 5036 (5033), 5047 (5045)
Rockart J F, **4**: 2501 (2499)
Rodabaugh S E, **3**: 1831 (1831)
Rodbell M, **2**: 1248 (1245)
Rödder W, **5**: 3455 (3452)
Rodenheaver G J, **1**: 482 (480)
Rodger W D, **3**: 1982
Rodin E Y, **4**: 2301 (2293, 2301)
Rodman L, **2**: 826
Rodrigo J J, **3**: 2174 (2173)
Rodrigo P, **3**: 1665 (1660)
Roels J A, **1**: 424
Roenfeld A, **4**: 2386 (2382)
Roesser R P, **5**: 3127 (3125)
Roff P A, **7**: 4956 (4955)
Rofman E, **3**: 2068 (2064), 2072 (2071)
Rofman R, **5**: 3480 (3478)
Roger F, **2**: 1577 (1575)
Roger P F, **4**: 2386 (2383)
Rogers A J, **5**: 3465 (3460, 3461)
Rogers E M, **7**: 4833 (4829, 4831)
Rogers G F C, **3**: 1995; **7**: 4585 (4580)
Rogers G J, **6**: 3802 (3796, 3797)
Rogers H, **1**: 720 (718, 719); **2**: 937 (937)
Rogers H H, **2**: 1521 (1518)
Rogers J, **3**: 1665 (1656)
Rogers J S, **4**: 2782 (2778)
Rogers K, **3**: 1979
Rogers M O, **4**: 2723 (2721)
Rogers P, **3**: 1665 (1661)
Rogers P I, **3**: 1611 (1609)
Rogers T A, **5**: 3390 (3386)
Rohn T, **2**: 1305 (1300)
Rohrbaugh J, **3**: 2061 (2060)
Röhrer F, **1**: 459 (457); **6**: 4066 (4063)
Rohrer R A, **6**: 4322 (4319)
Roithman D, **7**: 4833 (4828, 4829, 4830, 4831)
Rokeach M, **3**: 2203 (2195)
Rol N C, **5**: 3421 (3418)
Roland J J, **5**: 3471 (3466)
Rolewicz S, **5**: 3206
Rolf B, **7**: 4996
Rolf F J, **7**: 5008 (5006)
Rolfe A J, **2**: 1099 (1094); **6**: (3892)
Rolfe J M, **3**: 1669 (1666)
Rolfe P, **7**: 4856 (4853, 4855)
Roll Y, **3**: 2076
Röllinger H, **6**: 3834 (3833)
Rollne R A, **7**: 5113
Rolls B J, **3**: 2214 (2212, 2213)
Rolls E T, **3**: 2214 (2212, 2213)
Rom D, **6**: 3850 (3846)
Romanin-Jacur G, **1**: 493 (492)
Romanova L M, **3**: 2052 (2048)
Romeder J M, **2**: 879 (879)
Romeo A, **7**: 4833 (4828)
Romeo F, **7**: 4689 (4689)
Romer E G, **1**: 570 (568)
Romette J L, **1**: 433 (432)
Romslo K, **6**: 4252 (4250)
Ronco R J, **1**: 184 (183)
Rooks B W, **4**: 2444 (2441); **6**: 4108 (4102)
Rooney F, **7**: 4913 (4912)
Rooney J, **6**: 4108 (4106)
Roose A, **4**: 2517 (2516)
Rootselaar B, **6**: 3860 (3858), 4216 (4216)
Roper A T, **2**: 1566 (1560); **7**: 4827 (4826), 5177 (5171)
Rosario O, **4**: 2511 (2510)
Roscoe A H, **3**: 1669 (1665, 1666)
Rose A H, **3**: 1615 (1612)
Rose D, **1**: 312 (311)
Rose D G, **3**: 1982 (1980)
Rose D J, **6**: 4299
Rose E, **6**: 4172 (4171)
Rose J, **3**: 1878 (1877); **6**: 3768 (3766)
Rose J M, **4**: 2782 (2782)
Rose K A, **2**: 1552 (1549, 1550, 1551); **6**: 4229 (4228), 4235 (4233)
Rose L L, **4**: 2502 (2498); **7**: 4785
Rose M R, **2**: 1552
Rose N J, **5**: (3516)
Rose P E, **7**: 4601 (4601)
Rose P K, **2**: 1043 (1041)
Rosen C A, **6**: 4123
Rosen E M, **4**: 2517 (2516)
Rosen J, **7**: 4685 (4680)

Rosen J B, 2: 1099 (1094); 4: 2880 (2877)
Rosen K, 1: 769 (761)
Rosen M, 2: 963 (962)
Rosen R, 1: 294 (294); 2: 1348 (1346); 3: 1862 (1859, 1860); 5: 3440 (3440); 6: 4008 (4003), 4066 (4062), 4319
Rosenbaum J S, 2: 1248 (1246)
Rosenbeng H M, 2: 976
Rosenberg A E, 7: 4509 (4508)
Rosenberg B L, 3: 1669 (1668)
Rosenberg D, 4: 2488
Rosenberg G, 3: 1712 (1712)
Rosenberg J R, 5: 3242 (3240, 3241, 3242)
Rosenberg M, 5: 3235 (3234)
Rosenberg M J, 4: 2640 (2638)
Rosenberg R C, 2: 1329 (1327); 6: 4341 (4337, 4341)
Rosenblatt F, 6: 4168 (4166)
Rosenblatt M J, 3: 2076
Rosenbloom P, 5: 2924 (2921)
Rosenblueth A S, 2: (868), 873 (871, 872, 873)
Rosenbrock H H, 1: 580 (575, 578), 714 (713, 714), 736 (732, 734); 4: 2523 (2523), 2570 (2570), 2588 (2585, 2587), 2597 (2588, 2590), 2601 (2597, 2598), 2671 (2666, 2667), 2702 (2698), 2832 (2828, 2830), 2871 (2866, 2870); 5: 3197 (3189), 3201 (3197, 3200), 3215 (3206, 3207), 3224 (3220, 3221, 3223), 3228 (3224), 3319 (3319), 3383 (3380, 3381), (3387), 3406 (3402), 3436 (3432), 3441 (3441); 6: 3701, 3717 (3713), 3720 (3718), 3721 (3720, 3721), 3734 (3734), 3751 (3745, 3746, 3747, 3748, 3749, 3750), 3972 (3971); 7: 4457 (4456), 4547 (4541), 4554 (4553), 5192
Rosencher E, 1: 296 (294, 295)
Rosencrantz R, 6: 3695
Rosencwaig A, 4: 2511 (2510)
Rosenfalck P, 3: 1910 (1908, 1909)
Rosenfeld A, 1: 575; 3: 1781 (1781); 4: 2473 (2472), 2484 (2482); 6: 3644 (3641), 3653, 3654 (3646), 4160 (4157, 4158); 7: 4883
Rosenkrantz D J, 2: 884 (881, 882), 1155 (1145, 1150, 1152)
Rosenoer V M, 6: 3909 (3907)
Rosevear J W, 3: 2025 (2022)
Rosier D, 4: 2453 (2452)
Rosin R E, 2: 1451 (1451)
Roskam J, 1: 109; 7: 5170 (5166)
Rosner B S, 1: 281 (280)
Ross C W, 6: 3815 (3812)
Ross D T, 5: 3272 (3268); 6: 4355 (4353, 4354)
Ross D W, 6: 3820 (3817, 3819); 7: 4917 (4916), 4930 (4928)
Ross J, 1: 429 (428), 433 (431)
Ross J C, 1: 564 (561)
Ross J Jr, 1: 542 (541)
Ross N C, 2: 1214 (1213)
Ross N D H, 1: 633 (633)
Ross R, 4: 2901 (2899); 6: 4302 (4301)
Ross S, 3: 1601 (1600); 7: 4897 (4896)
Ross S M, 5: 2932
Rosser J B, 1: 333 (328); 5: 2921 (2918)
Rossi A, 1: 570 (569)
Rossi P H, 2: 1566 (1560)
Rossini F A, 2: 1566 (1560); 7: 4827 (4826), 5177 (5171)
Rossler J, 2: 1305 (1302)
Rossner H O, 5: 2943 (2935)
Rosswall T, 1: 142 (141)
Roth A E, 1: 687 (684); 2: 850 (849); 5: 2911 (2910)
Roth B, 4: 2386 (2376), 2412 (2393), 2456 (2453); 6: 4108 (4104, 4105)
Roth G J, 3: (1759)
Roth J, 3: 2174 (2172)
Roth P F, 6: 4302 (4301), 4303 (4300), 4306 (4303), 4323 (4321), 4403 (4402, 4403), 4406 (4404)
Roth Z S, 5: 3303 (3301)
Rothe C F, 1: 433 (432)
Rothenberg J R, 3: 2061 (2058)
Rothery R W, 7: 4922, 4927 (4925, 4926)
Rothnie J B, 2: 884 (881), 1154 (1148, 1153), 1155 (1145), 1160 (1159)
Rothschild M A, 6: 3909 (3907)
Rothstein J, 2: 1479 (1475)
Rothwell N J, 3: 2214 (2212)
Rotty R M, 2: 1344 (1343)
Rouahi A, 7: 4741 (4740)
Roubellat F, 3: 1648 (1644)
Roubens M, 3: 1847 (1847)
Roubens M C, 1: 608 (606, 607); 2: 1107 (1105)
Roucairol G, 6: 3670 (3667)
Rouchaleau Y, 5: 3293 (3290); 6: 3972 (3970)
Rouche N, 1: 675 (673); 5: 3402 (3399, 3400); 7: 4529 (4525)
Rouff M, 4: 2609 (2608)
Roumegoux J P, 3: 1761 (1759)
Rouse W B, 4: 2641 (2637)
Roussas G G, 1: 86 (83)
Rousset G, 7: 4749 (4745)
Roussias L, 1: 638 (638)
Routh E J, 3: (1745); 5: (3387); 7: 4517 (4517)
Roux J C, 1: 570 (569)
Roux M, 2: 1577 (1575)
Roveda C A, 4: 2827
Rovira A A, 3: 2113 (2111)
Rowe I H, 4: 2277 (2273)
Rowe R P, 4: (2351)
Rowe W D, 2: 1539 (1538)
Rowell C H F, 7: 5065 (5062)
Roxin E, 3: 1898 (1896); 5: 2963 (2963)
Roxin E O, 2: 1009
Roy B, 3: 1829 (1826); 5: 3124 (3122), 3152 (3152), 3165 (3157); 7: 4667 (4661, 4662, 4666)
Roy R J, 6: 4148 (4147)
Rozenberg C, 4: 2798 (2797)
Rozenberg G, 3: 2046 (2045); 4: 2798 (2795, 2796, 2797)
Rozencweig M, 2: 1479 (1478)
Rozenvasser E N, 5: (3387)
Rozin R, 3: 1910 (1910)
Rozonoev L I, 2: 1478 (1476)
Rubenstein L I, 5: 3099
Ruberti A, 1: 417 (414); 2: 1121 (1116, 1119), 1197; 4: 2527 (2526); 5: 3279 (3276); 6: 3961 (3959), 3977 (3973); 7: 5041 (5038)

Rubin A I, **6**: 4417
Rubin J, **7**: 4935 (4934)
Rubinstein A, **3**: 1921 (1919)
Ruch T C, **1**: 433 (432); **5**: 3253 (3248, 3250, 3252)
Ruckebusch G, **1**: 296 (294, 295)
Rückert H, **6**: 3890 (3889)
Ruddock K H, **4**: 2330 (2326, 2327, 2328), 2331 (2326, 2327, 2328)
Rudemo M, **3**: 2174 (2173)
Rudinger G, **3**: 1694 (1685)
Rudjord T, **5**: 3242 (3241)
Rudnianski J, **6**: (3862)
Ruelle D, **1**: (567), 570 (567); **5**: 3341 (3341)
Ruff L E, **2**: 1454 (1453); **3**: 1718 (1715)
Ruget G, **1**: 80 (77, 78, 79, 80), 91 (89, 90, 91)
Ruggeri A, **3**: 2025 (2024, 2025)
Rugh W J, **3**: 1779 (1769, 1770, 1776, 1778); **5**: 3376 (3374); **6**: 3962 (3957)
Ruhe A, **2**: 826
Ruighaver A B, **1**: 633 (632)
Rumelhart D E, **4**: 2641 (2637); **6**: 4167 (4167)
Runov J, **2**: 969
Ruppert R, **2**: 1080 (1079)
Ruschizka M, **6**: 4350 (4349)
Rushmer R F, **2**: 853
Ruska E A F, **7**: (4936)
Ruspini E H, **3**: 1604 (1602), 1838
Russak I B, **4**: 2846 (2843)
Russell A L, **2**: 973 (971, 972)
Russell C S, **2**: 1538 (1537); **4**: 2646 (2644, 2645)
Russell D L, **4**: 2807 (2807), 2861
Russell (Lord), **1**: 333 (330, 331); **2**: (870); **5**: (2918), 3455 (3450); **6**: 4008 (4001); **7**: 4996 (4994)
Russell P B, **6**: 4036 (4034, 4036)
Russo M, **6**: 4262 (4260)
Russwurm G, **1**: 177 (175)
Rust B W, **2**: 1326 (1324); **7**: 4454 (4453), 5008 (5005)
Rustagi J, **6**: 3997 (3996)
Rutherford D A, **3**: 1905 (1903); **4**: 2362 (2359, 2360)
Rutkiewic A F, **3**: 1669 (1666)
Rutledge R W, **3**: 1695 (1694)
Růžička J, **7**: 5122
Ryan E P, **5**: 3517 (3516, 3517); **6**: 3870, 4245 (4238, 4239, 4240, 4242, 4243)
Ryder B G, **2**: 827 (824)
Ryder E A, **7**: 4865
Ryder J, **1**: 177 (175)
Ryder W A, **5**: 2943 (2935)
Rydill L J, **6**: 4262 (4258)
Ryhming I L, **3**: 1992 (1992)
Rykiel E J, **2**: 1321 (1318), 1345 (1343)
Ryle G J A, **1**: 142 (139, 140)
Ryzin J, **3**: 1604 (1601)

S

Saad G H, **2**: 1099 (1098)
Saalfeld F E, **5**: 2943 (2935)
Saarimaa R, **7**: 4513 (4513)
Saaty T, **3**: 1847 (1847); **5**: 3164 (3159, 3160)
Saaty T L, **2**: 1355 (1352), 1539 (1538); **3**: 1829 (1829)
Sabadell A J, **1**: 184 (183)
Sabath J, **6**: 3806 (3805)
Sabatini R R, **6**: 4041 (4039)
Sabelli C, **6**: 3954 (3952, 3953)
Saberi A, **4**: 2560 (2559); **7**: 4439 (4436), 4534 (4533)
Sabir M S, **3**: 1694 (1687)
Sabri A, **2**: 923 (923)
Sacerdoti E D, **2**: 1573; **7**: 4785 (4783)
Sachdev M S, **6**: 3834 (3831, 3832)
Sachs J, **3**: 1985
Sachse H B, **7**: 4861 (4859)
Sackett D L, **2**: 1366 (1362)
Sacks A H, **1**: 486 (483)
Sacks G, **1**: 720 (720); **2**: 937 (937)
Sacomanno F, **6**: 3801 (3795)
Sacs I, **5**: 3554 (3551)
Sadamoto K, **6**: 4108 (4104)
Sadee W, **2**: 1248 (1246)
Sadeh E, **3**: 1887
Sadjadi F A, **6**: 4160 (4157)
Sadowski R P, **4**: 2453 (2451)
Sadowsky N L, **7**: 4856 (4855), 4871 (4870)
Saeki M, **4**: 2678 (2677)
Saeks R, **2**: 907 (907), 916 (916), 923 (921); **4**: 2559 (2559), 2570 (2570)
Saelid S, **6**: 3991 (3991), 4266 (4264)
Safabakhsh R, **4**: 2484 (2480)
Safe S, **3**: 1939
Safford F B, **6**: 4310 (4308)
Safir A, **2**: 1578 (1573)
Safonov M G, **2**: 827 (822), 916, 927 (924), 956 (952); **4**: 2560 (2559), 2846 (2843, 2845, 2846), 2872 (2872); **5**: 3197, 3228 (3225); **6**: 4176 (4173); **7**: 4822 (4821)
Safranko J, **7**: 5116
Safteruk K, **7**: 4661 (4660)
Sagasti F R, **2**: 1489
Sage A P, **1**: 618, 651 (648); **2**: 942 (941), 943 (938, 941), 947 (946), 987, 1003, 1566 (1560); **3**: 1999, 2060 (2053), 2192 (2186), 2202 (2196), 2203 (2194, 2196, 2199, 2202); **4**: 2338 (2337, 2338), 2501 (2498, 2499, 2500), 2517 (2515), 2522, 2628, 2640 (2639), 2641 (2639), 2686 (2684), 2697 (2692), 2706 (2702, 2703), 2713 (2706, 2708, 2709, 2710, 2711), 2714 (2706, 2709, 2710, 2711, 2712, 2713); **5**: 2917, 2979, 3115, 3116 (3104, 3114), 3144, 3568; **6**: 3655, 4155 (4154); **7**: 4495, 4762 (4755), 4771, 4785 (4781, 4783, 4784), 4793 (4791), 4809, 4810, 4828 (4826), 4948, 4952, 4986, 5005 (5002), 5164, 5177 (5171)
Saguez C, **2**: 808 (802), 1372 (1368); **3**: 1734 (1729, 1730, 1731, 1734); **7**: 5041 (5039, 5040)
Sågvall Hein A-L, **2**: 1577 (1577)
Saha J, **4**: 2842
Sahama Th G, **1**: 7 (1), 148 (144)
Sahn S A, **6**: 4061 (4060)
Sahni S, **1**: 725 (721, 722, 724, 725)
Said S M, **4**: 2253 (2249)
Saidman L, **1**: 281 (277)
Saila S B, **7**: 4646 (4645)
Sain M K, **1**: 18; **4**: 2601 (2597); **5**: 3220 (3215), 3224 (3220, 3223); **6**: 4133 (4129)
St Pierre R G, **2**: 1566
Saito S, **4**: 2511 (2510)
Saitta L, **1**: 769 (763, 764)
Sakairi T, **6**: 3889 (3889)
Sakakibara A, **7**: 4751 (4750)
Sakamaki F, **1**: 187 (187)
Sakamoto K, **1**: 511 (510)
Sakarovitch M, **4**: 2692
Sakata M, **5**: 2956 (2954)
Sakata S, **5**: 3131 (3128)
Sakawa M, **7**: 4709 (4708)
Sakawa Y, **5**: 3390 (3387); **6**: 4245 (4239)
Sakoe H, **7**: 4509 (4508)
Saksena V R, **7**: 4439, 4443 (4443), 4534 (4534)
Sakshaug E C, **6**: 3806 (3806)
Sakurai Y, **3**: 2021 (2017, 2018)
Sakuranaga M, **7**: 5070 (5066)
Salamon D, **3**: 2151 (2149)
Salanic G R, **7**: 4785 (4783)
Salcudean S, **6**: 4176 (4172)
Sale M J, **1**: 592 (590, 591)
Saleh B E A, **5**: 3131 (3129)
Saleh J, **4**: 2501 (2499)
Salehi S V, **6**: 3751 (3749); **7**: 4462 (4462)
Salinas D, **6**: 3626 (3625)
Salinas E, **6**: 3789
Salisbury J K, **4**: 2386 (2380); **6**: 4108 (4103, 4104, 4105)
Salkin H M, **1**: 630 (628); **4**: 2842
Sallet G, **1**: 124 (124); **2**: 1583; **4**: 2759 (2758); **5**: 2948 (2948), 2949 (2948)
Sällström P, **1**: 333 (332)
Salman M, **2**: 995 (995)
Salomaa A, **2**: 937 (934); **3**: 1712 (1712); **4**: 2798 (2795, 2797); **5**: 3276 (3273)
Salomonson V V, **6**: 4041 (4037)
Salter D R, **7**: 4920
Salut G, **4**: 2697 (2692, 2695), 2713 (2712)
Salvan A, **3**: 2025 (2022, 2024), 2174 (2173)
Saly S, **6**: 3829 (3829)
Salz J, **1**: 80 (76)
Salzman E W, **1**: 445 (443)
Sameh A H, **2**: 827 (818)
Samet P, **2**: 1410 (1409)
Saminaden V, **6**: 3954 (3951, 3953)
Sampson J R, **1**: 561 (559, 561); **6**: 4345 (4341, 4342, 4343, 4344)

Samson C, **2**: 1581 (1580); **5**: 3410 (3408)
Samuel A L, **4**: 2543 (2543); **6**: 4168 (4166)
Samuelson P A, **1**: 556 (554); **5**: 3535 (3533)
Sanathanan C K, **6**: 3780 (3779)
Sanchez E, **1**: 276 (272, 276), 575, 608 (607), 713 (710, 712, 713), 740 (736), 769 (762, 765); **2**: 776 (775), 1473, 1578 (1575); **3**: 1604 (1601, 1603), 1785 (1785), 1797, 1800 (1798, 1800), 1829 (1826), 1831 (1830), 1847 (1846), 1872 (1868), 1894 (1890), 1902 (1900, 1902), 1905 (1903); **4**: 2301 (2295, 2300); **5**: 2926 (2926), 2965 (2965), 3264 (3262), 3540 (3540); **6**: 3947 (3945), 4008 (4005); **7**: 4696 (4695), 4996 (4994, 4995)
Sanchez M, **4**: 2501 (2499)
Sanchez-Palencia E, **2**: 812 (809)
Sandberg I W, **3**: 1779 (1776); **4**: (2672); **5**: 3406 (3402)
Sande G, **3**: 1723
Sandell N R Jr, **2**: 827 (823), 916, 927 (924), (930, 931), 932 (931), 956 (952), 1388 (1388); **3**: 2163 (2153); **4**: 2560 (2559), 2622 (2621), 2846 (2843, 2844, 2845, 2846); **5**: 3197, 3228 (3225, 3226); **7**: 4469 (4466, 4468), 4822 (4821)
Sanders C W, **2**: 912 (910)
Sanders M S, **3**: 2186 (2180, 2185)
Sanders P, **1**: 482 (480)
Sanderson A C, **1**: 86 (83)
Sandewall E, **1**: 732 (726)
Sandford J, **5**: 2944
Sandhu P S, **5**: 3215 (3206)
Sandler G H, **6**: 4011 (4008, 4009)
Sandrin P, **2**: 1388
Sands P J, **1**: 142 (139)
Sandvik K L, **5**: 2993 (2991); **6**: 4252 (4250)
Sandys R C, **7**: 4930 (4928)
Sankar S, **2**: 1192 (1186)
Sanniti di Baja G, **7**: 4883
Sannuti P, **5**: 3082 (3081), 3502; **7**: 4439 (4435, 4438), 4443 (4439, 4441), 4534 (4530)
Sano A, **3**: 2021 (2017, 2018)
Sano H, **4**: 2723 (2721)
Sanoff S P, **1**: 63 (60); **6**: 3727 (3722, 3723, 3724), 3991 (3990), 4204 (4200)
Sanquinatti J W, **1**: 755 (753, 754)
Santagostino G, **6**: 3806 (3806)
Santamore W P, **2**: 853
Santiago J V, **3**: 2021 (2017)
Santos E S, **1**: 713 (712)
Sanz F, **2**: 1244 (1243)
Sapak L, **2**: 1192 (1191)
Saporta D, **6**: 3954 (3952)
Saposnik R, **2**: 777; **5**: 2979; **6**: 3655; **7**: 5164
Sara E, **7**: 4716 (4714)
Sarastano G, **2**: 1201 (1201)
Saravanamuttoo H I H, **3**: 1995
Sargent M, **7**: 4504
Sargent R G, **6**: 4359 (4357), 4396 (4395), 4415 (4415); **7**: 5011 (5008), 5018 (5015, 5017, 5018), 5019, 5106 (5105)
Sargent R W H, **6**: 3991 (3991), 4310 (4309)
Saridis G N, **1**: 126 (125, 126), 493 (492), 502 (499), 575, 608 (607), 664 (664), 713 (710, 712, 713); **3**: 1821 (1819, 1821), 1822 (1821), 1872 (1868), 1874, 2228 (2225); **4**: 2301 (2299), 2361; **5**: 3055 (3050), 3238 (3237), 3410 (3408), 3454 (3451, 3454); **6**: 3772 (3770), 3776 (3772), 3991 (3991), 4165 (4161, 4162, 4163, 4164), 4172 (4170, 4172), 4176 (4172), 4184 (4182), 4193 (4191), 4213 (4212, 4213); **7**: 4551 (4548), 4661 (4659), 4692 (4691), 4910 (4906, 4908, 4909, 4910)
Sarin R K, **3**: 2203 (2196); **4**: 2338 (2337); **5**: 3115 (3106, 3111, 3113), 3116 (3114); **6**: 4075; **7**: 4990
Sarma I G, **6**: 4183 (4182)
Sarna S K, **2**: 1036 (1033, 1034, 1035, 1036), 1040 (1036, 1037, 1038, 1039), 1043 (1040, 1041)
Sartorious H, **6**: 4146 (4145)
Sasaba T, **1**: 142 (141)
Sasano Y, **6**: 4037 (4035)
Sasscer D S, **2**: 1344 (1343)
Sassin W, **3**: 1719 (1715)
Sasso M, **3**: 1665 (1662)
Sasson A M, **6**: 3850 (3847, 3848, 3849)
Sastry D, **4**: 2706 (2705)
Sastry V A, **1**: 51 (50); **2**: 1116 (1110); **4**: 2714 (2712)
Sato K, **6**: 4117
Sato N, **6**: 3829 (3829)
Sato S, **7**: 5070 (5066)
Sato T, **4**: 2579 (2576, 2579)
Satterthwaite M, **7**: 4472 (4471, 4472)
Satyanarayanan M, **2**: 1132 (1128)
Sauer C G, **7**: 4490
Sauer C H, **4**: 2551 (2548, 2549)
Sauerbruch E F, **1**: (315)
Saunders B D, **5**: 3560 (3559)
Saunders D W, **1**: 486 (483)
Saunders K B, **4**: 2901 (2900, 2901); **6**: 4051 (4048, 4050)
Saunders M A, **4**: 2880 (2879); **6**: 4299
Saunders P T, **1**: 550 (548)
Saunders R A, **5**: 2942 (2937)
Savage L J, **3**: 1805 (1802, 1805); **7**: 4991
Savage P G, **4**: 2492 (2488, 2490)
Savage W E, **5**: 3421 (3419)
Savard P, **1**: 536 (534)
Savastano G, **3**: 2217; **6**: 4323 (4321), 4333 (4329)
Savich P, **7**: 4799 (4794)
Savitch W, **4**: 2798 (2797)
Savolainen V, **6**: 4204 (4200)
Savulescu S C, **6**: 3850 (3848)
Savvides L S, **5**: 3383 (3380)
Sawaragi T, **4**: 2325 (2325)
Sawaragi Y, **2**: 1497 (1494, 1497), 1500 (1499); **5**: 3390 (3385, 3388)
Sawers D, **7**: 4833 (4830)
Sawicki E, **1**: 177 (175); **7**: 5122
Sawyer C N, **7**: 5114
Saxner D, **5**: 2942 (2935)
Saxton K J H, **7**: 5143
Say M G, **2**: 1382, 1383
Saydam T, **6**: 4302 (4300, 4301)
Sayers B McA, **1**: 498 (496), 542 (539); **6**: 4055 (4052)
Sayers J F, **2**: 1214 (1213)
Sayre K M, **6**: 3644 (3635)

Scales J T, **3**: 1910 (1910)
Scalzo M, **5**: 3421 (3418)
Scandellari C, **3**: 2021 (2018), 2025 (2022, 2024)
Scarf H E, **1**: 671; **2**: 850 (842, 845); **5**: 3510 (3509); **6**: 3634
Scarf M, **3**: 1917 (1913)
Scattolini R, **7**: 4689 (4687, 4688, 4689)
Scavia D, **2**: 1272 (1269, 1270), 1552 (1549); **4**: 2646 (2645), 2652 (2650, 2651); **6**: 4235 (4231); **7**: 5008 (5007), 5135 (5130, 5131, 5132, 5133, 5134)
Schaake J C, **7**: 4929 (4928)
Schade D S, **3**: 2174 (2169)
Schade G, **2**: 1241
Schade O H, **4**: 2331 (2326)
Schaefer K, **5**: 2944 (2934)
Schaefer R E, **3**: 2192 (2186)
Schaefgen H W, **7**: 5170 (5166)
Schafer A W, **6**: 4281 (4280)
Schäfer O, **6**: 4211 (4211)
Schäfer R, **1**: 440
Schafer R E, **6**: 4029 (4027)
Schafer R M, **4**: 2601 (2597)
Schafer R W, **6**: 4289 (4285); **7**: 4566 (4560)
Schagen P, **7**: 4871 (4869)
Schamberger M L, **3**: 1862
Schanda E, **6**: 4045
Schank R C, **4**: 2640 (2638), 2641 (2637)
Schänzer G, **1**: 393 (393)
Scharf B, **1**: 326 (324)
Schatzoff M, **7**: 4651 (4651)
Schaude G R, **1**: 618
Schaufelberger W, **3**: 2228 (2225); **6**: 4204
Schawlow A L, **7**: 4504
Schedl H P, **2**: 1043 (1040)
Scheggi A M, **5**: 3465 (3461)
Schei A, **6**: 3806 (3806)
Scheinman V D, **4**: 2440 (2421), 2456 (2453); **6**: 4108 (4105)
Schellenberger R E, **7**: 5005 (5003)
Schelling T C, **3**: 1922 (1920)
Schenk G H Jr, **7**: 5110
Schepps J L, **1**: 450 (448, 449)
Scher A M, **1**: 542 (537, 540)
Scherer W T, **6**: 4155 (4154)
Scherlag B J, **2**: 1410 (1408, 1409)
Scherlis L, **1**: 333 (332)
Schertzer W M, **4**: 2794 (2789)
Scherzinger B, **6**: 4133 (4132)
Schetzen M, **1**: 489 (488); **3**: 1779 (1769); **5**: 3336 (3330); **6**: 3776 (3772), 3962 (3957)
Schiänzer G, **1**: 393 (392)
Schiavoni N, **7**: 4689 (4687, 4688, 4689)
Schief A, **4**: 2473
Schierau K, **5**: 3097
Schiesser W E, **2**: 1181 (1180, 1181)
Schiffer M, **4**: 2613 (2610), 2614 (2609)
Schildknecht C E, **6**: 3745 (3739)
Schiller P H, **4**: 2331 (2327)
Schilpp P, **1**: 333 (330)
Schindler P, **4**: 2782 (2778, 2779)
Schizas C, **2**: 952
Schkolnick M, **2**: 1174 (1170)
Schlageter G, **2**: 884, 1165
Schlaifer R, **2**: 1257 (1255); **7**: 4991
Schleter J C, **2**: 1521 (1517)
Schlick E O, **6**: 4275 (4271)
Schlincksupp H, **1**: 618
Schlünder E U, **3**: 2092 (2089)
Schmachtenberg H, **3**: 2092 (2091)
Schmandt W P, **7**: 4980 (4978)
Schmeidler D, **5**: 2911 (2908, 2909)
Schmeidler M, **2**: 850 (848)
Schmeiser B, **6**: 4396 (4395)
Schmeiser B W, **6**: 3950 (3948)
Schmelding D W, **4**: 2782 (2779)
Schmid C, **1**: 51 (50), 63 (60, 61, 62); **2**: 1093 (1091, 1092); **5**: 3044 (3040, 3044)
Schmid J P, **1**: 177 (176)
Schmid R M, **4**: 2827
Schmidlin D J, **5**: 3131 (3129)
Schmid-Schönbein G W, **1**: 478 (477)
Schmidt F W, **3**: 2097 (2094)
Schmidt G, **2**: 862 (862); **3**: 1741 (1737, 1738), 1990 (1986, 1987, 1989)
Schmidt G T, **4**: 2492 (2490, 2491)
Schmidt P, **5**: 3560 (3559)
Schmidt S F, **4**: 2618 (2618), 2622 (2619); **7**: 4540 (4535, 4536)
Schmidt W, **5**: 3139 (3139)
Schmith H, **6**: 3916 (3912)
Schmitt O H, **1**: 769 (760, 761)
Schmittendorf W, **1**: 687 (686)
Schmitz E, **6**: 4089 (4086)
Schmitz H H, **6**: 4370 (4369)
Schmugge T J, **3**: 2234 (2233)
Schnabel R B, **7**: 4975 (4974)
Schnakenberg J, **1**: 433 (432)
Schneider F, **7**: 4703 (4699)
Schneider H, **5**: 3197 (3190), 3464 (3461)
Schneider H C, **6**: 4333 (4329)
Schneider H J, **2**: 893 (892)
Schneider J, **7**: 4833 (4828, 4829, 4830, 4831)
Schneider J B, **6**: 4302 (4301)
Schneider W, **2**: 1577 (1577); **3**: 2192 (2187); **7**: 4793 (4791)
Schneiderman B, **3**: 2186 (2186); **5**: 3079 (3079)
Schnitzer M, **1**: 7 (2, 5), 8 (1, 3), 148 (143), 268 (263, 264)
Schoeffler J D, **3**: 2038; **5**: 3040 (3035), 3144 (3140); **6**: 3846 (3845)
Schoemaker P J, **3**: 2202 (2197, 2199), 2203 (2193, 2195)
Schoeman M E, **6**: 4155
Schoenheimer R, **3**: 2175 (2169)
Schoenstadt A L, **2**: 826 (822); **4**: 2846 (2843)
Schofield C L, **1**: 268 (264)
Schon D A, **7**: 4784 (4783), 4833 (4828)
Schonbein W R, **2**: 1479 (1478)
Schotland R M, **6**: 4037 (4036)
Schotter A, **5**: 2911 (2908)

Schotterer U, **1**: 531 (529)
Schour I, **2**: 973 (971)
Schoute F C, **7**: 4822 (4812)
Schouten H J A, **4**: 2582 (2582)
Schrack G F, **6**: 4302 (4300, 4301)
Schrader R, **1**: 628 (627)
Schraft R D, **4**: 2392; **6**: 4108 (4101)
Schrage L E, **6**: 3950 (3948); **7**: 4656 (4655, 4656)
Schrank W E, **4**: 2353 (2352)
Schreiber F A, **2**: 1160 (1157, 1159, 1160)
Schreiber J D, **2**: 1226 (1217)
Schreiber M M, **1**: 142 (140)
Schreiber P, **7**: 5186 (5180)
Schreiber R K, **7**: 4645 (4645)
Schreider Yu A, **6**: 4008
Schriber T J, **6**: 4302 (4300), 4369 (4369), 4370 (4368), 4396 (4395); **7**: 4935 (4933)
Schrijver A, **6**: 3739 (3738)
Schröder E, **6**: 4008 (4001)
Schroderbeck C G, **2**: 1003
Schroderbeck P O, **2**: 1003
Schroeder K R, **4**: 2901 (2898)
Schroeder M D, **2**: 1132 (1126)
Schroeder M R, **1**: 80 (78); **7**: 4509 (4506)
Schroeder R C, **3**: 2006 (2006)
Schroeder R G, **7**: 4793 (4792)
Schroer J A, **1**: 177 (176)
Schroeter H, **3**: 1756 (1750)
Schroter R C, **1**: 433 (432), 445 (441)
Schruben L, **4**: 2551
Schruben L W, **6**: 4396 (4395); **7**: 4654 (4652, 4653), 5008, 5011, 5019 (5016, 5017)
Schub C M, **2**: 894 (894); **6**: 4370 (4368)
Schuetzle D, **1**: 177 (176)
Schuler P, **5**: 3579 (3575, 3578)
Schulman J D, **4**: 2832
Schulman S G, **7**: 4504
Schultheis L W, **2**: 1595 (1594)
Schultz D G, **5**: (3387)
Schultz V, **6**: 3936
Schultz W C, **1**: 498 (496)
Schulz J, **2**: 867 (864)
Schum D A, **4**: 2338 (2335, 2337)
Schum G M, **4**: 2904 (2903, 2904)
Schumacher J M, **4**: 2816 (2807); **6**: 3734 (3733)
Schumann R, **1**: 63 (60), 72 (68); **3**: 2097 (2094, 2095); **5**: 3351 (3341, 3343, 3344, 3346); **6**: 3611 (3610), 4204
Schumitzky A, **2**: 1337 (1335, 1336)
Schur J, **1**: 495 (494)
Schuss Z, **5**: 3310 (3305, 3306)
Schuster G, **2**: 1454 (1452)
Schuster P, **6**: 4167 (4167)
Schützenberger M P, **3**: 1712 (1712), 1779 (1769); **5**: 3272 (3268), 3276 (3273, 3275)
Schutzer D, **1**: 651 (650)
Schuur J, **2**: 981 (980)
Schuy K D, **5**: 2944 (2934, 2935)
Schwab D J, **7**: 4889 (4885, 4887)
Schwab J L, **4**: 2386 (2383)
Schwan H P, **1**: 450 (446, 447, 448, 449), 468 (467)
Schwanke C O, **2**: 1109 (1109)
Schwarts E L, **1**: 505 (502)
Schwartz A, **6**: 3997 (3995)
Schwartz F, **3**: 2203
Schwartz J, **4**: 2343 (2342); **5**: 3455 (3450); **6**: 4223; **7**: 5047 (5043)
Schwartz J T, **1**: 529 (528); **5**: 3205 (3202)
Schwartz L, **1**: 529 (519); **3**: 1723, 1728 (1723, 1727); **4**: 2664 (2659, 2661); **6**: 3964 (3964)
Schwartz M, **6**: 4294
Schwartz S C, **1**: 755 (753)
Schwarz F, **5**: 3560 (3560)
Schwarz G, **4**: 2293 (2292); **7**: 4566 (4563)
Schwarz R J, **6**: 4108 (4103)
Schwarze G, **4**: 2325 (2321, 2322); **6**: 4306 (4303)
Schwarzer R R, **4**: 2608 (2607)
Schwede G W, **3**: 1888
Schwede H O, **1**: 281 (278)
Schweich D, **5**: 2944 (2934, 2935)
Schweizer B, **3**: 1781 (1780), 1809 (1809)
Schweizer M, **6**: 4123 (4121)
Schweppe F C, **1**: 490 (487, 488); **2**: 1099 (1096); **4**: 2257 (2256), 2618 (2615); **6**: 3829 (3826), 3834 (3832), 3846 (3845), 3850 (3846, 3847); **7**: 4469 (4467)
Schwerdtfeger K, **5**: 3139 (3136)
Schwerdtfeger P, **5**: 2973 (2972)
Schwertmann U, **1**: 8 (2, 5)
Schwiderski E W, **7**: 4889 (4887)
Schwinghamer P, **7**: 4454 (4453, 4454)
Schwodiauer G, **1**: 671; **5**: 2911 (2908)
Schwyhla W, **3**: 1802 (1801); **6**: 3710 (3707, 3709, 3710)
Sciacovelli D, **3**: 1665 (1661)
Scott B J, **5**: 3464 (3463)
Scott D, **1**: 333 (326, 331, 332), 720 (715, 719); **2**: 937 (933); **3**: 1712 (1712); **5**: 2921 (2920), 3116 (3105, 3108)
Scott E L, **7**: 4646 (4644)
Scott H A, **3**: 1648 (1645)
Scott K, **3**: 1958; **4**: 2775 (2772)
Scott M, **6**: 3806 (3804)
Scott P B, **6**: 4097
Scott R W, **5**: 3220 (3220)
Scott R W W, **3**: 1693 (1692)
Scott W B, **4**: 2453 (2452)
Scott W E, **5**: 3390 (3385)
Scott-Jones D F A, **1**: 580 (577)
Scott Morton M S, **2**: 947 (944)
Scribner B H, **1**: 310 (309)
Scrivens J H, **5**: 2943 (2934), 2944
Seaborn C, **5**: 2993 (2991)
Seals J, **1**: 450 (448, 449)
Seard C, **6**: 4051 (4050)
Searle J R, **4**: 2582 (2580)
Searle S R, **1**: 495 (494)
Sears T A, **6**: 4055 (4052)
Seavarda S, **6**: 4257
Sebestyen G S, **6**: 3645
Seborg D E, **1**: 51 (50); **2**: 1109 (1108, 1109), 1116 (1110), 1568 (1567, 1568); **6**: 3916 (3913), 4193 (4190)

Sechzer P H, 2: 963 (962)
Seckel E, 3: 2123
Sedgewick E M, 5: 3253 (3249, 3252)
Sedlak J, 2: 1143
Seed W A, 1: 433 (432), 445 (441)
Seegmüller G, 2: 1132 (1129), 1136 (1134, 1135), 1154, 1174 (1169, 1170, 1171)
Seemann R C, 2: 1109 (1108)
Seeram E, 6: 3940 (3936)
Segal M, 7: 4789 (4788)
Segall A, 5: 3303 (3301)
Segel L A, 1: 433 (429); 2: 1326 (1324); 5: 3082 (3082); 7: 4764 (4763)
Segen J, 1: 86 (83)
Segre G, 6: 3699 (3696)
Seguier G, 5: 3097
Seida T, 2: 1431 (1428)
Seidel H G, 3: 2014
Seif A, 3: 1604 (1603)
Seif F J, 1: 433 (432)
Seiffert U, 1: 380
Seigmurd D, 6: 3997 (3996)
Seinfeld J H, 1: 179 (179), 187 (187); 2: 1500 (1499); 6: 4244 (4238, 4240), 4245 (4238), 4364 (4363); 7: 4518 (4517)
Seitz W R, 5: 3465 (3463)
Sekiguchi T, 1: 713 (710, 712, 713); 3: 1902 (1901, 1902)
Sekuler R W, 4: 2331 (2329)
Selbuz H, 2: 995 (994); 7: 4822
Selby J D, 7: 4481
Self R E, 2: 840
Selfridge O G, 6: 4168 (4166)
Seligman N G, 1: 142 (138, 139, 140, 141)
Seligman P F, 4: 2608 (2607)
Selinger P G, 2: 1160 (1159), 1174 (1170)
Sel'kov E E, 1: 433 (431)
Selkurt E E, 1: 433 (432)
Selten R, 3: 1922 (1919); 5: 2911 (2910), 3256 (3255)
Seltzer S M, 5: 3012
Sempi C, 4: 2508 (2507)
Sen A, 7: 4472 (4469)
Sen P, 4: 2671 (2666, 2669)
Senda K, 2: 1361
Sengupta J K, 2: 1342 (1338)
Senn I A, 7: 4793 (4793)
Sentis R, 4: 2609 (2608); 7: 4944 (4942, 4943)
Seppa I, 5: 2944 (2935)
Seraji H, 4: 2671 (2671); 6: 3717 (3717)
Seraji M, 1: 703 (694)
Serazzi G, 1: 751 (749), 755 (752)
Serreault J Y, 7: 4716 (4714), 4726 (4726), 4749 (4748)
Sesonske A, 5: 3416
Sessin W, 2: 1454 (1452)
Sethi S, 2: 1099 (1095); 3: 1648 (1642); 6: 3706
Setterlund R H, 4: 2492 (2491)
Setzler E M, 2: 1345 (1343); 5: 3440 (3440)
Sevcik K, 2: 1154 (1153), 1165 (1161, 1165); 6: 3659
Severance D, 2: 1165
Seymour D F, 5: 3440
Seymour G, 6: 3916 (3913)
Sezer M E, 1: 703 (694); 2: 907 (907), 916, 923 (921), 952 (951); 4: 2559 (2559), 2560 (2557, 2558, 2559); 5: 3572; 7: 4685 (4683)
Shacham M, 4: 2517 (2516)
Shackle G L S, 3: 1822 (1822)
Shackshaft G, 2: 1383
Shaeffer D L, 2: 1551 (1551)
Shafer B A, 3: 2234 (2233)
Shafer G, 2: 1549 (1547); 3: 1802 (1801), 1805 (1802), 1822 (1821); 4: 2301 (2294), 2884 (2881); 6: 3710 (3707), 3990 (3984); 7: 4669 (4667)
Shafer K H, 4: 2511 (2510)
Shafer W, 1: 671
Shah P S, 3: 1611 (1609)
Shah S L, 6: 4193 (4190)
Shailer M J, 1: 326 (324, 325)
Shajenko P, 5: 3471 (3467)
Shaked U, 1: 580 (575, 578, 579), 581 (578, 579); 4: 2832 (2831)
Shakespeare W, 3: (2178)
Shamash Y, 1: 132 (127); 5: 3077 (3075, 3076)
Shamir U, 7: 5150
Shamos M I, 1: 347 (345)
Shankar S, 5: 3319 (3319)
Shanks C A, 5: 3238 (3236)
Shanno D F, 7: 4975 (4972, 4974)
Shannon C E, 1: 651 (649); 2: 869 (868), 1051 (1047), 1322 (1321), 1348 (1346), 1492; 4: 2497 (2492, 2494), 2501 (2498), 2508 (2504), 2543 (2542), (2543); 6: 3768, 4011 (4008), 4146 (4145), 4216 (4214); 7: 4696 (4693)
Shannon R E, 2: 1003; 3: 2076; 4: 2551 (2550); 6: 4313 (4312), 4370 (4369), 4402 (4396); 7: 5005 (5001, 5002, 5003, 5005), 5011, 5019
Shannon R W E, 3: 1985
Shanteau J, 3: 2203 (2196)
Shanthikumar J G, 3: 1648 (1645)
Shapiro H M, 1: 502 (499)
Shapiro J F, 1: 630 (629)
Shapiro L G, 6: 4160 (4157)
Shapiro R M, 2: 1155 (1153), 1165 (1165)
Shapley L S, 2: 849 (840, 845, 846), 850 (840, 841, 843, 844, 845, 846, 847, 848, 849); 3: 1917 (1913, 1917), 1926 (1923), 1929; 6: 3772 (3769)
Shappy R A, 7: 4646 (4645)
Sharan M, 5: 3496 (3495, 3496)
Sharma A, 7: 4766 (4765)
Sharma D D, 3: 1601 (1600); 7: 4897 (4896)
Sharma M, 5: 3465 (3462)
Sharpe D M, 7: 4645 (4645)
Sharpe F R, 6: (3751)
Sharpe R S, 7: 5186 (5184)
Shattuck S H, 2: 1154 (1150)
Shaver F A, 5: 2942 (2935)
Shaw A C, 6: 3654
Shaw C J, 6: 3898 (3894)
Shaw H J, 5: 3470 (3466, 3467)
Shaw J, 3: 2137
Shaw R, 6: 3744; 7: 5186 (5181, 5183)

She C Y, **2**: 1532 (1528)
Shearer J L, **3**: 1981; **6**: 4257
Shearer L, **5**: 2970
Sheble G B, **2**: 1388 (1387)
Sheehy J E, **1**: 142 (139, 140)
Sheem S K, **5**: 3470 (3469)
Sheer D, **5**: 3164 (3160)
Sheiner L B, **2**: 1248 (1247), 1249 (1247), 1366 (1365); **5**: 3238 (3236); **6**: 3699 (3698)
Sheingold D H, **6**: 4289
Sheirah M A, **2**: 1199 (1197)
Sheldon R W, **7**: 4454 (4453)
Sheldon S, **7**: 4905 (4902)
Shelton A K, **6**: 3751 (3747)
Shelton M, **4**: 2655 (2652)
Shelton R C, **6**: 4165 (4162)
Shen C M, **6**: 3790 (3789)
Shen L C, **3**: 2117 (2114)
Shen S W, **3**: 2025 (2022)
Shenk W, **6**: 4041 (4037)
Shepard J M, **1**: 372 (370)
Shepherd D, **2**: 1132 (1128)
Shepherd J T, **1**: 542 (540), 547 (544)
Shepkin A V, **1**: 37
Shepp L A, **5**: 3523 (3523)
Sheppard C W, **7**: 4906 (4902, 4903)
Sheppard L C, **1**: 498 (495, 496, 498), 502 (499)
Sheps S G, **2**: 1248 (1248)
Shere K, **1**: 651 (650)
Shere K L, **1**: 187 (185)
Sheridan T B, **1**: 372 (371); **2**: 1545 (1543); **3**: 1669 (1665, 1666, 1668), 2186 (2179, 2180)
Sherlock S, **7**: 4905 (4904)
Sherratt A, **1**: 556 (554)
Sherratt A F C, **1**: 167
Sherrill D L, **6**: 4051 (4048, 4049)
Sherwin R S, **3**: 2174 (2171)
Sherwood H, **3**: 1781 (1781)
Sheth P N, **5**: 2970 (2968)
Shettigar U K, **1**: 312 (311)
Shewchun J, **4**: 2511 (2510), 2723 (2721, 2722)
Shibata K, **5**: 3465 (3461)
Shibata R, **4**: 2293 (2292)
Shibata T, **3**: 1614 (1614)
Shichiri M, **3**: 2021 (2017)
Shieh L S, **5**: 3077 (3072, 3073); **7**: 4551 (4547, 4548, 4549, 4550)
Shiffrin R M, **3**: 2192 (2187); **7**: 4793 (4791)
Shigekawa I, **5**: 3336 (3332)
Shih Y-P, **6**: 3626 (3623)
Shik M L, **3**: 1910 (1909)
Shima T, **5**: 3568 (3567, 3568)
Shimano B, **4**: 2386 (2371, 2372, 2374, 2380), 2412 (2402), 2440 (2423, 2438); **6**: 4100
Shimizu H, **6**: 4037 (4035)
Shimizu K, **1**: 301 (298)
Shimomura T, **2**: 1361
Shimony A, **2**: 1478 (1475)
Shimura M, **1**: 665 (664), 746 (743); **3**: 1781 (1780), 1822 (1821), 1825 (1824), 1829 (1827), 1847 (1847), 1852 (1847, 1851), 1862 (1859), 1872 (1868), 1874, 1905 (1904); **4**: 2339 (2338); **6**: 3644 (3641), 3645, 3653 (3648)
Shin H, **1**: 445 (444)
Shindo J, **3**: 1936 (1934); **7**: 4518 (4518)
Shingledecker C A, **3**: 1669 (1666)
Shinji I, **2**: 1507
Shinners S M, **2**: 1321 (1318, 1320)
Shinohara K, **4**: 2723 (2721)
Shinohara M, **7**: 4518 (4518), 4751 (4750)
Shinokochi M, **5**: 3264 (3263)
Shinskey F G, **2**: 1109 (1108), 1116 (1109); **6**: 3683 (3681), 3688 (3687), 3691 (3690), 3701
Shiota M, **3**: 1611 (1609)
Shioya S, **3**: 1611 (1609), 1619 (1618)
Shipira A, **3**: 2202 (2196)
Shipley R A, **7**: 4906 (4903)
Shipman D W, **2**: 1154 (1144, 1145, 1146, 1148, 1153), 1165 (1164)
Shirai Y, **1**: 759 (759), 760 (755, 757, 758); **6**: 3766 (3762)
Shiratori T, **7**: 4980 (4977)
Shires D B, **2**: 1366 (1363, 1366)
Shiryaev A N, **5**: 3535 (3533); **7**: 4617
Shiryayev A N, **5**: 3303 (3293, 3294, 3299), 3336 (3333, 3334, 3335); **7**: 4637 (4631, 4633)
Shisha O, **5**: 2998 (2997)
Shitz D R, **2**: 1174 (1170)
Shneiderman B, **3**: 2178, 2179 (2178)
Shoda M, **3**: 1614 (1614)
Shoemaker W C, **4**: 2582 (2580)
Shoenfield J, **2**: 937 (937)
Shohat J A, **5**: 3131 (3129)
Shor N Z, **5**: 3546 (3544)
Shore J E, **6**: 3990 (3988); **7**: 4509 (4508)
Short C, **3**: 1862
Shorthill R W, **5**: 3471 (3466)
Shortliffe E H, **2**: 1332 (1332), 1366 (1365), 1578 (1573); **4**: 2640 (2637)
Shreve S E, **5**: 3523 (3518, 3519); **7**: 4612
Shrier A, **1**: 570 (568, 569)
Shrivastava S K, **2**: 1136 (1135, 1136)
Shriver B, **4**: 2681 (2680)
Shrum R, **1**: 142 (141)
Shub C M, **6**: 4302 (4301), 4303 (4300), 4306 (4303), 4323 (4321), 4333 (4329, 4330, 4331), 4369 (4368), 4370 (4369), 4403 (4402, 4403), 4406 (4404); **7**: 4654 (4652, 4653)
Shuba M F, **1**: 462 (459)
Shubik M, **1**: 687 (683); **2**: 849 (842), 850 (841, 845, 846, 848); **4**: 2353 (2352); **5**: 2911 (2907, 2911)
Shuder C B, **2**: 839
Shugan S M, **3**: 2192
Shugart H H, **2**: 1317 (1315, 1317), 1318, 1321 (1318), 1344 (1343), 1345 (1343), 1462; **7**: 4897 (4896), 4956 (4955), 5008 (5005)
Shuler K E, **6**: 4235 (4230)
Shulman B H, **1**: 301 (297)

Shulman L S, 2: 1577 (1574)
Shunta J P, 2: 1109 (1108)
Shupe M C, 7: 4799 (4794, 4798)
Shuster M D, 4: 2622 (2622)
Shuster W W, 3: 1601 (1600); 7: 4897 (4896)
Shutz L, 5: 2944
Shweder R A, 3: 2202 (2200), 2203 (2196)
Shyriaev A N, 7: 4613
Sibbald A, 2: 963 (962)
Sibley E H, 2: 893 (884, 888)
Sicherman A, 3: 2203 (2196)
Sideman S, 1: 312 (311)
Sidhu G S, 2: 785 (782, 783, 784); 7: 4469 (4466, 4468)
Siebert H, 4: 2253 (2253)
Siegal P, 1: 736 (735)
Siegel A, 7: 5123 (5122)
Siegel J H, 4: 2901 (2899)
Siegel M W, 5: 2943 (2934)
Siegel R, 1: 177 (175)
Siegenthaler U, 1: 531 (529)
Siegert H G, 2: 1132 (1126, 1132), 1136 (1136), 1174 (1170, 1171, 1172, 1173)
Siemens C W, 2: 1402 (1399, 1400); 7: (4514)
Siemens E W, 2: 1402 (1400); 7: (4514)
Sieverding C H, 7: 4585 (4582)
Siewarth J D, 4: 2747
Sifakis J, 6: 3670 (3668)
Sigel G H, 5: 3470 (3469)
Sigelle M, 7: 4749 (4745)
Sigurdsson G, 7: 4906 (4905)
Sikorski R, 5: 2924 (2922); 6: 3710 (3709)
Silberschatz A, 2: 884 (882), 1155
Silha R E, 2: 976
Silja D D, 2: 916
Šiljak D D, 1: 304, 675 (672, 673, 674), 703 (694, 696); 2: 777, 907 (907), 923 (918, 919, 920, 922), 952 (951), 1355 (1352); 4: 2556 (2551, 2555, 2556), 2559 (2557, 2558, 2559), 2560 (2557, 2558, 2559), 2675 (2674), 2702 (2697, 2702); 5: 3196 (3189), 3402 (3401), 3572 (3569, 3570, 3571); 6: 4011 (4008, 4009, 4010); 7: 4685 (4683)
Sillen L G, 1: 8 (4)
Sillness J, 2: 973 (971)
Silva J M, 7: 4749 (4745)
Silveira H M, 6: 3604 (3600)
Silver M T, 2: 1494
Silver W, 6: 3665 (3664, 3665)
Silver W M, 4: 2440 (2438)
Silverman B G, 3: 2203 (2201); 4: 2501 (2501), 2641 (2637, 2639)
Silverman L M, 1: 137 (133), 296 (294, 295), 410; 2: 785 (782, 784); 4: 2523 (2522), 2702 (2699), 2849 (2847); 5: 3220 (3215); 6: 3968 (3967), 4070 (4067)
Silvert W, 3: 1711 (1711), 1781 (1781); 6: 3621 (3620, 3621); 7: 4454 (4453, 4454), 4895 (4895)
Silvestri G J, 7: 4585 (4581)
Simaan M, 1: 687 (687), 703 (701); 2: 1269; 5: 3149 (3147, 3148)
Simazaki K, 2: 1522 (1521)
Sime M E, 3: 2179
Simek J G, 1: 575
Simkin A, 3: 1910 (1910)
Simmonard M, 2: 959; 4: 2839
Simmonet G, 6: 3954 (3952, 3953)
Simmons V P, 7: 4848 (4844)
Simmons W W, 2: 1545 (1543)
Simon B, 2: 1169 (1169); 7: 4892 (4891)
Simon B R, 1: 477 (477)
Simon H A, 1: 466 (466), 565 (564), 675 (674); 2: 947 (943), 992; 3: 2191, 2192 (2186); 4: (2579), 2641 (2636, 2637); 5: 3235 (3234); 7: 4785 (4782, 4783), 4789 (4786), 4793 (4791, 4792)
Simon J, 2: 1179 (1175)
Simon J C, 6: 3643 (3641)
Simon J D, 6: 3717 (3711)
Simon M K, 1: 80 (80), 91 (86, 87)
Simon Z, 1: 433 (432)
Simonnard M, 4: 2692
Simons G L, 4: 2469 (2467)
Simpson C F, 4: 2889
Simpson E H, 2: 1322 (1321)
Simpson H R, 6: 3898 (3895)
Simpson R J, 2: 1402
Simpson R W, 3: 1669 (1665, 1666, 1668)
Sims D, 4: 2640 (2637, 2638); 7: 4785 (4783)
Sims R B, 6: 4142 (4140)
Simson A, 1: 31 (21)
Sin K S, 1: 60 (60); 2: 1581 (1579); 5: 3410 (3407, 3408), 3475 (3473); 6: 4193 (4190); 7: 4611 (4608, 4610, 4611)
Sinai J C, 4: 2508 (2505, 2506, 2507)
Sinclair J D, 3: 2214 (2213)
Sincovec R F, 2: 1181 (1181)
Sindel D, 6: 4262 (4261)
Singer I M, 5: 3394 (3390, 3391, 3393, 3394)
Singer J R, 5: 3421 (3418, 3419, 3420)
Singer M, 5: 3560 (3559)
Singh B, 1: 450 (446, 447)
Singh H, 7: 4654 (4653)
Singh J S, 1: 142 (138, 139, 140)
Singh M G, 1: 132 (129), 703 (690, 692, 693, 695, 696, 697, 698, 699); 2: 862 (859, 861, 862), 916 (915), 923 (917, 922), 927 (924), 952 (951), 956 (952), 1192 (1191); 3: 1990 (1989), 2038, 2163 (2153, 2160, 2161), 2225; 4: 2560 (2557, 2559), 2697 (2692, 2695), 2706 (2702, 2704, 2705), 2713 (2709, 2712), 2714 (2709, 2710, 2713); 5: 3019 (3012, 3014, 3016, 3017, 3018, 3019), 3040 (3035, 3036, 3037, 3038), 3077 (3074), 3144, 3145 (3142), 3171 (3166, 3167, 3170); 7: 4434 (4433, 4434), 4571 (4568), 4689 (4688, 4689), 4751 (4751)
Singh R P, 5: 3060 (3059)
Singh Y P, 7: 4452 (4451)
Singleton W T, 4: 2361
Singpurwalla N D, 6: 4029 (4027)
Sinha N K, 1: 132 (128); 4: (2359); 5: 3077 (3077)
Sink L S, 3: 2163 (2153)
Sinko J W, 6: 3762 (3757)
Sion M, 2: 1262 (1259)
Sipilä A H, 6: 4066 (4064)
Sips H J, 1: 633 (632)

Siret J M, **1**: 132 (128, 129, 130, 131), 703 (698, 699)
Sirs J A, **1**: 445 (444)
Sisson R, **7**: 4771
Sisson R L, **4**: 2353 (2351); **6**: 4415 (4414)
Sivan R, **2**: 826 (814); **4**: 2245 (2240); **5**: 3475 (3472); **6**: 4011 (4011), 4165 (4162); **7**: 4554 (4552), 5170 (5169)
Sivaramakrishna Das P, **3**: 1781 (1781)
Sive M R, **6**: 3691 (3688)
Siy P, **3**: 1888
Sjöberg L, **1**: 276 (272)
Sjostrand F S, **7**: 4940
Skala H J, **5**: (2918), 2927 (2924); **7**: 4996 (4995)
Skarabis H, **2**: 1107
Skeat W C, **7**: 5102 (5100)
Skeen D, **2**: 1165
Skeist I, **6**: 3745 (3739)
Skellam J G, **7**: 4646 (4645)
Skelton R E, **1**: 133 (128)
Skene A M, **2**: 1107 (1104, 1105)
Skiles J W, **1**: 142 (139, 140)
Skinner F, **6**: 4108 (4103, 4104)
Skinner J B, **2**: 1366 (1364)
Skinner P, **4**: 2331 (2326)
Skirbekk G, **6**: 3865 (3861)
Sklansky J, **4**: 2386 (2382); **6**: 3645, 4165 (4163)
Sklar A, **3**: 1781 (1780), 1809 (1809)
Skogerboe G V, **1**: 142 (140, 141); **6**: 4229 (4228), 4230 (4229), 4235 (4233)
Skoglund V J, **6**: 4296
Skolem Th, **5**: (2918)
Skolnik M I, **5**: 2973 (2973)
Skorokhod A V, **6**: 3997 (3994); **7**: 4613, 4637 (4631, 4636), 4644 (4643)
Skwirzynski J K, **5**: 3329 (3320)
Skyum S, **4**: 2798 (2797)
Slate J B, **1**: 502 (499)
Slater J B, **1**: 633 (633)
Slater L E, **3**: 1706 (1701)
Slater M, **2**: 1262 (1258)
Slattery D, **4**: 2464 (2462)
Slattery P J, **2**: 963 (962)
Slatyer R O, **2**: 1318 (1314, 1315, 1316), 1321; **3**: 2046 (2045)
Slavan A, **3**: 2021 (2018)
Sleator W, **7**: 4980 (4978)
Slenger V A, **7**: 5116
Slinn W G N, **6**: 3933 (3931)
Sliwinska D, **3**: 1872
Sloan R E G, **7**: 4856 (4853)
Sloane N J A, **1**: 726 (722); **5**: 2956 (2955)
Sloman M, **2**: 1143
Slompson A B, **3**: 1791 (1789)
Slooff J W, **2**: 1179 (1176)
Slovic P, **3**: 2192 (2189), 2202 (2196), 2203 (2196, 2201); **4**: 2338 (2337); **5**: 3115 (3113), 3116 (3113); **6**: 4075
Slucki E, **6**: (3861)
Slud E, **1**: 86 (82)
Sly P G, **4**: 2782 (2780)
Smale S, **1**: (567); **2**: 1351 (1350); **7**: 4529 (4526), 4674 (4670), 5047 (5043)
Small H, **1**: 177 (175)
Smallwood R H, **1**: 462 (460, 461); **2**: 1040 (1036), 1043 (1042)
Smart D W, **6**: 4302 (4301)
Smeenk J H, **1**: 634 (632)
Smets P, **3**: 1604 (1603), 1802 (1801), 1805 (1803, 1805)
Smirnov V I, **7**: 4556
Smith A, **2**: 1033 (1029); **3**: (1607); **5**: 2944 (2935); **6**: 3940 (3939), 4303 (4302)
Smith A W M, **2**: 1043 (1041)
Smith B T, **2**: 827 (815, 818, 824); **5**: 3559 (3555)
Smith C A, **1**: 589 (585)
Smith C B, **3**: 1761 (1758)
Smith C F, **3**: 2113 (2111)
Smith C L, **3**: 2108 (2108)
Smith C P, **5**: 3464 (3459)
Smith C S, **4**: 2601 (2598)
Smith C W, **1**: 450 (446, 447)
Smith D J, **4**: 2794 (2789)
Smith F E, **1**: 643 (642)
Smith G P, **3**: 2214 (2211)
Smith H, **5**: 2993 (2992), 3030 (3027, 3028); **6**: 3621 (3620)
Smith H M, **6**: 3806 (3805)
Smith H T, **3**: 2179
Smith H W, **5**: 3390 (3385, 3388)
Smith J C, **1**: 504 (502)
Smith J H, **3**: 1998
Smith J M, **1**: 675 (672)
Smith J R, **6**: 3802 (3796, 3797)
Smith J T, **2**: 1445 (1441)
Smith K C, **3**: 1975
Smith M C, **1**: 581 (577); **4**: 2832 (2831, 2832); **5**: 3228
Smith M H, **2**: 1323 (1323), 1344 (1344), 1345 (1344)
Smith M J, **4**: 2659 (2656)
Smith M L, **2**: 1458
Smith N J, **4**: 2714 (2710, 2711, 2713)
Smith N T, **1**: 281 (277, 278), 289 (285)
Smith O J M, **6**: 4180 (4176); **7**: 4463 (4457)
Smith P D, **4**: 2794 (2789)
Smith R B, **3**: 1943
Smith R L, **7**: 4726 (4726)
Smith R M, **4**: 2782 (2779)
Smith R S, **5**: 3253 (3250, 3251, 3252)
Smith S W, **3**: 2175 (2174)
Smith T H, **3**: 2202 (2196)
Smith W M, **6**: 3745 (3739)
Smith W R, **6**: 3621 (3620), 3665 (3664, 3665)
Smithin T, **4**: 2640 (2639); **7**: 4785 (4783)
Smolitskiy K L, **3**: 1627
Smorodinsky M, **4**: 2508 (2506); **5**: (2910)
Šmuk K, **4**: 2277 (2277)
Smullin L D, **6**: 4036 (4033)
Smuts J C, **1**: 399 (394)
Smyth C J, **7**: 4856 (4856)
Snapper K, **4**: 2338 (2337)
Snaveley M K, **7**: 4504
Sneath P H A, **2**: 1107 (1104)

Snedecor G W, **7**: 5015 (5014)
Snell E S, **7**: 4856 (4852, 4853)
Snell F, **1**: 433 (432)
Snell F D, **7**: 5122
Snell F M, **5**: 3440 (3440)
Snell J L, **5**: 2929 (2927, 2928); **6**: 4356
Snodgrass W J, **4**: 2646 (2643), 2794 (2789, 2794)
Snoock T, **6**: 3898 (3895)
Snow P D, **4**: 2646 (2646)
Snyder A W, **7**: 5065 (5062)
Snyder D L, **5**: 3303 (3294, 3301, 3302); **7**: 4637 (4637)
Snyder L R, **4**: 2889
Snyder M F, **1**: 542 (537)
Snyder W E, **6**: 4160 (4157)
Snyman J A, **6**: 4346 (4346)
Sobel J, **5**: 2911 (2911)
Sobolev V V, **3**: 2052 (2046)
Sobolevskii P E, **6**: 4223 (4222)
Sochor A, **7**: 4996 (4995)
Socolar J E, **1**: 570 (568)
Söderström T, **1**: 80 (75, 76, 77); **4**: 2239 (2238), 2245 (2242), 2257 (2255), 2277 (2273, 2274, 2275, 2276, 2277), 2278 (2273, 2275, 2276), 2284 (2283), 2292 (2288), 2293 (2287, 2290), 2307 (2306), 2314 (2310, 2312, 2313), 2319 (2319); **5**: 3006 (3006), 3351 (3343); **6**: 3994 (3994); **7**: 4607 (4604), 4611 (4607, 4608), 5001 (5000)
Sodin L G, **5**: 3130 (3128)
Soeda T, **2**: 1497 (1494, 1497), 1500 (1499); **4**: 2325 (2325); **6**: 4245 (4238)
Soeldner J S, **3**: 2174 (2173)
Soete W, **2**: 952 (949, 950, 951)
Sohma H, **6**: 4261 (4260)
Sohnle R C, **1**: 561 (561)
Sohrab S, **6**: 4061 (4060)
Soisalon-Soininen E, **2**: 884 (883)
Soittola M, **4**: 2798 (2797); **5**: 3276 (3273)
Sojkz S A, **5**: 3421 (3418)
Sokal R R, **2**: 1107 (1104); **7**: 5008 (5006)
Sokolnikoff I S, **2**: 1186 (1183)
Sol H G, **2**: 893 (891), 947
Soland R M, **5**: 3115 (3114), 3165 (3161)
Solberg H H, **3**: 1648 (1645, 1646)
Solberg J J, **3**: 1648 (1644); **6**: 3904
Solheim O A, **2**: 927 (924); **7**: 4667 (4661)
Soliman H M, **2**: 923 (919, 920, 922)
Sollecito W E, **6**: 3870 (3868)
Solman A J, **7**: 4856 (4853, 4854)
Solo V, **1**: 81 (77, 78); **4**: 2314 (2313); **6**: 4209 (4207); **7**: 4469 (4467), 4611 (4607, 4608)
Soloman S, **7**: 4833 (4832)
Soloman S L, **6**: 4396 (4395)
Solomon A J, **5**: 2944 (2935)
Solomon H, **2**: 1106 (1103, 1104), 1107
Solomon S, **7**: 5011 (5008), 5019
Solomonoff R, **1**: 709 (708)
Solonnikov V A, **6**: 3596 (3592)
Solovay R, **5**: (2918)
Solso L, **3**: 2192 (2186)
Solt J, **4**: 2253 (2249)
Somerville M J, **5**: 3390 (3385, 3388)
Sommer G, **5**: 3455 (3454)
Sommerhoff G, **1**: 294 (293)
Sommers L E, **4**: 2782 (2781)
Sommerville I, **4**: 2501 (2500)
Somogyi A A, **5**: 3238 (3236)
Son S H, **2**: 1165 (1164)
Soncini-Sessa R, **5**: 3171 (3170)
Song J, **6**: 3755
Sonnenblick E H, **1**: 542 (541)
Sonnenschein H, **1**: 671; **5**: 2911 (2908, 2909)
Sontag E D, **4**: 2527 (2527); **5**: 3279 (3279), 3293 (3289, 3290, 3291, 3292), 3341 (3337, 3339), 3376 (3375); **6**: 3972 (3970, 3971)
Sorensen D C, **5**: 3546 (3544)
Sorensen M, **6**: 3626 (3625)
Sørensen S P L, **6**: 3691 (3688)
Sorenson H W, **4**: 2273 (2271), 2618 (2616), 2622, 2714 (2709); **6**: 4310 (4308)
Sorin S, **3**: 1926 (1923)
Sorine M, **3**: 1990 (1989); **4**: 2849 (2847); **6**: 4070
Sorum M J, **2**: 1107
Sosapavon F, **6**: 3789
Sosslav A B, **7**: 4948
Soubbotine A I, **2**: 1008 (1005, 1006), 1023 (1022); **4**: 2609 (2608)
Soudack A C, **2**: 808
Souder C W, **4**: 2453 (2451)
Souder W E, **7**: 4833 (4831)
Souganidis P E, **2**: 1017 (1015); **3**: 2068 (2066), 2072 (2069, 2072)
Soula G, **2**: 1578 (1575)
Soulas C, **1**: 340
Soulas M, **3**: 1990 (1989)
Sourander M, **6**: 3680 (3675, 3676)
Sousa W P, **2**: 1318
South R, **3**: 1769 (1766, 1768, 1769), 1969 (1968)
Southwell R V, **5**: 3140 (3135, 3136)
Sowell S, **3**: 1943
Spach M, **1**: 769 (760, 761)
Spain J D, **1**: 468 (466)
Spalding D B, **2**: 1416 (1413)
Spanbauer J P, **6**: 3917 (3911, 3915)
Spang H A, **4**: 2253 (2252); **6**: 4165 (4162)
Spangler W J, **4**: 2782 (2782)
Spanien E, **3**: 2169
Sparks J W, **7**: 4905 (4905)
Sparrow C T, **1**: 429 (428), 570 (568, 569); **7**: 4529 (4528)
Spear R, **2**: 1333 (1331)
Spearman C A, **2**: 840 (838); **3**: 1958 (1957)
Speckhart F H, **6**: 4311 (4311)
Speed H D M, **7**: 5150
Spence A M, **7**: 4822 (4819, 4820)
Spencer E A, **3**: 1693 (1687)
Spencer H, **6**: (3862)
Spencer J L, **3**: 1611 (1609)
Spencer M N, **4**: 2655 (2652)
Spencer W A, **1**: 542 (537)
Spenser R C Jr, **7**: 4585 (4581)

Sperb R P, **4**: 2613 (2613), 2614 (2613)
Sperry E, **1**: (341); **6**: 4275 (4271)
Sperry L, **1**: (341)
Spetzler C S, **2**: 943 (941); **3**: 2203 (2195)
Speyer J L, **2**: 932 (931), 1257 (1256); **6**: 4070; **7**: 4822 (4815)
Spicer E R, **3**: 1947, 1975
Spiegel M R, **1**: 751 (747)
Spiegethalter D J, **2**: 1107 (1104, 1105)
Spigarelli J L, **4**: 2782 (2782)
Spiguel C P, **6**: 4372 (4372)
Spilker J J, **2**: 1051 (1045)
Spillman W B, **5**: 3464 (3457), 3465 (3459)
Spingarn K, **4**: 2264 (2263)
Spirn J, **1**: 755 (752)
Spiteri P, **5**: 3480 (3478), 3481
Spittle P, **4**: 2356 (2354)
Spofford W O, **7**: 4897 (4897)
Sporton T M, **3**: 1958
Sprafka S A, **2**: 1577 (1574)
Sprague C H, **1**: 606 (606)
Sprague J M, **3**: 2213 (2210), 2214 (2211)
Sprague R H, **2**: 947 (945, 946); **4**: 2353 (2351), 2501 (2498), 2681 (2680); **5**: 3079 (3078); **6**: 4333 (4329, 4332), 4372 (4371)
Spratt E H, **1**: 315 (314)
Spreadbury P, **2**: 1393 (1391)
Spriet J A, **1**: 424 (421); **5**: 3088 (3086); **6**: 4313, 4360 (4359), 4372 (4371), 4388, 4392 (4388); **7**: 4692 (4691, 4692)
Spronk J, **5**: 3568 (3567)
Sproule M, **1**: 437
Sprugel D G, **2**: 1318
Sprules W G, **7**: 4454 (4453)
Sprung J L, **1**: 187 (187)
Spurr S H, **2**: 1321 (1318)
Sridhar R, **3**: 2025 (2022); **5**: 3389 (3385, 3388), 3390 (3385, 3388)
Srikantakumar P R, **4**: 2719 (2717); **7**: 4721 (4719, 4720), 4741 (4739)
Srinivasan M V, **7**: 5065 (5063, 5064)
Srinivasan R, **3**: 2025 (2022)
Staal J F, **6**: 3860 (3858), 4216 (4216)
Stacey W M Jr, **5**: 3416
Stachowicz M S, **3**: 1905 (1903)
Stack V T, **7**: 5114
Stackelberg H von, **1**: 703 (700); **5**: 3149 (3145)
Stackhouse T, **6**: 4108 (4102)
Stafford G D, **2**: 965, 969, 978
Stafford R A, **6**: 4168 (4166)
Stafford T G, **3**: 1943
Stagg D T, **6**: 4055 (4052)
Stagg G W, **6**: 3820 (3818)
Staite W E, **2**: 1402 (1399, 1400)
Staley S W, **1**: 177 (176)
Stalford H, **2**: 1269 (1264)
Stamm K, **1**: 380
Stampacchia G, **1**: 529 (524); **2**: 1261 (1259); **5**: 2997 (2997); **6**: 3596 (3594); **7**: 5036 (5031, 5033)
Stanat D F, **2**: 1087
Standefer J, **3**: 2174 (2169)
Standridge C R, **2**: 894 (894); **4**: 2551 (2550); **5**: 3035 (3034); **6**: 4372 (4371)
Staniewski P, **5**: 3454 (3454); **7**: 4661 (4658, 4660)
Stanley G M, **4**: 2517 (2516)
Stanley J C, **2**: 1565 (1561)
Stanley S M, **7**: 4856 (4852)
Stanski D R, **2**: 1249 (1247); **5**: 3238 (3236); **6**: 3699 (3698)
Stanton K N, **4**: 2310 (2309)
Starbuck W H, **6**: 4411 (4406); **7**: 4785 (4781)
Stark H L, **1**: 478 (477)
Stark L, **5**: 3242 (3241); **7**: 4876 (4873), 4877, 4880 (4880)
Starling E H, **1**: (543)
Starmer C, **3**: 1995
Staroswiecki M, **4**: 2547 (2547)
Starr A W, **2**: 1269 (1263)
Starr M K, **5**: 3121 (3121), 3165 (3157)
Starý P, **1**: 675 (672)
Stasko A B, **7**: 4848 (4848)
Stavroulakis P, **2**: 1192 (1186, 1187)
Staw B M, **7**: 4785 (4783)
Stea D, **4**: 2640 (2638)
Stearns R E, **2**: 850 (847), 884 (881, 882), 1155 (1145, 1150, 1152); **4**: 2497
Stechs H W, **3**: 2150 (2149), 2152; **5**: 3506
Stecke K E, **3**: 1648 (1642, 1644, 1645, 1646)
Stedinger J A, **2**: 1538 (1537)
Stedinger J R, **7**: 5140
Stedman D H, **1**: 184 (184)
Steel E W, **7**: 5099
Steel R A, **3**: 1982
Steele R, **7**: 4906 (4904)
Steemann Nielsen E, **6**: 4045
Steer M W, **7**: 4601 (4599, 4601)
Steer R W Jr, **2**: 1590 (1587, 1589), 1596 (1592)
Stefan P, **5**: 3366 (3366)
Stefanelli M, **3**: 1623 (1619, 1620, 1621, 1622)
Stefanescu A C, **5**: 3540 (3540)
Stefani G, **5**: 3184 (3181, 3182)
Stefik M, **2**: 1573
Stegemerten G J, **4**: 2386 (2383)
Stehfest H, **5**: 3171 (3170)
Steiger J, **4**: 2338 (2335)
Steiglitz K, **1**: 628, 630; **2**: 884 (882), 1083 (1080), 1125 (1123); **7**: 5170 (5166)
Stein E, **1**: 536 (532); **2**: 1410 (1408); **5**: 3538 (3537)
Stein E S, **3**: 1669 (1668)
Stein G, **1**: 736 (732, 735); **2**: 826 (822); **4**: 2846 (2845), 2872 (2872); **5**: 3196, 3228 (3225, 3227); **6**: 4165 (4162), 4227 (4224); **7**: 4554 (4552)
Stein K, **4**: 2356 (2355)
Stein P D, **2**: 853 (852)
Stein R B, **1**: 466 (465); **5**: 3248 (3245), 3253 (3250, 3251, 3252)
Stein R G, **7**: 4540 (4534, 4536)
Stein W E, **3**: 1831 (1830), 1845 (1845); **7**: 4661 (4660)
Steinberg D, **4**: 2412 (2397)

Steinberg S, **5**: 3311 (3307)
Steinbrunner J D, **7**: 4785 (4781, 4783, 4784)
Steiner J, **4**: (2610)
Steiner R E, **5**: 3425 (3425)
Steinhardt R, **1**: 142 (138)
Steinhorst R K, **6**: 4235 (4230, 4231)
Steinmetz G F, **3**: 1969
Steketee J, **7**: 4871 (4869)
Stell B, **3**: 1958
Stell W K, **7**: 5070 (5066)
Stella R, **4**: 2607 (2607)
Sten L, **5**: 3351 (3341)
Stenberg R, **7**: 5147 (5146, 5147)
Stengers I, **6**: 4168 (4167)
Stepanenko Y, **6**: 4128 (4125)
Stepanov V V, **2**: 1286 (1283)
Stephanopoulos G, **1**: 406 (403, 404); **4**: 2517 (2515)
Stephen R L, **1**: 312 (311)
Stephens A G, **3**: 1693 (1690)
Stephens H S, **7**: 5161
Stephenson F W, **6**: 4284 (4283)
Stephenson R P, **2**: 1249 (1245)
Sterling M J, **7**: 5106 (5104)
Sterling M J H, **2**: 1388 (1385); **6**: 3829 (3826, 3827); **7**: 5153 (5151, 5153)
Stern A C, **2**: 1528 (1528)
Stern H, **6**: 4257
Stern J, **1**: 486 (484)
Sternberg R J, **1**: 276 (272); **3**: 2203 (2201); **4**: 2641 (2639)
Sternberg S, **5**: 3394 (3390, 3391, 3393, 3394)
Sternby J, **6**: 4184 (4180)
Steuer R E, **5**: 3164 (3161), 3165 (3161)
Stevels I A L, **7**: 5187 (5181)
Stevens J C, **1**: 482 (480)
Stevens K, **5**: 2942 (2935, 2936)
Stevens M E, **1**: 575
Stevens P K, **6**: 3734 (3734)
Stevens R K, **1**: 177 (174, 175)
Stevens R T, **2**: 987
Stevens S S, **2**: 1527 (1523)
Stevens T H, **1**: 592 (589, 590)
Stevens T S, **1**: 177 (175)
Stevens W C, **1**: 289 (285)
Stevenson W D, **6**: 3855 (3851, 3852)
Steward A, **1**: 284 (281)
Steward D V, **7**: 4667 (4662, 4663)
Stewart C R, **7**: 4732
Stewart D, **6**: 4108 (4102, 4105, 4106)
Stewart D C, **5**: 3464 (3463)
Stewart G W, **2**: 826 (817, 818, 820, 823, 824), 827 (816, 817, 818, 819, 822), 1102 (1102); **4**: 2277 (2277), 2905; **5**: 2950 (2949), 2951 (2951), 3220 (3219); **6**: 3983 (3980)
Stewart I, **1**: 550 (548); **3**: 1788; **7**: 4674 (4671, 4672)
Stewart J C C, **5**: 3440
Stewart K C, **4**: 2453 (2452)
Stewart K H, **7**: 5106 (5103)
Stewart R A, **4**: 2412 (2393, 2397)
Stewart R M, **2**: 1078 (1076)
Stewart W J, **1**: 755 (752)
Stiefel E L, **5**: 3140 (3135)
Stiffler J J, **2**: 1051 (1049)
Stigum B P, **5**: 3104 (3104)
Stiles E W, **7**: 4645 (4645)
Stiles G F, **3**: 1700
Stiles J A, **6**: 3870 (3867)
Stillerman R, **7**: 4833 (4830)
Stimson K R, **3**: 1954 (1949); **7**: 5099, 5106 (5103, 5104, 5105)
Stinespring W F, **5**: 3131 (3129)
Stirgus C A, **6**: 4230 (4228)
Stiteler W M, **3**: 2045 (2045)
Stock M J, **3**: 2214 (2212)
Stock R, **1**: 594
Stockham J D, **1**: 179 (177)
Stockman G, **4**: 2473 (2472)
Stoddard C J, **1**: 462 (461)
Stodola A B, **3**: (1745); **7**: (4517)
Stoecker W F, **6**: 3918 (3918)
Stoelting R K, **1**: 289 (285)
Stoer J, **2**: 827 (816); **3**: 2068 (2064), 2072 (2071); **4**: 2827 (2817)
Stoffel J, **6**: 3815 (3811, 3813), 3846
Stohlman F Jr, **2**: 1318 (1316)
Stoica P, **4**: 2277 (2273, 2274, 2275, 2276), 2278 (2273, 2275, 2276), 2293 (2290), 2314
Stoker J J, **5**: 3390 (3384)
Stokes N W, **7**: 4799 (4794, 4797, 4798)
Stokey E, **2**: 1542 (1540); **5**: 3035 (3033)
Stokić D, **6**: 4086
Stoll A M, **7**: 4856 (4854)
Stoller D S, **2**: 1107
Stoltz K, **2**: 969
Stolwijk J A J, **1**: 643; **7**: 4876 (4875, 4876), 4877 (4873, 4874), 4880 (4879), 4881 (4880)
Stolzy J L, **1**: 142 (138)
Stone H S, **1**: 736, 740; **2**: 1282 (1281)
Stone H W, **5**: 3055
Stone J, **4**: 2331 (2325)
Stone L, **1**: 556 (554)
Stone M, **1**: 91 (89); **7**: 5015 (5014, 5015)
Stone M H, **3**: 1880 (1879); **5**: 2924 (2922)
Stonebraker M, **2**: 1155 (1150, 1151, 1153), 1160 (1159), 1165 (1161)
Stonert H, **6**: (3862)
Storah B, **3**: 1958
Store Y C, **7**: 5192
Storeton-West T J, **7**: 4848 (4845)
Storey C, **5**: 3398, 3402 (3399, 3400, 3401); **7**: 4457 (4456), 4547 (4541)
Storey J, **3**: 1756 (1751)
Stott B, **6**: 3790 (3789), 3829 (3828), 3846 (3845), 3850 (3849), 3855 (3854)
Stouffer S A, **7**: 4495 (4495)
Stout W F, **5**: 2934 (2933)
Stoy R D, **1**: 450 (448, 449)
Strahan V H, **5**: 3464 (3460)
Strahler A N, **4**: 2904 (2902)

Strain B R, **2**: 1521 (1518), 1522 (1521)
Strakosch G R, **2**: 1424 (1423); **7**: 5051 (5048)
Strang G, **2**: 812 (809), 1102 (1102), 1179 (1174); **3**: 1632; **5**: 2951 (2951); **6**: 3717 (3717)
Straškraba M, **2**: 1351 (1348)
Straszak A, **1**: 682, 703 (693); **5**: 3540 (3539)
Stratchey C, **3**: 1712 (1712)
Stratonovich R L, **4**: 2628 (2624); **5**: 3303 (3293)
Stratonovitch L, **5**: 3535 (3533)
Straus R W, **7**: 4593 (4588)
Strausfeld N J, **7**: 5065 (5062, 5064, 5065)
Strauss W, **6**: 4223 (4218, 4222)
Streeter D, **6**: 4165 (4163)
Streeter V L, **7**: 5161
Streibl N, **5**: 3130 (3129)
Streifer W, **6**: 3762 (3757)
Strejc V, **4**: 2284 (2280, 2282), 2314 (2314), 2319 (2317), 2325 (2321)
Stricker E M, **3**: 2213 (2210)
Strickland R, **1**: 340
Strickland S, **2**: 1249 (1245)
Striebel C, **5**: 3303 (3294), 3336 (3332); **7**: 4466 (4465), 4613 (4612)
Striker G, **2**: 1521 (1517), 1522 (1518, 1520); **4**: 2253 (2249)
Stringa L, **6**: 3645
Strobel H, **4**: 2325
Strobel H A, **1**: 13 (9)
Strodiot J J, **5**: 3546
Stromme S B, **7**: 4876 (4875), 4880 (4879)
Strong F C, **1**: 13 (8)
Stroock D W, **6**: 3596 (3594); **7**: 4617 (4615), 4627 (4623), 4637 (4632)
Stroock F, **5**: 3531 (3526)
Stroud H J F, **3**: 1961
Stroud N P, **2**: 1410
Struble G W, **1**: 575
Strutt J W *see* Rayleigh (Lord)
Stuart A, **4**: 2307 (2303, 2305); **6**: 3621 (3619); **7**: 5001 (4998, 4999)
Stubberman A W, **3**: 1958
Stubbs C W, **3**: 1982
Stuchly M A, **1**: 450 (445)
Stuchly S S, **1**: 450 (445, 449); **3**: 1694 (1687)
Stuckenberg N, **1**: 252 (250)
Štulí K, **7**: 5122
Stumm W, **1**: 8 (1, 2, 3, 4, 5, 7); **4**: 2782 (2779); **5**: 3266 (3264)
Sturgeon W, **7**: 5051 (5048)
Sturgis H, **2**: 1136 (1134, 1135), 1154 (1144), 1174
Sturm B, **5**: 3575
Sturt B, **3**: 1975
Styan G P H, **2**: 1105
Su R, **5**: 3197 (3196)
Subrahmanian E, **4**: 2551 (2548)
Subramanian A K, **4**: 2892
Subtleny S, **4**: 2798 (2795)
Suen C Y, **1**: 575 (573, 574); **6**: 3645
Suess M J, **1**: 185; **2**: 1536
Sugai S, **5**: 2943
Sugeno G, **4**: 2301 (2299)
Sugeno M, **1**: 713 (712); **2**: 1549 (1547); **3**: 1821 (1819, 1821), 1822 (1821), 1905 (1904); **5**: 3413 (3410); **6**: 3768 (3766), 3772 (3770), 3990 (3984)
Suggs J C, **1**: 177 (174)
Sugihara K, **1**: 760 (756, 757)
Sugimoto N, **6**: 4037 (4035)
Sugiura M, **2**: 1361
Sullivan A G, **6**: 3829 (3829)
Sullivan H G, **4**: 2582 (2580)
Suma K, **7**: 4856 (4854, 4855)
Summerfield M, **7**: 4490
Summers P, **4**: 2386 (2374)
Sumner F, **4**: 2484 (2480, 2484)
Sumpter C, **4**: 2368 (2367)
Sun F K, **7**: 4622 (4622)
Sundareshan M K, **2**: 923 (920, 921), 927 (924); **4**: 2560 (2559)
Sung K, **5**: 3568 (3568)
Sung R E, **4**: 2491
Suppan P, **1**: 281 (277, 278)
Suppe F, **3**: 2003 (2000)
Suppes P, **1**: 556 (555); **5**: 3115 (3105, 3107, 3109), 3116 (3105), 3455 (3450); **6**: 3860 (3859), 4216 (4215); **7**: 4991 (4987)
Surawicz B, **1**: 769 (761)
Surgenor D M, **1**: 445 (442, 443)
Suri R, **3**: 1648 (1646)
Suryanarayana M, **5**: 2963 (2958)
Suslin A A, **6**: 3972 (3971)
Sussmann H J, **1**: 413 (413), 434 (432); **2**: 1121 (1116); **4**: 2756 (2755, 2756), 2759 (2756, 2757), 2898 (2895, 2897, 2898); **5**: 2949 (2948), 3279 (3276), 3288 (3285), 3293 (3292), 3303 (3299), 3311 (3310), 3330 (3323, 3327), 3341 (3339), 3366 (3366), 3376 (3374), 3394 (3390, 3392), 3488 (3487), (3516); **6**: 3734 (3734), 3777 (3772), 3956 (3955, 3956), (3956), 3977 (3974, 3975, 3976), 4000 (3999, 4000); **7**: 4425 (4420), 4637 (4635), 4677 (4676, 4677), 4976 (4976)
Sutcliffe W H Jr, **7**: 4454 (4453)
Suter G W, **2**: 1539 (1538, 1539), 1552
Suter P, **7**: 5161
Sutherland V, **3**: 1985
Sutinen R, **6**: 3916 (3913)
Suverkrup R, **1**: 490 (488)
Suwa M, **1**: 760 (755)
Suzuki K, **6**: 3889 (3889)
Suzuki M, **7**: 4443 (4439)
Suzuki T, **5**: 3055 (3048); **7**: 4881 (4880)
Suzuki Y, **6**: 4033 (4031)
Svenson I, **7**: 4785 (4783)
Svenson O, **3**: 2192 (2186)
Sverdrup G M, **1**: 179 (179)
Svirezhev Yu M, **1**: 675 (674); **2**: 1351 (1348)
Svobodova L, **1**: 751; **2**: 1174 (1173)
Swain P H, **6**: 4033, 4160 (4158)
Swalm R O, **6**: 4075
Swan G W, **6**: 4066 (4062)

Swanick B H, **3**: 1611 (1609)
Swann C D, **7**: 4930 (4929)
Swann W H, **4**: 2601 (2598)
Swanson E B, **7**: 4785
Swanson G D, **6**: 4051 (4046, 4048, 4049, 4050), 4061 (4057, 4060), 4062 (4056, 4059, 4060, 4061)
Swartz J R, **3**: 1615 (1613)
Swartzman G L, **2**: 1552 (1550, 1551); **6**: 4229 (4228)
Sweedler M E, **3**: 1779 (1772)
Sweeting O J, **5**: 2942 (2936)
Sweetman J A, **7**: 5129 (5127)
Swerling P, **4**: 2628 (2624)
Swets J A, **3**: 1867 (1863)
Swierstra S, **1**: 196 (193); **3**: 1756 (1751, 1752, 1753, 1754)
Swift A C, **3**: 1982
Swift R S, **1**: 7 (3)
Swillens S, **2**: 1249 (1245)
Swindale N V, **6**: 4168 (4167)
Sworder D D, **6**: 4011 (4011), 4165 (4162)
Sydenham P H, **2**: 1559; **4**: 2533 (2528); **6**: 3884 (3882), 4289 (4288), 4294
Syding R, **1**: 51 (49, 50); **5**: 2993 (2990)
Sydov A, **6**: 4306 (4303)
Sykes B A, **3**: 1975
Symons S A, **1**: 450 (449)
Syrcos G P, **7**: 4534 (4530)
Syski R, **7**: 4835
Szalai K J, **1**: 252 (251)
Szaniawski K, **6**: (3862)
Szebehely V G, **7**: 4490
Szegö G, **4**: 2614 (2609, 2610, 2611)
Szegö G P, **2**: 1020 (1017, 1018); **3**: 1898 (1896); **4**: 2559 (2559)
Szent-Györgyi A, **1**: 450 (450)
Szeto E W, **2**: 981 (981); **7**: 4522 (4519)
Szidarovsky F, **3**: 1829 (1826)
Szilagyi J, **5**: 2942 (2935)
Szolovits P, **2**: 1577 (1577)
Szpirglas J, **5**: 3303 (3294)
Szücs E, **6**: 4296
Szurszewski J H, **2**: 1043 (1040)
Szybicki E, **7**: 4741 (4739), 4751 (4751)
Szymankiewicz J Z, **4**: 2551; **6**: 4393
Szyperski N, **6**: 4333 (4329, 4332)

T

Tabard N, **1**: 608 (607, 608)
Tabata Y, **5**: 2943
Tacke K H, **5**: 3139 (3136)
Tacker E C, **2**: 912 (910)
Tada K, **5**: 3464 (3461)
Tadian M, **1**: 633 (632), 634 (632)
Tadmor Z, **2**: 1586 (1585, 1586)
Tagasugi S, **2**: 1366 (1363, 1366)
Tager H, **3**: 2175 (2169)
Taggart W M, **4**: 2501 (2499, 2501)
Taguchi K, **3**: 1766 (1762)
Taguchi T, **6**: 4261 (4258)
Taha H A, **4**: 2842 (2839); **6**: 3904; **7**: 4648
Tahmessabi T, **6**: 4189 (4187)
Tai S, **5**: 3464 (3459)
Taillard J P, **6**: 4108 (4107)
Tai-Seng Chen, **3**: 2014
Taiwo O, **5**: 3197 (3189), 3215 (3206)
Takagi M, **6**: 3645
Takahar Y, **3**: 2038
Takahara I, **5**: 3088 (3083)
Takahara U, **2**: 1348 (1346)
Takahara Y, **1**: 93 (92), 294 (291), 405 (400, 404); **2**: 862 (861), 956 (952), 1286 (1283), 1486 (1480), 1549 (1547); **3**: 2043 (2039, 2041, 2042, 2043), 2163 (2153, 2155, 2159); **4**: 2713 (2710); **5**: 3019 (3015), 3035 (3033), 3040 (3035, 3036); **6**: 3631 (3629, 3631), 4322 (4319), 4355 (4354, 4355); **7**: 4571 (4570), 4666 (4661), 4779 (4775, 4776, 4777, 4778)
Takahashi K, **6**: 3850 (3850)
Takahashi M A, **7**: 4799 (4794, 4798)
Takahashi R, **2**: 1507
Takahashi S, **5**: 3390 (3385, 3388)
Takahashi Y, **2**: 1318 (1316), 1489 (1489); **5**: 3055 (3049)
Takahasi Y, **6**: 3703 (3702)
Takai Y, **2**: 1366 (1363, 1366)
Takamatsu T, **1**: 589 (585); **3**: 1611 (1609), 1619 (1618)
Takanashi K, **7**: 4593 (4588)
Takase K, **4**: 2386 (2374); **6**: 4086 (4085), 4117
Takashima S, **5**: 3055 (3048)
Takayama T, **2**: 1269
Takeda E, **3**: 1819
Takegaki M, **4**: 2386 (2376), 2412 (2393)
Takemura S, **1**: 360, 362, 367
Takens F, **1**: (567); **5**: 3341 (3341)
Takeuchi N, **6**: 4037 (4035)
Takeuchi S, **5**: 3464 (3461)
Takeuchi Y, **1**: 675 (673)
Takeyama I, **4**: 2440 (2431, 2432, 2439)
Taki K, **1**: 301 (297)
Takiuchi M, **2**: 1522 (1517, 1518)
Tal J, **5**: 3097
Talati K, **3**: 1831 (1830), 1845
Talavage J, **5**: 3165 (3157)
Talbot S A, **3**: 2175 (2173)
Talenti G, **4**: 2614 (2612)
Tallgren A, **2**: 965
Talliez B Y, **5**: 2944 (2934)
Tallman C R, **5**: 3465 (3458)
Talmon J L, **4**: 2319 (2319)
Talukdar S N, **6**: 3846
Tamarkin J D, **5**: 3131 (3129)
Tamburrini G, **7**: 4996 (4994)
Tamsen A, **2**: 963 (962)
Tamura H, **2**: 1305 (1304), 1497 (1494, 1497); **5**: 3104 (3099, 3101, 3102, 3103, 3104), 3116 (3107), 3156 (3154), 3171 (3166, 3167, 3169, 3170); **7**: 4883
Tamura S, **1**: 713 (712); **3**: 1604 (1602); **5**: 2927 (2924)
Tan M H, **3**: 2174 (2173)
Tanabe H, **6**: 4223 (4222)
Tanagho E A, **7**: 4980 (4976)
Tanaka H, **3**: 1603 (1603), 1802 (1801), 1819 (1816), 1838
Tanaka K, **1**: 665 (664), 713 (712), 746 (743); **3**: 1604 (1602), 1781 (1780), 1822 (1821), 1825 (1824), 1829 (1827), 1831 (1829, 1830), 1835, 1852 (1847, 1851), 1858, 1862 (1859), 1874, 1905 (1904); **4**: 2339 (2338); **5**: 2927 (2924); **6**: 3645, 3653 (3648)
Tanaka N, **2**: 1516 (1516)
Tanaka S, **4**: 2511 (2510)
Tanenbaum A S, **2**: 1132 (1130); **4**: 2368 (2365), 2686 (2684)
Tang D A, **6**: 4160 (4157)
Tang G Y, **6**: 4160 (4158)
Tang H S, **5**: 3070 (3070)
Tang W, **4**: 2556 (2552, 2555, 2556)
Tang Y S, **3**: 1778 (1776), 1779 (1776)
Tanimoto T T, **2**: 1479 (1475)
Tannenbaum A, **1**: 294; **4**: 2866 (2865)
Tanner C B, **2**: 1521 (1517, 1518)
Tanner M W G, **3**: 1958
Tanner R L, **1**: 177 (175)
Tanre D, **3**: 2052 (2050)
Tansley A G, **2**: 1356 (1356)
Tanur J M, **2**: 1106
Tao F T, **5**: 2944 (2935)
Tapia R A, **2**: 797 (795)
Tapiero C S, **2**: 1099 (1094, 1098), 1269
Tapley B D, **3**: 2006 (2005)
Tarapore E D, **2**: 1416 (1413)
Taras M J, **7**: 5122
Tardieu B, **1**: 310 (308); **7**: 4929 (4929)
Tarn T J, **5**: 3341 (3340)
Tarski A, **1**: 333 (326, 329); **2**: 937 (933), 938 (934); **5**: 3455 (3450)
Tartar J, **1**: 561 (561)
Tartar L, **1**: 675 (674); **2**: 812 (809, 810, 812), 1180; **4**: 2343 (2342)
Tarvainen K, **5**: 3568 (3567, 3568)
Tasai H, **5**: 2993 (2993)
Taschini A, **6**: 3806 (3806)

Tashker M G, **2**: 1244 (1244)
Tatjewski P, **2**: 1305 (1299, 1302, 1304); **7**: 4571 (4568, 4569, 4571)
Tatnall M L, **1**: 289 (285, 286, 289)
Taub F, **3**: 2032 (2032)
Taud R J, **2**: 1388 (1386, 1387, 1388)
Tauman Y, **3**: 1929 (1927)
Tausworthe R C, **5**: 2956 (2954)
Tavast R, **4**: 2517 (2516)
Tavill A S, **6**: 3909 (3908)
Tayler R J, **7**: 4892 (4889, 4891)
Taylor A E, **7**: 4566 (4557)
Taylor B W, **3**: 1648 (1647)
Taylor C L, **6**: 4108 (4103)
Taylor D J, **3**: 1693 (1690)
Taylor E O, **2**: 1383
Taylor G E, **4**: 2449 (2448, 2449)
Taylor H F, **5**: 3471 (3470)
Taylor I, **2**: 1040 (1036)
Taylor I W, **3**: 1947, 1975
Taylor J C, **5**: 3282 (3282)
Taylor J H, **5**: 3319 (3318), 3383 (3381), 3390 (3387); **7**: 4452 (4450)
Taylor L T, **4**: 2511 (2510)
Taylor P, **6**: 4133 (4132)
Taylor P M, **4**: 2449 (2448, 2449); **5**: 3383 (3381), 3389 (3386)
Taylor R, **2**: 1458
Taylor R H, **4**: 2386 (2373, 2374)
Taylor R M, **1**: 8 (2, 5)
Taylor R N, **3**: 2192 (2186); **7**: 4793 (4791)
Taylor S R, **2**: 1516
Taylor T D, **7**: 4498 (4496)
Taylor W K, **6**: 4108 (4107)
Tchon K, **7**: 4677 (4674)
Teasdale G, **4**: 2582 (2581)
Teener M D, **3**: 1736
Tehrani F T, **4**: 2901 (2899)
Teitelbaum T, **5**: 3079 (3078)
Teitelman W, **1**: 732 (726)
Tella R, **6**: 4123
Temam R, **1**: 405 (400), 529 (522, 524, 525, 526, 527); **2**: 1261 (1257, 1261); **3**: 1627; **5**: 3530 (3526); **7**: 5047 (5044, 5045)
Tempo R, **4**: 2257 (2256), 2635
Tenenbaum J M, **4**: 2386 (2382); **6**: 3653 (3652), 4160 (4157, 4159)
Teng P S, **1**: 142 (139, 140), 592 (590)
Tenland T, **5**: 3464 (3459)
Tenney R R, **2**: 932 (931)
Teo K L, **2**: 1211 (1210)
Teplitz P V, **7**: 4495
Teramae N, **4**: 2511 (2510)
Teranaka K, **2**: 1154
Terano T, **1**: 713 (710, 712, 713); **3**: 1822 (1821), 1902 (1902), 1905 (1903, 1904); **5**: 2927 (2925)
Teraoka H, **2**: 1516 (1516)
Terashita Y, **4**: 2655 (2652)
Tergny J, **5**: 3152 (3151), 3164 (3157)
Termini S, **2**: 1473 (1467, 1468, 1469, 1472, 1473), 1479 (1477); **3**: 1785 (1783), 1858 (1855), 1878 (1876); **5**: 2965 (2963, 2964, 2965); **7**: 4996 (4994, 4995)
Ter-Pogossian M M, **7**: 5187 (5180, 5181, 5182, 5183, 5184)
Terragni F, **5**: 3044 (3040, 3043, 3044)
Terrisson J C, **7**: 4842
Tertian R, **7**: 4504 (4500)
Testud J L, **6**: 3680 (3675)
Tetzlaff W H, **1**: 751 (751)
Teuber H L, **2**: 869 (868), 873 (872); **7**: 5077 (5075), 5078
Teutenberg J, **1**: 564 (561)
Thacker J G, **1**: 482 (480)
Thackray J E, **7**: 5155 (5155)
Thaler G J, **5**: 3390 (3383)
Thaler R, **3**: 2203 (2197, 2199)
Tham M T, **2**: 1116 (1115); **5**: 3188 (3188)
Thambynayagam R K M, **1**: 585
Tharp H S, **1**: 137 (134)
Tharp M O, **4**: 2501 (2499)
Thatcher J W, **3**: 1712 (1712)
Thatcher T W, **4**: 2464 (2461, 2463, 2464)
Thathachar M A L, **4**: 2719 (2716, 2717, 2718), 2737 (2731); **7**: 4721 (4717, 4718, 4719)
Thauer R, **7**: 4881 (4879)
Thayer W J, **6**: 4257
Thayse A, **3**: 1883 (1881)
Theil H, **4**: 2508 (2504, 2505)
Theocharous E, **5**: 3465 (3460)
Theodorescu R, **7**: 4721 (4717)
Thépot J, **2**: 1269 (1266, 1267)
Thevenet D, **5**: 2943 (2935)
Thewlis J, **7**: 4848 (4844)
Thibault C, **5**: 2944
Thiecke R J, **3**: 1933 (1931)
Thieme G, **3**: 1665 (1661)
Thies H, **6**: 4142
Thirurooran C, **2**: 1466 (1462, 1463), 1467 (1462)
Thissen W, **7**: 4762 (4762), 5005 (5002)
Thissen W A H, **7**: 4810
Thole U, **5**: 3455 (3452)
Thom N S, **7**: 5116
Thom R, **1**: 433 (432), 552; **2**: 867 (864); **3**: (1786, 1787)
Thoma A, **5**: 3523 (3521, 3523)
Thoma M, **2**: 1192 (1189); **4**: 2575 (2575); **6**: 4211 (4210)
Thomann R V, **4**: 2646 (2645), 2789 (2787); **7**: 5008 (5006)
Thomas B, **6**: 4089 (4088)
Thomas B E, **3**: 1975
Thomas B W, **5**: 2944 (2934)
Thomas C J, **4**: 2353 (2352)
Thomas C P, **3**: 2006 (2006)
Thomas D, **2**: 1186 (1183)
Thomas D J, **5**: 3425 (3425)
Thomas F, **2**: 1253 (1249); **5**: 2943 (2935)
Thomas H E, **1**: 13 (11)
Thomas H H, **3**: 2221
Thomas J, **6**: 3706
Thomas J D R, **4**: 2607 (2605)
Thomas J M, **3**: 1632

Thomas M C, **3**: 1642 (1638, 1640)
Thomas P B, **4**: 2464
Thomas R H, **2**: 1155 (1148, 1149, 1151)
Thomas R S, **1**: 177 (176)
Thomas Y, **6**: 3701
Thomason M G, **6**: 3644, 3654 (3648), 4160 (4157, 4158, 4159)
Thomassen K I, **6**: 4364 (4363)
Thomesse J P, **2**: 1143
Thompson B, **2**: 839
Thompson C W N, **4**: 2686 (2681, 2682, 2684, 2685)
Thompson D W, **7**: 4454 (4452)
Thompson G L, **2**: 1099 (1093, 1094, 1095); **6**: (3892)
Thompson J G, **3**: 2117
Thompson J H, **6**: 4061 (4061)
Thompson J M T, **1**: 550
Thompson J S, **1**: 80 (78)
Thompson K R, **5**: 2944 (2934)
Thompson L, **3**: 1975
Thompson M A, **5**: 3238 (3236)
Thompson P, **3**: 1943 (1942)
Thompson P W, **1**: 319
Thompson S P, **1**: 536 (532)
Thompson W, **7**: 5177 (5171)
Thompson W A, **7**: 4913 (4912)
Thompson W E, **4**: 2556 (2554)
Thompson W L, **4**: 2582 (2580)
Thompson W P, **2**: 1402 (1400)
Thomson E, **2**: 1402 (1402)
Thomson J, **4**: 2782 (2780)
Thooris B, **2**: 1179 (1174)
Thorell N E, **5**: 3351 (3341)
Thorley A R D, **7**: 5156, 5161 (5157)
Thorn R, **3**: 1693 (1691)
Thorneycroft J, **6**: (4260), 4275
Thorngate W, **4**: 2502 (2500)
Thornley J H M, **1**: 142 (140)
Thornton C G, **7**: 4540 (4535)
Thornton C L, **4**: 2892 (2891); **6**: 4204 (4202)
Thornton H L, **2**: 839; **3**: 1958 (1957)
Thornton K W, **6**: 4229 (4228), 4230 (4228, 4229)
Thorp J P, **4**: 2871 (2868)
Thorpe S A, **7**: 5070 (5066)
Thorson J, **4**: 2331 (2327)
Thouvenot P, **7**: 4856 (4856)
Thowsen A, **5**: 3135 (3133)
Thrall R M, **3**: 2060 (2058); **6**: 3772 (3769, 3771)
Thron H L, **1**: 547 (544)
Thum C H, **3**: 2175 (2170)
Thümmler C, **2**: 1305 (1302, 1303)
Thurber K, **4**: 2681 (2680)
Tibbals R, **5**: 3282 (3280)
Tibor V, **6**: 4311
Tickner E G, **1**: 486 (483)
Tiechroew D, **4**: 2502 (2500)
Tien J M, **2**: 987, 1565 (1559, 1564), 1566 (1559, 1561, 1562, 1563)
Tiengo A, **4**: 2631 (2629, 2631)
Tierney L, **7**: 4654 (4653)
Tignor S C, **3**: 1736
Tihomirov V, **1**: 529 (527)
Tikhonov A N, **7**: 4429 (4426), 4439 (4435)
Tillman C C, **7**: 4651 (4651)
Timmins C, **3**: 1961
Timpany P L, **4**: 2789 (2787, 2788)
Tinbergen J, **2**: 1342 (1340)
Tinney W F, **6**: 3855 (3853)
Tiran J, **3**: 2032 (2026, 2027, 2030, 2032), 2175 (2172)
Tirrell M, **6**: 3745 (3741)
Tisson J, **2**: 1244 (1244)
Titli A, **1**: 132 (128, 129), 703 (689, 690, 692, 693, 695, 696, 698, 699), 704 (692); **2**: 862 (862), 923 (922), 927 (924), 952 (951), 956 (952), 1192 (1190, 1191); **3**: 2038, 2163 (2153, 2155, 2160, 2161), 2225; **4**: 2697 (2692, 2695), 2706 (2705), 2713 (2712), 2714 (2709, 2710, 2713); **5**: 3019 (3012, 3014, 3016, 3017), 3040 (3035, 3036, 3037, 3038), 3144, 3145 (3142), 3171 (3166, 3167), 3481; **7**: 4571 (4568), 4666 (4665), 4667 (4661), 4741 (4736, 4737, 4738), 4751 (4751)
Titterington D M, **2**: 1107 (1104, 1105)
Titus J, **7**: 4897 (4896)
Tiwari J L, **6**: 4235 (4234); **7**: 4646 (4644)
Tjahjadi P I, **5**: 3071 (3070)
Toates F M, **3**: 2213 (2211, 2212), 2214 (2210, 2211, 2212, 2213)
Tobey G E, **1**: 272 (269, 270)
Tobias J V, **1**: 326 (324)
Tobin J D, **3**: 2025 (2022), 2174 (2171)
Tocher K D, **4**: 2353 (2352)
Tocher T K, **6**: 4334
Toda H, **1**: 301 (297)
Toda K, **3**: 1766 (1765)
Todd J M, **6**: 3634
Todd S, **4**: 2659 (2657)
Tödt F, **5**: 3579
Toffolo G, **3**: 2025 (2023, 2025); **4**: 2631 (2629, 2631)
Togai M, **1**: 713 (712)
Togawa T, **7**: 4856 (4853, 4854, 4855)
Toint P L, **7**: 4975 (4974)
Toivonen H, **1**: 564 (561)
Tokarev V V, **7**: 4490
Tokiwa Y, **1**: 176
Tokou K, **7**: 4593 (4588)
Tokumaru H, **2**: 1497 (1494)
Tokumary H, **1**: 675 (673)
Toland J F, **2**: 1262 (1261); **7**: 5047 (5046)
Tolimieri R, **3**: 1723; **7**: 4498 (4496)
Tolle J W, **6**: 3634
Tollet I H, **6**: 3916 (3914)
Tolliver T L, **2**: 1109 (1107)
Tolman R C, **4**: 2641 (2638)
Tolwinski B, **1**: 687 (685); **2**: 995 (995), 1192 (1191); **7**: 4631 (4630)
Tomita K, **1**: 570 (569)
Tomizuka M, **5**: 3004 (3004), 3055 (3049); **6**: 3703 (3702)
Tomlin J A, **6**: 4299
Tomović R, **2**: 1552 (1549); **6**: 4230 (4227, 4228), 4235 (4230); **7**: 4463 (4459)

Tomski G, **2**: 1023 (1022)
Tondl L, **6**: 3860 (3859, 3860), 4216 (4215)
Tong D D M, **6**: 3699 (3698)
Tong R, **2**: 1116 (1110)
Tong R M, **1**: 713 (712); **3**: 1847 (1846), 1852 (1849), 1894 (1894), 1898 (1895), 1905 (1903); **4**: 2362 (2359, 2360)
Tonkov E L, **7**: 4677 (4675)
Tooker J, **7**: 4833 (4829, 4830)
Topping J, **2**: 1559
Torasso P, **1**: 769 (763, 764)
Torborg R H, **2**: 1226 (1214, 1217, 1223, 1225)
Torgalkar A M, **1**: 482 (480)
Torgerson W S, **7**: 4986
Törn A A, **6**: 4303 (4300), 4350 (4349)
Tornatzky L G, **7**: 4833 (4828, 4829, 4830, 4831)
Törnebohm H, **6**: 3860 (3859)
Toro V, **4**: 2547 (2547)
Torvikoski T, **6**: 4204 (4200)
Toshiro M, **1**: 589 (585)
Toth H E, **3**: 1969 (1969)
Toth K C, **2**: 1160
Tou J, **1**: 709 (704, 707, 708)
Tou J T, **6**: 3645 (3636), 3653 (3646), 4033, 4160 (4157), 4165 (4161, 4163), 4168 (4166)
Touchard F, **5**: 2944
Tourassis V D, **5**: 3055, 3056
Toussaint G, **4**: 2636 (2635)
Toussaint G T, **6**: 3645 (3643)
Towe A L, **5**: 3253 (3248, 3250, 3252)
Townes C H, **7**: 4504
Townsend N A, **3**: 1985
Toynbee A, **1**: 294 (294)
Toyota H, **6**: 4033 (4031)
Tragenza P R, **7**: 5051
Trahan R, **5**: 3196 (3193), 3197 (3193)
Traiger I L, **2**: 883 (881, 882), 1154 (1144, 1145, 1147), 1174 (1170, 1171)
Trakhenbrot B A, **1**: 334
Tran V C, **3**: 1734 (1734)
Trannoy B, **7**: 5027
Tranter W H, **6**: 4281 (4278)
Trappl R, **1**: 333 (332), 399 (394); **2**: 1333 (1331); **3**: 1781 (1781), 1809 (1809), 2003 (2001); **5**: 3044 (3042); **6**: 4066 (4062), 4167 (4166), 4337, 4377 (4377), 4388
Travis C C, **1**: 494 (491)
Trayner C, **6**: 4007 (4007)
Trebst A, **2**: 1522 (1517, 1521)
Treichler J R, **4**: 2314 (2310)
Treleaven P C, **2**: 1136 (1133, 1134, 1136), 1174 (1169, 1170)
Treloar L R G, **1**: 486 (484)
Trémolières R, **2**: 1262 (1260); **3**: 1632; **7**: 4498 (4497), 5036 (5033)
Trentowski B, **1**: 399 (394)
Treu S, **5**: 3079 (3078)
Trevathan V L, **6**: 3683 (3681)
Trèves F, **3**: 1728; **4**: 2664
Tribolet J M, **1**: 80 (78); **6**: 4281 (4280)
Tribus M, **2**: 1479 (1475); **4**: 2508 (2507); **6**: 3989 (3988)
Triggs E J, **5**: 3238 (3236)
Trillas E, **2**: 1473 (1468, 1470, 1471); **3**: 1781 (1780), 1810 (1807, 1809), 1878 (1876); **6**: 4007 (4005); **7**: 4996 (4995)
Trinks W, **7**: 4517
Tripathi N, **6**: 4139 (4138)
Tripathi S K, **7**: 5018 (5017)
Trist E, **7**: 4833 (4828)
Troch I, **5**: 3430 (3426, 3428); **6**: 4311
Troianiello G, **4**: 2343
Trombetti G, **4**: 2614 (2612)
Troncale L, **1**: 508 (507); **6**: 3989 (3984), 4337
Tropf H, **4**: 2473 (2470, 2472); **6**: 4159 (4158)
Trotman D, **1**: 434 (430)
Trottenberg U, **5**: 3139 (3139)
Troutman P H, **6**: 3870 (3868)
Trowbridge E A, **1**: 478 (477), 486 (483, 484, 486)
Trucco E, **6**: 3762 (3756, 3758)
Truckenbrodt A, **4**: 2456 (2453); **6**: 4086 (4082), 4089 (4086, 4088)
True S, **4**: 2551
Truitt C J, **7**: 4726 (4722)
Trulsson E, **4**: 2277 (2273, 2275)
Trusty G L, **5**: 3470 (3469)
Truxal J G, **6**: 4284
Trybus L, **2**: 1192
Tryckare T, **1**: 109
Trygar T A, **3**: 2163 (2153)
Tsai W H, **6**: 3654 (3645, 3652)
Tsalighopoulos M G, **2**: 1590 (1587)
Tsao G T, **5**: 3579 (3578)
Tsao J W, **2**: 1522 (1518)
Tsay Y T, **7**: 4551 (4547, 4548, 4549, 4550)
Tscheschke P D, **7**: 4897 (4896)
Tse E, **1**: 133, 137 (134, 135, 137); **2**: 912 (908), 1257 (1256, 1257); **4**: 2257 (2255); **5**: 3035 (3033), 3055 (3048); **6**: 4165 (4162, 4163), 4193 (4190)
Tsetlin M L, **4**: 2719 (2715), 2737 (2731)
Tsichritzis D C, **2**: 893 (887)
Tsipis K, **5**: 3012
Tsivoglou E C, **5**: 3266 (3265)
Tsuchiya H M, **1**: 424 (422)
Tsuchiya S, **6**: 4108 (4104)
Tsuda I, **1**: 570 (569)
Tsui H T, **6**: 3645
Tsuji K, **5**: 3144 (3142)
Tsuji S, **6**: 3654 (3652), 4161 (4157)
Tsuji T, **7**: 4856 (4854, 4855)
Tsukahara N, **1**: 466 (465)
Tsukamoto Y, **1**: 713 (710, 712, 713); **3**: 1847 (1846), 1852 (1847), 1902 (1902), 1905 (1903); **5**: 2927 (2925)
Tsunogai S, **2**: 1516 (1516)
Tsuruoka K, **2**: 1154 (1153)
Tsutsumi S, **7**: 4751 (4750)
Tsypkin J A, **7**: 4452 (4451)
Tsypkin Ya Z, **1**: 81 (80); **3**: (1749); **4**: 2314 (2311), 2737 (2734); **5**: 3373 (3370, 3372), 3390 (3383, 3384, 3385, 3386, 3387), 3554 (3551, 3552, 3553, 3554); **6**: (4145), 4146 (4145), 4165 (4163), 4184 (4180)
Tu F, **6**: 3916 (3913)

Tuazon E C, **1**: 176 (175)
Tüchelmann Y, **4**: 2368 (2362)
Tucker A W, **2**: 849 (845, 846), 850 (846, 847, 848), 1262 (1257, 1258); **3**: 1917 (1913, 1917), 1926 (1922, 1923, 1926); **5**: 3163 (3159), 3164 (3159, 3161, 3163); **6**: 3739 (3736), 3820 (3818)
Tuckey J W, **5**: 3116 (3109)
Tuffal J, **7**: 4927 (4926, 4927)
Tuffs P S, **1**: 63 (63); **5**: 3188 (3188); **6**: 4172 (4172)
Tuggle F D, **3**: 2192 (2186)
Tukey J W, **2**: 879 (878); **3**: 1723 (1722); **4**: 2270 (2265, 2267); **7**: 4498 (4496), 5015 (5014, 5015)
Tullock E N, **2**: 965
Tully A E, **1**: 474 (469)
Tung F, **7**: 4466 (4465)
Tunis C J, **1**: 575
Tuomala M, **6**: 3680 (3675, 3676)
Turbabin A S, **3**: 2150 (2148), 2152
Turekian K K, **2**: 1516; **4**: 2782 (2780)
Turelli M, **1**: 675 (674)
Turgeon A, **3**: 2225
Turi A, **3**: 2010 (2008, 2010)
Turing A M, **1**: (304), (334), 720 (715, 719); **2**: (870, 871), 938 (933, 936); **7**: 4958
Turnbull D K, **3**: 1979
Turner R E, **4**: 2655 (2652)
Turner R H, **7**: 4539 (4536)
Turney J L, **4**: 2440 (2438)
Turoff M, **1**: 372 (360), 618; **2**: 1542 (1541), 1545 (1543); **3**: 2179; **7**: 4827 (4825)
Turquette A R, **5**: 2921 (2918)
Turton P S, **3**: 1979
Tushman M L, **3**: 2192 (2186)
Tustin A, **3**: 1749 (1749); **4**: 2575 (2571); **5**: 3390 (3385); **7**: 4452 (4443)
Tversky A, **3**: 2192 (2189), 2202 (2196), 2203 (2195, 2197, 2198, 2199, 2200); **4**: 2338 (2337); **5**: 3104 (3103, 3104), 3115 (3105, 3107, 3109, 3113); **7**: 4991 (4987)
Tweedle F, **3**: 1985 (1982, 1983)
Tweney R D, **7**: 4785 (4783)
Twenten A B, **5**: 3470 (3469), 3471 (3469)
Twiss B C, **7**: 4827
Twomey S, **2**: 1311 (1307)
Twort A C, **7**: 5099
Tydeman J, **7**: 5177 (5172)
Tye C, **5**: 3430 (3426, 3430); **6**: 4189 (4188, 4189)
Tygstrup N, **3**: 2143 (2141)
Tylee J L, **1**: 703 (695)
Tymoczko T, **1**: 333 (332)
Tyndall J, **2**: 1402 (1399)
Tyrrell Rockafellar R, **6**: 3734
Ty Smith N, **1**: 440
Tyson J J, **1**: 429 (424, 426, 428), 433 (432), 570 (569)
Tzafestas S G, **2**: 1192 (1186, 1187, 1189, 1191); **3**: 1990 (1989); **5**: 3135 (3132); **6**: 3626 (3623, 3625), 3703, 3972 (3970), 4364 (4360, 4361, 4363), 4365 (4361, 4363); **7**: 4547 (4546)

U

Uccellini L W, **6**: 4346 (4346)
Uchiyama M, **6**: 4086 (4085)
Udink ten Cate A J, **3**: 2097 (2094, 2096); **6**: 4277
Udupa S M, **4**: 2386 (2373)
Ueda K, **6**: 4033
Uematsu M, **2**: 1516 (1516)
Ueno S, **2**: 1311 (1307, 1309, 1310, 1311); **3**: 2052 (2046, 2047, 2050, 2052), 2053 (2047, 2048, 2049, 2050, 2051, 2052); **4**: 2655 (2652); **5**: 3465 (3458)
Uhley H, **1**: 769 (761)
Uhlmann E, **6**: 3825 (3821)
Uhr L, **6**: 3645
Uicker J J, **4**: 2386 (2378)
Ulanicki P, **7**: 4571 (4568)
Ulanowicz R E, **2**: 1355 (1355); **3**: 1695 (1694); **5**: 3044; **7**: 4956 (4955)
Ulianov N, **2**: 1080 (1080)
Ullman J D, **1**: 334, 725 (721, 722, 723); **2**: 884 (882), 1153 (1148), 1160 (1156, 1158, 1159); **3**: 2045 (2044); **4**: 2798 (2795, 2796); **6**: 3652 (3645, 3648), 3892, 3893, 4159 (4158)
Ullman J R, **1**: 575
Ullman S, **4**: 2331 (2330)
Ullmann J R, **6**: 3645 (3635, 3636)
Ulman J D, **2**: 1083 (1082); **5**: 3034 (3030)
Ulrich B, **5**: 3268 (3266)
Umano M, **3**: 1800 (1798, 1800)
Umbarger J E, **1**: 429 (424)
Umefuji O, **4**: 2559 (2559)
Umetani Y, **4**: 2391; **6**: 4108 (4107), 4113 (4109, 4112), 4128 (4124)
Umholtz F G, **6**: 4108 (4104)
Umino M, **5**: 3235 (3234)
Umpleby S, **1**: 333 (332)
Unbehauen H, **1**: 51 (50), 63 (60, 61, 62), 100; **2**: 1093 (1087, 1092), 1116 (1110); **3**: 2228 (2227); **4**: 2293 (2287); **5**: 3044 (3040, 3044); **6**: 4204, 4214 (4212); **7**: 5001 (4996)
Underwood E E, **7**: 4601 (4599)
Underwood E J, **1**: 268 (265)
Unger B W, **1**: 755 (752, 753, 754); **4**: 2550 (2547); **6**: 4402 (4398)
Unger K N, **5**: 3282 (3280)
Unny T E, **3**: 1862 (1859); **7**: 5140 (5138)
Upadhyay T N, **6**: 4165 (4162, 4163)
Ural'tseva N N, **6**: 3596 (3592)
Urdaibay C, **6**: 3789
Ure A, **7**: (4517)
Ure J W, **4**: 2769 (2767), 2777 (2775, 2776)
Uribe R B, **1**: 399 (394)
Uronen P, **5**: 2994 (2992), 3170 (3170); **6**: 3889 (3885, 3887, 3888), 3890 (3888, 3889), 3916 (3911, 3912, 3913, 3914)
Urquhart G C, **7**: 4848 (4847)
Urquhart J W, **2**: 1402 (1400)
Urwin R J, **6**: 3806 (3804)
Uselton S P, **1**: 347 (345)
Usher R E, **6**: 3829 (3828)
Ushijima T, **5**: 3082 (3081)
Utkin V I, **5**: (3517); **7**: 5031 (5027, 5028, 5030)
Utterback J M, **7**: 4832 (4830), 4833 (4828, 4829, 4831)
Uttley A M, **2**: 869 (868); **5**: 3390 (3386), (3516); **6**: 4168 (4166)
Uyeno D, **7**: 4935 (4934)
Uyttenhove H J J, **3**: 2003 (2000); **6**: 3989 (3984, 3986, 3987), 3990 (3984), 4372 (4371)
Uzawa H, **2**: 1261 (1257, 1260)

V

Vaca M V, **5**: 3303 (3294)
Vachira M, **2**: 1057 (1055)
Vaessen W, **7**: 4935 (4934)
Vagi F, **2**: 1116 (1115); **5**: 3188 (3188)
Vairavan K, **6**: 3631 (3631)
Vaishnav R N, **1**: 478 (477), 486 (484)
Vajda S, **1**: 494 (493)
Vajk I, **3**: 2010 (2008, 2010); **4**: 2892; **6**: 4184 (4181, 4182)
Vajta M, **2**: 1199 (1197); **3**: 2010 (2008); **4**: 2892
Valach M, **2**: 1489 (1486)
Valadier M, **2**: 1033; **5**: 3184 (3182)
Valavani L S, **1**: 51 (51), 60 (58, 59), 106 (102); **5**: 3060 (3056, 3059), 3061 (3061), 3065 (3062), 3070 (3070); **6**: 4209 (4206); **7**: 4611 (4607)
Valenca J M E, **5**: 3379 (3377)
Valenzuela H M, **5**: 3131 (3129)
Vali V, **5**: 3471 (3466)
Valk R, **6**: 3673 (3671)
Vallbona C, **1**: 542 (537)
Vallet C, **6**: 4336 (4335), 4337 (4335, 4336)
Valsamis D, **5**: 3383 (3382)
Valtorta M, **6**: 3790
Valverde L, **3**: 1781 (1780), 1810 (1807, 1809)
Vamos T, **1**: 372 (371)
van Amerongen J, **6**: 4277 (4275, 4276)
Van Atta D, **2**: 1458
van Belle F, **5**: 3030 (3027)
van Bemmel J H, **1**: 769 (762)
Van Bork L E, **2**: 1248 (1247)
Van Buskirk W C, **1**: 474 (471)
van Campenhout J M, **6**: 3645 (3643)
van Cauwenberghe A, **2**: 1586 (1585, 1586); **5**: 3430 (3426, 3428); **6**: 3744, 3889 (3887), 4172 (4170, 4171, 4172)
Van Cauwenberghe K, **1**: 176 (176), 177 (176)
Van Cott H P, **3**: 2186 (2181)
Van Cutsem T, **6**: 3846 (3845)
van Dam A, **3**: 2179; **6**: 4302 (4301)
van Damme E, **7**: 4622 (4622)
Van de Hulst H C, **3**: 2053 (2048); **4**: 2655 (2652)
Van de Riet R P, **2**: 1160
Van de Ven A H, **1**: 618; **3**: 2061 (2060); **4**: 2579 (2576)
Van de Wiele J P, **2**: 1372 (1368)
van Dedem G, **1**: 424 (423)
van Delft A, **5**: 3124 (3122)
Van den Berg J, **5**: 3248 (3244)
Van den Berg J H M, **3**: 1933 (1931)
van den Boom A J W, **4**: 2293 (2287), 2319 (2319); **7**: 5001 (4997)
van den Bosch P P J, **2**: 1388; **4**: 2800; **5**: 3071 (3070); **6**: 4148 (4147)
van den Brink G, **1**: 326 (325)
van den Burg A J, **5**: 3030 (3029); **7**: 4651 (4650)
van den Driessche P, **1**: 675 (674); **2**: 1080 (1079, 1080)
van den Enden A W M, **4**: 2293 (2287); **7**: 5001 (4997)
van der Enden J, **5**: 3421 (3418)
Van der Grinten P N E M, **6**: 3745 (3743, 3744)
van der Ham R T, **5**: 3030 (3029); **7**: 4651 (4650)
Van der Horst A, **1**: 184 (184)
Van Der Loos H F M, **6**: 4108 (4103)
van der Mark J, **1**: (535); **7**: 4980 (4978)
van der Ploeg R R, **1**: 142 (138)
van der Pol B, **1**: (535); **5**: 3024 (3020), 3390 (3383, 3384); **7**: 4980 (4978)
Van der Post J L, **3**: 1985 (1982, 1983)
Van der Valk A G, **2**: 1358
Van der Velde W E, **5**: 3373 (3369), 3389 (3385); **7**: 4452 (4451, 4452)
van der Ziel A, **2**: 1405
van Dijk J H M, **5**: 3253 (3249, 3250)
van Dixhoorn J J, **2**: 1329 (1327); **6**: 4359 (4358)
Van Dooren P, **2**: 826 (822), 827 (821, 822, 823); **4**: 2622 (2621), 2827 (2826)
Van Duijn J, **1**: 184 (184)
van Duyl W A, **7**: 4980 (4980), 4981 (4976, 4979)
Van Dycke M, **7**: 4429 (4427)
Van Dyne G M, **1**: 142 (141); **2**: 1538 (1537)
van Egmond A A J, **2**: 1596 (1592)
van Eijndhoven J H M, **4**: 2582 (2580)
Van Gigch J P, **2**: 1003; **7**: 4789 (4787)
Van Grieken R E, **1**: 177 (174)
Van Hall C E, **7**: 5116
Van Heijenoort J, **6**: 4008 (4001)
Van Hook R I, **2**: 1344 (1343)
Van Horn L D, **6**: 3680 (3675)
van Hulzen J A, **5**: 3560 (3559)
Van Loan C F, **2**: 826 (823), 827 (814, 821, 824), 828 (821)
Van Maarseveen M F A M, **3**: 1741 (1738)
van Mens P R, **2**: 967
Van Meteeren H P, **6**: 3954 (3952, 3953)
van Milaan J B, **1**: 535 (532)
Van Moerbecke P, **5**: 3535 (3533)
Van Nauta Lemke H R, **1**: 126 (125), 682 (678); **3**: 1990 (1986, 1987); **4**: 2361 (2359, 2360)
Van Ness H C, **7**: 4901 (4898)
Van Ostaeyen J, **6**: (4172)
van Overbeek A J M, **4**: 2293 (2288, 2291)
Van Putte K, **5**: 3421 (3418)
van Reeken A, **5**: 3030 (3029)
van Renesse R, **2**: 1132 (1130)
Van Rennes A, **7**: 4833 (4831)
van Rossum G J, **3**: 1969
van Rossum J M, **1**: 490 (488)
Van Ryzin J, **2**: 1105 (1105), 1106 (1105), 1107; **6**: 3645 (3641)
Van Sickle J, **4**: 2794 (2789)
Van Slyck L S, **6**: 3850 (3846, 3847, 3848)
Van Swieten A C M, **2**: 1020 (1020)
van Swieten A C W, **6**: 4148 (4147)

Van Trees H L, **1**: 664 (651, 664); **3**: 1999; **6**: 4062 (4057, 4060)
Van Vleck F S, **1**: 413 (410); **2**: 1099 (1096); **6**: 3972 (3972)
van Wijngaarden A, **3**: 1712 (1712)
Van Woerkom P T L M, **6**: 4148 (4147)
Van Zanten G A, **1**: 326 (325)
van Zomeren A H, **4**: 2582 (2582)
Vande Wiele R L, **3**: 2174 (2174)
Vander Brug G J, **6**: 4123
Vanderbrug G, **4**: 2386 (2382)
VanderWeele D A, **3**: 2213 (2210)
Vanhoutte P, **1**: 542 (540)
Vanier J, **3**: 1745
Vannelli A, **4**: 2678 (2678)
Vansteenkiste G C, **1**: 424 (421), 634 (632); **2**: 1201 (1201); **3**: 2217; **5**: 3088 (3086); **6**: 4313, 4323 (4321), 4333 (4329), 4360 (4359), 4372 (4371), 4388, 4392 (4388); **7**: 4692 (4691, 4692)
Vanthienen W, **3**: 2014
Varadarajan V S, **7**: 4425 (4423)
Varadhan S R S, **1**: 91 (89); **5**: 3531 (3526); **6**: 3596 (3594); **7**: 4617 (4615), 4627 (4623), 4637 (4632)
Varaiya P, **2**: 916, 927 (924), 956 (952), 1009 (1005); **4**: 2560 (2559); **7**: 4495 (4495), 4612 (4612), 4741 (4736, 4737, 4738), 4822 (4821)
Vardulakis A I G, **1**: 736 (735); **4**: 2832 (2830), 2871 (2869); **6**: 3717 (3712); **7**: 4457 (4455, 4456)
Vardy A E, **7**: 5161
Varela F J, **1**: 333 (332), 399 (394); **7**: 4996 (4995)
Varga A, **2**: 828 (821)
Varga R S, **2**: 927 (924); **3**: 1627, 2163 (2153, 2155, 2156)
Varju D, **4**: 2331 (2327)
Varley E, **5**: 3099
Varley S A, **2**: 1402 (1400)
Varna R, **7**: 4504 (4499)
Varrone P G, **1**: 301 (297)
Vârsan C, **4**: 2898 (2897)
Varshavskii V I, **4**: 2737 (2731)
Vary J A, **3**: 1975
Vasarhelyi M A, **7**: 4793 (4791, 4792)
Vasileva A B, **7**: 4429 (4427)
Vassiliou Y, **3**: 2179 (2178)
Vasudevan R, **3**: 2052
Vaughan D S, **2**: 1539 (1538)
Vazques J L, **4**: 2614 (2612)
Veda M, **6**: 4123
Veinott A F Jr, **6**: 3706, 3707
Vemuri V, **5**: 3165 (3160)
Vena P, **1**: 651 (650)
Vendrick A J H, **2**: 1595 (1592)
Vendrik J H, **1**: 482 (480)
Vendryes G A, **5**: 3440
Venells M V, **5**: 3464 (3463)
Venn J, **4**: (2577)
Veno S, **2**: 1311 (1307)
Ventre A G S, **3**: 1781 (1781), 1878 (1876, 1877)
Ventrello G, **1**: 301 (297)
Ventsel A D, **1**: 91 (89, 91)
Veran M, **3**: 1648 (1648)
Verbruggen H B, **2**: 1116 (1112); **3**: 1990 (1986, 1987)
Vercauter C, **6**: 3673 (3670, 3671, 3672)
Verdegay J L, **3**: 1819 (1817, 1818)
Verdin A, **5**: 3575
Vereecken R, **7**: 4981 (4977, 4978)
Verghese G C, **1**: 714 (714); **4**: 2618 (2617), 2832 (2829); **6**: 3751 (3751); **7**: 4466 (4465), 4469 (4467)
Verhagen C J D M, **6**: 3645 (3635)
Verhofstad J S M, **2**: 1137 (1134), 1165 (1161)
Verjus J P, **2**: 1209 (1208, 1209)
Vermillion J R, **2**: 973 (969, 970, 972)
Vermuri S, **6**: 3829 (3828)
Vernberg F J, **2**: 1311 (1307, 1309); **3**: 2053 (2048, 2049, 2052)
Veronda D R, **1**: 478 (475, 477), 486 (483)
Verrijn-Stuart A A, **2**: 893 (891)
Vertinski I, **7**: 5177 (5171)
Vesalius A, **1**: (315)
Veselý J, **7**: 5122
Vetter M, **4**: 2659 (2656)
Vial J P, **5**: 3550
Vichnevetsky R, **2**: 1181 (1181), 1192 (1186); **6**: 3626 (3623, 3625), 4302 (4300, 4301), 4303
Vicino A, **4**: 2257 (2256)
Vicker J J Jr, **5**: 2970 (2968)
Vickers G, **4**: 2588 (2584)
Vickers M D, **2**: 963 (962)
Vickson R G, **3**: 2202 (2196, 2200)
Vico G, **1**: 399 (394)
Vidal C, **1**: 570 (569)
Vidale M L, **2**: 1269 (1267)
Vidyasagar A, **2**: 808 (802)
Vidyasagar M, **1**: 132 (127); **3**: 2228 (2226, 2227); **4**: 2567 (2566), 2671 (2666, 2671), 2675 (2671, 2672, 2673, 2674, 2675), 2678 (2676, 2677, 2678), 2875; **5**: 3197 (3190), 3228 (3225), 3319 (3317), 3379 (3377), 3390 (3387); **6**: 4011 (4010), 4209 (4206); **7**: 4452 (4450), 4522, 4528
Vidyasagar S, **5**: 3441 (3441)
Vieira A, **2**: 1102 (1102)
Viemeister N F, **1**: 326 (325)
Viganotti C, **3**: 1623 (1619, 1620)
Viidik A, **1**: 478 (475, 477)
Viljanen A A, **6**: 4066 (4064, 4065)
Villani E, **7**: 4833 (4828)
Villanueva R, **6**: 4348 (4348), 4402, 4405 (4404)
Ville J, **5**: 2934 (2932)
Villermaux J, **5**: 2944 (2934, 2935)
Villier R, **2**: 832, 837
Villoria R L, **6**: 3904
Villwock R D, **3**: 1936 (1933)
Vince M A, **5**: 3465 (3462)
Vincelli S, **1**: 437
Vincent R, **2**: 1410
Vincent R A, **1**: 517 (516); **7**: 4925
Vincke P, **3**: 1847 (1847); **5**: 3152
Vinh N X, **7**: 4490
Vinogradov A P, **2**: 1516

Vinter R B, **3**: 2151, 2152; **5**: 3506 (3502, 3505), 3523 (3519)
Violante A, **1**: 7 (2, 5), 268 (263, 264)
Viot M, **3**: 1648 (1646); **5**: 3530 (3529), 3531 (3528); **7**: 4644 (4643)
Virkunnen J, **4**: 2756 (2756); **5**: 2927 (2925)
Virtue R W, **6**: 4051 (4048, 4049)
Visala A, **6**: 3682 (3681)
Viswanadham N, **4**: 2560 (2559), 2671 (2666, 2669, 2670, 2671), 2702 (2698, 2699, 2702); **6**: 4011 (4010)
Viswanathan R, **6**: 4165 (4163)
Vitale R A, **6**: 3619 (3618)
Vitányi P M B, **4**: 2798 (2795); **5**: 3035 (3033)
Viterbi A J, **1**: 80 (75), 91 (87); **2**: 1051 (1051)
Vitry G, **2**: 879 (878, 879)
Vitt A A, **5**: 3388 (3384)
Vizzini R W, **1**: 226
Vogel E F, **6**: 4214 (4212, 4213)
Vogel H G, **1**: 482 (480)
Vogel R M, **7**: 5051 (5048)
Vogelsaenger Th, **7**: 4958 (4957, 4958)
Vogland T P, **4**: 2601 (2598)
Vogt F B, **1**: 542 (537)
Vogt W, **4**: 2522 (2520)
Vogt W G, **7**: 4452 (4448)
Voisin P, **2**: 1342
Vollenweider R A, **4**: 2646 (2643, 2644), 2794 (2789)
Volman S F, **4**: 2331 (2327)
Volp C R, **2**: 967 (967)
Volta A G A A, **2**: 1402 (1398)
Volterra V, **2**: 1333 (1333), 1335 (1334), 1351; **3**: 1779 (1771, 1776, 1777); **5**: 3351 (3342), (3383, 3384)
Von Aspe B W, **4**: 2453 (2452)
Von Békésy G, **7**: 5065 (5062)
Von Bertalanffy L, **2**: 1348 (1346); **3**: 2164
von Eye A, **7**: 4833 (4830)
von Foerster H, **1**: 333; **2**: 869 (868), 873 (872), 1318 (1316); **6**: 3762 (3762), 4168 (4167)
von Holstein C A, **2**: 943 (941); **3**: 2203 (2195)
von Kollicker A, **1**: (532)
Von Mises R, **2**: 1107 (1103)
von Neumann J, **1**: 687 (682), 709 (704, 707); **2**: 850 (840, 843, 847), 869 (868), (870, 871, 872), 1269 (1262); **3**: 2061 (2053), (2175), 2203 (2193, 2196); **5**: 2911 (2907), 2921 (2918), 2998 (2995), 3104 (3099, 3103), 3116 (3105, 3112); **6**: 4011 (4008), 4075, (4343, 4375); **7**: 4799 (4794), 4991
von Rossum G J, **3**: 1969
von Uexküll J, **2**: 1486 (1481)
Vongpanitlerd S, **4**: 2835
Voorhis C C, **7**: 4871 (4868)
Vopěnka P, **7**: 4996 (4995)
Vorontsova I P, **4**: 2737 (2731)
Vouk V B, **1**: 148 (144)
Vovan L, **1**: 769 (762, 765); **3**: 1604 (1601, 1603)
Vozeh S, **2**: 1249 (1247); **5**: 3238 (3236); **6**: 3699 (3698)
Vranesic Z G, **5**: 2927 (2924)
Vranic M, **3**: 2175 (2171); **7**: 4906 (4904)
Vries J K, **4**: 2582 (2581)
Vroom V H, **7**: 4785 (4783)
Vukčević M B, **2**: 923 (919, 920, 922), 952 (951); **4**: 2560 (2559)
Vukobratović M, **2**: 1552; **4**: 2386 (2379); **6**: 4086, 4100, 4128 (4125), 4230 (4228)
Vuust J, **1**: 478 (475, 477)
Vyshnegradskii I A, **3**: (1745); **7**: (4517)

W

Wachpress E L, **5**: 3140 (3135)
Wada J, **1**: 570 (569)
Waddell J J, **4**: 2362
Waddington C H, **1**: 400
Waddington J, **3**: 1779 (1777); **7**: 4577 (4575)
Wade B W, **2**: 1174 (1170)
Wade W D, **6**: 4417
Wadey M D, **5**: 3373 (3372)
Waelbroeck J, **2**: 1342 (1340)
Waern B, **3**: 2225
Waeselynck J C, **7**: 4741 (4740)
Wafelmann H R, **4**: 2754
Waggoner R C, **2**: 1109 (1107)
Wagner H M, **2**: 1099 (1094); **6**: 3707 (3704), (3892); **7**: 4648
Wagner H N, **6**: 3940 (3937)
Wagner J G, **3**: 2143 (2140); **6**: 3698 (3698), 3699 (3697, 3698)
Wagner M L, **7**: 4981 (4978)
Wagner M M, **4**: 2583 (2583)
Wagner S, **7**: 4833 (4828)
Wagner Smitt L, **6**: 4262 (4259)
Wahba G, **2**: 1107
Wahdan A M, **7**: 4443 (4441)
Wahl E F, **2**: 1109 (1109)
Wahlström B, **4**: 2756 (2756); **5**: 2927 (2925)
Waide J B, **2**: 1323 (1323), 1345 (1343, 1344), 1356 (1353); **5**: 3440 (3440)
Waight J G, **2**: 1388 (1387)
Wait J V, **3**: 1910 (1910)
Wait R, **3**: 1632
Waite M, **6**: 4302 (4301)
Wajsberg M, **5**: (2919)
Wakeland W, **7**: 5177 (5171, 5174)
Walbaum H H, **2**: 781 (778)
Wald A, **1**: 252; **2**: 1107 (1103); **4**: 2287 (2285)
Walden M O, **2**: 1410 (1409)
Waldmann T, **6**: 3909 (3907)
Walisch W, **5**: 2944 (2935)
Walker A, **3**: 1979; **4**: 2798 (2796)
Walker A M, **7**: 4566 (4565, 4566)
Walker C, **2**: 1479 (1478)
Walker G, **7**: 4566 (4562)
Walker J K, **1**: 181
Walker J M, **1**: 312 (311)
Walker J W, **6**: 3855 (3853)
Walker M C, **7**: 4848 (4847)
Walker M W, **4**: 2386 (2369, 2378, 2379), 2412 (2392, 2401, 2402), 2440 (2434, 2438); **6**: 4086 (4083), 4100
Walker T M, **4**: 2681 (2680)
Walker W R, **1**: 142 (140, 141)
Wall C, **2**: 1596 (1594)
Wall H S, **5**: 3077 (3072)
Wall J E Jr, **7**: 4469 (4466, 4468)
Wall L S, **2**: 1532 (1528)
Wall P D, **6**: 4168 (4167)
Wall R, **5**: 2944 (2934)
Wall S M, **1**: 176
Wallace A R, **7**: (4954)
Wallace B W, **6**: 3917 (3911, 3915)
Wallace J N, **7**: 4577 (4575)
Wallen C C, **1**: 185
Wallenius J, **5**: 3124 (3123), 3165 (3157)
Wallenstein H M, **7**: 4703 (4702)
Waller A D, **1**: (532, 534)
Waller K V, **1**: 589 (586, 587); **2**: 1109 (1108, 1109)
Waller S A, **2**: 894
Wallin B G, **1**: 542 (539)
Walling J F, **1**: 177 (174)
Wallis J R, **7**: 4645 (4645)
Wallsten T S, **3**: 2191 (2186), 2192 (2186, 2190, 2191), 2202 (2200), 2203 (2202)
Wallstrom B, **7**: 4732
Walrand J, **7**: 4822 (4821)
Walsh F C, **4**: 2607 (2607)
Walsh J L, **7**: 4518 (4518)
Walsh P M, **4**: 2452 (2451, 2452)
Walsh T J, **1**: 456 (454, 455)
Walter B, **2**: 893 (892)
Walter E, **1**: 493 (492), 494 (492, 493)
Walter W, **4**: 2556 (2554)
Walter W G, **2**: 869 (868); **7**: 4775
Walters A A, **5**: 2979; **6**: 3655; **7**: 5164
Walters C, **2**: 1462
Walters D, **1**: 437
Walters J D, **1**: 643 (642)
Walters K, **7**: 5060 (5058)
Walters P, **4**: 2508 (2505, 2506)
Walters W J, **3**: 1975
Walton K, **7**: 4463 (4460, 4462)
Walton R E, **5**: 2911 (2911)
Waltz D, **1**: 347 (343)
Waltz M D, **6**: 4165 (4163)
Walzer P, **1**: 380
Wan F Y M, **6**: 4365 (4361)
Wan Y S, **5**: 3388
Wang C P, **4**: 2659 (2658)
Wang C T, **7**: 4551 (4548, 4550)
Wang C Y, **1**: 664 (664)
Wang D I C, **3**: 1611 (1609), 1619 (1618)
Wang E J, **6**: 3734 (3734)
Wang H Y, **3**: 1611 (1609), 1615 (1614), 1619 (1618)
Wang J, **2**: 808 (803)
Wang J S, **4**: 2782 (2778)
Wang P K C, **2**: 1186 (1182)
Wang P P, **1**: 664 (664); **2**: 776 (773, 774, 775), 1361; **3**: 1781 (1780, 1781), 1802 (1801), 1809 (1808, 1809), 1838 (1836, 1837); **4**: 2301 (2300); **5**: 3455; **6**: 3947 (3945), 4007; **7**: 4766
Wang Pei-zhuang, **4**: 2301 (2295, 2300); **6**: 3947 (3945)

Wang S H, **1**: 580 (578), 704 (694, 695), 736 (734); **2**: 902 (900, 901), 907 (902, 903, 904, 906, 907), 916 (913, 914, 915), 923 (917); **4**: 2560 (2559), 2702 (2699); **5**: 3220 (3215); **6**: 3717 (3713), 4133 (4130, 4131), 4139 (4135)
Wang S M, **4**: 2671 (2668)
Wang S S M, **6**: 4123 (4120)
Wang W H, **1**: 735 (734)
Wang Y, **1**: 542 (540), 547 (544)
Wang Y T, **1**: 580 (577); **4**: 2570 (2569, 2570), 2671 (2670); **5**: 3379; **6**: 4227 (4225)
Wanielista M P, **2**: 1553 (1552); **6**: 4229 (4228)
Wanner J C, **7**: 5170
Wanner M C, **4**: 2392
Ward D R, **5**: 3238 (3236)
Ward D S, **6**: 4062 (4057, 4059, 4061)
Ward I M, **1**: 478 (475)
Ward J D, **4**: 2582 (2580)
Ward J W, **6**: 4160 (4158)
Ward R C, **2**: 828 (824); **5**: 3268 (3266)
Ward R G, **4**: 2356 (2354)
Ward S A, **6**: 4051 (4049)
Wardi Y, **5**: 3197 (3189, 3193)
Wardrop J G, **7**: 4913 (4912), (4923), 4933 (4932)
Wareing A J, **2**: 1427 (1427), 1434 (1431)
Warfield J N, **1**: 372 (369); **4**: 2579 (2575, 2576, 2577, 2579), 2580 (2575, 2579); **7**: 4667 (4662)
Warga J, **5**: 2963 (2959)
Waring R H, **2**: 1321
Warman E A, **6**: 4302 (4301)
Warnecke H J, **4**: 2392; **6**: 3664 (3662), 4108 (4101)
Warner F, **4**: 2756 (2754); **5**: 2949 (2944)
Warner J C, **1**: 31 (21); **4**: 2686 (2684)
Warnock E H, **4**: 2338 (2338)
Warren A D, **7**: 4709 (4708)
Warren J V, **1**: 536 (532)
Warren M E, **4**: 2871 (2870); **5**: 3012
Warsahl D, **1**: 437
Warwick K, **5**: 3188; **7**: 4463 (4462)
Was G S, **2**: 1479 (1478)
Wasan M T, **5**: 3554 (3551)
Washburn R B, **1**: 567
Wasow W R, **3**: 1627; **7**: 5161 (5158)
Wasserman A I, **1**: 755 (754); **4**: 2685 (2681, 2685)
Wasserman K, **6**: 4051 (4049, 4050)
Watabe T, **3**: 1614 (1614)
Watanabe K, **7**: 4463 (4460)
Watanabe S, **2**: 1479 (1475, 1477); **3**: 1825; **5**: 3560 (3559); **6**: 3645 (3635), 3654, 3768 (3766); **7**: 4627 (4623), 4637 (4631), 4644 (4642)
Watanabe Y, **3**: 1614 (1614)
Waterfall R C, **7**: 4920
Waterfall W E, **2**: 1040 (1036)
Waterman D A, **2**: 1573
Waterman R H Jr, **7**: 4833 (4831)
Waterman T H, **2**: 1106, 1107
Waters G, **7**: 51865187 (5179, 5186)
Waterworth G, **3**: 1611 (1609)
Watkins D, **5**: 3465
Watsford R, **5**: 2993 (2990)
Watson A, **1**: 142 (141)
Watson H, **2**: 1410 (1408)
Watson H J, **6**: 4333 (4329)
Watson I B, **2**: 965
Watson J D, **1**: 559
Watson M, **6**: 3820 (3818)
Watson P F, **1**: 347 (346)
Watson S R, **3**: 1845 (1845), 1847 (1846)
Watson V J, **4**: 2646 (2645)
Watson W M, **4**: 2511 (2510)
Watt D G D, **2**: 1595 (1595)
Watt D M, **2**: 967
Watt J, **3**: (1607, 1745); **7**: (4513)
Watt K E F, **2**: 1329 (1327)
Watters G Z, **7**: 5161
Wattleworth J, **7**: 4913 (4913)
Watton R, **7**: 4871 (4869)
Watts D G, **2**: 1492, 1494; **4**: 2245 (2242), 2270 (2266, 2268), 2310 (2308, 2309)
Watts J, **5**: 3425 (3425)
Watts K E F, **3**: 2006 (2006)
Watts P, **6**: 4275 (4272)
Watts R D, **4**: 2533 (2529)
Waugh J S, **5**: 3425 (3422, 3425)
Wax M, **7**: 4466 (4465), 4469 (4467)
Way E C, **6**: 3990
Wayland H, **1**: 477 (475, 477)
Wayne K, **7**: 4832 (4832)
Ważewski T, **4**: 2566 (2560); **5**: 3406 (3404)
Wazzan A R, **6**: 4303 (4302)
Weall P, **4**: 2356 (2354)
Weaver E R, **3**: 1769 (1767)
Weaver J C, **5**: 2943 (2935, 2942), 2944 (2935, 2942)
Weaver J E, **2**: 1321 (1318, 1320)
Weaver W, **1**: 709 (704); **2**: 869 (868), 1051 (1047), 1322 (1321), 1348 (1346); **4**: 2497 (2492, 2494), 2501 (2498); **6**: 4146 (4145), 4216 (4214)
Webb G F, **2**: 1186 (1183)
Weber D L, **1**: 326 (324)
Weber J H, **7**: 4741 (4736)
Weber O, **1**: 352, 360, 362, 365, 367
Weber R J, **5**: 2911 (2909)
Weber S, **3**: 1810 (1809), 1821 (1819, 1821), 1822 (1822)
Webster D, **1**: 429 (428)
Webster F M, **6**: 3904 (3904)
Webster J R, **2**: 1323 (1323), 1326 (1324), 1345 (1344), 1356 (1353)
Webster N, **1**: 709 (704)
Wechsler J, **4**: 2588 (2584)
Wechsung R, **5**: 2943 (2935)
Weck M, **4**: 2368 (2362), 2469 (2465)
Wedde H, **6**: 4381, 4392 (4390)
Wedderburn J H M, **6**: 3751 (3749)
Wedeking H H, **4**: 2659 (2658)
Weed S B, **1**: 7 (1, 2), 8 (1, 2, 3, 5)
Weedy B M, **6**: 3809
Weekman V W III, **5**: 3082 (3079)
Weeks J K, **5**: 3030 (3029)

Wegman E J, **2**: 1107
Wegner P, **1**: 732 (726)
Wei C Z, **4**: 2283 (2279)
Wei J, **1**: 417 (416); **5**: 3082 (3079)
Weibel E R, **4**: 2904 (2902); **7**: 4601 (4599)
Weick K, **7**: 4833 (4830)
Weigand W A, **3**: 1611 (1609)
Weigert R G, **7**: 4454 (4453)
Weil J V, **6**: 4061 (4060)
Weiler P R, **4**: 2646 (2645); **7**: 4897 (4896)
Weiler R, **7**: 5065 (5062)
Weimann A, **3**: 1990 (1986, 1987)
Weinaug C F, **3**: 1975
Weinberg A M, **5**: 3416
Weinberg D, **2**: 992 (989, 991)
Weinberg G M, **1**: 93 (92); **2**: 992 (987, 991); **3**: 2179
Weinberg L, **6**: 4284 (4282)
Weinberg R, **6**: 4345 (4343)
Weinberger H, **4**: 2614 (2611, 2612)
Weinert H L, **7**: 4466 (4463, 4465), 4468 (4467)
Weinman J A, **3**: 2052 (2049); **4**: 2655 (2652)
Weinrich S D, **3**: 1611 (1609)
Weins J A, **5**: 3268 (3266)
Weinstein C J, **2**: 1078 (1077)
Weir M, **6**: 4388
Weisbin C R, **2**: 1552 (1549)
Weisner N, **6**: 4211 (4210, 4211)
Weiss A, **6**: 4252 (4250)
Weiss D, **7**: 5122
Weiss G H, **7**: 4646 (4644)
Weiss H K, **5**: 3390 (3385, 3386), (3516)
Weiss J J, **3**: 1845 (1845), 1847 (1846)
Weiss L, **4**: 2871 (2870, 2871); **6**: 3629, 3751 (3749)
Weiss M D, **5**: 2944 (2934)
Weiss M F, **3**: 2174 (2171)
Weiss P, **6**: 4007 (4001)
Weiss P A, **1**: 399 (394)
Weiss R M, **7**: 4981 (4976, 4978)
Weiss S, **5**: 2942 (2935)
Weiss S M, **2**: 1573, 1578 (1573)
Weissenberger S, **4**: 2566
Weissman C, **4**: 2386 (2374)
Weissmantel H, **6**: 4123 (4120)
Weist J D, **6**: 3904 (3904)
Weitzenfeld J, **3**: 2203 (2201)
Weitzman E D, **1**: 601 (601)
Weizenbaum J, **1**: 372 (370); **3**: (2178); **6**: 4411 (4406)
Welch B L, **2**: 1107 (1103)
Welch J D, **1**: 202 (201)
Welch P D, **4**: 2270 (2267), 2550 (2550); **6**: 4303, 4395 (4395), 4396 (4395); **7**: 4654 (4652, 4653), 4657 (4656)
Welch T R, **1**: 304 (301)
Weldon E J Jr, **1**: 80 (76)
Well E R, **1**: 445 (444)
Wells C H J, **7**: 4504
Wells D W, **2**: 1388 (1386); **6**: 3790 (3789)
Wells P N T, **7**: 4964
Wells R, **7**: 4577 (4575)
Wells W R, **1**: 252 (248)
Wellstead P E, **1**: 63 (60); **2**: 1466 (1463, 1464), 1467 (1462, 1463, 1465); **4**: 2253 (2251, 2252); **5**: 3188 (3184), 3232 (3229), 3410 (3407); **6**: 3727 (3721, 3722, 3723, 3724, 3725), 3991 (3990), 4176 (4175), 4193 (4190, 4193), 4204 (4200)
Welsch P D, **3**: 1723
Welsh D J A, **1**: 626 (624)
Welti D, **5**: 3425 (3423)
Wend H D, **2**: 952 (950, 951)
Wendel G J, **1**: 184 (184)
Wendel R E, **5**: 3165 (3161)
Wendlandt W W, **7**: 4504 (4499)
Wendt R E, **3**: 1693 (1688)
Wenk E Jr, **7**: 4828 (4824)
Wenstøp F, **1**: 126 (125); **5**: 3104 (3104)
Werblin F S, **7**: 5070 (5066, 5070)
Werkhoven-Goewie C E, **7**: 5129 (5129)
Werner E, **7**: 4703 (4699)
Werner R A, **6**: 4227 (4226)
Wertz J R, **6**: 4148 (4147)
Wertz V, **4**: 2292 (2288, 2291), 2293 (2288)
Weska J S, **6**: 4160 (4157)
Wesley M A, **4**: 2386 (2373, 2374), 2447 (2447), 2469 (2466)
Wesselink A F, **5**: 3260 (3260)
West D C, **2**: 1317 (1315, 1317), 1321 (1318)
West E J, **5**: 3470 (3469)
West J C, **5**: 3373 (3370), 3390 (3385); **7**: 4452 (4451)
West P G, **1**: 289 (284, 285, 286, 289)
West T, **5**: 3448
Westenskow D R, **1**: 289 (284)
Westerberg A W, **1**: 406 (403, 404)
Westerberg B, **5**: 3232 (3229), 3410 (3407)
Westerlund K T, **1**: 564 (561); **5**: 2994 (2993)
Westermeier T F, **1**: 252 (246)
Westesson P L, **2**: 976
Westmann R A, **1**: 478 (475, 477), 486 (483)
Weston C W, **7**: 5094 (5093)
Weston R H, **3**: 2014; **4**: 2368 (2367), 2464 (2461, 2463, 2464)
Westöö G, **4**: 2782 (2781)
Wetmore V, **6**: 3870 (3866, 3867)
Wets R J B, **5**: 3546 (3544)
Wetzel M C, **3**: 1910 (1910)
Wever R A, **1**: 601 (594, 595, 597, 598, 599, 600); **7**: 4880 (4877)
Wewerinke P H, **3**: 1669 (1666)
Weyl H, **1**: 294 (292)
Weyl S, **6**: 3653 (3652)
Whalen A D, **1**: 664 (659)
Wheatstone C, **2**: 1402 (1400)
Wheeler D J, **6**: 4271 (4268)
Wheeler G L, **1**: 592 (590, 591)
Wheeler J A, **1**: 294 (292)
Wheeler R M, **4**: 2719 (2718)
Wheelhouse K, **3**: 1992 (1991)
Wheelwright S C, **3**: 2192; **7**: 4827 (4826), 4833 (4832)
Whensch K D, **5**: 2944
Whicker F W, **6**: 3936
While C, **6**: 3690 (3690), 3691 (3690)

Whinston A B, **2**: 947; **6**: 3643, 3644
Whipp B J, **6**: 4051 (4046, 4048, 4049, 4050), 4061 (4057, 4060), 4062 (4059, 4061), 4066 (4062)
Whisler F D, **1**: 142 (140, 141)
Whistler W J, **5**: 2944 (2934)
Whitaker D A, **7**: 5129 (5128)
Whitaker H P, **5**: 3070 (3068)
Whitaker R, **2**: 1458
Whitby J D, **7**: 4856 (4853)
Whitby K T, **1**: 179 (178, 179)
Whitcomb W H, **2**: 1317 (1317)
White C C, **2**: 947 (946); **3**: 2203 (2202); **4**: 2338 (2338), 2501 (2499), 2714 (2706, 2709, 2711, 2712); **5**: 3116 (3114); **6**: 4155 (4154); **7**: 5005 (5002)
White D, **3**: 1985
White D C, **6**: 3680 (3675, 3676)
White D E, **4**: 2559 (2558, 2559), 2702 (2702); **5**: 3572 (3570); **6**: 4011 (4008, 4010)
White D J, **3**: 2191 (2186); **5**: 2932
White E B, **3**: 2203 (2199); **4**: 2641 (2639)
White G, **5**: 2944 (2934)
White G C, **1**: 417 (414); **2**: 1329 (1327), 1344 (1343)
White J, **2**: 1358; **3**: 1615 (1612)
White J W, **5**: 3282 (3280)
White K P Jr, **7**: 4495 (4495), 4762 (4756), 5177 (5172, 5173, 5174)
White P S, **2**: 1318 (1314), 1321 (1318), 1358
White R L, **4**: 2511 (2510)
White T, **1**: 310 (308)
White W, **6**: 4275 (4271)
White W D, **2**: 963 (962)
Whitehead A, **1**: 333 (330)
Whitehead R, **3**: (1745)
Whitehouse G E, **6**: 3904 (3904)
Whitelaw J H, **3**: 1693 (1686)
Whiteley A L, **3**: (1748)
Whitfield D, **1**: 726 (720)
Whitham G B, **7**: 4922 (4921)
Whitin T M, **4**: 2583 (2583)
Whiting B, **2**: 1248 (1247), 1249 (1247); **6**: 3699 (3698)
Whitmore G A, **3**: 2202 (2196, 2200); **5**: 3116 (3114)
Whitney D E, **4**: 2385 (2380), 2386 (2380), 2387 (2373, 2380), 2440 (2438), 2460 (2459); **6**: 4088 (4086), 4100, 4123 (4121)
Whittaker D K, **2**: 969
Whitten G Z, **1**: 188 (187)
Whitting I J, **3**: 1981 (1980), 1998
Whittington D, **2**: 1566 (1563)
Whittle P, **4**: 2270 (2266); **5**: 3006 (3006)
Whol M, **7**: 4933
Whorlow R W, **7**: 5060 (5057, 5058)
Whyte L L, **2**: 1348 (1346)
Wiberg D M, **4**: 2800 (2799); **5**: 3563; **6**: 4051 (4046, 4048, 4049), 4061 (4057, 4060, 4061), 4062 (4059), 4066 (4062)
Wickersheim K A, **5**: 3465 (3460)
Wickstrom G V, **7**: 4948
Wickstrom W A, **3**: 2113 (2112)
Widder D Y, **4**: 2664 (2659)
Widdicombe J G, **6**: 4061 (4057)
Widrow B, **1**: 81 (78, 79); **4**: 2314 (2311); **6**: 3645 (3636), 4160 (4157), 4168 (4166)
Wied G L, **3**: 1825 (1824)
Wiedey G, **5**: 3455 (3454)
Wiedner V R, **2**: 1521 (1517)
Wiegert R E, **2**: 1462
Wiegert R G, **2**: 1326 (1324); **7**: 5008 (5005, 5006)
Wiener A, **6**: 4155
Wiener N, **1**: (116), 294 (293), 372 (370), 490 (488); **2**: 869 (867, 868), 873 (869, 871, 872, 873), 1497 (1494); **3**: 1728 (1725), 1749 (1747, 1749), 1779 (1769); **4**: 2273 (2271), 2664; **5**: 3336 (3330), 3376 (3376); **6**: 4008 (4001); **7**: 4775 (4773)
Wiens J A, **1**: 142 (141)
Wierwille W, **3**: 1668 (1666), 1669 (1666)
Wierzbicki A, **5**: 3035 (3032)
Wierzbicki A P, **1**: 406 (403); **2**: 1276 (1273, 1274)
Wierzchon S T, **3**: 1822 (1821)
Wiesel D H, **4**: 2331 (2325, 2327)
Wiesendanger H V D, **5**: 2944 (2935)
Wieslander J, **4**: 2245 (2242); **5**: 3383 (3381)
Wiggins R A, **5**: 3131 (3130)
Wightman F L, **1**: 326 (325)
Wigner E P, **1**: 294 (290)
Wiig K M, **5**: 2994 (2992)
Wijn P F F, **1**: 482 (480)
Wilcox R H, **6**: 4168 (4166)
Wilcox T, **3**: 1810 (1807, 1809)
Wild C, **6**: 4223 (4221)
Wild K R, **3**: 1961
Wildberger A M, **2**: 1366 (1365)
Wilde C O, **6**: 3653 (3652)
Wildes J, **6**: 3850 (3846)
Wilensky R, **2**: 1573
Wilf H S, **2**: 1416 (1415)
Wilheit T, **6**: 4041 (4039)
Wilhelm R G, **6**: 3591 (3591), 3917 (3915), 4204 (4200)
Wilke C R, **1**: 424 (423)
Wilkins C L, **4**: 2511 (2510)
Wilkins R W, **1**: 547 (544)
Wilkinson G, **1**: 7 (4), 268 (262); **3**: 1958; **7**: 5094 (5090, 5092)
Wilkinson J F, **3**: 1992 (1991)
Wilkinson J H, **2**: 826 (817, 818, 820, 821), 827 (818), 828 (814, 815, 816, 818, 819, 821); **4**: 2827 (2817, 2819, 2823, 2826); **5**: 3180 (3176), 3228 (3226), 3559 (3555)
Wilkinson J R, **2**: 840
Wilkinson J W, **6**: 4235 (4231)
Wilkinson R, **1**: 482 (480)
Wilkinson R I, **7**: 4716 (4714), 4726 (4722), 4732 (4731), 4741 (4734)
Will H J, **6**: 4333 (4329, 4332)
Will P M, **6**: 3766 (3762), 4123 (4120)
Willeke K, **1**: 179 (178)
Willemain T R, **2**: 1566 (1564)
Willems J C, **2**: 981 (980); **4**: 2675 (2673), 2816 (2807), 2832 (2829, 2831), 2849, 2855 (2849); **5**: 3310 (3305), 3319 (3317), 3330 (3327); **6**: 3717 (3710, 3711, 3717), 3734 (3733); **7**: 4522 (4520), 4637 (4635)

Willems J H, **1**: 769 (762)
Willems J L, **1**: 769; **2**: 1080 (1078), 1351 (1348); **5**: 3319 (3319), 3398, 3402 (3399, 3400, 3401), 3441 (3441); **7**: 4529 (4527), 4554 (4551)
William M H, **4**: 2511 (2510)
Williams A, **2**: 1559; **7**: 4703 (4700)
Williams A E, **5**: 2942 (2937)
Williams A R, **7**: 4964
Williams B T, **6**: 4007 (4007)
Williams C, **2**: 992 (989)
Williams C S, **4**: 2314 (2311)
Williams D, **1**: 177 (175)
Williams D H, **5**: 2943 (2937, 2942)
Williams E C, **4**: 2353 (2351)
Williams G L, **7**: 4930 (4928)
Williams J, **1**: 532
Williams J C, **4**: 2533 (2528)
Williams J L, **2**: 916
Williams J L D, **7**: 5065 (5062)
Williams J W J, **1**: 347 (346)
Williams K J, **7**: 4637 (4631)
Williams L R, **7**: 5008 (5005, 5006), 5015 (5014)
Williams T J, **2**: 1568 (1567); **3**: 2014; **4**: 2517 (2511, 2513); **5**: 3145 (3142); **6**: 3890 (3888, 3889), 3916 (3911, 3913); **7**: 4593 (4589)
Williams T W C, **1**: 638 (638)
Williams W C, **7**: 4481
Williamson D, **5**: 3220 (3215); **6**: 3720 (3720)
Williamson D H, **4**: 2631 (2628)
Williges C, **3**: 1669 (1666)
Willman W, **2**: 932 (930)
Willman W W, **7**: 4466 (4466)
Willmott R, **3**: 1810 (1809); **6**: 4008 (4005)
Willoky A S, **3**: 1736
Willoner G, **6**: 4284 (4282)
Willshaw D J, **6**: 4168 (4167)
Willsky A S, **1**: 86 (82, 84), 251 (251), 252 (247, 249); **3**: 1741 (1738, 1739); **5**: 3272 (3268), 3303 (3301), 3336 (3336); **6**: 4281 (4278); **7**: 4466 (4465), 4469 (4466, 4468)
Wilnis P, **2**: 1174 (1171)
Wilser J E, **5**: 3165 (3157)
Wilshire F W, **1**: 177 (174)
Wilson A, **6**: 4275 (4273)
Wilson A G, **2**: 1348 (1346)
Wilson B L, **4**: 2608 (2607)
Wilson D, **2**: 1348 (1346)
Wilson E N, **5**: 2948 (2948)
Wilson E O, **5**: 2956 (2956)
Wilson H R, **1**: 434 (432)
Wilson I D, **2**: 956 (952)
Wilson J G, **3**: 1979
Wilson J M, **6**: 4388
Wilson J Q, **2**: 1565 (1563)
Wilson J R, **7**: 4654 (4652, 4653)
Wilson J W, **1**: 246; **3**: 1958
Wilson K R, **2**: 1497 (1494)
Wilson L, **4**: 2385 (2382)
Wilson R, **5**: 3256 (3255), 3440; **7**: 4622 (4622)
Wilson R G, **7**: 4577 (4573)
Wilson W, **6**: 4369 (4368)
Wiltshire J, **4**: 2640 (2639); **7**: 4785 (4783)
Winchester J W, **1**: 177 (174)
Windal G, **7**: 4703 (4700)
Windeknecht T G, **1**: 294 (292); **6**: 4355 (4354, 4355)
Windham M P, **1**: 608 (607)
Windsor H V, **5**: 3471 (3469)
Winefordner J D, **7**: 4504
Winer A M, **1**: 176 (175), 187 (187)
Winet M, **5**: 3116 (3105)
Winfield R P, **4**: 2646 (2645)
Winfree A T, **1**: 570 (568)
Wing P, **3**: 2099 (2099)
Wingate F P, **5**: 3465 (3458)
Wingate J R, **3**: 1958
Winge R K, **4**: 2347 (2345)
Wingfield A J, **4**: 2775 (2773)
Wings L, **1**: 187 (187)
Winkelman J R, **5**: 3180 (3176); **7**: 4439 (4439)
Winkler G, **6**: 4159 (4158)
Winkler K, **3**: 2143 (2141)
Winkler O, **5**: 3575
Winkler R, **7**: 5155 (5154)
Winkler R L, **5**: 2917
Winn C B, **3**: 2097 (2094)
Winograd T, **4**: 2641 (2637)
Winslow E H, **7**: 4504
Winsor C A, **6**: 4148 (4147)
Winston P A, **4**: 2640 (2638, 2639)
Winston P H, **1**: 347 (343); **4**: 2641 (2638, 2639, 2640); **6**: 4160 (4159), 4353 (4350)
Winter B, **6**: 4062 (4061)
Winter D A, **3**: 1913
Winter P, **1**: 585
Winter W K, **3**: 2006 (2006)
Winterbone D E, **1**: 583 (581); **2**: 1467 (1462); **4**: 2597 (2588); **6**: 4271 (4270)
Wintz P, **6**: 4160 (4157)
Wirgau K A, **6**: 3954 (3951, 3952)
Wirth N, **6**: 3898 (3896), 4093 (4089), 4289 (4288)
Wirtshafter D, **3**: 2213 (2212)
Wirz K, **1**: 459 (457)
Wise D A, **6**: 4266 (4262), 4267 (4262)
Wise S P, **1**: 466 (463)
Wisiniewska J, **6**: 4405 (4404)
Wismer D A, **1**: 638 (634, 636); **2**: 956 (952), 1192 (1188); **3**: 1990 (1989), 2038; **4**: 2714 (2710); **5**: 3019 (3016, 3017), 3040 (3035, 3039), 3164 (3158); **6**: 3846 (3845); **7**: 4709 (4705)
Wisneski D, **2**: 826
Wisnosky D E, **4**: 2453 (2450)
Wissler E H, **7**: 4877 (4873, 4875)
Witenhausen H, **5**: 3523 (3523)
Witherspoon P A, **3**: 2006 (2006)
Within T M, **6**: 3707 (3704)
Withington H W, **1**: 120
Witkamp M, **1**: 675 (674); **2**: 1344 (1343)

Witsenhausen H S, **2**: 932 (929, 930), 1269 (1263); **7**: 4822 (4816, 4817, 4821)
Witten I, **4**: 2550 (2547)
Witten J H, **2**: 1333 (1331)
Wittenberg B, **5**: 3410 (3407)
Wittenburg J, **3**: 1665 (1653)
Wittenmark B, **1**: 45 (39), 51 (49), 54 (54); **2**: 1116 (1110), 1199 (1197), 1253 (1250), 1257 (1257); **3**: 2010 (2010), 2021 (2018); **4**: 2892 (2891); **5**: 3006 (3006), 3188 (3184), 3232 (3229), 3351 (3341), 3410 (3407, 3409), 3430 (3426); **6**: 3591 (3590), 3727 (3721, 3722, 3723, 3724, 3725), 3871 (3870), 3991 (3990, 3991), 3994 (3994), 4164 (4164), 4172 (4168, 4171), 4176 (4172, 4175), 4180 (4179), 4183 (4181, 4182), 4186 (4185, 4186), 4193 (4190, 4191, 4193), 4197 (4194), 4204 (4200), 4209 (4207), 4211 (4209), 4213 (4211, 4212), 4214 (4212), 4267 (4265); **7**: 4607 (4603, 4606), 4611 (4607)
Wittgenstein E, **7**: 5122
Wittgenstein L, **1**: 709 (705)
Wittmann G T W, **4**: 2782 (2778, 2780)
Witzgall E, **6**: 3664 (3662)
Wobbe G, **3**: 1769 (1766)
Wobbe-Ohlenburg W, **6**: 4097
Wohl J G, **1**: 651 (648, 649)
Wohn F K, **5**: 2944
Wohnoutka R, **7**: 4930 (4928)
Wojcicki R, **6**: 3865 (3865)
Wold H, **1**: 556 (554)
Wolf E, **4**: 2723 (2720)
Wolf E M, **6**: 4356 (4356)
Wolf H, **2**: 1366 (1363, 1366)
Wolf R M, **2**: 1366 (1366)
Wolfe H B, **2**: 1269 (1267)
Wolfe J W, **2**: 1596 (1594, 1595)
Wolfe P, **1**: 405 (400); **2**: 956 (952), 958; **3**: 1926 (1926); **4**: 2880 (2877, 2879); **5**: 3170 (3167), 3171 (3169), 3545 (3544); **6**: 3739 (3736), 4299; **7**: 4666 (4661, 4664), 4969 (4966)
Wolfe T C, **5**: 2943 (2935)
Wolfe W L, **7**: 4871 (4869)
Wolff H S, **7**: 4856 (4853)
Wolff P H W, **5**: 3440
Wolfowitz J, **4**: 2508 (2506)
Wolovich G A, **4**: 2827 (2826)
Wolovich W A, **1**: 45 (40), 735 (734), 736 (734); **3**: 2108 (2108); **4**: 2523 (2522, 2523), 2570 (2568, 2569), 2671 (2666); **5**: 3060 (3059), 3220 (3216, 3220); **6**: 3727 (3724), 3969; **7**: 4554 (4553)
Wolter F, **3**: 1943
Wolthers H, **6**: 3699 (3697)
Womack B F, **2**: 1361
Wong D I C, **3**: 1615 (1614)
Wong E, **2**: 932 (930), 1160 (1159); **5**: 3303 (3293, 3294, 3295), 3336 (3334); **7**: 4637 (4631, 4633, 4634, 4635), 4822 (4818)
Wong F S, **7**: 4462 (4460)
Wong J K, **2**: 808
Wong J Tze-Fei, **1**: 424 (422)
Wong J W, **2**: 1087 (1084), 1201 (1200)
Wong K Y, **1**: 45 (42); **4**: 2278 (2273); **5**: 2994 (2992), 3232 (3230, 3232)
Wong P J, **7**: 4930 (4928)
Wong W S, **2**: 1154 (1145, 1146); **5**: 3303 (3301)
Wong Y T, **2**: 1078 (1076)
Wonham W M, **1**: 106 (102), 137 (134), 421 (419); **2**: 828 (814, 821, 822), 900 (894, 896); **4**: 2523 (2522, 2523), 2560 (2558), 2567 (2566), 2570 (2569, 2570), 2816 (2807), 2832 (2829), 2855 (2849, 2850, 2852, 2853, 2854), 2871 (2870); **5**: 3060 (3057), 3288 (3283, 3286), 3303 (3293, 3298); **6**: 3717 (3711), 3720 (3718), 3734 (3733), 4011 (4011), 4248 (4248); **7**: (4552), 4613, 4685 (4683), 4689 (4686)
Woo S L-Y, **1**: 478 (477)
Woo T C, **6**: 4161 (4157)
Wood A D, **1**: 552 (550); **7**: 4771
Wood B O, **4**: 2453 (2450)
Wood C F, **4**: 2601 (2598)
Wood D, **2**: 884 (883); **4**: 2798 (2797, 2798)
Wood D B, **3**: 1693 (1690)
Wood D E, **2**: 1388 (1387)
Wood D W, **1**: 725 (725)
Wood E F, **4**: 2239 (2236)
Wood E H, **1**: 546 (544)
Wood J E, **6**: 4108 (4104)
Wood J M, **4**: 2782 (2781)
Wood R, **3**: 1761 (1758)
Wood R K, **1**: 585; **2**: 1116 (1110, 1111, 1115); **5**: 3188 (3188)
Wood R T, **1**: 51 (50)
Wood W T, **2**: 1226 (1223)
Woodbury D A, **7**: (4516)
Woodbury J W, **5**: 3253 (3248, 3250, 3252)
Woodcock J P, **7**: 4964
Woodmansee R G, **1**: 142 (141)
Woods J W, **5**: 3134 (3131)
Woods R E, **6**: 4161
Woodside C M, **7**: 5177 (5171, 5173)
Woodson W E, **7**: 5086 (5082)
Woodward P M, **6**: 3898 (3894)
Woodwell G M, **1**: 532 (529)
Woody R, **5**: 3282 (3280)
Wooley R N, **6**: 4333 (4330)
Woollons D J, **2**: 1410
Woolvet G A, **5**: 3465 (3457)
Woon S K, **5**: 3383 (3381)
Worwood M, **3**: 1623 (1619)
Wotten I H, **4**: 2775 (2773)
Wozencraft J M, **2**: 1051 (1048, 1051)
Woźniak A, **2**: 1305 (1299, 1302, 1304); **7**: 4571 (4568, 4569, 4571)
Wrathall C, **2**: 827 (814)
Wright D A, **5**: 3421 (3418)
Wright E A, **7**: 4721 (4716)
Wright J D, **6**: 4133 (4132)
Wright J W, **4**: (2359)
Wright M H, **2**: 797 (797)
Wright O, **1**: (340)
Wright P, **1**: 634 (632); **3**: 2192 (2186)

Wright S, **3**: 2014; **5**: 3235 (3234)
Wright T, **2**: 1402 (1399)
Wright T M, **1**: 474 (469, 473)
Wright V, **1**: 482 (480)
Wright W, **1**: (340)
Wright W F, **3**: 2192 (2186)
Wright W J, **2**: 1388 (1387)
Wright W M, **7**: 4799 (4794, 4798)
Wronski G, **7**: 4577 (4573)
Wu C, **6**: 4303 (4302)
Wu C H, **4**: 2387 (2380)
Wu F F, **6**: 3846
Wu W W, **4**: 2511 (2510)
Wuehrer W, **3**: 2228 (2227)
Wuehrmann D H, **2**: 976
Wulf W A, **6**: 4402 (4398)
Wurtz R, **4**: 2331 (2325)
Wutke B, **2**: 1305 (1302)
Wuu T, **1**: 736 (735)
Wyard S J, **3**: 1910 (1908)
Wyatt J R, **5**: 2943 (2935)
Wyatt-Mair G, **5**: 2994 (2991)
Wylie E B, **7**: 5161
Wyman B F, **6**: 3972 (3970)
Wyman J, **1**: 429 (424)
Wymore A W, **2**: 1003; **5**: 3082 (3080), 3088; **6**: 4319, 4323 (4319), 4411 (4407)
Wyndham C H, **7**: 4856 (4855)
Wyper D J, **4**: 2582 (2580)
Wysowski C H, **6**: 4369 (4368)

X

Y

Yachida M, **6**: 4161 (4157)
Yachter M, **5**: 3389 (3385)
Yackel R A, **5**: 3502; **7**: 4439 (4438)
Yaged B Jr, **7**: 4716 (4714), 4726 (4725)
Yager R R, **1**: 126 (126), 664 (651), 769 (762, 765); **2**: 1473 (1471, 1472); **3**: 1781 (1781), 1786, 1802 (1802), 1819, 1823 (1822), 1829 (1827, 1829), 1831 (1830), 1835, 1838, 1845 (1845), 1852 (1847, 1848, 1850), 1858, 1867 (1864), 1872 (1868), 1874, 1878 (1876), 1888, 1894 (1889), 1902 (1900), 1905 (1903, 1904, 1905); **4**: 2301 (2294, 2295, 2300), 2883 (2881), 2884 (2883); **5**: 2965 (2963, 2964, 2965), 3540 (3539); **6**: 3619 (3617), 3710, 3772 (3769, 3770), 3947 (3945), 4008 (4005); **7**: 4661 (4657), 4695 (4693), 4789 (4788)
Yagi T, **3**: 2021 (2017)
Yagil G, **1**: 429 (426), 434 (429), 559
Yaglom A M, **6**: 3871 (3871)
Yagr R R, **3**: 1888
Yakimovsky Y, **4**: 2488
Yakubovich V A, **4**: 2566 (2563)
Yam E S, **4**: 2386 (2382)
Yamada H, **2**: 1522 (1518)
Yamada M, **2**: 1516 (1516)
Yamaguchi O, **7**: 4980 (4976)
Yamakami A, **7**: 4685 (4683)
Yamamoto H, **7**: 4881 (4880)
Yamamoto P S, **4**: 2901 (2900)
Yamamoto T, **1**: 450 (447); **2**: 1516 (1514, 1515, 1516)
Yamamoto W S, **4**: 2901 (2899, 2900); **6**: 4051 (4047, 4048, 4050)
Yamamoto Y, **1**: 450 (447)
Yamasaki Y, **3**: 1619 (1618), 2021 (2017), 2032 (2026, 2030, 2032)
Yamashiro S M, **6**: 4051 (4048), 4061 (4060), 4066 (4062, 4065)
Yamashita S, **3**: 1611 (1609)
Yamayee Z A, **2**: 1388 (1387)
Yamazaki H, **1**: 187 (187)
Yamazaki K, **1**: 713 (710, 713)
Yang A T, **4**: 2387 (2373)
Yang R T, **1**: 176 (175)
Yaniv I, **6**: 4252 (4251)
Yannakakis M, **2**: 884 (881, 882, 883)
Yao D D, **3**: 1649 (1646)
Yao J T P, **3**: 1802 (1801); **4**: 2883 (2881); **6**: 3772 (3770); **7**: 4669
Yao S B, **2**: 1160 (1159)
Yariv A, **5**: 3471 (3469)
Yarwood R I, **6**: 4252 (4251)
Yase Y, **1**: 268 (265)
Yasuda K, **4**: 2560 (2559)
Yasuda S, **5**: 3235 (3234)
Yasuhara A, **3**: 1936 (1934)
Yasui S, **6**: 3962 (3957)
Yates F E, **3**: 2032 (2026), 2175 (2174)
Yates R E, **7**: 4551 (4549, 4550)
Yau S S, **6**: 3653 (3648)
Ydstie B E, **2**: 1466 (1464); **5**: 3188 (3188), 3232 (3231); **6**: 3991 (3990, 3991)
Yeaple D, **4**: 2782 (2778)
Yee J P, **2**: 867 (865)
Yeh H C, **4**: 2904 (2903, 2904)
Yeh R T, **3**: 1883 (1881), 1888; **4**: 2681 (2678, 2679, 2680); **6**: 3898, 4402 (4398)
Yelowitz L, **3**: 1888
Yemm R, **5**: 3248 (3245)
Yen T, **5**: 3559 (3555)
Yetton P W, **7**: 4785 (4783)
Yeung L F, **5**: 3223 (3221, 3222)
Yevjevich V, **7**: 5015 (5013)
Yionoulis S M, **2**: 1244 (1244)
Ylinen R, **5**: 2994 (2991)
Yliniemi L, **5**: 2994 (2992)
Yntema D B, **7**: 4986
Yodzis P, **2**: 1318
Yonas Y, **1**: 651 (650)
Yonezawa K, **1**: 360, 362, 367
Yoon K, **5**: 3156 (3152)
Yor M, **5**: 3303 (3294)
Yorke J, **2**: 1351 (1349)
Yorke J A, **1**: 567 (565, 566)
Yoshiaki S, **1**: 589 (585)
Yoshika N, **5**: 2993 (2993)
Yoshikawa T, **7**: 4823 (4821)
Yoshimatsu M, **2**: 1522 (1518)
Yoshimura M, **2**: 1388
Yoshimura T, **2**: 1497 (1494, 1497)
Yoshino T, **5**: 3465 (3461)
Yoshitani Y, **5**: 2994
Yosida K, **2**: 1186 (1185); **6**: 4223 (4216, 4219, 4222)
Yost R A, **2**: 1174 (1170, 1171)
You K C, **6**: 3645 (3643), 3654 (3646, 3652), 4161 (4158)
Youla D C, **4**: 2570 (2568); **6**: 3972 (3971)
Youlal H, **2**: 1253 (1251)
Young A C, **1**: 433 (432), 542 (537, 540)
Young D A, **3**: 1979
Young D M, **3**: 2163 (2154, 2155)
Young I R, **5**: 3425 (3425)
Young J H, **2**: 1590 (1587, 1589)
Young K, **7**: 5031 (5028, 5029)
Young K-K D, **7**: 5031 (5029)
Young L C, **1**: 529 (526, 527); **3**: 2006 (2006); **5**: 2963 (2959)
Young L R, **2**: 1591 (1590), 1596 (1593, 1594)
Young M J, **1**: 437
Young P, **2**: 1333 (1331); **5**: 3248 (3246)
Young P C, **2**: 1311 (1307); **4**: 2277 (2276), 2278 (2273), 2284 (2283), 2293 (2290), 2314 (2313), 2320 (2318); **5**: 3188 (3188), 3242 (3241); **6**: 3717 (3711), 4172 (4172); **7**: 4692 (4690)

Young R J, **1**: 473 (473)
Young R M, **2**: 1342
Young S C K, **1**: 564 (561)
Young S J, **6**: 3898 (3894, 3896)
Young T Y, **6**: 3645
Younts S E, **1**: 7 (2, 5)
Yourdon E, **2**: 992 (987); **7**: 4472
Yovits M C, **2**: 869 (868), 1573; **4**: 2497 (2497), 2502 (2498); **6**: 4165 (4161), 4168 (4167); **7**: 4785
Ytterdahl S A, **2**: 1416 (1414, 1416)
Yu J Y, **6**: 3755
Yu P L, **5**: 3124 (3122), 3165 (3157, 3158, 3159, 3160, 3161)
Yu T K, **6**: 4245 (4238)
Yu T S, **6**: 3644 (3635, 3641)
Yuan J S C, **1**: 106 (102); **2**: 1099 (1094); **5**: 3060 (3057); **6**: (3892)
Yudin D B, **5**: 3546 (3544)
Yule G U, **7**: 4566 (4562)
Yum D Y Y, **2**: 1102 (1102)
Yuminaka T, **2**: 1431 (1428), 1434 (1431)
Yushkevich A A, **5**: 3523 (3519), 3535 (3533); **7**: 4612
Yusypchuk G J, **7**: 4466 (4463, 4465), 4468 (4467)
Yvon J P, **3**: 1734 (1734); **7**: 5041 (5038)

Z

Zabel E, **6**: 3707
Zaborsky J, **6**: 3806 (3804, 3805)
Zaborszky J, **4**: 2892
Zacks L, **2**: 1545 (1544)
Zacks S, **6**: 3616 (3614)
Zadeh L A, **1**: 126 (124, 125), 294 (291), 372 (360), 575 (573), 664 (651), 665 (664), 713 (711, 713), 740 (736, 738), 746 (743, 744); **2**: 1286, 1348 (1346), 1407 (1405), 1473 (1467), 1479 (1477), 1486 (1480), 1549 (1547); **3**: (1601), 1604 (1601), 1781 (1780), 1786 (1783), 1788 (1786), 1797, 1800 (1798), 1802 (1800, 1801), 1805 (1802, 1803, 1804, 1805), 1809 (1808), 1819 (1816), 1821 (1819, 1820, 1821), 1822 (1821), 1823 (1822), 1825 (1824), 1829 (1827), 1831 (1830), 1835, 1838 (1836), 1843 (1840, 1841, 1842), 1847 (1846), 1852 (1847, 1850, 1851, 1852), 1858 (1854), 1862 (1858, 1859), 1867 (1864, 1865, 1866, 1867), 1872 (1867, 1868, 1871), 1874, 1878 (1877), 1880 (1878), 1883 (1880), 1888, 1894 (1888), 1895 (1888, 1889, 1893), 1898 (1895, 1896), 1905 (1903, 1904); **4**: 2273 (2270, 2272), 2301 (2299), 2339 (2338), 2343 (2343), 2362 (2357), 2502 (2499), 2884 (2881, 2882, 2883); **5**: 2921 (2918, 2921), 2924 (2921), 3156 (3154, 3156), 3165 (3159, 3160), 3215 (3208, 3209), 3413 (3410), 3454 (3451), 3455 (3449, 3450, 3451, 3453), 3541 (3541); **6**: 3619 (3616), 3645, 3653 (3648), 3710 (3707), 3768 (3766), 3772 (3769, 3771), 3857 (3856), 3947, 3990, 4007 (4007), 4008 (4000), 4146 (4145), 4355 (4354); **7**: 4661 (4657, 4658), 4669, 4696 (4693), 4789 (4787, 4788), 4996 (4993)
Zadeh N, **7**: 4726 (4726)
Zagoruyko N G, **6**: 3645
Zagzebski J A, **7**: 4964
Zahler R S, **1**: 434 (432)
Zajonc R B, **3**: 2203 (2201)
Žak I E, **5**: 3135 (3131)
Zakai M, **5**: 3303 (3293, 3294, 3299), 3310; **7**: 4637 (4635)
Zakian V, **1**: 736 (734); **5**: 3197 (3189), 3215 (3206, 3207, 3210, 3211, 3213, 3214), 3383 (3381, 3382)
Zaks R, **6**: 4294
Zaldúa J M, **2**: 1193 (1186, 1187, 1188)
Zaman M, **1**: 736 (734)
Zamaria J F, **5**: 2944 (2934)
Zames G, **4**: (2672); **5**: 3228 (3225, 3228), 3319 (3317, 3318), 3379 (3377), 3389 (3387); **7**: 4529 (4528)
Zamierowski E, **4**: 2580 (2576, 2579)
Zamir S, **3**: 1926 (1925, 1926)
Zander A, **2**: 1542 (1541)
Zangemeister W, **2**: 1590 (1587, 1590)
Zangger P, **3**: 1910 (1909)
Zangwill W I, **2**: 1125 (1124); **4**: 2601 (2598); **5**: 3196 (3195); **6**: 3634, 3707, 4310 (4308)
Zanker P M, **2**: 1466 (1463, 1464), 1467 (1462, 1463, 1465); **5**: 3232 (3229), 3410 (3407); **6**: 3727 (3721, 3722, 3723, 3724, 3725), 4176 (4175), 4193 (4190, 4193)
Zanon G, **1**: 120
Zapol W M, **1**: 315 (314)
Zappa G, **6**: 4172 (4172)
Zarrop M B, **4**: 2264 (2263); **6**: 3727 (3724)
Zauderer J, **5**: 3389 (3385)
Zavalloni M, **3**: 1917 (1914)
Zavitz J, **7**: 4749 (4745)
Zecca P, **5**: 3184 (3181, 3182)
Zeckhauser R, **2**: 1542 (1540); **3**: 2061 (2060); **5**: 3035 (3033)
Zeeman E C, **1**: 434 (430, 431, 432), 550 (549), 552
Zehra G, **2**: 1372 (1371)
Zeidler H M, **7**: 4930 (4928)
Zeidler J R, **4**: 2314 (2311)
Zeiger H P, **6**: 3983 (3978)
Zeigler B P, **1**: 120, 561 (559, 560, 561), 755 (754); **2**: 1201 (1200), 1326 (1324), 1552 (1550); **3**: 1713, 2003 (2000, 2001), 2168 (2164, 2165, 2166, 2167); **4**: 2353 (2351, 2352); **5**: 3035 (3030, 3031, 3032, 3033), 3082 (3080, 3082), 3088 (3083, 3085, 3086, 3088); **6**: 3768, 3989 (3984), 4229 (4228), 4306 (4303), 4313 (4312), 4319 (4315, 4316), 4322 (4321), 4323 (4319, 4320, 4321), 4326 (4326), 4328 (4327, 4328), 4333 (4329), 4334, 4345 (4341), 4348 (4347, 4348), 4353 (4350, 4351, 4352, 4353), 4355 (4355), 4372 (4371, 4372), 4381, 4388 (4386), 4392 (4391), 4402 (4396, 4398, 4399), 4403 (4402), 4411 (4406, 4407, 4409, 4410), 4414; **7**: 4692 (4691), 4764 (4762, 4763), 4832 (4830), 4833 (4830), 4958 (4957), 5005 (5003), 5008 (5007), 5011 (5008)
Zeigler J G, **6**: 3701 (3700)
Zeigner A, **1**: 751 (749)
Zekan S, **7**: 4856 (4853)
Zeki S M, **4**: 2331 (2325)
Zelenski R E, **1**: 639 (639)
Zeleny M, **1**: 333 (332), 399, 400; **2**: 776 (773), 869 (869); **3**: 1819, 1829 (1826); **5**: 3035 (3033), 3116 (3115), 3121 (3118, 3121), 3124 (3122), 3152, 3164 (3161), 3165 (3157, 3158, 3159, 3160, 3161, 3163), 3568; **6**: 4377 (4376, 4377); **7**: 4709 (4708), 4789 (4786)
Zeleznik A J, **3**: 2174 (2172)
Zeller H, **4**: 2473
Zemanian A H, **4**: 2835
Zemany P D, **7**: 4504
Zenner K, **4**: 2368 (2362)
Zeppegno M, **3**: 1975
Zermeño-González R, **6**: 4097
Zettlemoyer A C, **5**: 2944 (2935)
Zettler F, **7**: 5065 (5062)
Zhang Nan-lun, **6**: 3947 (3945)
Zhao-Ying Z, **6**: 4176 (4172, 4174)
Zheng Y P, **2**: 995 (993, 994); **7**: 4631
Zhou Z, **2**: 1351 (1350)
Zhu D L, **1**: 405 (400, 403, 404, 405), 406 (405)
Zhu G, **2**: 1372 (1368)
Zhurbenko N G, **5**: 3546 (3544)
Ziebolz H, **6**: 3870 (3866)

Ziegler D K, **2**: 1551 (1550)
Ziegler J G, **3**: 2113 (2111); **6**: 3884 (3883)
Zieleniewski J, **6**: (3862)
Zielinski W L Jr, **1**: 177 (176)
Zielke W, **7**: 5161
Ziemer R E, **6**: 4281 (4278)
Ziemer R L, **3**: 2113 (2112)
Zienkiewicz O C, **2**: 1179 (1174); **3**: 1632; **6**: 3626 (3625)
Zierler N, **5**: 2956 (2952)
Zilles S, **3**: 1642 (1638); **4**: 2681 (2680)
Zimmels Y, **6**: 4252 (4251)
Zimmer H, **3**: 1979
Zimmer R, **7**: 4856 (4856)
Zimmerman W, **1**: 226
Zimmermann H J, **3**: 1809 (1808), 1819 (1816, 1818), 1852 (1848, 1851), 1867 (1863), 1872; **5**: 3124 (3122), 3455 (3450, 3451, 3452, 3454); **6**: 4007 (4007)
Zimmermann U, **2**: 867 (865, 866)
Zimskind P D, **7**: 4980 (4976)
Zingg W, **3**: 2032 (2030), 2174 (2170)
Zingraff J, **1**: 310 (308)
Zinnermann H, **4**: 2368
Zinober A S I, **7**: 5031 (5028, 5029, 5030, 5031)
Zionts S, **5**: 3124 (3123), 3164 (3157), 3165 (3157); **7**: 4709 (4709)
Zirino A, **4**: 2608 (2607)
Zirnweld J P, **5**: 2944
Zissis O O, **7**: 4871 (4869)
Zlobec S, **5**: 3362
Zloof M M, **3**: 2179 (2178)
Zmud R W, **2**: 947; **7**: 4793 (4791, 4792, 4793)
Znotinas N M, **3**: 1829 (1826, 1827, 1829)
Zollenkopf K, **3**: 1954 (1949)
Zopf G W, **2**: 869 (868)
Zotin A I, **7**: 4698
Zoutendijk G, **2**: 789 (786); **4**: 2412 (2393, 2397, 2399), 2880 (2879)
Zowe J, **5**: 3546 (3545)
Zraket C A, **1**: 651 (650)
Zubaly R B, **3**: 1766 (1766)
Zubov V, **2**: 1212 (1211); **7**: 4529 (4525)
Zuch E L, **6**: 4289
Zucker S W, **4**: 2387 (2382), 2484 (2482), 2488; **6**: 4160 (4157), 4161 (4157)
Zuev V E, **2**: 1311 (1307); **6**: 4037 (4034, 4035)
Zühlke D, **4**: 2368 (2362), 2469 (2465, 2467, 2468)
Zuidervaart J C, **1**: 634 (632); **2**: 1201 (1201); **3**: 2217; **6**: 4323
Zulli E, **5**: 3421 (3418)
Zurmühl R, **4**: 2827 (2817)
Zweifach B W, **1**: 478 (477)
Zweifel P F, **7**: 4944
Zwemmer E, **6**: 3933 (3932)
Zwicker E, **2**: 1527 (1523)
Zwietering T N, **6**: 3744 (3742)
Zygmund A, **3**: 1723 (1720)
Zysno P, **3**: 1867 (1863); **5**: 3455 (3452)
Zywiets Chr, **1**: 769

SUBJECT INDEX

The Subject Index has been compiled to assist the reader in locating all references to a particular topic in the Encyclopedia. Entries may have up to three levels of heading. For each entry the volume number, which appears in **bold** type, is followed by the relevant page number(s). Where there is a substantive discussion of the topic, the page numbers appear in ***italic bold*** type. As a further aid to the reader, cross-references have also been given to terms of related interest. These can be found at the bottom of the entry for the first-level term to which they apply.

(*A*,*B*)-invariant subspace **2**: 896
Abiotic control mechanisms **1**: ***1–7***
Absorption (light)
  atmospheric laser spectroscopy **4**: 2720, 2721
Abstract realization theory **1**: ***14–17***
Abstraction
  formalism **3**: 1713
AC generators **2**: 1380
AC motors **5**: ***3089–92***
Acceleration
  road traffic variables **7**: 4931
Acceleration measurement **7**: ***4509–13***
  aircraft **1**: 382
  Coriolis force **3**: 1680
  drag-free satellites **2**: 1242
  inertial navigation **4**: 2488
  optical fiber interferometric sensors **5**: 3469
  piezoelectric devices **7**: 4509
Acceptance/rejection
  random variate generation **6**: 3949
Accessibility
  automata theory **5**: 3270
  extension techniques **2**: 1581
Accessibility models *see* Gravity models
Accuracy
  instruments **2**: 1558
Acetamide herbicides **3**: 2145
Acetic acid
  nitrilotriacetic acid
    heavy metals limnology **4**: 2778
Acid rain
  ecological effects
    aluminum levels **1**: 264
Acidification
  limnology **4**: 2780
Acids
  herbicides **3**: 2145
  inorganic
    weathering **1**: 1
  organic
    weathering **1**: 1
Acoustic noise
  aerospace propulsion **1**: 230
  automated guideway transit **1**: 334
  control valves **2**: 829
  launchers **4**: 2730
  measurement **2**: ***1403–5***
    day–night average sound level **2**: 1525
    environment **2**: ***1522–7***
    loudness level **2**: 1523
    noise pollution level **2**: 1525
    rating sound level **2**: 1525
    sound intensity level **2**: 1522
    sound power level **2**: 1523
    sound pressure level **2**: 1522
    weighted sound pressure level **2**: 1523
  optical fiber interferometric sensors **5**: 3466
Acoustic wave propagation
  underwater telemetry **7**: ***4842–8***
Acoustoelectric transducers
  optical fiber interferometric sensors **5**: 3468
Actions
  praxiology **6**: 3861
Activation analysis **1**: ***18–21***
  airborne materials **1**: 20
  biological samples **1**: 21
  rocks **1**: 20
  soils **1**: 20
  water samples **1**: 20
Active control technology
  aerospace control **1**: 119
    control-configured vehicles **2**: 801
    history **1**: 343
  aircraft **1**: ***21–31***, 603
  helicopters **1**: 31; **3**: 2134
Active filters **6**: 4283
  state-variable filters **6**: 4283
Active guideway systems **1**: 335
Activity **1**: ***38***
  activity trees **7**: 4950
  critical activity **2**: ***867***
Actuators
  aircraft control systems **1**: 225
  autopilots **1**: 385
  control location problem **6**: ***4238–44***
  dynamics **3**: 2098
  electric generators **2**: 1376
  helicopter control systems **3**: 2132

hysteresis **3**: 2098
industrial robots
modular systems **4**: 2461
load **3**: 2098
location problems **2**: ***801–7***
modelling
HVAC processes **3**: ***2097–9***
valves **2**: 836
ADA **6**: 3896
Adams methods
ordinary differential equations **5**: 3557
Adaptation
thermal sensation **7**: 4872
Adaptive control **1**: ***38–45, 45–7***; **5**: 3473; **6**: ***4161***
algorithm convergence analysis **1**: 53, ***54–60***
positive-real condition **1**: ***51–4***
anticipatory systems **1**: 293
artificial satellites **6**: ***4146–8***
astronomical telescopes **6**: ***3927–9***
blood glucose **3**: ***2017–20***
blood pressure **1**: ***498–502***
bounded disturbances **1**: 105
cement industry **1**: ***561–4***
concrete mixing
cautious control **2**: ***771–3***
coordination level **1**: ***64–72***
design **1**: 103, ***60–3***
direct **1**: 104
distillation columns **2**: ***1109–15***
error models **1**: 52, 55
explicit
non-minium-phase systems **5**: 3407
explicit self-tuners
convergence analysis **2**: ***1578–80***
feedback control **5**: 3044
fermentation control **3**: 1610
fuzzy automation **1**: ***72***
fuzzy control systems **1**: ***72–3***
fuzzy controller **1**: ***73***
glass industry **3**: ***2008–10***
heat systems **3**: ***2092–6***
hybrid **6**: ***4194–7***
hyperstability **1**: ***93–7***
implementation **1**: 62
implicit
non-minimum-phase systems **5**: 3408
industrial applications **1**: ***47–51***
industrial robots **4**: 2412
learning automata **4**: 2714
learning systems **4**: 2731
linear time-invariant systems **5**: 3065
load frequency regulation **4**: ***2889–92***
minimum-phase systems **5**: ***2998–300***
minimum variance control **5**: ***3004–6***
model-following control
variable-structure systems **7**: 5029
model reference adaptive control systems
5: ***3056–60***; **6**: ***4184–6***
modelling **1**: 56
modified two-step approach **1**: 636
multilevel control systems **5**: 3141
muscle relaxation **5**: 3237
non-minimum-phase systems **5**: ***3406–9***
nuclear power stations **5**: ***3425–30***
paper industry **6**: ***3590–1***
parameter-adaptive systems **6**: ***3604–10***, 4161
performance-adaptive systems **6**: 4162
pH control **6**: 3681, 3687
positive-real condition
filtering **1**: 54
positive-real lemma **1**: 53
positive systems **1**: 53
positivity **1**: ***93–7***
recursive adaptive control
learning systems **4**: 2735
reduced-order models **1**: 105
robot arms **6**: 4084, 4099
self-adaptive near-time-optimal control
variable-structure systems **7**: 5030
self-selecting control
hydroelectric power stations **3**: 2225
self-tuning control **1**: 48; **6**: ***4184–6***
cement industry **1**: 561
modified minimum variance strategy **1**: 562
multivariable minimum variance **1**: 561
ships **6**: 4264
ships
dynamic positioning **6**: ***4262–6***
guidance **6**: ***4275–6***
stochastic adaptive control
ODE techniques **1**: 53
stochastic systems
algorithm convergence analysis **7**: ***4602–6***
martingale techniques **1**: 53
stability **7**: ***4607–11***
telephone networks **7**: 4737
titanium dioxide ovens **7**: ***4892–4***
asymptotic stability **7**: 4893
self-tuning algorithms **7**: 4892
variable-structure systems **7**: ***5027–31***
see also Self-tuning control, Parameter-adaptive control, Self-organizing systems, Model reference adaptive control systems
Adaptive systems **1**: ***91–3***
design **1**: 62
Lyapunov synthesis **1**: ***97–100***
recursive identification **4**: 2310
signal processing
abrupt changes **1**: ***81–5***
algorithms **1**: ***73–5, 75–80, 86–91***
Admissible solutions *see* Pareto optimality
Adsorption
anions
nutrients and pollutants **1**: 5
cations
nutrients and pollutants **1**: 5
molecular
nutrients and pollutants **1**: 6

nutrients and pollutants
soils and sediments **1**: 6
Advanced light rapid transit **1**: 338
Aerodynamic derivatives **1**: 216
Aerodynamic forces **1**: 107, 216
Aerodynamic moments **1**: 107
aircraft **1**: 216
Aerodynamics
aerospace control **1**: ***385***
combat aircraft **1**: 619
Aerosols
chemical analysis **1**: ***174–6***
particle size **1**: ***177–9***
wind velocity measurement **2**: 1528
Aerospace
industrial robots **4**: ***2449–52***
systems approaches **1**: ***112–20***
Aerospace computing
aerospace simulation **1**: 110
Aerospace control **1**: ***106***, ***112–20***, 388–93
active control technology **1**: ***21***, 119
aerodynamics **1**: 385
airships **1**: 252–7
assisted pilot **1**: 256
analog control **1**: 106
angle of attack **1**: ***290***
artificial control load **1**: ***207***
attitude control **1**: ***319***
autopilot modes **1**: ***380***
autopilots **1**: 116
civil aircraft **1**: 603
control-configured vehicles **2**: ***801***
control surfaces **1**: 112
digital control **1**: 106
drag-free satellites **2**: ***1242–4***
flexible structures **3**: ***1632–3***
flight envelopes **1**: 115; **3**: ***1669***
flight management systems **3**: ***1669***
fly-by-wire systems **3**: ***1700***
fuel economy **3**: 1750
full-authority control **3**: ***1769***
helicopters **3**: ***2118–23***, ***2124–33***, 2134
engine governors **3**: 2137
history **1**: ***340–3***
homing **3**: 2164
missiles **5**: 3010
instrument flying rules **7**: 5060
instrument landing systems **1**: 117
ministicks **1**: 602
missiles **5**: ***3006–12***
perturbation effects **1**: 116
redundancy
history **1**: 343
solar power satellites **7**: ***4475***
specifications **4**: 2842
trajectory control **3**: ***2080–9***
visual flying rules **7**: 5060
wing load alleviation **7**: ***5164***, 5164–70
Aerospace instrumentation
helicopters **3**: 2123
inertial navigation **4**: ***2488–91***
Aerospace propulsion **7**: 4482
afterburner control **1**: 231
air bleeds **1**: 231
air intake control design **1**: 231
combat aircraft **1**: 620
Concorde
air intake **1**: 232
control
motivators **1**: 226
objectives **1**: 226
physical principles **1**: ***226–31***
weapons effects **1**: 229
control system design **1**: ***231–6***
engine operating limitations
control **1**: 228
helicopter engine governors **3**: ***2137–9***
low speed control **1**: 229
measurable states **1**: 227
modelling **7**: 4482
nozzle control **1**: 228
peripheral systems **1**: 231
power offtakes **1**: 231
reheat **1**: 237
reversed thrust **1**: 231
rotor speed control **1**: 227
secondary nozzle control **1**: 231
starting **1**: 231
takeoff control **1**: 229
types **1**: ***236–8***
variable geometry air intake **1**: 237
variable geometry compressor **1**: 230
variable geometry exhaust **1**: 237
variable primary nozzle **1**: 237
variable secondary nozzle **1**: 238
Aerospace simulation **1**: ***106–8***, ***109–12***, 114; **6**: ***4393***
air traffic control **1**: ***202–7***
principles **1**: 202
real-time data processing **1**: 204
software **1**: 203
aircraft engines **1**: 223
flight simulators **1**: 109–12
helicopters **3**: 2123
motion **1**: 110
real-time **1**: 106
scenery **1**: 110
Aerospace systems
flight-crew workload **3**: 1665
Affine systems **1**: 123
see also Bilinear systems
Aggregate planning
linguistics **1**: ***124–6***
modelling
fuzzy algorithms **1**: 125
Aggregate subsystems
chained aggregation **1**: 133

Aggregation **1**: ***126–32***
  complex systems control **1**: 688
  control theory
    fuzzy set theory **4**: 2357
  decentralized stabilization algorithms **2**: 918
  definition **1**: 127
  ecological modelling
    errors **2**: ***1324–6***
  error measurement **2**: 1326
  large-scale systems **4**: 2561
  mathematical structure **2**: 1324
  model simplification **7**: 4762
  model structure simplification **5**: 3079
  parameter measurement errors **2**: 1326
  principle **1**: 127
  properties **1**: 127
  structural constraints **1**: 129
  theory **2**: 1325
Agriculture
  chemical stress
    models **1**: ***589–92***
  crop models
    air pollution **1**: 590
  energy systems **2**: 1461
  herbicides **3**: ***2143–6***
  modelling
    field applicability **1**: ***137–42***
  radioisotopes
    transport models **6**: ***3933–6***
Air conditioning **1**: ***148–50, 167–70***
  air-floor systems **1**: ***150–2***, 170
    ducted systems **1**: 151
    storage systems **1**: 151
  air-and-water systems **1**: 168
    fan-coil systems **1**: 169
    induction-unit systems **1**: 169
  all-air systems **1**: 167
  all-water systems **1**: 169
    fan-coil systems **1**: 169
    reversible heat pump systems **1**: 169
  comfort modelling **1**: 638
  control process interactions **3**: 2107
  control systems **1**: ***152–9***
  degree days **2**: 959
  dual-duct systems **1**: ***159–61***
    control **1**: 159
  economizer control **2**: ***1342***
  embedded panel systems **1**: 170
  fan-coil plenum systems **1**: ***161–2***
  floor distribution systems **1**: ***150–2***
  heat pumps **1**: ***165–7***; **3**: 2092
  humidity control **3**: 2206
  induction-unit systems **1**: ***162–5***
    changeover systems **1**: ***162***
    four-pipe systems **1**: 165
    nonchangeover systems **1**: 162
    three-pipe systems **1**: 165
    two-pipe systems **1**: 164
  laminar flow **1**: 170
  principles **1**: 148
  reset control **6**: ***4045***
  single-zone systems **1**: 168
    dual-duct systems **1**: 168
    variable-air-volume systems **1**: 168
  variable-air-volume systems **1**: ***171–3***
    duct work system regulation **1**: 172
    instability **1**: 173
    minimum fresh air **1**: 173
    plant start up **1**: 173
    terminal units **1**: 173
  window **1**: 169
Air flow
  fermentation measurement **3**: 1613
  respiration
    optimum pattern **6**: 4063
Air foils design **2**: 1177
Air frame configuration management **1**: 25
Air–gas ratio control
  industrial gas equipment **4**: 2353, 2356
Air pollution
  aerosol particle size **1**: ***177–9***
  agricultural crops
    models **1**: 590
  automatic analysis **1**: ***181–3***; **2**: ***1508–12***
  chemiluminescence **1**: 183
  control
    wind tunnels **1**: ***188–9***
  distributed parameter systems **2**: 1497
  estimation theory **2**: 1499
    Kalman filters **2**: 1499
  gas chromatography **3**: 1932
  hydrocarbons
    automatic analyzers **1**: ***179–81***
  information monitoring **2**: 1489
  laser spectroscopy **4**: 2721
  modelling **2**: ***1497–500***
    statistical analysis **2**: 1494
  monitoring **1**: ***184–5***
  numerical analysis
    spline interpolation **7**: ***4517–8***
  photoacoustic spectroscopy **6**: 3699
  photochemical smog **1**: ***185–7***
  plant (biological) **2**: 1518
  radiological modelling **6**: 3930
  spline interpolation **7**: ***4517–8***
  statistical analysis **2**: ***1497–500***
    Gaussian plume model **2**: 1498
  sulfur dioxide **2**: 1499
  urban areas **1**: 184
Air traffic control **1**: ***113, 196–201***
  aerospace simulation **1**: ***202–7***
    real-time data processing **1**: 204
    software **1**: 203
  aircraft flight modelling **1**: 193
  aircraft separation **1**: 193
  aircraft sequencing
    landing **1**: 191
  airports **1**: ***189–96***

automatic landing systems **1**: ***213***
conflict resolution **1**: 193
data communication systems **2**: 879
dynamic scheduling **1**: 190
fuel **3**: 1750
landing sequence **3**: 1751
mode of operation **1**: 197
radar beacon system **1**: 199
typical flight through **1**: 198, 196
USA **1**: 196
zones of convergence **3**: 1749
Aircraft
active control technology **1**: ***21–31***
airbus A300 **1**: 603
airbus A300-600
aircraft instrumentation **1**: 241
airbus A300B
computers **1**: 381
airbus A310 **1**: 602
airbus A320
fly-by-wire systems **3**: 1700
aircraft instrumentation **1**: 241
analytic redundancy **1**: 247
Boeing 707 **1**: 601
inertial platform **1**: 381
tabs **1**: 385
Boeing 757
aircraft instrumentation **1**: 241
Boeing 767 **1**: 602
aircraft instrumentation **1**: 241
Caravelle **1**: 601, 602
autopilots **1**: 381
civil
long-term future **1**: ***601–4***
combat **1**: ***618–23***
air superiority **1**: 619
electronic systems **1**: 621
equipment **1**: 621
materials **1**: 620
multirole capability **1**: 619
stealth penetration **1**: 619
STOL **1**: 619
structures **1**: 620
sustained supersonic cruise fighter **1**: 619
Comet **1**: 602
Concorde **1**: 603
control-configured vehicles **2**: 801
fly-by-wire systems **3**: 1700
propulsion units **1**: 232, 233, 237
vibration control **1**: 386
configuration
propulsion control **1**: 229
costs **3**: 1750
Curtis seaplane **1**: 380
descent procedure
air traffic control **1**: 190
Douglas C54
autopilots **1**: 381
dynamics **1**: 107
aeroelastic model **7**: 5164–70
history of development **1**: 340
state equations **1**: ***213–20***
electric sensing devices **1**: 250
engine control systems **1**: ***220–6***
extreme attitude
propulsion control **1**: 229
F-18
aircraft instrumentation **1**: 244
flight-crew workload **3**: ***1665–8***
flight modelling
air traffic control **1**: 193
fuel **3**: ***1749–56***
inertial systems **4**: ***2488***
landing **3**: 2085
control–pilot communication **1**: 192
landing sequencing
air traffic control **1**: 191
Lockheed 1011
flight management system **1**: 381
Lockheed 1011–500 **1**: 604
Lockheed Vega 5B **1**: 380
Mirage 2000
aircraft instrumentation **1**: 244
Mirage 4000 **1**: 244
Mirage III
control-configured vehicles **2**: 801
modelling
analytic redundancy **1**: 247
motion
differential equations **1**: 215
modelling **1**: 213
NASA/Langley TC 127
control-configured vehicles **2**: 801
Nimrod
aircraft instrumentation **1**: 244
noise measurement **2**: 1525
propulsion units **1**: 231
relaxed static stability **1**: 24
Saab Viggen
aircraft instrumentation **1**: 244
safety **1**: 115
state equations
singular perturbations **1**: ***252***
Tornado
aircraft instrumentation **1**: 244
tracking precision
air traffic control **1**: 193
trajectory prediction
air traffic control **1**: 193
trajectory tracking
air traffic control **1**: 193
wing load alleviation **7**: 5164–70
see also Helicopters
Aircraft data communication systems **2**: ***879***
Aircraft instrumentation
artificial horizons **1**: 113
cathode ray tubes **1**: ***238–45***
civil aircraft

display systems **1**: 241
electromechanical devices **1**: 238
head-down displays **1**: 239, 244
head-up displays **1**: 238, 239, 244; **3**: ***2080–9***
history **1**: 341
instrument landing systems **1**: 117
military aircraft
display systems **1**: 244
multimode
combat aircraft **1**: 621
propulsion units
Concorde **1**: 236
Aircraft motion
pitch **1**: 382
roll **1**: 382
yaw **1**: 382
Airports
air traffic control **1**: ***189–96***
automatic landing systems
bad weather landing **1**: ***207–13***
control towers **1**: 197
instrument landing systems **1**: 190
Airships
aerostatic balance **1**: 253
aerostatic problems **1**: 253
Alcyon I **1**: 252
performance **1**: 253
architecture **1**: 252
attitude control **1**: 253
center of gravity
vertical variations **1**: 253
control and guidance **1**: ***252–7***
servomechanisms **1**: 256
deformability **1**: 254
flight mechanics **1**: 255
gas motion **1**: 253
piloting laws **1**: 256
Airworthiness
automatic landing systems
bad weather landing **1**: 210
propulsion units **1**: 235
Albedo
ground albedo
atmosphere–ground systems **2**: 1307
Landsat imaging **4**: 2652
mapping
invariant embedding **3**: 2046
single scattering
atmosphere–ground systems **2**: 1307
terrestrial atmosphere **3**: 2046
Aldehyde photolysis
air pollution **1**: 185
Algae
biomass determination **7**: 5112
cultivation **7**: 5112
growth modelling
Cladophora **1**: ***257–61***
model components **1**: 258
growth potential
determination **7**: ***5112–3***
Microcystis aeruginosa **7**: 5112
tests **7**: 5112
Algebra
fuzzy set theory **3**: ***1780–1,*** 1883
linear quadratic regulation **4**: 2848
Algebraic many-valued logics **5**: ***2921–3***
Algebraic topology
homotopy **3**: ***2168–9***
Algorithms **1**: ***262;*** **2**: 1082
adaptive algorithms **4**: 2310
large deviations **1**: ***86–91***
rare events **1**: ***86–91***
scalar **1**: 89
signal processing **1**: ***73–5,*** 75–80
approximation algorithms
computational complexity **1**: 725
approximation theory
combinatorial optimization **1**: ***626–8***
worst-case analysis **1**: 626
branch-and-bound methods
maintenance scheduling **6**: 3836
complexity **1**: 705, 707
computability and decidability **2**: 934
convergence analysis
explicit self-tuners **2**: 1578
criterial algorithms **5**: 3553
digital control design **2**: ***1057–66***
distributed algorithms
partition **3**: 2154
duality (mathematics) **2**: 1260
efficiency
analysis and design **1**: 720–3
fast algorithms
constant parameter systems **2**: ***781–5***
fast square-root algorithms
normalized Chandrasekhar equations **2**: 783
fuzzy set theory **5**: 3452
independence
combinatorial optimization **1**: 625
iterative algorithms
convergence **2**: 1121
minimization
optimal control **5**: 3486
nonlinear systems
self-tuning regulators **5**: 3344
optimal algorithms **5**: 3551
optimization **5**: 3551
functional equations **5**: ***3546–7***
simulation problems **6**: 4307
output feedback
pole assignment **6**: 3713
parameter estimation **5**: 3343
instrumental variable techniques **4**: 2276
path-following methods **6**: 3632
power systems computer control **6**: 3840
probabilistic algorithms
computational complexity **1**: 725
recursive identification methods **4**: 2281

routing algorithms **2**: 1123
self-tuning regulators
nonlinear systems **5**: ***3341–51***
semi-infinite optimization
multivariable control systems design **5**: 3193
smoothing algorithms **7**: ***4463–6***
state feedback
pole assignment **6**: 3711
stochastic approximation **5**: 3551
Aliasing
digital signals **6**: 4286
identification **4**: 2269
pseudo-random signals **4**: 2308
Alkaline earth metals
ecology **1**: 267
α,α′-bi-toluidine, *N,N′*-bis(*p*-phenylbenzylidine)
aerosols
chemical analysis **1**: 176
α-cut
fuzzy set theory **3**: ***1872***; **4**: 2299
Alternatives **1**: ***262***
Altimeters *see* Radioaltimeters
Altitude control
airships **1**: 253
Altitude measurement *see* Height measurement
Aluminum
ecological effects **1**: 264
environment
chemistry **1**: 262
effects on nutrients and toxic substances **1**: ***262–8***
soil **1**: 143
toxic effects **1**: 264
toxic substances **1**: 144
Aluminum compounds
adsorption
weathering **1**: 2
Aluminum production
electrolysis **2**: 1413
Amino acids
protein synthesis control **1**: 557
water concentration **7**: 5122
Ammonia
chemical analysis
water pollution **7**: 5117
ion-selective electrodes
water pollution **4**: 2606
Ammonium compounds
ecology **1**: 266
Amplification
artificial intelligence **4**: 2542
voltage amplification **1**: 269
see also Gain
Amplifiers
bandwidth **1**: 269
chopper stabilization **1**: 272
common mode rejection ratio **1**: 269
commutating autozero amplifiers **1**: 272
instrumentation **1**: ***268–72***
operational amplifiers **1**: 270
spectroscopy
biology **1**: 9
telephony
frequency response techniques **3**: 1745
voltage amplifier specifications **1**: 268
Amplitude modulation
signal transmission **6**: 4289
Anabaena flos-aquae
algal growth potential **7**: 5112
Analgesia
demand analgesia
postoperative pain **2**: ***960–3***
Analog computers **3**: 2214
similarity theory **6**: 4295
Analog control
ship propulsion **6**: 4269
Analog–digital conversion **2**: 1066; **6**: 4287
digital communication systems **2**: 1045
Analog signal processing *see* Signal processing
Analogical inference **1**: ***272–6***
Analogy
similarity theory **6**: 4295
Analysis of correspondences **2**: 877
hot blast cupola **2**: 877
rotary cement kiln **2**: 878
Anemometers **5**: 2972
flow measurement **3**: 1677
laser Doppler anemometers **2**: 1529
optical fiber sensors **5**: 3459
Anesthesia
depth **1**: 285
feedforward control **1**: ***281–4***
educational computer models **2**: 1364
feedback control **1**: ***277–81***
inhalational agent anesthesia
identification and control **1**: ***284–9***
modelling **1**: 281
uptake characteristics **1**: 284
Angle of attack **1**: ***290***; **3**: 2081
Angular velocity measurement **7**: 4511
Aniline herbicides **3**: 2146
Animal behavior
grammars **3**: 2045
Anisotropy
biological materials **1**: 483
Antennas
artificial satellites
digital communication systems **2**: 1052
directional
air traffic control **1**: 199
Anthracene
aerosol chemical analysis **1**: 176
Anticipating control
thermostats **1**: ***290***
Anticipatory systems **1**: ***290–4***
degrees of freedom **1**: 291
Anticollision systems
civil aircraft **1**: 604

Antirolling tanks
  ships **6**: 4272
Antisymmetric relations **1**: ***294***
Approach controller
  air traffic control **1**: 198
Approximate programming
  direct approximate programming
    industrial robots **4**: 2399
  industrial robots **4**: 2397
Approximate reasoning **1**: ***296***; **3**: 1872
  patient diagnosis **2**: 1574
Approximation analysis
  penalty methods **2**: 797
Approximation theory
  algorithms
    combinatorial optimization **1**: ***626–8***
  costate coordination method **2**: 859
  decentralized successive approximation
    hierarchical control **3**: 2153
    optimal control **3**: 2157
  discrete-time control problems **2**: 1097
  discrete two-timescale systems **7**: 4434
  Hamilton–Jacobi–Bellman equation **3**: 2069
  impulse responses **1**: 294
  quality measurement **1**: 127
  realization theory
    optimal methods **1**: ***294–6***
Arc lamps
  control systems history **2**: 1399
ARIMA models
  system identification **4**: 2304
Arithmetic relators
  simulation modelling formalism **6**: ***4334–6***
ARMA processes
  spectral estimation
    stationary time series **7**: 4565
  stationary time series **7**: ***4559***
Armaments *see* Weapons
ARMAX models
  dynamic systems modelling **4**: 2315
  self-tuning control **6**: 4173
  stochastic systems
    state-space methods **7**: 4547
  system identification **4**: 2305
Aromatic compounds
  distillation control **6**: 3676
Arrow–Hurwicz algorithm **1**: 403
Arsenic
  aerosol chemical analysis **1**: 174
  ecology **1**: 267
  soil **1**: 143
  toxic substances **1**: 144
Arsonate herbicides **3**: 2145
Artificial control load
  aerospace control **1**: ***207***
Artificial intelligence **1**: ***304–7***; **2**: 870
  applications **1**: 307
  cybernetics **2**: 868
  dialectical inquiry system **7**: 4780
  expert systems **2**: ***1569–73***
    medical diagnostic computing **2**: ***1573–7***
    medical reasoning **2**: 1365, 1573
    taxonomy **2**: 1569
  history **7**: 4773
  inference engines **4**: ***2492***
  intelligence amplification **4**: ***2542–3***
  intelligent control systems **6**: 4161
  meanings **1**: 304
  modelling **1**: 304
  self-organizing systems **6**: 4165
  systems analysis **7**: 4767
  technology **1**: 304
  theorists **1**: 306
Artificial organs
  kidneys **1**: ***307–9***
  lungs **1**: ***312–5***
  pancreas
    blood glucose control **3**: 2017
  portable kidneys **1**: ***310–2***
  respiratory systems **1**: ***315–9, 315–9***
    high-frequency positive-pressure ventilation
        1: 315
    ventilation **1**: 315, 316, 317, 318
Artificial satellites
  adaptive control **6**: ***4146–8***
  drag-free satellites **2**: ***1242–4***
    control laws **2**: 1243
    error analysis **2**: 1244
  geostationary satellites **6**: 4037
  geosynchronous orbit **2**: 1051
  heavy
    launch vehicles **4**: 2724
  image sensors **6**: 4038
  inertial navigation **4**: 2489
  modelling **6**: 4147
  orbits **6**: 4037
  remote sensing **6**: ***4037–40***
  solar power satellites **7**: ***4472–6***
  telemetry **7**: 4837
  see also Satellite relay systems, Satellite links
ASCII *see* Code standards
ASHRAE scale
  thermal sensation **7**: 4871
Assembling
  industrial robots **4**: 2379, 2419
    aerospace industry **4**: 2451
    programming **4**: ***2347–51***
  industrial vision systems **4**: 2469
Assessment
  multiple attribute utility theory **5**: 3112
Astronomical telescopes
  adaptive control **6**: ***3927–9***
Asymptotic analysis **4**: 2235
  nonlinear estimation theory **5**: ***3303–10***
Asymptotic stability
  Lyapunov methods **5**: 3399
Asymptotic tracking
  linear systems **4**: 2569

Asynchronous fixed-point algorithms **2**: 1123
  convergence **2**: 1121
Asynchronous generators
  power systems **6**: 3795
Asynchronous machines
  variable speed induction machines
    numerical control **7**: ***5019–27***
Atmosphere
  landsat imaging **4**: ***2652–5***
    atmospheric effects correction system **4**: 2653
  modelling **1**: 187
  optical thickness **2**: ***1307–11***; **4**: 2653
  particle size **1**: 179
  remote sensing
    optical radar **6**: ***4033–6***
  see also Terrestrial atmosphere
Atmosphere–ground systems
  optical parameters **2**: 1307
  radiation **4**: 2652
Atmosphere–water systems
  optical parameters **2**: 1310
Atmospheric analysis
  laser spectroscopy **4**: ***2719–23***
Atmospheric humidity
  measurement **5**: 2973
  physiological temperature control **7**: 4867
  thermal comfort **1**: 639, 644; **7**: 4866
Atmospheric motion
  physiological temperature control **7**: 4867
Atmospheric optics
  laser Doppler anemometry **2**: 1528
Atmospheric pollution *see* Air pollution
Atmospheric precipitation
  hydrologic forecasting **3**: 2228
  nutrients and pollutants
    environment **1**: 3
  rainfall-runoff modelling **6**: ***3940–4***
  trace elements
    water pollution **7**: 5125
Atmospheric temperature
  measurement **5**: 2972
  physiological temperature control **7**: 4867
  thermal comfort **7**: 4866
Atmospheric turbulence
  aerospace control **1**: 116, 247
  airships **1**: 255
  civil aircraft **1**: 604
  modelling **7**: 5166
Atomic absorption spectroscopy
  aerosols **1**: 174
  biology **1**: 11
Atomic actions
  distributed database systems **2**: 1170
Atomic emission spectroscopy
  inductively coupled plasma method **4**: ***2343–7***
Atrioventricular node **1**: 451
Attenuators **2**: 1392
Attitude control **1**: ***319***; **3**: 2085
  artificial satellites **6**: 4146
  space vehicles **1**: ***319–23***
    Kalman filters **4**: 2622
Attribute trees **7**: 4952
Attributes **1**: ***323***; **5**: 3117
  attribute measures **1**: ***323***
Auctions
  noncooperative games **5**: 2909
Auditory perception *see* Hearing
Auger effect
  electron probes **2**: 1420
Augmented error
  model reference adaptive control systems
      5: 3059, ***3061–5***
Autocorrelated processes
  random variate generation **6**: 3949
Autocorrelation functions **4**: 2246
  analytical computation **4**: 2246
  signal analysis **6**: 4279
Autology **1**: ***326–32***
  complexity **1**: 706
Automata theory **1**: ***333–4***, 559; **6**: 4375
  cellular automata **6**: ***4341–5***
  finite automata **1**: 333; **5**: ***3268–72***
  formal languages **5**: 3270
  internal model principle **4**: 2566
  learning systems **6**: 4166
  Lindenmayer systems **4**: 2795
  memoryless automata **1**: 333
  nerode equivalence construction **1**: 17
  operative automata **6**: 4166
  systems homomorphism **3**: 2165
  telecommunication networks **7**: 4717
  Turing machines **1**: 334
Automated guideway transit **1**: ***334–40***
  breaking **1**: 336
  command and control **1**: 336
  cost **1**: 335
  electrical auxiliaries **1**: 336
  guidance **1**: 335
  guideway **1**: 335
  motorization **1**: 335
  switching **1**: 336
  technical characteristics **1**: 335
Automatic analyzers
  air pollution **1**: ***181–3***
    hydrocarbons **1**: ***179–81***
Automatic generation control
  electric generators **6**: 3813
  power system computer control **6**: 3827
  power systems **6**: ***3809–15***, 3815, 3845
    dynamic response **6**: 3813
Automatic landing systems **1**: ***213***
  all-weather landing categories **1**: 207
  bad weather landing **1**: ***207–13***
  components **1**: 213
  design **1**: 207
  monitoring **1**: 210
  regulations **1**: 207

Automatic programming
  simulation program generators **4**: 2547
Automatic vehicle operation **1**: 336
Automatic vehicle protection **1**: 336
Automatic vehicle supervision **1**: 336
Automation
  anesthesia **1**: 281
  buildings
    air conditioning **1**: 157
  employment **1**: 369
  manufacturing industries
    washing machines **3**: ***1597–600***
  manufacturing processes **3**: ***1633–6***
  personnel aspects **6**: ***3659–63***
  ship control **6**: ***4257–61***
  social effects **1**: 367
Automobile electronics **1**: ***372–80***
  engine and transmission electronics center **1**: 376
Autonomous control
  biofeedback **1**: 436
Autopilots **1**: 116, ***380–8***
  automatic landing systems **1**: 213
  design **1**: 381
  feedback **1**: ***388–93***
  flight management systems **3**: 1669
  heading control loop **1**: 392
  history **1**: 341, 380
  lateral mode **1**: 391
  longitudinal modes **1**: 392
  modes **1**: ***380***
  pilots and **1**: 385
  pole assignment **1**: 391
  radio beam control **1**: 392
  trajectory control **3**: 2088
Autopoiesis **1**: ***393–9***; **2**: 869
  autology **1**: 332
  modelling
    computers **1**: 396
    simulation **6**: ***4374–7***
Autoregressive moving-average proceses *see* ARMA processes
Autoregressive processes
  spectral estimation **7**: 4562
  stationary time series **7**: ***4559***
Autothrottles
  feedback control **1**: 393
Auxiliary problem principle
  decomposition
    coupling through cost only **1**: 400
  duality
    coupling constraints **1**: 402
  optimization **1**: ***400–5***
Availability
  reliability of systems **6**: 4024
  systems evaluation **6**: 4018
Average energetic consumption
  brains as fuzzy systems **1**: 505
Averaging
  nonlinear estimation theory **5**: 3308
Axiology **1**: ***406***
Azinphosmethyl
  agricultural models **1**: 590

B

Backtracking
  computational complexity **1**: 725
Bang–bang control **1**: ***410–3***
  history **5**: 3511
  Lie brackets **4**: 2756
  linear systems
    singular solutions **1**: 411
    time-varying case **1**: 411
  multivariable systems **5**: 3204
  polymerization process control **6**: 3741
  real analyticity **6**: 3955
  see also On–off control
Bargaining sets
  cooperative game theory **2**: 845
Baroreflexes modelling **1**: ***536–42***
BASIC **6**: 3894
Basin system **1**: 507
Batch processing
  inorganic chemicals industry
    control **4**: 2512
Batch processing (industrial)
  aerospace industry
    industrial robots **4**: 2450
  glass industry
    control **3**: 2012
  industrial robots **4**: 2465
Bayes methods **1**: ***414***
  entropy minimax theory **2**: 1474
  fuzzified **3**: 1603
  fuzzy causes **3**: 1788
  identification **4**: 2285
  impact analysis **4**: ***2331***
  pattern recognition **6**: 3640
  pH control **6**: 3690
  probabilistic inference **4**: ***2336***
  recursive identification **4**: 2310
  self-organizing control **6**: 4163
  two-filter smoothing algorithms **7**: 4467
  water resource management **7**: 5139
Beacon collision avoidance system **1**: 201
Bedford scale
  thermal sensation **7**: 4871
Beer–Lambert law **7**: 4502
Behavior systems **1**: ***414***; **2**: 1547
Benes problem
  invariance **5**: 3324
Benzo [*a*] pyrene
  aerosol chemical analysis **1**: 176
Benzo [*e*] pyrene
  aerosol chemical analysis **1**: 176
Bias
  modelling
    Kalman filters **4**: 2619
  parameter estimation **6**: 3613
Bias current
  amplifiers **1**: 270
Bifurcation
  Hopf bifurcation
    describing functions **2**: 981
  multistate systems **5**: 3490
  see also Catastrope theory
Bilinear equations
  ecology **1**: ***414–7***
  solution **1**: 416
Bilinear models
  biological
    structural identifiability **1**: 492
  distillation columns **2**: ***1116–21***
  reduced-order models
    distillation columns **2**: 1118
Bilinear systems
  approximation **5**: 3374
  realization theory **5**: ***3276–9***; **6**: 3973
  waiting function
    binary pseudo-random signals **4**: 2251
  see also Affine systems
Binary pseudo-random signals
  analytical computation **4**: 2247
  linear systems
    waiting function **4**: 2249
Binary random signals **1**: ***418–21***
  analytical computation **4**: 2247
Biochemical oxygen demand **7**: ***5113***
  measurement **7**: 5113
  reaction characteristics **4**: 2783
Biochemistry
  distributed parameter systems **2**: 1183
  educational computer models **2**: 1362
  electrical activity modelling
    gastrointestinal tract **5**: 3019
  multistate systems **5**: 3489
  oscillators
    dynamic behavior **1**: ***424–8***
    modelling **1**: 427
    regulation **1**: 428
  simulation **1**: ***421–4***
  switching behavior **1**: ***429–33***
Biocontrol
  lipoproteins **4**: ***2884–7***
Biodegradability
  determination **1**: ***434***
Bioelectric phenomena
  digestive tract
    bioelectric potential **1**: ***459–62***
  large intestine **2**: ***1036–40***
  small intestine **2**: ***1040–2***
  stomach
    modelling **2**: ***1033–6***
  ureter

bioelectric potential **7**: 4976
modelling **7**: ***4976–80***
visual system
feedback lateral inhibition **7**: 5060
Bioenergetics
toxic substances modelling **3**: 1600
Biofeedback **1**: 435–7
dentistry **2**: 967
Biogeochemical cycling
ecosystems **5**: 3440
Biogeocoenosis **2**: 1356
Biological materials
stereology **7**: ***4598–601***
Biological measurement
allometric laws
particle size **7**: 4452
body size **7**: ***4452–4***
electrocardiography **1**: ***450–6***
gas chromatography–mass spectrometry **3**: 1938
Biological oscillators
periodically forced hysteresis transitions **1**: 432
Biological specimen preparation
scanning electron microscopy **6**: 4150
transmission electron microscopy **7**: 4938
Biological systems
hysteresis **1**: 431
Biology
causality **1**: 553
cellular automata **6**: 4343
holism **3**: 2163
models
cusp catastrophe **1**: 432
physiological models
optimal control **6**: 4062
simulation **1**: ***466–7***
spectroscopy **1**: ***8–13***
subjectivity **7**: 4697
Biomass
Cladophora growth modelling **1**: 258
power generation **2**: 1457
Biomechanics **1**: ***482–6***
biped locomotion
mode decomposition **6**: 4126
connective tissue **1**: ***474–7***
gait analysis **3**: ***1907–10, 1910–3***
skin **1**: ***478–82***
Biomedical engineering
intracranial pressure regulation
feedback **4**: ***2580–2***
Biomedical NMR **1**: 13; **5**: ***3421–5***
imaging
echo-planar **5**: 3424
Fourier zeugmatography **5**: 3423
Biomedical phenomena
modelling **1**: ***486–9***
simulation
computer graphics **6**: 4302
Biomedicine
identification errors **4**: ***2253–6***
Biosphere
carbon cycle modelling **1**: 531
Biothermics
physiological control **7**: ***4877–80***
Biotransport
lipids **4**: ***2884–7***
Bioturbation
limnologically toxic substances **4**: 2781
Birefringence
optical fiber sensors **5**: 3460
Birth control
optimization **6**: 3754
Black-box models
radioecological modelling **6**: 3932
radioisotopes **6**: 3934
Blending
cement industry adaptive control **1**: 561
polymerization process control **6**: 3744
Block-diagonalization *see* Block-triangularization
Block Gerschgorin theorem **2**: 924
decentralized stabilization **2**: ***923–7***
Block matrix formulae *see* Matrix algebra
Block-triangularization **5**: 3171
Blood
cell packing **1**: 444
constitutive relations **1**: 443
extracorporeal
purification **1**: 310
gases
continuous *in situ* measurement **1**: 439
in vitro measurement **1**: 439
measurement **1**: ***437–40***
glucose
adaptive control **3**: ***2017–20***
hunger **3**: 2210
mechanical properties **1**: ***440–5***
plasma **1**: 442
red cells
deformation **1**: 444
viscosity **1**: 442
determination **1**: 441
Blood flow *see* Hemodynamics
Blood pressure
coronary **2**: 851
feedback control
hypertension **1**: ***495–8***
regulation **1**: 536
surgery
adaptive control **1**: ***498–502***
Blood volume
drinking control **3**: 2212
Bode diagrams **4**: ***2664–5***
control system analysis computing **4**: 2798
pseudo-asymptotic Bode diagrams **4**: 2665
Body temperature measurement **7**: 4852
Boilers
efficiency **2**: ***1226–7***
flame safeguard control **3**: ***1632***
nuclear power station boilers

multivariable control systems **5**: ***3431–6***
process control **7**: ***4572–7***, ***4577–80***
space heating **2**: ***1227–33***
steam thermal units **7**: ***4580–5***
structure **7**: 4572
see also Steam plants
Bolometers
detectors in spectroscopy **1**: 9
Boltzmann equations **4**: 2536, 2537
linear Boltzmann equations **7**: ***4940–4***
control **7**: 4944
Bond graphs
causality **6**: 4340
ecological modelling **2**: ***1326–9***
elements **6**: 4338
junction elements **6**: 4339
simulation modelling formalism **6**: ***4337–41***
sources **6**: 4339
variables **6**: 4338
Bondavera–Shapley theorem
cooperative game theory **2**: 844
Bone mechanical properties **1**: ***468–73***
Boreholes
water supply simulation **7**: 5107
Boron
ecology **1**: 267
Bottom-up methods
economic modelling **2**: 1340
Boundary-layer problems
singular perturbations **7**: ***4425–9***
Boundary-layer systems
singular perturbations **7**: 4435
Boundary-value problems
distributed parameter systems **2**: 1182
Legendre polynomials
spectral analysis **7**: 4497
multigrid methods **5**: 3136
numerical methods
modelling **5**: ***3097–9***
optimal stopping **5**: 3534
singular perturbations **7**: 4430, 4435
variational inequalities
spectral analysis **7**: 4497
Bounded external noise
model reference adaptive control systems **5**: 3059
Bounded rationality
heuristic programming **1**: ***502***
Boundedness
adaptive control
deterministic case **1**: 57
stochastic case **1**: 59
convergence analysis
adaptive control **1**: 59
disturbances
adaptive control **1**: 105
parameter-adaptive control **6**: 3602
Bourdon tube
process control history **6**: 3882
Brain
activity during language perception **1**: ***502–4***
fuzzy learning strategies **5**: ***3261–4***
fuzzy systems **1**: ***505–8***
Brainstorming **1**: ***612***; **7**: 4802
Brainwriting **1**: ***610***; **4**: 2576; **7**: 4802
Branch-and-bound
branching strategies **1**: 629
combinatorial mathematics
optimization **1**: ***628–30***
computational complexity **1**: 725
fathoming criteria **1**: 629
relaxation **1**: 629
separation **1**: 628
Branch-and-bound methods
air traffic control
landing sequencing **1**: 192
maintenance scheduling
power systems **6**: 3836
Bridge circuits **2**: ***1402–3***
dc transmission networks **6**: 3821
Brittle lacquer
dentistry **2**: 968
Brownian motion **1**: ***511–2***
existence **1**: 511
Markov property **1**: 512
martingale properties **1**: 512
nonlinear estimation **5**: 3298
pathwise continuity **1**: 511
realization theory **6**: 3964
stochastic calculus **7**: 4631
stochastic integrals
iterated integrals **5**: 3330
see also Wiener processes
Buildings
automation
air conditioning **1**: 157
HVAC systems
analog simulation **3**: 2099
management systems **1**: ***512–4***
elevators **2**: 1424
space heating **2**: 1217
structural damage assessment
fuzzy set theory **7**: ***4667–9***
Bulk materials
level measurement **4**: ***2743–7***, ***2747–54***
Burners
industrial gas equipment **4**: 2353
Busbars
infinite busbars
power systems **6**: 3796
Butterfly catastrophe **1**: 551
Butterfly valves **2**: 834

C

CAD
  aerospace industry **4**: 2450
  computer-aided process planning
    aerospace industry **4**: 2451
  control systems **2**: 825
    design environments **1**: ***726–32***
    design languages **1**: 727
    optimization languages **1**: 729
    user interfaces **1**: 730
  elevator systems **2**: ***1427–31***
  facilities
    control system studies **1**: 733
  interactive terminals
    human factors **1**: 730
  multivariable systems **1**: ***732–5***
  nonlinear systems **5**: ***3379–83***
  simulation **3**: 2216
Cadmium
  soil **1**: 144
  toxic substances **1**: 144
Calcium
  ion-selective electrodes
    water pollution **4**: 2607
Calculus
  first-order predicate
    artificial intelligence **1**: 306
Calculus of variations **1**: ***519–29***; **7**: ***5043–5***
  duality methods **1**: 524
  geodesics **7**: 5045
  multivalued differential equations **5**: 3183
  optimal control **5**: 3512, 3513
Calibration
  mass spectrometers **5**: 2942
  thermometers **7**: 4851
CAMAC **6**: 4292
Cameras
  microprocessor chips **5**: 2986
CAMS *see* Mass spectrometry
Canonical analysis **2**: 879
Capacitance
  angular velocity measurement applications **7**: 4512
  level measurement applications **4**: 2740
  power systems **6**: 3809
Capnometry
  blood carbon dioxide **1**: 440
CAPS (simulation language)
  simulation
    activity-based models **6**: 4392
Capturability
  linear systems **2**: 1007
Carbon
  carbon cycle modelling **1**: 529–31; **2**: 1322
    validation **1**: 531
  total organic carbon
    water pollution indices **7**: ***5115–6***
Carbon dioxide
  blood measurement **1**: 437
  ocean modelling **1**: 529
  respiration
    cats **6**: 4053
    regulation **6**: 4046
  terrestrial atmosphere
    modelling **1**: 529
Carbon monoxide
  air pollution
    automatic analyzers **1**: 182
Carboxylic acids
  aerosol chemical analysis **1**: 176
Cardiac action potential **1**: 451
Cardiac simulation
  exercise **1**: 541
Cardinals **1**: ***536***
Cardiology
  baroreflexes modelling **1**: ***536–42***
  cardiac arrhythmias **1**: 455
  cardiac muscle contraction **1**: 452
  cardiopulmonary bypass
    artificial lungs **1**: 312
  coronary hemodynamics
    modelling **2**: ***850–3***
    regulation **2**: 852
  educational computer models **2**: 1363
  electrocardiography **1**: 760–1
  hemodynamics modelling **1**: ***542–6***
  His bundle electrocardiography **2**: 1407–10
  stroke volume control **1**: 540
  thermoregulation **7**: 4879
Cardiophysiology
  dysrhythmias
    random processes **1**: 569
Cardiovascular state
  fuzzy set theory **2**: 1406
Carnot cycle **7**: 4580
Cassini ovals **5**: 3177
Casting
  optimal control **3**: 1729
  pressure die
    industrial robots **4**: 2419
Catalysis
  environmental nutrients and pollutants **1**: 7
Catalytic cracking
  process control **6**: 3678
Catastrophe theory **1**: ***547–50***
  butterfly catastrophe **1**: 551
  cusp catastrophe **1**: 550
  ecosystem analysis **1**: ***550–2***
  fold catastrophe **1**: 550
  fuzzy set theory **3**: ***1786–8***
  structural stability **7**: 4671

Cathode-ray tubes
  aircraft instrumentation **1**: ***238–45***
    helicopters **3**: 2123
  beam index tubes **1**: 240
  cockpit displays **1**: 602
  penetron tubes **1**: 240
  phosphor screens **1**: 239
  raster scanning **1**: 241
  shadow mask tubes **1**: 240
  stroke scanning **1**: 241
  technology **1**: 239–41
Cathodoluminescence
  electron probes **2**: 1417
Cationic herbicides **3**: 2144
Cats
  electromyography
    neuromuscular control **6**: 4052
Causal loop diagram
  system dynamics modelling **7**: 4757
Causal operators **4**: 2677
Causality **1**: ***552–6;*** **6**: 4317
  anticipatory systems **1**: 290
  bond graphs **6**: 4340
  fuzzy causes **3**: 1788–91
  fuzzy relations **5**: 2925
Cautious control
  concrete mixing **2**: ***771–3***
Cavitation
  flow **3**: 1697
  valves **2**: 828
CCV *see* Control configured vehicles
Cells (biological)
  Coulter counters
    cytometers **2**: 866
  deformation
    Coulter counters **2**: 865
  growth
    fermentation **3**: 1609
  modelling
    tumor treatment **7**: 4957
  stereology **7**: 4599
  surface area measurement
    stereology **7**: 4599
  volume measurement
    stereology **7**: 4599
Cells (electric)
  standard cells **2**: 1390
Cellular automata *see* Finite automata
Cellular connectivity
  brains as fuzzy systems **1**: 506
Cellular spaces
  autopoiesis **1**: 393
  cellular automata **6**: 4343
  simulation modelling **1**: ***559–61***
Cement industry
  adaptive control **1**: ***561–4***
  process control **5**: ***2989–93***
Center controller
  air traffic control **1**: 198
Center of gravity
  vertical variations
    airships **1**: 253
Central nervous system
  motor systems **1**: 462
Centrality **2**: 927
  ecological modelling **2**: 1330
Centralization
  incentive systems **1**: 33
Centrifugation
  trace elements
    water pollution **7**: 5123
Cerebral motor cortex
  motor systems **1**: 463
Cerebrum
  muscle movement function **1**: 463
Certainty equivalence **1**: ***564–5***
  control **2**: 1256
  decision theory and analysis **3**: 2197
  multiattribute utility theory **5**: 3100
  self-tuning control **6**: 4190
Chained aggregation **1**: ***133–7***
  geometric interpretation **1**: 134
  geometry **1**: 134
Chandrasekhar equations
  constant parameter systems **2**: 781
  distributed parameter systems **2**: 1191
  normalized
    fast square-root algorithms **2**: 783
Channel capacity
  information theory **4**: 2505
Chaos
  ecological modelling **2**: 1331
  observability **5**: 3341
  see also Random processes
Character recognition **1**: ***570–4***
  contextual analysis **1**: 573
  geometrical and topological feature extraction **1**: 572
  global feature extraction **1**: 572
  hand-printed **1**: 574
  linguistic description **1**: 573
  pattern recognition **1**: 572
  statistical distribution of points **1**: 572
  techniques **1**: 573
  thinning transformations **1**: 571
Character representation
  computers **1**: ***575***
Character sets
  Chinese **1**: 574
  Indian
    character recognition **1**: 574
  Japanese
    character recognition **1**: 574
Characteristic-locus method
  design study **1**: ***581–3***
  feedback control
    design **1**: ***575–80***
  multivariable control systems **5**: 3197

Characteristic polynomials **5**: 3172
Chebyshev approximation
  spectral analysis **7**: 4497
Checklist paradigm
  computer security **1**: 741
Chelates *see* Coordination complexes
Chemical analysis **1**: 174
  aerosols **1**: ***174–6***
  atomic emission spectroscopy
    inductively coupled plasma method **4**: 2346
  automatic analysis **2**: ***1508–12***
    hydrocarbons **1**: 179–81
  chemical oxygen demand **7**: 5114
  concentration measurement
    flow measurement **3**: 1690
  distillation
    process control **2**: 1108
  environmental measurements **3**: ***1936–8***
    infrared spectrometry **4**: ***2509–11***
    ion-selective electrodes **4**: ***2604–7***
  fingerprinting of oil **7**: 5110
  liquid chromatography **4**: 2887
  mass spectrometry **5**: 2934
  process stream composition
    microprocessor chips **5**: 2988
  water pollution **2**: 1535; **7**: ***5109***
    inorganic nitrogen compounds **7**: ***5117–22***
Chemical equilibrium
  fugacity **7**: 4898
Chemical oxygen demand **7**: ***5114***
Chemical potential *see* Thermodynamic properties
Chemical reactions
  characteristics **4**: ***2782–9***
  periodic **1**: 424
Chemical reactors
  modelling **1**: 585
    parameter estimation **1**: 587
    partial differential equations **6**: 4362
  process control **1**: ***585–8***
    fuzzy set theory **4**: 2359
  process controller design
    Nyquist diagrams **4**: 2590
Chemical systems
  hysteresis **1**: 431
  random processes **1**: ***567–9***
Chemical technology
  optical fiber sensors **5**: 3462
  simulation **1**: ***583–5***
Chemical variables control **6**: ***3680–2***
Chemical variables measurement **6**: 3681
  elements concentration
    marine systems **2**: 1513
  organic nitrogen compound concentration
    water **7**: ***5122–3***
  oxygen concentration
    gases **5**: ***3572–5***
    liquids **5**: ***3575–9***
  trace element concentration
    water pollution **7**: ***5123–7***
Chemiluminescence **1**: ***183–4***
  environmental analysis **2**: 1511
  nitrogen oxides
    air pollution **1**: 182
  ozone
    air pollution **1**: 183
Chewing
  jaw movement
    measurement **2**: 978
  strain **2**: 968
Chlorides
  ion-selective electrodes
    water pollution **4**: 2605
Chlorine
  ecology **1**: 267
  soil **1**: 146
Chlorine compounds
  toxic substances **1**: 145
Chlorophyll
  water quality
    remote sensing **6**: 4044
Choice functions
  fuzzy set theory **4**: 2295
Cholecystokinin
  hunger control **3**: 2211
Choppers (circuits)
  amplifiers **1**: 272
  dc motors **5**: 3094
  variable speed induction machines **7**: 5022
Chromatography **1**: ***592–4***
  chemical variables control **6**: 3682
  efficiency **1**: 594
  environmental analysis **3**: ***1931–2***
  fingerprinting of oil **7**: 5111
  gas chromatography **3**: ***1929–31***; **5**: 2934, 2988
    air pollution **1**: 182
    capillary columns **3**: 1930
    carrier gases **3**: 1930
  gas–liquid chromatography **3**: 1929
  gas–solid chromatography **3**: 1929
  high-pressure liquid chromatography **4**: 2887
  liquid chromatography **4**: ***2887–9***
    nuclear magnetic resonance **5**: 3419
  see also Gas chromatography–mass spectrometry
Chromium
  soil **1**: 144
  toxic substances **1**: 144
Church's thesis
  computability and decidability **1**: 715
Circadian rhythms
  biological results **1**: 594
  forced Duffing's oscillator **1**: 433
  physiological models **1**: ***594–601***
    stability conditions **1**: 600
  threshold-level hypothesis **1**: 595
Circle criterion
  nonlinear systems **5**: ***3376–9***
  see also Stability criteria

Circuit breakers
  power systems **6**: 3808
Cladophora
  growth modelling **1**: ***257–61***
  loss modelling **1**: 259
Clark sensors **1**: 437; **5**: 3577
Class concepts
  process-based models **6**: 4365
Classification
  decision-theoretic methods **6**: 3637
  electrocardiography
    fuzzy set theory **1**: 764
  fuzzy set theory **3**: ***1601–3***
  hierarchical scene analysis **6**: 4157
  linguistics **3**: 1603
  multispectral images **6**: ***4029–33***
    algorithm accuracy **6**: 4031
  Thom's theorem **1**: 548
  see also Pattern recognition
Classification maps
  Landsat imaging
    atmosphere **4**: 2655
Climate
  space heating **2**: 1215
Clo units **1**: 605
Clocks
  microprocessor chips **5**: 2986
  water clock
    feedback control **3**: 1604
Closed loop systems
  discrete time control problems **2**: 1098
  frequency response **7**: 4451
  fuzzy set theory **7**: 4658
  glucose regulation **3**: 2026
  internal stability **4**: 2568
  loop stabilizer **4**: 2568
  multivariable control systems
    semi-infinite optimization **5**: 3189
  optimal control **5**: ***3496–7***
  state feedback control
    slow–fast decomposition **7**: 4530
  see also Feedback
Closed sets **1**: ***605***
Closed systems **1**: ***605***
Clothing
  area factor **1**: 605
  insulation **1**: ***605–6***
  manufacture
    industrial robots **4**: ***2447–9***
Cluster analysis
  discriminant analysis **2**: 1103
  electrocardiography **1**: 761
  multispectral images **6**: 4030
  see also Pattern recognition
Coal **2**: 1461
  coal gas manufacturing **3**: 1958
  energy resources **3**: 1714
Coalition function form games
  cooperative game theory **2**: 840
Coating technology
  industrial robots **4**: 2416
  tablets
    process control **6**: 3694
Cobalt
  soil **1**: 145
  toxic substances **1**: 144
Cockpits
  civil aircraft **1**: 602
  display of traffic information **1**: 201
Code-division multiplexing *see* Multiplexing
Code length
  brains as fuzzy systems **1**: 506
Code standards
  ASCII **1**: 575; **6**: 4291
  EBCDIC **1**: 575
  signal transmission **6**: ***4291***
Coding *see* Encoding
Cognition **1**: ***608***
Cognitive maps
  knowledge representation **4**: 2638
Cognitive styles
  systems management **7**: ***4790–3***
  taxonomies **7**: 4790
Cognitive systems
  expert systems **2**: 1572
  neurophysiology **1**: 502
Coherence function
  signal analysis **6**: 4280
Coherent anti-Stokes Raman spectroscopy
  atmospheric analysis **4**: 2722
Cohesion
  interactive systems **7**: 4777
Cold rolling **6**: 4139
Collagen
  biomechanics **1**: 475, 479
Collective enquiry **1**: ***608–18***; **7**: ***4802***
  interpretive structure modelling **4**: ***2575–9***
  workshop dynamic modelling **7**: ***5171***
Collision avoidance systems **1**: 201
  ships **6**: 4261
Colorimetry
  nitrogen oxides
    air pollution **1**: 182
Combat aircraft **1**: ***618–23***
  missions **1**: 618
Combinatiorial mathematics
  optimization
    independence **1**: ***624–6***
Combinatorial mathematics **6**: 3919
  optimization **2**: ***1080–3***
    approximate algorithms **1**: ***626–8***
    branch-and-bound methods **1**: ***628–30***
    linear programming **6**: 3735
    linear systems **6**: 3737
  polyhedral combinatorial mathematics
    cutting plane methods **6**: ***3735–9***
Comfort
  air conditioning **1**: 149

body temperature **7**: ***4865–6***
comfort equation **1**: ***639–41***
control systems
modelling **1**: ***638***
effective temperature **1**: 642
indices **1**: ***642–3***
comfort equation **1**: 639
predicted mean vote **1**: 641
predicted percentage dissatisfied **1**: 641
skin wettedness **1**: 644
standard effective temperature **1**: ***644–5***
new effective temperature **1**: 642
operative temperature **1**: 643
resultant temperature **1**: 643
Command and control systems
military computing **1**: ***645–51***
Comminution
mineral processing **5**: 2989
Common mode rejection ratio **1**: 269
voltmeters **2**: 1397
Communication
air traffic control **1**: 197
control–pilot
air traffic control **1**: 192
control system applications **2**: 1140
data-link
air traffic control **1**: 197
M-sequences **5**: 2955
operations research
simulation **4**: 2352
subjectivity **7**: ***4696–8***
Communication channels
distributed processing **2**: ***1202–9***
Communication command
military computing **1**: ***645–51***
Communication networks
combat aircraft **1**: 623
industrial robots **4**: 2365
Communities
ecology
disturbances **2**: 1316
Commutating autozero amplifiers **1**: 272
Commutative rings
realization theory **6**: 3970
Commutators
thyristors
dc motors **5**: 3093
Compact sets **1**: ***665***
Comparators
electric current measurement **2**: 1373
Comparison principle
Lyapunov methods **4**: 2554
Comparison systems
vector norms
nonlinear systems **5**: 3402
Compartment modelling
biological
structural identifiability **1**: 490
catenary **1**: 492
ecosystem stability **2**: 1349
ecosystems **2**: 1327, ***1343–4***
environ analysis **2**: 1483
hepatobiliary kinetics **3**: 2140
hormone regulation **3**: 2171
iron metabolism **3**: 1619
ketone body metabolism **4**: 2629
lake response time **4**: ***2789–94***
mamillary
structural identifiability **1**: 492
pharmacokinetics **2**: 1245
nonlinear models **6**: ***3695–8***
phosphorus kinetics
lakes **5**: 3043
radioisotopes **6**: 3932
water quality **7**: 5130
Compensation
state estimation
pole assignment **6**: 3730
Competitive equilibrium **1**: ***665–71***
Competitive hierarchy
ecological modelling **2**: 1315
Complementarity
autology **1**: 326
linguistic **1**: 329
Completeness **2**: 915
Complex systems
analysis **1**: 689; **4**: 2543
block lower triangular form **4**: 2552
characterization **1**: 687
control **1**: 688, 693
decentralized deterministic design techniques **1**: 695
decentralized fixed modes **1**: 694
multicriteria optimization **1**: 689
reduction methods **1**: 688
tracking methods **1**: ***675–81***
coordination **1**: 690
hierarchical overlapping coordination **5**: ***3564***
data analysis **1**: 689
dissimilarity indices **1**: 689
informational analysis **1**: 689
statistical methods **1**: 689
decomposition **1**: 688; **4**: 2543
horizontal decomposition **1**: 688, 690
spatio temperal decomposition **1**: 688
temporal vertical decomposition **1**: 697
vertical decomposition **1**: 688, 696
description **1**: 677
dynamic systems
partition **4**: 2545
management
game theory **1**: ***682–7***
methodology **1**: ***687–703***
multicriteria formulation **1**: 699
Pareto equilibrium **1**: 702
quasi-optimal control
temporal vertical decomposition **1**: 697

regular perturbation techniques **1**: 692
size reduction **1**: 698
stability **4**: 2551
structural analysis **1**: 690
suboptimal control
aggregated model **1**: 699
transinformation **4**: 2544
see also Large-scale systems, Interconnected systems
Complexation
heavy metals
limnology **4**: 2778
metal complexes
environment **1**: 4
nutrients and pollutants
environment **1**: 7
Complexity **1**: ***704–9***
autology **1**: 706
von Neumann's hypotheses **1**: 707
Compliance
industrial robots **4**: 2379
Composite fuzzy relations **1**: ***710***
equations
solution algorithms **1**: ***710***
Composite materials
aerospace industry **4**: 2452
Composite systems **1**: ***713–4***
see also Large-scale systems
Composition
random variate generation **6**: 3948
Compressibility
biological materials **1**: 483
skin **1**: 479
Compressors
gas turbines **3**: 1994
natural gas technology **3**: 1946, 1977
Compromise programming **5**: 3157
multicriteria decision making **5**: 3119
Computability and decidability **1**: ***715–20***; **2**: ***932–7***
cellular automata **6**: 4342
complexity **1**: 704
noncomputable functions **1**: 719
programming languages **1**: 715
self-applicability **1**: 717
Turing machines **1**: 715
Computational complexity **1**: 705, ***720–5***; **2**: 1082
axiomatization **1**: 708
multivariable control system design **5**: 3208
Computational geometry
three-dimensional vision
modelling **1**: 343
Computed tomography
plants (biological) **2**: 1518
radiology **7**: 5182
Computer-aided analysis
flexible spacecraft
control system components **3**: ***1649–64***
gas distribution systems **3**: 1950
linear continuous systems **4**: ***2798–800***
reconstructability analysis **6**: 3983
see also Modelling, Simulation
Computer-aided design *see* CAD
Computer architecture **1**: ***736***
architecture extension **2**: 1451
distributed database systems **2**: 1144
emulation **2**: 1450
microprocessors
discrete-event simulation **2**: ***1083–7***
simulation models **1**: ***630–3***
see also Microprogramming
Computer graphics
computerized instrumentation
correlators **5**: 2981
gas distribution systems
steady-state analysis **3**: 1954
interpretive structure modelling **4**: 2577
simulation **6**: ***4299–302***
translatable graphics **4**: 2577
water supply network analysis **7**: 5105
Computer metatheory
automata theory **1**: ***333–4***
Computer networks
design
simulation program generators **4**: 2548
distributed systems **2**: ***1125***
learning algorithms **7**: 4716
Computer software
concurrency **6**: ***3596***
correctness
large-scale systems **4**: 2679
distance measurement **4**: 2487
event-based simulation **6**: ***4403–5***
large-scale systems
reliability **4**: ***2678–81***
microprogramming **5**: ***2989***
multiprogramming **5**: ***3165***
operating systems **5**: ***3448***
parallel processing **6**: ***3596***
program compilers **6**: ***3892***
program interpreters **6**: ***3893***
robustness
large-scale systems **4**: 2679
simulation
activity-based models **6**: ***4392–3***
databases **2**: 893
optimization **6**: 4309
partial differential equations **2**: ***1180–1***
process-based models **6**: ***3878–9***
transaction flow-based models **7**: ***4933–5***
systems software **7**: ***4810***
utility programs **7**: ***4986***
Computer systems
measurement **1**: ***746–51***
performance modelling **6**: 3655
Computer–user interfaces
ergonomics **3**: 2179
modelling **5**: 3077
Computers **3**: 2215
aerospace control **1**: 383

aircraft engines **1**: 224
autopilots **1**: 384
Concorde propulsion units **1**: 234, 235
flexible spacecraft **3**: ***1649–64***
helicopters **3**: 2132
space shuttle computer redundancy **7**: 4477
character codes **1**: ***575***
distributed systems
reliability **2**: ***1132–6***
fermenter control **3**: 1618
information systems **4**: 2502
memory hierarchy **1**: ***740***
modelling
autopoiesis **1**: 396
power system component identification **6**: 3777
security
fuzzy set theory **1**: ***741–3***
possibility theory **1**: ***743–6***
simulation **1**: ***751–4***; **6**: 4311
synchronization **7**: ***4751–2***
telemetry **7**: 4839
Computing time **1**: 721
Concentration control *see* Chemical variables control
Concrete mixing
cautious control **2**: ***771–3***
Concurrency (computers) **6**: ***3596***
database management systems **2**: ***880–3***
see also Multiprocessing programs
Conditional gain **4**: 2674
Conductimetry
environmental analysis **2**: 1508
sulfur dioxide
air pollution **1**: 181
Confidence regions
ecological modelling
parameter estimation **6**: 3619
Confirmation *see* Validation, Verification
Conflict analysis **7**: ***4793–8***
Conflict resolution
air traffic control **1**: 193, 197
decision theory and analysis
fuzzy set theory **2**: 775
fuzzy set theory **2**: ***773–5***
Conjugate-point theory **6**: ***4068***
Connectance
ecological modelling
stochastic processes **7**: 4644
Connective tissue
anisotropic **1**: 474
biomechanics **1**: ***474–7***
compressibility **1**: 477
preconditioning **1**: 475
time-dependent features **1**: 474, 477
Connectives
fuzzy sets
general class **3**: ***1792***
Connectivity
ecosystem network structure **2**: 1352
graphs **7**: 4662, 4666
simple connectivity **7**: 4664
stability **2**: ***776–7***
strong connectivity **7**: 4662
Consensus methodologies *see* Collective enquiry
Conservation laws
simulation modelling
discrete arithmetic-based models **6**: 4345
Consistency
distributed processing **2**: 1131, 1161, 1165, 1171, 1203
parameter estimation **6**: 3614
process control **2**: ***777–81***
Constant parameter systems
fast algorithms **2**: ***781–5***
see also Optimization
Constraint theory **2**: ***801***
optimization
feasible direction methods **2**: ***786–8***
methods of multipliers **2**: ***789–93***
penalty methods **2**: ***797–801***
quadratic programming **2**: ***793–7***
systems analysis **4**: 2492
Constructability
linear systems **4**: 2863
Consumers
welfare economics **7**: 5163
Contents measurement **4**: 2737
Contextual relations **2**: ***801***
Continuity
Hamilton–Jacobi–Bellman equation **3**: 2063
Continuous control **4**: 2341
Continuous mathematics **2**: ***1087***
Continuous time dynamic simulation **5**: 3520
Continuous time systems
Gauss–Markov processes
realization theory **6**: 3963
Continuously stirred tank reactor
modelling **4**: 2647
Contouring control
autonomous **1**: 297
articulated robot arms **1**: ***296–301***
principles **1**: 297
mixed **1**: 297
programmed **1**: 296
robot arms
experimental results **1**: 300
Contractibility
overlapping decentralized control **5**: 3570
Contraction theory
convergence analysis **3**: 2155
Contrapositive symmetry
fuzzy implication operators **3**: 1808
Control in coefficients **2**: ***808–12***
Control configured vehicles
aerospace control **2**: ***801***
helicopters **3**: 2134
Control equipment
control location problem **6**: ***4238–44***
electrohydraulic servovalves **6**: ***4252–7***

sensor location problem **6**: ***4238–44***
valves **2**: ***833–7***
see also Actuators, Electric sensing devices
Control information systems
military computing **1**: ***645–51***
Control laws
decentralized **2**: 924
local **2**: 895
Control rooms
ergonomics **3**: 2181
Control rules
competition
fuzzy systems **2**: ***812***
fuzzy logic controller **3**: 1811
generation
fuzzy systems **2**: ***813***
Control system analysis
flexible spacecraft **3**: ***1649–64***
multivariable control systems
nuclear power station boilers **5**: ***3431–6***
nonlinear systems **5**: ***3366–73***
singular perturbations **7**: ***4435–9***
state-space methods **7**: ***4551–4***
structural stability **7**: 4669
Control system analysis computing
linear continuous systems **4**: ***2798–800***
Control system computing
linear algebra **2**: 824
numerical analysis **2**: ***813–25***
Control system synthesis
aircraft wing load alleviation **7**: 5168
flexible spacecraft **3**: 1649
multivalued differential equations **5**: 3183
multivariable control systems **5**: 3197
frequency-domain synthesis **5**: ***3431–6***
geometric approach **4**: ***2807–16***
nonlinear decoupling **5**: ***3282–8***
multivariable servomechanism problem **6**: ***4129–32***
decentralized control **6**: ***4133–8***
nonlinear systems **5**: ***3366–73***
state-space methods **7**: ***4551–4***
see also Design
Control systems
affine systems **1**: ***120–4***
air conditioning **1**: ***152–9***
elements **1**: 154
aircraft engines **1**: ***220–6***
aircraft systems
independence **1**: 233
autonomous systems
structural stability **7**: 4675
biological systems **1**: 436
CAD **1**: 733; **2**: 825
design environments **1**: ***726–32***
chemical engineering
simulation **1**: 585
comfort
modelling **1**: ***638***
decentralized systems
distributed data processing **2**: ***1137***
geometric approach **2**: ***894–9***
stability **2**: 923
design **5**: 3376
design computations
numerical analysis **2**: ***813–25***
discrete time systems **2**: ***1093–8***
deterministic systems **2**: 1093–5
stochastic systems **2**: 1095–8
distributed systems **2**: 1138
domestic gas appliances **2**: ***1212–3***
evolution **4**: 2856
feedback
stability **7**: 4524, 4527
fuzzy set theory
education **2**: ***1358–61***
gas turbines
design **1**: 581
industrial gas equipment **4**: ***2353–6***
interconnected systems
stability **2**: 925
Kalman systems **1**: 121
lighting **4**: ***2775–7***
linear systems **1**: 122
abstract **4**: 2860
multiple control systems **6**: 4009, 4010
nonautonomous systems
structural stability **7**: 4675
parameter optimization **7**: ***4678–84***
realibility **6**: ***4008–11***
regulation
viability **2**: 1028
saturating control systems **5**: ***3516–7***
structural stability **7**: ***4674–7***
see also Controllers
Control theory
biological motor systems
modelling **1**: 466
dynamic systems **2**: ***1276–8***
extension techniques **2**: ***1581–3***
feedback
fuzzy set theory **4**: 2356
fuzzy set theory
applications **4**: 2358
geometric control theory **5**: 3181
group theory **5**: ***2944–8***
Lie brackets **4**: 2754
linear hyperbolic systems **4**: ***2800–7***
linear systems **4**: ***2855–61***; **7**: ***4617–21***
controllability **2**: 821
observability **2**: 821
reconstructability **2**: 821
multivalued differential equations **5**: ***3180–4***
real analyticity **6**: ***3954–6***
universal inputs **7**: ***4975–6***
see also Controllers
Control valves *see* Valves
Controllability **1**: ***120–4***; **5**: 3513
control systems **4**: 2856

discrete time control systems **2**: 1093
extension techniques **2**: 1581
global controllability
space vehicles (rigid) **1**: 321
Lie brackets **4**: 2756
Lie groups **4**: ***2756–9***
linear hyperbolic systems **4**: 2801
linear systems **4**: 2862
local controllability **4**: ***2893–8***
multivalued differential equations **5**: 3181
space vehicles (rigid) **1**: 321
multivariable control systems **4**: 2808, 2811, ***2816–26***
controllability subspaces **4**: 2850
nonlinear decoupling **5**: 3286
nonlinear discrete time systems **5**: ***3288–92***
nonlinear systems
geometry **5**: 3366
numerical analysis
linear control theory **2**: 821
perturbed systems **2**: 1022
polynomial systems
matrix algebra **6**: 3747
semigroup theory **6**: 4222
space vehicles (rigid)
Poisson stability **1**: 320
Riemannian geometry **1**: 320
state-space methods **7**: 4553
Controllers
air conditioning **1**: 154
final control elements **1**: 156
proportional controllers **1**: 154
artificial satellites **6**: 4147
blood pressure
adaptive control **1**: 499
domestic appliances
gas appliances **2**: 1212
space heating **2**: ***1239–41***
electric generators **2**: 1376
furnaces
Nyquist diagrams **4**: 2594
gas equipment **2**: 837
turbines **4**: 2589
HVAC systems **2**: 1225; **3**: 2092
tuning **3**: ***2108–13***
learning automata **4**: 2716
lighting **4**: 2771
maintenance **6**: 4009
positive feedback
respiration **6**: 4047
respiration **6**: 4046
integral controllers **6**: 4046
optimal controllers **6**: 4046
sampled-data systems
Nyquist diagrams **4**: 2595
structural stability **4**: 2566
weighted minimum variance **1**: 39
see also Control systems, Control theory, Regulation
Convergence **1**: 58
adaptive control **1**: ***54–60***; **5**: 3406
boundedness **1**: 57, 59
positive-real condition **1**: ***51–4***
asynchronous iteration
decomposition–coordination **5**: 3477
contraction theory **3**: 2155
costate coordination algorithms **2**: 860
distributed asynchronous fixed-point algorithms **2**: ***1121–5***
explicit self-tuners **2**: ***1578–80***
fuzzy control process **3**: ***1793–4***
global convergence
quadratic programming **2**: 795
identification **4**: 2236
martingale methods **5**: 2932
stochastic adaptive systems **7**: 4607
methods of multipliers **2**: 791
nonlinear estimation theory **5**: 3310
numerical methods
inequality-constrained problems **4**: 2879
optimization
unconstrained **7**: 4965
parameter-adaptive control **6**: 3602, 3606
quasi-Newton methods **7**: 4973
self-tuning control **6**: 4182, 4192, 4207
stochastic adaptive control algorithms
ODE method **7**: ***4602–6***
weak convergence
recursive algorithms **6**: ***3994–7***
Convex programming
decomposition
price **1**: 400
Convexity
convex functions
nonlinear programming **5**: 3357
convex sets
nonlinear programming **5**: 3353
K convexity **5**: 3509
nondifferentiable optimization **5**: 3541
optimization **1**: 404
variational techniques **7**: 5046
Conveyance
structural stability **7**: 4673
Conveyors
industrial robots **4**: 2430
Convolution **4**: 2250
Fourier transforms **3**: 1726
Cooling
degree days **2**: 959
hydronic systems **3**: ***2234***
Cooling systems *see* HVAC systems
Cooperative game theory **1**: 684; **2**: ***840–9***; **7**: ***4811–22***
see also Game theory
Coordinate systems
aircraft dynamics **1**: 213

Coordinated planning laws **1**: 36
Coordinating variable
  optimization **5**: 3013, 3036
Coordination
  asynchronous iteration **5**: ***3475–80***
  costate coordination method **2**: ***859–62***
  distributed parameter systems **2**: ***1186–92***
  dynamic systems **2**: ***1297–305***
  global feedback **7**: 4567
  goal coordination method **3**: 2032
  hierarchical control **3**: ***2153***
  hierarchical overlapping coordination **5**: ***3564–8***
  local feedback **7**: 4568
  mixed coordination
    hierarchical control **3**: 2160
    large-scale optimization **5**: ***3012–9***
  multilevel control systems **5**: 3140
  robot arms **6**: 4084
  steady-state systems **7**: ***4566–71***
Coordination complexes
  metals
    periodic classification **1**: 4
  nutrients and pollutants
    environment **1**: 6
Copper
  ion-selective electrodes
    water pollution **4**: 2607
  soil **1**: 146
  toxic substances **1**: 144
CORAL-66 **6**: 3894
Cores
  cooperative game theory **2**: 844
Coriolis force
  acceleration measurement **3**: 1680
  flowmeters **3**: 1692
Coronary hemodynamics
  coronary impedance **2**: 851
  epicardial vessels **2**: 853
  modelling **2**: ***850–3***
  regulation **2**: 852
  see also Hemodynamics
Correctness
  database management systems **2**: 881
Correlation **2**: ***853***
Correlation functions
  data
    determination **4**: 2248
Correlation methods **7**: 4656
  equipment **4**: 2249
  identification **4**: 2241, ***2245–53***, 2317
Correlators **4**: 2249
  computerized instrumentation **5**: 2980
  polarity correlation
    zero-crossing-time algorithm **5**: 2980
Corrosion of pipes
  natural gas technology **3**: 1939, 1955
Cost–benefit analysis **2**: ***854–9***; **7**: 4809
  environmental modelling **2**: 1537
  systems acquisition **7**: 4753
Cost–benefit ratios **2**: ***859***
Cost benefits
  identification **2**: 854
Cost–effectiveness analysis **2**: ***854–9***; **7**: 4809
  identification **2**: 855
  quantification **2**: 855
  systems acquisition **7**: 4752
Cost functions
  feedforward self-tuning control **6**: 4187
  gradient **5**: 3482
  linear quadratic Gaussian **1**: 39, 43
  self-tuning control **6**: 4173
  weighted minimum variance **1**: 39, 43
Costate coordination method **2**: ***859–62***
  three-level algorithm **2**: 861
  two-level algorithm **2**: 860
Costate prediction approach
  parameter estimation
    large-scale systems **4**: 2702
Costs
  air traffic control **1**: 189
  identification **2**: 854
  project management **6**: 3901
Coulometry
  environmental analysis **2**: 1509
  sulfur dioxide
    air pollution **1**: 182
Coulter counters **2**: ***862–6***
  basic principles **2**: 862
  cells (biological)
    dieletric breakdown **2**: 865
  focused flow **2**: 864
  nonfocused flow **2**: 864
  sweep flow **2**: 865
Counting
  microscopic particles
    Coulter counters **2**: 862, 863
Counting processes
  stochastic calculus **7**: 4636
Coupling **2**: ***867***
  interactive systems **7**: 4777
Covariance matrices **2**: 1270
CPM **6**: 3899
Craftsman robots **4**: 2412
Cramér–Rao bound
  identification **4**: 2237
  parameter estimation **6**: 3614
Creaons
  ecosystem digraphs **2**: 1346
Creativity **1**: ***508–11***
  artificial intelligence **1**: 307
  characteristics **1**: 509
  creatron **1**: 509
Creatron
  creativity **1**: 509
Credibility
  simulation **6**: ***4406–11***

Crisp controller **2**: ***867***
Critical activity **2**: ***867***
Critical clearing time
  power systems **6**: 3796
Critical path analysis **2**: ***867***; **6**: 3900
Critical time override
  load regulation **2**: ***867***
Cross-correlation function
  flowmeters **3**: 1689
  signal analysis **6**: 4281
Cross-impact analysis *see* Impact analysis
Cross-spectral density
  signal analysis **6**: 4280
Crushing
  mineral processing **5**: 2990
  process control **5**: 3280
Cubic sensor problem **5**: 3325
Cumulative sum test
  dynamical systems
    abrupt changes **1**: 81
Current comparators
  alternating-current **2**: 1374
  direct-current **2**: 1374
  electric current measurement
    2: 1373, 1374
Current transformers
  alternating-current **2**: 1373
  direct-current **2**: 1373
  electric current measurement **2**: 1373
Cusp catastrophe **1**: 551
  biological models **1**: 430, 432
Cutset in graphs **7**: 4664
Cutting (fabrics)
  industrial robots **4**: 2447, 2448
Cyanide
  ion-selective electrodes
    water pollution **4**: 2605
Cybernetics **2**: ***867, 867–9***
  artificial intelligence **2**: 868, 870
  automata theory **1**: ***333–4***
  general systems theory **2**: 869
  history **2**: 868, ***869–73***; **4**: 2271; **7**: 4772
  industries **2**: 869
  praxiology **6**: 3862
  self-organizing systems **6**: 4165
Cycloconverters
  ac motor control **5**: 3091
Cytometers
  cells (biological)
    Coulter counters **2**: 866

D

Damping
  simulation **5**: 2967
Dams
  hydroelectric power stations **3**: 2219
  safety
    Bayesian decision analysis **7**: 5139
Dantzig–Wolfe decomposition principle **2**: 956
Data **2**: ***875***
Data access
  optimization with distributed database systems **2**: 1158
Data acquisition
  power systems **6**: 3846
Data analysis
  interaction evaluation **2**: ***875–9***
Data communication systems
  aircraft communication systems **2**: ***879***
  buildings
    management **1**: 513
  packet switching **6**: 3581
  telemetry **7**: ***4835–42***
Data compression
  environmental information **2**: ***1492–4***
  environmental testing **2**: ***1489–92***
  industrial vision systems **4**: 2471
Data conversion **6**: 4286
Data dictionaries *see* Database management systems
Data handling
  CAD
    multivariable systems **1**: 734
  optimization
    distributed database systems **2**: 1160
Data prediction
  self-tuning control **6**: ***4168–72***
Data privacy **2**: 887
Data processors *see* Computers
Data reduction and analysis
  industrial vision systems **4**: 2471
Data smoothing
  environmental information **2**: ***1492–4***
  environmental testing **2**: ***1489–92***
  see also Signal processing
Data structures **2**: 875, ***879***
  multidimensional scaling **5**: ***3124***
Data synchronization **2**: 875
Data systems **2**: ***880***, 1546; **3**: 2001
Data transfer
  industrial robots **4**: 2364
Database management systems **2**: ***884–92***, 945; **4**: 2503
  concurrency **2**: ***880–3***
    correctness **2**: 881
    locking **2**: 882
    recovery **2**: 883
    reliability **2**: 883
    transactions **2**: 881
  consistency of data **2**: 885
  external redundancy
    modelling **4**: 2656
  fuzzy set theory **3**: ***1795–7***
  integrity of data **2**: 885
  languages **4**: ***2655–8***
  library search systems
    gas chromatography–mass spectrometry **3**: 1935
  relational approach **4**: 2656
  simulation **2**: 893
  synchronization **2**: 1203
Databases **2**: ***893***
  environmental databases
    INFOTERRA **2**: 1503–6
  fuzzy databases **3**: ***1795–7***
    information retrieval **3**: ***1798–800***
  simulation **2**: ***893–4***
  see also Distributed database systems
Dataless systems **3**: 2001
  see also Primitive systems
DC generators **2**: 1380
  history **2**: 1401
  regulators
    history **2**: 1401
DC motors **5**: ***3092–7***
  control **5**: 3095
  digital control **5**: 3096
  fuzzy control **2**: ***1071–2***
  fuzzy model **2**: ***1072–3***
  position control **5**: 3096
  rolling-mill control **6**: 4140
  speed control **5**: 3096
DC transmission networks **6**: ***3820–5***
Deadband control
  temperature **2**: ***894***
Deadlock
  distributed processing **2**: 1203
Deburring
  industrial robots **4**: 2416
Decentralization
  distributed parameter systems **2**: ***1186–92***
  regulation
    viability **2**: 1029
Decentralized control **2**: 912, 928; **4**: 2557
  algebraic approach **4**: ***2666–71***
    matrix fraction descriptions **4**: 2666
    polynomial matrix descriptions **4**: 2666
  cooperative method **1**: 679
  coordinated method **1**: 677
  decentralized feedback
    large linear systems **4**: 2666
  decomposition–coordination **2**: 952
  design **1**: 131
  dynamic systems

on-line coordination **2**: 1302
fixed modes **2**: ***902–7***
geometric approach **2**: ***894–9***
large-scale systems **4**: ***2557–9***
linear quadratic problems **2**: 954
linear systems
algebraic approach **4**: ***2666–71***
overlapping decentralized control **5**: ***3568–72***
pole assignment **2**: ***900–2***
robust controller design **6**: ***4133–8***
structural constraints **7**: 4685
stochastic control **2**: ***927–32***
Decentralized control systems
geometric approach **2**: ***894–9***
Decentralized disturbance decoupling problem **2**: 895
output feedback **2**: 897
Decentralized estimation **2**: 931
Decentralized feedback
large linear systems **4**: 2666
unassignable modes **4**: 2666
Decentralized filters **4**: 2695
Decentralized stabilizing compensator **2**: 900
Decentralized systems **2**: 902
stability **2**: ***916–23***, 951
Decidability *see* Computability and decidability
Decision analysis *see* Decision theory and analysis
Decision making
automated systems **6**: 3661
deterministic incentives **2**: 993
holistic approaches **3**: 2193
human information processing **3**: 2186
incentives **7**: 4627
information analysis **4**: 2499
minimax techniques **5**: 2996
operations research
simulation **4**: 2352
team theory **7**: 4811
see also Collective enquiry, Decision theory and analysis
Decision rooms
office automation **5**: 3448
Decision support systems *see* Management information systems
Decision tables
electrocardiography
medical diagnostic computing **1**: 762
fuzzy controllers **4**: ***2898***
Decision theory and analysis **2**: 933, ***938–42***; **5**: 3104; **7**: 4806
additive difference **3**: 2200
cognitive styles **7**: 4790
conflict analysis **7**: ***4794***
conflict resolution
fuzzy set theory **2**: 775
conjunctive **3**: 2200
disjunctive **3**: 2199
dominance **3**: 2200
entropy measures
fuzzy set theory **2**: 1472
fuzzy random variables **3**: 1844
fuzzy set theory **3**: 1873
group decision making **3**: ***2053–60***
heuristics **3**: 2199
hierarchical inference **4**: 2338
hierarchical scene analysis **6**: 4157
human judgment **3**: ***2193–202***
impact analysis **4**: 2338
lexicographic theory **3**: 2200
Markov decision processes **5**: 2930
methodology **5**: 3112
minimum difference lexicographic theory **3**: 2200
multiattribute utility theory **5**: ***3099–103***, 3104
multicriteria decision making **5**: ***3116–21***, ***3121–4***
surrogate worth trade-off method **7**: 4704
multilevel systems **5**: ***3145–9***
multiobjective decision making **5**: ***3150–2***
multiobjective optimization **5**: 3152
noncooperative games **5**: 2910
pattern recognition **6**: ***3634–43***
philosophical aspects **7**: 4782
praxiology **6**: 3860
principles
characterization **3**: 2042
formulation **3**: 2040
scenarios **6**: 4153
self-tuning control
data prediction **6**: 4171
Stackelberg strategies **5**: ***3145–9***
stochastic games **7**: 4621
subjective entropy **7**: 4692
systems analysis **7**: 4767
trees (mathematics) **7**: 4950
utility measurement **7**: 4981
utility theory **7**: 4986
wholistic decision theory **3**: 2201
see also Game theory
Decision trees **7**: 4950
Decoding
channel decoding **2**: 1049
Decomposition **2**: ***947***
asynchronous iteration **5**: ***3475–80***
block decomposition **3**: 2155
block matrix formulae **1**: 495
complex systems
estimation theory **4**: 2692
coordination methods **1**: 400
coupling through cost only
auxiliary problem principle **1**: 400
Dantzig–Wolfe decomposition **4**: 2691
decentralized stabilization **2**: 917
directed graphs
ecosystems **2**: ***1345–6***
distributed parameter systems **2**: ***1186–92***
dual decentralized method **3**: 2160
dynamic systems
on-line coordination **2**: 1302
function space **3**: 2156
functional decomposition **5**: 3140

goal coordination method **3**: 2032
hierarchical control **3**: ***2153***
hierarchical decomposition
simulation modelling **6**: ***4353–5***
hierarchical overlapping coordination **5**: 3564
indirect methods
hierarchical control **3**: 2153
interaction prediction method **3**: 2159
interconnected systems **4**: 2552
linear programming **4**: 2690
local decomposition
nonlinear decoupling **5**: 3284
multilevel decomposition **3**: 2158
multistage decomposition algorithms **5**: ***3166–70***
multi-timescale methods
singular perturbations **7**: 4432
nested decomposition **5**: 3166
optimization **1**: ***400–5***; **2**: ***956–8***
overlapping decentralized control **5**: 3569
parallel decomposition **3**: 2154
algorithms **1**: 402
price
augmented Lagrangians **1**: 403
Lagrangians **1**: 402
procedures **2**: 951
sequential decomposition **1**: 402; **3**: 2154
singular value decomposition **2**: 910
numerical analysis **2**: 819
slow–fast decomposition **7**: ***4529–34***
structural properties **2**: ***948–52***
system decomposition
generalized dependence **3**: ***2003–5***
temporal decomposition **5**: 3142
time decomposition
singularly perturbed systems **7**: 4439
two-level Newton relaxation **3**: 2162
two-level parallel decomposition **3**: 2158
two-level sequential decomposition **3**: 2158
vertical decomposition **5**: ***3140–4***
Decomposition–coordination **2**: ***952–6***; **5**: 3014
structural analysis **7**: 4661
Decomposition models
multiattribute utility measurements **7**: 4983
multiattribute utility theory **7**: 4989
Decoupling
disturbance decoupling **4**: ***2522–3***
input–output decoupling **4**: ***2522–3***
input–output systems **5**: 3287
multivariable control systems **4**: 2853
nonlinear multivariable systems **5**: ***3282–8***
Deduction **2**: ***959***
Deformation
cells (biological) **1**: 444
Coulter counters **2**: 865
connective tissue **1**: 474, 475
Defuzzification operators **2**: ***959***
fuzzy logic controller **3**: 1812
Degeneracy
distributed parameter systems
delays **2**: 1210
Degeneration
nonlinear oscillators
mode analysis **5**: 3020
Degree days
meteorology **2**: ***959***
natural gas load forecasting **3**: 1980
Delauney triangulation
three-dimensional vision **1**: 345
Delay differential systems
hereditary systems **3**: 2147
stability **5**: 3133
Delays
distributed parameter systems **2**: ***1209–10***
Delphi method **1**: ***613***; **2**: ***959***; **4**: 2579; **7**: 4802
fuzzy numbers **3**: 1835
philosophical aspects **7**: 4779
technological forecasting **7**: 4825
see also Collective enquiry
Delta modulation **2**: 1046
Demand control
load shedding **2**: ***963***
Demodulators
digital communication systems **2**: 1048
Denaturation
proteins
cusp catastrophe **1**: 432
DENDRAL **1**: 307
Density
control **2**: ***777–81***
sugar industry **7**: 4701
mass flow measurement **3**: 1677
measurement
natural gas technology **3**: 1962
Density function
probability
random signals **1**: 418
Dental caries
measurements **2**: 964
Dental materials **2**: ***968–9***
Dental plaque
measurement **2**: 968
Dentistry
measurements **2**: ***963–5***, ***968–9***, ***969–72***, ***976–8***
equipment and devices **2**: ***965–7***
radiography **2**: ***973–6***
tactile **2**: ***969–72***
visual **2**: ***969–72***
Dentures
measurements **2**: 966
Departure controller
air traffic control **1**: 198
Dependence
set theory
generalized dependence **3**: 2003
Dependency digraph **4**: 2678
Depth of anesthesia *see* Anesthesia
Dermis
biomechanics **1**: 478

Descent methods
feasible direction methods **2**: 787
optimal control
numerical solution **5**: 3481
unconstrained optimization **7**: ***4964–9***
Describing functions **7**: ***4443–52***
accuracy **7**: 4450
evaluation **7**: 4444
fuzzy logic controllers **3**: 1812
nonlinear systems **3**: 1749; **5**: 3368, 3385
reliability **2**: ***978–81***
Descriptive (systems terminology) **2**: ***981***
Design
adaptive control **1**: ***60–3***, 103
adaptive systems **1**: 62
aircraft engine control systems **1**: 223
articulated robot arm control
electrohydraulic servosystems **1**: 301–4
characteristic locus method
gas turbines **1**: ***581–3***
control systems **5**: 3376
compensation **7**: 4451
controller ergonomics **3**: 2181
industrial robots **4**: ***2421–40***
digital control **2**: ***1066–71***
algorithms **2**: ***1057–66***
direct digital control **2**: ***1073–8***
distributed elliptic systems **2**: ***1174–8***
dominance design
multivariable systems **5**: ***3220–3***
elevator systems **2**: ***1435–40***
experiment design **4**: 2257
instruments **4**: ***2533–6***
microprocessor chips **5**: 2984
systems approach **4**: 2534
manufacturing computer control **3**: 1644
model reference adaptive control systems **2**: ***1087–92***; **5**: 3047
model reference adaptive systems **5**: ***3060–1***
multivariable control systems **5**: 3208
method of inequalities **5**: ***3206–14***
robustness **5**: ***3224–8***
semi-infinite optimization algorithms **5**: 3193
Nyquist diagrams **4**: ***2584–8, 2588–96***
road vehicles
fuel consumption **3**: 1758
robot operating systems **4**: ***2456–9***
classical controller design **4**: ***2421–40***
syntax graphs **6**: ***4089–93***
self-tuning control **5**: 3231
predictive control **6**: ***4176–80***
simulation
experiments **7**: ***4648–51***
programs **6**: ***4396–402***
stabilization control
inverse Nyquist array methods **4**: ***2584–8***
top down methods
interpretive structure modelling **4**: 2576
see also Control system synthesis
Desorption
toxic substances
limnology **4**: 2778
Detectability
control systems **4**: 2858
Detection algorithms
automatic incident detection **3**: 1735
Detectors
air conditioning **1**: 154
static pressure detectors **1**: 172
freeways
automatic incident detection **3**: ***1734–6***
gas chromatography **3**: 1930
industrial gas equipment **4**: 2355
spectroscopy
biology **1**: 9
Determinants
block matrix formulae **1**: 494
Deterministic control
learning automata **4**: 2714
Deterministic systems **2**: ***995***
Developmental systems
heterarchical systems
simulation modelling formalism **6**: 4351
Lindenmayer systems **4**: ***2795–8***
Dewatering
process control **2**: ***1249–53***
Dewpoint control
air conditioning **1**: 157
Diabetes
educational computer models **2**: 1363
glucose
adaptive control **3**: 2017
glucose regulation
modelling **3**: 2021
modelling and control **3**: ***2025–32***
Diagnostic inference **2**: ***1003***
Diagonal dominance
dyadic expansion **5**: 3200
Dialog generation
management systems **2**: 946
Dielectric properties of substances
soft tissue **1**: ***445–50***
Diesel engines
speed control self-tuning regulators **2**: 1462
Diet
optimization by linear programming **4**: 2836
Difference equations
hereditary systems **3**: 2147
multi-timescale systems **7**: 4431
ordinary difference equations **2**: ***1003***
sampled-data systems **6**: 4143
simulation modelling **6**: ***4356–9***
singular perturbations **7**: 4430
Differential control
governors
history **7**: 4516
Differential equations **2**: ***1003–4***; **5**: ***3560–3***
finite difference method **3**: 1623

functional differential equations
hereditary systems **3**: 2147
graph theory **2**: 1078
hereditary systems **3**: 2151
hydraulic systems **1**: 546
initial-value problems
integration **5**: ***3555–8***
linear systems
Laplace transforms **4**: 2664
memory models **2**: 1004
multivalued differential equations
control theory **5**: ***3180–4***
Nagumo theorem **2**: 1024
path-following methods **6**: 3633
Peano theorem **2**: 1024
simulation **6**: 4307
modelling **6**: ***4356–9***
modelling formalism **5**: 3086
symbol manipulation **5**: ***3559–60***
solution **6**: 4310
functional expansions **3**: 1772
stochastic differential equations **7**: 4633
Girsanov transformation **3**: 2007
stochastic processes **7**: 4623
stochastic control **7**: 4639
stochastic systems
optimal control **5**: 3520
viable trajectories **2**: 1004
see also ODE methods, Partial differential equations, Linear differential equations
Differential games **2**: ***1009–10***
closed-loop games **2**: ***1004–8***
complex systems management **1**: 682
dynamic duopoly **2**: 1262
Isaacs equations **2**: ***1010–6***
linear quadratic games **2**: ***1017–20***
modelling
management problems **1**: 686
open-loop games **2**: ***1020–3***
stochastic games **7**: ***4613–7***
stopping-time problems **7**: 4616
Differential inclusions **2**: 1004
memory **2**: 1032
viability theory **2**: ***1023–32***
Differential inequalities
isoperimetric problems **4**: 2611
Differentiation
control in coefficients **2**: 811
stationary points **7**: ***4555–6***
Diffraction patterns
red blood cell deformability **1**: 444
Diffusion
control **7**: 4638
diffusion equation with jumps **4**: 2540
distributed parameter systems **2**: 1184
linear Boltzmann equations **7**: 4943
linear diffusion
distributed parameter systems **2**: 1182
nonlinear filtering **5**: 3299
optimal stochastic control **5**: 3523
toxic substances
limnology **4**: 2780
Diffusion pumps
mass spectrometer apparatus **5**: 2939
Digestive tract
bioelectric potential **1**: ***459–62***
electrical activity modelling **2**: ***1036–40***
large intestine
bioelectric potential **1**: 1462
electrical activity modelling **2**: ***1036–40***
small intestine
bioelectric modelling **2**: ***1040–2***
stomach
bioelectric modelling **2**: ***1033–6***
Digital–analog conversion **2**: 1067; **6**: 4287
Digital communication systems **2**: ***1043–51***
output signal-to-noise ratio **2**: 1051
satellite links **2**: ***1051–7***
Digital control
aircraft engines **1**: 223
redundant full authority **1**: 224
computers **2**: 1067
aircraft engines **1**: 224
dc motors **5**: 3096
design **2**: ***1066–71***
algorithms **2**: ***1057–66***
disturbances **2**: 1070
sample rate selection **2**: 1070
separation processes **6**: 4249
ship propulsion **6**: 4269
sugar industry **7**: 4703
three-term control **6**: ***3701–3***
Digital filters **6**: 4288
multidimensional systems stability **5**: 3131
random signals **1**: 419
Digital phase-locked loop
cycle-slipping phenomenon **1**: 86
Digital signals
signal processing **6**: ***4284–8***
signal transmission **6**: 4289
see also Signal processing, Signal transmission
Digital simulation
discrete-event simulation
service time **6**: ***4252***
HVAC systems **3**: ***2113–7***
software **3**: 2116
see also Simulation
Digital storage
hierarchy **1**: ***740***
Digital voltmeters **2**: 1390, 1395
dual slope systems **2**: 1395
Digitization
digital control algorithm design **2**: 1061
Digraphs *see* Directed graphs
Dilatancy *see* Thixotropy
Dimension
generalized dependence **3**: 2005

Dimensional analysis
  similarity theory **6**: 4295
Diophantine equations
  predictor models **6**: 3870
  self-tuning control **6**: 4173
  solutions **1**: 42
Direct digital control
  air conditioning **1**: 157
  boilers **7**: 4573
  design **2**: ***1073–8***
    sampling frequency **2**: 1074
  digital notch filters **2**: 1076
  discrete controllers **2**: 1077
  glass industry **3**: 2013
  multirate controllers **2**: 1077
  resonance **2**: 1076
  subrate controllers **2**: 1077
Directed graphs **2**: 915, 948, ***1071***
  ecological modelling **2**: 1326
  ecosystem decomposition **2**: ***1345–6***
  ecosystem models **2**: 1327
  ecosystem network structure **2**: 1352
  ecosystem stability **2**: 1349
  elementary components **2**: 1345
  Petri nets **6**: ***3665–70***
  project management **6**: 3899
  signed graphs **2**: 1079
  stability **2**: 1078–80
  transportation modelling **7**: 4947
Directed relations **2**: ***1080***
Directed systems **2**: ***1080***, 1545, 1548; **3**: 2001
Direction control
  robot arms **1**: 300
Discontinuities
  catastrophe models **1**: 552
  controlled discontinuities
    populations **2**: 1335
Discrete-event models
  cell space simulation
    heterarchical systems **6**: 4351
Discrete-event systems
  simulation **2**: 1199
Discrete-interval random binary signals
  analytical computation **4**: 2246
Discrete mathematics **2**: ***1087***
Discrete systems
  digital control algorithm design **2**: 1064
  optimization **2**: ***1080–3***
  simulation modelling **6**: ***4345–6***
Discrete-time systems
  aggregation forms **4**: 2565
  control systems **2**: ***1093–8***
    deterministic systems **2**: 1093–5
    linear deterministic case **2**: 1095
    linear quadratic deterministic case **2**: 1095
    linear quadratic stochastic case **2**: 1097
    stochastic systems **2**: 1095–8
  distributed-lag models **5**: 3166
  fuzzy systems **3**: ***1895***
  Gauss–Markov processes
    realization theory **6**: 3963
  innovations **2**: ***1099–102***
  Kalman filters
    state-space models **4**: ***2622–7***
  Lyapunov methods **4**: 2565
  nonlinear systems
    algebraic theory **5**: ***3288–92***
    input–output maps **4**: ***2523–7***
  optimal stochastic control **5**: ***3471–5***, 3518, 3526
  random processes
    dynamics **1**: ***565–7***
  simulation modelling formalism **5**: 3087
  singular perturbations **7**: ***4430***
  stability **4**: 2556; **7**: 4528
Discrete variables **2**: ***1102***
Discretization
  discrete-time control problems **2**: 1097
  distributed-parameter models **2**: ***1295–7***
  finite difference method **3**: 1623
  goal coordination method **3**: 2036
  model structure simplification **5**: 3081
  multigrid methods **5**: 3136
  shape design **2**: 1176
  stochastic control **5**: 3526
Discriminant analysis **2**: ***1102–5***
  membership criteria **2**: 1103–5
  multispectral images **6**: 4030
  pattern space representation **2**: 1103
  sample space representation **2**: 1103
  see also Pattern recognition
Disintegration
  automata theory **6**: 4376
Dispersion
  abiotic dispersion
    limnology **4**: 2778
  biotic dispersion
    limnology **4**: 2781
  toxic substances **4**: ***2778–82***
Dissipativeness **4**: 2674
  large-scale system stability **7**: ***4519–22***
Dissolution *see* Dissolving
Dissolving
  nutrients and pollutants
    environment **1**: 3
  toxic substances
    limnology **4**: 2778
Distance measurement
  edge method **1**: 756
    junction dictionary **1**: 756
  fuzzy set theory **3**: 1869
  region method **1**: 757
  structured light **4**: 2486
  three-dimensional vision
    computers **1**: ***755–9***
    industrial robots **4**: 2484
  triangulation **1**: 755
Distillation
  crude oil

process control **6**: 3678
disturbances **2**: 1108
models **2**: 1109
petrochemical industry
process control **6**: 3676
Distillation columns
adaptive control **1**: 50; **2**: ***1109–15***
bilinear models **2**: ***1116–21***
binary columns
adaptive control **2**: 1110
continuous columns
control **2**: 1107
process control **2**: ***1107–9***
Distributed computer systems **2**: ***1126–32***
reliability **2**: ***1132–6***
task allocation **2**: 1281
Distributed control
glass industry **3**: 2014
learning automata **4**: 2717
Distributed data processing
control systems
implementation **2**: 1141
management **2**: 1141
reliability **2**: 1141
safety **2**: 1141
decentralized control systems **2**: ***1137***
synchronization **2**: ***1202–9***
Distributed database systems
efficiency **2**: ***1155–60***
failure recovery procedures **2**: ***1160–5***
multiprocessing programs **2**: ***1143–53***
schedules **2**: 1147
multiprocessing systems **2**: ***1165–9***
performance modelling **6**: 3655
reliability **2**: ***1169–74***
see also Distributed processing
Distributed-lag models
multistage decomposition algorithms **5**: 3166
Distributed parameter models **2**: ***1292–7***
discretization **2**: 1295
Distributed parameter networks
power systems **6**: 3802
Distributed parameter systems **2**: ***1182–6***
air pollution **2**: 1497
control **2**: ***1186–92***
delays **2**: ***1209–10***
estimation **2**: ***1186–92***
identification **2**: ***1193–7***
mathematical formulation **2**: 1184
multivariable systems **5**: 3205
optimization
calculus of variations **1**: 519, 521
self-tuning regulators **2**: ***1197–9***
sensor location problem **6**: 4238
stochastic systems **7**: ***4617–21***
Distributed processing **2**: ***1125***
discrete-event simulation
dedicated processors **2**: 1085
microprocessor architecture **2**: ***1083–7***
distributed database systems **2**: ***892***
synchronization **2**: ***1202–9***
see also Distributed database systems
Distributed systems
iterative algorithm convergence **2**: ***1121–5***
Distribution networks
power systems **6**: 3780
Distributions
conditional
random variate generation **6**: 3949
Laplace transforms **4**: 2660
District governors
gas equipment **2**: 837, 838
Disturbance decoupling
multivariable control systems **4**: 2522
nonlinear multivariable systems **5**: 3286
Disturbance rejection
linear systems **4**: ***2567–70***
nuclear power station control **5**: 3425
Disturbances
boundedness
adaptive control **1**: 105
ecological succession **2**: 1314
ecology **2**: ***1311–7***, ***1318–21***
communities **2**: 1316
control **2**: 1317
diversity **2**: 1315
excitation **2**: 1312
investigations **2**: 1314
nature **2**: 1312
propagation **2**: 1312
ecosystems **2**: 1356–8
endogenous disturbances
ecology **2**: 1313
exogenous disturbances
ecology **2**: 1313
impulse
ecology **2**: 1313
multivariable control systems design **5**: 3192
nonstationary disturbances
system identification **4**: 2304
process disturbances
ecology **2**: 1313
Diversions
road traffic control **6**: 4079
Diversity
ecological disturbances **2**: 1315
DNA
synthesis control **1**: 556
Document retrieval *see* Information retrieval
Documentation *see* Information science, Program and system documentation
Domains of attraction **2**: ***1211–2***
Domestic appliances
boiler and furnace efficiency **2**: ***1226–7***
gas appliance control systems **2**: ***1212–3***
space heating **2**: ***1214–26***
Dominance
diagonal dominance

multivariable control systems **5**: 3200
dominance sharing
Nyquist diagrams **4**: ***2597–600***
optimization
multivariable systems **5**: ***3220–3***
Dominance (of alternatives) **2**: ***1241***
multiple attribute utility theory **5**: 3111
noncooperative games **3**: 1917
Dominance matrices **3**: 1828
Doppler effect
fission reactors **5**: 3438
laser Doppler anemometry **2**: 1529
laser Doppler flowmeters **3**: 1677
laser Doppler velocimeters **7**: 4513
radar road traffic detectors **7**: 4918
DRAFT
activity-based simulation models **6**: 4393
Dredging
limnology
toxic substances **4**: 2780
Drilling
industrial robots
aerospace industry **4**: 2451
ships
adaptive control **6**: 4262
Drinking control **3**: 2212
Driver information center
automobile electronics **1**: 374
Drugs *see* Pharmacology
Drum boilers *see* Boilers
Drying
process control **2**: ***1249–53***
sugar industry
process control **7**: 4702
Dual control **2**: ***1253–7***; **6**: ***4163***
Duality **2**: ***1257–61***; **7**: 5045
auxiliary problem principle
coupling constraints **1**: 402
calculus of variations **1**: 524
convex optimization **7**: 5046
coupling constraints **1**: 400
generalized dependence **3**: 2005
linear multivariable systems **4**: 2828
linear programming **4**: ***2836–9***
linear systems **4**: 2863
methods of multipliers **2**: 791
multistage decomposition algorithms **5**: 3169
nonconvex optimization **7**: 5046
nonlinear programming
surrogate worth trade-off method **7**: 4709
optimization
goal coordination method **3**: 2037
state-space methods **7**: 4553
Duffing's oscillator
forced oscillator models **1**: 432
Duty cycling
load regulation **2**: ***1262***
minimum on-time **5**: ***2998***

Dyadic expansions
multivariable control systems **5**: ***3197–201***
semi-infinite optimization **5**: 3190
Dyadic transfer matrices **5**: 3197
Dynamic compensation **2**: 913
model reference adaptive control systems **5**: 3056
Dynamic control **2**: 1276–8
petrochemical industry **6**: 3674
robot arms **6**: ***4082–6***
Dynamic duopoly **2**: ***1262–8***
Dynamic games **1**: 684; **5**: ***3116–21***
deterministic incentives **2**: 993
moods of play **1**: 686
Stackelberg strategies **5**: ***3145–9***
see also Game theory
Dynamic modelling
finite element models
flexible spacecraft **3**: ***1650–3***
flexible spacecraft
multibody modelling **3**: 1653
polymerization **6**: 3740
Dynamic optimization **2**: ***1276–8***
complex systems **1**: 675–81
Dynamic positioning of ships
adaptive control **6**: ***4262–6***
Dynamic processors
local **2**: 898
Dynamic programming **2**: 1277, ***1278–9***; **7**: 4623
backward dynamic programming **1**: ***407–8***
computational complexity **1**: 724
deterministic control
perturbation techniques **5**: 3499
dynamic duopoly **2**: 1263
enumeration procedure **1**: ***407***
forward dynamic programming **3**: ***1713–4***
load dispatching **6**: 3820
fuzzy numbers **3**: 1833
fuzzy set theory **5**: 3538
Hamilton–Jacobi–Bellman equation **3**: ***2063–8***
Isaacs equation **2**: 1014
load dispatching
power systems **6**: 3819
Markov decision processes **5**: 2930
optimal stochastic control **5**: 3518, 3520, 3524
optimal stopping **5**: 3532
planning horizons **6**: 3704
separation principle **6**: 4247
stochastic control
discrete-time systems **2**: 1097
fuzzy set theory **7**: 4658
perturbation techniques **5**: 3501
stochastic maximum principle **7**: 4643
stock control **5**: 3509
surrogate worth trade-off method **7**: 4709
water resource management **7**: 5137
water supply pump optimization **7**: ***5152***
Dynamic simulation
workshop dynamic modelling **7**: ***5170***

Dynamic systems **2**: 1277, ***1282–6***
classification **2**: 1283
complex system partitions
Conant's procedure **4**: 2545
Richetin's procedure **4**: 2546
control **2**: 1277
definition **2**: 1283
fundamental structure **1**: 133
fuzzy set theory **4**: 2300
interconnected systems **2**: 923
stability **4**: ***2551–6***
large-scale systems
hierarchical control **3**: 2153
modelling **2**: ***1286–91***, ***1292–7***
multilayer control **2**: 1297
multilevel–multilayer control **2**: 1299
on-line coordination **2**: ***1297–305***
optimization
discretization **5**: 3015, 3038
goal coordination method **3**: 2036
interaction prediction approach **5**: 3016
model coordination method **5**: 3038
partial minimal realization **6**: 3627
problems **2**: 1285
simulation
chemical engineering **1**: 583
single systems
stability **7**: 4523
stability **2**: 912; **7**: 4523
structural stability **7**: 4670
stochastic models
simulation **7**: 4651
structure **2**: 948
time-domain modelling **4**: 2314
Dynamical systems **5**: 3080
abrupt changes detection **1**: ***81–5***
ecosystem stability **2**: 1348
information theory **4**: 2505
modelling **4**: 2240
likelihood functions **4**: 2286
probabilistic models **4**: 2286
Dynamics
agroecosystems **1**: 591
aircraft
state equations **1**: ***213–20***
continuous time dynamics
random processes **1**: 566
launchers **4**: 2728, 2729
multi-timescale systems **5**: ***3171***
muscle relaxation **5**: 3236
oscillators
biochemistry **1**: ***424–8***
physiological models
muscles **5**: ***3243–8***
random processes
discrete-time processes **1**: ***565–7***
Dynkin formula
stochastic calculus **7**: 4632

E

Earth orbital parameters
  launchers **4**: 2728
EBCDIC *see* Code standards
Echo suppression
  adaptive algorithms **1**: 78
Ecological modelling **2**: ***1323–4, 1330–2***
  aggregation errors **2**: ***1324–6***
  bond graphs **2**: ***1326–9***
  comparison **2**: 1332
  conceptualization **2**: 1330
  credibility **2**: 1332
  developmental modelling **2**: 1331
  environ analysis **2**: 1480
  epidemics **7**: 4645
  errors **2**: 1332
  flow models **5**: 2927
  flow networks **3**: ***1694–5***
  forecasting **3**: 1711
  gnotobiotic systems **3**: 2032
  grammars **3**: ***2043–5***
  heterarchical systems simulation **6**: 4353
  law of the minimum **2**: ***1329***
  Markov chains **5**: ***2927–9***
  Michaelis–Menten models **2**: ***1330***
  mixing **4**: ***2782–9***
  model-order estimation **5**: ***3040–4***
  neutral modelling **2**: 1331
  nitrification modelling
    lotic systems **5**: 3264
  nitrogen cycle modelling **5**: ***3266–8***
  parameter estimation **6**: ***3619–21***
  perturbation techniques **6**: ***3664–5***
  prediction **2**: 1332
  predictive models
    toxic chemicals assessment **7**: 4895
  radioecological modelling **6**: ***3929–33***
  rainfall-runoff modelling **6**: ***3940–4***
  sensitivity analysis **2**: 1549; **6**: ***4227–9, 4230–4, 4235–7***
  simplification **2**: 1332
  simulation experimentation **6**: ***4227–9***
  stochastic processes **7**: ***4644–5***
  temperature **2**: ***1333***
  time-scale estimation
    limmology **7**: ***5090–4***
  top-down modelling **7**: ***4894–5***
  toxic chemicals assessment **7**: ***4895–7***
  toxic substances
    fate **3**: 1600
    fugacity models **7**: ***4897–901***
  trophic structures **7**: ***4954–6***
  validation
    phytoplankton model **7**: ***5005–7***
  water quality **7**: ***5129–34***
  see also Compartment modelling, Ecosystem modelling, Population modelling
Ecological succession
  disturbances **2**: 1314
Ecology
  aluminum **1**: 264
  bilinear equations **1**: ***414–7***
  control
    disturbances **2**: 1317
  disturbances **2**: ***1311–7, 1318–21***
    excitation **2**: 1312
    propagation **2**: 1312
  diversity **2**: ***1321***
  ecosphere **2**: ***1342***
  energy systems **2**: ***1458–62***
  environ theory **2**: ***1479–85***
  error analysis **2**: ***1549–51***
  food chains **3**: ***1700***
  food webs **3**: 1700
  fuzzy set theory
    systems theory **3**: 1859
  gnotobiotic systems **3**: ***2032***
  information theory **2**: ***1321***
  maximum-power principle **5**: ***2956***
  nominal system behavior
    definition **2**: 1312
  organization of ecological systems
    modelling **2**: 1330
  random variations
    modelling **2**: 1331
  sensitivity analysis **2**: ***1549***
  size dependence **7**: ***4452–4***
  toxic substances **2**: ***1322–3***
  see also Ecological modelling, Ecosystems, Populations
Econometrics *see* Economic cybernetics
Economic cybernetics **7**: 4769, 4804
  history **4**: 2272
  modelling **2**: 1337
Economic dispatching *see* Load dispatching
Economics
  certainty equivalents **1**: ***565***
  cost–benefit ratios **2**: ***859***
  duality
    linear programming **4**: 2838
  dynamic duopoly **2**: ***1262–8***
  economic equilibrium **2**: ***1337***; **5**: 2977
    tatonnement **7**: ***4811***
  economic rationality **7**: 4781
  exchange **1**: 665
  fossil fuels **3**: 1715
  industrial robots **6**: 4096
  market clearing equations **5**: ***2927,*** 2974
  microeconomic modelling **7**: 4769, 4804
  microeconomic systems analysis **5**: ***2973–9***

microeconomics
technological innovation **7**: 4831
modelling **2**: ***1337–42***; **5**: 2975
normative economics **7**: ***5162–4***
oligopolistic competition
noncooperative games **5**: 2908
perfect competition **6**: ***3654–5***
power stations **3**: 1716
private ownership **1**: 665
production control
process industry **6**: 3888
production functions **5**: 2974; **6**: ***3890***
public goods **6**: ***3909***
revealed preference theory **5**: 2974; **6**: ***4066***
shadow prices **6**: ***4257***
steel industry
hierarchical control **7**: 4588
substitution **7**: ***4699***
team theory **7**: 4818
theory of the consumer **5**: 2975
theory of the firm **5**: 2975; **7**: ***4865***
time horizons **5**: 2978; **7**: 4892
welfare functions **7**: ***5162***
Economizer control
air conditioning **2**: ***1342***
Ecosphere **2**: ***1342***
Ecosystem modelling **2**: 1324
compartment modelling **2**: 1327, ***1343–4***
sensitivity analysis **6**: 4235
flow modelling
Markov chains **5**: 2927
hierarchical modelling **2**: ***1329***
lake modelling **4**: ***2643–6, 2650–1***
lake response time **4**: 2789
nitrogen cycle modelling **5**: 3266
parameter estimation **6**: 3620
qualitative stability **2**: ***1348–51***
see also Compartment modelling, Ecological modelling
Ecosystems **2**: ***1356***; **7**: 4775
abiotic control mechanisms **1**: ***1–7***
agriculture
herbicides **3**: ***2143–6***
analysis
catastrophe theory **1**: ***550–2***
fuzzy set theory **3**: ***1858–61***
directed graphs
decomposition **2**: ***1345–6***
disturbed ecosystems **2**: 1356–8
dynamics
compartment models **2**: 1344
elementary components **2**: 1346
grammars **3**: 2044
network structure **2**: ***1352–5***
nutrient cycling **5**: ***3440***
primary productivity **6**: ***3877***
stability **1**: ***672–4***
terrestrial carbon cycle **1**: 531
toxic chemicals assessment **7**: ***4895–7***
see also Ecosystem modelling, Ecology
ECSL
activity-based simulation models **6**: 4392
Edge detection
gray-scale image sensing **4**: 2481
Edges
graph theory **2**: ***1358***
Education
fuzzy set theory **2**: ***1358–61***
Educational aids
interpretive structure modelling **4**: 2579
Educational computing
interpretive structure modelling **4**: 2579
medical models **2**: ***1361–6***
Effective systems **2**: ***1366***
Effective temperature **1**: 642
Effectiveness
praxiology **6**: 3863
Efficiency
distributed database systems **2**: ***1155–60***
parameter estimation **6**: 3614
praxiology **6**: 3863
Efficient frontiers **2**: ***1367***
Efficient solutions
multicriteria decision making **5**: 3121
see also Pareto optimality
Efficient systems **2**: ***1367***
Eigen fuzzy sets **2**: ***1367***
Eigenvalues and eigenfunctions **2**: 818
assignment of energy levels
state-space methods **7**: 4552
control **2**: ***1367–72***
isoperimetric problems **4**: 2611
linear Boltzmann equations **7**: 4943
Einthoven triangle
electrocardiography **1**: 532
Elastic deformation
bone **1**: 470
connective tissue **1**: 475
Elastin
biomechanics **1**: 475
Electric control equipment
history **2**: ***1398–402***
Electric current measurement **2**: ***1372–5***
four terminal resistances **2**: 1372
power dissipation **2**: 1373
Electric field measurement
optical fiber sensors **5**: 3461
Electric generators **2**: ***1380–2***
automatic generation control **6**: 3813
control equipment **2**: 1401
load and frequency control **2**: ***1375–80***
parallel operation **2**: 1378
power systems **6**: 3780, 3796
speed control
dynamic response **2**: 1379
voltage control
reactive power control **6**: 3951

Electric motors *see* AC motors, DC motors, Stepping motors
Electric power generation **6**: ***3806–9***
  capacity **6**: 3790
  gas turbines **3**: ***1992–5***
  interconnected systems **6**: 3791
  see also Power systems
Electric sensing devices
  aircraft **1**: 250
    analytic redundancy **1**: 247
  aircraft engines
    control systems **1**: 225
  autopilots **1**: 381
  blood gas measurement **1**: 437
  drag-free satellites **2**: 1243
  failure detection
    analytic redundancy **1**: ***246–51***
  flowmeters **3**: 1689
  geometrical sensors
    three-dimensional vision **1**: 343
  helicopters
    automatic flight control systems **3**: 2132
  instrumentation **4**: 2530
  lighting control **4**: 2771
  location problems **2**: ***801–7***; **6**: ***4238–44***
  mechanoception **6**: ***4118–23***
  oxygen sensors
    applications **5**: 3578
    calibration **5**: 3578
    dissolved **5**: 3577
    errors **5**: 3578
  propulsion units
    Concorde **1**: 234, 235
  tactile sensors
    snakelike robots **6**: 4109
Electric variables measurement
  bridge circuits **2**: ***1402–3***
Electrical activity
  gastrointestinal tract **5**: 3019
  stomach **2**: 1033
Electrical conductivity
  level measurement applications **4**: 2740
Electrical engineering
  control systems history **2**: ***1398–402***
Electrical faults
  power system transients **6**: 3803
  power systems **6**: 3807
Electricity **2**: 1461
  modelling
    partial differential equations **6**: 4362
Electrocardiography **1**: ***450–6***
  cardiology **1**: 760–1
  classification
    fuzzy set theory **1**: 764
  clinical electrocardiography **1**: 455
  diagnosis
    fuzzy set theory **2**: ***1405–7***
  His bundle electrocardiography **2**: ***1407–10***
  medical diagnostic computing
    fuzzy set theory **1**: ***760–8***
  modelling **1**: ***532–5***
  physical basis **1**: 453
Electrochemical analysis
  oxygen concentration determination **5**: 3573
    dissolved oxygen **7**: 5115
  trace metals
    water pollution **7**: 5127
Electrodes
  extracellular electrodes
    digestive tract **1**: 460
  ion-selective electrodes
    environmental analysis **4**: ***2604–7***
  pH measurement **6**: ***3683–7***
  surface electrodes
    digestive tract **1**: 460
Electroencephalography **2**: ***1410–2***
  language perception **1**: 502
Electrohydraulics
  servomechanisms
    articulated robot arm control **1**: ***301–4***
  servovalves **6**: ***4252–7***
Electrolysis
  environmental analysis **2**: 1510
  modelling **2**: 1413
  parameter estimation **2**: 1415
  process control **2**: ***1413–6***; **5**: 3282
Electromagnetic devices
  separation process control **6**: 4251
Electromagnetic waves **7**: 4498
  spectra **7**: 4500
Electromagnets
  design **2**: 1177
Electromechanical devices
  adaptive control **1**: 47
Electromyography
  gait analysis **3**: 1907, 1908, 1913
  neuromuscular control in cats **6**: 4052
Electron device noise
  optical fiber interferometric sensors **5**: 3466
Electron lenses
  focusing **7**: 4938
  transmission electron microscopy **7**: 4937
Electron microscopy
  cell stereology **7**: 4599
  electron optics
    resolution **7**: 4939
  scanning electron microscopy **6**: ***4149–52***
  scanning transmission electron microscopy **6**: 4151
  transmission electron microscopy **7**: 4936, ***4936–40***
Electron probes **2**: ***1416–20***
Electron spectra **7**: 4501
Electron spectroscopy **7**: 4499
  chemical analysis **1**: 175
Electron spin resonance *see* Paramagnetic resonance
Electronic mail **5**: 3444
  voice mail boxes **5**: 3444
Electronics
  automobiles **1**: ***372–80***

Electrooptical devices
  missile guidance systems **5**: 3011
Electrophysiology *see* Bioelectric phenomena
Elements (chemical)
  distribution in marine environment **2**: ***1513–6***
Elevators
  breakdown rate **2**: 1449
  CAD **2**: ***1427–31***
  computer control **2**: ***1431–4***
    software structure **2**: 1433
  control algorithms **2**: 1426
  control systems **2**: ***1420–4***
  data logging **2**: 1443
  design **2**: ***1435–40,*** 1448
  door operation **2**: 1448
  group control **2**: 1422
  handling capacity **2**: 1435, 1436
  history **7**: ***5047–51***
  maintenance **2**: 1425
  management systems **2**: ***1424–7***
  mean system response time **2**: 1449
  multiple-car control algorithms **2**: 1422
  noise **2**: 1448
  operation **2**: 1446
  passenger arrival distribution **2**: 1440
  passenger comfort **2**: 1447
  passenger data **2**: 1442, 1448
  performance **2**: 1425, ***1441–5***
  reliability **2**: 1448
  round trip time **2**: 1435
  servicing **2**: 1424, 1448
  simulation **2**: ***1427–31***
  specifications **2**: 1425, 1449
  standards **2**: 1449
  supervision **2**: 1449
  traffic **2**: 1425
  transit time **2**: 1449
  user requirements **2**: ***1446–9***
Emergencies
  steel industry
    hierarchical control **7**: 4592
Emergent properties **2**: ***1450***
Empirical results **2**: ***1450***
Employment
  automation **1**: 369
Emulation
  computer system simulation **2**: ***1450–1***
Encoding
  channel encoding **2**: 1047
  coding errors
    digital communication systems **2**: 1044
    digital signals **6**: 4285
  digital encoding
    digital signals **6**: 4285
  information theory **4**: 2507
  signal transmission **6**: 4290
  source encoding **2**: 1046
End effectors **4**: 2368; **6**: ***4082***
  aerospace industry **4**: 2451
  control **4**: 2393
  robot arms **1**: 296
  robots **2**: ***1451***
  snakelike robots **6**: 4112
  vacuum grippers **4**: 2452
Endogenous variables **2**: ***1451***
Energy flow **2**: 1458
Energy hierarchies **2**: 1459
Energy loss spectroscopy **2**: 1419
Energy resources **2**: ***1452–4***
  consumption
    aircraft **3**: ***1749–56***
    road vehicles **3**: ***1756–61***
    ships **3**: ***1761–6***
  evaluation **2**: 1461
  fossil fuels **3**: ***1714–8***
  renewable **2**: ***1454–7***
  solar power satellites **7**: ***4472–6***
Energy savings
  lighting control systems **4**: 2775
Energy systems
  ecology **2**: ***1458–62***
  kinetics **2**: 1460
Engineering
  identification
    history **4**: 2272
  simulation **4**: ***2351–2***
Engines
  velocity control
    self-tuning regulators **2**: ***1462–6***
Entropy
  nutrients and pollutants
    environment **1**: 7
  see also Thermodynamic properties
Entropy (information) **6**: ***4257***
  conditional entropy **4**: 2543
  definition **3**: 1876
  fuzzy set theory **2**: ***1467–73;*** **3**: 1797, 1869
    brains **1**: 506
    decision processes **2**: 1472
    measures of fuzziness **5**: 2963
    ordering **3**: ***1875–8***
  Hartley entropy **4**: 2505
  information theory **4**: 2505, 2543; **7**: 4696
  maximum-entropy principle
    reconstructability analysis **6**: 3988
  possibilistic uncertainty **6**: 3988
  Shannon entropy **6**: ***4257***
  subjective entropy **7**: ***4692–5***
  subjectivity
    natural languages **7**: 4697
Entropy minimax theory **2**: ***1473–8***
Environ analysis **2**: ***1479–85***
  trophic structures **7**: 4956
Environ theory **2**: ***1479–85***
  Markov chains **5**: 2929
Environment **2**: ***1486–9***
  aluminum chemistry **1**: 262
  automated guideway transit **1**: 334

chemical analysis
ion-selective electrodes **4**: ***2604–7***
comfort indices **1**: 639
data
compression **2**: 1489–92
smoothing **2**: 1492–4
statistical analysis **2**: ***1494–7***
databases **2**: ***1500–3***
environ theory **2**: 1480
freshwater environment
abiotic control mechanisms **1**: ***1–7***
aluminum **1**: ***262–8***
green space
evaluation **2**: ***1506–7***
influenced environment **2**: 1487
influencing environment **2**: 1487
information monitoring **2**: ***1489***
input **2**: 1487
marine environment
distribution of elements **2**: ***1513–6***
modelling
partial differential equations **6**: 4363
noise measurement **2**: ***1522–7***
odor measurement **2**: ***1527–8***
output **2**: 1487
pollutants
equilibrium distribution **1**: 143
pollution
plants **2**: ***1516–21***
problem solving **2**: ***1536–7***
risk analysis **2**: ***1538–9***
risk management
systems management **7**: 4788
socioeconomic aspects **2**: 1537
solution chemistry
nutrients and pollutants **1**: 3
surface chemistry
nutrients and pollutants **1**: 5
terrestrial environment
abiotic control mechanisms **1**: ***1–7***
aluminum **1**: ***262–8***
trace metals
activation analysis **1**: ***18–21***
see also Ecology
Environmental control
heat systems
adaptive control **3**: 2095
Environmental databases
INFOTERRA **2**: ***1503–6***
Environmental measurements
automatic systems **2**: ***1508–12***
chemical analysis
infrared spectrometry **4**: ***2509–11***
data smoothing **2**: ***1492–4***
gas chromatography **3**: ***1931–2***
gas chromatography–mass spectrometry **3**: ***1933–6***, ***1936–8***
photoacoustic spectroscopy **6**: ***3699***
Environmental policy
systems analysis **2**: ***1539–41***
ELMES support system **2**: ***1542–4***
Environmental testing
data compression **2**: ***1489–92***
mass spectrometry **5**: 2936
Enzymes
allosteric regulation **1**: 424
modelling **1**: 422
repression
oscillators **1**: 425
Epidemics
modelling **7**: 4645
Epidermis
biomechanics **1**: 478
Epistemology
pragmatic information
epistemic gain **6**: 3859
Equal-loudness contours
hearing **1**: 323
Equality
fuzzy set theory **3**: ***1875***
Equality-constrained problems
Lagrange multipliers **2**: 793
Equalizers
adaptive signal processing
algorithms **1**: 75
Equations of motion
anticipatory systems **1**: 291
space vehicles (rigid) **1**: 319
Equivalence classes
linear multivariable systems **4**: 2867
Equivalence relations **2**: ***1549***; **3**: 1836
fuzzy set theory **4**: 2295
Ergonomics **3**: ***2179–86***
computing **3**: 2179
head-up displays **3**: 2087
Error analysis
drag-free satellites **2**: 1244
ecological models **2**: ***1549–51***; **6**: 4231
first-order error analysis **2**: ***1552–3***
simulation model validation **7**: 5013
Error correction
pattern recognition **6**: 3648
Error signals
anticipatory systems **1**: 293
Error statistics
aggregation
ecological modelling **2**: ***1324–6***
bias **2**: 1325
relative error **2**: 1325
total error **2**: 1325
total relative error **2**: 1325
Errors
augmented errors
adaptive control **1**: 104
model reference adaptive control systems **5**: 3063
auxiliary errors
model reference adaptive control systems **5**: 3063

classification **2**: 1554
continuous errors
modelling **1**: 101
definition **2**: 1554
degrees of error
many-valued logics **5**: 2920
equations
adaptive control **1**: 101
hybrid errors
modelling **5**: 3066
identification errors
analytical evaluation **4**: 2255
biomedicine **4**: ***2253–6***
influence errors
time variation **2**: 1557
input errors
system identification **4**: 2304
instruments **4**: 2532
intrinsic instrumental errors **2**: 1555
intrinsic errors
time variation **2**: 1557
measurements **2**: ***1553–9***
modelling
adaptive control **1**: 55
model reference adaptive control systems **5**: ***3065–7***
random errors **2**: 1557
reliability
industrial robots **4**: 2445
statistical analysis
simulation approaches **4**: 2256
systematic errors **2**: 1557
Erythrocytes
Coulter counters **2**: 864
Escalators
vertical transportation
history **7**: 5048
Escherichia coli
tryptophan operon
modelling **1**: 426
Esterline recorder
elevator performance measurement **2**: 1441
Esthetics
pragmatic fuzzy systems **6**: 3856
Estimation theory **2**: ***1559***
air pollution **2**: 1499
ARMA spectral estimation
stationary time series **7**: 4565
autoregressive spectral estimation
stationary time series **7**: 4562
control system computing **2**: ***813–25***
data smoothing
environmental data **2**: 1492
errors
large stochastic systems **4**: 2706
fuzzy set theory **6**: ***3616–8***
large stochastic systems **4**: 2706
linear systems **7**: ***4617–21***
maximum-entropy spectral estimation
stationary time series **7**: 4563
maximum likelihood methods **4**: 2284
model-order estimation
ecological models **5**: ***3040–4***
moving-average spectral estimation
stationary time series **7**: 4564
nonlinear estimation **5**: ***3293–302***
asymptotic analysis **5**: ***3303–10***
stochastic control **5**: ***3311–6***
nonlinear filtering **5**: 3293–302
geometric theory **5**: 3323
stochastic calculus **7**: 4631
stochastic integrals **5**: 3330
optimal control **5**: 3472
self-tuning control **6**: 4173
smoothing algorithms **7**: 4463
spectral estimation **4**: 2266
stationary time series **7**: 4559
systems analysis **7**: 4805
transfer functions **4**: 2268
two-filter smoothing algorithms **7**: ***4466–9***
see also Identification, Filtering and prediction theory, State estimation, Parameter estimation
Ethanol
fermentation measurement **3**: 1613
Ethernet *see* Networks (circuits)
Euclidean norm **2**: 924
Euler angles **1**: 107
aircraft
measurement **1**: 382
Euler equations
aircraft motion **1**: 215
optimization
calculus of variations **1**: 522
European Space Agency
flexible spacecraft
computer-oriented control analysis **3**: 1656
Eutrophication **7**: 4952
algal growth potential **7**: 5112
dynamic models
variance estimates **2**: ***1269–72***
lake modelling **4**: 2646; **5**: 3040
indeterminacy **4**: 2650
modelling
variance estimates **2**: 1272
toxic substances
limnology **4**: 2779
Evaluation **2**: ***1559***
Evaluation design **2**: ***1559–65***
Evaporation
control **2**: ***1566–8***
heat loss
physiology **7**: 4867
industrial plants **2**: 1566
modelling **2**: 1567
sugar industry
process control **7**: 4701
trace organic compound determination **7**: 5127

Evaporators **2**: 1566
Events **2**: ***1568***
Evolution
  control systems **4**: 2856
  Laplace transforms **4**: 2661
  semigroup theory **6**: 4221
  variational inequalities **7**: 5034
Exact penalty technique **1**: 405
Excitation
  ecological disturbances **2**: 1312
Exercise
  respiration feedback control **6**: ***4045–51***
Exogenous variables **2**: ***1568***
Experiment design
  frequency-domain analysis **4**: 2262
  identification **4**: ***2257–63***
Expert systems *see* Artificial intelligence
Explanation in systems science **2**: ***1578***
Exposure analysis model system **2**: 1323
Extended Kalman filters **2**: ***1581***; **4**: 2619
  large stochastic systems **4**: 2708
  see also Kalman filters
Extended Lebesgue space **4**: 2676
Extension principle **2**: ***1581***
  linguistic probabilities **4**: 2882
Extension techniques **2**: ***1581–3***
  Lie groups
    controllability **4**: 2758
External normalization
  database management languages **4**: 2658
Externality **2**: ***1583***
Extrapolation **2**: ***1583***
  systems analysis **7**: 4804
  technological forecasting **7**: 4826
Extremal control *see* Optimal control
Extrusion (polymers)
  process control **2**: ***1583–6***
Eye
  movement **7**: 5084
  nonlinear modelling **7**: 5065
  retinocortical visual pathways **4**: 2325
  structure and function **7**: ***5078–85***
  vestibular system modelling **2**: ***1591–5***
    slow-phase caloric nystagmus **2**: 1586

F

Factor analysis methods **2**: 876
Fail operational–fail safe
  space shuttle **7**: 4477
Failure analysis **6**: 4011
  distributed database systems **2**: ***1160–5***
  electric sensing devices
    analytic redundancy **1**: 247
  gas transmission networks **3**: 1989
Failure location
  innovation test
    electric sensing devices **1**: 248
Fair play **1**: 36
Fans
  air conditioning
    variable-air-volume **1**: 172
    extract fans **1**: 173
    fan coil **1**: 161
Fast algorithms
  constant parameter systems **2**: ***781–5***
Fast Fourier transforms **3**: ***1719–23***; **4**: 2265
  spectral analysis **7**: 4496
Fast modes **5**: 3171, 3174
Fast subsystems **5**: 3174
Fatigue
  aircraft active control technology **1**: 29
  pipes
    natural gas transmission systems **3**: 1983
Fault diagnosis
  many-valued logics **5**: 2925
Fault location
  estimation filters
    electric sensing devices **1**: 248
  space shuttle computers **7**: 4481
Fault trees **3**: ***1601***
  system performance **6**: 4013
Feasible decomposition method *see* Model coordination method
Feasible direction methods
  optimization **2**: ***786–8***
Feature extraction
  discriminant analysis **2**: 1103, 1105
  fuzzy set theory **3**: ***1601–3***
  see also Pattern recognition
Feedback **3**: ***1607***
  adaptive control **5**: 3044
  articulated robot arm control
    electrohydraulic servosystems **1**: 303
  autopilots **1**: ***388–93***
  blood pressure control **1**: ***495–8***
  control systems
    history **3**: 1606
    stability **7**: 4524, 4527
  control theory
    fuzzy set theory **4**: 2356
  cybernetics **2**: 871
  decomposition–coordination **2**: 952
  design
    characteristic-locus method **1**: ***575–80***
    multivariable-root-locus method **1**: ***575–80***
  distillation **2**: 1109
  dynamic output **2**: 912
  dynamic systems **2**: 1277
  environment
    problem solving **2**: 1536
  external respiration **6**: ***4051–5***
  fermentation control **3**: 1608
  float valves **3**: 1604
  frequency domain analysis **6**: 4224
  general anesthesia **1**: ***277–81***
  global feedback
    on-line coordination **7**: 4567
  glucose regulation
    modelling and control **3**: 2025
  history **3**: ***1604–7***
  intracranial pressure regulation **4**: ***2580–2***
  irregular optimal control
    generalized trajectories **4**: ***2608***
  linear quadratic regulation **5**: ***3513–4***
  local feedback
    on-line coordination **7**: 4568
  local static state **2**: 895
  loops
    autopilot functions **1**: 386
    lateral dynamics **1**: 390
    longitudinal dynamics **1**: 389
    pilot and autopilot **1**: 385
  modelling
    system dynamics **7**: 4759
  motivation
    hunger and thirst **3**: 2209–13
  multivariable control systems **4**: 2808, 2842; **5**: ***3197–201***
  muscle relaxation **5**: ***3235–8***
  nuclear power station control **5**: 3425
  optimal feedback **5**: 3497
    linear time-optimal problem **5**: ***3514–5***
    piecewise analytic feedback **6**: ***3998–4000***
    stock control **5**: 3509
  optimal stochastic control **5**: 3471
  output feedback control **1**: 131
    singular perturbations **7**: 4439
  parameter-adaptive control **6**: ***3596–604***
  pH control **6**: 3681, 3690
  respiration
    exercise **6**: ***4045–51***
  robot arms **6**: 4083
  self-tuning control
    input–output analysis **6**: 4205
  singularly perturbed systems
    output feedback control **7**: ***4439–43***

spatiotemperal feedback systems **5**: 3129
state feedback control **1**: 130, 388; **2**: 912; **7**: ***4529–34***
static output **2**: 897, 913
system dynamics
modelling **7**: 4757
telephony **3**: 1745
temperature **3**: 1604
ventilation
modelling **6**: 4055
vestibular system **2**: ***1591–5***
visual system neural networks **7**: ***5060–5***
see also Closed loop systems
Feedback strategies
game theory **1**: 685
Feedback synthesis
bang–bang theorems **1**: 413
Feedforward
adaptive control
telephone networks **7**: 4738
anesthetic depth **1**: ***281–4***
local feedforward **2**: 896
robot arms **6**: 4083
self-tuning control **6**: ***4187–9***
visual system neural networks **7**: ***5060–5***
Fermentation
gas analysis **3**: 1612
instrumentation **3**: 1615
measurement **3**: ***1611–4***
indirect parameter measurement **3**: 1614
modelling **1**: 422
process control **3**: ***1607–11, 1615–8***
Ferromanganese nodules
dispersion
limnology **4**: 2779
Filtering and prediction theory **2**: 871; **5**: ***3293–303***
data smoothing
environmental data **2**: 1492
distributed parameter systems **2**: 1190
experiment design **4**: 2257
finite-dimensional nonlinear filters **5**: 3327
Girsanov transformation **3**: 2007
Kalman filters
variations of formulae **4**: ***2615–8***
multidimensional systems **5**: 3129
multivariate analysis **5**: 3234
nonlinear filtering **5**: ***3293–302***
geometric theory **5**: 3323
iterated Ito integral expansions **5**: ***3330–6***
quantum theory **5**: ***3319–29***
stochastic calculus **5**: ***3330***; **7**: 4631
predictive control
fast-model **6**: ***3866–70***
self-tuning control
data prediction **6**: ***4168–72***
sensor location problem **6**: 4239
smoothing algorithms **7**: 4463
square root algorithms **7**: ***4534–9***
traffic simulation **7**: 4927
two-filter smoothing algorithms **7**: ***4466–9***
see also Estimation theory, Kalman filters, Prediction
Filters
analog prefilters **2**: 1067
image filtering
visual perception **4**: ***2325–30***
power system computer control **6**: 3839
spatiotemporal filters
visual perception **4**: 2326
state-variable filters **6**: 4283
see also Active filters, Digital filters, Passive filters
Filtration
natural gas transmission systems **3**: 1984
trace elements
water pollution **7**: 5123
Final control elements
air conditioning **1**: 156
Finite automata **5**: ***3268–72***
cellular automata **6**: ***4341–5***, 4351
arithmetic relators **6**: 4335
distributed system simulation **2**: 1200
Finite difference methods **3**: ***1623–7***
discretization **2**: 1295
heat transfer analysis **3**: 2114
partial differential equations **6**: 3623
Finite element analysis **3**: ***1627–32***
heat transfer analysis **3**: 2114
optimal stochastic control **5**: 3524
partial differential equations **6**: 3625
Finite element models
flexible spacecraft **3**: ***1650–3***
modal representation **3**: 1651
representation **3**: 1650
Finite memory
definitions **6**: 3629
past determinancy **6**: ***3629–31***
properties **6**: 3631
Fins
ship stability **6**: 4260, 4273
Firmware *see* Microprogramming
First-order variance propagation
lake modelling
eutrophication **2**: 1269
indeterminacy **4**: 2650
Fisher's information matrix
biomedical phenomena
modelling **1**: 486, 487
maximum likelihood identification **4**: 2285
see also Information theory
Fission reactors **5**: ***3436–40***
bidirectional fuelling **5**: 3439
breeder reactors **5**: 3415
control rods
neutron absorber **5**: 3438
coolant
density changes **5**: 3438
core power-density **5**: 3439
delayed neutrons **5**: 3438

fast neutron **5**: 3415
feedback effects **5**: 3438
fission-product poisons **5**: 3438
neutron kinetics **5**: 3438
reactivity **5**: 3436
reactor control **5**: 3438
thermal reactors **5**: 3414
see also Nuclear power stations
Fixed-lag smoothing algorithms **7**: 4466
Fixed modes **2**: 913
algebraic characterization **1**: 694
complex systems
decentralized control **1**: 694; **2**: ***902–7***
decentralized fixed modes **2**: 900, 902, 903
alternative characterizations **2**: 903
calculations **2**: 903
interconnected systems **2**: 906
properties **2**: 903
large linear systems **4**: 2667
structural fixed modes **1**: 694
Fixed-point algorithms
distributed algorithms
convergence **2**: 1122
smoothing algorithms **7**: 4465
Fixed systems
Lyapunov stability **7**: 4524
Flame ionization detectors
hydrocarbons **1**: 179
Flame photometry
sulfur dioxide **1**: 182
Flame protection
boilers **3**: ***1632***
domestic gas appliances **2**: 1212
industrial gas equipment **4**: 2355
thermoelectric devices
gas **2**: 1212
Flame traps
industrial gas equipment **4**: 2356
Flare
aerospace control **3**: 2085
Flashing
flow **3**: 1697
Flexible aerospace vehicles **3**: ***1632–3***
computer-oriented control analysis **3**: ***1649–64***
ATS-F model **3**: 1659
DCAP-2 **3**: ***1657***
ERS-I model **3**: 1660
Hipparcus models **3**: 1662
IPS model **3**: 1660
L-SAT model **3**: 1662
control analysis **3**: ***1659–63***
control models **3**: ***1656–9***
control synthesis **3**: 1649
finite element models **3**: 1650–3
augmented modal form **3**: 1652
base excitation **3**: 1652
modal representation **3**: 1651
modelling **3**: ***1659–63***
multibody modelling **3**: ***1653–6***
performance predictor **3**: 1649
Flexible manufacturing systems *see* Manufacturing computer control
Flexible programming
fuzzy set theory **5**: 3539
Flicker noise *see* Random noise
Flight management systems **1**: 387, 602; **3**: ***1669***
air traffic control **1**: 193
aircraft instrumentation **1**: 243
combat aircraft **1**: 622
flight director
autopilots **1**: 386
flight economy **3**: 1749–56
four-dimensional navigation **3**: 1719
fuel **3**: 1754
Flight path
control **1**: 386
differential equations **1**: 215
flight envelopes **3**: ***1669***
limiting **1**: 26
Flight plans
air traffic control **1**: 202
aerospace simulation **1**: 204
radar correlation **1**: 205
Flight simulation *see* Aerospace simulation
Floats
project engineering **3**: ***1669***
Flotation
process control **5**: 3280; **6**: 4249
trace elements
water pollution **7**: 5127
Flow
flashing **3**: 1697
fluid dynamics **3**: 1696
gases **3**: 1697
natural gas technology **3**: 1940
road traffic variables **7**: 4931
systemic veins **1**: 544
Flow characteristics
valves **2**: 829
Flow coefficients
gases **2**: 828
liquids **2**: 828
Flow control
heat exchangers **3**: 2089
natural gas **3**: 1962
process control **3**: ***1695–700***
valves **3**: 1696
characteristics **2**: 829; **3**: 1697, 1698
Flow imaging
two-phase flow **3**: 1692
Flow measurement **3**: 1695
capacitance methods **3**: 1691
correlation techniques
microprocessor chips **5**: 2987
natural gas **3**: ***1962–5***, 1984
process control
history **6**: 3882
quantity flow **3**: ***1681–5***

signal noise **3**: 1696
two-phase flow **3**: ***1685–93***
volumetric flow **3**: ***1670–7***
Flow rate
fermentation measurement **3**: 1613
Flowcharting **3**: ***1694–5***
Flowmeters
anemometers **3**: 1677
Coriolis force meters **3**: 1692
cross-correlation function **3**: 1689
cross-correlation meters **3**: 1673
differential pressure meters **3**: 1670
Doppler flowmeters **3**: 1686
electric sensing devices **3**: 1689
electromagnetic meters **3**: 1675
flow quantity meters **3**: 1672
fluidic meters **3**: 1673
inferential meters **3**: 1684
laser Doppler flowmeters **3**: 1677
natural gas technology **3**: 1962
nuclear magnetic resonance **3**: 1688
open channel meters **3**: 1677
positive displacement meters **3**: 1681
quantity meters **3**: 1681
radar meters **3**: 1676
semipositive meters **3**: 1684
tracers **3**: 1673
ultrasonic meters **3**: 1676
variable area meters **3**: 1672
vortex shedding meters **3**: 1673
Fluid dynamics **3**: 1696
mixing models **4**: ***2782–9***
Fluidized beds
manufactured fuel gas **3**: 1960
Fluoranthene
aerosol chemical analysis **1**: 176
Fluorescence
atmospheric laser spectroscopy **4**: 2720, 2721
chemical technology
optical fiber sensors **5**: 3462
environmental analysis **2**: 1509
time-decay fluorescence
optical fiber sensors **5**: 3460
wavelength-dependent fluorescence
optical fiber sensors **5**: 3460
Fluorescence spectra
fingerprinting of oil **7**: 5111
Fluorides
ion-selective electrodes
water pollution **4**: 2606
Fluorine
ecology
aluminum compounds **1**: 267
toxic substances **1**: 144–5
Fluorometry
sulfur dioxide
air pollution **1**: 182
Flutter mode
aircraft active control technology **1**: 31
Flutter suppression system
aircraft **1**: 31
Fly-through piloting
helicopters **3**: 2130
Fly-by-wire controls **1**: 603; **3**: ***1700***
space shuttle **7**: 4477
Fold catastrophe **1**: 550
Food chains
ecology **3**: ***1700***
radioecological modelling **6**: 3931
radioisotopes
radioecological modelling **6**: ***3933–6***
trophic levels **7**: 4954
Food processing industry
process control **3**: ***1700–5***
quality control **3**: 1701
sugar industry
optical rotation **1**: 12
process control **7**: ***4699–703***
viscosity control **2**: 781
Food webs **3**: 1700; **7**: 4955
Force control
industrial robots **6**: ***4113–7***
Force measurement **3**: ***1706–10***
gait analysis **3**: 1911
industrial robots **4**: 2379
Forecasting theory **3**: ***1711***
effectiveness **3**: 1711
evaluation **3**: ***1710–1***
hydrologic forecasting **3**: ***2228–33***
impact analysis **4**: ***2331–8***
information theory **4**: 2505
load forecasting
power systems **6**: 3830
planning horizons **6**: ***3704***
power **3**: 1711
systems analysis **7**: 4767
systems methodology **7**: 4803
time series **7**: 4804
validity **3**: 1711
water demand **7**: ***5154–5***
see also Technological forecasting
Forests
gap dynamics
hurricanes **2**: 1315
Forgetting factors
variable forgetting factors
self-tuning control **6**: ***3990–1***
velocity control
diesel engines **2**: 1466
Forging
industrial robots **4**: 2418
Formal languages **3**: ***1712***
automata theory **5**: 3269
definition **6**: 3645
Lindenmayer systems **4**: 2795
linguistic relativization
complexity **1**: 705
rational generating series **5**: 3272

rational languages **5**: 3271
stochastic languages
pattern recognition **6**: 3646
Formal logic
algebraic many-valued logics **5**: ***2921–3***
fuzzy Łukasiewicz logic **3**: ***1815–6***
interpretive structure modelling **4**: 2577
Łukasiewicz logic **5**: 2921–4
many-valued logics **5**: ***2918–21***
Formalism **3**: ***1712–3***
FORTRAN **6**: 3894
simulation languages
transaction flow-based models **7**: 4934
Fossil fuels **2**: 1461
energy **3**: ***1714–8***
life cycle **3**: 1714
Fourier analysis
discrete Fourier transforms
pseudo-random signals **4**: 2308
fast-Fourier-transform algorithm
pseudo-random signals **4**: 2308; **6**: 3777
identification **4**: 2242
signal analysis **6**: 4277
spectral analysis **7**: 4496
Fourier transform infrared spectroscopy **1**: 175
environmental analysis **4**: 2510
Fourier transforms **3**: ***1723–8***; **4**: 2265
applications **3**: 1727
fast Fourier transforms **3**: 1722
signal analysis **6**: 4277
Fracture toughness
bone **1**: 471
center-notched cylindrical specimens **1**: 472
compact tension specimens **1**: 472
single-edge-notched specimens **1**: 471
Frame representation
knowledge representation **4**: 2639
Frames **3**: ***1729***
Free-boundary problems
optimal stopping **5**: 3534
Free energy
fugacity **7**: 4898
Free surfaces control **3**: ***1729–34***
Freeways
automatic incident detection **3**: ***1734–6***
traffic control **7**: 4908, ***4910–3***
modelling **3**: 1736
multilayer control **3**: ***1736–41***
ramp assignment **7**: ***4913–5***
ramp metering **7**: ***4915–7***
traffic state estimation **3**: 1738
Freeze drying
pharmaceutical industry
process control **6**: 3692
trace elements
water pollution **7**: 5124
Frequency
accuracy **3**: 1742
characteristic frequency
feedback control **1**: 575
definition **3**: 1741
spectral purity **3**: 1742
stability **3**: 1742
standard transmission **3**: 1742
Frequency control
ac motors
stators **5**: 3089
electric generators **2**: ***1375–80***
power systems **6**: 3809
adaptive control **4**: 2891
see also Load-frequency control
Frequency converters
variable speed induction machines **7**: 5019
Frequency discrimination
hearing **1**: 323–5
Frequency-division multiplexing *see* Multiplexing
Frequency-domain analysis
experiment design **4**: 2262
feedback control **6**: 4224
identification **4**: 2241, ***2264–70***
large-scale systems
model simplification **5**: ***3071–7***
multivariable control systems
nuclear power station boilers **5**: 3431
nonlinear systems **5**: 3372
CAD **5**: 3380
stability **5**: 3318
Frequency-domain synthesis
direct digital control design **2**: 1074
multivariable control systems **5**: 3197
nuclear power station boilers **5**: ***3431–6***
Frequency matrix
feedback control **1**: 575
Frequency measurement **3**: ***1741–5***
frequency meters
accuracy **3**: 1743
gating error **3**: 1743
time-base errors **3**: 1744
trigger error **3**: 1743
precision **3**: 1744
stability
long-term **3**: 1744
short-term **3**: 1744
Frequency modulation
signal transmission **6**: 4289
telemetry **7**: 4835
Frequency response
calculations
linear control theory **2**: 822
closed-loop systems **7**: 4451
history **3**: ***1745–9***
identification **4**: ***2320–5***
linear systems **4**: 2264
sampled-data systems **6**: 4143, 4145
Frequency shift keying **6**: 4290
telemetry **7**: 4837
Frequential interpretation
dynamic behavior

global system **5**: 3174
Fuel
conservation
aircraft **3**: ***1749–56***
road vehicles **3**: 1756, 1759
ships **3**: ***1761–6***
fuel flow control
helicopters **3**: 2137
gas equipment **4**: ***2353–6***
controllers **2**: 837
gas interchangeability **3**: ***1766–9***
prediction **3**: 1766
gas turbines **3**: 1994
manufactured gas **3**: ***1958–61***
manufacturing principles **3**: 1959
space heating **2**: 1216
Fugacity
fugacity capacity **7**: 4898
toxic substances modelling **7**: ***4897–901***
Full-authority control
aerospace **3**: ***1769***
helicopter governors **3**: 2139
Functional analysis
calculus of variations **1**: 519
multivariable systems **5**: ***3201–5***
nonlinear systems
CAD **5**: 3382
optimal control **5**: 3511
stability **7**: 4528
Functional equations
analytic causal functionals **3**: 1770
critical points **1**: 527
functional differential equations
hereditary systems **3**: 2147, 2152
optimization algorithms **5**: ***3546–7***
Functional expansions **3**: ***1769–78***
iterated integrals **5**: 3330
Functioning mechanisms
optimal design **1**: 36
regular mechanisms
incentive systems **1**: 34
theory of active systems **1**: 31
Furnaces
control
burner location **2**: 801
Nyquist diagrams **4**: 2593
efficiency **2**: ***1226–7***
glass industry
adaptive control **3**: 2008
process control **3**: 2013
pressure control **6**: 3875; **7**: 4578
self-tuning control
data prediction **6**: 4171
space heating **2**: ***1227–33***
steel industry
system waiting function **4**: 2251
Fuzzification operators **3**: ***1779***
Fuzziness **3**: ***1780,*** 1867
energy-type measures **5**: 2965
entropy measures **2**: 1472
entropy orderings **2**: 1471
partition
measures **3**: 1603
scientific theories **7**: 4993, 4995
see also Fuzzy set theory
Fuzzy algorithms **3**: ***1782***
ISODATA algorithms **3**: 1602
Fuzzy cardinals **3**: ***1782–5***
elementary operations **3**: 1784
fuzzy equipotence class **3**: 1783
fuzzy numbers **3**: 1783
fuzzy sets
numbers of elements **3**: 1783
links **3**: 1785
notation **3**: 1783
ordinary quasiorderings **3**: 1784
Fuzzy clustering
classification **3**: 1601
Fuzzy coalitions **3**: 1913
Fuzzy concept learning
analogical inference **1**: 272
characteristics **1**: 276
Fuzzy conditional inference **3**: ***1791–2***
compositional rules **3**: ***1792***
fuzzy logic controller **3**: 1811
Fuzzy control
adaptive fuzzy control system **1**: ***72–3***
adaptive fuzzy controller **1**: ***73***
algorithm completeness **3**: ***1792–3***
applications in industry **4**: ***2356–61, 2362***
control theory **4**: 2358
convergence of fuzzy control process **3**: ***1793–4***
Czogała's and Pedrycz's controller **3**: ***1839***
dc series motor **2**: ***1071–2***
expert fuzzy controller **2**: ***1568–9***
fuzzy logic controller **3**: ***1811–2;*** **5**: 3453
algebraic analysis **3**: ***1812***
good-mapping property **3**: ***1812***
implementation of parameters **3**: ***1812***
interactivity **3**: ***1812–3***
linguistic synthesis **5**: 3155
self-improvement **3**: ***1813***
sensitivity **3**: ***1813***
stability **3**: ***1814***
structure **3**: ***1814***
fuzzy Łukasiewicz controller **3**: ***1815***
fuzzy possibilistic controller **3**: ***1838***
fuzzy probabilistic controller **3**: ***1839***
industry **4**: ***2356–61, 2362***
learning fuzzy controller **4**: ***2730–1***
linguistic analysis **3**: ***1794***
linguistic synthesis **3**: ***1794–5***
optimal fuzzy target control **5**: ***3515–6***
reproducibility property **3**: ***1795***
self-organizing fuzzy logic controller **3**: ***1813***
Fuzzy databases **3**: ***1795–7***
data manipulation language **3**: 1798
definition **3**: 1798

information retrieval 3: ***1798–800***
Fuzzy equations 3: 1852
Fuzzy events 3: ***1800–1,*** 1840, 1871, 1878
independence 3: 1804, 1841
probability 3: ***1802–5,*** 1819, ***1839–42***
conditional probability 3: 1803
Fuzzy games 3: ***1913–7***
Fuzzy ideals 2: ***773–5***
dynamics 2: 775
Fuzzy identification 3: ***1805–6***
Fuzzy implication 3: ***1806***
fuzzy logic controller 3: 1811
multidimensional 5: ***3124***
Fuzzy implication operators 3: ***1806–9***
checklist paradigm 3: 1807
Fuzzy integers 3: 1784
fuzzy ordering 3: 1785
fuzzy structure 3: 1785
Fuzzy logic 3: ***1811***
algebraic many-valued logics 5: 2921
fuzzy Łukasiewicz logic 3: ***1815–6***
many-valued logics 4: 2293; 5: 2918, 2925
algebraic logics 5: 2921
modus ponens
generalized 5: ***3089***
modus tollens
generalized 5: ***3089***
Fuzzy matrices 3: ***1853***
Fuzzy measures 3: ***1819–21, 1821–2, 1878–80***
epistemological hierarchy of systems 2: 1547
fuzzy integral approach 3: 1821
fuzzy orthogonal random measures 5: 3412
fuzzy random measures 5: ***3410–3***
information theory 6: 3766
Markov kernels 5: 3411
normal fuzzy measures 5: ***3410–3***
reconstructability analysis 6: 3984
see also Fuzzy set theory
Fuzzy models 3: ***1825***
accuracy 3: ***1825***
dc series motor 2: ***1072–3***
derivation 3: ***1825–6***
evaluation 3: ***1826***
linguistic relations 3: ***1825***
Fuzzy numbers 3: 1820, ***1829–31,*** 1855
applications 3: ***1831–5***
Delphi method 3: 1835
dynamic programming 3: 1833
fuzzy cardinals 3: 1783
fuzzy integers 3: 1784
hybrid numbers 3: 1856
interval of confidence 3: 1831
level of presumption 3: 1831
operations 3: 1831
ranking 3: 1846
trapezoidal numbers 3: 1832
triangular numbers 3: 1832
Fuzzy partitions 3: ***1835–8***
Fuzzy random variables 3: ***1843–5***
Fuzzy reasoning 1: ***296;*** 3: ***1847–52***
Baldwin's method 3: 1848
fuzzy learning strategies 5: 3263
fuzzy logic 3: ***1811***
Mizumoto's method 3: 1852
Tsukamoto's method 3: 1850
Yager's method 3: 1851
Zadeh's method 3: 1847
Fuzzy recurrency 3: 1785
Fuzzy relational equations 3: ***1852–3,*** 1889
Fuzzy relations 3: ***1853,*** 1857; 5: 2925
binary relations 6: 4004
causality 5: 2925
composite relations 1: ***710–3***
equations 1: 711
composition 1: 711
eigen fuzzy sets 2: ***1367***
fuzzy binary relations 3: 1836
fuzzy logic controller 3: 1811
inverse operation 5: 2925
*n*ary relations 6: 4006
ordering 3: 1846
verbalization 7: ***5047***
Fuzzy set theory 3: ***1853–8, 1867–71, 1872–4***
algebra 3: ***1780–1,*** 1880
signal detection 1: 652
α-cut 3: ***1872***
applications 3: 1858
cardiovascular state 2: 1406
Cartesian product 3: ***1786***
castastrophe theory 3: ***1786–8***
character recognition 1: 573
classification 3: ***1601–3***
compositional rules of inference 3: ***1792***
computer security 1: ***741–3***
conflict resolution 2: ***773–5***
control theory 3: 1893
applications 4: 2358
feedback 4: 2357
decision theory and analysis 3: 1873
discriminant analysis 2: 1105
dynamic programming 5: 3538
dynamical systems 3: ***1895–8***
ecology
systems theory 3: 1859
ecosystem analysis 3: ***1858–61***
electrocardiography
classification 1: 764
diagnosis 1: ***760–8;*** 2: ***1405–7***
entropy 2: ***1467–73;*** 3: 1797, 1855, 1869; 5: 2963
ordering 3: ***1875–8***
equality 3: ***1875***
extension principle 2: ***1581***
finite universe 3: 1874
foundations 3: 1859
fuzzy cardinals 3: 1783
fuzzy derivatives 3: 1897

fuzzy integrals **3**: ***1810,*** 1821; **4**: 2299; **5**: 3413
fuzzy learning strategies **5**: ***3261–4***
fuzzy linear programming
  multicriteria decision making **5**: 3121
fuzzy mapping **3**: ***1816***
  fuzzy sets **3**: ***1816***
  ordinary sets **3**: ***1816***
fuzzy matrices **3**: ***1853***
fuzzy numbers **3**: 1829–31
fuzzy partitions **3**: ***1835–8***
  combinatorial results **3**: 1837
  metric spaces **3**: 1837
fuzzy power set theory **3**: 1809
fuzzy random variables **3**: ***1843–5,*** 1871
fuzzy restrictions **3**: ***1853***
fuzzy sets and random sets **4**: ***2293–301***
fuzzy variables **3**: ***1905***
human reasoning **1**: 124
inclusion **3**: ***1875***
intersection **3**: ***1792***
intersection operators **3**: 1780
invariance **3**: 1897
lattice ordering **3**: 1877
linguistic approximation **4**: ***2880***
linguistic control **1**: 125
linguistic hedges **4**: ***2880–1***
linguistic probabilities **4**: 2882
linguistic variables **4**: ***2884***
mathematical programming **3**: ***1816–9***
max–min fuzzy relational operator **3**: ***1852–3***
measurement
  education **2**: ***1358–61***
measures of fuzziness **5**: ***2963–5, 2965–6***
medical diagnostic computing **1**: ***736–40,*** 762; **2**: 1574
membership evaluation **3**: 1787, ***1823–4***
modelling
  multicriteria modelling **3**: ***1826–9***
  operations research **5**: 3450
multidimensional fuzzy implication **5**: ***3124***
multiobjective optimization **5**: 3154
negation-distance measures of fuzziness **5**: 2964
nonmetric cluster analysis **1**: 606
operations **3**: ***1875***
operations research **5**: ***3449–54***
optimal stochastic control **5**: 3471
optimization **5**: ***3538–40***
ordering
  definition **3**: 1876
  linear **3**: 1876
  semantics **3**: 1875
ordinary mapping
  fuzzy sets **5**: ***3563***
parameter estimation **6**: ***3616–8***
possibility **3**: ***1862–7***
possibility distribution functions **6**: ***3769***
possibility distributions **6**: ***3769***
possibility measures **6**: 3770
pragmatic fuzzy systems **6**: ***3855–7***
prediction **3**: 1892
principle of incompatibility **4**: ***2343***
probabilistic fuzzy sets **3**: 1858
probability **3**: ***1862–7,*** 1870
  fuzzy causes **3**: ***1788–91***
queries **3**: 1796
random sets **6**: ***3945–7***
ranking **3**: ***1845–7***
relational algebra **3**: 1795; **6**: 4004
relational assignment equations **6**: ***4000***
representation theorem **3**: 1817
scientific theories
  vagueness **7**: ***4993***
σ-fields **3**: ***1878–80***
signal detection and estimation **1**: ***651–64***
singletons **3**: ***1880***
special systems **4**: 2293
stochastic control **7**: ***4657–61***
structural damage assessment **7**: ***4667–9***
subjective entropy **7**: 4692
switching functions **3**: ***1880–3, 1883–7***
  applications **3**: 1882
  canonical forms **3**: 1881
  minimal forms **3**: 1882
synthesis problems
  ordering **3**: 1877
system performance **7**: ***4764–6***
systems management **7**: 4787
theoretical operators **3**: 1780
topology **3**: 1870
union **3**: ***1792***
Fuzzy singletons **3**: ***1880***
Fuzzy systems **3**: ***1888–94, 1895***; **5**: 3538
  adaptive fuzzy automation **1**: ***72***
  brains **1**: ***505***
  computer implementation
    fuzzy discretization **3**: ***1800***
  control with constraints **3**: ***1895***
  control rules
    competition **2**: ***812***
    generation **2**: ***813***
  convergence **3**: 1793
  description **3**: 1898
  dynamical systems **3**: ***1895–8***
  fuzzy algorithms **3**: ***1782***
  fuzzy automata **3**: ***1782***
    identification **3**: ***1782***
  identification **3**: ***1843,*** 1891, ***1898–902***
  optimal control **5**: ***3488–9***
    fixed horizons **5**: 3488
  optimization **3**: 1893
  polynomial form **3**: 1901
  pragmatic fuzzy systems **6**: ***3855–7***
  prediction of future states **6**: ***3866***
  stability **3**: 1814, 1897; **7**: ***4529***
Fuzzy transformations **3**: ***1902–5***
  extrapolating **3**: 1903
  relational **3**: 1903

G

Gain
  Bode diagrams **4**: 2665
  characteristic gain
    feedback control **1**: 575
  gain matrix
    feedback control **1**: 575
  scheduling **1**: 45
  see also Amplification
Gait analysis **3**: ***1907–10***
  measurement **3**: ***1910–3***
Galvanometers
  string galvanometers **1**: 532
    electrocardiography **1**: 532
Game theory **3**: ***1913***
  bargaining solutions **1**: 684
  characteristic functions **1**: 682
  coalition analysis **7**: 4798
  complex systems management **1**: ***682–7***
  conflict analysis **7**: ***4794***
  cooperative games **1**: 684; **2**: ***840–9***; **7**: 4798, ***4811–22***
    applications **5**: 2910–1
    fuzzy games **3**: 1914
  cooperative solutions **1**: 684
  core of a game **1**: 684
  deterministic incentives **2**: 993
  differential games **2**: ***1010***
    closed-loop games **2**: ***1004–8***
    dynamic duopoly **2**: ***1262–8***
    Isaacs equation **2**: ***1010–6***
    linear quadratic games **2**: ***1017–20***
    open-loop games **2**: ***1020–3***
    stochastic games **7**: ***4613–7***
    stopping-time problems **7**: 4616
  dominance
    noncooperative games **3**: 1917
  dynamic duopoly **2**: ***1262–8***
  dynamic games **1**: 684; **5**: ***3116–21***
    Stackelberg strategies **1**: 684; **5**: ***3145–9***
  extensive form **1**: 682
  feedback strategies **1**: 685
  feedback structures
    dynamic duopoly **2**: 1263
  finite games
    Nash equilibrium **5**: 3255
  fuzzy games **3**: 1913
  games with denumerably many players **3**: 1929
  games with infinitely many players **3**: ***1926–9***
  hypergames **7**: 4798
  information structure **1**: 682
  learning automata **4**: 2718
  management **1**: 685; **5**: ***2907–11***
  martingales
    stochastic calculus **7**: 4631
  memory strategies **1**: 685
  metagame analysis **7**: 4794
  minimax techniques **5**: ***2994–7***
  mixed games **3**: 1928
  multiattribute utility theory **5**: ***3099–103***
  multicriteria decision making **5**: ***3121–4***
  multiperson stochastic games **7**: 4622
  Nash equilibrium **1**: 35, 683, 700; **5**: 2908, ***3255***
  nonatomic games **3**: 1926
  noncooperative games **1**: 683; **3**: ***1917–21***
    applications **5**: 2908–10
  noncooperative solutions **1**: 683
  normal form **1**: 682
  one-person stochastic games **7**: 4621
  Pareto optimality **1**: 684; **5**: ***3156–63***; **6**: ***3621–3***
  prisoners' dilemma **3**: 1918
  repeated games **3**: ***1922–6***
  semidifferential games **1**: 686
  Shapley value **2**: 848; **3**: 1926
  side-payments **2**: 844
  social choice theory **7**: ***4469–72***
  Stackelberg strategies **1**: 684; **5**: ***3145–9***
  static games **2**: 1263
    Nash equilibrium **5**: 3255
    repeated games **3**: ***1922–6***
  stochastic differential games **7**: ***4613–7***
  stochastic games **3**: 1923; **7**: ***4621–2***
  systems analysis **7**: 4804
  trees (mathematics) **1**: 682
  two-person games **5**: ***2994–7***
  see also Decision theory and analysis, Multicriteria decision making
Gamma cameras **6**: 3938
Gamma rays
  detection and measurement
    optical fiber sensors **5**: 3462
  medical diagnostic imaging **6**: 3936
Gantt charts
  project management **6**: 3899
Garments *see* Clothing, Textile industry
Gas analysis
  fermentation measurement **3**: 1612
Gas chromatography *see* Chromatography
Gas chromatography–mass spectrometry **1**: 176
  computerized search systems **3**: 1933
  environmental analysis **3**: 1932, ***1933–6***, ***1936–8***, ***1936–8***
Gas distribution systems *see* Natural gas
Gas equipment
  ionization detectors **4**: 2355
Gas (fuel) *see* Fuel
Gas (natural) *see* Natural gas
Gas turbines **1**: 236
  aerospace propulsion **1**: 226
  closed-cycle gas turbines **3**: 1994
  control **3**: 1995

design **1**: 581
improvement
Nyquist diagrams **4**: 2588
industrial gas turbines **3**: 1993
modelling **1**: 581
power generation **3**: ***1992–5***
ship propulsion control **6**: 4268
Gases
blood gases
continuous in situ measurement **1**: 439
in vitro measurement **1**: 439
measurement **1**: ***437–40***
flow **3**: 1697
coefficients **2**: 828
Gauss–Markov processes
realization theory **6**: ***3962–4***
see also Markov processes, Random processes, Stochastic processes
Gaussian distribution **3**: 1998
Gaussian noise *see* Random noise
Gaussian random variables **3**: ***1998–9***
Gelfand equation
isoperimetric problems **4**: 2612
Gene synthesis control **1**: 556
General Purpose Simulation System *see* GPSS
General systems **3**: ***1999–2000***
praxiology **6**: 3860
reconstructability analysis **6**: 3984
self-organizing systems **6**: 4165
General systems problem solver **3**: ***2000–3***
Generalized dependence **3**: ***2003–5***
Generalized gradients **7**: 5040
Generalized Hessenberg representation **1**: 133
Generalized likelihood ratio tests
dynamical systems
abrupt changes **1**: 84
Generating series
nonlinear systems **4**: 2525
Generative systems **2**: 1546; **3**: 2001
Generators *see* Electric generators
Genons **2**: 1346
Geodesics
calculus of variations **7**: 5045
Geometry
computational geometry
three-dimensional vision **1**: 343
similarity theory **6**: 4294
Geophysics
modelling **3**: ***2005–6***
partial differential equations **6**: 4363
Geothermal power **2**: 1456, 1461
Gerschgorin set **2**: 925
Gerschgorin theorem **2**: 923; **5**: 3176
multivariable control systems **5**: 3200
Girsanov transformation **3**: ***2007–8***
Givens transformations **5**: ***2949–50***
square root factors **7**: 4535
Glass
composition **3**: 2011
forming
process control **3**: 2013
manufacture **3**: 2011
Glass industry
adaptive control **3**: ***2008–10***
process control **3**: ***2011–4***
Global positioning satellite system
inertial navigation **4**: 2491
Globe thermometers **3**: ***2016–7***
Glomerular filtration **3**: 2204
GLR tests *see* Generalized likelihood ratio tests
Glucose regulation
adaptive control **3**: ***2017–20***
closed-loop control **3**: 2026
hunger **3**: 2210
modelling **3**: ***2021–5***
diabetics **3**: ***2025–32***
open-loop control **3**: 2026
Glycolysis
oscillators
modelling **1**: 426
Gnathosonics **2**: 966
Gnotobiotic systems
ecology **3**: ***2032***
Goal coordination method **3**: ***2032–8***
Goal-driven approaches
problem solving **3**: ***2038***
Goal-oriented systems **1**: 31–7; **3**: ***2038–43***
modelling **3**: 2039
stochastic control
fuzzy set theory **7**: 4657
teleology **7**: ***4848***
workshop dynamic modelling **7**: 5174
Goal programming **5**: 3119, 3151, 3157
multicriteria decision making **5**: 3121
Goals **5**: 3119
Gödel-enumeration
self-applicability **1**: 326
Gödel sentences
decidability **2**: 935
Goniometers
gait analysis **3**: 1907, 1912
Governors
domestic gas appliances **2**: 1212
electric generators **2**: 1376
helicopter engines **3**: ***2137–9***
history **7**: ***4513–7***
industrial gas equipment **4**: 2356
proportional–integral governors
helicopter engines **3**: 2137
proportional governors
helicopter engines **3**: 2137
GPSS
process-based models **6**: 3878
transactions-flow models **7**: 4933
Gradient methods
conjugate-gradient methods **5**: 3485; **7**: ***4969–71***
control in coefficients **2**: 812
optimal control

implementation **5**: 3484
numerical solution **5**: ***3481–7***
Grammars
definition **6**: 3645
ecological modelling **3**: ***2043–5***
fuzzy relational grammars **1**: 507
programming languages **7**: 4752
Granular materials
pharmaceutical industry
process control **6**: 3692
Graph theory **3**: ***2046***; **7**: 4666
combinational optimization **2**: 1082
differential equations **2**: 1078
ecosystem network structure **2**: 1352
edges **2**: ***1358***
nodes **5**: 3268
nondirected relations **5**: ***3279***
Petri nets **6**: ***3665–70***
structural analysis **7**: ***4661–6***
transitivity **7**: ***4935***
trees (mathematics) **7**: 4949
see also Directed graphs, Graphs
Graphic matroid
combinatiorial optimization **1**: 624
Graphitization
natural gas distribution systems
leakage **3**: 1955
Graphs **2**: ***1080***
connectivity **7**: 4662
elementary components **2**: 1345
hierarchical structures **7**: 4662
Nyquist diagrams
design applications **4**: 2586
simulation **6**: 4349
valued graphs
partition **7**: 4665
see also Directed graphs, Graph theory
Gravity models
spatial allocation **7**: ***4491–5***
Great Lakes
algal control modelling **1**: 257
Greediness
combinatorial optimization
independence **1**: 625
Green's function
isoperimetric problems **4**: 2609
Grinding
mineral processing **5**: 2990
process control **2**: 780; **5**: 3280
Grippers *see* End effectors
Gross specific growth rate
Cladophora modelling **1**: 258
Ground controller
air traffic control **1**: 198
Groundwater
hydrologic forecasting **3**: 2230
Group activity *see* Collective enquiry
Group rapid transit **1**: 336
Group theory
control theory **5**: 2947
Lie groups
control theory **5**: ***2944–8***
controllability **4**: ***2756–9***
Guidance
combat aircraft **1**: 622
launchers **4**: 2728–9
missiles **5**: ***3006–12***
road traffic **6**: ***4075–9***
ship adaptive control **6**: ***4275–6***
see also Navigation
Guided missiles *see* Missiles
Gusts
aircraft active control technology **1**: 27
airships **1**: 254
Gut *see* Digestive tract
Gyroscopes
inertial navigation **4**: 2488
ship stability **6**: 4259, 4271

H

H-bahns **1**: 338
Hadamard matrix **5**: 2956
Hall effect devices
  tachometers **7**: 4512
Halothane
  anesthesia modelling **1**: 281
Halting problems
  Turing machines
    decidability **2**: 935
Hamilton–Jacobi–Bellman equation **2**: 1011; **3**: ***2063–8***
  continuity **7**: 4624
  numerical methods **3**: ***2068–72***
  optimality principle **5**: 3537
  smooth solutions **5**: 3537
  stochastic differential games **7**: 4614
  stochastic processes **7**: ***4622–6***
  uniqueness **7**: 4626
  viscosity **7**: 4624
Hamiltonian cycles
  combinatiorial optimization **1**: 624
Hammerstein models
  calculation **5**: 3349
  self-tuning optimal control **6**: 4180
Hamming window *see* Smoothing windows
Hand-off
  air traffic control **1**: 198
Hankel matrix
  realization theory **6**: ***3965–8***
  singular values **1**: 409
Harbors *see* Ports
Harmonic analysis
  generalized
    history **4**: 2271
  nonlinear systems **5**: 3384
    CAD **5**: 3380
Harmonic balance
  reliability **2**: 979
Hazard detection
  fuzzy switching functions **3**: 1882
Head-up displays **3**: ***2080–9***
Heading control loop
  autopilots **1**: 392
Health hazards
  industrial robots **6**: 4095
Hearing **1**: ***323–5***
  binaural processing **1**: 325
  monaural localization **1**: 325
  sound localization **1**: 325
  temporal resolution **1**: 325
Heat
  acclimatization
    physiological models **7**: 4876
  launchers **4**: 2730
Heat conduction
  distributed parameter systems **2**: 1182
Heat engines
  domestic gas appliances **2**: 1213
Heat equations **6**: 3591
  Fourier transforms **3**: 1728
  Laplace transforms **4**: 2663
  semigroup theory **6**: 4221
  stochastic systems **7**: 4617
Heat exchangers
  control theory
    fuzzy set theory **4**: 2359
  process control **3**: ***2089–92***
Heat losses
  evaporative
    physiology **7**: 4867
  physiology **7**: ***4866–8***
Heat pumps **3**: 2092
  air conditioning **1**: 169, ***165–7***
Heat transfer
  HVAC systems
    digital simulation **3**: 2113
  modelling
    numerical methods **5**: 3097
  Stefan problem **5**: 3097
  temperature control
    physiology **7**: 4867
  see also Radiative transfer
Heating systems *see* HVAC systems
Heavy metals
  desorption
    sediments **4**: 2778
  dispersion **4**: 2780
    microorganisms **4**: 2782
  ecology
    aluminum compounds **1**: 267
Height measurement
  feedback **1**: 393
Helicopters **1**: 114; **3**: ***2118–23***
  active control technology **1**: 31
  aerospace control **3**: ***2124–33***
  control configured vehicles **3**: 2134
  controls **3**: 2126
  dynamics **3**: 2120
  engine governors **3**: ***2137–9***
  fly-by-wire controls **3**: 2134
  long-term prospects **3**: ***2133–6***
  principles of flight **3**: 2118–22
  vibration control **3**: 2135
  see also Aircraft
Hematopoiesis
  bifurcations
    random processes **1**: 569
  feedback control
    modelling **1**: 569

Hemodialysis **1**: 307, 310
  dialyzate regeneration **1**: 311
Hemodynamics
  coronary hemodynamics modelling **2**: ***850–3***
  educational computer models **2**: 1362
  gravity **1**: 544
  hepatic uptake **3**: 2139
  modelling **1**: ***542–6***
  nuclear magnetic resonance **5**: 3419
  organs
    measurement **7**: 4902
  thermoregulation **7**: 4872
Hemofiltration **1**: 310
  hemofiltrate regeneration **1**: 311
Hemoperfusion **1**: 311
Hepatobiliary transport
  kinetics modelling **3**: ***2139–43***
Herbicides
  agroecosystems **3**: ***2143–6***
  s-triazine **3**: 2145
Hereditary systems **3**: ***2147–50, 2151–2***
  optimal control **5**: ***3502–6***
Heterarchical systems
  simulation modelling formalism **6**: ***4350–3***
Heuristic programming
  history **7**: 4774
  medical diagnostic computing **2**: 1575
Heuristics **3**: ***2153***
  see also Hypotheses
Hierarchical control **1**: 688
  dynamic systems
    on-line coordination **2**: 1302
  large-scale systems **1**: 135
  large stochastic systems **4**: 2710
  manufacturing computer control
    Petri nets **6**: ***3670–3***
  partition **3**: ***2153–62***
  power systems **6**: 3814
Hierarchical inference **4**: ***2331–8***
Hierarchical structures
  graphs **7**: ***4662***
Hierarchical systems
  biofeedback **1**: 435
  decomposition
    simulation modelling **6**: ***4353–5***
  ecological modelling **2**: ***1329***
  parameter estimation
    large-scale systems **4**: 2702
  simulation program design **6**: 4396
  steel industry **7**: ***4585–93***
    personnel **7**: 4590
  structural analysis
    graphs **7**: 4662
  walking robots **6**: 4127
Hierarchy **3**: ***2163***
Hierarchy of systems
  epistemology **2**: ***1545–8***
  general systems problem solver **3**: 2000
High-level languages
  FORTRAN **6**: 3894
  simulation
    computer graphics **6**: 4301
High-pressure liquid chromatography **1**: 176
High-voltage engineering
  dc transmission networks **6**: ***3820–5***
Hilbert space
  Wiener integral **7**: 4617
Hilbert transforms
  multidimensional systems **5**: 3128
His bundle
  electrocardiography **1**: 451
HOCUS
  activity-based simulation models **6**: 4392
Holism **3**: ***2163–4***; **6**: 3998
  systems complexity **1**: 704
Holography
  combat aircraft **1**: 622
Homeostatic mechanisms **2**: 871
Homing
  aerospace **3**: ***2164***
    passive homing **3**: 2164
    semiactive homing **3**: 2164
  missiles **5**: 3010
Homogenization
  nonlinear estimation theory **5**: 3308
  optimal control
    control in coefficients **2**: 809
  polymerization
    process control **7**: 3744
Homomorphism **3**: ***2164–8***
  hierarchy **3**: 2165
Homotopy **3**: ***2168–9***
  homotopic maps **3**: 2168
  path-following methods **6**: 3631
Hopf bifurcation theorem **7**: 4526
Horizontal situation indicator
  aircraft instrumentation **1**: 242
Hormonal regulation **3**: ***2169–74***
Hot blast cupola
  data analysis **2**: 876
Hot rolling **6**: 4139
Householder transformations
  matrix algebra **5**: ***2950–1***
  square root factors **7**: 4535
Hulls
  ship fuel conservation **3**: 1761
Human factors
  aerospace **1**: 116
  flight-crew workload **3**: 1666–8
  task analysis **3**: 2183
Human factors engineering *see* Ergonomics
Human information processing **3**: ***2186–91***
  cognitive styles **7**: 4790
Humanistic systems **6**: 3856
Humic substances
  adsorption
    minerals **1**: 3

Humidistats **3**: ***2206***
Humidity control **1**: 160; **3**: ***2206–9***
  air conditioning **1**: 148, 154, 168
  comfort modelling **1**: 639
  humidistats **3**: ***2206***
  see also Moisture control
Hunger
  feedback control **3**: ***2209–13***
  thirst **3**: 2213
Hurricanes
  forest gap dynamics **2**: 1315
HVAC systems
  actuator modelling **3**: ***2097–9***
  adaptive control **3**: ***2092–6***
    energy-saving potential **3**: 2094
    storage capacity effects **3**: 2094
  analog simulation **3**: ***2099–104***
  control process interactions **3**: ***2105–8***
  controller tuning **3**: ***2108–13***
  digital simulation **3**: ***2113–7***
  heat pumps **3**: 2092
  hydronic systems **3**: ***2234***
  parameter-adaptive control **3**: 2093
  pumps **6**: 3917
  solar power **2**: 1455
  see also Space heating
Hybrid computers **3**: ***2214–7***
  applications **3**: 2216
  automatic control
    simulation **6**: 4311
Hybrid numbers
  fuzzy set theory **3**: 1856
Hydration
  weathering of minerals **1**: 1
Hydraulic phenomena
  limnology **4**: 2780
Hydraulic systems
  differential equations **1**: 546
  hydraulic turbines
    nonlinear modelling **6**: 3774
Hydrocarbons
  air pollution
    automatic analyzers **1**: ***179–81***
  saturated aliphatic hydrocarbons
    aerosols **1**: 176
Hydrocyclones
  separation process control **6**: 4250
Hydrodynamics
  aircraft engine control systems **1**: 223
  road traffic simulation **7**: ***4920–2***
Hydroelectric power stations **2**: 1461; **3**: ***2217–21, 2221–5***
  components
    transfer functions **6**: 3777
  self-selecting control **3**: ***2225–7***
Hydrogen bonds
  nutrients and pollutants
    environment **1**: 6
Hydrogen economy **3**: 1959
Hydrogenerators **2**: 1382
Hydrology
  forecasting **3**: ***2228–33***
  groundwater **7**: 5141
  hydroelectric power stations **3**: 2218
  hydrological cycle **7**: 5140
  mixing models **4**: ***2782–9***
  nitrogen cycle modelling **5**: 3266
  rainfall-runoff modelling **3**: 2228; **6**: ***3940–4***
  streamflow forecasting **3**: 2228
Hydrolysis
  nutrients and pollutants
    environment **1**: 3
  weathering
    minerals **1**: 1
Hydronic systems **3**: ***2234***
Hydrophones
  optical fiber interferometric sensors **5**: 3468
Hydrostatics
  level measurement **4**: 2739
Hyperbolic systems
  Hamilton–Jacobi–Bellman equation **3**: 2071
  linear control theory **4**: ***2800–7***
Hypercycles
  self-organizing systems **6**: 4167
Hyperfinite model theory
  many-valued logics **5**: 2921
Hypergames *see* Game theory
Hyperstability
  adaptive control **1**: ***93–7***
  model reference adaptive systems **2**: 1087
Hypertension
  patient treatment
    feedback control **1**: ***495–8***
  see also Blood pressure
Hyperthermia *see* Biothermics
Hypothalamus
  thermoregulation **7**: 4872
Hypothermia *see* Biothermics
Hypotheses **3**: ***2234***
  management **5**: ***2911–7***
  military decision making **1**: 649
  see also Heuristics
Hysteresis
  biological systems **1**: 431

I

IDEF$_2$
  simulation databases **2**: 894
Identifiability **4**: 2236, 2303
  distributed parameter systems **2**: 1196
  parameter estimation
    simulation **7**: 4691
Identification **2**: ***1559***; **4**: ***2239–45***
  asymptotic identification **4**: ***2235–9***
  Bayes methods **4**: 2285
  black-box models **4**: 2288
  control in coefficients **2**: 809
  correlation methods **4**: ***2245–53***, 2317
  Cramér–Rao inequality **4**: 2285
  distributed parameter systems **2**: ***1193–7***
  experiment design **4**: 2242, ***2257–63***
  Fisher information matrix **4**: 2285
  frequency-domain methods **4**: ***2264–70***
  frequency response **4**: ***2320–5***
  fuzzy automata **3**: ***1782***
  fuzzy identification **3**: ***1805–6***
  fuzzy sets with random sets **4**: ***2293–301***
  fuzzy systems **3**: ***1843***, 1891, 1899
  history **4**: ***2270–2***
  identification errors
    analytical evaluation **4**: 2255
    biomedicine **4**: ***2253–6***
  input–output models
    hydrologic forecasting **3**: 2231
  input–output sequences **5**: 3041
  instrumental variable techniques **4**: ***2273–7***
  least squares method **4**: ***2278–83***
  linear continuous systems
    control system analysis computing **4**: 2800
  maximum likelihood method **4**: ***2284–7***
  model reference adaptive control systems **6**: 3596, 4196
  model structure determination **4**: ***2287–92***
  model validation **4**: 2242; **7**: ***4996–500***
  nonlinear systems
    power plants **6**: 3772
    state-space models **5**: ***3374–6***
  on-line methods **4**: ***2310–4***
  overlapping model structures **4**: 2288
  parameter-adaptive algorithms **6**: 4162
  phosphorus kinetics **5**: 3043
  procedure **4**: 2241
  pseudo-random signal method **4**: ***2307–10***
  real-time methods **4**: ***2310–4***
  reconstructability analysis **6**: 3985
  recursive methods **4**: ***2310–4***
  sensor location problem **6**: 4241
  simultaneous identification and control
    fuzzy systems **3**: ***1902***
  steepest-descent methods
    lung parameters **1**: 457
  step response **4**: 2320
  stochastic control **2**: 1253
  system identification **5**: 3473
  systems analysis **7**: 4769
  time-domain methods **4**: ***2314–9***
    least squares method **4**: 2278
  transient response **4**: ***2320–5***
  Volterra series **6**: 3772
  see also Estimation theory, Parameter estimation, System identification
IFAC *see* International Federation of Automatic Control
IFR *see* Instrument flying rules
Ignition
  domestic gas appliances **2**: 1213
  industrial gas equipment **4**: 2354
Image processing *see* Picture processing
Image sensors
  artificial satellites **6**: 4038
  materials **6**: 4038
  medical thermography **7**: 4869
  plants (biological)
    environmental pollution **2**: ***1516–21***
Imaging
  gray scale imaging **4**: ***2479–84***
  industrial vision systems **4**: 2470
  medical diagnostic imaging
    radioisotopes **6**: ***3936–40***
    thermography **7**: ***4868–71***
  nuclear magnetic resonance **5**: ***3421–5***
  radiology **7**: ***5179–86***
  ultrasonics **7**: ***4959–64***
Immunofluorescence
  Coulter counters **2**: 866
Impact analysis **4**: ***2331–8***; **7**: 4767, 4803
  systems methodology **7**: 4804
  workshop dynamic modelling **7**: 5171
Impedance
  amplifier impedances **1**: 269
Impulse control **4**: ***2339–42***
Impulse response **1**: 14
  random signals **1**: 419
Impulsive function **2**: 1335
Incentive systems
  degree of centralization **1**: 33
  design **1**: 34
Incentives
  deterministic incentives **2**: ***993–5***
  game theory **1**: 685
  stochastic processes
    decision making **7**: ***4627–30***
Incidence angle
  measurement **1**: 381
Inclusion
  fuzzy set theory **3**: ***1875***

Inclusion principle
  overlapping decentralized control **5**: 3570
Inclusion regions **2**: 923
Independence **4**: ***2343***
  multiattribute utility theory **5**: 3105
  optimization
    combinatiorial mathematics **1**: ***624–6***
Independent variables **4**: ***2343***
Indeterminacy
  modelling **2**: 1551
    lake ecosystems **4**: ***2650–1***
Inductance measurement *see* Electric variables measurement
Induction machines *see* Asynchronous machines
Induction motors **5**: 3089
Inductively coupled plasma atomic emission spectroscopy **1**: 174
Industrial dynamics modelling **7**: 4756
Industrial plants
  adaptive control **1**: ***47–51***
  automation **6**: 3659
  cybernetics **2**: 869
Industrial robots **4**: ***2368–85, 2392***
  aerospace industry **4**: ***2449–52***
  anatomy **4**: 2368
  applications **4**: ***2416–21, 2440–4***
  arms **4**: ***2387–91***
    electrohydraulic servosystems **1**: ***301–4***
    flexible **6**: ***4086–8***
    mechanical structure **6**: 4101
    snake robots **4**: 2391
    spine robots **4**: 2391
    walking robots **4**: 2391
  automatic adjusting **4**: ***2412–5***
  compliance **4**: 2379
  contouring control **1**: 296
  control **4**: 2392, ***2456–9***
    classical design **4**: ***2421–40***
  coordinate systems **4**: 2369
  design **4**: ***2440–4***
  end effectors
    mechanical structure **6**: 4102
  garment manufacture **4**: ***2447–9***
  hand-level control **4**: ***2393–401***
  hand trajectories **4**: 2394
  intelligent interfaces **4**: 2363
  joints **6**: 4105
    control **4**: ***2401–9***
    disposition **6**: 4105
    trajectories **4**: 2402
  kinematic design **6**: 4104
  manufacturing computer control **3**: 1637
  mechanical structure **6**: ***4100–8***
  modular systems **4**: ***2460–4***
    actuation systems **4**: 2462
  off-line programming **4**: 2364
  optimal control **4**: 2375, ***2392–411***
  path planning
    optimal control **4**: ***2396***
  path tracking **4**: 2372, 2392
  performance index **4**: 2382
  position control **4**: 2372
    lasers **4**: ***2453–5***; **6**: 3762
  production **4**: ***2362–7***
  programming languages **4**: ***2347–51***, 2374, ***2464–8***
    motion **4**: 2466
  sensory feedback **4**: 2450
  social aspects **6**: ***4093–7***
  system failure and recovery **4**: ***2444–7***
  tactile and force control **6**: ***4113–7***
  time and motion studies **4**: 2383
  vision
    gray-scale images **4**: ***2479–84***
    three-dimensional vision **4**: ***2484–7***
  wrists
    mechanical structure **6**: 4101
  see also Robots, Robot arms
Industrial vision systems **4**: 2381, ***2469–72, 2473–9***
  aerospace industry **4**: 2451
  laster tracking techniques **6**: ***3762–5***
  three-dimensional objects
    distance measurement **1**: ***755–9***
    modelling **1**: ***343–7***
Inequalities
  multivariable control systems design **5**: 3190, 3206
Inequality-constrained problems **2**: 793
  penalty methods **2**: 799
  quadratic programming **2**: 796
Inertial navigation systems **4**: ***2488–91***
  airborne systems **1**: 384
  components **4**: 2488
  gimbal systems **4**: 2488
  helicopters **3**: 2123
  mechanization **4**: 2489
  missiles **5**: 3007
  strapdown systems **4**: 2488
  see also Navigation
Inference **4**: ***2492***
  causal inference
    multivariate analysis **5**: 3234
  diagnostic inference **2**: ***1003***
  hierarchical inference **4**: ***2331–8***
  probabilistic inference
    Bayesian updating **4**: 2336
Inference engines **4**: ***2492***
Infinite-dimensional systems
  stability **7**: 4528
Infinite-valued logics **5**: 2921
Information
  interaction analysis **4**: ***2543–7***
Information analysis
  environment **2**: ***1489***
  human information processing **3**: ***2186–91***
  on-line information **2**: 928
  philosophical aspects **7**: ***4779***
  requirements determination **4**: ***2497–501***
  semantic data **6**: 4214
  systems methodology **7**: 4803

Information display
  autopilots **1**: 386
Information patterns
  classical patterns
    linear stochastic systems **2**: 931
  linear stochastic systems **2**: 929
  one-step delay **2**: 931
  partially nested patterns **2**: 930
Information retrieval **5**: 3445
  analogical inference **1**: 275
  fuzzy databases **3**: ***1798–800***
  modelling **6**: 4323
Information science
  simulation model documentation **6**: ***4372–4***
Information sources in systems and control
    8: ***5575–5585***
Information systems **4**: ***2502–4***
  control units **4**: 2503
  input units **4**: 2502
  management **4**: 2503
  organization **4**: 2502
  output units **4**: 2503
  processing units **4**: 2503
  storage **4**: 2502
    modelling **6**: ***4323–6***
Information theory **4**: ***2504–8***, 2543
  Bernoulli systems **4**: 2506
  coding theorems **4**: 2507
  decomposition **4**: 2545
  digital commuication systems **2**: 1044
  ecology **2**: ***1321***
  Fisher information matrix
    maximum likelihood identification method
      4: 2285
  flow networks **3**: 1694
  Hartley information **6**: 3766
  incentives
    stochastic processes **7**: 4628
  law of requisite variety **4**: 2495, ***2730***
  maximum entropy principle **4**: 2507; **5**: ***2951***
  model validation **7**: 5000
  N-dimensional theory **4**: 2493
  picture processing **4**: 2475
  possibility measures **6**: ***3766–8***
  pragmatic information **6**: ***3858–60***
  semantics **6**: ***4214–6***
  Shannon entropy **6**: ***4257***
  signal detection
    fuzzy set theory **1**: ***651–64***
  signalling **7**: 4817
  source entropy **4**: 2506
  structure **7**: 4811–22
    game theory **1**: 682
  subjectivity **7**: 4697
  systems analysis **4**: ***2492–7***
  team theory **7**: ***4811–22***
  transmission **4**: ***2509***; **6**: 3858
    semantics **6**: 4216
  see also Entropy (information)
INFOTERRA **2**: ***1503–6***
Infrared absorption analysis
  environmental analysis **2**: 1509
Infrared detectors
  lighting control **4**: 2773
  medical thermography **7**: 4869
  remote sensing **6**: 4039
Infrared imaging
  medical applications **7**: ***4868–71***
Infrared spectroscopy
  chemical analysis
    environmental analysis **4**: ***2509–11***
  fingerprinting of oil **7**: 5111
  organic molecules and substances **1**: 11
Initial-transient techniques
  stochastic simulation **7**: ***4651–4***
Initial-value problems
  ordinary differential equations
    integration **5**: ***3555–8***
  singular perturbations **7**: 4430, 4435
Injection molding
  industrial robots **4**: 2440
Innovation
  research **6**: 3660
  technological innovation **7**: ***4828–32***
Innovations processes
  discrete-time systems **2**: ***1099–102***
  iterated Ito integrals **5**: 3333
  Kalman filters
    discrete-time systems **4**: 2622
  stationary time series **7**: 4558
Inorganic chemicals production
  control of variables **4**: 2514
  modelling **4**: 2516
  process control **4**: ***2511–6***
Input–output analysis **7**: 4769, 4804
  ecological modelling
    Markov chains **5**: 2929
  economic modelling **2**: 1341
  ecosystems
    network structure **2**: 1355
    trophic structures **7**: 4956
  flow networks **3**: 1694
  Leontief inverse **4**: ***2737***
  self-tuning control **6**: ***4205–9***
  sensitivity analysis **6**: ***4235–7***
  transactions tables **7**: 4935
Input–output decoupling
  multivariable control systems **4**: 2522
Input–output models **4**: ***2517–22***; **7**: 4776
  applications **4**: 2520
  hepatobiliary kinetics **3**: 2140
  hydrologic forecasting **3**: 2231
  lake ecosystems **4**: ***2646–9***
  mathematics **4**: 2519
  nonlinear discrete-time systems **4**: ***2523–7***
  rainfall-runoff modelling **6**: 3942
  results **4**: 2518
  stability

nonlinear feedback systems **5**: 3317
systems homomorphism **3**: 2164
use **4**: 2517
water quality **4**: 2649
Input–output sequences
identification **5**: 3041
Input–output systems
dynamical systems **6**: 3629
nonlinear decoupling **5**: 3287
Input signals
tracking
multivariable control systems **5**: 3192
Input voltage offset
amplifiers **1**: 270
Inputs **4**: ***2527***
INS
simulation databases **2**: 894
Inspection
industrial vision systems **4**: 2469, ***2473–9***
Instability theory **5**: 3399
see also Stability
Instrument flying rules **7**: ***5060***
head-up displays **3**: 2087
helicopters **3**: 2122
navigation **3**: 1719
Instrument landing systems **1**: 381
bad weather landing **1**: 213
helicopters **3**: 2122
Instrumental variable techniques **4**: ***2273–7***
basic method **4**: 2273
extended method **4**: 2273
Instrumentation
amplifiers **1**: ***268–72***
computerized instrumentation **5**: 2980–4
mass spectrometers **5**: 2940
design **4**: ***2533–6***
fermenter control **3**: 1615
mass spectrometers **5**: ***2934–42***
production control
process industry **6**: 3885
spectroscopy **1**: 8
radiation sources **1**: 8
structure of systems **4**: 2532
see also Aerospace instrumentation
Instruments **4**: ***2527–33***
design **4**: ***2533–6***
errors **2**: ***1553–9***; **4**: 2532
classification **2**: 1554
instumental influence errors **2**: 1556
intrinsic instrumental errors **2**: 1555
information flow **4**: 2528
level measurement **4**: 2747
microcomputers **5**: ***2980–4***
microprocessor chips **5**: ***2984–8***
performance **4**: 2531
power flow functions **4**: 2529
signal flow functions **4**: 2528
uncertainty **2**: ***1553–9***
Insulation
clothing insulation **1**: ***605–6***
Insulin
adaptive control **3**: 2017
glucose regulation
modelling **3**: 2021, 2025
hunger **3**: 2210
Integer programming **7**: 4554
Integral control
governors
history **7**: 4514
Integral equations
hereditary systems **3**: 2147
Integrated circuits
flight simulators **1**: 110
Integration
initial-value problems
ordinary differential equations **5**: ***3555–8***
Integrodifferential equations **4**: ***2536–41***
hereditary systems **3**: 2147
linear Boltzmann equations **7**: 4940
Intelligence amplification
artificial intelligence **4**: ***2542–3***
Intelligent control systems **6**: ***4161***
Interacting groups **4**: ***2543***
Interacting systems
HVAC systems **3**: 2105
Interaction analysis
data analysis **2**: ***875–9***
information theory **4**: ***2543–7***
Interaction graphs **2**: 1280
Interaction matrices **4**: ***2547***
Interaction prediction principle **1**: 400
dynamic systems **5**: 3016
Interactive methods **4**: ***2614***
process methods **4**: ***2547***
Interactive programming
simulation program generators **4**: 2547
Interactive systems **7**: ***4775–9***
adaptive control
design **1**: 62
Interactive terminals **4**: ***2551***
CAD
human factors **1**: 730
Interconnected systems
chained aggregation **1**: 136
decentralized fixed modes
characterization **2**: 906
decomposition **4**: 2552
dynamical systems
optimization and control **2**: 859
stability **4**: ***2551–6***
stabilization **2**: 916
see also Large-scale systems, Complex systems
Interconnection matrices
large-scale systems **4**: 2672
Interference (signal)
series mode
voltmeters **2**: 1397

voltmeters **2**: 1397
  guard techniques **2**: 1397
Interferometers
  Mach–Zehnder interferometer **5**: 3467
  multimode interferometers **5**: 3469
  optical fiber interferometric sensors **5**: ***3465–70***
  Sagnac effect interferometer **5**: 3465
Internal model principle **4**: ***2566–7, 2567–70***
International Federation of Automatic Control **4**: ***2571–5***
International Federation for Systems Research **4**: ***2570***
Interpolation
  splines (mathematics)
    air pollution pattern analysis **7**: ***4517–8***
Interpretation
  hierarchical scene analysis **6**: 4159
Interpretive structure modelling **4**: ***2575–9***
Interrogation signal
  air traffic control **1**: 199
Intracranial pressure
  feedback control **4**: ***2580–2***
Intuition
  control theory **4**: 2357
Invariance
  decision theory **3**: 2041
  extended invariance principle
    ground albedo mapping **3**: 2048
  poles and zeros **4**: 2832
Invariant embedding
  ground albedo mapping **3**: 2046
Invariants
  linear multivariable systems **4**: ***2866–71***
Inventory control *see* Stock control
Inverse Nyquist array methods **6**: 3700
  articulated robot arm control design **1**: 304
  design engineering **4**: ***2584–8***
  multivariable control systems **5**: 3197
  see also Nyquist array methods
Inverse problems
  parameter estimation
    distributed parameter systems **2**: 1193
Inverse transformation **6**: 3948
Inverters
  ac motors **5**: 3090
  current-fed inverters
    ac motors **5**: 3090
    synchronous motors **5**: 3092
  dc motors **5**: 3093
  dc transmission networks **6**: 3821
  variable speed induction machines **7**: 5019
  voltage-fed inverters
    synchronous motors **5**: 3092
Iodine
  ecology
    aluminum compounds **1**: 267
Ion chromatography
  chemical analysis **1**: 175
  inorganic nitrogen compounds
    water pollution **7**: 5122
Ion dipole
  nutrients and pollutants
    environment **1**: 6
Ion-exchange reactions
  weathering **1**: 2
Ion pumps
  mass spectrometer apparatus **5**: 2939
Iron compounds
  adsorption and weathering **1**: 2
Isaacs–Breakwell theory **2**: 1006
Isaacs equation **2**: ***1010–6***
Ischemia
  electrocardiography **1**: 455
Isomorphic systems **4**: ***2609***
Isoperimetric problems **4**: ***2609–13***
Iterative methods
  asynchronous iteration
    decomposition–coordination **5**: ***3475–80***
  distributed algorithms
    convergence **2**: 1122
  gas distribution systems
    steady-state analysis **3**: 1948
  stochastic calculus **5**: ***3330–6***
Ito integrals **5**: ***3330–6***; **7**: 4631

J

Jaws
  measurements **2**: ***963–5***, 976
Join procedure
  reconstructability analysis **6**: 3988
  relational join **6**: ***4000***
Joints (industrial robots) *see* Industrial robots
Judgment (discernment)
  decision theory and analysis **3**: ***2193–202***
    holistic approaches **3**: 2193
  see also Decision theory and analysis

K

K convexity **5**: ***3509–10***
Kalman filters
  aerospace control **1**: 116
  air pollution **2**: 1499
  aircraft wing load alleviation **7**: 5168
  applications **4**: ***2618–22***
  design **4**: 2620
    software **4**: 2621
  discrete-time systems
    state-space models **4**: ***2622–7***
  divergence **4**: 2620
  ecological modelling **2**: 1332
  global
    computational requirements **4**: 2696
  hydrologic forecasting **3**: 2231
  iterated Ito integrals **5**: 3335
  large stochastic systems **4**: 2708
  least squares identification **4**: 2281
  linear feedback **2**: 930
  load forecasting **6**: 3832
  missile guidance systems **5**: 3008
  multivariable control systems **4**: 2845
  nonlinear estimation **5**: 3300
  nonlinear filtering **5**: 3326
  performance analysis **4**: 2620
  pH control **6**: 3690
  recursive filters **4**: 2693
  separated stochastic control **5**: 3474
  ship positioning **6**: 4259, 4264
  smoothing algorithms **7**: 4463
  square root algorithms **7**: 4534
  state estimation **4**: 2692
  traffic simulation **7**: 4929
  variations of formulae **4**: ***2615–8***
  see also Extended Kalman filters, Filtering and prediction theory
Kata thermometers **3**: 2017
Ketone bodies
  metabolism modelling **4**: ***2628–31***
Kidneys **3**: ***2204–6***
  acid–base regulation **3**: 2205
  disorders **3**: 2206
  electrolyte control **3**: 2205
  metabolic waste excretion **3**: 2205
  tubular function **3**: 2205
  water reabsorption **3**: 2205
Kilns *see* Ovens
Kinematics
  snake locomotion **6**: 4109
Kinetic data
  patient diagnosis **4**: ***2631–5***
  phosphorus kinetics
    identification **5**: 3043
  physiological models **4**: 2632
    classification **4**: 2634
Knocking
  automobile electronics **1**: 376
Knowledge bases **4**: ***2636***
Knowledge representation **4**: ***2636–40***
  artificial intelligence **1**: 306
  expert systems **2**: 1569
  production systems **4**: 2637
Knowledge simulation
  process control **6**: 4317
Kolmogoroff's forward equation
  ecological population modelling **2**: 1327
Kronecker canonical forms
  linear multivariable systems **4**: ***2866–71***
Krypton
  fission reactors **5**: 3437
KSIM modelling *see* Workshop dynamic modelling
Kullback J-divergence
  dynamical systems
    abrupt changes **1**: 85

L

L-systems *see* Lindenmayer systems
Lag window **4**: 2267
Lagrange multipliers **2**: 789–93, 1274
  calculus of variations **1**: 524
  constraint theory **2**: ***789–93***
  hypothesis testing **5**: 2914
  quadratic programming **2**: ***793–7***
Lagrangians **2**: 1274
  augmented Lagrangians **1**: 400
    price decomposition **1**: 403
  nonlinear programming **5**: 3036
  optimization **5**: 3013
  price decomposition **1**: 402
  saddle point **2**: 1274
Lakes
  ecosystems
    input–output models **4**: ***2646–9***
    modelling **4**: ***2643–6***
    uncertainty **4**: ***2650–1***
  eutrophication modelling **5**: 3040
    variance estimates **2**: 1269
  Lake George
    modelling **4**: 2645
  modelling
    history **4**: 2643
  phosphorous modelling **4**: 2789
  redox reactions **1**: 3
  response time
    compartment modelling **4**: ***2789–94***
  tide modelling **7**: 4887
  toxic substances
    dispersion **4**: ***2778–82***
  trophic structure **7**: 4952
  water quality modelling **7**: 5129
  see also Limnology
Land use modelling **7**: 4491
Landing
  automatic landing systems **1**: 393
Landsat imaging—*see* Remote sensing
Langer's lines
  skin
    biomechanics **1**: 479
Language (natural)
  metalanguage **1**: 328
  subjectivity **7**: 4696
Language perception
  brain activity **1**: ***502–4***
  fuzzy languages **1**: 507
Language translation
  artificial intelligence **1**: 307
Languages
  nonprocedural languages
    database management systems **4**: 2656
Languages (formal) *see* Formal languages
Languages (programming) *see* Programming languages
Laplace transforms **4**: ***2659–64, 2664–5***
  sampled-data systems **6**: 4145
Large intestine *see* Digestive tract
Large-scale integration
  microprocessors **5**: 2980
Large-scale systems **1**: 687, ***713–4***; **4**: 2557
  centrality **2**: 927
  computer software
    reliability **4**: ***2678–81***
  continuous-time systems **4**: 2561
    aggregation forms **4**: 2561
    stability **4**: 2561
  control **1**: 688
  decentralized control **2**: 902; **4**: ***2557–9***
    algebraic approach **4**: ***2666–71***
    design **1**: 131
    robust controller design **6**: ***4133–8***
  decentralized stabilization **2**: ***916–23***; **4**: 2559
  decomposition **1**: 688
  decomposition–coordination **2**: 952–6
    asynchronous iteration **5**: ***3475–80***
  difference equations **7**: 4429
  discrete singular pertubations **7**: 4429
  discrete-time systems **4**: 2564
    aggregation forms **4**: 2565
    stability **4**: 2565
  dynamic systems
    on-line coordination **2**: 1297
  hierarchical control **1**: 135; **3**: 2153
  hierarchical overlapping coordination **5**: ***3564–8***
  horizontal decomposition **1**: 688
  input–output setting
    well posedness **4**: 2677
  input–output stability **4**: ***2671–5***
  interaction inputs **2**: 908
  interaction measurement condition **2**: 910
  linear programming **4**: ***2686–92***
  Lyapunov methods **4**: ***2560–5***
  minimax techniques
    matrix algebra **7**: 4684
  model simplification
    frequency-domain methods **5**: ***3071–7***
  multicriteria optimization **1**: 689
  multilevel control **5**: 3140
  multilevel filters **4**: 2692
  multistage decomposition algorithms **5**: ***3166–70***
  multi-timescale methods
    singular perturbations **7**: 4431
  nonlinear systems stability **5**: 3402
  observers **4**: ***2697–702***
  on-line coordination **7**: ***4566–71***
  optimization **4**: 2704
    goal coordination method **3**: 2032
    mixed coordination method **5**: 3016
    model coordination method **5**: ***3035–9***

order reduction **5**: 3071
state feedback control **7**: 4439
overlapping decentralized control **5**: ***3568–72***
parameter estimation **4**: ***2702–5***
parameter optimization **7**: 4678
real systems development **4**: ***2681–5***
reduction methods **1**: 688
spatiotemporal decomposition **1**: 688
stability
decentralized feedback **4**: 2666
dissipativeness **7**: ***4519–22***
state estimation
decentralized observers **2**: ***907–12***
power systems **6**: 3846
state feedback control
slow–fast decomposition **7**: ***4529–34***
statistical models
entropy minimax theory **2**: 1473
stochastic systems **4**: ***2706–13***
control **4**: 2712
hierarchical control **4**: 2710
learning automata **4**: ***2714–9***
structural analysis
graphs **7**: ***4661–6***
technological innovation models **7**: 4829
vertical decomposition **1**: 688; **5**: 3140
well posedness **4**: ***2675–8***
see also Complex systems, Composite systems
Laser beam applications
distance measurement **4**: 2486
industrial robots
position control **4**: 2453–5
laser Doppler anemometers **5**: 3459
laser Doppler flowmeters **3**: 1677
laser Doppler velocimeters **3**: 1686; **7**: 4513
solar power satellites **7**: 4474
tracking systems
position measurement **6**: ***3762–5***
wind velocity measurement **2**: ***1528–32***
Laser spectroscopy
atmospheric analysis **4**: ***2719–23***
infrared spectrometry
environmental analysis **4**: 2510
long-path absorption
atmospheric analysis **4**: 2722
Latest completion **4**: ***2723***
Latest start
project engineering **4**: ***2723***
Launchers
Ariane **4**: 2725
Atlas **4**: 2726
B1 launchers **4**: 2724
Black Arrow **4**: 2724
China **4**: 2724
Delta **4**: 2726
Diamant **4**: 2724
Europe **4**: 2724
flight sequence **4**: 2729
heavy satellites **4**: 2724
Japan **4**: 2726
launch bases **4**: 2728, 2730
performance **4**: 2730
light payload **4**: 2724
missions **4**: 2728
multistage launchers **4**: ***2723–7***
performance **4**: 2728
problems **4**: ***2728–30***
propellants **4**: 2728
Satellite Launch Vehicle **4**: 2724
Saturn **4**: 2726
Scout **4**: 2724
Space Shuttle **4**: 2727
Titan **4**: 2726
USA **4**: 2726
USSR **4**: 2727
Lauvent series **1**: 15
Law of large numbers **3**: 1841
Law of the minimum
ecological modelling **2**: ***1329***
Law of potential violation **1**: 742
Law of requisite variety *see* Information theory
LCR multiports **1**: 714
uncontrollability **1**: 714
unobservability **1**: 714
Lead
aerosol chemical analysis **1**: 174
agricultural models **1**: 591
soil **1**: 146
toxic substances **1**: 144
Lead vectors
electrocardiography **1**: 532
Leak detection **7**: 5148
natural gas technology **3**: ***1954–8***
water supply **7**: ***5143–7***
Leakage *see* Leak detection
Learning control
learning automata **4**: 2714
Learning systems **4**: ***2731–7***; **6**: ***4161***, 4166
algorithms
switching networks **7**: ***4716–21***
analogical inference **1**: ***272–6***
artificial intelligence **1**: 307
dual control **2**: 1253
fuzzy learning systems **3**: ***1810***
intelligence amplification **4**: 2543
large stochastic systems **4**: ***2714–9***
learning automata
controllers **4**: 2716
pattern recognition **6**: 3637
philosophical aspects **7**: 4780
Least order systems
polynomial systems
matrix algebra **6**: 3748
Least squares approximations **6**: 4206
extended method **2**: ***1581***
generalized **3**: ***2005***; **5**: 3028
identification **4**: ***2278–83***, 2317
identification errors

biomedicine **4**: 2255
minimum variance control **5**: 3004
model reference adaptive control systems **5**: 3051
nuclear power station control
parameter estimation **5**: 3428
numerical analysis **2**: 817
regression analysis
simulation models **5**: 3026
state estimation
linear systems **4**: 2865
power systems **6**: 3848
two-filter smoothing algorithms **7**: 4467
Leaves (plants)
leaf damage **2**: 1520
image instrumentation **2**: 1519
temperature
image instrumentation **2**: 1518
Leontief inverse
input–output analysis **4**: ***2737***
Leslie's equations
population modelling **2**: ***1333***; **6**: 3760
Level control **4**: 2742
boilers **7**: 4576, 4578
glass industry
adaptive control **3**: 2010
sugar industry **7**: 4701
Level measurement
bulk materials storage **4**: ***2743–7***
instruments **4**: 2747
intermediate storage **4**: ***2737–43***
liquids **4**: ***2747–54***
Level meters
bulk materials **4**: 2743
Level of significance **4**: ***2754***
Lie algebras **1**: 121; **4**: 2757; **5**: 3327
bilinear equations **1**: 416
group theory
control theory **5**: 2945
invariance **5**: 3324
nonlinear estimation theory **5**: 3306
nonlinear filtering **5**: 3323
nonlinear realization theory **5**: 3390
space vehicles (rigid)
attitude control **1**: 320
Lie brackets **4**: ***2754–6***
functional expansions **3**: 1773
nonlinear systems **5**: 3364
Lie groups **5**: 2945
Abelian lie groups **1**: 416
bilinear equations **1**: 414
control theory **5**: ***2944–8***
controllability **4**: ***2756–9***
Life-cycle costing **2**: 996; **4**: ***2760–4***
Life cycles
simulation models **6**: 4328
Lifts *see* Elevators
Light absorption
chemical technology
optical fiber sensors **5**: 3462
flow measurement **3**: 1692
wavelength-dependent absorption
optical fiber sensors **5**: 3460
Light scattering
flow measurement **3**: 1692
optical fiber sensors **5**: 3463
Lighting
control **4**: ***2770–4, 2775–7***
costs **4**: 2775
energy savings **4**: 2775
gray-scale image sensing **4**: 2480
industrial vision systems **4**: 2470
switching **4**: ***2764–9***
Likelihood function
parameter estimation **6**: 3613
Likelihood ratios **4**: ***2777***
hypothesis testing **5**: 2915
Likeness relations **3**: 1602
Limit cycles
closed-loop systems **7**: 4445
multidimensional systems
rounding multiplier outputs **5**: 3126
nonlinear systems **5**: 3384
Limited purpose models
hormone regulation **3**: 2170
Limits to growth
modelling **7**: 4756
Limnology
lake trophic structure **7**: 4952
mixing models **4**: 2782
reaction characteristics **4**: 2782
timescale estimation **7**: ***5090–4***
toxic substances
dispersion **4**: ***2778–82***
fugacity models **7**: 4901
water quality modelling **7**: 5129
see also Lakes, Rivers
Lindenmayer systems **4**: 2795, ***2795–8***
cellular automata **6**: 4344
heterarchical systems
simulation modelling formalism **6**: 4351
Linear aggregation **1**: 698
Linear algebra
complete computer programs
control system computing **2**: 824
numerical analysis **2**: ***817***
Linear differential equations **5**: 3560
numerical analysis **2**: 821
see also Differential equations
Linear dynamic modelling
multivariable control systems
rolling mills **4**: 2602
Linear dynamic optimization **5**: 3016
Linear ordering **4**: ***2832***
Linear predictive coding
speech **7**: 4506
Linear programming **4**: ***2835–6, 2839–41***; **7**: 4554
Dantzig–Wolfe decomposition method **4**: 2691
decomposition methods **2**: 956; **4**: 2690

duality **4**: ***2836–9***
fuzzy linear programming **3**: ***1810–1***
multicriteria decision making **5**: 3121
large-scale systems **4**: ***2686–92***
Markov decision processes **5**: 2931
multicriteria decision making **5**: 3122
optimal solutions
algebraic approach **4**: 2841
graphical solution approach **4**: 2840
optimization
combinatorial mathematics **6**: 3735
nondifferentiable optimization **5**: 3543
power systems **6**: 3786
revised simplex method **6**: 4297
Rosen partition method **4**: 2691
simplex method **4**: ***2686,*** 2841; **6**: ***4296–9***
stochastic programming **7**: 4646
Linear quadratic Gaussian techniques *see* LQG techniques
Linear quadratic problems **4**: ***2846–9***
decentralized control **2**: 954
multivariable systems control
robustness **5**: 3226
optimal control **5**: ***3513–4***; **6**: 3701, ***4066–70***
singular perturbations **7**: 4438
Riccati equation and conjugate-point theory **6**: ***4066–70***
separation principle **6**: 4248
singular perturbations **7**: 4438
Linear systems **4**: ***2861–5***
algebraic realization **4**: 2865
bang–bang control **1**: 410
Bode diagrams **4**: 2664
classical control **5**: 3473
closed-loop capture **2**: 1007
combinatorial optimization **6**: 3737
complex systems control
reference variables **1**: 676
control
quadratic criteria **1**: 691
control system analysis computing **4**: ***2798–800***
control theory **4**: ***2855–61***
frequency response calculations **2**: 822
numerical analysis **2**: 821–4
decentralized control
algebraic approach **4**: ***2666–71***
decentralized feedback **4**: 2666
decentralized feedback
disturbance rejection **4**: 2670
tracking **4**: 2670
disturbance rejection **4**: ***2567–70***
duality **4**: 2863
dynamic systems
parameter estimation **1**: 487
estimation **7**: ***4617–21***
filtering
iterated Ito integrals **5**: 3335
frequency response **4**: 2264
geometric realization **4**: 2865
hereditary systems **3**: 2150
hierarchic calculations
quadratic criteria **1**: 691
hyperbolic systems
control theory **4**: ***2800–7***
identification **4**: 2265
internal model principle **4**: ***2567–70***
Laplace transforms **4**: 2665
Lie brackets **4**: 2756
local controllability **4**: 2893
Lyapunov methods **5**: 3400
minimal design problems **5**: ***3215–20***
modelling
nonparametric **1**: 488
multidimensional systems **5**: 3129
multigrid methods
algorithms **5**: 3138
multivariable control systems **4**: ***2827–32***
controllability **4**: ***2816–26***
design **5**: 3206
functional analysis **5**: 3201
geometric synthesis **4**: ***2807–16***
invariants **4**: ***2866–71***
Kronecker canonical forms **4**: ***2866–71***
linear quadruples **4**: ***2849–55***
observability **4**: ***2816–26***
optimal control **1**: 38
numerical methods **4**: ***2875–9***
output tracking **4**: ***2567–70***
partition **3**: 2161
pole-shifting **4**: 2864
polynomial matrices **6**: 3745
realization **4**: 2862
robustness **4**: ***2871–2***
stability **4**: ***2872–5***
decentralized feedback **4**: 2666
state feedback control
slow–fast decomposition **7**: 4529
stationary dynamical
balanced realization **1**: ***409–10***
stochastic systems **2**: 929
time-invariant systems
decentralized stabilization **2**: 917
structural perturbations **2**: 922
time-varying systems
decentralized stabilization **2**: 922
waiting function
binary pseudorandom signals **4**: 2249
Linearization techniques
ecological modelling **6**: 3665
lumped-parameter models **2**: 1288
Lyapunov methods **5**: 3400
Linesearch algorithms
optimization **7**: 4966
Linguistic analysis
fuzzy control systems **3**: ***1794***
Linguistic approximation **4**: ***2880***
Linguistic controllers
process control **3**: 1811; **4**: ***2880***

Linguistic hedges **4**: ***2880–1***; **5**: 3155
  multiobjective optimization **5**: 3156
Linguistic models **4**: ***2881***
Linguistic probabilities **4**: ***2881–3***
Linguistic synthesis
  fuzzy control systems **3**: ***1794–5***
Linguistic variables **4**: ***2884***
  aggregate planning **1**: ***124–6***
  conflict resolution
    fuzzy set theory **2**: 774
  medical diagnostic computing
    electrocardiography **1**: ***762***
Lipids
  biotransport **4**: ***2884–7***
Lipoproteins
  classification **4**: 2884
  dynamics and control **4**: ***2884–7***
Liquefaction of gases
  level measurement **4**: 2745
  natural gas storage **3**: 1970
Liquid chromatography *see* Chromatography
Liquid crystal devices
  medical thermography **7**: 4868
Liquid–liquid extraction
  trace elements
    water pollution **7**: 5124
Liquids
  flow coefficients **2**: 828
  level measurement **4**: ***2747–54***
List processing
  discrete-event simulation **2**: 1085
Load (electrical)
  classification **6**: 3808
  electric power generation
    daily peak load variation curve **6**: 3791
    load duration curve **6**: 3791
    load probability **6**: 3790
Load alleviation
  aircraft **1**: 26
Load capacitance
  power systems **6**: 3809
Load cells
  gait analysis **3**: 1911
  gyroscopic cells **3**: 1708
  hydraulic cells **3**: 1707
  pneumatic cells **3**: 1708
  strain gauges **3**: 1709
  weighing **3**: 1706
Load dispatching
  power system computer control **6**: 3827
  power system control **6**: 3781, ***3815–20***, 3845
Load distribution
  natural gas technology **3**: 1942
Load expectation
  electric power generation **6**: 3790
Load flow
  automatic control **6**: 3854
  power systems
    state estimation **6**: 3846
  steady-state modelling **6**: ***3850–5***
Load forecasting
  natural gas supply systems **3**: 1977, ***1979–81***
  power system computer control **6**: 3827
  power systems **6**: ***3830–4***
Load–frequency control
  adaptive control **4**: ***2889–92***
  power system computer control **6**: 3827
  power systems **6**: 3815
  see also Load regulation
Load regulation
  critical time override **2**: ***867***
  duty cycling **2**: ***1262***
    minimum on-time **5**: ***2998***
  electric generators **2**: ***1375–80***
  see also Load–frequency control
Load shedding
  demand control **2**: ***963***
  power system computer control **6**: 3827
Local area networks *see* Networks (circuits), Telecommunication networks
Local consistency
  reconstructability analysis **6**: 3985
Local models
  pole assignment **2**: 901
Local output **2**: 912
Locking
  database management systems
    concurrency **2**: 882
Locomotion
  control **3**: ***1907–10***
  measurement **3**: 1910
  movement patterns **3**: 1907
  snakes **6**: 4109
Logic (formal) *see* Formal logic
Logistic growth equation
  population dynamics **1**: 415
Logistics
  launch bases **4**: 2730
Look-up tables *see* Decision tables
Loop analysis
  system dynamics modelling **7**: 4757
Lotic systems
  nitrification modelling **5**: ***3264–6***
Lotka equations
  population modelling **6**: 3759
Lotka–Volterra models
  ecosystems **1**: 673
  population modelling **2**: ***1334***
Lotteries
  multiattribute utility theory **5**: 3099
LQG problem
  cost function **1**: 39
  optimal control **1**: 38
LQG techniques
  aircraft wing load alleviation **7**: 5169
  discrete-time control **5**: 3472
  minimum variance control **5**: 3005
  multivariable control systems

robustness **4**: ***2842–6***
process control
petrochemical industry **6**: 3674
self-tuning control **6**: ***4172–6***; **7**: 4547
Luenberger observer
pole assignment **6**: ***3717–20***
see also State estimation
Łukasiewicz logic **5**: 2922
Luminescence **7**: 4499
Lumped-parameter models **2**: ***1286–91***
mechanical systems **5**: 2966
nonlinear systems **2**: 1288
Lumped-parameter networks
power systems **6**: 3802
Lumped-parameter systems
HVAC systems
digital simulation **3**: 2113
Lumping *see* Aggregation
Lungs
connectivity **4**: 2904
educational computer models **2**: 1362
mechanical properties **1**: 457
modelling **4**: ***2898–901, 2902–4***
trees
classification **4**: 2902
see also Respiration
Lyapunov functions **4**: 2551
differential inclusions **2**: 1030
domains of attraction **2**: 1211
metabolic control **1**: 427
scalar functions **4**: 2552
vector functions **4**: 2554
weak coupling conditions **4**: 2553
Lyapunov methods
adaptive systems
Lyapunov synthesis **1**: ***97–100***
convergence of numerical methods
distributed asynchronous fixed-point algorithms **2**: 1124
direct method **4**: 2551
fixed systems **7**: 4524
large-scale systems **4**: ***2560–5***
linear control theory **2**: 822
Lyapunov equation **4**: ***2904–5***
M-matrices **4**: 2553
model reference adaptive control systems **6**: 4184
stability **5**: 3068
nonlinear systems **5**: ***3394–8, 3398–401***, 3402
pneumodynamics
modelling **1**: 457
power system analysis computing **6**: 3799
Lyapunov stability **7**: 4520, 4673
multivariable systems **5**: 3203
nonlinear systems **5**: ***3394–8***
populations **6**: 3753
singular perturbations **7**: 4436
vector Lyapunov functions
decentralized stabilization algorithms **2**: 918
Lyophilization *see* Freeze drying

M

M-array *see* Pseudo-random pattern
M-matrices
  Lyapunov methods **4**: 2553
M-sequences **5**: ***2951–6***
  binary random signals **1**: 418
  characteristic M-sequences **5**: 2953
  characteristic polynomials **5**: 2952
  delayed version **5**: 2953
  generating function **5**: 2952
  generation **5**: 2951
  properties **5**: 2952
    shift and add property **5**: 2953
  system identification **5**: 2955
Machine tools
  handling
    industrial robots **4**: 2419
  manufacturing computer control **3**: 1642
Machine vision systems *see* Industrial vision systems
Mackereth sensor
  oxygen concentration determination **5**: 3577
Macroscopic models
  transportation **7**: ***4923–5***
MACSYMA **1**: 307
Magnesium
  ion-selective electrodes
    water pollution **4**: 2607
Magnetic field measurement
  optical fiber sensors **5**: 3461
Magnetic recording
  telemetry **7**: 4837
Magnetic resonance
  biology **1**: 13
  spectroscopy **7**: 4499
  see also Paramagnetic resonance, Nuclear magnetic resonance
Magnetohydrodynamics
  modelling
    partial differential equations **6**: 4363
Magnetometers
  optical fiber interferometric sensors **5**: 3469
  road traffic detectors **7**: 4918
Maintenance engineering
  aircraft engines
    control systems **1**: 225
  mass spectrometers **5**: 2942
  natural gas technology
    scheduling **3**: 1976
  power systems
    scheduling **6**: ***3834–7***
Man–machine systems **3**: ***2175–8***
  aircraft **1**: 116
    combat aircraft **1**: 622
  flight-crew workload **3**: 1665
  future developments **3**: 2178
  history **3**: 2175
  input–output devices
    data tablets **3**: 2177
    light pens **3**: 2177
    mice **3**: 2177
  multicriteria decision making **5**: 3121
  robot operating systems **6**: 4089
  steel industry
    hierarchical control **7**: 4588, 4592
  task allocation **3**: 2177
Management
  complex systems
    game theory **1**: 685
  conflict analysis **7**: ***4793–8***
  control decisions **2**: 943
  firms
    differential games **1**: 686
  game theory **5**: ***2907–11***
  hypothesis testing **5**: ***2911–7***
  interpretive structure modelling **4**: 2579
  large-scale systems **4**: 2681
  model validation **7**: ***5001–5***
  ports **3**: 2072
  simulation model management **6**: ***4328–32***
  steel industry
    hierarchical control **7**: 4590
Management information systems **2**: ***943–7***
  cognitive styles **7**: 4790, 4792
  ecological modelling **2**: 1332
  environmental policy making **2**: ***1542–4***
  evaluation **2**: 947
  expert systems **2**: 1569
  philosophical aspects **7**: 4784
  requirements
    analysis **4**: 2499
  scenarios **6**: 4153
  simulation model documentation **6**: 4374
  simulation model management **6**: 4332
Management science
  complexity **7**: 4786
  reducibility **7**: 4786
  simulation **4**: 2351
  systems management **7**: 4787
  technological forecasting **7**: 4823
  technological innovation **7**: 4831
Management systems
  screening methods **5**: 3114
Maneuvering
  aircraft
    active control technology **1**: 25
    load control **1**: 27
  ships **5**: 3256
Manifolds
  geometry **5**: 3364
  group theory **5**: 2944
  Lie groups **4**: 2757

Manipulators *see* Industrial robots
Manufacturing computer control **3**: ***1633–6, 1637–42***
  design **3**: 1644
  industrial robots **4**: 2362
  modelling
    analytical modelling **3**: 1645
    simulation **3**: ***1642–8***
  Petri nets **6**: ***3670–3***
  simulation **3**: 1646
Manufacturing processes
  automation
    washing machines **3**: ***1597–600***
  industrial robots **4**: 2379
  industrial vision systems **4**: ***2473–9***
Many-valued logics **3**: 1815; **5**: ***2918–21***
  algebraic logics **5**: ***2921–3***
  applications **5**: 2918, ***2924–6***
  fuzzy implication operators **3**: 1806
  fuzzy set theory **4**: 2293
  vagueness **7**: 4995
Mapping
  ordinary mapping
    fuzzy sets **5**: ***3563***
Marine systems
  distribution of elements **2**: ***1513–6***
Market clearing equations
  economics **5**: ***2927***
  microeconomic systems analysis **5**: 2974
Markov kernels
  fuzzy measures **5**: 3411
Markov parameters **1**: 14
  realization theory
    numerical methods **6**: 3977
Markov processes **2**: 1276; **5**: ***2932***; **7**: ***4632–3***
  absorbing chains **5**: 2928
  Brownian motion **1**: ***511–2***
  controlled processes
    learning systems **4**: 2734
  decision models **7**: 4806
  ecological modelling
    forest succession **7**: 4645
    Markov chains **5**: ***2927–9***
  ecosystems
    trophic structures **7**: 4955
  learning automata **4**: 2716
  Markov decision processes **5**: ***2930–2***
  optimal stochastic control **5**: 3471
  realization theory **6**: ***3962–4***
  reliability of systems **6**: 4024
  simulation modelling **6**: ***4355–6***
  telecommunication networks **7**: 4719
  telephone traffic **7**: 4735
  two-filter smoothing algorithms **7**: 4467
  uncontrolled processes
    learning systems **4**: 2733
  water resource management **7**: 5136
Martingale methods **5**: ***2932–4***
  optimal stochastic control **5**: 3521
  stochastic adaptive systems **1**: 53
    stability **7**: ***4607–11***
  stochastic calculus **7**: 4631
Mass balance
  algal growth modelling **1**: 257
Mass flow measurement **3**: ***1677–81***, 1692
  momentum change **3**: 1679
  pressure difference **3**: 1678
  sonic velocity **3**: 1680
  thermal transfer **3**: 1679
  vibrating vane **3**: 1680
Mass measurement **3**: ***1706–10***
Mass spectrometry **5**: ***2934–42***
  gas chromatography–mass spectrometry **3**: 1930
    environmental measurement **3**: ***1933–6***
  microprocessor chips **5**: 2988
  plasma mass spectrometry **5**: 2935
  quadrupole analyzers **5**: 2938
Mass spectrometry–gas chromatography *see* Gas chromatography–mass spectrometry
Mass transport
  algal growth modelling **1**: 258
Matching
  combinatiorial optimization **1**: 624
Materials handling
  automation **3**: 1643; **4**: 2368
  industrial robots **4**: 2444
    fabrics **4**: 2448
  industrial vision systems **4**: 2469
  ports **3**: 2072
  sugar industry **7**: 4700
  warehouse automation **7**: ***5087–9***
Mathematical analysis
  history **4**: 2270
Mathematical models
  competitive hierarchy **2**: ***1315***
Mathematical programming
  fuzzy set theory **3**: ***1816–9***
  matrix algebra
    large-scale systems **7**: 4678
  parameter optimization **7**: 4678
  simulation problems **6**: 4307
  surrogate worth trade-off method **7**: 4709
  systems analysis **7**: 4770
  systems methodology **7**: 4806
Mathematics
  hypothesis testing **5**: 2913
  relations **6**: ***4000–7***
  vagueness **7**: 4993
Matrices, polynomial *see* Polynomial matrices
Matrix algebra
  block diagonalization **7**: 4433
  block matrix formulae **1**: ***494–5***
  factorization **1**: 494
  finite matrices
    singular values **1**: 409
  inversion **1**: 494
  matrix groups
    control theory **5**: ***2944–8***
  matrix pencils **4**: 2867

poles and zeros **4**: 2828
multivariable systems **4**: 2820
polynomial systems
least order systems **6**: 3748
linear systems **6**: 3745
rational matrices
Smith–McMillan canonical forms **7**: ***4455–6***
sign stability **2**: 1078
square root factors **7**: 4535
stability **2**: 1078
state-space methods **7**: 4540
triangularization **1**: 494
Givens transformations **5**: ***2949–50***
Householder transformations **5**: ***2950–1***
see also Polynomial matrices
Matrix fraction representation **1**: 14
realization theory **6**: ***3968–9***
Matrix representation for digraphs
structural analysis **2**: 950
Maximization
subsolutions
Hamilton–Jacobi–Bellman equation **3**: 2071
see also Optimization
Maximum *a posteriori* approach **4**: 2702
Maximum-entropy principle **5**: ***2951***
entropy minimax theory **2**: 1474
information theory **4**: 2507
reconstructability analysis **6**: 3988
spectral estimation
stationary time series **7**: 4563
Maximum length sequences *see* M-sequences
Maximum-likelihood methods
autoregressive spectral estimation
stationary time series **7**: 4562
classification
multispectral images **6**: 4031
identification **4**: 2317
Maximum-power principle
model validation **7**: 4999
Maximum principle **2**: 1277; **5**: ***2956–63***
deterministic control
perturbation techniques **5**: 3498
higher-order conditions **7**: ***4419–25***
history **5**: 3511
matrix maximum principle
decentralized observers **2**: 910
multivalued differential equations **5**: 3182
multivariable systems **5**: 3204
piecewise analytic feedback **6**: 3998
regular synthesis **6**: 3998
Riccati equations **6**: 4069
singular control **5**: 3506
singular perturbations **7**: 4436
singular solutions **7**: ***4419–25***
stochastic maximum principle **5**: 3315
transversality **7**: 4948
Means–ends analysis **5**: ***2963***
Measurement
active measurement **2**: 1559
computer systems **1**: ***746–51***
fermentation **3**: ***1611–4***
fuzzy set theory
education **2**: ***1358–61***
indirect measurement **2**: 1558
indirect parameter measurement
fermentation **3**: 1614
sensor location problem **6**: 4238
uncertainty **2**: ***1553–9***
Measurement errors **2**: ***1553–9***
biomedicine
statistical error description **4**: 2254
classification **2**: 1554
identification errors
biomedicine **4**: 2254
Measurement standards
voltage measurement **2**: ***1390–1***
Measurement systems
active measurement **4**: 2532
instruments **4**: ***2527–33***
passive measurement **4**: 2532
Mechanical properties of substances
blood **1**: ***440–5***
bone **1**: ***468–73***
Mechanics
modelling **5**: 2968
simulation **5**: ***2966–70***
Mechanoception
electric sensing devices **6**: ***4118–23***
snakelike robots **6**: 4109
industrial robots **6**: ***4113–7***
Medical computing
educational models **2**: ***1361–6***
Medical diagnostic computing
electrocardiography
fuzzy set theory **1**: ***760–8***
expert systems **2**: ***1573–7***
fuzzy set theory **1**: 762, ***736–40***; **2**: ***1405–7***
modelling **1**: 761
pattern recognition
electrocardiography **1**: 761
Meetings
interpretive structure modelling **4**: 2576
Membership functions **3**: 1835
fuzzy set theory **3**: 1787, ***1823–4***, 1874
Membrane problem
finite element analysis **3**: 1627
Membranes
electrical properties **1**: 505
neuronal membranes
fuzzy encoding **1**: 506
Memory (physiological)
associative memory **6**: 4166
human information processing **3**: 2187
Mercury
dispersion
chlorides **4**: 2782
microorganisms **4**: 2781
ion-selective electrodes

water pollution **4**: 2607
soil **1**: 146
toxic substances **1**: 144
Metabolism
dynamics
tracers **7**: ***4901–5***
glucose
hunger **3**: 2210
ketone bodies modelling **4**: ***2628–31***
metabolic rate **5**: ***2970***
activity **5**: 2970
protein dynamics and control **6**: ***3905–9***
Metagames *see* Game theory
Metalanguage *see* Language (natural)
Metallurgical industries
nonferrous metallurgy
process control **5**: ***3280–2***
see also Steel industry
Metalogics
self-applicability **1**: ***326–32***
Metalworking
industrial robots
aerospace industry **4**: 2451
Metarules **5**: ***2971***
Metasystems **2**: 1548; **3**: 2001; **5**: ***2971***
Meteorological instruments
laser Doppler anemometry **2**: 1528
Meteorology
degree days **2**: ***959***
measurement techniques **5**: ***2971–3***
Meter provers **3**: 1682
Metering
water supply **7**: 5145, 5148
Method of lines
partial differential equations
simulation **2**: 1180
Method of multipliers *see* Lagrange multipliers
Methodology **5**: ***2973***
praxiology **6**: 3861
systems metamethodology **5**: ***2971***
systems methodology **5**: ***2971***
Metric spaces
fuzzy partitions **3**: 1837
Metros
automatic driving control **1**: ***362–5***
automatic line supervision **1**: 353, ***347–52***
automatic train operation **1**: 353
automatic train protection **1**: 352, ***365***
communication **1**: 364
operation quality monitoring **1**: 351
passenger information **1**: 351
power distribution regulation **1**: 349
routing control **1**: 349
signalling **1**: 365
technical failure management **1**: 351
traffic control **1**: 348, ***352–60***
see also Rail traffic
Michaelis law **7**: 5040
Michaelis–Menten models
ecology **2**: ***1330***
pharmacokinetics **6**: 3697
Microbiology
simulation **1**: ***421–4***
Microcomputers **5**: ***2980***
instrument systems **5**: ***2980–4***
model reference adaptive control systems **5**: ***3047–55***
power systems **6**: ***3837–42***
Microeconomics *see* Economics
Microelectrodes
bioelectric potential
digestive tract **1**: 460
Microfiltration
red blood cells
deformability **1**: 445
Microorganisms
dispersion
limnology **4**: 2781
Micropipette aspiration
red blood cells
deformability **1**: 444
Microprocessor chips **5**: ***2980***
architectures
discrete-event simulation **2**: ***1083–7***
instruments **5**: ***2980–4, 2984–8***
state of the art **2**: 1084
Microprogramming **5**: ***2989***
computer system simulation **2**: ***1450–1***
simulation models **1**: ***630–3***
parallel structure **1**: 633
see also Computer architecture
Microscopic particles
Coulter counters **2**: 862
Microwave Doppler velocity meters **3**: 1686
Microwave imaging
medical applications **7**: 4870
Microwave landing systems
helicopters **3**: 2122
Microwave measurement
remote sensing **6**: 4039
snow water equivalent
hydrologic forecasting **3**: 2233
Microwave spectroscopy **7**: 4499
Microwaves
radiometers
clinical **7**: 4855
solar power satellites **7**: 4473
Mie scattering
atmospheric laser spectroscopy **4**: 2720
Milestone charts
project management **6**: 3899
Military computing
communication command and control information systems **1**: 645
control and coordination **1**: 647
planning and commitment **1**: 646

MIMO systems
  self-tuning control **5**: 3184
Min–max pessimistic strategy **1**: 700
Mineral processing industry **5**: ***2989–93***
  adaptive control **1**: 50
  density control **2**: 778
    sedimentation **2**: 780
  process control **5**: 2989, 3280
  separation control **6**: 4248
Minerals
  complexes
    formation **1**: 3
  weathering **1**: 1
    neoformation **1**: 2
Minimal design problems
  multivariable control systems **5**: ***3215–20***
Minimax rule **6**: 3613
  linguistic probabilities **4**: 2882
Minimax techniques **5**: ***2994–7***; **7**: 5041
  duality (mathematics) **2**: 1259
Minimization **7**: 5041
  algorithms
    optimal control **5**: 3486
  switching functions
    fuzzy set theory **3**: 1885
  see also Optimization
Minimum alveolar concentration
  anesthesia **1**: 285
Minimum audible pressure
  hearing **1**: 323
Minimum-entropy principle
  entropy minimax theory **2**: 1475
Minimum-norm control
  multivariable systems **5**: 3204
Minimum-phase systems
  adaptive control **5**: ***2998–300***
Minimum prediction error methods
  system identification **4**: 2304
Minimum principle
  Pontryagin theory
    optimal control **5**: 3521
  see also Maximum principle
Minimum on-time
  duty cycling **5**: ***2998***
Minimum-time control
  multivariable systems
    functional anlaysis **5**: 3203
Minimum variance control **5**: ***3004–6***
  distillation columns **2**: 1110
  nonlinear systems **5**: 3345
    coefficient calculation **5**: 3350
  self-tuning control **6**: 4191
    feedforward control **6**: 4187
    optimal control **6**: 4181
    pole assignment **6**: 3725
    predictive control **6**: 4176
Mining
  process control **5**: 3280
Missiles
  guidance **1**: 114; **5**: ***3006–12***
Mixed coordination method **5**: ***3012–9***
Mixing
  concrete mixing
    cautious control **2**: ***771–3***
  modelling **4**: ***2782–9***
Mode analysis
  oscillators
    coupled nonlinear oscillators **5**: ***3019–24***
    two-dimensional array oscillators **5**: 3020
Mode S
  air traffic control **1**: 197, 200
Model-base management systems **2**: 946
Model coordination method **5**: ***3035–9***
Model error
  first-order error analysis **2**: 1552
Model-following control
  predictive self-tuning control **6**: 4176
Model management systems **6**: 4328
Model order selection
  instrumental variable techniques **4**: 2277
Model reference adaptive control systems **1**: 45, 48;
    5: ***3044–7, 3056–60***; **6**: ***4163, 4184–6***
  astronomical telescopes **6**: 3927
  augmented error **5**: 3059
  augmented error signals **5**: ***3061–5***
  bounded external noise **5**: 3059
  comparative studies **5**: ***3067–70***
  continuous time **6**: 3596, 4194
  design **5**: 3047
  discrete-system design **2**: ***1087–92***
  dynamic compensation **5**: 3056
  error models **1**: 56; **5**: 3057, ***3065–7***
  evaporators **2**: 1567
  first-order parallel systems
    Lyapunov synthesis **1**: 98
  global stability **5**: 3056
  higher-order state vector models
    Lyapunov synthesis **1**: 99
  identification-error controllers **5**: 3048
  input-error controllers **5**: 3048
    Lyapunov methods **6**: 4184
  microcomputer implementation **5**: ***3047–55***
  model updating **5**: 3070
  output-error controllers **5**: 3048
  relative degrees **5**: 3056
  ships
    guidance **6**: ***4275–6***
  stability
    Lyapunov methods **5**: 3068
  stochastic systems **5**: 3046
  variable-structure systems **7**: 5029
  see also Adaptive control
Model reference adaptive systems **1**: ***100–5***
  direct control **5**: ***3060–1***
  indirect control **5**: ***3060–1***

Model structure determination **4**: ***2287–92***
Model validation *see* Validation, Modelling, Validation
Modelling **5**: ***3083***; **6**: ***4311–3***; **7**: ***5011–5***
activity-based simulation **6**: ***4333–4, 4392–3***
actuators
HVAC processes **3**: ***2097–9***
aerosol particle size **1**: 179
aerospace propulsion **7**: 4482
age-dependent models
ecological populations **1**: 415
agriculture
air pollution **1**: 590
chemical stress **1**: ***589–92***
field applicability **1**: ***137–42***
air pollution **2**: ***1497–500***
aircraft
analytic redundancy **1**: 247
motion **1**: 213
aircraft dynamics
aeroelastic model **7**: 5165
analytic model **7**: 5166
aircraft engine control systems **1**: 222
aircraft flight
air traffic control **1**: 193
aircraft wing load alleviation
aeroelastic **7**: 5164–70
algal growth
Cladophora **1**: ***257–61***
model segmentation **1**: 258
analytical modelling
manufacturing computer control **3**: 1645
anesthesia **1**: 281
artificial intelligence **1**: 304
artificial satellites **6**: 4147
astronomical telescopes
adaptive control **6**: 3928
atmospheric modelling **1**: 187
turbulence **7**: 5166
autopoiesis
computers **1**: 396
simulation **6**: ***4374–7***
autoregressive models
environmental data **2**: 1494
baroreflexes
cardiology **1**: ***536–42***
bilinear equations
ecological population dynamics **1**: 415
biochemical oscillators **1**: 427
bioelectric phenomena
digestive tract **1**: 460
small intestine **2**: ***1040–2***
stomach **2**: ***1033–6***
ureter **7**: 4978
biological modelling **1**: 466
cusp catastrophe **1**: 432
motor control theory **1**: 466
structural identifiability **1**: ***490–3***
biomedical phenomena **1**: ***486–9***
black-box models
radioecology **6**: 3934
blood glucose
adaptive control **3**: 2017
blood pressure
adaptive control **1**: 499
boilers **7**: 4575
boundary-value problems **5**: ***3097–9***
carbon cycle **1**: 529–31
validation **1**: 531
cardiac interval reflex **1**: 537
catastrophe **1**: 550
cellular automata **6**: ***4341–5***
cellular spaces **1**: 559–61
chemical reactors **1**: 585
Nyquist diagrams **4**: 2591
parameter estimation **1**: 587
comfort
control systems **1**: ***638***
compartment models
bilinear models **1**: 414
environ analysis **2**: 1482
radioecology **6**: 3934
computer–user interfaces **5**: ***3077–9***
computers
environment **2**: 1536
security **1**: 741
conceptualization
ecology **2**: 1330
environment **2**: 1536
continuous error
adaptive control **1**: 101
continuous linear systems **4**: ***2798–800***
continuous models
microprogramming **1**: ***630–3***
control system applications **2**: 1140
control theory
fuzzy set theory **4**: 2357
coronary hemodynamics **2**: ***850–3***
decentralized observers **2**: 908
decision theory and analysis **3**: 2193
deterministic modelling
ecology **7**: 4644
deterministic models
population control **2**: 1336
discrete error
adaptive control **1**: 102
discrete-event simulation **6**: 4397; **7**: 4763
discrete models
microprogramming **1**: ***630–3***
distributed computer systems **2**: 1129
performance index **6**: ***3655–9***
synchronization **2**: 1167
distributed parameter systems **2**: 1193
distributed systems
distributed simulation **2**: ***1199–201***
drinking **3**: 2213
drying **2**: 1249
dynamic duopoly **2**: 1267

dynamic models
distillation **2**: 1109
polymer extrusion **2**: 1585
dynamic systems **2**: ***1286–91, 1292–7***; **4**: 2240
likelihood functions **4**: 2286
probabilistic models **4**: 2286
dynamics
food chains **6**: 3934
ecology **2**: ***1323–4, 1330–2***
aggregation errors **2**: ***1324–6***
bond graphs **2**: ***1326–9***
comparison **2**: 1332
credibility **2**: 1332
developmental **2**: 1331
errors **2**: 1332
grammars **3**: ***2043–5***
law of the minimum **2**: ***1329***
Michaelis–Menten models **2**: ***1330***
parameter estimation **6**: ***3619–21***
perturbation techniques **6**: ***3664–5***
populations **2**: 1324
prediction **2**: 1332
sensitivity analysis **2**: 1549
simplification **2**: 1332
temperature **2**: ***1333***
economics **2**: ***1337–42***; **5**: 2795
ecosystems **2**: 1324
compartment models **2**: ***1343–4***
environ analysis **2**: 1483
qualitative stability **2**: ***1348–51***
stability **1**: 672
electric power generation
reliability **6**: 3791
electrocardiography **1**: ***532–5***
electrolysis **2**: 1413
environment **2**: 1536
environmental data **2**: ***1494–7***
equilibrium models
food chains **6**: 3934
errors
model reference adaptive control systems **5**: ***3065–7***
eutrophication
variance estimates **2**: ***1269–72***
evaluation design **2**: ***1559–65***
evaluative modelling
toxic organic materials fate **3**: 1600
evaporators **2**: 1566
external respiration **6**: ***4051–5***
flexible spacecraft
multibody modelling **3**: ***1653–6***
system components **3**: 1649
fuel consumption
road traffic **3**: 1759
furnaces
Nyquist diagrams **4**: 2594
fuzzy algorithms
aggregate planning **1**: 125
fuzzy relations
databases **3**: 1800
gas transmission networks **3**: 1986
gas turbines **1**: 581
Nyquist diagrams **4**: 2588
glass industry
process control **3**: 2013
global models **3**: ***2015–6***
glucose regulation **3**: ***2021–5, 2025–32***
glycolysis
oscillators **1**: 426
goal-oriented systems **3**: 2039
hemodynamics **1**: ***542–6***
hepatobiliary kinetics **3**: ***2139–43***
hereditary systems **3**: 2147–50
hierarchical scene analysis **6**: 4155
hierarchical systems
ecology **2**: ***1329***
history **4**: 2270
hormone regulation **3**: 2169
human information processing **3**: 2186
hybrid errors **5**: 3066
hydroelectric power stations **3**: 2222
hydrology **3**: 2230
identification **4**: 2235, 2320
black-box models **4**: 2241
identification errors
biomedicine **4**: 2253
in vivo drug effect prediction **2**: 1244
indeterminacy **2**: 1551
lake ecosystems **4**: ***2650–1***
industrial robots
articulated arm control **1**: 302
flexible arms **6**: 4086
industrial vision systems **4**: 2472
information gathering **7**: 4783
information storage **6**: ***4323–6***
inorganic chemicals production **4**: 2516
Kalman filters **4**: 2619
ketone body metabolism
noncompartmental models **4**: 2629
lakes
ecosystems **4**: ***2643–6, 2650–1***
history **4**: 2643
response time **4**: ***2789–94***
land use **7**: 4491
large intestine electrical activity **2**: ***1036–40***
relaxation oscillators **2**: 1037
large-scale systems
hierarchical overlapping coordination **5**: 3564
model simplification **5**: ***3071–7***
life cycle models **6**: 4328
linear models
environmental data **2**: 1494
linear programming **4**: 2839
lipoprotein dynamics **4**: 2887
load flow **6**: ***3850–5***
load forecasting **6**: 3830
lungs **4**: ***2898–901, 2902–4***
management models

differential games **1**: 686
validation **7**: ***5001–5***
manufacturing computer control **3**: ***1642–8***
Markov decision models **7**: 4806
mean-variance
decision theory and analysis **3**: 2196
mechanical systems **5**: 2968
medical computing
medical education **2**: ***1361–6***
medical diagnostic computing **1**: 761
microbiology **1**: 421
microeconomic systems analysis **5**: 2975; **7**: 4804
mixing **4**: ***2782–9***
model behavior
taxonomy **5**: ***3030–4***
multiattribute models
decision theory and analysis **3**: 2196
multicriteria modelling
fuzzy set theory **3**: ***1826–9***
multi-timescale systems **5**: 3176
natural gas demand **3**: 1979
natural gas system planning **3**: 1996
natural gas transmission systems **3**: 1990
network modelling
project management **6**: ***3898–903***
neutral models
ecology **2**: 1331
nitrification modelling
lotic systems **5**: 3264
nitrogen cycle modelling **5**: 3266
nonlinear models
air pollution **2**: 1494
power stations **6**: ***3772–5***
vision **7**: ***5065–70***
nonlinear systems **5**: 3341
nonparametric models
biomedical phenomena **1**: 488
normative models
risk aversion **6**: 4070
on–off control
many-valued logics **5**: 2926
operations research **4**: 2351; **5**: 3450
fuzzy set theory **5**: 3450
optimization **6**: 4306
linear programming **4**: 2839
order
aggregation **1**: 128
order estimation
ecological models **5**: ***3040–4***
ordinary differential equation models
symbol manipulation **5**: ***3559–60***
organic compounds fate
PEST **3**: ***1600–1***
parameter estimation
sparse data systems **1**: 486
supervision **1**: 64
partial differential equations
examples **6**: 4361
linear partial differential equations **6**: 4360
nonlinear partial differential equations **6**: 4361
simulation **2**: ***1180–1***
partial modelling
approximate similarity **6**: 4295
peripheral resistance reflex **1**: 537
pharmacokinetics **2**: 1245
nonlinear models **6**: ***3695–8***
phosphorus kinetics **5**: 3043
phytoplankton models
validation **7**: ***5005–7***
pneumodynamics **1**: 457
population models **6**: 3751, ***3756–62***
ports **3**: 2076
power systems **6**: 3782, 3796
state estimation **6**: 3847
praxiology **6**: 3863
predictive modelling **3**: 1710
anticipatory systems **1**: 292
risk aversion **6**: 4070
pressure control
furnaces **6**: 3875
probability distributions
reliability **6**: ***4027–9***
process-based models **6**: ***3878–9***
simulation modelling formalism **6**: ***4365–6***
protein synthesis **1**: 558
pumps
HVAC processes **6**: 3918
queuing theory **6**: ***3922–5***
radioecology **6**: ***3929–33***
agricultural food chains **6**: ***3933–6***
rainfall–runoff modelling **3**: 2230; **6**: ***3940–4***
rational actor
decision theory and analysis **3**: 2194
reduced-order models **1**: 127
adaptive control **1**: 105
generalized Hessenberg representation **1**: 134
regional models **3**: ***2015–6***
respiration **1**: 568; **6**: ***4055–61***
road traffic **7**: 4912
freeways **3**: ***1736–41***
microscopic models **7**: ***4925–7***
rolling mills **6**: 4141
sampled-data systems
Nyquist diagrams **4**: 2595
scenarios **6**: 4153
self-tuning control **6**: ***4197–9***
explicit delay **6**: 4199
implicit delay **6**: 4199
semantic data **2**: 891
separation processes **6**: 4249
simulation
arithmetic relators **6**: ***4334–6***
computer systems **1**: ***751–4***
databases **2**: 893
differential equations **6**: ***4356–9***
discrete arithmetic-based models **6**: ***4345–6***
event-based models **6**: ***4346–8***
extended Petri net graphs **6**: ***4348–50***

hierarchical decomposition **6**: ***4353–5***
Markov chains **6**: ***4355–6***
partial differential equations **6**: ***4360–4***
symbol manipulation **6**: ***4377–80***
system dynamics **6**: ***4366–8***
systems control **6**: 4315
transaction flow-based models **6**: ***4368–9***
verification **6**: 4402
simulation model documentation **6**: ***4372–4***
simulation model taxonomy **6**: ***4381–7***
singular dose–response models
agriculture **1**: 589
spatial allocation **7**: 4491
speech **7**: 4505
state-space modelling **7**: ***4540–7***
steady-state modelling
polymer extrusion **2**: 1585
stochastic models
populations **1**: 415; **2**: 1335
stochastic processes
dynamic systems **7**: 4651
ecology **7**: ***4644–5***
structural modelling
systems methodology **7**: 4803
structure simplification **5**: ***3079–82***
order reduction **5**: 3081
support systems **6**: ***4370–2***
switching
artificial lighting **4**: 2765
switching networks **7**: ***4726–31***
system components
flexible spacecraft **3**: 1649
system dynamics **7**: ***4755–62***
system identification **4**: 2301
system specifications **5**: ***3083–8***; **7**: ***4762–4***
system theory **3**: 1872
systems analysis **7**: 4769, 4803
validation **6**: 4407
systems methodology **7**: 4803
team theory **7**: 4812
technological forecasting **7**: 4826
technological innovation **7**: 4829
three-dimensional vision **1**: ***343–7***
tides **7**: ***4883–8***
top-down modelling
ecology **7**: ***4894–5***
toxic chemicals assessment **7**: ***4895–7***
toxic substances
ecology **2**: 1322
transaction flow-based models
simulation **7**: ***4933–5***
transport models
radioecology **6**: 3933
transportation **7**: ***4944–7***
trophic structures **7**: ***4954–6***
tryptophan
operon **1**: 426
tumors **7**: 4957
updating
model reference adaptive control systems **5**: ***3070–1***
ureter
bioelectric phenomena **7**: ***4976–80***
utility
ecology **2**: 1331
validation **4**: 2242; **7**: ***4996–500, 5008–10***
operations research **4**: 2382
phytoplankton **7**: ***5005–7***
statistical analysis **7**: ***5015–8***
systems analysis **6**: 4407
vestibular system **2**: ***1591–5***
slow-phase caloric nystagmus **2**: ***1586–90***
water demand **7**: 5154
water distribution networks **7**: 5106
water quality **7**: ***5129–34***
water resources **7**: 5142
water supply **7**: 5099
network analysis **7**: 5102
transients **7**: 5157
welfare economics **7**: 4805, 5163
see also Simulation
Modems **6**: 4290
Moderators
nuclear power **5**: 3414
Modes
selection
aggregation **1**: 128
MODULAR **6**: 3896
Modularity
manufacturing computer control **3**: ***1633–6***
Modulation
delta modulation **2**: 1046
digital communication systems **2**: 1048
signal transmission **6**: 4289
Modules
programming **7**: ***4472***
Modus ponens
fuzzy logic **5**: ***3089***
medical diagnostic computing **2**: 1575
Modus tollens
generalized
fuzzy logic **5**: ***3089***
medical diagnostic computing **2**: 1575
Moisture control
nuclear magnetic resonance **5**: 3418
see also Humidity control
Molecular spectra
nuclear magnetic resonance **5**: 3416
see also Spectra
Molybdenum
ecology
aluminum **1**: 267
soil **1**: 146
toxic substances **1**: 144
Monitoring
air pollution **1**: ***184–5***
spline interpolation **7**: 4517–8
building services **1**: 513

electric sensing devices
failure detection **1**: 249
environmental information **2**: 1489
industrial robots **4**: 2364
natural gas technology **3**: 1976
power systems
microcomputers **6**: 3837
state estimation **6**: 3847
ships **7**: ***5051–5***
technological forecasting **7**: 4825
water quality **2**: ***1532–5***
Monochromators
spectroscopy
biology **1**: 9
Monte Carlo methods **5**: ***3089***
conditional **7**: 4655
ecological modelling **2**: 1551; **7**: 4644
lakes **4**: 2650
optimal stochastic control **5**: 3529
parameter estimation
ecological modelling **6**: 3620
partial differential equations **6**: 3623
sensitivity analysis **6**: 4228
system failure and recovery
probability **6**: 4012
Morphisms
model structure simplification **5**: 3080
system specifications **6**: 4326–8
Mössbauer effect
spectroscopy **7**: 4500
Motion
aircraft
lateral motion **1**: 218
longitudinal motion **1**: 218
Motivation
hunger and thirst
feedback control **3**: ***2209–13***
Motor systems
biological **3**: 1908
nervous control **1**: ***462–6***
Motorways *see* Freeways
Mouth
satiety
hunger control **3**: 2210
Moving-average processes
spectral estimation
stationary time series **7**: 4564
stationary time series **7**: ***4559***
see also ARMA processes
Moving ramps
vertical transportation
history **7**: 5048
MPL **6**: 3895
MRAC *see* Model reference adaptive control systems
MRAS *see* Model reference adaptive systems
MS/OR *see* Management science, Operations research
Multiattribute utility theory **5**: ***3099–103***, ***3104–15***, 3117, 3157; **7**: 4806
decision theory and analysis **5**: 3104
independence **5**: 3105
multicriteria decision making **5**: 3121
trees (mathematics) **7**: 4952
water resource management **7**: 5137
Multichannel systems
decentralized stabilization **2**: 917
Multicriteria decision making **5**: ***3116–21***, ***3121–4***, ***3150–2***
fuzzy mathematical programming **3**: 1816
social choice theory **7**: 4469
Stackelberg strategies **5**: ***3145–9***
surrogate worth trade-off method **7**: ***4704–9***
water resource management **7**: 5135
see also Game theory
Multicriteria formulation
complex systems **1**: 699
Multidimensional scaling **5**: ***3124***
Multidimensional systems **5**: ***3127–30***
oscillations **5**: ***3124–7***
stability **4**: 2874; **5**: ***3131–4***
Multigoal programming **5**: 3119
Multigrid methods **5**: ***3135–9***
algorithms
linear systems **5**: 3138
Multi-input, multi-output systems *see* MIMO systems
Multi-input systems
correlation methods **4**: 2252
local controllability **4**: 2895
Multilayer control
dynamic systems **2**: 1297
freeway traffic **3**: ***1736–41***
Multilevel control **2**: 952
coordination **5**: 3013
dynamic systems **2**: 1297
functional hierarchy **5**: 3140
temporal hierarchy **5**: 3142
vertical decomposition **5**: ***3140–4***
Multilevel filters **4**: ***2692–7***
algebraic structure **4**: 2695
computational requirements **4**: 2696
multiplication requirements **4**: 2696
structure **4**: 2694
Multiloop design
suboptimal control **1**: 131
Multimodel control
tracking methods
complex systems **1**: 679
Multiobjective control
dynamic multistage processes **2**: ***1272–6***
multiobjective coordination
hierarchical overlapping coordination **5**: ***3564***
Multiobjective decision making **5**: 3118
hepatobiliary kinetics modelling **3**: 2143
multiattribute utility theory **5**: ***3099–103***
optimization **5**: 3150
Multiobjective optimization **5**: 3114, ***3150–2***, ***3152–6***, ***3156–63***
information gathering **7**: 4781
interactive methods **5**: 3152

surrogate worth trade-off method **7**: ***4703–9***
Multipass processes
multidimensional systems **5**: 3129
stability **5**: 3134
Multiple controllers
blood pressure
surgery **1**: 499
structure **6**: 4008
Multiple regression **2**: 878
Multiple systems
dynamic systems
stability **7**: 4524
Multiplexing **6**: 4290
code-division multiplexing
satellite links **2**: 1056
frequency-division multiplexing **6**: 4290
satellite links **2**: 1055
packet switching **6**: ***3581–90***
satellite links **2**: 1055
time-division multiplexing **6**: 4290
time random multiplexing
satellite links **2**: 1056
Multiplier methods *see* Lagrange multipliers
Multiprocessing programs
distributed database systems **2**: ***1143–53***
see also Concurrency (computers)
Multiprocessing systems
distributed database systems **2**: ***1165–9***
Multiprogramming **5**: ***3165***
synchronization **7**: 4751
Multiprojection **4**: 2695
multilevel filters **4**: 2692
Multirate control systems
self-tuning control **5**: ***3184–8***
Multispace processes
dynamic multispace processes
multiobjective control **2**: ***1272–6***
Multispectral images
landsat imaging
atmosphere **4**: 2652
plants (biological)
environmental pollution **2**: 1517
Multistate systems
optimal control **5**: ***3489–96***
Multi-timescale systems **5**: ***3171–6***
difference equations **7**: 4431
dynamic location **5**: ***3176–80***
linear systems **5**: ***3176–80***
slow–fast decomposition **7**: 4529
Multivariable control systems
applications
consistency control **2**: 779
control system synthesis
geometric approach **4**: ***2807–16***
controllability **4**: ***2816–26, 2827–32***
controllability subspaces **4**: 2850
decentralized robust controller design **6**: ***4133–8***
decoupling problem **4**: 2853
design **5**: 3206, 3208
method of inequalities **5**: ***3206–14***
semi-infinite optimization **5**: ***3189–96***
digital three-term control **6**: 3702
disturbance decoupling **4**: ***2522–3***
dyadic expansions **5**: ***3197–201***
feedback
evaporators **2**: 1567
functional analysis **5**: 3201
geometric synthesis **4**: ***2807–16***
geometric theory **2**: 822
input–output decoupling **4**: ***2522–3***
internal model principle **4**: 2566
invariants **4**: 2866–71
Kronecker canonical forms **4**: 2866–7
linear quadruples
geometric approach **4**: ***2849–55***
linear systems
LQG design
robustness **4**: ***2842–6***
minimal design problems **5**: ***3215–20***
nonlinear decoupling **5**: ***3282–8***
nuclear power stations **5**: ***3431–6***
pole-assignment self-tuning **6**: 3723
observability **4**: 2816–26
optimal control **1**: 38
robust controller design **6**: ***4129–32***
structural constraints **7**: 4685
rolling mills
Nyquist diagrams **4**: ***2601–4***
self-tuning control **5**: ***3184–8***
stability design **5**: 3377
see also Multivariable systems
Multivariable-root-locus method
feedback control
design **1**: ***575–80***
Multivariable systems
adaptive control **1**: 104
aircraft motion **1**: 388
CAD **1**: ***732–5***
design
parameter optimization **1**: 391
robustness **5**: ***3224–8***
dominance design
optimization **5**: ***3220–3***
functional analysis **5**: ***3201–5***
Hessenberg form **4**: 2820
least squares identification **4**: 2280
linear systems
controllability **4**: ***2816–26***
invariants **4**: ***2866–71***
Kronecker canonical forms **4**: ***2866–71***
observability **4**: ***2816–26***
poles and zeros **4**: ***2827–32***
Nyquist diagrams
design applications **4**: 2584
observers **4**: 2701
self-tuning control **5**: ***3229–32***
state-space methods **7**: 4547
variable-structure systems **7**: 5028

see also Multivariable control systems
Multivariate analysis **5**: ***3232–5***
Multivariate control systems
  boilers **7**: 4575
Multivariate realization theory **6**: ***3969–72***
Muscle
  dynamics
    physiological models **5**: ***3243–8***
  movement
    nervous control **1**: 462
  neurophysiology **5**: ***3248–52***
  relaxation
    dynamics **5**: 3236
    feedback control **5**: ***3235–8***
    measurement **5**: 3235
  spindles
    anatomy **5**: 3240
    identification **5**: 3241
    output signals **5**: 3240
    physiological models **5**: ***3238–42***
    physiology **5**: 3240
  striated muscle
    characteristics **5**: 3248
MYCIN system
  medical diagnostic computing **2**: 1574
Myoelectricity *see* Bioelectric phenomena

N

Nash equilibrium **1**: 35, 700; **5**: 2908, ***3255***
  dynamic duopoly **2**: 1263
  game theory **1**: 683
  noncooperative games **3**: 1919–21
  stochastic differential games **7**: 4613
Natural gas **2**: 1461; **3**: ***1966–9***
  distribution systems **3**: ***1939–43***
    leakage **3**: 1955
    steady-state analysis **3**: ***1947–54***
  energy resources **3**: 1714
  interruptible loads **3**: 1946
  liquefaction
    storage **3**: 1970
  production **3**: 1966
  properties **3**: 1968
  salt-cavity storage **3**: 1972
  seasonal demand **3**: 1944
  seasonal storage **3**: 1969
  storage **3**: 1944, ***1969–75***
  substitute natural gas **3**: ***1958–61***
  supply and demand **3**: ***1943–7***
  supply systems
    control **3**: ***1975–9***
    load forecasting **3**: ***1979–81***
    transient analysis **3**: 1976
  system planning **3**: ***1995–8***
  transmission systems **3**: ***1982–5***
    supervision and control **3**: ***1985–9***
    transient analysis **3**: ***1990–2***
Natural gas technology
  compressors **3**: 1946
  density measurement **3**: 1962
  flow control **3**: 1962
  flow measurement **3**: ***1962–5***
    flowmeters **3**: 1962
  leak detection **3**: ***1954–8***
  leakage control **3**: 1956
  materials
    ductile iron **3**: 1939
    gray cast iron **3**: 1939
    polyethylene **3**: 1939
  pipe joints
    leak detection **3**: 1955
  pipeline design **3**: 1991
  pressure measurement **3**: 1962
  telemetry **3**: 1962, ***1975–9***
  temperature measurement **3**: 1962
Navigation
  air traffic control
    aerospace simulation **1**: 205
  aircraft **1**: 113
    automatic direction finders **1**: 113
    combat aircraft **1**: 623
  artificial satellites **2**: 1243
  display
    aircraft instrumentation **1**: 242
    cockpits **1**: 602
  errors
    inertial systems **4**: 2490
  four-dimensional navigation **3**: ***1719***
    civil aircraft **1**: 604
  helicopters **3**: 2122
  missiles **5**: 3007
  ships **5**: ***3256–60***; **6**: 4260
    traffic control **7**: ***5051–5***
  see also Inertial navigation systems
Needs **5**: ***3260***
Nerves
  excitation
    hysteresis **1**: 432
Net present value **5**: ***3260***
Network analysis
  gas distribution systems **3**: 1941, 1947
    steady-state analysis **3**: ***1947–54***
  network ascendancy **3**: ***1694–5***
  nonlinear optimization problems **2**: 1279
  systems methodology **7**: 4807
  water supply **7**: 5097, 5102
    simulation **7**: ***5106–9***
    steady-state analysis **7**: ***5102–5***
Network control
  road traffic **6**: 4076
Network methods **5**: ***3261, 3261***
  project management **6**: ***3898–903***
Network synthesis
  ISO model **6**: 4293
  linear passive synthesis **4**: ***2832–5***
Network topology
  ecosystem networks **2**: ***1352–5***
  gas distribution systems **3**: 1949
  power system control **6**: 3826
  signal transmission **6**: 4293
Networks (circuits)
  ethernet
    signal transmission **6**: 4293
  local area networks
    signal transmission **6**: 4293
  stability **5**: 3132
  wide area networks
    signal transmission **6**: 4293
Neural nets **6**: 4165
  cellular automata **6**: 4344
  cybernetics **2**: 872
  fuzzy learning strategies **5**: 3261
  fuzzy systems **1**: 507
  modelling **1**: 569
  perceptrons **6**: 4166
  visual system
    feedback lateral inhibition **7**: ***5060–5***

Neurophysiology
  biological motor systems **1**: ***462–6***
  blood pressure
    adaptive control **1**: 502
  cognitive systems **1**: 502
  fuzzy learning strategies **5**: ***3261–4***
  locomotion control **3**: 1909
  muscles **5**: ***3248–52***
    electromyography **6**: 4052
  random processes **1**: 569
  self-organizing systems **6**: 4167
  visual perception **7**: ***5081***
Neutral systems **2**: 1545; **3**: 2001; **5**: ***3264***
Neutron activation analysis **1**: 174
  flow measurement **3**: 1688
New effective temperature **1**: 642
Newton methods **2**: 792
  unconstrained optimization **7**: 4968, ***4971–4***
NGT *see* Nominal group technique
Nickel
  soil **1**: 146
  toxic substances **1**: 144
Nitrates
  chemical analysis
    aerosols **1**: 175
    water pollution **7**: 5119
  ecology
    aluminum **1**: 265
  ion-selective electrodes
    water pollution **4**: 2606
Nitric oxide
  air pollution
    automatic analyzers **1**: 182
    chemiluminescence **1**: 183
Nitrification
  lotic systems **5**: ***3264–6***
Nitrites
  chemical analysis
    water pollution **7**: 5119
  ion-selective electrodes
    water pollution **4**: 2606
Nitrogen
  ecology
    aluminum **1**: 265
  ion-selective electrodes
    water pollution **4**: 2606
  organic nitrogen
    aluminum compounds with nitrogen **1**: 266
  soil **1**: 146
  toxic substances **1**: 145
Nitrogen compounds
  inorganic nitrogen compounds
    water pollution **7**: ***5117–22***
  organic nitrogen compounds
    water pollution **7**: 5121
  water
    concentration **7**: ***5122–3***
Nitrogen cycle modelling **2**: 1322; **5**: ***3266–8***
Nitrogen dioxide
  air pollution
    automatic analyzers **1**: 182
  photolysis
    air pollution **1**: 185
Nitrogen oxide
  ion-selective electrodes
    air pollution **4**: 2606
Nitrous acid
  photolysis
    air pollution **1**: 185
Nodes
  graph theory **5**: ***3268***
Noise **2**: 1403
  amplifiers **1**: 270
  modelling
    Kalman filters **4**: 2619
  optical fiber interferometric sensors **5**: 3466
  radiology
    x rays **7**: 5183
Noise (acoustic) *see* Acoustic noise
Nominal group technique **1**: ***612***; **4**: 2576; **7**: 4802
Nonautonomous systems **7**: 4528
Noncommutative rings
  realization theory **6**: 3970
Noncommuting variables
  finite automata **5**: ***3268–72***
  rational generating series **5**: ***3272–6***
  realization theory
    bilinear systems **5**: ***3276–9***
Noncooperative games **1**: 683; **3**: ***1917–21***
  applications **5**: 2908–10
Nondirected relations **5**: ***3279***
Nondispersive infrared analyzers
  air pollution
    hydrocarbons **1**: 179
Nondominated solutions *see* Pareto optimality
Nonelectric sensing devices
  air conditioning **1**: 154
  optical fiber sensors **5**: ***3455–64***
    interferometric devices **5**: ***3465–70***
Noninferior solutions *see* Pareto optimality
Noninteractive control
  articulated robot arm control
    electrohydraulic servosystems **1**: 304
  multivariable control system synthesis **4**: 2812
Nonlinear coupling
  transinformation **4**: 2543
Nonlinear differential equations **5**: 3561
  nonlinear control systems **5**: ***3383–8***
Nonlinear estimation theory **5**: ***3293–302***
  asymptotic analysis **5**: ***3303–10***
  stochastic control **5**: 3311–6
  see also Estimation theory
Nonlinear filtering **5**: ***3293–302***
  geometric theory **5**: 3323
  iterated Ito integral expansions **5**: ***3330–6***
  quantum theory **5**: ***3319–29***

stochastic calculus **7**: ***4631–7***
see also Filtering and prediction theory
Nonlinear modelling
power stations **6**: ***3772–5***
Nonlinear optimization
spatial dynamic programming **2**: 1279
Nonlinear programming **5**: ***3351, 3352–62***; **7**: 4554
approximate programming
industrial robots **4**: 2397
convex analysis **5**: 3353
convex functions **5**: 3357
discrete-time control systems
deterministic case **2**: 1093
goal coordination method **3**: 2032
locally Lipschitz functions **5**: 3359
model coordination method **5**: 3035
nondifferentiable optimization **5**: ***3541–5***
optimal control **5**: 3512
smooth case **5**: 3355
surrogate worth trade-off method **7**: 4709
Nonlinear systems **5**: ***3366–73***
approximation **5**: ***3374–6***
automata theory **5**: 3268
bang–bang control **1**: 412
blood pressure control
surgery **1**: 499
CAD **5**: ***3379–83***
characteristic controllers **5**: 3344
circle criterion **5**: ***3376–9***
correlation methods **4**: 2252
decentralized stabilization **2**: 922
decoupling **5**: ***3282–8***
discrete-time systems
algebraic theory **5**: ***3288–92***
input–output maps **4**: ***2523–7***
dynamic systems
multilevel optimization **2**: 859
feedback systems stability **5**: ***3316–9***
functional expansions **3**: 1769
geometric approach **5**: ***3362–6***
geometry **5**: ***3362–6***
hereditary systems **3**: 2150
history **5**: ***3383–8***
invertible systems **5**: 3291
local controllability **4**: 2893
local realization **5**: ***3390–4***
Lyapunov methods **5**: ***3394–8, 3398–401***, 3402
modelling **2**: 1288; **5**: 3341
nonparametric **1**: 488
multi-timescale systems **5**: ***3176–80***
Lurie–Postnikov type **5**: 3179
multivariable control systems
design **5**: 3206
observability **5**: ***3336–41***
global results **5**: 3339
piecewise linear systems **5**: 3292
realization theory **6**: ***3973–7***
sampling **5**: 3292
self-tuning control
algorithms **5**: ***3341–51***
optimal control **6**: 4180
sensitivity analysis **6**: 4225
stability
vector norms **5**: ***3402–6***
state feedback control
slow–fast decomposition **7**: 4532
state-space methods **7**: 4540
universal inputs **5**: 3339
Non-minimum-phase systems
adaptive control **5**: ***3406–9***
pole-assignment self-tuning **6**: 3723
Non-Newtonian fluids
viscosity measurement **7**: 5056
Nonproper systems
polynomial systems
matrix algebra **6**: 3750
Normative methods **5**: ***3413***
Notch filters
digital filters
direct digital control **2**: 1076
Nozzles
design **2**: 1177
NP-complete problems
combinational optimization **2**: 1082
computational complexity **1**: 724
NP problems
computational complexity **1**: 724
Nuclear fission **2**: 1461
nuclear power **5**: ***3414–6***
see also Fission reactors
Nuclear fusion
nuclear power **5**: 3416
Nuclear instrumentation
mass spectrometers **5**: 2935
Nuclear magnetic resonance **5**: ***3416–21***
flowmeters **3**: 1688
imaging
plants (biological) **2**: 1518
Nuclear physics
optical fiber sensors **5**: 3462
Nuclear power **5**: ***3413–6***
fission reactors **5**: ***3436–40***
Nuclear power stations **5**: 3439
economics **3**: 1716
fast reactors **5**: 3439
multivariable control systems **5**: ***3431–6***
nonlinear modelling **6**: 3774
self-tuning control **5**: ***3425–30***
feedforward control **6**: 4189
thermal reactors **5**: 3439
see also Fission reactors, Power stations
Nuclear spin–lattice relaxation
biomedical NMR **5**: 3424
Nuclear spin–spin relaxation **5**: 3425
Nucleolus
cooperative game theory **2**: 848

Null hypothesis **5**: ***3440***
Numerical analysis
  control system computing **2**: ***813–25***
  linear algebra **2**: ***817***
  partial differential equations **6**: ***3622–6***
  problem conditioning **2**: 816
  splitting
    hierarchical control **3**: 2153
Numerical control
  variable speed induction machines **7**: ***5019–27***
Numerical methods
  boundary-value problems **5**: ***3097–9***
  Hamilton–Jacobi–Bellman equation **3**: ***2068–72***
  identification **4**: 2318
  integration **5**: ***3555–8***
  linearly constrained problems **4**: ***2875–9***
  multivariable control system design
    method of inequalities **5**: 3213
  optimal control
    multistate systems **5**: 3494
  optimal stochastic control **5**: ***3523–30***
  parameter estimation **4**: 2238
  realization theory **6**: ***3977–83***
  water distribution network simulation **7**: 5107
Nutrients
  adsorption
    soils and sediments **1**: 6
  cycling
    ecosystems **5**: ***3440***
  environment
    aluminum **1**: ***262–8***
    solution chemistry **1**: 3
    surface chemistry **1**: 5
Nyquist array methods
  multivariable control systems **5**: 3197
  see also Inverse Nyquist array methods
Nyquist criterion **4**: 2264; **5**: ***3440–1***
  generalization
    multivariable control systems **1**: 577
Nyquist diagrams **5**: 3441; **6**: 3700
  describing functions **2**: 979
  design applications **4**: ***2584–8, 2588–96***
  dominance sharing
    optimization **4**: ***2597–600***
  multivariable systems **5**: 3224
    rolling mill control **4**: ***2601–4***
Nystagmus
  slow-phase caloric nystagmus modelling **2**: ***1586–90***

O

Objectives **5**: 3118, ***3443***
Objectives trees **5**: ***3443***; **7**: 4949
Observability
  automata theory **5**: 3270
  control systems **4**: 2857
  discrete-time systems **2**: 1093
  linear multivariable systems **4**: ***2816–26***
  linear systems **4**: 2863
  nonlinear discrete-time systems **5**: ***3288–92***
  nonlinear estimation
    stochastic control **5**: 3311
  nonlinear systems **5**: ***3336–41***
    geometry **5**: 3365
    global results **5**: 3339
  numerical analysis
    linear control theory **2**: 821
  observability map **1**: 16
  polynomial systems
    matrix algebra **6**: 3747
  power system control **6**: 3826
  state-space methods **7**: 4553
Observers *see* State estimation
Obstacle problem **7**: 5037
Occupancy
  lighting control **4**: 2766, 2774
Oceans
  carbon cycle modelling **1**: 529
  distribution of elements **2**: 1513
  tide modelling **7**: 4886
Octanol–water partition coefficient
  toxic substances
    limnology **4**: 2779
ODE methods **6**: 3994
  adaptive algorithms **1**: 76, 87
  convergence analysis
    stochastic adaptive control algorithms **7**: ***4602–6***
  recursive algorithms **6**: ***3991–4***
  stochastic adaptive control **1**: 53
Odontology
  forensic **2**: 968
Odor
  environment **2**: ***1527–8***
Office automation **5**: ***3443–8***
  history **7**: 4774
Oil **2**: 1461
  energy resources **3**: 1714
Oil technology
  fingerprinting **7**: 5110
    active tagging method **7**: 5110
    remote sensing method **7**: 5111
On-line operations
  dynamic systems
    coordination **2**: ***1297–305***
  power system control **6**: ***3842–6***
  steady-state systems
    coordination **7**: ***4566–71***
  viscosity measurement **7**: 5059
On–off control
  modelling
    many-valued logics **5**: 2926
  self-tuning control **6**: ***4209–11***
  see also Bang–bang control
One-dimensional systems
  time-invariant systems
    stability **4**: 2872
  time-varying systems **4**: 2873
Ontology
  pragmatic fuzzy systems **6**: 3855
Open-loop control
  differential games **2**: ***1020–3***
  discrete time systems **2**: 1098
  dynamic systems **2**: 1277
  fuzzy set theory **7**: 4658
  optimal control **5**: ***3496–7***
  perturbed systems **2**: 1022
Open-loop systems
  glucose regulation **3**: 2026
Open questions **5**: ***3448***
Operating points file **2**: 876
Operating systems
  robots
    syntax graph design method **6**: ***4089–93***
Operating systems (computers) **5**: ***3448***
Operational amplifiers
  amplifiers **1**: 270
  analog signal processing **6**: 4283
Operational control decisions **2**: 943
Operational performance decisions **2**: 943
Operations research **5**: ***3449***; **7**: ***4801–6***
  distributed processing **2**: 1204
  fuzzy set theory **5**: ***3449–54***
    algorithms **5**: 3452
  history **7**: 4772
  incentives
    decision making **7**: 4629
  limitations **7**: 4786
  modelling **5**: 3450
  optimization **5**: 3450
  planning horizons **6**: ***3703–6***
  power systems **6**: 3785
  simulation **4**: ***2351–2***; **6**: 4311
  stock control **7**: 4773
  time and motion studies
    industrial robots **4**: 2383
Operative temperature **1**: 643
Optical character recognition **1**: 571
Optical fibers
  combat aircraft **1**: 622
  lighting control **4**: 2773
  optical fiber sensors **5**: ***3455–64***

interferometric devices **5**: ***3465–70***
Optical heterodyne spectrometry
atmospheric analysis **4**: 2722
Optical instruments
road traffic detectors **7**: 4918
Optical modulation
optical fiber sensors **5**: 3455
Optical parameters
earth–atmosphere systems **2**: ***1307–11***
Optical radar
atmospheric analysis **4**: 2722
remote sensing **6**: ***4033–6***
differential absorption **6**: 4036
equation **6**: 4034
fluorescence **6**: 4036
Mie laser radar **6**: 4034
Raman scattering **6**: 4036
resonance scattering **6**: 4036
Optical rotation
biology **1**: 12
Optima **5**: ***3471***
Optimal control **5**: ***3513***, 3516
casting **3**: 1729, 1732
closed-loop control **5**: ***3496–7***
complex systems
tracking methods **1**: 675
conjugate points **6**: ***4066–70***
decentralized approximation **3**: 2157
decomposition–coordination
asynchronous iteration **5**: ***3475–80***
differential equations
dynamic programming **2**: 1278
discrete optimal control problem
multistage decomposition algorithms **5**: 3166
discrete-time systems **2**: ***1093***; **5**: ***3471***, 3474
duality (mathematics) **2**: 1260
eigenvalues and eigenfunctions **2**: 1367
existence
control in coefficients **2**: 809, 810
feedback
linear time-optimal problem **5**: ***3514–5***
feedforward
evaporators **2**: 1567
finite-time linear quadratic control
Riccati equations **5**: 3502
Fuller's problem **5**: 3487
fuzzy systems **5**: ***3488–9***
fixed horizons **5**: 3488
hierarchical control
partition **3**: 2153
history **5**: ***3510–2***
impulse control **4**: 2339–42
populations **2**: ***1335–7***
industrial robots **4**: 2375, ***2392–411***
infinite-time linear quadratic control
Riccati equations **5**: 3502
irregular feedback
generalized trajectories **4**: ***2608***
K convexity **5**: ***3509–10***
linear quadratic problems **5**: ***3513–4***; **6**: 3701, ***4066–70***
singular perturbations **7**: 4438
Lyapunov equation **4**: 2904
maximum principle **5**: ***2956–63***
multistate systems **5**: ***3489–96***
nondifferentiable optimization **5**: 3542
nonexistence
control in coefficients **2**: 809
nonlinear systems
singular perturbations **7**: 4436
nuclear power stations **5**: 3428
numerical solution
finite-dimensional approximation **5**: 3483
gradient methods **5**: ***3481–7***
open-loop control **5**: ***3496–7***
optimal fuzzy target control **5**: ***3515–6***
optimal stopping **5**: ***3531–5***
impulse control **4**: 2339
pertubation techniques **5**: ***3497–502***
pH control **6**: 3690
physiological models **6**: 4062
piecewise analytic feedback **6**: ***3998–400***
planning horizons **6**: ***3703–6***
power systems **6**: 3787
predictive control **7**: 3866
production smoothing **6**: 3890–2
real analyticity **6**: 3955
regular synthesis **6**: ***3998–400***
respiration **6**: ***4062–6***
Riccati equations **5**: ***3502–5***; **6**: ***4066–70***
singular control **5**: 3508
s, *S* policy **5**: ***3509–10***
second-variation theory **6**: 4069
self-tuning control **6**: ***4180–3***
separation process control **6**: 4250
singular control **5**: ***3506–8***; **7**: ***4419–25***
singular perturbations **7**: 4436
slow–fast decomposition **7**: 4531
state-space methods **7**: 4552
stochastic control
discrete-time systems **5**: ***3471–5***
fuzzy set theory **7**: 4657
numerical methods **5**: ***3523–30***
stochastic systems **1**: ***38–45***; **5**: ***3517–23***
stock control **4**: 2583
suboptimal solutions **4**: 2559; **7**: ***4699***
stability **4**: 2557
switching
regular synthesis **5**: ***3487***
systems analysis **7**: 4770
systems methodology **7**: 4806
time-delay systems
Smith predictors **7**: 4461
time horizons **7**: 4892
transversality **7**: ***4948–9***
see also Optimal feedback, Optimization

Optimal estimators **4**: 2693
Optimal feedback **5**: 3497
  linear time-optimal problem **5**: ***3514–5***
  optimal stochastic control **5**: 3527
  piecewise analytic feedback **6**: ***3998–400***
  see also Optimal control
Optimal pair **7**: 5037
Optimal relay **5**: ***3516–7***
Optimal sampling
  biomedical phenomena
    modelling **1**: 487
Optimality **1**: 700; **5**: ***3535–8***
  control in coefficients **2**: 811
  differential games **2**: 1004
  learning systems **4**: 2732
  Lie brackets **4**: 2756
  stochastic systems **2**: 1255
Optimality principle
  Hamilton–Jacobi–Bellman equation **3**: 2063
Optimization **1**: ***634–8***
  air traffic control
    landing sequencing **1**: 192
  aircraft
    fuel **3**: 1753
  approximate realization theory **1**: ***294–6***
  bundle methods **5**: 3544
  calculus of variations **1**: 519–29
  combinatorial mathematics **2**: ***1080–3***
    approximate algorithms **1**: ***626–8***
    branch-and-bound methods **1**: ***628–30***
    independence **1**: ***624–6***
    linear programming **6**: 3735
    linear systems **6**: 3737
  complex systems **1**: ***675***
  computational complexity **1**: 725
  constrained optimization
    canonic equations **6**: 4068
  constraint theory
    feasible direction methods **2**: ***786–8***
    methods of multipliers **2**: ***789–93***
    penalty methods **2**: ***797–801***
    quadratic programming **2**: ***793–7***
  control location problem **2**: 802; **6**: 4242
  convexity **1**: 404
    duality (mathematics) **2**: 1257
    duality methods **7**: 5046
  cutting-plane methods **5**: 3545
  data access
    distributed database system **2**: 1158
  data manipulation
    distributed database system **2**: 1160
  decentralized observers **2**: 909
  decentralized robust controller design **6**: 4135
  decision making
    deterministic incentives **2**: 993
  decomposition–coordination **2**: 953
  decomposition methods **2**: ***956–8***
  descent methods **5**: 3541
  differential games **2**: ***1009–10***
  discrete systems **2**: ***1080–3***
  dominance design **5**: ***3220–3***
  dominance sharing
    Nyquist diagrams **4**: ***2597–600***
  dynamic multistage processes **2**: 1276
  dynamic optimization
    on-line coordination **2**: 1297
  dynamic programming **1**: 407; **2**: 1277, 1278
    forward dynamic programming **3**: ***1713–4***
  dynamic systems **2**: ***1276–8***
  ϵ-constraint approach **1**: 636
  Euler equations
    calculus of variations **1**: 522
  functional equations
    algorithms **5**: ***3546–7***
  fuzzy environment **5**: ***3540–1***
  fuzzy mathematical programming **3**: 1816
  fuzzy set theory **5**: ***3538–40***
  fuzzy systems **3**: 1893
  glucose regulation in diabetics **3**: 2030
  goal coordination method **3**: ***2032–8***
  gradient methods **5**: 3541
    goal coordination method **3**: 2035
  hierarchical decomposition **7**: 4664
  hierarchical optimization **2**: 859
  impulse control **4**: 2339–42
  integer programming
    maintenance scheduling **6**: 3834
  Kiev methods **5**: 3544
  large-scale systems **4**: 2704
    goal coordination method **3**: 2032
    mixed coordination method **5**: 3016
    model coordination method **5**: ***3035–9***
  linear programming **4**: 2835, 2839; **5**: 3543; **6**: 3735
  linear quadratic differential games **2**: 1017
  linearly constrained problems
    numerical methods **4**: 2875
  load dispatching
    power systems **6**: 3816
  maintenance scheduling
    power systems **6**: 3834
  Markov decision processes **5**: 2930
  microeconomic systems analysis **5**: 2974
  modelling
    environment **2**: 1537
  modified two-step approach **1**: 636
  Morse theory **7**: 5042
  multilevel optimization
    nonlinear dynamic systems **2**: 859
  multiobjective decision making **5**: 3150
  multiobjective optimization **5**: 3114, ***3150–2***, ***3152–6***, ***3156–63***
    information gathering **7**: 4781
    interactive methods **5**: 3152
    surrogate worth trade-off method **7**: ***4703–9***
  natural gas supply systems **3**: 1977
  nonconvex optimization
    duality methods **7**: 5046
  nondifferentiable optimization **5**: ***3541–5***

nonlinear dynamic systems **2**: 859
nonlinear programming **5**: ***3352–62***
nonsmooth optimization
auxiliary problem principle **1**: ***400–5***
numerical methods
descent methods **4**: 2877
equality-constrained problems **4**: 2875
inequality-constrained problems **4**: 2878
operations research **5**: 3450
optimal stochastic control
local feedback **5**: 3527
optimum Stackelberg strategies **5**: 3145
output feedback control
singularly perturbed systems **7**: 4442
overlapping decentralized control **5**: 3570
parameter optimization **7**: 4678
robust controller design **6**: 4132; **7**: 4685
parametric approach **1**: 636
process control **6**: 3885
relaxation theory
calculus of variations **1**: 526
respiration modelling **6**: 4057
robot hand trajectories **4**: 2394
robot joint trajectories **4**: 2404
semi-infinite optimization
multivariable control systems design **5**: ***3189–96***
sensitivity analysis **5**: ***3547–50***
sensor location problem **6**: 4240
shape design **2**: 1175
simulation models
regression analysis **5**: 3029
simulation problems **6**: ***4306–10***
smooth case
auxiliary problem principle **1**: ***400–5***
nonlinear programming **5**: 3355
space vehicle trajectories **7**: ***4481–90***
static optimization **7**: ***4554–5***
mixed coordination method **5**: 3014
stochastic processes
Girsanov transformation **3**: 2007
stochastic programming **7**: ***4646–7***
surrogate worth trade-off method **7**: 4704
systems analysis **7**: 4770
systems methodology **7**: 4803
two-level optimization
multistage decomposition algorithms **5**: 3170
two-step approach **1**: 634–8
uncertainty **5**: ***3551–4***
unconstrained optimization
conjugate-gradient methods **7**: ***4969–71***
descent methods **7**: ***4964–9***
quasi-Newton methods **7**: 4974
Riccati equations **6**: 4067
variational inequalities **7**: 5036
variational techniques **7**: ***5041–6***
voltage
reactive power control **6**: 3952
water supply pumps **7**: 5147–9, ***5150–3***
see also Constant parameter systems, Maximization, Minimization, Optimal feedback, Optimal control
Optimum start control
space heating **5**: ***3554***
Options **5**: ***3555***
Optoelectronic devices
angular velocity measurement **7**: 4512
industrial vision systems **4**: 2473
Order determination
modelling **7**: 4996
Ordering
fuzzy set theory
definition **3**: 1876
entropy (information) **3**: ***1875–8***
lattices **3**: 1877
synthesis problems **3**: 1877
linear ordering
fuzzy set theory **3**: 1876
Ordinals **5**: ***3555***
Ordinary differential equation method *see* ODE methods
Ordinary differential equations *see* Differential equations, ODE methods
Organic compounds
fate modelling **3**: ***1600–1***
water pollution **7**: ***5127–9***
Organization charts **5**: ***3564***
Orsat method
oxygen concentration determination **5**: 3572
Orthodontics **2**: 965
Orthogonal bases
innovations
discrete-time systems **2**: 1100
Oscillations
multidimensional systems **5**: ***3124–7***
Oscillators
biochemistry
dynamic behavior **1**: ***424–8***
modelling **1**: 427
regulation **1**: 428
coupled nonlinear
mode analysis **5**: ***3019–24***
genetic **1**: 424
glycolipids
modelling **1**: 426
metabolic **1**: 424
two-dimensional array
mode analysis **5**: 3020
Ostrowski bands
Nyquist diagrams **4**: 2586
Outages *see* Power systems
Output-error models
dynamic systems modelling **4**: 2315
Output feedback
algorithms
pole assignment **6**: 3713
Output feedback control *see* Feedback

Output impedance
  amplifiers **1**: 269
Output tracking
  linear systems **4**: ***2567–70***
Outputs **5**: ***3564***
Ovens
  titanium dioxide ovens
    adaptive control **1**: 50; **7**: ***4892–4***
Overhead lines *see* Power overhead lines
Overrelaxation **1**: 401
Overvaluing technique
  nonlinear systems stability **5**: 3402
Overvoltage
  power system transients **6**: 3803
Oxidation
  nutrients and pollutants
    environment **1**: 3
  weathering
    minerals **1**: 1
Oximetry
  blood oxygen **1**: 440
Oxygen
  biochemical oxygen demand
    water pollution indices **7**: 5115, 5116
  blood oxygen measurement **1**: 437
  chemical oxygen demand
    water pollution indices **7**: 5115, 5116
  concentration determination in gases **5**: ***3572–5***
  concentration determination in liquids **5**: ***3575–9***
  depletion
    limnology **4**: 2779
  dissolved oxygen
    determination **7**: 5113
    electric sensing devices **5**: 3577
    water pollution indices **7**: ***5114–5***
  electric sensing devices
    applications **5**: 3578
    calibration **5**: 3578
    errors **5**: 3578
  fermentation **3**: 1612
    fermenter control **3**: 1617
    indirect measurement **3**: 1614
  liquids
    partial pressure **5**: 3576
  total oxygen demand
    water pollution indices **7**: ***5116–7***
Ozone
  agricultural models **1**: 590
  air pollution
    automatic analyzers **1**: 182
    chemiluminescence **1**: 183
    photolysis **1**: 185

P

Pacemakers
  electrocardiography **1**: 452
Packaging
  industrial robots **4**: 2449
  pharmaceutical industry
    process control **6**: 3694
Packet switching **6**: 3581–90
  see also Multiplexing
Padé approximants
  model simplification **5**: 3074
Pain
  postoperative pain
    demand analgesia **2**: ***960–3***
Pancreas
  blood glucose control
    artificial pancreas **3**: 2017
Paper industry
  adaptive control **1**: 50; **6**: ***3590–1***
  chemical variables control **6**: 3682
  consistency control **2**: 777
  humidity control **3**: 2207
  process control **6**: ***3909–16***
Parabolic systems **6**: ***3591–5***
  linear parabolic systems **6**: 3592
Parallel processing **6**: ***3596***
  distributed system simulation **2**: 1201
  hardware
    cellular automata **6**: 4343
Parallel transmission
  signal transmission **6**: 4291
Paramagnetic resonance
  biology **1**: 13
  oxygen analysis **5**: 3574
Parameter adaptation algorithm
  minimum phase systems
    adaptive control **5**: 3001
Parameter-adaptive control **6**: ***4161***
  algorithms **6**: ***3604***
    design **6**: 3607
  applications **6**: ***3604–10***
  feedback systems **6**: ***3596–604***
  stochastic systems **6**: 3603
  time-varying adaptations gains **6**: 3601
  see also Self-tuning control
Parameter estimation **1**: 634–8; **6**: ***3611–6***
  algorithms **5**: 3343
  Bayesian approach **6**: 3611
  Bayesian linear least-squares estimation **6**: 3615
  best invariant estimators **6**: 3613
  bilinear models
    distillation columns **2**: 1119
  chemical reactor modelling **1**: 587
  deterministic least-squares estimation **6**: 3616
  distributed parameter systems **2**: 1193
  ecological modelling **6**: ***3619–21***
  electrolytic cells **2**: 1415
  ϵ-constraint approach **1**: 638
  filters **1**: 68
  fuzzy identification **3**: 1805
  fuzzy set theory **6**: ***3616–8***
  history **4**: 2271
  identifiability **4**: 2236
  identification errors
    biomedicine **4**: 2253
  instrumental variable techniques **4**: ***2273–7***
  large-scale systems **4**: ***2702–5***
  large stochastic systems **4**: 2706
  linear systems
    dynamic systems **1**: 487
  loss function **6**: 3611
  maximum likelihood estimators **6**: 3613
  minimax rule **6**: 3613
  modelling **1**: 486
  modified two-step approach **1**: 636
  non-Bayesian approach **6**: 3612
  nuclear power station control **5**: 3428
  numerical methods **4**: 2238
  optical parameters
    atmosphere–ground systems **2**: 1308
    atmosphere–water systems **2**: 1310
  orthogonality principle **6**: 3615
  parametric approach **1**: 636
  phosphorus kinetics models
    lakes **5**: 3043
  physiological models
    ketone body metabolism **4**: 2629
    kinetic data **4**: 2634
  properties **6**: 3613
  pseudo-random signals **4**: 2307
  rainfall-runoff modelling **6**: 3944
  recursive estimation
    learning systems **4**: 2733
  respiration modelling **6**: 4057
  risk function **6**: 3611
  self-tuning control **5**: 3230; **6**: 4202
    data prediction **6**: 4169
    state-space methods **7**: 4548
  sequential methods **4**: ***2310–4***
  signal phase estimation
    fuzzy set theory **1**: 659
  simulation **7**: ***4690–2***
  stochastic control **2**: 1253
  supervision
    modelling **1**: 64
  systems analysis **7**: 4769
  two-step approach **1**: 635
  see also Identification, Estimation theory
Parameter optimization
  control systems **7**: ***4678–84***
  large-scale systems **7**: 4678

robust controller design **7**: 4685
Parameter sensitivity
digital control **2**: 1070
Parameters **6**: ***3621***
Parametric models
estimation theory **5**: 3304
Parametric programming
linear programming **4**: 2841
Pareto optimality **1**: 701; **2**: 1273; **5**: 3152, ***3156–63***; **6**: ***3621–2***
decision theory **3**: 2041
fuzzy games **3**: 1914
game theory **1**: 684
multicriteria decision making **5**: 3120, 3121
noncooperative games **3**: 1918
proper Pareto optimality **5**: 3163
solution **2**: 1273
surrogate worth trade-off method **7**: 4704
Partial differential equations
classification **6**: 4360
elliptic equations
numerical solution **6**: 3623
finite difference method **3**: 1623
Fourier transforms **3**: 1724, 1727
hyperbolic equations
numerical solution **6**: 3622
hyperbolic systems **4**: 2800
linear equations
modelling **6**: 4360
linear stochastic equations **7**: 4618
modelling **6**: 4361
multigrid methods **5**: 3135
nonlinear equations
modelling **6**: 4361
nonlinear stochastic equations **7**: 4619
numerical solution **6**: ***3622–6***
parabolic equations
numerical solution **6**: 3622
semigroup theory **6**: 4217
simulation
modelling **6**: ***4360–4***
software **2**: ***1180–1***
spectral analysis **7**: 4495
stochastic control **5**: 3313
Partial minimal realization **6**: ***3627–9***
Partial ordering **6**: ***3629***
reconstructability analysis **6**: 3986
Partially known systems
adaptive control **5**: 3060
Particle size
aerosol particles **1**: ***177–9***
allometric laws **7**: 4453
atmosphere **1**: 179
distribution
aerosols **1**: 177
dynamic model **1**: 179
Partition **6**: ***3629***
function space **3**: 2156
hierarchical control **3**: ***2153–62***
set theory **6**: ***3629***
valued graphs **7**: 4665
PASCAL
simulation languages
transaction flow-based models **7**: 4934
Passive filters **6**: 4282
Passive guideway systems **1**: 335
Passive networks
linear network synthesis **4**: ***2832–5***
Passive Stackelberg strategy **1**: 700
Passive systems
dissipativeness **7**: 4519
Passivity
nonlinear systems stability **5**: 3317
Past determinacy
definitions **6**: 3629
finite memory **6**: ***3629–31***
properties **6**: 3631
Paternosters
vertical transportation
history **7**: 5049
Path-following methods **6**: ***3631–4***
Patient diagnosis
characteristics **2**: 1573
electrocardiography
fuzzy set theory **2**: ***1405–7***
kinetic data **4**: ***2631–5***
Patient monitoring
microcomputers **5**: 2987
Pattern discovery
entropy minimax theory **2**: 1475
Pattern recognition **1**: 307; **6**: ***3634***
Bayes classification **6**: 3640
character recognition **1**: 572
classification algorithms
kinetic data **4**: 2634
cluster center techniques **6**: 3651
decision making
industrial vision systems **4**: 2479
decison-theoretic approaches **6**: ***3634–43***
discriminant analysis **2**: ***1102–5***
discriminant functions **6**: 3638
distance measures **6**: 3636
electrocardiography **1**: 761
feature extraction **6**: 3635
feature selection **6**: 3635
fuzzy learning strategies **5**: 3262
grammars **6**: 3645
hierarchical scene analysis **6**: ***4155–9***
human information processing **3**: 2188
industrial robots
laser tracking techniques **6**: 3763
industrial vision systems **1**: 758; **4**: 2381, 2469–72, 2471, 2474, ***2484–7***
learning by showing **1**: 758
linear classifiers **6**: 3639
many-valued logics **5**: 2924
medical diagnostic computing
fuzzy set theory **1**: 761

multispectral images **6**: ***4029–33***
multivariate analysis **5**: 3234
nearest neighbor classification **6**: 3638
nonmetric cluster analysis **1**: ***606–8***
fuzzy clustering algorithms **1**: 606
validity functionals **1**: 607
patient diagnosis
kinetic data **4**: 2632
primitive selection **6**: 3645
remote sensing **6**: 4040
sentence-to-sentence algorithms **6**: 3650
sequential decision model **6**: 3642
syntactic methods **6**: ***3645–52***
cluster analysis **6**: 3650
nearest neighbor rule **6**: 3651
stochastic languages **6**: 3646
template-matching **6**: 3635
thinning transformations **7**: ***4881–3***
vision
spatial frequency channels **7**: 5074
see also Classification, Cluster analysis, Feature extraction
Penalized equations **7**: 5038
Penalty methods **2**: ***797–801***
methods of multipliers **2**: 790
optimal control
multistate systems **5**: 3494
Pendulums
inverted pendulum
dynamic control **6**: 4124
Peptide
concentration **7**: 5122
Perceptrons
neural nets **6**: 4166
Performance-adaptive systems **6**: ***4162***
Performance–cost ratio
computer systems **1**: 746
Performance index **6**: ***3655***
air conditioning control systems **1**: 157
classification algorithms
kinetic data **4**: 2635
distributed database systems **2**: 1166
modelling **6**: ***3655–9***
elevator systems **2**: ***1441–5***
fuzzy set theory **7**: 4764
learning automata **4**: 2715
multivariable control systems **5**: 3208
Pericardium
biomechanics **1**: 482
Periodic distributions
Fourier series **3**: 1720
Periodograms
spectral estimation
stationary time series **7**: 4560
Periodontal disease
measurements **2**: 964
Peristaltic flow
ureter
electrical activity modelling **7**: 4976
Permittivity
soft tissue **1**: 446
temperature effects **1**: 449
Peroxidase–oxidase reaction
random processes
physiological systems **1**: 569
Persistently exciting inputs
adaptive control **1**: 105; **5**: 3061
Personality
systems management **7**: 4790
Personnel
automation **6**: ***3659–63***
flight-crew workload **3**: ***1665–8***
steel industry
hierarchical control **7**: 4590, 4591, 4592
Personnel rapid transit systems **1**: 337, ***334–40***
PERT **6**: ***3892***, 3899
systems acquisition **7**: 4754
Perturbation techniques
ecosystem modelling **6**: ***3664–5***
nonlinear systems **5**: 3384
open-loop control
controllability **2**: 1022
optimal control **5**: ***3497–502***
optimal stochastic control **5**: 3530
regular perturbations
deterministic control **5**: 3498
stochastic control **5**: 3500
singular perturbations **5**: 3502
Perturbations
ecology **2**: 1318
Perturbations (singular) *see* Singular perturbations
Pesticides
agricultural models **1**: 591
ecology
aluminum **1**: 267
soil **1**: 148
toxic substances **1**: 145
Petri nets **2**: 1140; **6**: ***3665–70***
extended Petri nets
simulation modelling **6**: ***4348–50***
flexible manufacturing systems **6**: ***3670–3***
marked graphs **6**: 3668
structured adaptive nets **6**: 3671
structured colored adaptive nets **6**: 3672
Petrochemical industry
process control **6**: ***3673–9***
Petroleum industry
distillation control **2**: 1107
manufactured fuel gas **3**: 1960
reservoirs
simulation **3**: 2005
pH **6**: 3683, 3688
fermentation measurement **3**: 1612
pH control **6**: ***3680–2, 3688–90***
sugar industry **7**: 4700
pH measurement **6**: ***3683–7***
temperature effects **6**: 3686

Pharmaceutical industry
  process control **6**: ***3691–5***
Pharmacodynamics modelling **2**: 1244
Pharmacokinetics
  demand analgesia
    postoperative pain **2**: 961
  modelling **2**: 1244
    linear models **6**: 3696
    nonlinear models **6**: ***3695–8***
Pharmacology
  drug interactions **2**: 1246
  educational computer models **2**: 1362
  in vivo drug effect prediction **2**: ***1244–8***
Phase
  Bode diagrams **4**: 2665
Phase modulation
  signal transmission **6**: 4289
Phase-plane techniques **5**: 3384
  nonlinear systems **5**: 3367
Phase shift keying **2**: 1048; **6**: 4290
  telemetry **7**: 4837
Phase-space methods
  structural stability **7**: 4670
Phasic encoding
  brains
    fuzzy systems **1**: 506
Phenanthrene
  aerosol chemical analysis **1**: 176
Phosphofructokinase
  oscillators **1**: 426
Phosphorus
  algal growth **1**: 258, 259, 260
  kinetics
    identification **5**: 3043
  nutrient cycling
    modelling **2**: 1322
  phosphorus management
    algal growth modelling **1**: 257
  soil **1**: 147
    aluminum phosphates **1**: 266
  toxic substances
    agricultural soils **1**: 145
Photoacoustic spectroscopy
  environmental analysis **6**: ***3699***
    atmosphere **4**: 2721
Photochemistry
  smog formation **1**: ***185–7***
Photoelasticity
  dentistry **2**: 968
Photogrammetry
  dentistry **2**: 967
Photographic emulsions
  image instrumentation **2**: 1517
Photometry
  distance measurement **4**: 2485
  environmental analysis **2**: 1509
  microprocessor chips **5**: 2986
Photomultipliers
  remote sensing
    satellites **6**: 4038
  spectroscopy
    biology **1**: 9
Photosynthesis
  plant stress
    leaf light emission **2**: 1521
Photovoltaic cells
  detectors
    spectroscopy **1**: 9
  solar power **2**: 1456
Physics fundamentals
  causality **1**: ***552–6***
    philosophical analyses **1**: 554
    state-space approach **1**: 555
  vagueness **7**: 4993
Physiological models
  albumin dynamics **6**: 3905
  circadian rhythms **1**: ***594–601***
    stability conditions **1**: 600
  ferrokinetic models **3**: 1619
  ketone body metabolism **4**: ***2628–31***
  kinetic data **4**: 2632
    patient diagnosis **4**: 2631
  muscle dynamics **5**: ***3243–8***
  muscle spindles **5**: ***3238–42***
  optimal control **6**: 4062
  postoperative pain
    demand analgesia **2**: 961
  protein dynamics **6**: 3905
  respiration
    exercise **6**: 4048
  respiratory system **4**: ***2898–901, 2902–4***
  thermoregulation **7**: ***4873–6***
  visual perception
    spatiotemporal filters **4**: 2326
Physiology
  educational computer models **2**: 1362
  metabolic dynamics
    tracers **7**: 4901
  random processes **1**: ***567–9***
  switching behavior **1**: ***429–33***
  thermoregulation **7**: ***4866–8, 4873–6, 4877–80***
Phytoplankton
  lake phytoplankton modelling **4**: 2650
Pi-bonding
  nutrients and pollutants
    adsorption mechanisms **1**: 7
Picture processing **1**: 508
  gray-scale image processing
    industrial vision systems **4**: 2476
  image representation
    industrial vision systems **4**: 2474
  medial axis transformation **7**: 4881
  multidimensional systems theory **5**: 3130
  processing stages **1**: 508–11
  remote sensing **6**: 4040
  thinning transformations **7**: ***4881–3***
    algorithms **7**: 4882
  three-dimensional vision

modelling **1**: 343
PID control *see* Three-term control
Piezoelectric devices
accelerometers **7**: 4509
pressure transducers **6**: 3876
spark ignition
domestic gas appliances **2**: 1213
Pilot plants
pharmaceutical industry
process control **6**: 3691
Piloting
airship piloting laws **1**: 256
pilot workload **3**: ***1665–8***
targeted piloting **1**: 603
see also Autopilots
Pipe flow
gas distribution systems
steady-state analysis **3**: 1948
water supply transients **7**: 5157
Pipes
natural gas transmission systems **3**: 1939, 1982
water distribution networks
simulation **7**: 5106
Pitch angle
autopilot stabilization **1**: 389
total control
civil aircraft **1**: 603
PL/1
simulation languages
transactions-flow-based models **7**: 4934
PL/M **6**: 3895
PL/Z **6**: 3895
Plan-independent penalties **1**: 33
Plane radiant temperature
physiological temperature control **7**: 4868
Planning (environmental)
traffic control
freeways **7**: ***4910–3***
ramp assignment **7**: ***4913–5***
ramp metering **7**: ***4915–7***
Planning horizons **6**: ***3703–6***
fuzzy set theory **7**: 4658
Markov decision processes **5**: 2930
production smoothing **6**: ***3890–2***
Planning law equations
design **1**: 32
Plant saturation
multivariable control systems design **5**: 3192
Plants (biological) **2**: 1516–21
detection of plant stress
leaf light emission **2**: 1521
pollution effects on plants
image sensors **2**: 1516–21
Plasma applications
inductively coupled plasma emission spectrometry **4**: ***2343–7***
plasma mass spectrometry **5**: 2935
Plasma diagnostics
optical fiber sensors **5**: 3462
Plausibility functions
fuzzy set theory **3**: 1822
Plausibility measures **6**: ***3707–10***
axioms **6**: 3707
mean values **6**: 3708
measure-theoretical representation **6**: 3708
reconstructability analysis **6**: 3984
Pneumatic control equipment
signal transmission **6**: 4289
sugar industry **7**: 4703
Pneumodynamics **1**: ***457–9***
airway resistance **1**: 458
parameter identification
steepest-descent methods **1**: 457
tidal volume **1**: 458
Point of equilibrium **1**: 700
Point processes
random variate generation **6**: 3949
Poisson processes
stochastic calculus **7**: 4637
Poisson ratio
connective tissue **1**: 474
Poisson stability
space vehicle controllability **1**: 320
Polarimeters
biology **1**: 12
optical fiber interferometric sensors **5**: 3469
Pole assignment **2**: ***900–2***; **6**: ***3710–7***, ***3727–34***
decentralized systems **2**: 900, 916
solution **2**: 900
implicit pole assignment
non-minimum-phase systems **5**: 3407, 3409
linear multivariable systems **4**: ***2827–32***
polynomial matrices **6**: ***3720–1***
self-tuning control **6**: ***3721–7***, 4185, 4193; **7**: 4547
PID control **6**: ***4211–3***
state estimation **6**: 3729
state feedback **6**: 3727
Luenberger observer **6**: ***3717–20***
Poles and zeros
invariance properties **4**: 2832
large linear systems **4**: 2667
linear systems **4**: 2864
matrix algebra **4**: 2828
model structure determination **4**: 2290
multivariable control system synthesis **4**: 2809
self-tuning control **6**: ***3721–7***
stationary time series **7**: 4559
system algebraic functions **4**: 2831
transfer functions **4**: 2830
Policy capture
decision making **7**: 4807
systems methodology **7**: 4799–4810
Pollution
adsorption
soils and sediments **1**: 6
environment **2**: 1516–21
databases **2**: 1503
equilibrium distribution **1**: 143

environmental solution chemistry **1**: 3
launchers **4**: 2730
surface chemistry
environment **1**: 5
see also Air pollution, Water pollution
Polyhedra **6**: ***3734***
convex polyhedra **6**: 3734
regular polyhedra **6**: 3734
three-dimensional vision
modelling **1**: 343
Polymerization
process control **6**: ***3739–44***
batch reactors **6**: 3740
reactors **6**: 3739
Polymers
extrusion **2**: 1583
melts
viscosity measurement **7**: 5058
Polynomial matrices
factorization
stochastic optimal control **1**: 38
pole assignment **6**: ***3720–1***
realization theory **6**: 3971
see also Matrix algebra
Polynomial systems **6**: ***3745–51***
general systems **5**: 3290
internally bilinear systems **5**: 3289
multilinear systems **5**: 3289
Polynomials
irreducible polynomials
multidimensional systems **5**: 3128
Polynuclear aromatic hydrocarbons
aerosol chemical analysis **1**: 176
Polysaccharides
adsorption
minerals **1**: 3
Pontryagin theory **2**: 1007, 1021
minimum principles
optimal control **5**: 3521
see also Maximum principle
Popov criterion **5**: 3318
hydropower system stability **3**: 2226
Lyapunov methods **5**: 3396
Population modelling **2**: 1324; **6**: 3756, ***3756–62***
bilinear models **1**: 415
ecology **6**: 3756
epidemic modelling **7**: 4645
Kolmogoroff's forward equation **6**: ***3758***
Leslie's equations **2**: ***1333, 1333***
Lotka–Volterra models **2**: ***1334, 1334***
Malthus' model **6**: 3761
timescales **6**: 3757
Volterra's equations **2**: ***1334, 1334***
von Foerster–M'Kendrick equation **6**: ***3758***
Populations **6**: ***3756***
continuous equations **6**: 3752
control
impulsive control **6**: ***3751–5***
discrete equations **6**: 3752
dynamics **6**: ***3751–5, 3756–62***
distributed parameter systems **2**: 1183
ecology **6**: ***3756***
information theory **2**: ***1321***
logistic growth **6**: ***3755–6***
stochastic processes **7**: 4645
growth
age-dependent **2**: 1328
indices **6**: 3753
logistic growth **6**: ***3755–6***
Lyapunov stability **6**: 3752
optimal impulsive control **2**: ***1335–7***
stochastic equations **6**: 3752
Ports
management **3**: ***2072–6***
radar traffic control **7**: 5051
ship monitoring **7**: 5051
Position control
industrial robots **1**: 300; **4**: 2423
lasers **4**: ***2453–5***
Position measurement
laser tracking techniques **6**: ***3762–5***
Positive end expiratory pressure
artificial respiratory systems **1**: 315
Positive-real transfer functions
adaptive control **1**: 51–4
Positive systems
continous linear time-invariant systems **1**: 95
discrete linear time-invariant systems **1**: 96
discrete linear time-varying systems **1**: 96
dynamic systems **1**: 94
Positivity
adaptive control **1**: ***93–7***
Possibility
computer security **1**: ***743–6***
fuzzy set theory **3**: 1858, ***1862–7***
inference **6**: 3771
information theory **6**: 3766
possibilistic uncertainty
reconstructability analysis **6**: 3988
possibility distribution functions **6**: ***3769***
possibility distributions **6**: ***3769***
possibility measures **6**: ***3769–72***
epistemological hierarchy of systems **2**: 1547
fuzzy random variables **3**: 1844
fuzzy set theory **3**: 1822
fuzzy transformations **3**: 1903
information theory **6**: ***3766–8***
reconstructability analysis **6**: 3984
probabilistic interpretation **6**: 3770
systems management **7**: 4787
Post optimality analysis
linear programming **4**: 2841
Potassium
ecology
aluminum **1**: 266
Potassium iodide
measurement of oxidants
air pollution **1**: 183

Potentiometers **2**: ***1389–90***
digital voltmeters **2**: 1395
pH measurement **6**: 3687
Potter's algorithm *see* Square root algorithms
Power cepstrum
signal analysis **6**: 4279
Power density spectrum **4**: 2251
signal analysis **6**: 4278
stationary time series **7**: 4556
Power overhead lines
dc transmission networks **6**: 3821
power systems **6**: 3808
Power series
bilinear systems **5**: 3268
rational generating series **5**: ***3272–6***
nonlinear systems **5**: 3375
Power station computer control **7**: 4573
load frequency regulation **4**: 2890
Power station load **6**: 3781, 3808
Power stations
economics **3**: 1716
nonlinear modelling **6**: ***3772–5***
steam thermal units **7**: ***4580–5***
see also Nuclear power stations
Power system analysis **6**: 3796–3801
Power system control **6**: 3782, 3837
automatic generation control **6**: ***3809–15***, 3845
computer control
integrated software **6**: ***3825–9***
data acquisition **6**: 3844
dc transmission networks **6**: 3823
frequency control **6**: 3809
load dispatching **6**: ***3815–20***, 3845
load flow **6**: 3854
network topology **6**: 3826
on-line control **6**: ***3842–6***
security control **6**: 3842
state estimation **6**: 3826
Power systems **6**: ***3806–9***, ***3820–5***
components **6**: 3808
identification **6**: 3777
disturbances **6**: 3796
fault levels **6**: 3798
hydroelectric power stations
self-selecting control **3**: 2225–7
load dispatching **4**: 2892
load flow
steady-state modelling **6**: ***3850–5***
load forecasting **6**: ***3830–4***
load frequency control
adaptive control **4**: ***2889–92***
long-term planning **6**: 3786
maintenance scheduling **6**: ***3834–7***
microcomputers **6**: ***3837–42***
network structure **6**: 3807
operation **6**: ***3780–4***
outages **6**: 3834
computer protection **6**: 3828
per-unit representation **6**: 3851
planning **6**: ***3784–9***
power overhead lines **6**: 3808
protection
microcomputers **6**: 3838
pseudo-random signals **6**: ***3777–80***
reactive power loss minimization **6**: 3950
reinforcement strategy **6**: 3785
reliability **6**: ***3790–5***
security of supply **6**: 3781
simulation **6**: 3828
stability **6**: ***3795–801***, 3799
state estimation **6**: 3845, ***3846–50***
transients **6**: ***3802–6***
transmission network calculations **6**: 3852
unit commitment **2**: ***1383–8***; **6**: 3827
see also Electric power generation
Power transmission
reliability **6**: 3793
transmission lines **6**: 3808
unit commitment **2**: 1386
Pragmatic fuzzy systems **6**: ***3855–7***
Pragmatic information **6**: ***3858–60***
Pragmatics **6**: 4223
Praxiology **6**: ***3860–5***
modelling **6**: 3863
Precipitation (atmospheric) *see* Atmospheric precipitation
Precision
control theory **4**: 2357
Predation
ecology **6**: 3756
Leslie's equations **2**: 1333
Lotka–Volterra models **2**: 1334
modelling **6**: 3756
Volterra's equations **2**: 1334
ecosystem stability **2**: 1349
Prediction
computers
history **4**: 2271
fuzzy systems **6**: ***3866***
operations research
simulation **4**: 2352
systems science **6**: ***3865***
see also Filtering and prediction theory
Predictive control
digital three-term control **6**: 3702
fast-model **6**: ***3866–70***
applications **6**: 3869
switching **6**: 3867
on–off control **6**: 4209
optimal control **6**: 3866
road traffic
predictor models **7**: 4927
self-tuning control **6**: 4168, 4173
design **6**: ***4176–80***
Smith predictors **7**: ***4458–62***
Predictor–corrector methods
ordinary differential equations **5**: 3556
path-following methods **6**: 3633

water distribution network simulation **7**: 5108
Predictor models **6**: ***3870–1***
anticipatory systems **1**: 292
dynamic systems **4**: 2286
identification **4**: 2257, 2315
minimum variance control **5**: 3004
self-tuning control **5**: 3229
Preference
group decision making **3**: 2057
risk aversion **6**: 4071
Preferential independence **5**: 3117; **6**: ***3871***
Prescriptive methods **6**: ***3872***
Preservation
model structure simplification **5**: 3081
Pressing
industrial robots **4**: 2418
workpiece handling **4**: ***2418–20***
Pressure
absolute pressure **6**: 3876
gauge pressure **6**: 3876
natural gas technology **3**: 1940
Pressure control
boilers **7**: 4575, 4578
control loop
properties **6**: 3874
domestic gas appliances **2**: 1212
furnaces **6**: 3875
gas equipment **2**: 838
industrial gas equipment **4**: 2355
physical aspects **6**: 3872
process control **6**: ***3872–5***
properties **6**: 3873
retroaction **6**: 3873
self-regulation **6**: 3873
sugar industry **7**: 4701
Pressure measurement **6**: ***3876–7***
differential pressure measurement **6**: 3877
natural gas technology **3**: 1962
Pressure transducers **6**: 3876
capacitive devices **6**: 3876
differential transformer **6**: 3876
inductive devices **6**: 3876
optical fiber sensors **5**: 3458
interferometric devices **5**: 3468
piezoelectric devices **6**: 3876
potentiometric devices **6**: 3876
Price
decomposition–coordination methods
convex programming **1**: 400
water supply **7**: 5148
Primary flight display
aircraft instrumentation **1**: 242
cockpits **1**: 602
Primary productivity
ecosystems **6**: ***3877***
Primary surveillance radar
air traffic control **1**: 197
Primitive polynomials **5**: 2952
Primitive systems **2**: 1545
see also Dataless systems, Source systems
Principal component analysis **2**: 876
Principle of incompatibility
fuzzy set theory **4**: ***2343***
Printing problems
Turing machines
decidability **2**: 936
Priority
distributed processing **2**: 1203
Probabilistic analysis
algorithms
approximation theory **1**: 627
Probabilistic sets **6**: ***3878***; **7**: 4692
distribution function **6**: ***3878***
Probability
betting behavior **3**: 1804
computational complexity **1**: 720
conditional probability **3**: 1841
density function
random signals **1**: 418
dynamic systems modelling **4**: 2286
frequency **3**: 1804
fuzzy causes **3**: ***1788–91***
fuzzy events **3**: ***1802–5***, 1819, ***1839–42***
fuzzy set theory **3**: 1858, ***1862–7***
Gauss–Markov processes
realization theory **6**: 3962
hyperfinite model theory **5**: 2921
hypothesis testing **5**: 2913
impact analysis **4**: ***2331–8***
likelihood ratio **4**: 2334, ***2777***
linguistic probabilities **4**: ***2881–3***
probability measures **2**: 1547
stochastic system analysis **6**: 4027
subjective entropy **7**: 4692
subjective probability **7**: ***4696***
fuzzy set theory **4**: 2300
system failure and recovery **6**: 4011
Probability measures
epistemological hierarchy of systems **2**: 1547
reconstructability analysis **6**: 3984
Probability models
identification **4**: 2260
Probes
electron probes
x-ray analysis **2**: ***1416–20***
enzyme probes
fermentation **3**: 1611
Problem solving
artificial intelligence **1**: 306
environmental problems **2**: ***1536–7***
goal-driven approaches **3**: ***2038***
means–ends analysis **5**: ***2963***
satisficing **6**: ***4149***
top-down methods **7**: ***4894***
Process control
adaptive control **1**: 47

automation **4**: 2368
boilers **7**: ***4572–7, 4577–80***
chemical reactors **1**: ***585–8***
  Nyquist diagrams **4**: 2592
chemical variables **6**: 3681
consistency **2**: 777–81
crisp controller **2**: ***867***
density **2**: 777–81
dewatering **2**: ***1249–53***
distillation columns **2**: ***1107–9***
drying **2**: ***1249–53***
electrolysis **2**: ***1413–6***
elements **6**: 3881
expert fuzzy controller **2**: ***1568–9***
fermentation **3**: ***1607–11, 1615–8***
  measurement **3**: 1611
flow **3**: ***1695–700***
food processing industry **3**: ***1700–5***
fuzzy control language **3**: ***1793***
fuzzy logic controller **3**: ***1811–2***
  structure **3**: ***1814***
fuzzy set theory **4**: 2357
glass industry **3**: ***2011–4***
heat exchangers **3**: ***2089–92***
history **6**: ***3879–84***
inorganic chemicals production **4**: ***2511–6***, 2512, 2513
level measurement **4**: ***2747–54***
linguistic controller **4**: ***2880***
manual control **5**: ***2918***
manufacturing computer control
  Petri nets **6**: 3671
mineral processing **5**: 2989
multivariable self-tuning control **5**: 3184
nonferrous metallurgical industries **5**: ***3280–2***
nuclear magnetic resonance **5**: 3418
optimization **6**: 3885
petrochemical industry **6**: ***3673–9***
  advanced systems **6**: 3675
pH control **6**: 3680
pharmaceutical industry **6**: ***3691–5***
polymer extrusion **2**: ***1583–6***
polymerization **6**: ***3739–44***
pressure control **6**: ***3872–5***
process control systems **4**: 2513
  function **4**: 2514
  state identification **4**: 2516
process state **4**: 2512
process state estimation **4**: 2515
pulp industry **6**: ***3909–16***
rolling mills **6**: ***4139–42***
self-tuning control **5**: 3229
separation **6**: ***4248–52***
sugar industry **7**: ***4699–703***
three-term control **5**: 3229; **6**: 3699
verbal reporting **7**: ***5047***
viscosity control **2**: 777
viscosity measurement **7**: 5059
see also Pressure control, Production control

Process industry
  production control **6**: ***3884–9***
Process interaction
  simulation modelling formalism **6**: 4365
Production
  industrial robots **4**: ***2362–7***
Production control
  history **7**: 4774
  industrial vision systems **4**: 2470
  process industry **6**: ***3884–9***
  production smoothing **6**: ***3890–2***
    modelling **6**: 3891
  structure **6**: 3886
  tasks **6**: 3885
Production functions **6**: ***3890***
  microeconomic systems analysis **5**: 2974
Production rules **6**: ***3890***
Program compilers **6**: ***3892***
Program diagnostics
  simulation model documentation **6**: 4374
Program interpreters **6**: ***3893***
Program and system documentation
  simulation models **6**: ***4372–4***
Programming
  characteristics
    evaluation design **2**: 1561
  flowcharting **3**: ***1695***
  methodology **6**: ***3898***
  modules **7**: ***4472***
  off-line programming
    industrial robots **4**: 2364, 2450
  simulation programs
    design **6**: ***4396–402***
    generators **4**: ***2547–50***
    operations research **4**: 2352
  water supply
    network analysis **7**: 5105
Programming languages **6**: ***3893***
  database management systems **4**: ***2655–8***
    algebraic systems **4**: 2656
  expert systems **2**: 1572
  fuzzy languages **3**: ***1843***
  industrial robots **4**: ***2347–51***, 2374, ***2464–8***
  manufacturing computer control **3**: 1638
    ADA **3**: 1638
    LMAC **3**: ***1637–42***
  real-time languages
    RTL/2 **6**: 3895
  real-time systems **6**: ***3893–7***
  robot operating systems **6**: 4092
  semantics **6**: ***4216***
  simulation
    operations research **4**: 2352
  syntax **7**: ***4752***
Project engineering **6**: ***3898–903***
  floats **3**: ***1669***
  latest completion **4**: ***2723***
  latest start **4**: ***2723***
  net present value **5**: ***3260***

Projected gradient algorithms **1**: 402
Projection
  set theory **6**: ***3904***
PROM *see* Read-only storage
Propagation
  ecological disturbances **2**: 1312
Propellants
  cryogenic propellants **4**: 2728
  launchers **4**: 2728–30
  liquid propellants
    pressurized feed systems **4**: 2728
    turbopump-fed systems **4**: 2729
  solid propellants **4**: 2728
  storable propellants **4**: 2728
Propellers
  pitch angle
    ships **6**: 4269
Proportional-integral control *see* Two-term control
Proportional-integral-derivative control *see* Three-term control
Propulsion
  drag-free satellites **2**: 1243
  launchers **4**: 2728
  ships
    control **6**: ***4267–71***
    fuel conservation **3**: 1764
  see also Aerospace propulsion, Propellants
Prospect theory
  decision theory and analysis **3**: 2198
Proteins
  albumin dynamics modelling **6**: 3905
  cellular protein synthesis control **1**: ***556–8***
  denaturation
    cusp catastrophe **1**: 432
  metabolism
    dynamics and control **6**: ***3905–9***
  structure **1**: 557
  synthesis modelling **1**: 558
  water
    concentration **7**: 5122
Proton-induced x-ray emission spectra
  aerosols **1**: 174
Pseudo bond graphs
  ecological modelling **2**: 1327
Pseudo-random binary sequences **4**: 2309
  power system component identification **6**: 3777
  transient analysis
    natural gas transmission systems **3**: 1991
Pseudo-random pattern **5**: 2955
Pseudo-random signals
  identification **4**: ***2307–10***
  power system component identification **6**: ***3777–80***
Psychology
  causality **1**: 553
  visual perception
    image filtering **4**: 2325
Psychophysical ontology
  pragmatic fuzzy systems **6**: 3855
Psychophysics
  flight-crew workload **3**: 1667
  vision **7**: ***5070–7***
Public goods **6**: ***3909***
Pulp industry *see* Paper industry
Pulse amplitude modulation
  telemetry **7**: 4836
Pulse code modulation
  telemetry **7**: 4836
Pulse–reset systems
  ecological modelling **7**: 4645
Pulse stability **7**: ***4522–3***
Pulverized fuels
  boilers **7**: 4572
Pumps
  characteristics
    flow control **3**: 1699
  control applications **6**: ***3917–8***
  controllers
    history **7**: 4514
  on–off control
    many-valued logics **5**: 2926
  water supply **7**: 5095
    optimization **7**: ***5147–9, 5150–3***
    simulation **7**: 5106
    surge protection **7**: 5155
Purification
  extracorporeal blood **1**: 310
Pyrene
  aerosol chemical analysis **1**: 176
Pyroelectric devices
  medical thermography **7**: 4869
Pyrometers **7**: 4849
  optical fiber sensors **5**: 3459
  temperature measurement **7**: 4860

Q

Q-analysis **6**: ***3919–22***
QSIM modelling *see* Workshop dynamic modelling
Quadratic problems
nonlinearly constrained optimization **4**: 2875
Quadratic programming
optimization **4**: 2879
sequential quadratic programming
Lagrange multipliers **2**: *793–7*
Quadratic-in-state equations
bilinear equations **1**: 414
Quadratic variation
stochastic calculus **7**: 4632
Quality parameters
switching networks **7**: ***4732–3***
Quantum theory
many-valued logics **5**: 2918
nonlinear filtering **5**: ***3319–29***
Quasilinearization
optical parameters
atmosphere–ground systems **2**: 1308
Quasi-Newton methods
equality-constrained problems **2**: 795
unconstraied optimization **7**: ***4971–4***
Quasiorderings **6**: ***3922***
Quasi-variational inequalities **4**: 2340
Query processing
distributed database systems **2**: 1156
Questionnaires
information theory **4**: 2508
Queuing theory **7**: 4805
distributed database systems **6**: 3657
modelling **6**: ***3922–5***
server occupancy **6**: ***4252***
traffic control **6**: 4076

R

Radar
  Doppler effect
    road traffic detectors **7**: 4918
  flight plan correlation **1**: 205
  frequency response techniques **3**: 1748
  position-fixing systems
    inertial navigation **4**: 2491
  primary radar **1**: 197
    air traffic control **1**: 202
  secondary radar **1**: 197
    air traffic control **1**: 202
  ship monitoring **7**: 5051
  synthetic aperture radar
    remote sensing **6**: 4039
Radar equipment
  ports **7**: 5052
Radiation
  atmosphere–ground systems **4**: 2652
Radiation detection and measurement
  medical diagnostic imaging **6**: 3936
    scintillation counters **6**: 3936
Radiative transfer
  ground albedo mapping **3**: 2046
  see also Heat transfer
Radioactive tracers
  compartment models **2**: 1343
  fingerprinting of oil **7**: 5110
Radioactivity measurement
  flow measurement **3**: 1690
Radioaltimeters **1**: 384
Radioecology
  modelling **6**: ***3929–33***
    agricultural food chains **6**: ***3933–6***
Radiofrequency spectroscopy **7**: 4499
Radiofrequency techniques
  Coulter counters **2**: 866
  lighting control signalling **4**: 2773
Radiography
  dentistry **2**: 973–6
  imaging **7**: ***5179–86***
  x rays **7**: ***5179–86***
Radioisotopes
  agricultural transport models **6**: ***3933–6***
  ecological modelling **6**: ***3929–33***
  medical diagnostic imaging **6**: ***3936–40***
  terrestrial systems **6**: 3931
Radiometers
  net radiometers **5**: ***3260–1***
  remote sensing **6**: 4039
Radiopharmaceuticals
  ferrokinetic modelling **3**: ***1619–23***
Radiowave propagation
  underwater telemetry **7**: 4843
Rail traffic
  automatic control **1**: ***365–7***
    reliability **1**: 360
    safety **1**: 360
  electronic coupling **1**: 338
Railroads
  automatic control **1**: ***365–7***
  automatic driving control **1**: ***362–5***
  automatic line supervision **1**: ***347***
  see also Metros, Rail traffic
Railways *see* Railroads
Rain
  hydrologic forecasting **3**: 2228
  rainfall-runoff modelling **6**: ***3940–4***
Rainfall-runoff modelling **6**: ***3940–4***
  model classification **6**: 3941
Raman spectroscopy **7**: 4499
  atmospheric laser spectroscopy **4**: 2721
Ramp metering
  freeway traffic control **6**: 4076
Random-access storage
  digital control **2**: 1067
Random binary signals *see* Binary random signals
Random noise **2**: 1403
  optical fiber interferometric sensors **5**: 3466
  physiological models
    circadian rhythms **1**: 597
  realization theory **6**: 3964
Random number generation **6**: ***3945, 3948–9***
  discrete-event simulation **2**: 1086
Random numbers
  Gaussian **5**: 2954
  uniform **5**: 2954
Random processes
  chemical systems **1**: ***567–9***
  combined randomness and uncertainty **3**: 1856
  description **1**: 567
  dynamical systems **1**: ***565–7***
  dynamics
    discrete time **1**: ***565–7***
  fuzzy set theory **3**: 1853
  nonlinear control systems **5**: 3387
  physiology **1**: ***567–9***
  random simulation **5**: 3026
  random walk theory
    seed dispersal modelling **7**: 4645
  stability **7**: 4528
  stationary random processes
    identification **4**: 2271
  see also Markov processes, Stochastic processes
Random sampling
  ecological simulation modelling **6**: 4232
Random sets **4**: 2294
  fuzzy set theory **4**: ***2293–301***; **6**: ***3945–7***
  parameter estimation **6**: 3617
  see also Set theory

Random signals **1**: ***418–21***
  amplitude characteristics **1**: 419
  autocorrelation **1**: 421
  binomial distribution **1**: 421
  characteristic function **1**: 419
  dual distribution **1**: 420
  frequency characteristics **1**: 421
  generation
    M-sequences **5**: 2954
  normal distribution **1**: 420
  Poisson distribution **1**: 421
  triangular distribution **1**: 420
  uniform distribution **1**: 420
  white signals **1**: 421
Random variables **6**: ***3948***
Randomness **1**: 704; **3**: 1867
  algorithmic complexity **1**: 707
Rangeability
  valves **2**: 830
Rankine cycle **7**: 4580
Ranking
  fuzzy set theory **3**: ***1845–7***
Raster scanning
  cathode ray tubes **1**: 241
Rate distortion theory
  data compression **2**: 1490
Rating matrices
  fuzzy multicriteria modelling **3**: 1826
Rational generating series
  bilinear systems **5**: ***3272–6***
Rationality **6**: ***3950***
  bounded rationality **7**: 4783
    decision theory and analysis **3**: 2195
  decision theory and analysis **3**: 2194
  information gathering **7**: 4781
Rayleigh scattering
  atmospheric laser spectroscopy **4**: 2720
Reachability
  control systems **7**: 4676
  ecosystem network structure **2**: 1353
  linear systems **4**: 2862
  reachability map **1**: 16
Reaction kinetics
  ecological modelling **7**: 5090
  toxic substances modelling **3**: 1600
Reactive power
  loss minimization **6**: ***3950–4***
Read-only storage
  digital control **2**: 1067
    PROM **2**: 1067
Real analyticity **6**: ***3954–6***
  Lie brackets **4**: 2755
Real-time systems **2**: 1137
  programming languages **6**: ***3893–7***
  simulation **6**: ***4418***
  telephone networks **7**: ***4733–40, 4749–51***
Realization theory **1**: ***14–17***
  algorithms
    system identification **4**: 2306
  anticipatory systems **1**: 291
  approximate theory
    optimal methods **1**: ***294–6***
  balanced realization
    linear stationary dynamical system **1**: ***409–10***
  bilinear models
    distillation columns **2**: 1119
  bilinear systems **5**: ***3276–9***
  extension techniques **2**: ***1581–3***
  finite Volterra series **6**: ***3956–61***
  Gauss–Markov processes **6**: ***3962–4***
  Hankel matrix **6**: 3627, ***3965–8***
  input–output models **3**: 2165
  linear systems **4**: ***2861–5***
  local realization
    nonlinear systems **5**: ***3390–4***
  matrix fraction representation **6**: ***3968–9***
  nonlinear systems **6**: ***3973–7***
  numerical methods **6**: ***3977–83***
  partial minimal realization **6**: ***3627–9***
  polynomial matrices **6**: 3971
  recursion theorems **6**: 3965
  recursive algorithms **6**: 3979
  state-space methods **6**: ***3969***
  stochastic differential equations **6**: 3964
  time invertability **6**: 3973
  transform domain methods **6**: ***3971***
  two-filter smoothing algorithms **7**: 4467
  see also Control theory
Receptor binding modelling
  pharmacodynamics **2**: 1245
Reconstructability
  numerical analysis
    linear control theory **2**: 821
Reconstructability analysis **6**: ***3983–9***
  relational join **6**: 4000
Reconstruction *see* Reconstructability analysis
Recovery management
  database management systems
    concurrency **2**: 883
  distributed database systems **2**: 1169
Rectifier substations
  dc transmission networks **6**: 3821
Rectifying circuits
  dc motors **5**: 3093
Rectilinear scanners
  medical diagnostic imaging **6**: 3937
Recursion theorems
  computability and decidability **1**: 718
  realization theory **6**: 3965
Recursive algorithms
  adaptive control algorithms
    convergence analysis **7**: 4602
  ordinary differential equation method **6**: ***3991–4***
  realization theory **6**: 3979
  weak convergence **6**: ***3994–7***

Recursive functions **1**: 716
Recursive identification **4**: ***2310–4***
Recursive least squares methods
  discrete-time case **6**: 3597
Redox reactions
  toxic substances
    limnology **4**: 2779
Reduction
  aggregation **1**: 127
  nutrients and pollutants
    environment **1**: 3
  weathering
    minerals **1**: 1
Reductionism **6**: ***3997–8***
Redundancy
  aerospace control
    history **1**: 343
  analytic redundancy
    aircraft **1**: 247
    sensor failure detection **1**: ***246–51***
  automatic landing systems **1**: 209
  computers
    space shuttle **7**: 4479
  distributed database systems
    failure recovery procedures **2**: 1161
  information theory **4**: 2505
  reliability of systems **6**: 4021
  space shuttle control systems **7**: ***4476–81***
Reflex control
  exercise
    simulated response **1**: 541
Reflexivity
  graphs **7**: 4666
  relations **7**: ***4751***
Refraction
  level measurement applications **4**: 2741
Refractive index
  detectors
    liquid chromatography **4**: 2888
  optical fiber sensors **5**: 3460, 3462
Regression **6**: ***3998***
Regression analysis
  linear regression analysis
    identification **4**: 2280
  load forecasting **6**: 3832
  multivariate analysis **5**: 3233
  simulation models **5**: ***3024–9***
  systems analysis **7**: 4805
Regular perturbation techniques
  complex systems **1**: 692
Regular synthesis
  optimal control **6**: ***3998–4000***
    switching **5**: ***3487***
  real analyticity **6**: 3956
Regular systems *see* Bilinear systems
Regulation
  biochemical oscillators **1**: 428
  control systems
    viability **2**: 1028
  coronary hemodynamics **2**: 852
  decentralization
    viability **2**: 1029
  glucose
    diabetics **3**: ***2025–32***
    modelling **3**: ***2021–5***
  hormones **3**: ***2169–74***
  minimum-phase systems
    adaptive control **5**: 2999
  speed
    governors **7**: ***4513–7***
  see also Control systems, Control theory, Controllers
Relational algebra
  fuzzy set theory **3**: 1795
Relational join **6**: ***4000***
Relations (mathematics) **6**: ***4000–7***
  binary relations **7**: 4666
  crisp binary relations **6**: 4000
  interpretive structure modelling **4**: 2575
  *n*ary relations **6**: 4006
  operations **6**: 4001
  relation maps **4**: 2576
Relaxation oscillators
  coupled oscillators
    electrocardiography **1**: 534
    large intestine electrical activity modelling **2**: 1036
  stomach electrical activity modelling **2**: 1034
  synthesized relaxation oscillators
    large intestine electrical activity modelling **2**: 1037
  van der Pol equation
    circadian rhythms **1**: 594
Relaxation (physiology)
  muscle
    dynamics **5**: 3236
    feedback control **5**: ***3235–8***
    measurement **5**: 3235
Relaxation theory **2**: 810
  calculus of variations **1**: 526
  decomposition–coordination
    asynchronous iteration **5**: 3476
  differential equations **2**: 1026
  hierarchical control **3**: 2153
  multigrid methods **5**: 3136
Relay control *see* Bang–bang control
Relays
  lighting **4**: 2772
  nonlinear control systems **5**: 3370, 3386
Reliability
  applications **6**: ***4020–6***
  control systems **6**: ***4008–11***
  database management systems **2**: 883
  describing-function method **2**: ***978–81***
  distributed computer systems **2**: ***1132–6***
  distributed database systems **2**: ***1169–74***
  industrial robots **4**: 2444
  large-scale systems
    computer software **4**: ***2678–81***
  power systems **6**: ***3790–5***

computer control **6**: 3841
rail traffic
automatic control **1**: ***360–2***
space shuttle control system **7**: ***4476–81***
statistical design and control **6**: ***4027–9***
systems reliability **6**: ***4011–5***
evaluation **6**: 4017
theory **6**: ***4015–20***
see also System failure and recovery
Remote sensing **6**: ***4037–40***
atmosphere **4**: ***2652–5***
landsat imaging **4**: 2652
optical radar **6**: ***4033–6***
atmospheric effects **6**: 4040
buildings
management systems **1**: 513
ground albedo mapping
invariant embedding **3**: ***2046–52***
laser spectroscopy **4**: 2722
meteorological measurement **5**: 2973
multispectral scanning **6**: 4038
image classification **6**: ***4029–33***
pattern recognition **6**: 4040
snow-cover area
hydrologic forecasting **3**: 2233
water pollution **6**: ***4041–5***
fingerprinting of oil **7**: 5111
wind velocity
optical lasers **2**: ***1528–32***
Repeatability
instruments **2**: 1558
Repeated games *see* Game theory
Repeaters
regenerative transponders
satellite links **2**: 1052, 1054
satellite links **2**: 1051, 1054
Repolarization
electrocardiography **1**: 452
Representating analogies
knowledge representation **4**: 2639
Representation theory
algebraic many-valued logics **5**: 2923
multiattribute utility theory **5**: 3107
Reproducibility
instruments **2**: 1558
Research and development management
scenarios **6**: 4154
technological innovation **7**: 4831
Reservoirs **7**: 5141
hydroelectric power stations
modelling **3**: 2222
petroleum
simulation **3**: 2005
water supply **7**: 5095
leak detection **7**: 5143
multiattribute utility theory **7**: 5137
simulation **7**: 5107
Reset control
air conditioning **6**: ***4045***
reset ratio **6**: ***4045***
Residual subsystems
chained aggregation **1**: 133
Resistance thermometers **7**: 4850
conduction errors
self-heating **7**: 4862
Resolution
digital signals **6**: 4285
Resonance
direct digital control systems **2**: 1076
nonlinear oscillators
mode analysis **5**: 3020
Respiration
educational computer models **2**: 1362
electromyography **6**: 4052
feedback control
exercise **6**: ***4045–51***
mechanics **6**: 4062
modelling **1**: 568; **4**: ***2898–901, 2902–4***; **6**: ***4055–61***
external respiration **6**: ***4051–5***
optimal control **6**: ***4062–6***
steady-state **6**: 4046
Respiration (artificial) *see* Artificial organs
Resultant temperature **1**: 643
Retina *see* Eye
Retrodiction
systems science **6**: ***4066***
Revealed preference theory **6**: ***4066***
microeconomic systems analysis **5**: 2974
Rheology
blood **1**: 441
viscosity measurement **7**: 5056
Riccati equations
constant parameter systems **2**: 784
hereditary systems **6**: ***4066–70***
linear control theory **2**: 822
linear quadratic differential games **2**: ***1017–20***
optimal control **5**: ***3502–5***
Riccati differential equations
hierarchical control **3**: 2161
Riemannian geometry
space vehicles (rigid)
controllability **1**: 320
Right-invariant systems
group theory
Lie groups **4**: 2757
Rigidity (functions)
real analyticity **6**: 3955
Rings, commutative *see* Commutative rings
Rings, noncommutative *see* Noncommutative rings
Risk
aircraft landing
automatic landing systems **1**: 207
ecosystems uncertainty analysis **2**: 1539
environmental risk analysis **2**: ***1538–9***
extrapolation error **2**: 1538
fault tree analysis **2**: 1538
global risk attitude **6**: 4071
group decision making **3**: 2053

local risk attitude **6**: 4072
monotonic risk attitude **6**: 4072
multivariate risk attitude **6**: 4073
patterns **6**: 4075
quotient method **2**: 1538
relative risk attitude **6**: 4073
toxic chemicals assessment **7**: 4895
water resource management **7**: 5138
Risk aversion **6**: ***4070–5***
multiattribute utility theory **5**: 3100
Rivers
reservoirs **7**: 5142
toxic substances
dispersion **4**: ***2778–82***
fugacity models **7**: 4901
see also Limnology
Riveting
industrial robots
aerospace industry **4**: 2451
Road traffic
acceleration **7**: 4931
communication **6**: ***4081–2***
concentration **7**: 4931
density **7**: 4930
detectors **7**: 4919
detectors **7**: ***4917–20***
diversion control **6**: ***4079–81***
flow **7**: 4917, 4931
freeway corridor control
ramp assignment **7**: ***4913–5***
ramp metering **7**: ***4915–7***
fuel consumption modelling **3**: 1759
fundamental diagram **7**: 4932
guidance systems **6**: ***4075–9***
headway **7**: 4930
intelligent control systems **7**: ***4906–10***
occupancy **7**: 4931
detectors **7**: 4919
predictive models **7**: 4928
route control **6**: 4076
simulation
car-following models **7**: 4925
hydrodynamic theory **7**: ***4920–2***
macroscopic models **7**: ***4923–5***
microscopic models **7**: ***4925–7***
predictive models **7**: ***4927–9***
spacing **7**: 4930
speed **7**: 4930
variables **7**: ***4930–3***
vehicle classification **7**: 4919
see also Traffic, Traffic control
Road vehicles
fuel consumption **3**: ***1756–61***
Roasting
process control **5**: 3281
Robot arms **4**: 2369
adaptive control **6**: 4099
articulated arms
autonomous contouring control **1**: ***296–301***
axes **4**: 2387
contouring control
experimental results **1**: 300
coordination control **6**: ***4097–100***
design **4**: 2389
direction control **1**: 300
dynamic control **6**: ***4082–6***, 4099
flexible arms **6**: ***4086–8***
industrial robots **4**: ***2387–91***
manipulation variables
noninteractive control **1**: 299
mathematical models **6**: 4099
position control **1**: 300
resolved-motion rate control **6**: 4098
see also Industrial robots, Robots
Robot hands *see* End effectors
Robotics **6**: ***4093***
Robots **6**: ***4097***
applications
washing machine manufacture **3**: ***1597–600***
biped robots
dynamics **6**: 4125
hierarchical control **6**: 4127
coordination control **6**: ***4097–100***
end effectors **2**: ***1451***
operating system design
syntax graphs **6**: ***4089–93***
snakelike robots **6**: ***4109–13***
tactile sensors **6**: ***4118–23***
teach function **6**: ***4123***
three-dimensional vision **1**: 755
modelling **1**: 343
walking robots **6**: ***4123–8***
see also Industrial robots, Robot arms
Robustness
closed loop systems **4**: 2569
decentralized control **4**: 2558
decentralized servomechanism design **6**: 4133
decentralized stabilization algorithms **2**: 919
ecosystem modelling **1**: 674
fuzzy optimization **5**: 3539
linear systems **4**: ***2871–2***
decentralized feedback **4**: 2670
multivariable control system design **5**: 3192, ***3224–8***; **6**: ***4129–32***, ***4133–8***; **7**: ***4685–9***
LQG design **4**: ***2842–6***
self-tuning control
input–output analysis **6**: 4208
servomechanism design **6**: 4129
zero-error regulation **7**: 4686
Rolling
autopilots **1**: 390
ship control **6**: 4259, 4271
Rolling mills **6**: 4139
modelling **6**: 4141
multivariable control systems
Nyquist diagrams **4**: ***2601–4***
process control **6**: ***4139–42***
scheduling **6**: 4140

ROM *see* Read-only storage
Root loci
control system analysis computing **4**: 2799
frequency response techniques **3**: 1748
see also Stability criteria, Characteristic-locus method, Multivariable-root-locus method
Rotary cement kilns
data analysis **2**: 876
Rotors
helicopters **3**: 2118
Routh approximants
model simplification **5**: 3073
Runge–Kutta methods
ordinary differential equations **5**: 3557
Runoff
rainfall-runoff modelling **6**: ***3940–4***

S

Saccharin
  drinking control **3**: 2212
Saddle points **1**: 402
Safety
  aircraft **1**: 115, 603
    automatic landing in bad weather **1**: 210
    sensor failure detection **1**: 246
  aircraft separation **1**: 190
  airports
    air traffic control **1**: 189
  automated guideway transit **1**: 334, 336
  domestic gas appliances **2**: ***1212–3***
  helicopters
    automatic flight control systems **3**: 2125
  industrial gas equipment **4**: ***2353–6***
  industrial robots **6**: 4096
  launch bases **4**: 2730
  rail traffic
    automatic control **1**: ***360–2***
  road traffic
    control **6**: 4081
    simulation **7**: 4927
Sampled-data systems
  design
    Nyquist diagrams **4**: 2594
  history **6**: ***4143–6***
  z transforms **7**: 5189
Sampling **6**: ***4146***
  aliasing **6**: 4286
  data compression
    environmental testing **2**: 1491
  digital communication systems **2**: 1045
  digital control **2**: 1070
  digital signals **6**: 4286
  ecological simulation modelling **6**: 4231
  experiment design **4**: 2257
  history **6**: ***4143–6***
  identification errors
    biomedicine **4**: 2254
  importance sampling **7**: 4654
  nonlinear discrete-time systems **5**: 3292
  optimal sampling schedule
    biomedical models **4**: 2254
  power system computer control **6**: 3839
  stratified sampling **7**: 4655
  systematic sampling **7**: 4656
  water pollution **2**: 1534
    chemical analysis **7**: 5109
Satellite imaging *see* Remote sensing
Satellite links
  digital communication systems **2**: ***1051–7***
  transmission losses **2**: 1052
Satellite relay systems
  digital communication systems **2**: 1051
Satellites (artificial) *see* Artificial satellites
Satiety
  hunger and thirst **3**: ***2210–2***
Satisficing **6**: ***4149***
Saturating control **5**: ***3516–7***
Saunders valve **2**: 834
Scalarizing function
  Pareto optimal solution **2**: 1273
Scale formation
  evaporation control **2**: 1568
Scanning electron microscopy **6**: ***4149–52***
Scanning transmission electron microscopy **6**: 4151
Scattering
  ground albedo mapping **3**: 2046
  optical parameters **2**: 1307
Scenarios **6**: ***4152–4***; **7**: 4803
  technological forecasting **7**: 4827
Scene analysis
  descriptors **6**: 4157
  hierarchical **6**: ***4155–9***
Scheduling
  air traffic control
    dynamic scheduling **1**: 190
  distributed processing **2**: 1203
  power systems
    maintenance scheduling **6**: ***3834–7***
  process industry **6**: 3887
Schrödinger equation
  Fourier transforms **3**: 1728
  nonlinear filtering **5**: 3320
  semigroup theory **6**: 4222
Scientific theories
  vagueness **7**: ***4993–5***
Scintillation counters **6**: 3936
Scripts **6**: ***4161***
Searching
  artificial intelligence **1**: 306
Seas
  tide modelling **7**: 4888
Seasons
  modelling
    lakes **4**: 2644
Second guessing phenomenon **2**: 930
Secondary surveillance radar
  air traffic control **1**: 197, 199
Sectors
  air traffic control **1**: 198
Security of data
  computers
    fuzzy set theory **1**: ***741–3***
    possibility theory **1**: ***743–6***
  database management systems **2**: 886
Sedimentation
  modelling
    lakes **4**: 2643

Sediments
  adsorption
    nutrients and pollutants **1**: 6
Segmentation **2**: 878
  algal growth modelling **1**: 258
  gray-scale image sensing **4**: ***2479–84***
  scene analysis
    hierarchical systems **6**: 4156
Selenastrum capricornutum
  algal growth potential **7**: 5112
Selenium
  ecology **1**: 267
  soil **1**: 147
  toxic substances **1**: 145
Self-adjusting systems
  adaptive fuzzy control **1**: ***72–3***
  fuzzy logic controller **3**: ***1813***
  learning fuzzy controller **4**: ***2730–1***
  see also Self-organizing systems
Self-applicability
  metalogics **1**: ***326–32***
  partial self-applicability
    Gödel-enumeration **1**: 326
  self-reference **1**: 330
Self-organizing control **6**: ***4161–4***
  multilevel control systems **5**: 3141
Self-organizing systems **6**: ***4165–7***
  ecology **2**: 1330
  flow networks **3**: ***1694–5***
  fuzzy logic controller **3**: ***1813***
  see also Self-adjusting systems
Self-production *see* Autopoiesis
Self-reproducing automata **6**: 4375
Self-selecting control
  hydroelectric power stations **3**: ***2225–7***
Self-tuning control **1**: 46; **5**: 3060, 3343; **6**: ***4164***, ***4184–6***, ***4190–3***
  applications **6**: ***3604–10***
  blood glucose control **3**: 2017
  certainty equivalence principle **6**: 4190
  data prediction **6**: ***4168–72***
  distributed parameter systems **2**: ***1197–9***
  engine speed control **2**: ***1462–6***
  error models **1**: 56
  feedforward control **6**: ***4187–9***
  glass industry **3**: 2009
  HVAC systems **3**: 2112; **5**: 3347
  hybrid **6**: 4194
  implementation **6**: ***4199–204***
  input–output analysis **6**: ***4205–9***
  linear quadratic Gaussian **6**: ***4172–6***
  load frequency control **4**: 2891
  modelling **6**: ***4197–9***
  multirate control **5**: ***3184–8***
  multivariable control **5**: ***3184–8***
  non-minimum-phase systems **5**: 3408
  nonlinear systems
    algorithms **5**: ***3341–51***
  nuclear power stations **5**: ***3425–30***
  on–off control **6**: ***4209–11***
  optimal control **6**: ***4180–3***
  pH control **6**: 3690
  pole assignment **6**: ***3721–7***, 4193
  predictive control **5**: ***3229–32***
    design **6**: ***4176–80***
  predictor models **6**: 3870
  state-space methods **7**: ***4547–51***
  three-term control
    digital control **6**: 3702
    pole assignment **6**: ***4211–3***
  variable forgetting factors **6**: ***3990–1***
  see also Adaptive control, Parameter-adaptive control
Self-tuning filters
  ship positioning **6**: 4265
Semantics **6**: 4223
  programming languages **6**: ***4216***
  semantic data
    modelling **2**: 891
  semantic information **6**: 3858, ***4214–6***
  semantic nets
    artificial intelligence **1**: 306
    knowledge representation **4**: 2637
Semicircular canals *see* Vestibular system
Semiconductor devices
  pressure transducers **6**: 3876
  switches
    lighting control **4**: 2772
Semicontinuous functions
  topology **6**: ***4216***
Semidifferential games **1**: 686
Semigroup theory **6**: ***4216–22***
  compact semigroups **6**: 4218
  contraction **6**: 4218
  distributed-parameter systems
    functional analysis **5**: 3205
  dual semigroups **6**: 4220
  Feller's semigroups **6**: 4220
  holomorphic semigroups **6**: 4219
  linear Boltzmann equations **7**: 4941
  nonlinear estimation **5**: 3313
  nonlinear semigroups **6**: 4220
Semimartingale methods **5**: ***2932–4***
Semiotics **6**: ***4223***
  pragmatic information **6**: 3858
Sensitivity **6**: ***4223***
  digital control **2**: 1070
  fuzzy logic controller **3**: ***1813***
  muscle relaxation **5**: 3237
  optical fiber interferometric sensors **5**: 3466
Sensitivity analysis **6**: ***4223–7***
  agricultural modelling **1**: ***137–42***
  conflict analysis **7**: 4798
  cost effectiveness **2**: 857
  ecological modelling **2**: ***1549–51***; **6**: ***4230–4***
  food-chain models
    radioisotopes **6**: 3935
  fuzzy multicriteria modelling **3**: 1826

input–output analysis **6**: ***4235–7***
large models **6**: 4234
linear programming **4**: 2841
management models **7**: 5003
multivariable systems design **5**: 3224
natural gas system planning **3**: 1996
optimization **5**: ***3547–50***
radioecological modelling **6**: 3933
simulation experimentation **6**: ***4227–9***
simulation models
regression analysis **5**: 3029
system dynamics modelling **7**: 4757
technological forecasting **7**: 4824
workshop dynamic modelling **7**: 5173
Sensors *see* Nonelectric sensing devices, Electric sensing devices
Separation
density control **2**: 780
electromagnetic devices **6**: 4251
process control **5**: 3280; **6**: ***4248–52***
Separation principle **6**: ***4245–8***
control location problem **6**: 4242
large stochastic systems **4**: 2712
stochastic control **5**: 3474
Sequential circuits
boiler control **7**: 4573
Sequential probability ratio test **1**: 249
Serial transmission
signal transmission **6**: 4292
Serializability theory
distributed database systems
multiprocessing programs **2**: 1143
Series (mathematical)
Fourier series **3**: ***1719–23***
truncated Fourier series **3**: 1722
see also Time series
Server occupancy
queuing theory **6**: ***4252***
Service time **6**: ***4252***
Servocompensators
closed loop systems **4**: 2567
robust controller design **6**: 4130
decentralized controllers **6**: 4135
Servomechanisms
accelerometers **7**: 4510
autopilots **1**: 385
electrohydraulics **6**: ***4252–7***
articulated robot arm control **1**: ***301–4***
linear systems **4**: 2567
multivariable servomechanism problem
decentralized robust controller design **6**: 4133
robust controller design **6**: 4129
self-tuning control
data prediction **6**: 4172
pole assignment **6**: 3722
Set covers **6**: ***4257***
reconstructability analysis **6**: 3985
set-covering problems **6**: ***4257***
Set theory
alternative set theory
vagueness **7**: 4995
antisymmetric relations **1**: ***294***
closed sets **1**: ***605***
compact sets **1**: ***665***
convex sets
nonlinear programming **5**: 3353
disjoint x-bases
generalized dependence **3**: 2004
disjoint x-span
generalized dependence **3**: 2004
equivalence relations **2**: ***1549***
fuzzy mapping **3**: ***1816***
generalized dependence **3**: ***2003–5***
linear ordering **4**: ***2832***
many-valued logics **5**: 2918
observable sets
generalized dependence **3**: 2003
partial ordering **6**: ***3629***
partition **6**: ***3629***
projection **6**: ***3904***
quasiordering **6**: ***3922***
random sets **4**: 2294
reflexivity of relations **7**: ***4751***
relational join **6**: ***4000***
relations (mathematics) **6**: 4003
set-covering problems **6**: ***4257***
set covers **6**: ***4257***
simplicial complexes **6**: ***4299***
simulation **6**: 4312
subjectivity **7**: 4697
total relations **7**: ***4895***
transitivity of relations **7**: ***4935***
see also Fuzzy set theory
Severinghaus sensor systems **1**: 439
Sewing
industrial robots **4**: 2449
Shadow prices **6**: ***4257***
Shannon entropy *see* Entropy (information)
Shape control
rolling mills **6**: 4141
Shape design **2**: 1174
Shape representation
three-dimensional vision
polyhedra **1**: 343
Shapley value *see* Game theory
Shear flow
viscosity measurement **7**: 5056
Ships
control **6**: ***4257–61***
dynamic positioning
adaptive control **6**: ***4262–6***
fuel conservation **3**: ***1761–6***
guidance
adaptive control **6**: ***4275–6***
maneuvering **5**: ***3256–60***; **6**: 4257
monitoring **7**: ***5051–5***
motion estimates

Kalman filters **6**: 4264
motions
dynamic positioning **6**: 4262
propulsion control **6**: ***4267–71***
stability **6**: ***4271–4***
Shivering
mechanism **7**: 4879
thermoregulation **7**: 4871–3
Shock waves
road traffic simulation **7**: ***4920–2***
Shortest-path problem
routing algorithms **2**: 1123
Shuttle loop transit **1**: 336
Side-lobe suppression signal
air traffic control **1**: 199
Sideslip
aircraft
measurement **1**: 381
σ-hyperfields
fuzzy set theory **6**: 3945
Signal analysis
bioelectric potential
digestive tract **1**: 460
estimation
classical method **1**: 661
fuzzy set theory **1**: ***651–64***
fundamentals **6**: ***4277–81***
harmonics **3**: ***2076–80***
mean values **3**: 2078
peak values **3**: 2077
root-mean-square values **3**: 2077
Signal detection
fuzzy set theory **1**: ***651–64***
signal known exactly **1**: 651
signal known except for phase **1**: 651
Signal flow graphs
multivariable control system **4**: 2808
state-space equations **1**: 388
Signal processing
active filters **6**: 4283
adaptive algorithms **1**: ***73–5, 75–80, 81–5, 86–91***
adaptive systems
abrupt changes **1**: ***81–5***
analog signals **6**: ***4281–4***
digital signals **6**: ***4284–8***
His bundle electrocardiography **2**: 1409
instruments **4**: 2528
mechanoception
electric sensing devices **6**: 4119
multidimensional systems theory **5**: 3129
nonlinear analog signal processing **6**: 4284
passive filters **6**: 4282
pneumatic systems **6**: 4282
see also Signal transmission
Signal transmission **6**: ***4289–94***
analog transmission **6**: 4289
digital transmission **6**: 4289
Signalling
control **2**: 930
information theory **7**: 4817
metros **1**: 353, 365
nonlinear signalling **2**: 930
Significance feedback
self-organizing systems **6**: 4166
Significance level
hypothesis testing **5**: 2912
Signorini problem **7**: 5037
Similarity **6**: ***4294–6***
approximate similarity
partial modelling **6**: 4295
geometry **6**: 4294
relations **3**: 1795, 1836
analogical inference **1**: 272
decision theory **3**: 2041
fuzzy relations **3**: 1602
transformation
biological models structural identifiability **1**: 492
Simplex method **4**: ***2686***, 2841; **6**: ***4296–9***
revised simplex method **6**: 4297
see also Linear programming
Simplicial algorithms
path-following methods **6**: 3632
Simplicial complexes **6**: ***4299***
Simplicity principle
complexity **1**: 705
Simulation **6**: ***4311–3***; **7**: ***5011–5***
activity-based models **6**: ***4333–4***
CAPS **6**: 4392
computer software **6**: ***4392–3***
aerospace simulation **6**: ***4393***
air pollution control
wind tunnels **1**: ***188–9***
automatic control **6**: ***4310–1***
automatic landing systems **1**: 210
biochemistry **1**: ***421–4***
biology **1**: ***466–7***
blood glucose
adaptive control **3**: 2018
blood pressure
adaptive control **1**: 499
bond graphs **6**: 4337–41
CAD **3**: 2216
chemical engineering **1**: ***583–5***
class concepts **6**: 4365
computer graphics **6**: ***4299–302***
computer systems
emulation **2**: 1450
computer–user interfaces **5**: ***3077–9***
credibility **6**: ***4406–11***
databases **2**: ***893–4***
deterministic simulation **5**: 3026
discrete-event simulation
dedicated processors **2**: 1085
microprocessor architecture **2**: ***1083–7***
systems analysis **7**: 4805
distillation columns
bilinear models **2**: 1119
distributed system models **2**: ***1199–201***

dynamic simulation
  workshop dynamic modelling 7: 5170
ecological simulation modelling
  sensitivity analysis 6: ***4230–4***
elevator systems 2: ***1427–31***
ergonomics 3: 2184
event-based simulation 6: ***4346–8***
  computer software 6: ***4403–5***
experiment design 7: ***4648–51***
experimental frames 6: 4389
flow-based models
  computer software 7: ***4933–5***
geometry-dependent simulation
  structural graphics 6: 4301
geophysics 3: ***2005–6***
hardware
  data-oriented hardware 1: 632
  system-oriented hardware 1: 632
HVAC systems 3: ***2099–104,*** 2113
industrial robots
  aerospace industry 4: 2451
Lindenmayer systems 4: 2795
linear continuous systems
  control system analysis computing 4: 2799
manufacturing computer control 3: ***1642–8***
mechanical systems 5: ***2966–70***
metamodels 5: 3025
methodology
  model manipulation 6: ***4311–3***
  top-down view 6: ***4319–22***
  systems management 6: ***4313–9***
microprogramming 1: ***630–3***
  parallel structure 1: 633
model behavior
  regression summarization 5: 3024
model documentation 6: ***4372–4***
model equivalence 6: ***4326–8***
model life cycle 6: 4328
model management 6: ***4328–32***
model taxonomy 6: ***4381–7***
modelling
  activity-based models 6: ***4333–4***
  autopoiesis 6: ***4374–7***
  cellular spaces 1: ***559–61***
  computer systems 1: ***751–4***
  differential equations 6: ***4356–9***
  information storage 6: 4323
  model simplification 7: ***4762–4***
  partial differential equations 6: ***4360–4***
  support systems 6: ***4370–2***
  symbol manipulation 6: ***4377–80***
modelling formalism
  arithmetic relators 6: ***4334–6***
  bond graphs 6: ***4337–41***
  cellular automata 6: ***4341–5***
  discrete arithmetic 6: ***4345–6***
  event-based formalism 6: ***4346–8***
  extended Petri nets 6: ***4348–50***
  heterarchical systems 6: ***4350–3***
  hierarchical decomposition 6: ***4353–5***
  Markov chains 6: ***4355–6***
  ordinary differential equations 6: ***4356–9***
  partial differential equations 6: ***4360–4***
  process-based formalism 6: ***4365–6***
  system specifications 5: 3086
  systems dynamics 3: 2015; 6: 4366–8
  transactions-flow techniques 6: ***4368–9***
multiple objectives
  modelling 6: 4388
natural gas transmission systems 3: 1990
operations research 4: ***2351–2***
optimization 6: ***4306–10***
  regression analysis 5: 3029
ordinary differential equations 6: ***4356–9***
  symbol manipulation 5: 3559
parameter-dependent simulation
  interactive graphics 6: 4300
parameter estimation 7: ***4690–2***
partial differential equations 6: ***4360–4***
  numerical solution 6: ***3622–6***
  software 2: ***1180–1***
ports 3: 2076
power systems 6: 3828
process-based models 6: ***4365–6***
  computer software 6: ***3878–9***
program design 6: ***4396–402***
program generators 4: ***2547–50***
radioecological modelling 6: 3932
rainfall-runoff modelling 6: 3940
random simulation 5: 3026
real-time systems 6: ***4418***
regression analysis 5: ***3024–9***
road traffic 7: 4924
  car-following models 7: 4925
  hydrodynamic theory 7: ***4920–2***
  macroscopic models 7: ***4923–5***
  microscopic models 7: ***4925–7***
  predictive models 7: ***4927–9***
scenarios 6: 4153
sensitivity analysis 6: ***4227–9***
  regression analysis 5: 3029
ships
  dynamic positioning 6: 4265
  maneuvering 6: 4261
  traffic control 7: 5053
statistical analysis of output 6: ***4393–5***
statistical error analysis 4: 2256
steady-state simulation
  chemical engineering 1: 583
stochastic processes
  initial-transient techniques 7: ***4651–4***
  variance-reduction techniques 7: ***4654–6***
system dynamics modelling 3: 2015; 6: ***4366–8***; 7: 4755
system specification morphisms 6: ***4326–8***
systems analysis 7: 4769
systems control 6: ***4313–9***
systems homomorphism 3: 2164

systems methodology 7: 4803
taxonomy 6: ***4411–3***
time-advance methods 6: ***4414–5***
time-dependent simulation
animation graphics 6: 4300
toxic chemicals assessment 7: ***4895–7***
training 6: ***4415–7***
transaction flow-based models 6: ***4368–9***
tumors
treatment 7: ***4956–8***
validation 6: ***4388–92***
documentation 6: 4374
philosophical aspects 7: ***5011–5***
regression analysis 5: 3029
statistical analysis 7: ***5011–5***
verification 6: ***4402–3***
"wait until" concepts 6: 4365
water distribution networks 7: ***5106–9***
see also Modelling
Simulation data language
simulation databases 2: 894
Simulation languages 4: 2548
discrete-event simulation 6: 4398
HVAC systems
digital simulation 3: 2116
SIMON
activity-based models 6: 4393
SIMSCRIPT
simulation databases 2: 894
SIMSCRIPT II. 5
process-based models 6: 3878
SIMULA
process-based models 6: 3878
transactions-flow-based models 7: 4934
SLAM
transactions-flow-based models 7: 4934
taxonomy 6: 4303–4
Single-input, single-output systems *see* SISO systems
Single-loop systems
input–output stability 4: 2673
Singular control 7: ***4419–25***
optimal control 5: ***3506–8***
see also Optimal control
Singular perturbations 7: ***4435–9***, 4529
boundary-layer problems 7: ***4425–9***
boundary-value problems 7: 4430
complex systems control 1: 696
discrete perturbations 7: ***4429–34***
distributed parameter systems 2: ***1186–92***
initial-value problems 7: 4430
output feedback control 7: ***4439–43***
Singularly perturbed models 5: 3173
Singularly perturbed systems 5: 3174
SISO systems
controllability 4: 2817
design
robustness 4: 2842
minimum variance control 5: 3004
observability 4: 2817
Sizing
control valves 2: ***828–32***
in situ sizing 2: 830
microscopic particles
Coulter counters 2: 862, 863
Skidding
automobile electronics 1: 375
Skill-based behavior
expert systems 2: 1570
Skin
anisotropy 1: 478
biomechanics 1: ***478–82***
compressibility 1: 479
load–deformation response
time dependence 1: 478
mechanical testing 1: 480
in vitro 1: 479
in vivo 1: 479
temperature
comfort indices 1: 639
temperature measurement 7: 4854
wettedness
comfort 1: 644
Slack
systems engineering 7: ***4454***
Slip power recovery
ac motors 5: 3091
Sloughing
Cladophora modelling 1: 260
Slow modes 5: 3174
Slow subsystems 5: 3174
Sludge
gas chromatography–mass spectrometry 3: 1938
Small intestine *see* Digestive tract
Smelting
process control 5: 3282
Smith predictors 7: ***4457–62***
Smith–McMillan canonical forms 7: ***4445–7***
Smoothing 7: ***4463–6***
multigrid methods 5: 3136
operators 4: 2677
square root algorithms 7: 4539
two-filter smoothing algorithms 7: ***4466–8***
Smoothing windows
stationary time series 7: 4560–2
Snakes
locomotion 6: 4109
Snow
hydrologic forecasting 3: 2228
Sobolev spaces
Fourier transforms 3: 1727
Social aspects of automation 1: ***367–71***; 6: ***4093–7***
Social and behavioral sciences
autopoiesis 1: 394
causality 1: 554
computing
interpretive structure modelling 4: 2575
flight-crew workload 3: 1666
holism 3: 2163

workshop dynamic modelling **7**: 5171
Social choice theory **7**: ***4469–72***
Social judgment theory **7**: 4807
Sodium
drinking control **3**: 2212
Sodium nitroprusside
blood pressure
surgery **1**: 498
Soft tissue
dielectric properties **1**: ***445–50***
Software engineering
methodology **6**: ***3898***
Soil
adsortion
nutrients and pollutants **1**: 6
gas chromatography–mass spectrometry **3**: 1938
radioecological modelling **6**: 3932
toxic substances **1**: ***143–8***
Solar power **2**: 1455, 1461
satellites **7**: ***4472–6***
Solids
heterogeneous solids
stereology **7**: ***4598–601***
Solubility product
toxic substances
limnology **4**: 2779
Solutions
environmental chemistry
nutrients and pollutants **1**: 3
Solvent extraction
trace organic compounds **7**: 5127
Sorbent extraction
trace organic compounds **7**: 5128
Sorption
artificial kidneys **1**: 310
trace elements
water pollution **7**: 5126
Source systems **3**: 2001; **7**: ***4476***
data systems **2**: 880
see also Primitive systems
Space heating
boilers **2**: ***1227–33***
degree days **2**: 959
domestic appliances **2**: ***1214–26***
controllers **2**: ***1239–41***
furnaces **2**: ***1227–33***
optimum start control **5**: ***3554***
self-tuning regulators **5**: 3347
solar power **2**: 1455
temperature control **2**: ***1234–9***
see also HVAC systems
Space shuttle
control systems
redundancy and reliability **7**: ***4476–81***
Space vehicles
attitude control
Kalman filters **4**: 2622
computer-oriented control analysis **3**: ***1656–9***
flexible spacecraft
control analysis **3**: ***1649–64***
finite element models **3**: ***1650–3***
impulsive control **2**: 1335
rigid spacecraft
attitude control **1**: ***319–23***
trajectory optimization **7**: ***4481–90***
Spark ignition
domestic gas appliances **2**: 1213
Sparse data systems
modelling
parameter estimation **1**: 486
Sparsity **2**: 1279
Spatial allocation
gravity models **7**: ***4491–5***
Spatial decomposition methods **2**: ***1279–82***
Spatial dynamic programming **2**: ***1279–82***
Special position interrogation pulse
air traffic control **1**: 200
Spectra **7**: ***4498–503***
molecular spectra
nuclear magnetic resonance **5**: 3416
Spectral analysis **4**: 2264; **7**: ***4495–7***
correlation methods **4**: 2251
ecosystem modelling **6**: 3664
identification **4**: 2242
multidimensional systems theory **5**: 3128
stationary time series **7**: ***4556–66***
time series
tides **7**: 4891
water quality **6**: 4041
Spectral factorization
self-tuning control **6**: 4173
Spectral window **4**: 2267
Spectrophones
optical fiber interferometric sensors **5**: 3469
Spectroscopy **7**: ***4498–503***
biology **1**: ***8–13***
environmental analysis **2**: 1509
photoacoustic spectroscopy **6**: 3699
inductively coupled plasma atomic emission spectrometry **4**: ***2343–7***
quantitative measurements **7**: 4502
Speech **7**: ***4504–9***
automatic recognition **7**: 4507
synthesis **7**: 4507
Speed measurement *see* Velocity measurement
Speed regulation
governors
history **7**: ***4513–7***
SPHINX system
medical diagnostic computing **2**: 1575
Spike nutrients
algal growth potential **7**: 5112
Splines (mathematics)
interpolation
air pollution pattern analysis **7**: ***4517–8***
Spray coating technology
industrial robots **4**: 2416

Spreadsheets **3**: 2178
Springs
  simulation **5**: 2967
Square root algorithms
  state-space models **7**: ***4534–9***
Square root factors
  state-space models **7**: 4534
Stability **7**: ***4518, 4523–8***
  absolute stability
    nonlinear control systems **5**: 3387
  adaptive control **5**: 3061, 3406
    non-minimum-phase systems **5**: 3408
  asymptotic stability **7**: 4673
  bounded stability
    Lyapunov methods **5**: 3400
  closed-loop systems **4**: 2568; **7**: 4445
    nonlinear systems **5**: ***3316–9***
  conflict analysis **7**: 4795
  connective stability **4**: 2674
    ecosystems **1**: 672
  connectivity **2**: ***776–7***
  control systems **7**: ***4674–7***
    air conditioning **1**: 156
    design **5**: 3376
  convergence analysis
    explicit self-tuners **2**: 1579
  cooperative decentralized control **1**: 679
  decentralized feedback **4**: 2666
  decentralized robust controller design **6**: 4136
  decentralized systems **2**: 951; **4**: 2700
  differential equations **5**: 3563
  direct digital control
    design **2**: 1074
  directed graphs **2**: 1078–80
  dissipativeness **7**: ***4519–22***
  ecosystem models
    qualitative stability **2**: ***1348–51***
  ecosystems **1**: ***672–4***
  fuzzy logic controller **3**: ***1814***
  fuzzy systems **3**: 1897
  helicopters **3**: 2121
    automatic flight control systems **3**: 2124
  hereditary systems **3**: 2149
  hydropower systems **3**: 2226
  input–output stability **7**: 4520
    large-scale systems **4**: ***2671–5***
  instability theorems **4**: 2674
  interconnected dynamical systems **4**: ***2551–6***
  Lagrange stability **4**: 2556
  large-scale systems
    input–output stability **4**: ***2671–5***
    Lyapunov methods **4**: 2560
  linear systems **4**: ***2872–5***
  Lyapunov methods **4**: 2904; **5**: 3399; **7**: 4520
    bounded stability **5**: 3400
    fixed systems **7**: 4524
    large-scale systems **4**: 2560
    model reference adaptive control systems **5**: 3068
    nonlinear systems **5**: ***3395***
  model reference adaptive control systems **2**: 1088; **5**: 3056
    Lyapunov methods **5**: 3068
  multidimensional systems **5**: ***3131–4***
  multiple control systems **6**: 4010
  multistate systems **5**: 3493
  multivariable systems **4**: 2843; **5**: 3224
    functional analysis **5**: 3203
  non-minimum-phase systems **5**: 3408
  nonlinear feedback systems **5**: ***3316–9***
  nonlinear systems **5**: 3372
    Lyapunov methods **5**: ***3395***
    vector norms **5**: ***3402–6***
  numerical stability **2**: 816
  output feedback control
    singularly perturbed systems **7**: 4441
  populations **6**: 3752
  power systems **6**: ***3795–801***
  relaxed static stability
    aircraft **1**: 24
  self-tuning control **6**: 4192
    input–output analysis **6**: 4206
    parameter-adaptive control algorithms **6**: 3606
  ships
    history **6**: ***4271–4***
  singular perturbations **7**: ***4435***
  Smith predictors **7**: 4460
  stochastic adaptive systems **7**: 4604
    martingale methods **7**: ***4607–11***
  structural stability **7**: 4525, ***4669–74***
    internal model principle **4**: 2566
  suboptimality **4**: 2557
  systems stability
    history **3**: 1745
  values and pulses **7**: ***4522–3***
  walking robots **6**: 4124
  weighted minimum variance controllers **1**: 41
Stability criteria
  characteristic-locus method **1**: 575
  circle criterion **5**: ***3376–9,*** 3398
  describing-function method **2**: 978
  Nyquist criterion **5**: ***3440–1***
  root loci **3**: 1748
Stabilization
  control systems **4**: 2858
  decentralized stabilization **2**: 900, ***912–6, 916–23***
    algorithms **2**: 917
    block Gerschgorin theorem **2**: ***923–7***
    hierarchical algorithms **2**: 921
    large-scale systems **4**: 2559
  exponential stabilization **2**: 924
  fuzzy systems **7**: ***4529***
  interconnected control systems **2**: 925
  manipulation
    industrial robots **4**: 2457
  multivariable control system synthesis **4**: 2809
Stabilizers
  aircraft control **1**: 386

Stable sets
  combinatorial optimization **1**: 624
  cooperative game theory **2**: 847
Stackelberg strategies **5**: ***3145–9***
  dynamic duopoly **2**: 1263
  feedback Stackelberg strategies **5**: 3148
  game theory **1**: 684
Standard effective temperature **1**: ***644–5***
Standards
  natural gas technology
    flow measurement **3**: 1963
  signal transmission **6**: 4292
  temperature scale **7**: 4849
  see also Measurement standards, Code standards
State-affine systems **5**: 3289
  approximation **5**: 3375
  nonlinear identification **6**: 3772
State equations **1**: 388
  aircraft dynamics **1**: ***213–20***
    linear equations **1**: 219
  anticipatory systems **1**: 290
  bond graphs **2**: 1328; **6**: 4341
State estimation **4**: 2516
  adaptive control **1**: 102
  continous-time systems
    adaptive control **1**: 102
  control location problem **6**: 4241
  decentralized observers **2**: ***907–12***
    complete interaction estimate exchange **2**: 911
  decoupled methods
    power systems **6**: 3848
  discrete-time systems
    Kalman filters **4**: 2624
  freeway traffic control **3**: 1738
  gas transmission networks **3**: 1987
  inorganic chemicals production **4**: 2515
  Kalman filters **4**: 2624, 2692
  large stochastic systems **4**: 2706
  linear systems **4**: 2864
    hyperbolic systems **4**: 2806
  nonlinear state estimation **4**: 2310
  optimal decentralized observers
    complete interaction measurements **2**: 910
  parallel simulation
    gas transmission networks **3**: 1987
  pole assignment **6**: 3729
  power systems **6**: 3845, ***3846–50***
    control **6**: 3826
  sensor location problem **6**: 4241
  see also Identification, Estimation theory, Filtering and prediction theory, Luenberger observer
State feedback **1**: 388
  pole assignment **6**: 3711, ***3717–20***, 3727
  stochastic control theory **2**: 928
  see also Feedback
State reduction
  model structure simplification **5**: 3080
State-space methods **2**: 1289; **7**: ***4551–4***
  aerospace control **1**: 116
  biological models
    structural identifiability **1**: 490
  consistency control **2**: 779
  control system analysis computing **4**: 2799
  decomposition
    linear systems **4**: 2863
  differential equation models **6**: 4357
  discrete-time processes
    Kalman filters **4**: 2622
  distillation columns
    bilinear models **2**: 1117
  Gaussian models
    GLR tests **1**: 84
  group theory **5**: 2944
  HVAC systems
    digital simulation **3**: 2114
  large-scale systems
    model simplification **5**: 3071
  linear multivariable systems
    transformations **4**: 2867
  matrix triangularization
    Givens transformations **5**: 2950
    Householder transformations **5**: 2950
  model simplification **7**: 4762
  nonlinear identification **6**: 3772
  nonlinear systems **5**: 3346
    approximation **5**: ***3374–6***
  realization theory **6**: ***3969***
  self-tuning control **6**: 4199; **7**: ***4547–51***
  smoothing algorithms **7**: 4463
  square root algorithms **7**: ***4534–9***
  state-space modelling **7**: ***4540–7***
  state-space transformations **7**: ***4540–7***
  well posedness **4**: 2678
State systems
  systems homomorphism **3**: 2165
State-transition systems **7**: ***4554***
State variables **7**: ***4554***
  aircraft motion **1**: 217
Static optimization *see* Optimization
Stationary points
  differentiation **7**: ***4555–6***
Stationary time series **7**: ***4556–66***
Statistical analysis
  air pollution **2**: ***1497–500***
  environmental data
    modelling **2**: ***1494–7***
  errors
    simulation approaches **4**: 2256
  microprocessors
    discrete-event simulation **2**: 1086
  model structure determination **4**: 2289
  model validation **7**: 4997, ***5011–5***, ***5015–8***
  modelling **5**: 3024
  multivariate analysis **5**: ***3232–5***
  simulation output **6**: ***4393–5***
  telecommunication traffic **7**: 4834

Statistical mechanics **4**: 2246
Statistical models
  entropy minimax theory **2**: 1473
Statistics
  fuzzy statistics
    random sets **6**: 3945
  history **4**: 2271
  hypothesis testing **5**: 2912
  level of significance **4**: ***2754***
  regression analysis **6**: 3998
  sampling
    ecological simulation modelling **6**: 4231
Stators
  dc motors **5**: 3092
  frequency control
    ac motors **5**: 3089
  voltage control
    ac motors **5**: 3089
Steady-state analysis
  aggregation **1**: 129
  gas distribution systems **3**: ***1947–54***
Steady-state optimization
  process control
    petrochemical industry **6**: 3674
Steady-state systems
  on-line coordination **7**: ***4566–71***
Steady states
  multiple steady states **1**: 429
    biological systems **1**: 431
Steam boilers *see* Boilers
Steam cycle
  Carnot cycle **7**: 4580
  Rankine cycle **7**: 4580
Steam generators *see* Boilers
Steam plants
  nonlinear modelling **6**: 3774
  see also Boilers
Steam power stations
  boilers **7**: 4572
Steam thermal units **7**: ***4580–5***
Steam turbines
  control theory
    fuzzy set theory **4**: 2359
  process control **7**: 4573
Steel industry
  hierarchical control **7**: ***4585–93***
    personnel **7**: 4592
  mass spectrometry **5**: 2935
  strip steel
    production control **7**: 4588
  tracking
    hierarchical control **7**: 4590
Steering *see* Guidance
Stefan problem *see* Heat transfer
Stepping motors **7**: ***4593–8***
Stereology
  biological materials **7**: ***4598–601***
Stereometry
  distance measurement **4**: 2485
Sterilization
  pharmaceutical industry
    process control **6**: 3691
Stiff differential systems
  stability **5**: 3133
Stirring
  fermentation measurement **3**: 1613
Stochastic calculus **5**: 3330–6; **7**: ***4631–7***, 4638
Stochastic control **1**: ***38–45***; **2**: 1253; **5**: ***3524–30***; **7**: ***4611–2***
  Bayesian framework **2**: 1253
  control location problem **6**: 4242
  differential equations
    stochastic processes **7**: 4639
  discrete-time systems **2**: 1095–8; **5**: ***3471***
  discretization **5**: 3526
  fuzzy set theory **7**: ***4657–61***
  learning automata **4**: 2714
  methods **7**: 4612
  model reference adaptive control systems **5**: 3046, ***3056–60***
  nonlinear estimation **5**: ***3311–6***
  nonlinear stochastic control **1**: 46
  optimal stochastic control **5**: ***3517–23***
    discrete-time systems **5**: ***3471–5***
    nuclear power stations **5**: 3428
    numerical methods **5**: ***3523–30***
  parameter-adaptive control **6**: 3603, 4162
  performance-adaptive control **6**: 4163
  problem formulation **7**: 4611
  regular perturbations **5**: 3500
  self-tuning control **6**: 4190
    data prediction **6**: 4168
  separated stochastic control **5**: 3474
  separation principle **6**: ***4245–8***
  see also Stochastic systems, Stochastic processes
Stochastic games **7**: ***4621–2***
  differential games **7**: ***4613–7***
Stochastic gradient method
  power systems **6**: 3788
Stochastic integrals **5**: 3294; **7**: ***4631–7***
  iterated integrals **5**: ***3330–6***
Stochastic maximum principle **5**: 3315; **7**: ***4638–43***
Stochastic processes **4**: 2245, 2265; **7**: ***4631–2, 4646***
  block matrix formulae **1**: 495
  Brownian motion **1**: ***511–2***
  computer systems **6**: 3655
  differential equations
    stochastic control **7**: 4639
  ecological modelling **7**: ***4644–5***
  formal languages
    pattern recognition **6**: 3646
  Girsanov transformation **3**: 2007
  Hamilton–Jacobi–Bellman equation **7**: ***4622–6***
  hydrologic forecasting **3**: 2232
  incentives **7**: ***4627–30***
  linear Boltzmann equations **7**: 4940
  load forecasting **6**: 3830
  Markov decision processes **5**: 2930

nonlinear estimation **5**: 3294
optimal stopping **5**: 3531–5
realization theory **7**: 3962
recursive identification **4**: 2310
simulation
initial transient techniques **7**: ***4651–4***
modelling **6**: 4355
variance-reduction techniques **7**: ***4654–6***
stationary time series **7**: ***4556–66***
stochastic adaptive control algorithms
convergence analysis **7**: ***4602–6***
stochastic games **3**: 1923; **7**: 4621–2
stochastic integrals **5**: 3294; **7**: ***4631–7***
iterated integrals **5**: ***3330–6***
stochastic maximum principle **5**: 3315; **7**: ***4638–43***
zero–one processes
fuzzy set theory **4**: 2295
see also Markov processes, Random processes
Stochastic programming **7**: ***4646–7***
Stochastic signals
frequency response techniques **3**: 1749
history **4**: 2271
Stochastic systems
adaptive control **5**: 2999
stability **7**: ***4607–11***
analysis
probability distributions **6**: 4027
ARMAX models
state-space methods **7**: 4547
discrete-time control systems **2**: 1095–8; **5**: ***3471***
distributed parameter systems **7**: ***4617–21***
model reference adaptive control systems **5**: 3046, ***3056–60***
optimal control **5**: ***3517–23***
optimality **2**: 1255
parameter-adaptive control **6**: 3603, 4162
ship positioning **6**: 4263
stability **4**: 2556
see also Stochastic control, Stochastic processes
Stochastic transition matrices **2**: 1276
Stock control **4**: ***2583***
backward dynamic programming **1**: 408
dynamic programming **5**: 3509
forward dynamic programming **3**: 1714
fuzzy set theory **7**: 4661
impulse control **4**: 2339
K convexity **5**: ***3509–10***
operations research **7**: 4773
planning horizons **6**: ***3703***
s, *S* policy **5**: ***3509–10***
Stomach *see* Digestive tract
Stomata
air pollution effects
image instrumentation **2**: 1518
movement
image instrumentation **2**: 1521
Stomatognathic physiology **2**: ***976–8***
Stopping rule
optimal impulsive control
populations **2**: 1335
Stopping-time problems
differential games **7**: 4616
optimal stopping **5**: ***3531–5***
Storage
level measurement **4**: 2737
bulk materials **4**: ***2743–7***
natural gas **3**: ***1969–75***
polymerization
process control **6**: 3744
tank distortion **4**: 2743
water pollution
chemical analysis **7**: 5109
Storage function
dissipativeness **7**: 4520
Storage requirements
convergence behavior **2**: 860
Strain gauges
dentistry **2**: 968
optical fiber interferometric sensors **5**: 3469
pressure measurement **6**: 3876
Strain measurements
connective tissue **1**: 475
Strange attractors **1**: 565, 567
Strategic planning decisions **2**: 943
Strategies **7**: ***4661***
Stratification
lake modelling **4**: 2643
Stratum corneum
biomechanics **1**: 478
Strength of couplings **2**: 951
Stress
chemical stress
agricultural models **1**: ***589–92***
Stress–strain relations
biomechanics **1**: 480
blood **1**: 443
bone
macroscopic scale **1**: 468
microscopic scale **1**: 469
Stroboscopes
angular velocity measurement **7**: 4511
Stroke scanning
cathode ray tubes **1**: 241
Structural-(A,B)-invariant subspace **2**: 895
Structural analysis **6**: 4353
graph representation **7**: ***4661–6***
matrix representation for digraphs **2**: 950
Structural complete systems
structural observability **2**: 950
Structural constraints
robust controller design **7**: ***4685–9***
Structural controllability
graph-theoretical criteria **2**: 949
Structural identifiability
biological models **1**: ***490–3***
Structural modelling **5**: 3268; **7**: ***4692***
contextual relations **2**: ***801***
edges **2**: ***1358***

transitivity **7**: ***4935***
Structural modes
aircraft active control technology **1**: 29
Structural observability
structural complete systems **2**: 950
Structural operators
hereditary systems **3**: 2149
Structural perturbations
linear time-invariant systems
decentralized stabilization **2**: 922
Structural properties
decomposition **2**: ***948–52***
Structural stability **7**: 4525, ***4669–74***
catastrophe theory **7**: 4671
dynamic systems **7**: 4670
Structure **2**: 948
ecosystem networks **2**: ***1352–5***
Structure graphs **2**: 948
Structure systems **2**: 1547; **3**: 2001; **7**: ***4692***
reconstructability analysis **6**: 3985
Structured programming
programming methodology **6**: 3898
Subgradient algorithms **1**: 405
Subjective entropy **7**: ***4692–5***
Subjective probability **7**: ***4696***
Subjectivity
communication **7**: ***4696–8***
control theory **4**: 2357
natural languages **7**: 4696
pragmatic fuzzy systems **6**: 3855
Suboptimal control **1**: ***126–32***
design **1**: 130
Suboptimal solutions **7**: ***4699***
Substations
power systems **6**: 3807
Substitution
economics **7**: ***4699***
Subways *see* Metros
Sufficiency
parameter estimation **6**: 3613
Sugar industry *see* Food processing industry
Sulfates
aerosol chemical analysis **1**: 174
ion-selective electrodes
water pollution **4**: 2607
Sulfides
ion-selective electrodes
water pollution **4**: 2607
Sulfur
ecology
aluminum compounds **1**: 267
ion-selective electrodes
water pollution **4**: 2606
Sulfur dioxide
agricultural models **1**: 589
air pollution
automatic analyzers **1**: 181
Sulfuric acid
aerosol chemical analysis **1**: 174
Supervision and executive programs
power system computer control **6**: 3841
Supervisory levels
multilevel control systems **5**: 3140
Supply rate
dissipativeness **7**: 4519
Support
fuzzy set theory **3**: 1874
Support systems **6**: 4370
Surface chemistry
nutrients and pollutants
environment **1**: 5
Surgery
blood pressure
adaptive control **1**: ***498–502***
cardiac surgery
patient monitoring **5**: 2987
pain
demand analgesia **2**: ***960–3***
Surrogate worth trade-off method **7**: ***4703–9***
Surveillance
freeway traffic control **3**: ***1736–41***
Surveys
systems science **7**: ***4709***
Swashplates
helicopters **3**: 2119
Sweating
comfort indices **1**: 639
thermoregulation **7**: 4868, 4871–3, 4879
Switches
lighting **4**: 2770
low-pressure cut off **4**: 2355
power systems **6**: 3808
Switching
artificial lighting **4**: 2775
biochemistry **1**: ***429–33***
lighting **4**: ***2764–9***
optimal control
regular synthesis **5**: ***3487***
power system control **6**: 3782
power system transients **6**: 3805
Switching functions
fuzzy set theory **3**: ***1880–3, 1883–7***
applications **3**: 1882
canonical forms **3**: 1881
minimal forms **3**: 1882
minimization algorithm **3**: 1885
Switching networks
control policies
learning algorithms **7**: ***4716–21***
modelling **7**: ***4726–31***
packet switching **6**: ***3581–90***
quality parameters **7**: ***4732–3***
reactive power control **6**: 3951
telecommunication networks **7**: ***4710***
long-term planning **7**: ***4721–6***
short-term planning **7**: ***4741–8***
see also Telecommunication networks, Telephone networks

Switching surfaces
variable-structure systems **7**: 5028
Symbol manipulation
ordinary differential equation models **5**: ***3559–60***
simulation **6**: ***4377–80***
Symmetrization
isoperimetric problems **4**: 2610
Synapses **1**: 506
Synchronization
computers **7**: ***4751–2***
control applications **2**: 1139
distributed computer systems **2**: 1130
distributed processing **2**: 1166, ***1202–9***
modelling **2**: 1167
distributed system simulation **2**: 1200
simulation
time-advance methods **6**: 4414
Synchronous generators **2**: 1382
control **2**: 1375
power systems **6**: 3796
Synchronous motors **5**: 3091
Synectics **7**: 4802
Syntactic methods
pattern recognition **6**: ***3645–52***
cluster analysis **6**: 3650
nearest neighbor rule **6**: 3651
stochastic languages **6**: 3646
Syntax **6**: 4223
programming languages **7**: ***4752***
syntax graphs
robot operating systems **6**: ***4089–93***
System acquisition **7**: ***4752–5***
System digraphs **4**: 2677
System dynamics
diagrams **7**: 4758
land use **7**: 4495
modelling **7**: ***4755–62***
simulation modelling **6**: ***4366–8***
workshop dynamic modelling **7**: 5171
System failure and recovery
catastrophic failure
probability **6**: 4013
distributed computer systems **2**: 1131
industrial robots **4**: ***2444–7***
mean time to failure **6**: 4013
probability **6**: 4011
see also Reliability
System identification **4**: ***2239–45***
model structure determination **4**: 2302
practical aspects **4**: ***2301–7***
see also Identification
System performance **2**: 996
fuzzy set theory **7**: ***4764–6***
System procurement **7**: ***4752–5***
System specifications
hierarchy **5**: 3083
modelling **5**: ***3083–8***
morphisms **6**: 4326–8
System theory **7**: ***4766***
automata theory **5**: 3269
environ theory **2**: 1479
fuzzy set theory
ecology **3**: 1859
general systems problem solver **3**: 2000
history **7**: ***4771–4***
modelling **3**: 1872
Systemic processes **7**: ***4766***
Systems **7**: ***4767***
see also Control systems
Systems analysis **2**: ***982–7***, ***987–92***; **7**: ***4767–70***, 4799, ***4803–6***
complexity **2**: 990
computer numerical analysis **2**: ***813–25***
constraints **2**: 989
environmental policy making **2**: ***1539–41***
ELMES support system **2**: ***1542–4***
evaluation **2**: 985
history **7**: 4772
indigenous **2**: 989
information theory **4**: ***2492–7***
methodology **2**: 983
microeconomic systems analysis **5**: ***2973–9***
model validation **6**: 4407
operational employment **2**: 986
optimizing **2**: 989
reconstructability analysis **6**: ***3983–9***
satisficing **2**: 989
structure determination **7**: 4767
Systems control
simulation **6**: ***4313–9***
Systems design *see* Systems analysis, Systems synthesis
Systems ecology **7**: ***4775***
Systems engineering **2**: ***995–1003***; **7**: ***4767–70***, ***4775***, ***4800–1***
definition **2**: 996
ergonomics **3**: 2180
large-scale systems **4**: 2681
praxiology **6**: 3862
process **2**: 997
slack **7**: ***4454***
water resources **7**: ***5135–40***
see also Systems management
Systems hierarchy
epistemology **2**: ***1545–8***
general systems problem solver **3**: 2000
Systems homomorphism **3**: ***2164–8***
Systems knowledge
philosophical aspects **7**: ***4779–84***
Systems management **7**: ***4785–9***
cognitive styles **7**: ***4790–3***
conflict analysis **7**: ***4793–8***
see also Systems engineering
Systems methodology **7**: ***4799–809***
Systems philosophy **7**: ***4779–84***
holism **3**: 2164
Systems science **7**: ***4810***
behavior systems **1**: ***414***

closed systems **1**: ***605***
coupling **2**: ***867***
data systems **2**: ***880***
deterministic systems **2**: ***995***
directed systems **2**: ***1080***
dynamic systems **2**: ***1282***
effective systems **2**: ***1366***
efficient systems **2**: ***1367***
isomorphic systems **4**: ***2609***
metasystems **5**: ***2971***
neutral systems **5**: ***3264***
nominal group technique **7**: ***4810***
source systems **7**: ***4476***
state-transition systems **7**: ***4554***
structure systems **7**: ***4692***
surveys **7**: ***4709***
Systems software **7**: ***4810***
Systems synthesis **7**: ***4810***

T

Tablets
  coating techniques
    process control **6**: 3694
  pharmaceutical industry
    process control **6**: 3692
Tachometers **7**: 4511
Tannic acid
  aluminum compounds **1**: 264
Task allocation
  distributed computer systems **2**: 1281
Tatonnement **7**: ***4811***
Taylor series expansion
  error analysis **2**: 1552
Teach function
  robots **6**: ***4123***
Team theory **2**: 931
  information theory **7**: ***4811–22***
Technological forecasting **7**: ***4823–7***, 4831
  systems acquisition **7**: 4752
  systems management **7**: 4788
  workshop dynamic modelling **7**: ***5170–7***
Technological innovation **7**: ***4828–32***
  research **6**: 3660
Teeth
  ageing effects **2**: 965
  extraction effects **2**: 965
  measurements **2**: ***963–5***
    mobility **2**: 966
  surface loss **2**: 964
Telecommunication networks **7**: ***4710–5***
  modelling **7**: ***4726–31***
  multiplexing **7**: 4726
  planning
    long-term planning **7**: ***4721–6***
    short-term planning **7**: ***4741–8***
  routing **7**: 4716
  satellite links **2**: ***1051–7***
  variables measurement **7**: ***4833–5***
  see also Switching networks, Telephone networks
Telecommunication traffic
  configuration **7**: 4721
  variables
    measurement **7**: ***4833–5***
  see also Telephone traffic
Telecommunications
  office automation **5**: 3446
  spacecraft **7**: ***4835–42***
Teleconferencing **5**: 3446
Telecontrol
  power systems **6**: 3782
Telemetry **7**: ***4835–42***
  data communication systems **7**: ***4835–42***
  dental materials **2**: 968
  digital coding **7**: 4837
  natural gas technology **3**: 1962, ***1975–9***
  satellites **7**: 4837
  underwater telemetry **7**: ***4842–8***
  water supply **7**: 5098
Teleology **1**: 290; **2**: 871; **7**: ***4848***
Telephone exchanges **7**: 4710
Telephone networks **7**: 4712
  dimensioning **7**: 4713
  organization **7**: 4734
  quality parameters **7**: ***4732–3***
  queuing theory **6**: 3924
  real-time control **7**: ***4733–40***, ***4749–51***
  see also Telecommunication networks
Telephone traffic **7**: ***4726***
  routing **7**: 4730, 4734, ***4749–51***
  see also Telecommunication traffic
Telephony
  frequency response techniques **3**: 1745
Telesimulation
  air traffic control **1**: 203
Temperature
  ecological modelling **2**: ***1333***
  fermentation measurement **3**: 1612
  leaf temperature
    image instrumentation **2**: 1517
  skin temperature
    comfort indices **1**: 639
Temperature control **7**: ***4872–3***
  acclimatization **7**: 4876
  air conditioning **1**: 148, 154
  boilers **7**: 4575, 4578
  deadband control **2**: ***894***
  feedback control **3**: 1604
    history **3**: 1606
  heat exchangers **3**: 2089
  physiological control **7**: ***4866–8***, ***4877–80***
  physiological models **7**: 4873–6
    comfort modelling **1**: 639
  polymerization **6**: 3741
  space heating **2**: ***1234–9***
Temperature measurement **7**: ***4848–51***, ***4851–6***, ***4856–61***
  internal organs **7**: 4855
  medical thermography **7**: 4869
  natural gas technology **3**: 1962
  optical fiber sensors **5**: 3459
  radiation effects **7**: 4864
  see also Thermometers
Temperature scale **7**: 4849
Tempered distributions
  Fourier transforms **3**: 1725
Tensile strength
  biological materials **1**: 482
Terrestrial atmosphere
  carbon cycle modelling **1**: 529
  ground albedo mapping **3**: 2046

optical parameters **2**: ***1307–11***; **4**: 2653
see also Atmosphere
Test equipment
aircraft propulsion units
Concorde **1**: 235
Textile industry
industrial robots **4**: 2444
Theorem proving
vagueness **7**: ***4993–5***
Theory of the firm **5**: 2975; **7**: ***4865***
Thermal comfort **7**: ***4865–6***
air conditioning control systems **1**: 152
Thermal imaging *see* Infrared imaging
Thermal insulation
clo units **1**: 605
tog units **1**: 605
Thermal noise **2**: 1403
Thermal sensation **7**: ***4871–2***
Thermal variables measurement
net radiometers **5**: 3260
Thermistors **7**: 4859
medical thermography **7**: 4869
Thermocouples **7**: 4850
detectors
spectroscopy **1**: 9
location problems **2**: 801
process control
history **6**: 3882
thermometers **7**: 4857
types **7**: 4858
Thermodynamic properties **7**: 4898
Thermodynamics
fugacity **7**: 4898
nonequilibrium thermodynamics
ecosystems **2**: 1330
self-organizing systems **6**: 4167
Thermoelectric devices
domestic gas equipment **2**: 1212
industrial gas equipment **4**: 2355
Thermography *see* Infrared imaging
Thermometers **7**: 4849
atmospheric temperature measurement **5**: 2972
clinical thermometers
radiation devices **7**: 4854
single-use devices **7**: 4852
zero-heat-flow devices **7**: 4853
conduction errors **7**: 4861
expansion thermometers **7**: 4856
gas thermometers **1**: 9
globe thermometers **3**: ***2016–7***
installation **7**: ***4861–5***
Kata thermometers **3**: 2017
optical fiber interferometric sensors **5**: 3469
resistance thermometers **7**: 4859
thermocouples **7**: 4857
types **7**: ***4856–61***
vapor pressure **7**: 4856
see also Temperature measurement
Thermoregulation *see* Temperature control
Thermostats
anticipating control **1**: ***290***
control **7**: 5040
multistage thermostats **7**: ***4881***
Thickness control
rolling mills **6**: 4141
Thinning transformations **7**: ***4881–3***
Thiocarbamate herbicides **3**: 2145
Thiocyanates
ion-selective electrodes
water pollution **4**: 2607
Thiolcarbamate herbicides **3**: 2145
Thirst
feedback control **3**: ***2209–13***
hunger **3**: 2213
Thixotropy **7**: 5056
negative thixotropy **7**: 5056
Threat alert collision avoidance system **1**: 604
Three-dimensional vision
computers
distance measurement **1**: ***755–9***
industrial robots **4**: ***2484–7***
modelling **1**: ***343–7***
see also Industrial vision systems
Three-level structure
dynamic multistage processes **2**: 1275
Three-term control **6**: ***3699–701***
airship automatic pilot **1**: 256
boilers **7**: 4575
consistency control **2**: 778
digital control **6**: ***3701–3***
direct digital control **2**: 1073
pH control **6**: 3681, 3687
process control **5**: 3229
robust controller design **6**: 4131
self-tuning control
pole placement **6**: ***4211–3***
self-tuning digital control **6**: 3702
Three-way valves **2**: 834
Threshold systems
combinatiorial optimization **1**: 624
Thresholding
gray-scale image sensing **4**: 2480
Thrust
aircraft engine control systems **1**: 220
Thyristor applications
dc motors **5**: 3093
dc transmission networks **6**: 3821
lighting control **4**: 2772
see also Choppers (circuits)
Thyroid-stimulating hormone
secretion
cusp catastrophe **1**: 432
Tidal maps **7**: 4884
Tidal power generation **2**: 1457
Tides
modelling **7**: ***4883–8***
time series **7**: ***4889–92***

Time
  ecological timescales **7**: 5090
  measurement
    frequency measurement **3**: 1741
    microprocessor chips **5**: 2986
  simulation modelling **6**: 4368
    nonanticipatory time **6**: 4354
    transition time **6**: 4354
  static time **6**: 4354
Time-advance methods
  simulation **6**: ***4414–5***
Time decomposition *see* Decomposition
Time-delay systems
  Smith predictors **7**: ***4457–62***
    optimal control **7**: 4461
Time-division multiplexing *see* Multiplexing
Time-domain analysis
  identification **4**: 2241, ***2314–9***
    least squares method **4**: 2278
  nonlinear systems
    CAD **5**: 3381
Time-flow mechanism
  simulation **6**: 4414
    activity-based modelling **6**: 4333
    modelling formalism **6**: 4365
Time horizons **7**: ***4892***
Time-optimal control **5**: 3513
  optimal feedback **5**: ***3514–5***
Time random multiplexing *see* Multiplexing
Time series
  analysis
    water demand forecasting **7**: 5154
  environmental-data smoothing **2**: 1492
  forecasting theory **7**: 4804
  history **4**: 2271
  load forecasting models **6**: 3832
  random variate generation **6**: 3949
  road traffic **7**: 4928
  stationary time series **7**: ***4556–66***
  tides **7**: ***4889–92***
Time-sharing systems **3**: 2176
Timoshenko equations
  linear hyperbolic systems **4**: 2800
Titanium dioxide ovens
  adaptive control **1**: 50; **7**: ***4892–4***
Tobacco industry
  humidity control **3**: 2207
Tog units **1**: 605
Tomography
  dentistry **2**: 973
  emission computerized axial tomography **6**: 3939
  radiography
    x rays **7**: 5182
Tonic encoding
  brains as fuzzy systems **1**: 506
Top-down methods
  design
    interpretive structure modelling **4**: 2576
  economic modelling **2**: 1340
  hepatobiliary kinetics modelling **3**: 2140
  problem solving **7**: ***4894***
  simulation methodology **6**: ***4319–22***
Topical magnetic resonance
  biology **1**: 13
Topology
  algebraic topology
    homotopy **3**: ***2168–9***
  closed sets **1**: ***605***
  compact sets **1**: ***665***
  finite topology
    minimax techniques **5**: 2997
  Q-analysis **6**: 3919
  semicontinuous functions **6**: ***4216***
Total organic carbon
  water pollution indices **7**: ***5115–6***
Total oxygen demand
  water pollution indices **7**: ***5116–7***
Total relations **7**: ***4895***
Tower controller
  air traffic control **1**: 198
Toxic substances
  classification **1**: 143
  deleterious effects **1**: 143
  dispersion **4**: ***2778–82***
  ecology **2**: ***1322–3***
  environment
    aluminum compounds **1**: ***262–8***
  fate modelling **3**: 1600
  soil **1**: ***143–8***
  transformation **1**: 143
Trace atmospheric gas analysis **3**: 1938
Trace elements
  water pollution
    concentration **7**: ***5123–7***
  see also Trace metals
Trace metals
  activation analysis **1**: 18–21
Trace organic compounds
  water pollution **7**: ***5127–9***
Tracers
  compartment models **2**: 1343
    dynamics **2**: 1343
  ketone body metabolism **4**: 2628
  metabolic clearance rate
    glucose **7**: 4903
  metabolic dynamics **7**: ***4901–5***
  turnover rates
    glucose **7**: 4902
  velocity measurement
    two-phase flow **3**: 1687
Tracking
  adaptive algorithms **1**: 78
  complex systems control **1**: ***675–81***
    multimodel control **1**: 679
    reference system **1**: 677
  input signals
    multivariable control systems **5**: 3192
  Kalman filters **4**: 2621

linear systems
decentralized feedback **4**: 2670
minimum phase systems
adaptive control **5**: 2998
multivariable control system synthesis **4**: 2813
sampling
digital control **2**: 1070
Tracking systems
laser position measurement **6**: ***3762–5***
Tractors
warehouse automation **7**: 5088
Trade-offs **7**: ***4906***
surrogate worth trade-off method **7**: 4703
trade-off functions **7**: 4705
Traffic
elevators **2**: 1425; **7**: 5050
noise measurement **2**: 1524
transportation modelling **7**: 4945
see also Air traffic, Rail traffic, Road traffic
Traffic alert and collision avoidance system **1**: 201
Traffic control **6**: 4077, ***4081–2***
bus routes **1**: ***514–7***
computer implementation **3**: 1740
control centers **1**: 197
diversion control **6**: ***4079–81***
freeways **7**: 4908, ***4910–3***
automatic incident detection **3**: ***1734–6***
modelling **3**: 1736
multilayer control **3**: ***1736–41***
ramp assignment **7**: ***4913–5***
ramp metering **7**: ***4915–7***
metros **1**: ***352–60, 365–7***
automatic line supervision **1**: ***347–52***
multilayer control
freeway traffic **3**: 1739
pretimed control
ramp metering **7**: 4916
radar
ports **7**: 5051
responsive control
ramp metering **7**: 4916
road traffic **7**: 4917
state estimation **3**: 1738
strategies
modelling **7**: 4912
Traffic lights
bus priority **1**: 514
freeway ramp metering **7**: 4915
modelling **7**: 4924
Traffic recording **6**: ***4075–9***
road traffic **7**: ***4917–20***
Training
simulation **6**: ***4415–7***
operations research **4**: 2352
Trains *see* Railroads, Rail traffic, Metros
Trajectories
aerospace control **3**: 2081
generalized trajectories
irregular optimal feedback **4**: ***2608***
industrial robots
programming **4**: 2347
random processes
dynamics **1**: 565
space vehicles
optimization **7**: ***4481–90***
Transactions
database management systems **2**: 881
distributed database systems **2**: 1156, 1161, 1170
Transactions tables **7**: ***4935***
Transcription
proteins **1**: 557
Transducers
electric generators **2**: 1376
industrial vision systems **4**: 2474
instrumentation **4**: 2529; **5**: 2985
power system computer control **6**: 3839
Transfer functions **1**: 14–15
approximate realization theory **1**: 295
bilinear systems **5**: 3268
control system analysis computing **4**: ***2798***
estimation **4**: 2268
frequency response techniques **3**: 1748
Laplace transforms **4**: 2665
measurement **6**: 3777
modelling **2**: 1289
multivariable control systems
inverse closed-loop transfer function **4**: 2843
poles and zeros **4**: 2830
Transform domain methods
realization theory **6**: ***3971***
Transformation matrices
aircraft dynamics **1**: 213–4
Transformations
state-space transformations **7**: ***4540–7***
Transformers
electric current measurement **2**: 1373
models **6**: 3852
power systems **6**: 3808
Transforms **4**: 2265
Transient analysis
aggregation **1**: 128
gas distribution systems **3**: 1948, 1976
gas transmission networks **3**: 1986, ***1990–2***
history **3**: 1745
identification **4**: 2241
power systems **6**: ***3802–6***
water supply systems **7**: ***5155–6, 5156–61***
Transient response **7**: ***4935***
identification **4**: ***2320–5***
Transients
power systems **6**: ***3802–6***
water supply **7**: ***5156–61***
suppression and control **7**: ***5155–6***
Transinformation **4**: 2544
applications **4**: 2547
nonlinear coupling **4**: 2543

Transition probability (systems) **7**: ***4935***
Transitivity **7**: ***4935***
  graphs **7**: 4666
  relations **7**: ***4935***
Translation
  ecological modelling **2**: 1331
  proteins **1**: 557
Transmission
  ground albedo mapping **3**: 2046
Transmission electron microscopy **7**: ***4936–40***
  particle-size measurement **7**: 4940
Transmission networks
  dc transmission networks **6**: 3821
  load flow calculations
    steady-state modelling **6**: 3850
  natural gas networks
    supervision and control **3**: ***1985–9***
  nodal equations **6**: 3852
Transport equations *see* Boltzmann equations
Transport processes
  fugacity **7**: 4898
  linear Boltzmann equations **7**: 4940
  mixing models **4**: ***2782–9***
  models
    radioisotopes **6**: 3933
Transportation
  energy consumption
    aircraft **3**: ***1749–56***
    road vehicles **3**: ***1756–61***
    ships **3**: ***1761–6***
  modelling **7**: ***4944–7***
  planning **7**: 4944
  traffic zones **7**: 4945
  urban transportation
    new modes **1**: 334
  vertical transportation
    history **7**: ***5047–51***
Transposed systems
  hereditary systems **3**: 2150
Transstockers
  warehouse automation **7**: 5088
Transversality **7**: ***4948–9***
Travelling-wave tubes
  amplifiers
    satellite links **2**: 1052
Trees (mathematics) **7**: 4802, ***4949–52***
  decision trees
    classification algorithms **6**: 4032
  fault trees **3**: ***1601***
  game theory **1**: 682
  graphs **7**: 4664
  hierarchical inference **4**: 2335
  objectives trees **5**: ***3443***
Triangular covariance factorization
    discrete-time systems innovations **2**: 1101
Trophic structures
  lakes **7**: ***4952–3***
  modelling **7**: ***4954–6***
Trotter formula
  Hamilton–Jacobi–Bellman equation **3**: 2070
Trucks
  warehouse automation **7**: 5088
Tryptophan
  operon
    modelling **1**: 426
Tumors
  tissue dielectric properties **1**: 449
  treatment
    simulation **7**: ***4956–8***
Tuning
  HVAC system controllers **3**: ***2108–13***
Turbidity
  fermentation measurement **3**: 1613
  optical fiber sensors **5**: 3463
Turbine meters **3**: 1680
Turbines
  aircraft engine control systems **1**: 220
  helicopter engine governors **3**: 2137
Turbofans **1**: 236
  variable compressor geometry **1**: 230
Turbogenerators **2**: 1382
Turbojets **1**: 236
Turboshafts **1**: 236
  fixed shaft engine **1**: 236
  free turbine engine **1**: 237
Turbulence
  nonlinear observability **5**: 3341
  see also Atmospheric turbulence
Turing machines **1**: 334; **2**: 871; **7**: ***4958***
  computability and decidability **1**: 715
  computational complexity **1**: 723
  halting problems **2**: 935
  predicate **1**: 717
Two-dimensional random pattern
  M-sequences **5**: 2955
Two-filter smoothing algorithms **7**: ***4466–8***
Two-level structure
  dynamic multistage processes **2**: 1274
Two-phase flow
  flow measurement **3**: ***1685–93***
  velocity measurement **3**: 1686
Two-term control
  boilers **7**: 4575
  pH control **6**: 3690
Two-timescale systems
  approximate models **7**: 4434
  discrete-time systems **7**: 4433

U

U-uncertainty **7**: ***4959***
Ultrametricity **3**: 1836
  fuzzy partitions **3**: 1835–6
Ultrasonic devices
  Doppler flowmeters **3**: 1687
  road traffic detectors **7**: 4918
Ultrasonic materials testing
  industrial robots
    aerospace industry **4**: 2452
Ultrasonics
  acoustic microscopy **7**: 4962
  biological measurement
    biological tissues **7**: 4960
  concentration measurement
    two-phase flow **3**: 1691
  distance measurement **4**: 2486
  Doppler methods **7**: 4962
  level measurement applications **4**: 2741
  lighting control **4**: 2773
  medical applications
    safety **7**: 4964
  pulse–echo methods **7**: 4961
  theory **7**: 4959
  ultrasonic imaging
    medical applications **7**: 4959–64
    transmission methods **7**: 4961
Ultraviolet detectors
  liquid chromatography **4**: 2888
Ultraviolet spectroscopy
  inorganic nitrogen compounds
    water pollution **7**: 5118
  ozone
    air pollution **1**: 183
Uncertainty
  analysis **6**: 4228
  ecosystems risk analysis **2**: 1539
  information theory **6**: 3766
  instruments **2**: ***1553–9***
  measurements **2**: ***1553–9***
  measures
    fuzzy set theory **4**: 2294
  modelling
    lakes **4**: ***2650–1***
  optimization **5**: ***3551–4***
  possibilistic uncertainty **6**: 3988
  pragmatic information **6**: 3858
  specification **2**: 1557
  U-uncertainty **7**: ***4959***
Unconstrained optimization *see* Optimization
Underrelaxation **1**: 401
Undersealing
  automobiles
    industrial robots **4**: 2416
Underwater telemetry **7**: ***4842–8***
Universal inputs **7**: ***4975–6***
Universal unfolding
  catastrophe theory **1**: ***547–50,*** 548
Universality
  observability **5**: 3339
Unobservable subspace
  chained aggregation **1**: 134
Updating
  model updating
    model reference adaptive control systems **5**: ***3070–1***
Uracil herbicides **3**: 2145
  ecosystems
    agricultural soils **3**: ***2143–7***
Uranium
  enriched uranium
    nuclear power **5**: 3414
  soil **1**: 147
  toxic substances **1**: 145
Urban dynamics
  land use **7**: 4495
  modelling **7**: 4756
Urea
  herbicides **3**: 2145
  water
    concentration **7**: 5122
Uremia
  artifical kidneys **1**: 310
Ureter **7**: 4976
  bioelectric phenomena
    modelling **7**: ***4976–80***
User manuals *see* Program and system documentation
Utility programs **7**: ***4986***
Utility theory **2**: 940; **7**: 4802, ***4981, 4986–90***
  decomposition models **7**: 4989
  expected utility **5**: 3099; **7**: 4988
    decision theory and analysis **3**: 2193
  group decision making **3**: 2056
  microeconomic systems analysis **5**: 2974
  multiattribute utility theory **5**: ***3099–103***; **7**: 4989
  multiobjective decision making **5**: 3150
  ordered value differences **7**: 4987
  Pareto optimality **2**: 1273
  risk aversion **6**: 4071
  subjective expected utility
    decision theory and analysis **3**: 2195
  subjective utility
    decision theory and analysis **3**: 2197
  utility independence **5**: 3100, 3117
  utility measurements **7**: ***4981–5***
  water resource management **7**: 5138
Uzawa algorithm **1**: 403

V

Vacuum pumps
  mass spectrometer components **5**: 2939
Vagueness
  control theory **4**: 2357
  scientific theories **7**: ***4993–5***
Validation **7**: ***4996***
  carbon cycle modelling **1**: 531
  global models **3**: 2015
  hepatobiliary kinetics modelling **3**: 2143
  identified models **4**: 2242, 2303; **7**: ***4996–500***
  information theory **7**: 5000
  management modelling **7**: ***5001–5***
  model structure determination **4**: 2287
  modelling **4**: 2242; **7**: ***5008–10***
    operations research **4**: 2352
    phytoplankton **7**: ***5005–7***
    statistical analysis **7**: ***5015–8***
    systems analysis **6**: 4407
  optimal tests **7**: 4998
  physiological models
    ketone body metabolism **4**: 2629
  radioecological models
    agricultural food chains **6**: 3933
  simulation models **6**: ***4388–92***
    documentation **6**: 4374
    philosophical aspects **7**: ***5011–5***
    regression analysis **5**: 3029
    statistical analysis **7**: ***5011–5***
  simulation output **6**: 4393
  statistical analysis **7**: 4997
  systems analysis **7**: 4769
  workshop dynamic modelling **7**: 5173
  see also Verification
Validity
  simulation **6**: ***4388–92***
  threats to validity **2**: 1561
Value stability **7**: ***4522–3***
Value systems **7**: ***5019***
Valves **2**: ***833–7***
  electrohydraulic servovalves **6**: ***4252–7***
  flow control **3**: 1696
    characteristics **3**: 1697
  gas equipment **2**: 1213, ***837–9***
    low-pressure cut-off valves **2**: 1213; **4**: 2355
    natural gas technology **3**: 1962
    natural gas transmission systems **3**: 1984
  guiding **2**: 834
  noise **2**: 829
  nonreturn valves
    industrial gas equipment **4**: 2356
  positioners **2**: 837
  relays
    gas equipment **2**: 1213
  seating **2**: 834
  shut-off valves
    industrial gas equipment **4**: 2354
  sizing **2**: ***828–32***
    in situ sizing **2**: 830
  solenoid valves
    gas equipment **2**: 1213
  spool valves **6**: 4252
  water supply valves **7**: 5097
    simulation **7**: 5107
    surge protection **7**: 5155
van der Waals forces
  adsorption of nutrients and pollutants
    environment **1**: 6
Vanadium
  soil **1**: 147
  toxic substances **1**: 145
Vanishing viscosity method
  Hamilton–Jacobi–Bellman equation **3**: 2068
Vaporization
  trace elements
    water pollution **7**: 5124
Variable-speed gear
  dc motors **5**: 3092
  induction machines
    numerical control **7**: ***5019–27***
Variable-structure systems
  adaptive control **7**: ***5027–31***
  bilinear equations **1**: 414
Variables **7**: ***5031***
  discrete variables **2**: ***1102***
  endogenous variables **2**: ***1451***
  exogenous variables **2**: ***1568***
  independent variables **4**: ***2343***
  random variables **6**: ***3948***
  state variables **7**: ***4554***
Variance
  dynamic eutrophication models **2**: ***1269–72***
  fuzzy set theory **3**: 1897
Variance-reduction techniques
  stochastic processes
    simulation **7**: ***4654–6***
Variational inequalities **1**: ***523***; **7**: ***5031–6***
  boundary control **7**: 5040
  control **7**: ***5036–41***
  elliptic systems **7**: 5032
    control **7**: 5037
  parabolic systems **6**: 3594
    distributed control **7**: 5039
  quasi-variational inequalities **4**: 2340
Variational techniques **7**: ***5041–6***
  calculus of variations **1**: ***519–29***; **7**: ***5043–5***
  duality methods **7**: 5045
  optimal stopping **5**: 3535
Vasomotor control
  thermoregulation **7**: 4873

Vector fields
  control systems **5**: 3363
Vector maximum problem **5**: 3122
Vector norms
  comparison systems **5**: 3402
  nonlinear systems stability **5**: ***3402–6***
Vector radiant temperature
  physiological temperature control **7**: 4868
Vectorcardiograms
  modelling **1**: 534
Vee-ball valves **2**: 834
Vehicles
  automated guideway transit
    vehicle size **1**: 334
  automobile electronics **1**: 375
Veins (systemic)
  venous flow **1**: 544
Velocity control
  aircraft **1**: 386
  engines
    self-tuning regulators **2**: ***1462–6***
  road traffic variables **7**: 4930
Velocity measurement **7**: ***4509–13***
  road traffic detectors **7**: 4919
  two-phase flow **3**: 1686
    tracers **3**: 1687
Ventilation
  control
    anesthesia **1**: 285
    modelling **6**: 4055
Ventilation systems *see* HVAC systems
Verbal reporting
  process control **7**: ***5047***
Verbalization **7**: ***5047***
Verhulst–Pearl equation
  populations
    logistic growth **6**: 3755
Verification **7**: ***5047***
  hydrologic forecasting **3**: 2233
  modelling
    management **7**: 5001
  radioecological models
    agricultural food chains **6**: 3933
  simulation **6**: ***4402–3***
  see also Validation
Verisimilitude
  scientific theories **7**: 4995
Vestibular system
  caloric **2**: 1587
  eye movement **7**: 5085
  eye movement control **2**: 1586
  eye movement modelling **2**: ***1591–5***
VFR *see* Visual flying rules
Viability theory
  control systems
    regulation **2**: 1028
  decentralization **2**: 1029
  differential inclusions **2**: ***1023–32***
  state-dependent theorem **2**: 1027
  time-dependent theorem **2**: 1027
Vibrating systems
  distributed parameter systems **2**: 1182
Vibration control
  aircraft **1**: 386
  helicopters **3**: 2135
  launchers **4**: 2729
Vibration measurement
  optical fiber sensors **5**: 3458
Videotelephony **5**: 3446
Virtual circuits
  packet switching **6**: 3583
Viscoelasticity **7**: 5056
  coronary hemodynamics **2**: 851
  equations **4**: 2537
  long memory equations **4**: 2539
  nonlinear viscoelasticity
    connective tissue **1**: 477
Viscometers **7**: 5057
  capillary tube **1**: 441
  concentric cylinder **1**: 441
  industrial systems **7**: 5059
  rotational **1**: 441
Viscosity
  blood viscosity **1**: 442
    determination **1**: 441
  control **2**: 777
    food processing industry **2**: 781
  fermentation measurement **3**: 1613
  Hamilton–Jacobi–Bellman equation **7**: 4624
  Isaacs equation **2**: 1014
  units **7**: 5056
Viscosity measurement **7**: ***5056–60***
Viscosity solutions
  Hamilton–Jacobi–Bellman equation **3**: 2066
Visible and ultraviolet spectroscopy
  organic molecules and substances **1**: 10
Vision
  color vision **7**: 5075
  industrial robots
    gray-scale images **4**: ***2479–84***
  nonlinear modelling **7**: ***5065–70***
  visual system function **7**: ***5078–85***
  visual system neural networks
    feedback lateral inhibition **7**: ***5060–5***
  visual system structure **7**: ***5078–85***
  see also Visual perception
Vision systems *see* Industrial vision systems
Visual flying rules **7**: ***5060***
  head-up displays **3**: 2085
Visual perception **7**: 5079
  computational approach **4**: 2329
  electrophysiological studies **4**: 2327
  feedback lateral inhibition **7**: 5060
  image filtering **4**: ***2325–30***
  light/dark adaptation **7**: 5070
  psychophysics **7**: ***5070–7***
  spatial vision
    contrast sensitivity function **7**: 5074

see also Vision
Voltage amplifiers
  specifications **1**: 268
Voltage control
  ac motor stators **5**: 3089
  reactive power control **6**: ***3950–4***
Voltage dividers **2**: 1392
  inductively coupled devices
    electric current measurement **2**: 1374
  voltmeters **2**: 1395
Voltage drift
  amplifiers **1**: 272
Voltage measurement **2**: ***1389–93***
  alternating signals **2**: 1396
  commercial systems **2**: ***1393–8***
  see also Electric variables measurement
Volterra equations
  population modelling **2**: ***1334***
Volterra series
  bilinear equations **1**: 414
  finite Volterra series
    realization **6**: ***3956–61***
  functional expansions **3**: 1776
  nonlinear systems **4**: 2525
    approximation **5**: 3374
Volterra systems
  correlation methods **4**: 2252
Voltmeters
  ac/dc transfer systems **2**: 1398
  differential voltmeters **2**: 1390
  electronic analog meters **2**: 1394
  input impedance **2**: 1394
  interference (signal) **2**: 1397
  passive analog meters **2**: 1393
Volumetric control
  high pressure
    gas equipment **2**: 839
  low pressure
    gas equipment **2**: 838
Voluntary control
  biofeedback **1**: 436
von Neumann's hypotheses
  complexity **1**: 707
Voting **3**: ***2053–60***; **7**: 4807
  voting approaches **3**: 2053
  voting paradox **3**: 2054; **7**: ***5086***

W

Warehouse automation 7: ***5087–9***
Washing machines
  industrial robots 3: ***1597–600***
Wastewater 7: 5155
  biodegradability determination 1: 434
  chemical analysis 7: 5109
Water
  organic nitrogen compounds
    concentration 7: ***5122–3***
  water resources 7: ***5140–3***
  weathering
    minerals 1: 1
Water demand
  forecasting 7: ***5154–5***
Water pollution
  chemical analysis 7: ***5109***
    ion-selective electrodes 4: 2604
  data smoothing 2: 1493
  environmental solution chemistry 1: 3
  fingerprinting of oil 7: ***5110–2***
    active tagging method 7: 5110
    remote sensing method 7: 5111
  gas chromatography 3: 1932
  gas chromatography–mass spectrometry 3: 1938
  inorganic nitrogen compounds
    chemical analysis 7: ***5117–22***
  temperature
    ion-selective electrodes 4: 2605
  toxic substances modelling
    fugacity models 7: 4901
  trace elements
    concentration 7: ***5123–7***
  trace organic compounds 7: ***5127–9***
  see also Water quality, Pollution
Water pollution indices
  chemical oxygen demand 7: ***5114***
  dissolved oxygen 7: ***5114***
  total organic carbon 7: ***5115–6***
  total oxygen demand 7: ***5116–7***
Water pressure
  leak detection 7: 5143
  water supply 7: 5148
Water quality 7: 5098, 5142
  lakes
    input–output models 4: 2649
  modelling 7: ***5129–34***
    ecological systems 1: 257
  monitoring 2: ***1532–5***
  remote sensing 6: ***4041–5***
  trophic structure 7: 4952
  see also Water pollution
Water resources 7: ***5140–3***
  Markov decision processes 7: 5136
  multiattribute utility theory 7: 5137
  multicriteria decision making 7: 5135
  systems engineering 7: ***5135–40***
  waterfront appraisal 7: ***5161–2***
  see also Water supply
Water supply 7: ***5095–8, 5140–3,*** 5154
  chemical analysis 7: 5109
  design
    simulation 7: 5109
  leakage
    leak detection 7: ***5143–7***
    simulation 7: 5108
  load demand
    simulation 7: 5107
  modelling 7: ***5099–102***
  network simulation 7: ***5106–9***
  planning
    simulation 7: 5108
  pumps 7: 5147
    optimization 7: ***5147–9, 5150–3***
  steady-state network analysis 7: ***5102–5***
  system analysis 7: 5151
  transients 7: ***5156–61***
    suppression and control 7: ***5155–6***
  see also Water resources
Water treatment 7: 5142, 5148
  mixing models 4: 2788
Wave equations
  Fourier transforms 3: 1728
  semigroup theory 6: 4222
Wave power generation 2: 1457
Weakly Lipschitz operators 4: 2677
Weapons
  combat aircraft 1: 622
Weather forecasting
  meteorological measurement 5: 2972
Weathering
  minerals 1: 1
Weighing 3: 1706
  direct comparison method 3: 1706
  electronic devices 3: 1706
  force balance technique 3: 1707
  indirect comparison method 3: 1706
  pneumatic devices 3: 1708
Weighted minimum variance problem
  controller 1: 39
  cost function 1: 39
Weighting function
  linear systems
    binary pseudorandom signals 4: 2249
Weighting vector
  fuzzy multicriteria modelling 3: 1829
Weights
  moving weights
    ships 6: 4273
Welding
  industrial vision systems 4: 2470

X

X-ray analysis
  electron probes **2**: 1417
X-ray emission spectra
  proton-induced x-ray emission
    aerosols **1**: 174
X-ray fluorescence analysis
  aerosols **1**: 174
X-ray spectroscopy **7**: 4500
X rays
  radiology **7**: ***5179–86***
Xenon
  fission reactors **5**: 3437
XIMP modelling *see* Workshop dynamic modelling

Y

Yawing
aircraft dynamics
state equations **1**: 216
aircraft motion **1**: 382
autopilots **1**: 390
Yeast
concentration
fermentation control **3**: 1608
fermentation measurement **3**: 1612
oxygen supply **3**: 1613
fermenter control **3**: 1615
Yield
agriculture
chemical stress modelling **1**: 589
Young–Helmholtz theorem
visual systems **7**: 5076

Z

z transforms **7**: 5189–92
  digital control
    algorithm design **2**: 1057
  frequency-response techniques **3**: 1749
  sampled-data systems **6**: 4143
Zakai equation
  nonlinear estimation
    stochastic control **5**: 3312
Zeiger's lemma **1**: 16
Zeros *see* Poles and zeros
Zinc
  soil **1**: 147
  toxic substances **1**: 145
Zirconium
  oxygen sensor **5**: 3573
Zooplankton
  lakes
    modelling **4**: 2650

INFORMATION SOURCES IN SYSTEMS AND CONTROL

Spyros G Tzafestas
National Technical University
Athens, Greece

This article gives a review of the various sources of information in the field of systems and control. The broad sense of the field has been adopted here, as is done throughout the Encyclopedia, and so information sources are included that refer to mathematical, managerial, electrical, electronic, chemical, mechanical, computer, communication, transportation, aerospace, energy, environmental and urban systems, etc.

The available information sources vary in both type and extent. In order for the information to be useful it must be pertinent, new to the user, appropriate, comprehensive and concise. The purpose of looking for information in the available sources may be to learn something new, to extract from the universal knowledge those elements needed in the case at hand, to find data for computations or data for individuals, companies and institutions, to study the evolution of the methodology or technology in a particular field, and so on.

The fundamental steps for obtaining and processing information from various sources are: awareness of the existence of information, obtaining access to it, extraction of relevant pieces of information from the source, filtering out the information in order to keep the useful one, and finally organization, compilation, evaluation, and dissemination of the acquired knowledge. Of course, for an effective, productive and intelligent use of the information sources, the user must have a deep understanding of his or her problem and sufficient experience of how to identify the specific information needed and select the most appropriate source(s) in each case.

1. Classification of Information Sources

Information sources for science and technology have been reviewed, classified and compiled by many authors and editors (Grattidge and Creps 1978, Westbrook 1983). The information transfer process in engineering has been studied by Schuchman (1981), who has observed that engineers have a marked preference for books over journal articles, for personal contacts over printed sources, and for nearby, easily accessible or interpretable sources over the remote or difficult-to-use ones (Westbrook 1986).

The information sources on systems and control presented here follow closely the classification adopted by Westbrook (1986) in his similar article on Materials Information Sources. The differences are mainly due to the different nature of the two fields.

Thus our discussion will be made under the following general headings:

Guides, sourcebooks and services
Encyclopedias and handbooks
Dictionaries and glossaries
Abstract and review sources
Technical journals
International book series and conference proceedings
Sources on people
Other information sources
Continuing education courses

Owing to space limitation it is not possible to provide an exhaustive listing of all the sources in each class. An effort has been made to include sufficient sources to identify and differentiate each category. More sources can be easily located and identified using the material in this article.

2. Guides, Sourcebooks and Services

There is a large number of reference guides, sourcebooks and services for information and documentation related to science and technology, including electrical, computer, system and control engineering. These guides are devoted to various aspects such as literature, world meetings, manufacturers, products, academic and research institutions, industrial research, service bureaux, etc. Some of them of a rather general nature are listed below.

A Bibliography of Computing, 1974. Infotech, Maidenhead, UK
A Guide to New Technical Horizons, IEEE Technical Activity Guide (TAG). IEEE, New York
A Guide to Today's CAD/CAM/CAE Systems. North-Holland, Amsterdam
Abdelmalek N 1981 *Robotics Bibliography 1970–80.* National Research Council, Ottawa

ACM Guide to Computing Literature (annual index). ACM/Kinokumya, Tokyo

Aleksander I (ed.) 1985 *The World Yearbook of Robotics Research and Development*. Kogan Page, London

Automation and Computer Technology for Manufacturing (NTIS-PR-612). US Department of Commerce, Washington, DC

Ball I D L (ed.) 1968 *Industrial Research in Britain*. Harrap, London

Bauly J A, Bauly C B 1981 *World Energy Directory*. Longman, London

Benedetto F D (ed.) 1977 *Institutes and Scientific and Technological Research Items—Italy*. DGCCST-Ministero Degli Affari Esteri (Officio VII), Rome

Bundy A (ed.) 1986 *Catalogue of Artificial Intelligence Tools*. Springer, Berlin

Chen C C 1977 *Scientific and Technical Information Sources*. MIT Press, Cambridge, Massachusetts

Cumulative index 1956–1980: Control Systems Society 25th anniversary index. *IEEE Trans. Autom. Control* **26**(4)

Current Bibliographic Directory of the Arts and Sciences (Annual). Institute for Scientific Information, Philadelphia, Pennsylvania

Directory of Scientific and Technical Associations and Institutes in Israel (Directory No. 2), 1966. National Council for Research and Development, Jerusalem

Directory of Water Pollution Research Laboratories, 1965. Organization for Economic Cooperation and Development/Bureau de Publications, Paris

East European Research Index: Medical, Agriculture, Engineering, 1977. Hodgson, London

Electronic Engineers Master (EEM): Manufacturers' Catalogs, 1968–69. United Technical Publications, New York

Environmental Pollution: A Guide to Current Research, 1971. CCM Information Corporation, New York

Eurodata Foundation Yearbook, 1984. Eurodata Foundation, London

European Research Index: Medical, Agriculture, Engineering, 1977. Hodgson, London

FAO Documentation: Current Bibliography. Food and Agriculture Organization/United Nations, Rome

Foundyller C M (ed.) 1982 *CAD/CAM Computer Graphics: Survey Review and Buyer's Guide*. North-Holland, Amsterdam

Gianniny O A Jr 1982 *Mechanical Engineering Education in America: Its First Century*. ASME Publications, New York

Gough B E (ed.) 1977 *World Environmental Directory: United States and Canada*. Business Publishers, Silver Spring, Maryland

Gough B E (ed.) 1978 *World Environmental Directory: Africa–Asia–Australia–New Zealand–Europe–Middle America–South America*. Business Publishers, Silver Spring, Maryland

Guide Technique de l'Electronique Professionnelle (four languages), 1965. Publicité and Editions Techniques/L'AIE, Paris

Guide to World Science, 1975. Hodgson, London

Harvey A P, Pernet A (eds.) 1981 *European Sources of Scientific and Technical Information* (reference on research). Hodgson, London

Henderson G P, Henderson S P A (eds.) 1980 *Directory of British Associations and Associations in Ireland*. Gale, Detroit, Michigan

Incentives for Industrial Research, Development and Innovation. Kogan Page/EEC Publications Office, London

Index to Scientific Reviews, 1979. Institute for Scientific Information, Philadelphia, Pennsylvania

International Directory of Computer and Information System Services. Europa, London

International Directory of New and Renewable Energy Information Sources and Research Activities, 1982. UNESCO/SERI, Paris

International Scientific Organizations, 1965. Organization for Economic Cooperation and Development/Bureau de Publications, Paris

Kelly's Directory of Manufacturers and Merchants (with accompanying index of services), 1967–68. Kelly's Directories, East Grinstead, UK

Knopman D S (ed.) 1976 *Scientific Research in Israel* (NCRD 4-76). National Council for Research and Development, Jerusalem

L'Annuaire Industriel (I: Fabrications, II: Firmes), 1968–69. Kompass, Paris

McGraw-Hill 1984 *Yearbook of Science and Technology* (updated yearly). McGraw-Hill, New York

MIND—The Meeting Index, SEMT Series. InterDock Corporation, New York

National Historic Mechanical Engineering Landmarks, 1979. ASME Publications, New York

Netherlands Research Guide, 1963. Netherlands Institute for Documentation and Filing, The Hague

OECD Guide to European Sources of Technical Information, 1964. Organization for Economic Cooperation and Development, Paris

OECD Nuclear Safety Research Index, 1982. Organization for Economic Cooperation and Development, Paris

Palmer A M *Research Centers Directory (USA and Canada)*. Gale, Detroit, Michigan

Pernet A (ed.) 1976 *Nuclear Research Index: A Guide to World Nuclear Research*. Hodgson, London

Rhodes A (ed.) 1975 *Industrial Research Laboratories of the United States*. Bowker, New York

Saur K G 1968 *Handbuch der Technischen Dokumentation und Bibliographie*. Verlag Dokumentation, Munich

Shotwell R 1985 *The Ethernet Sourcebook*. North-Holland, Amsterdam

The World of Learning (Institutes and Names), 1975. Europa, London

Wedgwood C G (ed.) *International Electronics Directory '86: The Guide to European Manufacturers and Agents*. North-Holland, Amsterdam

Weisman H 1967 *Technical Correspondence: A Handbook and Reference Source for the Technical Professional*. Wiley, New York

Williams C H 1970 *Guide to European Sources of Technical Information*. Hodgson, London

World Directory of National Science Policy-Making Bodies, 1967. Unipub, New York

World Meetings: Outside United States and Canada. Macmillan, Riverside, New Jersey

World Meetings: United States and Canada. Macmillan, Riverside, New Jersey

Yearbook of International Organizations (UIA 235), 1977. Union of International Associations, Brussels

Zils M (ed.) 1981 *World Guide to Scientific Associations and Learned Societies*, 3rd edn. Saur, Ridgewood, New Jersey

3. Encyclopedias and Handbooks

In general, encyclopedias consist of a large number of articles written by individual authors. They are usually multivolume works having well-defined scopes, and are useful for quickly obtaining background information, or when entering a new scientific area. Most of the handbooks are of the same nature as the encyclopedias in that they consist of collections of individual articles, but as the scope of handbooks is usually narrower, their articles tend to cover topics in greater depth. In the following we give some examples of encyclopedias and handbooks in the electrical, electronic, control and information engineering fields. To facilitate the access to information, encyclopedias and handbooks present their topics in alphabetic sequence and have extensive cross-referenced indexes.

(*a*) *Encyclopedias*

Concise Encyclopedia of Science and Technology, 1984. McGraw-Hill, New York

Considine D M (ed.) 1971 *Encyclopedia of Instrumentation and Control.* McGraw-Hill, New York

Encyclopedia of Electronics and Computers, 1983. McGraw-Hill, New York

Encyclopedia of Engineering, 1983. McGraw-Hill, New York

Encyclopedia of Science and Technology, 1982. McGraw-Hill, New York

Meetham A R, Hudson A 1969 *Encyclopedia of Linguistics Information and Control.* Pergamon, Oxford

Ralston A, Chester L M (eds.) 1976 *Encyclopedia of Computer Science.* Van Nostrand-Reinhold, New York

Rota G-C (ed.) 1979 *Encyclopedia of Mathematics and its Applications.* Addison-Wesley, Reading, Massachusetts

(*b*) *Handbooks*

Abramowitz M, Stegun I A (eds.) 1965 *Handbook of Mathematical Functions.* Dover, New York

Barr A, Feigenbaum E A 1980 *The Handbook of Artificial Intelligence.* Heuristech Press/William Kaufmann, Stanford/Los Altos, California

Baumeister T (ed.) 1978 *Standard Handbook for Mechanical Engineers.* ASME Publications, New York

Cawkell A E (ed.) 1986 *Handbook of Information Technology and Office Systems.* North-Holland, Amsterdam

Chang S S L 1983 *Fundamentals Handbook of Electrical and Computer Engineering.* Wiley, Chichester

Considine D (ed.) 1974 *Process Instruments and Control Handbook.* McGraw-Hill, New York

Dean C, Whitlock Q 1984 *A Handbook of Computer-Based Training.* Kogan Page, London

Fink D G, Beaty H W (eds.) 1978 *Standard Handbook for Electrical Engineers.* McGraw-Hill, New York

Fink D G, Christiansen D (eds.) 1982 *Electronics Engineers Handbook.* McGraw-Hill, New York

Hamilton T D S 1977 *Handbook of Linear Integrated Electronics for Research.* McGraw-Hill, New York

Harper C A 1977 *Handbook of Components for Electronics.* McGraw-Hill, New York

Harper C A 1979 *Handbook of Electronic Systems Design.* McGraw-Hill, New York

Hnate E 1979 *A User's Handbook of D/A and A/D Converters.* Wiley, Chichester

Hunt V D 1983 *Industrial Robotics Handbook.* Industrial Press, New York

Kraus A, Bar-Cohen A 1983 *Thermal Analysis and Control of Electronic Equipment.* Hemisphere, Washington, DC

Koslov B A, Ushakov I A 1970 *Reliability Handbook.* Rinehart and Wilson, New York

Machol R (ed.) 1965 *System Engineering Handbook.* McGraw-Hill, New York

Mardiguian M 1985 *Interference Control in Computers and Microprocessor-Based Equipment*, ICT-DWCI. EMC Handbooks, Gainsville, Virginia

Mark's Standard Handbook for Mechanical Engineers, 1958. McGraw-Hill, New York

Markus J 1980 *Modern Electronic Circuits Reference Manual.* McGraw-Hill, New York

Mathematical Handbook for Scientists and Engineers, 1968. McGraw-Hill, New York

Maynard D (ed.) 1971 *Industrial Engineering Handbook.* McGraw-Hill, New York

Moschytz G S 1981 *Active Filter Design Handbook.* Wiley, Chichester

Singh M G, Titli A (eds.) 1979 *Handbook of Large Scale Systems Engineering Applications.* North-Holland, Amsterdam

Smeaton R (ed.) 1977 *Switchgear and Control Handbook.* McGraw-Hill, New York

Sydenham P H 1983 *Handbook of Measurement Science.* Wiley, Chichester

Truxal J 1959 *Control Engineers Handbook.* McGraw-Hill, New York

4. Dictionaries and Glossaries

Dictionaries and glossaries provide in alphabetical sequence spelling, definitions and descriptions of terms. Some bilingual and multilingual dictionaries exist in the electrical, electronic and systems engineering field. The following list provides some examples.

Alford M H T, Alford V L 1970 *Russian–English Scientific and Technical Dictionary.* Pergamon, Oxford

Amos S W 1981 *Dictionary of Electronics.* Butterworth, London

Belle I 1977 *Dictionnaire Technique General Anglais–Français.* Dunod, Paris

Broadbent D T, Masubuchi M 1981 *Multilingual Glossary of Automatic Control Technology.* Pergamon, Oxford

Burger E, Schuppe W 1970 *Four Language Technical Dictionary of Data Processing, Computers and Office Machines.* Pergamon, Oxford

Chernukhin A E 1977 *English–Russian Polytechnical Dictionary.* Pergamon, Oxford

Dictionary of Scientific and Technical Terms, 1983. McGraw-Hill, New York

IEC Multilingual Dictionary of Electricity, 1980. International Electrotechnical Commission, Geneva

IEEE Standard Dictionary of Electrical and Electronics Terms, 1984. IEEE, New York

Iyanaga S, Kawada Y 1977 *Encyclopedic Dictionary of Mathematics.* MIT Press, Cambridge, Massachusetts

Kuznetzov B 1981 *Russian–English Polytechnical Dictionary.* Pergamon, Oxford

Macura P 1971 *Russian–English Dictionary of Electrotechnology and Allied Sciences.* Wiley–Interscience, New York

Markus J 1978 *Electronics Dictionary.* McGraw-Hill, New York

Meadows A J, Gordon M, Singleton A 1984 *Dictionary of Computing and New Information Technology.* Kogan Page, London

Paenson I 1970 *Systematic Glossary of the Terminology of Statistical Methods.* Pergamon, Oxford

Sneddon I N 1976 *Encyclopaedic Dictionary of Mathematics for Engineers and Applied Scientists.* Pergamon, Oxford

Tver D, Boltz R (eds.) 1983 *Robotics Sourcebook and Dictionary.* Industrial Press, New York

Wenrich P 1978 *Anglo-American and German Abbreviations in Science and Technology.* Verlag Documentation, Munich

5. Abstract and Review Sources

The activity of paper and book abstracting and reviewing services of scientific societies and publishers in the systems, computer and control fields is currently very intensive and is continually increasing. Each year many new abstract and review sources are launched in the technical literature. An exhaustive listing of all these sources is impossible in an article of this extent, but an interesting sample of the more significant ones is given below.

Applied Mechanics Reviews
Artificial Intelligence
Automation Express
Computer Abstracts
Computer and Control Abstracts
Computing in Electronics and Power
Computing Reviews
Current Papers in Computers and Control
Current Papers in Electrical and Electronic Engineering
Cybernetics Abstracts
Electrical and Electronics Abstracts
Electronic Circuits
Electronics and Communications Abstracts
Engineering Management Review
International Abstracts in Operations Research
International Aerospace Abstracts
Japan Annual Reviews in Electronics, Computers and Telecommunications
Mathematical Reviews
New Literature on Automation
Online Review
Power Systems and Applications
Robotics and Control
Software Engineering
Surveys in Computer Science
Zentralblatt für Mathematik

A good source of information, complementary to standard abstracting and reviewing sources, is formed by the bulk of review papers written by independent authors either by their own volition or for their employers, or sometimes for some contracting agency. According to the 1979 Index to Scientific Reviews of ISI (Westbrook 1986), review articles are estimated to constitute about 3–4% of the total literature, and an average of about 70 references are cited in each review. Although review papers are beneficial, especially for those entering a new field, their appearance is more or less random. As a result of this, many topics are reviewed very frequently while others are not reviewed at all. Thus, this activity needs some kind of coordination, and an Index of Scientific Review Papers would be very useful. Very frequently, review and tutorial papers are included in multiauthored research books published by most publishers.

Examples of books of this type are given below.

Cellier F E (ed.) 1982 *Progress in Modelling and Simulation.* Academic Press, London

Chen K (ed.) 1972 *Urban Dynamics: Extensions and Reflections.* San Francisco Press, San Francisco, California

Mass N J (ed.) 1974 *Readings in Urban Dynamics*, Vol. 1. Wright-Allen, Cambridge, Massachusetts

Saridis G N (ed.) 1985 *Advances in Automation and Robotics.* JAI Press, Greenwich, New York

Schwarz L B (ed.) 1981 *Multi-Level Production/Inventory Control Systems: Theory and Practice, TIMS Studies in the Management Sciences.* North-Holland, Amsterdam

Sinha N K (ed.) 1986 *Microprocessor-Based Control Systems.* Reidel, Dordrecht

Tzafestas S G (ed.) 1982 *Distributed-Parameter Control Systems: Theory and Application.* Pergamon, Oxford

Tzafestas S G (ed.) 1986 *Multidimensional Systems: Techniques and Applications.* Dekker, New York

Zeigler B P, Elzas M S, Klir G J, Ören T I (eds.) 1979 *Methodology in Systems Modelling and Simulation.* North-Holland, Amsterdam

6. Technical Journals

Technical journals constitute the main body of the primary literature on systems and control. Other primary literature sources are the proceedings of countless national and international conferences, technical reports, and dissertations.

To help the reader of the present article a sufficient number of journals are listed under the following headings.

Mathematics for engineering and system computations
System modelling, simulation and computer-aided design
General systems
Systems control and optimization
Information and computer systems
Artificial intelligence and knowledge-based systems
Robotics and manufacturing systems
Electrical and electronic engineering systems
Measurement, signal processing and communication systems
Energy and power systems

6.1 Journals on Mathematics for Engineering and System Computations

Advances in Applied Probability
Applied Numerical Mathematics
Computational Mechanics
Computational Statistics and Data Analysis
Computer Methods in Applied Mechanics and Engineering
Computers and Chemical Engineering
Computers and Fluids
Computers in Mechanical Engineering
Discrete Applied Mathematics
Engineering Computations Journal
Finite Elements in Analysis and Design
Interfaces in Computing
International Journal for Computation and Mathematics in Electrical and Electronic Engineering (COMPEL)
International Journal for Numerical Methods in Engineering
Journal of Applied Probability
Journal of Computational and Applied Mathematics
Journal of Computational Chemistry
Journal of Computational Mathematics
Journal of Computational Physics
Mathematical Engineering in Industry
Mathematical Programming Study
Mathematics of Computation
Mathematics of Operations Research
Moscow University Computational Mathematics and Cybernetics
SIAM Journal of Applied Mathematics
SIAM Journal of Numerical Analysis
SIAM Journal of Scientific and Statistical Computation
Stochastic Process Applications

6.2 Journals on System Modelling, Simulation and Computer-Aided Design

Applied Mathematical Modelling
Computer-Aided Control Engineering
Computer-Aided Design
Computer-Aided Design of Microwave Circuits
Der Versuchs- und Forschungsingenieur (Simulation Journal)
IEEE Transactions on Computer-Aided Design of Integrated Circuits and Systems
International Journal of Modelling and Simulation
Journal of Computer-Aided Engineering
Mathematics and Computers in Simulation
Simulation
Simulation and Games
Systems Analysis Modelling Simulation

6.3 Journals on General Systems

(*a*) *Circuits and systems*

Circuits, Systems and Signal Processing
IEEE Circuits and Systems Magazine
IEEE Transactions on Circuits and Systems
International Journal of Circuit Theory and Applications
Networks
Proceedings of the IEE, Part G: Electronic Circuits and Systems

(*b*) *Cybernetics, management and economic systems*

American Economic Review
American Journal of Economics and Sociology
Behavioral Science
Cybernetica
Cybernetics and Systems
Decision Support Systems
Econometrica
Engineering Cybernetics
Engineering Management Review
European Journal of Operational Research
Human Systems Management
IEEE Transactions on Systems, Man and Cybernetics
Interfaces
International Journal of Production Research
Journal of Economics
Journal of Marketing
Journal of Mathematical Economics
Journal of Operations Research of Japan
Journal of the Operational Research Society
Kybernetes
Kybernetik
Management Science
Operations Research
Operations Research Letters
Operations Research Spektrum
Proceedings of the IEE, Part A: Physical Sciences, Measurement and Instrumentation, Management, Education and Systems Engineering
Review of Economic Studies
Revue Belge de Statistique, d'Informatique et de Recherche Operationelle
Sloan Management Review
Zeitschrift für Operations Research

(*c*) *Industrial and systems engineering*

AIAA Journal
Chemical Engineering Progress
Chemical Engineering Science
Fuzzy Sets and Systems
IEEE Transactions on Aerospace and Electronic Systems
IEEE Transactions on Industry Applications
IEEE Transactions on Reliability
Industrial Engineering and Chemistry
Industrial Research and Development
International Journal of Engineering Science
International Journal of Mechanical Sciences
International Journal of Solid Structures
Japanese Railway Engineering
Journal of Aerospace Sciences
Journal of Aircraft
Journal of Applied Systems Analysis
Journal of Basic Engineering
Journal of Industrial Engineering
Journal of the Franklin Institute
Microelectronics Reliability
Nuclear Science and Engineering
Nuclear Technology
Production Engineer
Reliability Engineering
Siemens Research and Development
Soviet Journal of Automation and Information Sciences
Systems Research
Systems Science
Traffic Engineering
Traffic Engineering and Control
Transportation Research
Transportation Research Record
Transportation Science

(*d*) *Environmental, agricultural, urban and related systems*

Agronomical Journal
Die Naturwissen Schaften
Environment and Planning
Environmental Science and Technology
Experimental Agriculture
International Regional Science Review
Journal of Agricultural Science
Journal of Applied Ecology
Journal of Applied Meteorology
Journal of Atmospheric Science
Journal of Environmental Quality
Journal of Fermentation Technology
Journal of Hydraulics
Journal of Hydrology
Policy Sciences
Population Studies
Quarterly Journal of the Royal Meteorological Society
Regional Science and Urban Economics
Regional Studies
Science
Society of Petroleum Engineering Journal
Town Planning Review
Transactions of the British Geographers
Water Resource Systems
Water Resources Bulletin
Water Resources Research

(*e*) *Biomedical and bioengineering systems*

Acta Radiologica
Biological Cybernetics
Biomedical Research
Biometrica
Biometrics
Biophysics and Bioengineering
Biotechnology and Bioengineering
Computers in Biomedical Research
IEEE Transactions on Biomedical Engineering
Journal of Mathematical Biology
Journal of Theoretical Biology
Medical and Biological Engineering and Computing
Medical Physics
Neuroscience
Radiology
Theoretical Population Biology

6.4 Journals on Systems Control and Optimization

Automatic Control & Computer Science (Translation)
Automatica
Automation and Remote Control (Translation of *Automatika i telemekhanika*)
Automatisierungstechnik
Automatisierungstechnische Praxis
Automatisme
Control and Computers
Control & Instrumentation
Control Engineering
Control Systems Magazine
Control: Theory and Advanced Technology
Decision Informatique
Foundations of Control Engineering
IEEE Transactions on Automatic Control
IMA Journal of Mathematical Control
Industrial and Process Control Magazine
Information and Control
International Journal of Control
International Journal of Systems Science
Journal of Dynamic Systems, Measurement and Control
Journal of Economic Dynamics and Control
Journal of Guidance Control Dynamics
Journal of Optimization Theory and Applications
Large Scale Systems: Theory and Applications
Measurement and Control
Optimal Control Applications and Methods (OCAM)
Proceedings of the IEE, Part D: Control Theory and Applications
Ricerche di Automatica
SIAM Journal on Control and Optimization
Systems and Control Letters
Wissenschaftlich-technische Zeitschrift für die Automatisierungstechnik

6.5 Journals on Information and Computer Systems

ACM Computing Surveys
ACM Transactions on Data Base Systems
Acta Informatica
Algorithmica
Communications of the ACM
Computer
Computer Age
Computer Bulletin
Computer Compacts
Computer Design
Computer Economic Reports
Computer Graphics and Applications Magazine
Computer Journal
Computer Magazine
Computer Speech and Language
Computer Survey
Computers and Electrical Engineering
Computers and Biomedical Research
Computers in Industry
Computing
Data Dynamics
Data Processing
Data Processing Digest
Database
Datamation
Design and Test of Computers Magazine
Distributed Computing
Education and Computing
Future Generations Computer Systems
IBM Journal of Research and Development
IBM System Journal
IEEE Transactions on Computers
Infomac
Information and Management
Information and Word Processing Report
Information Processing and Management
Information Processing Letters
Information Sciences
Informationstechnik
Infosystems
Japan Data Processing Letter
Journal of Computer and Information Sciences
Journal of Computer and System Sciences
Journal of Information Science
Journal of Information Technology
Journal of Parallel and Distributed Computing
Journal of Symbolic Logic

Journal of Systems and Software
Journal of the ACM
Lettre de l'Industrie Informatique
Micro & Personal Computer
Microcomputer Applications
Microcomputer Review
MicroMagazine
Microprocessing and Microprogramming
Microprocessors & Microsystems
Mini & Micros
Mini Micro Software
Mini Micro Systems
Mitteilungen aus dem Frauenhoferinstitut für Informations- und Datenverabeitung
New Generation Computing
Online
Online Libraries & Microcomputers
Parallel Computing
P.C. & Compatibles
Performance Evaluation
Pergamon Infotech: Computer State of the Art Reports
Peripherals Digest
Personal Computer Industry Views
Personal Computerworld
Proceedings of the IEE, Part E: Computers and Digital Techniques
RAIRO Computer Science
SIGMICRO Newsletter
Technique et Science Informatique

6.6 Journals on Artificial Intelligence and Knowledge-Based Systems

AI Magazine
Artificial Intelligence
Data and Knowledge Engineering
Expert Magazine
Expert Systems
IEEE Transactions on Pattern Analysis and Machine Intelligence
International Journal of Expert Systems Research and Applications
International Journal of Knowledge Engineering
International Journal of Man–Machine Studies
Journal of Algorithms
Journal of Artificial Intelligence
Journal of Automated Reasoning
Journal of Computer and System Sciences
Journal of Intelligent and Robotic Systems: Theory and Applications
SIGART Newsletter

6.7 Journals on Robotics and Manufacturing Systems

Assembly Automation
FMS Magazine
IEEE Transactions on Components, Hybrids and Manufacturing Technology
Industrial Robot Journal
International Journal of Robotics and Automation
International Journal of Robotics Research
Journal of Manufacturing Systems
Journal of Robotic Systems
Journal of Robotics and Automation
Lettre de la Robotique
Manufacturing Engineering
Precision Engineering
Robotics
Robotics Age
Robotics and Computer Integrated Manufacturing
Robotics Today
Robots
Sensor Review

6.8 Journals on Electrical and Electronic Engineering Systems

Archiv für Elektrotechnik
Brown Boveri Technik
Canadian Electrical Engineering Journal
Computer & Elektronik
Electrical Engineering in Japan
Electronic Components and Applications
Electronic Design
Electronic Engineering
Electronic Modelling
Electronique Industrielle
Electronics Letters
Electronics Week
Elektronische Rechenanlagen
Electrotechnik und Maschinenbau
GTE Automatic Electrical Journal
IEEE Transactions on Geoscience Electronics
IEEE Transactions on Industrial Electronics
Industrie-Elektronik
International Journal of Electronics
Journal of Electrical and Electronics Engineering
Proceedings of the IEEE
Proceedings of the IREE
Radio Electronic Engineering
Revue Générale de l'Electricité
Soviet Journal of Communications Technology and Electronics

6.9 Journals on Measurement, Signal Processing and Communication Systems

Annales des Télécommunications
Bell System Technical Journal
Communications Magazine
Computer Vision Graphics and Image Processing
IEEE Transactions on Acoustics, Speech and Signal Processing
IEEE Transactions on Communications
IEEE Transactions on Electroacoustics
IEEE Transactions on Electromagnetic Compatibility
IEEE Transactions on Information Theory
IEEE Transactions on Instrumentation and Measurement
IEEE Transactions on Medical Imaging
Instrumentation Technology
Instruments and Experimental Techniques (Translation of *Pribori i Technika Eksperimenta*)
International Journal of Remote Sensing
ISA Transactions
Journal on Selected Areas in Communications
Network Magazine
Proceedings of the IEE, Part F: Communications, Radar and Signal Processing
Radiological Imaging
Signal
Signal Processing
Speech Communication

Transactions of the Society of Instruments and Control Engineering

6.10 Journals on Energy and Power Systems

Electric Machines and Power Systems
Electric Power Systems Research
Electronics and Power
Energie und Automation
Energy
IEEE Transactions on Energy Conversion
IEEE Transactions on Power Apparatus and Systems
IEEE Transactions on Power Delivery
IEEE Transactions on Power Electronics
IEEE Transactions on Power Systems
International Journal of Energy Systems
International Journal of Heat and Mass Transfer
Journal of Engineering for Power
Journal of Heat Transfer
Journal of Nuclear Energy
Journal of Power Sources
Power Reactor Technology
Proceedings of the IEE, Part B: Electric Power Applications
Proceedings of the IEE, Part C: Generation, Transmission and Distribution
Solar Energy
Zeitschrift für Elektrische Energietechnik

7. International Book Series and Conference Proceedings

International book series and conference proceedings are published by various publishers and scientific societies. Since it is impossible to list all of them here, only a small subset of book series on systems and control, along with the conference proceedings of the main existing societies, are listed.

7.1 Book Series on Systems and Control

Academic Press Books in Electrical Engineering
Advances in Automation and Robotics
Alternative Energy Sources
Annals of Discrete Mathematics
Annual Review of Information Science and Technology
Applied Information Technology Book Series
Benchmark Papers in Electrical Engineering and Computer Science
Computer Science and Applied Mathematics Series
Control and Dynamic Systems Book Series (formerly *Advances in Control Systems*)
Electrical Engineering Book Series
Holt-Saunders International Editions: Engineering-Computer Science
IEEE Press Selected Reprint Series
International Series on Computer Science
International Series on Systems and Control
International Series on Systems and Control Engineering
Lecture Notes in Computational Mathematics
Lecture Notes in Computer Science
Lecture Notes in Control and Information Sciences
Lecture Notes in Economics and Mathematical Systems
Lecture Notes in Mathematics
Lecture Notes in Statistics
Marcel Dekker Electrical Engineering and Electronics Series
Marcel Dekker Manufacturing Engineering and Material Processing Series
Mathematical Programming Studies Series
Mathematical Systems in Economics Book Series
Mathematics and its Applications Series
Methods of Operations Research Book Series
Microelectronics and Signal Processing Series
Microprocessor-Based Systems Engineering Series
MIT Press Series in Artificial Intelligence
North-Holland Systems and Control Series
North-Holland Systems Science and Engineering Series
Notes and Reports in Computer Science and Applied Mathematics Series
Springer Book Series on Communications and Control Engineering
Springer Book Series on Information Sciences

7.2 Regular Conference Proceedings

ACC Proceedings
ACTA IMEKO
AFCET Proceedings Publications
AFIPS Conferences Proceedings Series
ASME Symposia Proceedings Series
EURASIP Conference Proceedings Series
IAEA Proceedings Series
IASTED Symposia Series
IEEE/CSS CDC Proceedings Series
IEEE Symposia Proceedings Series
IFAC Publications
IFIP Publications
IMA Conference Proceedings Series
IMACS Symposia Proceedings Series
SCS Symposia Proceedings Series

8. Sources on People

Information on individual scientists and on "who knows what" can be found in scientific directories and biographical volumes. This information may vary from the simple form of citation index or current address, degrees and affiliation, to a full biographical note of a shorter or longer type. It is useful to mention that currently many new directories are under preparation, either in printed form or in on-line computer form.

Some representative sources on people are listed below.

Community Leaders of the World. American Biographical Institute, Raleigh, North Carolina
Current Contents Address Directory, 1985 (Journal Lists, Author Index, Organization Index). Institute for Scientific Information, Philadelphia, Pennsylvania
Dictionary of International Biography. International Biographical Centre, Cambridge, UK
Dove J C (ed.) 1985 *Who's Who in European Institutions, Organizations and Enterprises*. Who's Who Sutter's International Red Series/Who's Who in Italy, Milan
International Book of Honor. American Biographical Institute, Raleigh, North Carolina
International Directory of Distinguished Leadership. American Biographical Institute, Raleigh, North Carolina

International Directory of Research and Development Scientists (annual). Institute for Scientific Information, Philadelphia, Pennsylvania
International Engineering Directory (annual). Consulting Engineering Council of the US, Washington, DC
International Register of Biographies. Universal Intelligence Data Bank of America, Independence, Missouri
International Who's Who in Engineering. International Biographical Centre, Cambridge, UK
ISI's Who is Publishing in Science (annual). Institute for Scientific Information, Philadelphia, Pennsylvania
McGraw-Hill Modern Men of Science, 1966. McGraw-Hill, New York
Mechanical Engineers in America Born Prior to 1861: A Biographical Dictionary, 1980. ASME Publications, New York
Men of Achievement. International Biographical Centre, Cambridge, UK
Roysdon C, Khatri L A (eds.) 1978 *American Engineers of the Nineteenth Century: A Biographical Index*. Garland, New York
Science Citation Index. Institute for Scientific Information, Philadelphia, Pennsylvania
Source Index. Institute for Scientific Information, Philadelphia, Pennsylvania
The International Who's Who, 1974–75. Europa, London
Who Knows and What, 1954. Marquis, Chicago, Illinois
Who's Who in America, 1980–81. Marquis, Chicago, Illinois
Who's Who in Consulting, 1973. Gale, Detroit, Michigan
Who's Who in Science in Europe, 1978. International Publications Service, New York
Who's Who in Technology Today, 1981. Technology Recognition Corporation, Pittsburgh, Pennsylvania
Who's Who in the East. Marquis, Chicago, Illinois
Who's Who in the South and Southwest. Marquis, Chicago, Illinois
Who's Who in the West. Marquis, Chicago, Illinois
Who's Who in the World, 1980–81. Marquis, Chicago, Illinois
Who's Who in Western Europe. International Biographical Centre, Cambridge, UK
World Biographical Hall of Fame. American Biographical Institute, Raleigh, North Carolina

9. Other Information Sources

Under this heading we include special information sources such as standards and specifications, patents, symbols, abbreviations, safety regulations, and mathematical and engineering tables.

9.1 Standards and Specifications

The main body that coordinates the making of standards and specifications in USA is the American National Standards Institute (ANSI). The corresponding body in the UK is the British Standards Institution (BSI). The body that deals with the standards affairs on an international basis is the International Organization for Standardization (ISO).

Some principal guides to standards and specifications are listed below.

British Standards Yearbooks. British Standards Institution, London
Catalogue des Normes Françaises. Association Française de Normalisation, Paris
Chumas S I 1975 *Directory of U.S. Standardization Activities*. National Bureau of Standards (NBS), Washington, DC
Code of Federal Regulations (Parts 0-199), 1984. Office of the Federal Register, US Government Printing Office, Washington, DC
English Translations of German Standards, 1970. DNA/Beuth-Vertieb, Cologne
Normblatt-Verzeichnis, 1969. DNA/Beuth-Vertieb, Cologne
Struglia E J 1965 *Standards and Specifications–Information Sources*. Gale, Detroit, Michigan

A small sample of available volumes on safety, instrumentation, software, communications, energy, and other standards is given below.

ASME Codes and Standards. ASME, New York
Ferson L M 1982 *ISA Electrical Safety Standards*. Wiley, Chichester
Folts H C (ed.) 1981 *Compilation of Data Communications Standards*. McGraw-Hill, New York
Graphic Symbols for Logic Functions Standards. ANSI/IEEE, New York
IAEA Safety Codes and Guides, 1984. International Atomic Energy Agency, Vienna
IAEA Safety Standards: Safety Series. International Atomic Energy Agency, Vienna
IEEE Instrumentation Interface Standards. ANSI/IEEE, New York
IEEE Recommended Practice for Energy Conservation and Cost Effective Planning in Industrial Facilities. IEEE, New York
IEEE Standards for Local Area Networks. ANSI/IEEE, New York
IEEE Software Engineering Standards. ANSI/IEEE, New York
Klein M (ed.) 1965 *Einführung in die Din-Normen*. Teubner, Stuttgart
Park W T 1978 *Robot Safety Suggestions*, SRI Intl. Artificial Intelligence Center, Menlo Park, California
Quak K 1961 *Bergiffslexikon: Benennungen und Definitionen aus den Deutschen Normen*. DNA/Beuth-Vertieb, Cologne
Standards for Computer-Aided Manufacturing, 1977. National Bureau of Standards, Washington, DC
Study on Standardization of Industrial Robots, 1976. Japan Industrial Robot Association, Tokyo
Workshop on Standards for Image Pattern Recognition. National Bureau of Standards, Washington, DC

9.2 Patents and Symbols

A few guides and sources on patents, symbols and abbreviations are now given (Westbrook 1986).

(a) Patents

Central Patents Index. Document Publications, London
Haughton B (ed.) 1972 *Technical Information Sources: A Guide to Patent Specifications, Standards and Technical Report Literature*. Shoestring, Hamden, Connecticut
Official Gazette of the U.S. Patent Office (weekly since 1872). US Patent Office, Washington, DC

(b) Symbols and abbreviations

Armstrong-Lowe D 1975 *A Guide to International Recommendations on Names and Symbols.* World Health Organization, Geneva

Arnell A (ed.) 1963) *Standard Graphical Symbols—A Comprehensive Guide for Use in Industry, Engineering and Science.* McGraw-Hill, New York

ASME Graphics Symbols Standards Package (ASME/NSI). ASME Publications, New York

ASME Letter Symbol Standards Package (ASME/NSI). ASME Publications, New York

Buttress F A 1966 *World List of Abbreviations.* Hill, London

Dreyfus H 1972 *Symbol Sourcebook.* McGraw-Hill, New York

Hasegawa K, Kaneko T 1978 *Study on the Standardization of Terms and Symbols Relating to Industrial Robots in Japan.* Springer, New York

Mitchell J H 1968 *Writing for Professional and Technical Journals.* Wiley, New York

Moser R C 1969 *Space-Age Acronyms: Abbreviations and Designations.* IFI/Plenum, New York

Polon D D (ed.) 1965 *Encyclopedia of Engineering Signs and Symbols.* Odyssey, New York

Shepard W 1971 *Shepard's Glossary of Graphic Signs and Symbols.* Dent, London

Style Manual, 1967. US Government Printing Office, Washington, DC

9.3 Mathematical and Engineering Tables

There exists a vast quantity of reference books providing comprehensive tables of mathematical and engineering functions and data.

Listed below is a sample of literature sources of this kind.

Bateman H 1954 *Tables of Integral Transforms.* McGraw-Hill, New York

Belousov S L 1962 *Tables of Normalized, Associated Legendre Polynomials.* Pergamon, Oxford

Beyer W H (ed.) 1966 *Handbook of Tables for Probability and Statistics.* Chemical Rubber Co., Cleveland, Ohio

Boll M 1964 *Tables Numériques Universelles.* Dunod, Paris

Cayther D B (ed.) 1975 *UK Nuclear Data Progress Report: April 1974–March 1975* (INDC (UK)-25/U). Nuclear Physics Division, AERE Harwell

Christova E A 1959 *Table of Bessel Functions of the True Argument and of Integrals Derived from Them.* Pergamon, Oxford

Coceva C, Panini G C 1981 *Neutron Cross Sections of Fission Product Nuclei (Proc. Specialists Meeting)*, NEANDC(E)-209-L. Comitato Nazionale Energia Nucleare, Dipartmento Ricerca Technologica di Base ad Avanzata, RIT/FIS-LDN(80)1

Compilation of Actinide Neutron Nuclear Data (KDK-35), 1979. Swedish Nuclear Data Committee, Stockholm

Fort E (ed.) *Rapport des Activités de Recherche dans le domaine de données Nucléaire en France*, NEANDC(E)-212-U. Départment de Réacteurs à Neutrons Rapides, Section de Physique de Neutrons Rapides

Foster R M 1975 *Fourier Integrals for Practical Applications.* Prentice-Hall, Englewood Cliffs, New Jersey

Hansen E R 1975 *A Table of Series and Products.* Prentice-Hall, Englewood Cliffs, New Jersey

Neutron Cross Sections of 28 Fission Product Nuclides Adopted in JENDL-1, 1981: JAARI 1268, NEA-NDC(J)-68/U, INDC (JAP)-55/L. Japan Atomic Energy Research Institute, Tokyo

Progress Report on Nuclear Data Activities in Sweden for 1982. Swedish Nuclear Data Committee, Stockholm, Sweden

Progress Report on Nuclear Data Research in the European Community, Jan.–Dec. 1973: Nuclear Energy Agency Nuclear Data NEANDC/(E) 161/U. CBNM-Euratom, Geel, Belgium

Qaim S M (ed.) *Progress Report on Nuclear Data Research in the FR Germany*, April 1978–March 1979: NEA-NDC(E)-202U, INDC (Ger)-21/L + Special. Institut für Chemie, Kernforshungsanlage Julich

10. Continuing Education Courses

A number of academic institutes and associations provide professional education and specialized information in engineering and applied sciences. Intensive courses, including those in the general systems and control field, are held throughout each year in various countries and cities.

Some representative, but nonexhaustive, examples of such Continuing Engineering Education Programs are the following.

Advanced Science and Technology Education Program. Continuing Education Institute–Europe, Rörstopsvägen 5, S-61200 Finspång, Sweden

Berlin Continuing Engineering Education Program (BECEEP). AMK Berlin, Messedamm 2, D-1000 Berlin 19, FRG

European Association for Signal Processing Short Course Series. EURASIP, PO Box 134, CH-1000 Lausanne 13, Switzerland

George Washington University Continuing Engineering Education Program, Washington, DC 20052, USA

IEEE Standards Seminar Series. Standards Board of the IEEE, New York

Short Course Programs. Ecole Polytechnique Fédérale de Lausanne, EPFL, CH-1007 Lausanne, Switzerland

Training Courses Program, ICT-DWCI. Interference Control Technologies Inc. (Don White Consultants, Subsidiary), Gainsville, Virginia

11. Conclusions

This article provides an extensive outline of the present state of the art in systems and control information. An attempt was made to cover the major types of information sources on the whole field in the broad sense, that is, including analysis and design methodology, computing, and applications. Of course, the sources discussed in the article belong to the class of sources that are open to the public. A vast amount of information, however, is classified or confidential (for military or commercial reasons), which implies that a lot of useful existing knowledge does not become available to the world at

large. A direct consequence of this is that much redundant work is carried out and that many existing discrepancies are never studied and resolved. Today, a great effort is being made to automate the documentation and processing of bibliographic and other scientific and technical information through the use of computers. Among the issues included in this automation are the access to machine readable files, the electronic storage and retrieval of information, and the interactive processing of data using dialoguing and expert system methods (Bundy 1986, Hayes-Roth *et al.* 1983, Waterman 1985).

Bibliography

Bundy A 1986 *Catalogue of Artificial Intelligence Tools*. Springer, Berlin

Grattidge W, Creps J E 1978 Information systems in engineering. *Ann. Rev. Inf. Sci. Technol.* **13,** 972

Hayes-Roth F, Waterman D A, Lenat D B (eds.) 1983 *Building Expert Systems*. Addison-Wesley, Reading, Massachusetts

Schuchman H L 1981 *Information Transfer in Engineering*, Report for Futures Group. Futures Group, Glastonbury, Connecticut

Waterman D A 1986 *A Guide to Expert Systems*. Addison-Wesley, Reading, Massachusetts

Westbrook J H 1983 *Extraction and Compilation of Numerical and Factual Data. Development and Use of Numerical and Factual Data Bases*, AGARD Lecture Series No. 130. NATO Advisory Group for Aerospace Research and Development, Neuilly-sur-Seine, France

Westbrook J H 1986 Materials information sources. In: Bever M B (ed.) *Encyclopedia of Materials Science and Engineering*. Pergamon, Oxford

LIST OF ACRONYMS AND ABBREVIATIONS

This list has been designed as a reference source to the wide variety of acronyms and abbreviations now in common use in the field of systems and control. Specific computer languages have been indicated by an asterisk.

The list was compiled by Professor P. Borne of Institut Industriel du Nord, Villeneuve d'Ascq, France, and while every attempt has been made to make it as comprehensive as possible, it is recognized that no list can be complete in such a rapidly developing field.

1D	One-dimensional (space)
2D	Two-dimensional (space)
3D	Three-dimensional (space)
AACC	American Automatic Control Council
AAS	Advanced administrative system
ABP	Actual block processor
AC	Analog computer
ac	Alternating current
ACAP	Advanced composite airframe program
ACAS	Airborne collision-avoidance system
ACB	Access control block; adapter control block; application control block
ACC	Accumulate; accumulator; application control code
ACCA	Asynchronous communications control attachment (feature)
ACCT	Account
ACD	Automatic call distributor
ACE	Automated computing engine; automated cost estimating; area control error
ACF	Advanced communications function
ACH	Automated clearing houses
ACI	Automatic car identification
ACIA	Asynchronous communications interface adapter
ACK	Acknowledge character
ACL	Application control language
ACM	Association for Computing Machinery
ACR	Alternate recovery; audio cassette recorder; automatic call recording
ACRE	APAR control remote entry
ACS	Application customizer service
ACSL	Advanced continuous simulation language
ACT	Abend control table
ACU	Address control unit; automatic calling unit
ACV	Address control vector
A–D	Analog-to-digital
ADA	Automatic data acquisition
ADAC	Automated direct analog computer
ADAM	Advanced data management
ADAPT	Adaptation of automatically programmed tools
ADC	Analog-to-digital converter; air data computer
ADCCP	Advanced data communications control procedure
ADDDS	Automatic direct distance dialling system
ADE	Automatic design engineering; automatic drafting equipment
ADES	Automatic digital encoding system
ADF	Application development facility
ADI	Altitude direction indicator
ADM	Adaptive delta modulation
ADP	Automatic data processing
ADPE	Automatic data-processing equipment
ADPS	Automatic data-processing system
ADR	Address; application definition record; applied data research
ADT	Active disk table; application-dedicated terminal
ADU	Automatic dialling unit
ADX	Automatic data exchange
AE	Asymptotic expansion
AEC	Architectural, engineering and construction (applications)
AED	Automated engineering design
AEIMS	Administrative engineering information management system
AEW	Airborne early warning
AF	Arrow form
af	Audio frequency
AFCET	Association Francaise pour la Cybernetique Electronique et Technique; Association Francaise pour la Cybernetique Economique et Technique
AFF	Automatic frequency follower
AFIPS	American Federation of Information Processing Societies
AFP	Automatic floating point
AFR	Application function routine
AFT	Active file table
AGC	Automatic generation control
AGR	Advanced gas-cooled reactor
AHL	A hardware programming language
AHP	Analytic hierarchy process
AI	Artificial intelligence
AIA	Aerospace Industries Association
AIC	Akaike's information criterion
AID	Attention identifier
AIEE	American Institute of Electrical Engineers (now IEEE)
AILS	Advanced integrated landing systems
AIM	Application interface module; avalanche-induced migration
AIMS	Automated industry management services
AIP	Average instructions per second

AIRS	Advanced inertial reference sphere
ALD	Automatic logic diagram
ALGOL*	Algorithmic oriented language
ALL	Application load list
ALS	Automatic line supervision
ALU	Arithmetic logic unit
AM	Average magnitude; amplitude modulation
AMA	Automated message accounting
AME	Angle-measuring equipment; asymptotic matched expansion
AMFC	Adaptive model-following control
AMH	Application message handler
AMIS	Automated management information system
AML	Approximate maximum likelihood
AMR	Automatic meter reading
AMS	American Mathematical Society
AMT	Amount
AMVSB	Amplitude modulation with vestigial side band
ANA	Automatic network analyzer
ANSI	American National Standards Institute
AO	Automated operator
A–O	Answer–originate
AOI	Automated operator interface
AOQ	Average outgoing quality
AOQL	Average outgoing quality limit
AP	Accounts payable; attached processor
APAR	Authorized program analysis report
APCR	Air pollution control regulation
APF	Authorized program facility
APG	Automatic priority group
API	Application program interface
APICS	American Production and Inventory Control Society
APL*	A programming language
APR	Alternate path retry; automatic passbook recording
APS	Assembly programming system
APSE	ADA programming support environment
APT	Actual parameter area; automatic programming tool; automatically programmed tool; automatic picture transmission
AQAM	Air-quality assessment model
AQL	Acceptable quality level
AQS	Automated quotation systems
AR	Accounts receivable; address register; air-release (valve)
ARCS	Advanced reconfigurable computer system
ARDI	Analysis requirements, determination, design and development
ARM	ACRE/RETAIN merge; asynchronous response mode; availability, reliability, maintainability; autoregressive modelling
ARMA	Autoregressive moving average
ARMAX	Autoregressive moving average with exogeneous inputs
ARO	After receipt of order
ARP	Angle-resolved photoemission
ARQ	Automatic repeat request
ARS	Adaptive random search
ARSR	Air route surveillance radar
ART	Algebraic reconstruction algorithm
ARTCC	Air route traffic control center
ARTS	Automatic radar terminal system
ARU	Audio response unit
ARUPS	Angle-resolved ultraviolet photoelectron spectroscopy
AS	Asymptotic stability; address syllable
ASA	Accelerated storage adapter
ASC	Advanced scientific computer
ASCII	American Standard Code for Information Interchange
ASDE	Airport surface detection equipment
ASID	Address space identifier
ASHRAE	American Society of Heating, Refrigerating, and Air Conditioning Engineers
ASK	Amplitude-shift keying
ASKA	Automatic system for kinematic analysis
ASM	Auxiliary storage manager; all-sky monitor
ASME	American Society of Mechanical Engineers
ASP	Asymmetric multiprocessing system; attached support processor
ASR	Automatic send–receive; airport surveillance radar
ASTRA	Automatic scheduling with time-integrated resource allocation; advanced structural analyzer
ATC	Air traffic control
ATCRBS	Air traffic control radar beacon system
ATDM	Asynchronous time-division multiplexing
ATE	Automatic (or automated) test equipment
ATL	Automated tape library
ATLAS	Automated tape lay-up system
ATM	Automated teller machine
ATMS	Advanced text management system
ATN	Augmented transmission network
ATO	Automatic train operation
ATP	Alternative term plan; automated train protection
ATPG	Automatic test pattern generation
ATQ	Active transition queue
ATR	Advanced test reactor
ATS	Administrative terminal system; automated telemetry system
ATTN	Attention

ATTP	Aptitude test for programmer personnel
AUTODIN	Automatic digital network
AV	Audiovisual; atrioventricular
AVD	Alternate voice and data service
AVOID	Airfield vehicle obstacle indication device
AVR	Automatic volume recognition
AWT	Advanced wastewater treatment
BAL	Basic assembly language; branch linkage
BAM	Block allocating map
BAP	Basic assembly program
BAR	Base address register
BARSA	Billing, accounts receivable, sales analysis
BART	Bay area rapid transit
BASIC*	Beginners all-purpose symbolic instruction code
BB	Begin bracket indicator
BBD	Bucket-brigade devices
BC	Basic control
BCAS	Beacon collision avoidance system
BCB	Bit control block; buffer control block
BCC	Block check character; body-centered cubic
BCD	Binary-coded decimal
BCH	Block control header
BCL	Base-coupled logic
BCM	Bound control module
BCO	Binary coded octal
BCP	Byte-controlled protocol
BCS	Basic control system; British Computer Society
BCU	Basic counter unit; block control unit
BDAM	Basic direct-access method
BDE	Batch data exchange
BDES	Batch data exchange services
BDP	Bulk data processing
BDU	Basic device unit
BEL	Bell character
BEMA	Business Equipment Manufacturers, Association (USA)
BER	Bit error rate
BERT	Bit error rate test
BEST 3	Bending evaluation of structures
BEX	Broadband exchange
BF	Body-fixed
BFS	Basic fault-tolerant system
BG	Bond graph; bond graph model
BHJ	Bellman–Hamilton–Jacobi (equation)
BIBO	Bounded-input, bounded-output
BICARSA	Billing, inventory control, accounts receivable, sales analysis
BIFET	Bipolar field effect transistor
BINET	Bicentennial information network
BIOMED	Biomedical engineering
BIOS	Basic input–output supervisor
BIPS	Billion instructions per second
BISAM	Basic indexed sequential access method
BIT	Binary digit
BIU	Basic information unit; bus interface unit
BKF	Blocking factor
BLERT	Block error rate test
BLK	Blink
BLP	Boundary-layer problem
BLU	Basic link unit
BMC	Bulk media conversion
BMD	Bubble memory device
BMP	Batch message processing (program)
BMS	Bit mark sequencing
BNA	Burrough's network architecture
BNF	Backus normal form
BNN	Boundary network node
BO	Branch office
BOC	Basic operating company
BOD	Biochemical oxygen demand
BOE	Beginning of extent
BOF	Beginning of file
BOMP	Bill of material processor
BOP	Basic operator panel; bit-oriented protocol
BORAM	Block-oriented random access
BORS(C)HT (functions)	Battery feed, overload protection, ringing, supervision codec and channel filters, hybrid and test
BOS	Basic operating system
BOT	Beginning of tape
BPAM	Basic partitioned access method
BPDE	Biharmonic partial differential equation
BPI	Bits per inch; bytes per inch
BPM	Batch processing monitor
BPOS	Batch processing operating system
BPS	Basic programming support; bits per second
BPSS	Bell packet switching system
BR	Base register
BS	Backspace character; bounded set
BSAM	Basic sequential access method
BSC	Binary synchronous communication(s)
BSCA	Binary synchronous communications adapter
BSE	Bilinear strict equivalence
BSI	British Standards Institution
BSTAT	Basic status register
BTAM	Basic telecommunications access method
BTL	Beginning tape label
BTR	Behind tape reader
BTS	Bound task set; burster–trimmer–stacker
BTU	Basic transmission unit; British thermal unit
BVP	Boundary-value problem
BW	Bandwidth
BWR	Boiling water reactor
C3	Communications, command, control

CA Capacity assignment; computer-aided; channel adapter; communications adapter; continue-any (mode); composants acoustiques
C & C Command & control
CACSD Computer-aided control system design
CAD Computer-aided design; computer-assisted drafting
CADAM Computer-augmented design and manufacturing
CADD Computer-aided design and drafting
CAE Computer-aided engineering; computer-assisted engineering
CAI Computer analog input; computer-aided instruction; computer-assisted instruction
CAL Common assembler language; computer-assisted learning; conversational algebraic language
CAM Computer-aided manufacturing; computer-aided modelling; content-addressed memory; computer-assisted makeup
CAMAC Computer-automated measurement and control
CAN Cancel character
CANTRAN Cancel transmission
CAR Computer-assisted retrieval; check authorization record
CARR Carrier
CARS Coherent anti-Stokes–Raman scattering
CAS Computer-aided simulation; computer accounting system; collision avoidance system
CASS Common address space section
CAT Computer-aided testing; computer-aided translation; computerized axial tomography
CATV Community antenna television
CAU Command/arithmetic unit
CAW Channel address word
CAX Community automatic exchange
CAZ Commutating auto-zero
CB Citizen band; conduction band
CBA Cost–benefit analysis
CBCT Customer–bank communications terminal
CBE Concave branch elimination (algorithm)
CBMS Computer-based message system
CBT Computer-based terminal
CC Card code; card column; communications computer; control computer
CCA Central Computer Agency; common communications adapter; communications control area; current cost accounting
CCB Character control block; command control block
CCCB Completion code control block
CCD Charge-coupled device; circumscribing circle diameter
CCE Channel command entry
CCF Cobol communications facility; controller configuration facility; communications control field
CCH Channel-check handler
CCHS Cylinder–cylinder-head sector
CCIS Common-channel interoffice signalling
CCL Communications control language
CCLS Constant-coefficient linear systems
CCM Charge-coupled memory
CCP Character-controller protocol; communications control program
CCPT Controller creation parameter table
CCR Channel control routine
CCROS Card capacitor read-only store
CCSA (Bell system) common control switching arrangement
CCT Carriage control tape; computer compatible tape
CCTV Closed-circuit television
CCU Central control unit; channel control unit; communications control unit
CCW Channel command word
CD Communication device; compact disk
CDA Copper Development Association
CDB Corporate database
CDC Call directing code; call direction code
CDF Cummulative distribution function
CDP Certificate in data processing
CDRM Cross-domain resource manager
CDROM Compact disk read-only memory
CDT Communications display terminal; command definition table
CDTI Cockpit display of traffic information
CDTS Centralized digital telecommunications system
CDV Check-digit verification
CDW Charge density wave
CE Customer engineer; customer engineering; current efficiency
CEGL Cause–effect graph language
CEO Comprehensive electronic office
CEP Civil engineering package; circular error of probability
CESD Composite external symbol dictionary
CFA Capacity of flow assignment
CFIA Component failure impact analysis
CFM Cubic feet per minute
CFMS Chained file management system
CFP Creation facilities program
CFR Commercial fast-breeder reactor; coprime fractional representation
CFS Continuous forms stacker; cubic feet per second

CG	Channel grant
CGI	Computer-generated imagery
CGIC	Compressed-gas-insulated cable
CGL	Charge generation layer
CGM	Coarse-grained material
CGMID	Character generation module identifier
CHAR	Character
CHAT	Computer-harmonized, application-tailored
CHCV	Channel control vector
CHF	Critical heat flux
CHIO	Channel I–O
CHIPS	Clearing house interbank payment system
CHM	Chemical machining
CHP	Channel pointer
CHPS	Characters per second
CIA	Computer Industry Association; computer interface adapter
CIB	Command input buffer
CIC	Communications intelligence channel
CICP	Communication interrupt control program
CICS	Customer information control system
CID	Communication identifier; connection ID; current-image diffraction
CIDA	Channel indirect data addressing
CIDF	Control interval definition field
CIE	Commission Internationale de l'Eclairage
CIF	Central information file
CIG	Computer image generation
CIL	Condition-incident log; core image library
CIM	Communication interface module; computer input microfilm; computer-integrated manufacturing
CINDAS	Center for Information and Numerical Data Analysis and Synthesis
CIOCS	Communication input–output control system
CIR	Carrier-to-interference ratio
CIU	Communications interface unit; computer interface unit
CKD	Count-key-data
CKDS	Cryptographic key data set
CKT	Circuit
CLA	Computer Law Association; custom logic array; centerline average
CLAT	Communication line adapter
CLC	Closed-loop control; communications link controller
CLHP	Closed left-half plane
CLIST	Command list
CLK	Clock
CLM	Commutatorless motor
CLOC	Closed-loop optimal control
CLOP	Closed-loop optimal problem
CLP	Current line pointer
CLS	Closed-loop system
CLT	Central limit theorem; communications line terminal
CMA	Computer monitor adapter
CMC	Communications mag card; concurrent media conversion
CMC7	Caractère magnetique code 7
CMD	Command
CMI	Computer-managed instruction
CML	Current mode logic
CMM	Communications multiplexer module
CMND	Command
CMOD	Customer module
CMOS	Complementary metal–oxide–semiconductor
CMS	Conversational monitor(ing) system; computerized manufacturing system
CNA	Communication network architecture
CNC	Communications network controller; computer numerical control
CNI	Communications navigation–identification
CNM	Communication network management
CNP	Communications network processor
CNR	Carrier-to-noise ratio
CNRS	Centre National de la Récherche Scientifique (France)
CNTL	Control
COAM	Customer-owned and maintained
COBOL*	Common business-oriented language
COC	Computer on the chip
COD	Chemical oxygen demand
CODASYL	Conference on data systems languages
COG	Center of gravity
COGO	Coordinate geometry program
COL	Computer-oriented language
COLA	Cost of living adjustment
COM	Computer output microfilm; computer output microform
COMSTAT	Competitive statistical analysis
COP	Character-oriented protocol; communications output printer; coefficient of performance
CORS	Canadian Operational Research Society
COS	Class of service; commercial operating system
COVER	Covariance equivalent realization
CP	Card punch; central processor; command processor; continuous path; control program; communications processor; charge conjugation parity; cross-ply; characteristic polynomial
CPA	Channel program area; critical path analysis
CPAA	Charged particle activation analysis
CPAB	Computer programmer aptitude battery
CPB	Channel program block
CPBX	Computerized private branch exchange

CPC Card programmed calculator; computer process control; computer program component
CPE Computer performance evaluation; cross-program editor; central processing element; circular probable error
CPF Control program facility
CPH Characters per hour
CPI Changes per inch; characters per inch
CPL Characters per line; conversational programming language
CPM Cards per minute; characters per minute; critical path method; cycles per minute
CPMC Continuous parameter Markov chain
CPN Colored Petri net
CPO Concurrent peripheral operations
CPOL Communications procedure-oriented language
CPS Card programming system; characters per second; cycles per second
CPT Customer provided terminal
CPU Central processing unit; control processing unit
CQ Commercial quality
CQMS Circuit quality monitoring system
CR Card reader; carriage return; credit
CRA Computer Retailers' Association; catalogue recovery area
CRAM Card random-access memory
CRBE Conversational remote batch entry
CRC Cyclic redundancy check
CRCC Cyclic redundancy check character
CRE Carrier return character
CRHP Closed right-half plane
CRJE Conversational remote job entry
CRN Common random numbers
CROM Control read-only memory
CROS Card capacitor read-only store
CRP Channel request priority
CRT Cathode ray tube
CRTL Control
CRV Cryptography verification request
CS Control system; communication services
CSB Communication scanner base
CSDT Continuous-space, discrete-time
CSECT Control section
CSMA Carrier sense multiple access
CSMA/CD Carrier sense multiple access with collision detection
CSMP Continuous system modelling program
CSN Computer service network
CSO Computer service office
CSR Console send–receive
CSRS Coherent Stokes–Raman spectroscopy
CSSL Continuous system simulation language
CST Channel status message
CSTR Continuous stirred-tank reactor
CSU Customer service unit; customer setup
CSW Channel status word
CT Computed tomography
ct Continuous time
CTAK Cipher text auto key
CTB Concentrator terminal buffer
CTCA Channel-to-channel adapter
CTD Chemical transport and deposition; conductivity–temperature–depth
CTL Charge transport layer
CTMT Combined thermomechanical treatment
CTR Controlled thermonuclear reactor
CTS Carriage tape simulator; conversational terminal system; clear-to-send; communications technology satellite
CUE Computer utilization efficiency; configuration utilization evaluator
CUR Cost per unit requirement
CUTS Computer' users tape system
CV Coefficient of variation; calculus of variations
CVD Chemical vapor deposition
cw Continuous wave
CWP Computer word processing
DA Data administrator; device adapter; differential analyzer; direct access; discriminant analysis; data available
D–A Digital-to-analog
DAA Data access arrangement
DABS Discrete address beacon system
DAC Data acquisition and control; design augmented by computer; digital-to-analog converter
DAF Destination address field
DAL Data-access line; digital access line
DAM Direct access method
DAMA Demand-assigned, multiple-access
DAP Data-access protocol
DAPS Direct access programming system
DAR Damage assessment routine
DARS Digital altitude reference system
DARSS Diode array rapid-scan spectrometer
DAS Data acquisition system; data automation system; digital–analog system
DASD Direct-access storage device
DAT Dynamic address translation; disk allocation table
DAV Data above voice
DB Debit; database
DB/DC Database/data communications
DBA Database administrator
DBC Database computer
DBD Database description
DBDA Database design aid
DBDL Database definition language
DBF Database facility

DBMS	Database management software; database management system
DBS	Direct broadcast satellite
DBTG	Database task group
DC	Derivated causality; data center; data communication; data conversion; device control; digital computer; direct control
dc	Direct current
DCA	Distributed communications architecture
DCB	Data control block; device control block
DCC	Digital communications console
DCE	Data communications equipment; data circuit-terminating equipment
DCF	Data communication facility; data count field; discounted cash flow
DCL	Data control list; depth of cut line
DCQ	Double cascade quench patenting
DCS	Data collection system; data communications subsystem; defense communications system; distributed computer system; decentralized control system
DCSP	Direct-current straight polarity
DCT	Device characteristics table; dispatcher control table
DCTL	Direct-coupled transistor logic
DCU	Data control unit; display control unit
DCW	Data control word
DD	Data definition
DDA	Digital differential analyzer; demand deposit accounting; direct disk attachment
DDB	Device descriptor block
DDBMS	Distributed database management system
DDC	Direct digital control
DDCMP	Digital data communications message protocol
DDD	Direct distance dialling
DD/D	Data dictionary/directory
DDDL	Dictionary data definition language
DDG	Digital display generator
DDL	Data definition language; data description language
DDM	Device descriptor module
DDP	Distributed data processing
DDPS	Digital data-processing equipment
DDR	Dynamic device reconfiguration
DDS	Dataphone digital service; data distribution system; digital data system; dynamic dispatch system
DDSA	Digital data service adapter
DDT	Data description table
DE	Data entry; dictating equipment
DEA	Data encryption algorithm
DEB	Data extent block
DEDB	Data entry database
DED/D	Data element dictionary/directory
DEF	Destination element field
DEL	Delay; delete character
DES	Data encryption standard
DETAB	Decision table
DEU	Data encryption unit
DFC	Data flow control; disk file controller
DFLD	Device field
DFT	Diagnostic function test; discrete fourier transform
DFU	Data file utility
DHCF	Distributed host command facility
DI	Dependant inertia
DIAL	Data interchange at the application level
DIB	Data integrity block
DIBOL*	Digital equipment business-oriented language
DIC	Data interchange code; differential interference contrast
DID	Direct inward dialling
DI–DO	Digital input–digital output
DIF	Device input format
DINA	Distributed information processing network architecture
DIOCB	Device I–O control block
DIP	Dual-in-line package
DITTO	Data interfile transfer, testing and operations utility
DIV	Divisor
DJ	Dependant inertia moment
DL	Dependant inductance
DL/1*	Data language/1
DLA	Data link adapter
DLAT	Directory look-aside table
DLC	Data link control
DLE	Data link escape
DLGS	Doppler landing guidance system
DLPSC	Disturbance-localization problem with state dead-beat control
DLT	Data loop transceiver
DLU	Data line unit
DM	Dependant mass; data management; data manager; decision maker
DMA	Direct memory access
DMF	Disk management facility
DMH	Device message handler
DML	Data manipulation language; device media language
DMLS	Doppler microwave landing system
DMM	Digital multimeter
DMOS	Diffused MOS
DMS	Data management system
DMY	Day, month, year
DNA	Digital network architecture; distributed network architecture
DNC	Direct numerical control
DNIC	Data network identification code

DO	Dissolved oxygen
DOC	Documentation; direct operating cost
DOD	Direct outward dialling
DOF	Degree of freedom; device output format; depth of field
DOM	Dissolved organic matter
DOR	Digital optical recording
DOS	Disk operating system
DOSF	Distributed office support facility
DOT	Domain tip
DP	Distributed parameters; data processing; double precision; dynamic programming
DPA	Display/printer adapter
DPAGE	Device page
DPB	Dynamic pool block
DPC	Data processing center; disk pack controller
DPCM	Differential pulse-code modulation; predictive compression
DPCX	Distributed processing control executive
DPDT	Double-pole, double-throw
DPF	Dual polarized frequency
DPH	Disk pack handler
DPM	Data plant management; data processing machine; data processing manager; distributed presentation management; documents per minute
DPMC	Discrete-parameter Markov chain
DPP	Digital picture processing
DPPX/BASE	Distributed processing programming executive base
DPS	Distributed-parameter system; data processing system; disk programming system; distributed presentation services; document processing system
DPSK	Differential phase-shift keying
DPU	Display processor unit
DQCB	Disk queue control block
DRAM	Dynamic RAM
DRAW	Digital read-after-write disk
DRC	Data recording control
DRD	Data recording device
DRDS	Dynamic reconfiguration data set
DRO	Destructive readout
DRQ	Data ready queue
DRS	Data rate selector
DS	Data set; dynamical system; distributed systems
DSAC	Data set authority credential
DSAF	Destination subarea field
DSC	Disk storage controller; direct satellite communications
DSCA	Default system control area
DSCB	Data set control block
DSCP	Data services command processor
DSD	Data set definition
DSDS	Dataphone switched digital service
DSDT	Data set definition table
DSE	Data set extension; data switching exchange; distributed system environment
DSECT	Dummy control section
DSI	Data stream interface; digital speech interpolation
DSID	Data set identification
DSL	Data set label; development support library
DSLO	Distributed systems license option
DSM	Data services manager
DSN	Distributed systems network
DSO	Data set optimizer; direct system output
DSP	Digital signal processor
DSPP	Doubly stochastic Poisson process
DSRB	Data services request block
DSS	Digital subset; disk storage subsystem; dynamic support system
DST	Data services task; device service task
DSU	Data service unit; digital service unit; disk storage unit
DSW	Device status word
DSX	Distributed systems executive
DT	Data transmission; display terminal
dt	Discrete time
DTC	Data transmission channel; desk-top computer
DTE	Data terminal equipment
DTF	Define the file
DTL	Diode–transistor logic
DTM	Dynamic transient master control block
DTMF	Dual-tone multifrequency (receiver)
DTMS	Database and transaction management system
DTPA	Dynamic transient pool area
DTPM	Dynamic transient pool management
DTR	Distribution tape reel; data terminal reader
DTS	Data transfer sequence; dynamic transient segment register save; digital tandem switch
DTT	Domain tip technology
DTU	Display terminal unit
DTV	Digital television display
DTW	Dynamic time warping
DUART	Dual universal asynchronous receiver–transmitter
DUT	Device under test
DUV	Data under voice
DVB	Device base control block
DVMS	Digital voice messaging system
DVST	Direct-view storage tube
DVT	Device vector table
DWA	Daily weighted average
DXAM	Distributed indexed access method

DXS	Data exchange system
EA	Effective address
EAM	Electrical accounting machine
EAROM	Electrically alterable read-only memory
EAS	Exponential asymptotic stability; entity attribute set
EAU	Extended arithmetic unit
EAX	Electronic automatic exchange
EB	Electronic beam; event block
EBAM	Electron-beam-accessed memory
EBCDIC	Extended binary-coded decimal interchange code
EBCE	Electron-beam-control electronics
EBIC	Electron-beam-induced current
EBR	Electron-beam recorder; electron-beam recording; experimental breeder reactor
EC	Engineering change; extended control; electrical conductor grade; erosion corrosion
ECB	Event control block
ECC	Error correction code; error check(ing) and correction
ECCCS	Emergency command, control and communications system
ECD	Electron capture detector
ECG	Electrocardiogram
ECL	Emitter-coupled logic; execution control language
ECM	Extended core memory; electronic countermeasures; electrochemical machining; electronically commutated motor
ECOM	Electronic computer-originated mail
ECP	Electron channelling pattern
ECR	Electronic cash register
ECS	Expanded character set
ECS	Electrochemical Society
ECSL	Extended control and simulation language
ECT	Environment control table
ECTA	Error correcting tree automation
ECU	Electronic control unit
EDC	Error detection and correction; energy distribution curve
EDD	Electronic document distribution
EDM	Electrical discharge machining
EDOS	Extended disk operating system
EDP	Electronic data processing
EDPE	Electronic data-processing equipment
EDPM	Electronic data-processing machine
EDPS	Electronic data-processing system
EDS	Engineering design system; exchangeable disk storage
EDSAC	Electronic delay storage automatic computer
EDST	Elastic diaphragm switch technology
EDVAC	Electronic discrete-variable automatic computer
EDX	Energy-dispersive x-ray analysis
EEG	Electroencephalogram
EEPROM	Electrically erasable programmable read-only memory
EEROM	Electrically erasable read-only memory
EFP	Expanded function operator panel
EFTS	Electronic funds transfer system
EIN	European informatics network
EIOS	Extended input–output system
EIRV	Error interrupt request vector
EIS	Environmental impact statement; extended instruction set
EJCC	Eastern Joint Computer Conference
EKF	Extended Kalman filter
elint	Electronic intelligence
ELM	Error log manager; extended length message
ELS	Extended least-squares; emitter location system
ELT	Emergency locator transmitter
EM	Electromotor; end of medium character
EMC	Electromagnetic compatibility
EMG	Electromyogram
EMH	Expedited message handling
EMI	Electromagnetic interference
EMM	Extended matrix method
EMP	Electromagnetic pulse
EMR	Electromagnetic radiation
ENQ	Enquiry character
ENS	Extended network service
ENSIP	Engine structural-integrity program
EOA	End of address
EOB	End of block
EOC	End of card; end of chain
EOD	End of data
EOE	End of extent
EOF	End of file; end of form
EOG	Electrooculogram
EOI	End of inquiry; end of item
EOJ	End of job
EOL	End of line; end of list
EOM	End of message
EOP	End of program
EOQ	Economic order quantity
EOR	End of record; end of reel, end of report; end of run
EOS	Extended operating system
EOT	End of tape; end of text; end of transmission
EOV	End of volume
EP	Emulator program
EPA	Environmental Protection Agency
EPDE	Elliptic partial differential equation
EPIC	Executive planning information and communication system
EPL	European Program Library
EPMA	Electron probe microanalysis
EPNL	Effective perceived noise level
EPO	Emergency power-off

EPROM	Erasable programmable read-only memory
EPSS	Experimental packet-switched service
EPST	Extended partition specification table
EQ	Equal to
ER	Explicit route
ERB	Execution request block
ERC	Error retry count
ERCC	Error checking and correction
EREP	Environmental recording, editing and printing
ERG	Electroretinogram
ERM	Error recovery manager
ERMA	Electronic recording method of accounting
ERP	Error recovery procedure
ERR	Error
ERTS	Earth resources technology satellite
ES	Exponential stability
ESC	Escape character
ESCA	Electron spectroscopy for chemical analysis
ESD	External symbol dictionary
ESL*	European system language
ESMR	Electrically scanning microwave radiometer
ESN	Effective segment number
ESR	Early storage reserve
ESS	Electronic switching system
ESTV	Error statistics by tape volume
ESV	Error statistics by volume
ET	Electrical typewriter
ETB	End of transmission block
ETC	Extended text compositor
ETOS	Extended tape operating system
ETP	Extended tape processing; extended term plan
ETX	End of text
EUROCRA	European OCR Association
EVA	Error volume analysis
EVN	Error variance norm
EXCP	Execute channel program
EXLST	Exit list
EXR	Exception request
EXTRN	External reference
F/L	Fetch/load
FA	Flow assignment; final address
FAA	Federal Aviation Administration (USA)
FAC	Function authority credential
FAMOS	Floating gate avalanche injection MOS
FAMS	Forecasting and modelling system
FAPS	Financial analysis and planning system
FAX	Facsimile
FBA	Fixed block architecture
FBT	Facility block table
FC	Font change character
FCB	Forms control buffer; function control block
FCFS	First-come, first-served
FCI	Flux changes per inch
FCS	Frame check sequence
FD	File description; full duplex
FDA	Frequency-domain approach; finite-difference analysis, Food and Drug Administration (USA)
FDM	Frequency-division multiplexing; finite-difference method
FDMA	Frequency-domain multiple access
FDP	Form description program
FDT	Function data table
FDX	Full duplex transmission
FE	Field engineer; field engineering; format effector
FEA	Finite-element analysis
FEC	Forward error correction
FECB	File extent control block
FEFO	First-ended, first-out
FEM	Finite-element method
FEP	Front-end processor
FESS	Flywheel energy storage system
FET	First exit time; field-effect transistor
FF	Flip flop; form feed
FFA	Function-to-function architecture
FFT	Fast Fourier transform
FFTF	Fast flux test facility
FHD	Fixed-head disk
FHP	Fixed header prefix
FIC	First in chain
FICB	File identification control block
FID	Format identification
FIFF	First-in, first-fit
FIFO	First-input, first-output
FIGS	Figures shift
FINAC	Fast interline nonactivate automatic control
FIP	Finance image processor
FIR	Finite impulse response
FIT	Failure unit
FJCC	Fall Joint Computer Conference
FLAD	Fluorescence-activated display
FLIR	Forward-looking infrared
FLOPS	Floating point operations per second
FM	Facilities management; frequency modulation
FMCW	Frequency-modulated continuous wave
FMD	Function management data
FMH	Function management header
FMS	Flexible manufacturing system; file management supervisor
FNC	First normal canonical form; function generator
FND	Function regenerator
FNP	Front-end network processor
FOC	Fiber-optics communications
FOCAL	Formula calculator
FOCUS	Forum of control data users

FORMAT	FORTRAN matrix abstraction technique
FORTRAN*	Formula translation
FOSDIC	Film optical scanning device for input to computers
FOV	Field of view
FPE	Final prediction error
FPH	Floating point hardware
FPL	File parameter list
FPLA	Field programmable logic array
FPM	Fixed-point method
FPP	Floating point processor
FPROM	Field programmable read-only memory
FPS	Financial planning system
FPT	File parameter table
FR	Fractional representation
FROM	Fusible read-only memory
FRR	Functional recovery routine
FRU	Field replaceable unit
FS	File separator; future series; fuzzy systems
FSA	Field search argument
FSCB	File system control block
FSE	Full screen editor
FSK	Frequency-shift keying
FSN	File sequence number
FSP	Full screen processing
FSR	Full-scale range
FST	File status table
FSV	Floating-point status vector
FT	Fourier transform
FTA	Fast turnaround
FTAB	Field tab
FTE	Frame table entry
FTIR	Fourier-transform infrared (spectroscopy)
FTLP	Fixed-term lease plan
FTP	File transfer protocol; fixed-term plan
FV	Foreline valve
FVT	Final value theorem
FW	Fiscal week
FWDC	Full-wave-rectified three-phase alternating current
FWH	Flexible working hours
FWHM	Full width at half maximum
FX	Flexible exchange
FY	Fiscal year
GA	Go-ahead sequence
GADDR	Group address
GAI	Gain
GAM	Graphics access method
GAO	General accounting office
GAP	General accounting package
GAS	Global asymptotic stability
GBG	Generalized bond graph
GCFR	Gas-cooled fast reactor
GCR	Group coded recording
GD	Gaussian distribution
GDG	Generation data group
GDP	Gross domestic product
GE	Greater than or equal to
GERT	Graphical evaluation and review technique
GFD	Geophysical fluid dynamics
GFT	Generalized Fourier transform
GIGO	Garbage in, garbage out
GIS	Generalized information system
GJP	Graphic job processor
GM	Group mark
GML	Generalized mark-up language
GNA	Gauss–Newton algorithm
GND	Ground
GNP	Gross national product
GNR	Guest name record
GP	General purpose
GPC	General peripheral controller; general-purpose computer
GPDC	General-purpose digital computer
GPIA	General-purpose interface adapter
GPIB	General-purpose interface bus
GPS	General problem solver; graphic programming services; global positioning system
GPSS	General-purpose simulation system
GPT	General-purpose terminal
GPU	Graphic processing unit
GS	Group separator
GSA	General services administration
GSAM	Generalized sequential access method
GSE	Ground support equipment
GSO	Geosynchronous orbit
GSP	Graphic subroutine package
GSPS	General system problem solver
GSR	Global shared resources
GSVC	Generalized supervisor call
GT	Greater than
GTF	Generalized supervisor call
GUUB	Global uniform ultimate boundedness
GWAM	Gross words per minute
GY	Gyration; gyrator
HAM	High-availability manager
HAR	Harbor advisory radar
HASP	Houston automatic spooling program
HC	Hydraulic cylinder; hierarchical control
HCF	Host command facility
HCM	Hard-core monitor
HCP	Host command processor; host communication processor
HCR	Hybrid communication routine
HCT	Hard-copy task
HD	Half duplex
HDA	Head disk assembly
HDAM	Hierarchical direct access method
HDLC	High-level data link control
HDR	Header
HDX	Half duplex
HEL	High-energy laser
HEM	Heat-exchange method

HEMT	High-electron-mobility transistor
HERF	High-energy-rate forming
HEX	Hexadecimal
hf	High frequency
HFC	High-frequency control
HHC	Hand-held computer
HIC	Hybrid integrated circuit
HIDAM	Hierarchical indexed direct access method
HIDM	High-information delta modulation
HIPO	Hierarchical input, process output
HIRS	High-resolution infrared radiation sounder
HISAM	Hierarchical indexed sequential access method
HJB	Hamilton–Jacobi–Bellman (equation)
HJBT	Heterojunction bipolar transistor
HLC	High-level center
HLL	High-level language
HLW	High-level (radioactive) waste
HM	High modulus
HMOS	High-performance MOS
HN	Host-to-network
HNR	Handwritten numeral recognition
HOF	Head of form
HOL	High-order language
HP	Hydraulic pump; high pressure
HPDE	Hyperbolic partial differential equation
HPLC	High-pressure liquid chromatography
HRS	High-resolution spectrometer
HS	High speed; hierarchic sequential
HSAM	Hierarchic sequential access method
HSM	High-speed memory
HSP	High-speed printer
HSR	High-speed reader
HT	Horizontal tabulation character
HTFFR	High-temperature fast-flow reactor
HTGR	High-temperature gas-cooled reactor
HTL	High threshold logic
HV	High voltage
HVDC	High-voltage direct current
HY	High yield
I2L	Integrated injection logic
IAC	International Association of Cybernetics
IAL*	International algebraic language
IAP	Industry applications programs
IAR	Instruction address register
IAS	Interactive application system
IASTED	International Association of Science and Technology for Development
IBG	Interblock gap
IBN	Indexed by name
IBRO	Inter-bank computer bureau
IBT	Interrupt bit table
IC	Integrated circuit; instruction counter
ICA	Integrated communications adapter
ICAM	Integrated communications access method; integrated computer-aided manufacturing program
ICAO	International Civil Aviation Organization
ICB	Interrupt control block
ICC	Integrated communications controller; intercomputer coupler
ICCF	Interactive computing and control facility
ICDS	Input command data set
ICE	In-circuit emulation; in-circuit emulator
ICR	Independent component release
ICS	Interactive counting system
ICV	Initial chaining value
ID	Identifier
IDA	Integrated disk adapter
IDAPS	Image data-processing system
IDCB	Immediate device control block
IDF	Internal distribution frame
IDM	Intelligent database machine
IDMS	Integrated database management system
IDN	Integrated digital network
IDP	Integrated data processing
IDR	Instrumental determinant ratio
IDS	Integrated data store
IDT	Interdigital transducer
id(z)	Input decoupling (zero)
IEC	International Electrotechnical Commission
IEE	Institution of Electrical Engineers (UK)
IEEE	Institute of Electrical and Electronic Engineers (USA)
IEO	Integrated electronic office
IERE	Institution of Electronic and Radio Engineers (USA)
IES	Inverted echo sounder
IFA	Integrated file adapter
IFAC	International Federation of Automatic Control
IFB	Invitation for bid
IFCB	Interrupt fan control block
IFES	Image feature extraction system
IFIP	International Federation of Information Processing
IFIPS	International Federation of Information Processing Societies
IFORS	International Federation of Operational Research Science
IFR	Instrument flying rules
IFT	Inverse Fourier transform
IGES	Initial graphic exchange standard
IGFET	Insulated-gate field-effect transistor
IHI	Interactive hybrid interpreter
IHX	Intermediate heat exchanger
ILBT	Interrupt level branch table
ILT	Inverse Laplace transform

IMA	Institute of Mathematics and its Applications
IMACS	International Association for Mathematics and Computers in Simulation
IMC	Internal model control; Institute of Measurement and Control (UK)
IMEKO	International Measurement Confederation
IMH	Intermodal message handler
IMIS	Integrated management information system; integrated municipal information system
IML	Initial microprogram load
IMM	Intelligent memory manager
IMMA	Ion microprobe mass analyzer
IMP	Interface message processor
IMPL	Initial microprogram load
IMR	Iterative modified refined; interruption mask register
IMS	Information management system
IMVC	Input media conversion
IN	Information network (IBM)
INA	Inverse Nyquist array
INFO	Information
INP	Intelligent network processor
INRIA	Institut National de Recherche en Informatique et Automatique (France)
INS	Inertial navigation system
INT	Integrator
INX	Index character
IO	Intermediate output
I–O; i–o; I/O	Input–output
IOAU	Input–output access unit
IOB	Input–output block
IOC	Input–output controller
IOCR	Input–output control routine
IOCS	Input–output control system
iod(z)	Input–output decoupling (zero)
IOM	Input–output multiplexer
IOP	Input–output processor
IOQ	Input–output queue
IOS	Input–output supervisor
IP	Index of performance; information processing; information provider
IPA	Information processing architecture
IPC	Integrated program for commodities; illustrated parts catalog; industrial process control; integrated peripheral channel
IPCS	Interactive problem control system
IPL	Information processing language; initial program load; initial program loader; initial program loading
IPM	Instrumental product matrix
IPN	Instant private network
IPO	Installation productivity option
IPR	Isolated pacing response
IPS	Information processing system; installation performance specification
IPSS	International packet-switched service
IPT	Improved programming technologies
IPTS	International practical temperature scale
IQF	Interactive query facility
IQL	Incoming quality level
IR	Iterative refined; indicator register; industrial relations; industrial robot; information retrieval; interrupt request; infrared
IRA	Individual retirement account
IRBT	Intelligent remote batch terminal
IRC	International record carrier
IRG	Interrecord gap
IRIS	Instantaneous retrieval information
IRISA	Institut de Recherche en Informatique et Systemes Aléatoires (France)
IRL	Information retrieval language
IRM	Information resource management
IRMS	Information retrieval management system
IRR	Internal rate of return
IRS	Inquiry and reporting system; Internal Revenue Service
IRSS	Intelligent remote station support
IRT	Index return character
IS	Information separator
ISA	Instrument Society of America
ISAM	Indexed sequential-access method
ISC	Integrated storage control
ISDN	Integrated digital switching and transmission network; integrated services digital network
ISE	Integral square error
ISI	International Standards Institute; intelligent standard interface
ISL	Initial system load; integrated Schottky logic
ISO	International Organization for Standardization (International Standards Organization); independent sales organization
ISR	Information storage and retrieval; input select and reset
IST	Input stack tape; interrupt service task
IT	Indent tab character; information technology; intelligent terminal
ITACS	Integrated tactical air-control system
ITB	Intermediate text block
ITC	Investment tax credit
ITDM	Intelligent time-division multiplexing
ITF	Interactive terminal facility
ITNS	Integrated tactical navigation subsystem
ITR	Isolation test routine
ITS	Invitation to send

ITU	International Telecommunications Union
IUP	Installed user procedure
IV	Instrumental variables
IVD	Ion vapor deposition
IVNE	Instrumental variables normal equations
IVP	Installation verification procedure; initial-value problem
IVT	Initial-value theorem
JCB	Job control block
JCFI	Job control file (internal)
JCFS	Job control file (source)
JCL	Job control language
JCP	Job control program
JECS	Job entry central services
JEPS	Job entry peripheral services
JES	Job entry subsystem; job entry system
JFET	Junction field-effect transistor
JIS	Job information station
JJL	Josephson junction logic
JPA	Job pack area
JS	Junction structure
KAK	Key-auto-key
KAU	Key station adapter unit
KBD	Keyboard
KBM	Klylov and Bogolinbov Method
KBPS	Kilobits per second
KCS	Kilocharacters per second
KDS	Keyboard display station
KF	Kalman filter
KOPS	Thousands of operations per second
KSR	Keyboard send/receive
KTR	Keyboard typing reperforator
KWIC	Keyword in context
KWOC	Keyword out of context
KY(L)	Kalman–Yakubovich (lemma)
LA	Line adapter
LAN	Local area network
LAP	Link access protocol
LAPB	Link access protocol balanced
LASAR	Logic-automated stimulus and response
LASP	Local attached support processor
LASR	Letter writing with automatic send–receive
LB	Lower bound
LBG	Load balancing group
LBN	Logical block number
LBR	Laser beam recording
LBT	Listen before talk
LC	Limit cycle; local control; liquid crystal
LCB	Line control block
LCC	Life-cycle cost
LCD	Liquid crystal display
LCDDS	Leased circuit digital data service
LCFS	Last-come, first-served
LCH	Logical channel queue
LCN	Local computer network
LCP	Language conversion program
LCS	Large-capacity storage
LCU	Loop control unit
LDA	Logical device address
LDM	Limited-distance modem
LDO	Logical device order
LDT	Logical device table
LE	Less than or equal to
LED	Light-emitting diode
LESS	Least cost estimating and scheduling
LF	Lyapunov function; line feed
lf	Low frequency
LFC	Low-frequency control
LGN	Logical group number
LH	Left hand
LHP	Left-half plane
LIC	Last-in-chain; linear integrated circuit
LIFO	Last-input, first-output
LIP	Lasalle invariance principle
LISP*	List processing
LIT	Load initial table
LKY(L)	Lefschetz–Kalman–Yakubovich (lemma)
LLG	Logical line group
LLL	Low-level language
LM	Load module
LMFC	Linear model-following control
LMSE	Linear mean square estimation
LO	Line occupancy
LOCMOS	Local oxidation complementary MOS
LODESS	Linear ordinary differential equation system specification
LP	Linear programming; line printer; load point; longitudinal parity
LPA	Link pack area
LPB	Load program block
LPC	Linear predictive coding
LPM	Lines per minute
LPN	Logical page number
LPS	Linear programming system; lines per second
LQ	Linear quadratic
LQG	Linear quadratic Gaussian
LQP	Linear quadratic problem
LQSF	Linear quadratic state feedback
LRC	Longitudinal redundancy check
LRCC	Longitudinal redundancy check character
LRPDS	Long-range position-determining system
LRU	Line replaceable units
LS	Least significant; low speed; least squares
LSB	Least significant bit
LSC	Least significant character; loop station connector
LSD	Least significant digit; line-sharing device; line signal detector
LSI	Large-scale integration
LSID	Local session identification

LSP	Loop splice plate
LSQ	Least squares
LSQA	Local system queue area
LSR	Local shared resources
LSS	Large-scale system; loop surge suppressor
LSSS	Large-scale systems stability
LSSTA	Large-scale systems theory and applications
LSU	Local store unit; library store unit
LT	Laplace transform; less than
LTERM	Logical terminal
LTI	Linear time-invariant
LTM	Long-term memory
LTPL	Long-term procedural language
LTRS	Letters shift
LTU	Line termination unit
LTV	Linear time-varying
LTWA	Log tape write ahead
LU	Logical unit
LUT	Local user terminal
LVDT	Linear variable differential transformer
LWC	Loop wiring concentrator
LWR	Light water reactor
MA	Maintenance agreement
MAC	Machine-aided cognition; multiple-access computer; multiple-access computing
MACS	Modular application customizing system
MAD	Mean absolute deviation
MAGIC	Matrix analysis via generative and interpretive computations
MAGLEV	Magnetic levitation
MAI	Multiple access interface
MAL	Macro assembly language
MAP	Maximum *a posteriori*; macro assembly program; maintenance analysis procedures; mean arterial pressure; memory allocation and protection
MAR	Minimum angle of resolution
MAT	Medial axis transformation
MAU	Media access unit
MB	Megabyte
MBO	Management by objectives
MBX	Management by exception
MBZ	Must be zero
MC	Management committee; monitor call instruction; memory controller
MCAR	Machine check analysis and recording
MCBF	Mean characters between failures
MCC	Magnetic card code; master control code
MCH	Machine check handler
MCI	Machine check interruption
MCP	Master control program; message control program
MCRR	Machine check recording and recovery
MCS	Management control system; message control system; multiple console support
MCU	Magnetic card unit; microprocessor control unit
MD	Machine direction
MDF	Main distribution frame
MDR	Magnetic document reader; mark document reader
MDS	Microprocessor development system; multiple data set system
MDTS	Modular data transaction system
MDY	Month, day, year
ME	Manufacturing engineering
MES	Miscellaneous equipment specification; mapping and earth science (applications)
MESE	Maximum-entropy spectral estimator
MEW	Microwave early warning
MF	Multifrequency; microfiche
mf	Medium frequency
MFCM	Multifunction card machine
MFCU	Multifunction card unit
MFG	Manufacturing
MFH	Magnetic film handler
MFLD	Message field
MFM	Modified frequency modulation
MFMR	Multifrequency microwave radiometer
MFR	Multifrequency receiver
MFS	Message format service
MFT	Multiprogramming with a fixed number of tasks
MGA	Multiple gas analyzer
MGY	Modulated gyrator
MH	Message handler
MHD	Magnetohydrodynamic; moving-head disk
MHS	Magnetic hand scanner
MIB	Member information bank
MIC	Message identification code; middle-in-chain; monolithic integrated circuit
MICOS	Medical information and communication area
MICR	Magnetic ink character recognition
MICS	Management information and control system; microprocessor inertia and communication system
MID	Message input descriptor
MIH	Missing interruption handler
MIMD	Multiple instruction, multiple data stream
MIMO	Multi-input, multi-output
MIP	Mixed integer programming
MIPS	Million instructions per second
MIS	Management information system; manufacturing information system; medical information system; multistate information system
MISD	Multiple instruction, single data stream

MIT	Master instruction tape
MKY(L)	Meyer–Kalman–Yakubovich (lemma)
ML	Maximum likelihood; machine language
MLC	Maximum likelihood covariance; magnetic ledger card; medium-level center; multiline controller
MLE	Microprocessor language editor
MLP	Multivariable Lurie–Postnikov-type (system); multiline printing
mLP	Monovariable Lurie–Postnikov-type (system)
MLR	Multiline reading
MLS	Multilayered structure
MLT	Monolithic logic technology
MMS	Manufacturing monitoring system
MMU	Memory management unit
MNC	Multinational company
MNCS	Multipoint network-control system
MNOS	Metal, nitride, oxide, silicon
MOC	Mathematics of control
MOD	Message output descriptor
MOS	Metal–oxide semiconductor
MOSFET	Metal–oxide–semiconductor field-effect transistor
MP	Multiprocessor; multipurpose
MPC	Multiprocessor controller
MPP	Message processing program
MPPS	Message processing procedure specification
MPR(F)	Minimal positive real (function)
MPS	Mathematical programming system; multiprogramming system; multiple partition support
MPSP	Multiparameter singular perturbation
MPSX	Mathematical programming system extended
MPU	Microprocessor unit
MR	Memory reclaimer
MRA	Multiple regression analysis
MRAC	Model reference adaptive control; meter-reading access circuit
MRAS	Model reference adaptive system
MRM	Machine-readable material
MRO	Multiregion operation
MRP	Materials requirements planning; manufacturing resource planning
MRR	Material removal rate
MRT	Multiple requestor program
MS	Mass spectroscopy; mark sensing; master scheduler
MSB	Most significant bit
MSC	Mass storage control; mass storage controller; multiple systems coupling (feature)
MSD	Most significant digit
MSDB	Main storage database
MSE	Mean square error
MSF	Mass storage facility
MSG	Message
MSHP	Maintain system history program
MSI	Medium-scale integration
MSNF	Multisystem networking facility
MSP	Most significant position
MSPR(F)	Minimal strict positive real (function)
MSR	Magnetic stripe reader
MSS	Multispectral scanner; magnetic slot scanner; mass storage system
MSSF	Monitoring and system support facility
MST	Monolithic systems technology; medium-scale technology
MSU	Modem sharing unit
MSVC	Mass storage volume control
MT	Machine translation; magnetic tape; mechanical translation; measured time
MTA	Multiple terminal access
MTBF	Mean time between failures
MTBM	Mean time between maintenance
MTBO	Mean time between overhauls
MTC	Magnetic-tape controller
MTF	Modulated transformer
MTH	Magnetic-tape handler
MTL	Merged transistor logic
MTNS	Metal, thick oxide, nitride, silicon
MTS	Multi-timescale; message toll service
MTTA	Mean time to arrive
MTTF	Mean time to failure
MTTR	Mean time to repair
MTU	Magnetic-tape unit
MUL	Multiplier
MUSE	Modcom users exchange
MUST	Message user service transcriber
MUT	Module under test
MUX	Multiplexer
MVS	Multiple virtual storage
MVT	Multiprogramming with a variable number of tasks
MW	Molecular weight
MY	Man-year
NA	Numerical aperture; not applicable
NAK	Negative acknowledge character
NAM	Network analytic machine
NAS	National Academy of Sciences (USA); network administration station
NASA	National Aeronautics and Space Administration (USA)
NASTRAN	NASA structural analysis
NAU	Network addressable unit
NBS	National Bureau of Standards (USA)
NBS	Numeric backspace character
NC	Numeric computer; numerical control; necessary condition; network control; normally closed; numerically controlled
NCC	Network control center
NCCF	Network communications control facility

NCI	Noncoded information
NCP	Network control program
NCR	No carbon required
NCS	Network control station
NDB	Nondirectional beacon
NDC	Normalized device coordinates
NDE	Nondestructive evaluation
NDF	No defect found
NDI	Nondestructive inspection
NDL	Network definition language
NDPS	National data-processing service
NDR	Nondestructive read character
NDRO	Nondestructive readout
NDS	Network development system
NDT	Nondestructive testing; net data throughput
NE	Not equal to
NEIO	New international economic order
NEP	Never-ending program
NFF	No fault found
NFP	Network facilities package
NIB	Node initialization block
NIC	Network information center
NIH	Not invented here
NIP	Nucleus initialization program
NIS	Network information services
NIT	Nearly intelligent terminal
NIU	Network interface unit
NJE	Network job entry
NJI	Network job interface
NL	Nonlinear; new line
NLKF	Nonlinear Kalman filter
NLP	Nonlinear programming
NLS	Non-Lyapunov stability
NLTI	Nonlinear time-invariant
NLTV	Nonlinear time-varying
NMC	Network measurement center; network management center
NMO	National member organization
NMOS	*n*-Channel MOS
NMR	Nuclear magnetic resonance
NO	Normally open
NOS	Network operating system
NOSP	Network operation support program
NP	Nonprint
NPC	Nonprinting character
NPDA	Network problem determination application
NPE	Normally projective equivalent
NPR	Numerical position readout
NPS	Negatively Poisson stable; network processing supervisor
NRP	Newton–Raphson procedure
NRT	Nonrequestor terminal
NRZ	Nonreturn to zero
NRZI	Nonreturn to zero I
NS	Network services
NSC	Necessary and sufficient condition
NSD	Negative semidefinite
NSF	National Science Foundation
NSP	Network services protocol; numeric space character
NSPE	Network services procedure error
NTF	No trouble found
NTIS	National Technical Information Service
NTO	Network terminal option
NTPF	Number of terminals per failure
NTR	Nine thousand remote (protocol)
NTU	Network terminating unit
NUA	Network users association
NUL	Null character
O & M	Organization and methods
OAF	Origin address field
OAR	Office of Aerospace Research
OAR	Operator authorization record
OAS	Organizational accounting structure
OBC	On-board computer
OBI	Omnibearing indicator
OBR	Outboard recorder
Obs	Observable
OCC	Operator control command
OCCF	Operator communication control facility
OCDS	Output command data set
OCF	Operator console facility
OCL	Operator control language
OCP	Order code processor
OCR	Optical character recognition; optimal character recognition
OCR-A	Optical character recognition font A
OCR-B	Optical character recognition font B
OCRIT	Optical character recognizing intelligent terminal
ODE	Ordinary differential equation
ODESS	Ordinary differential equation system specification
ODF	Orientation distribution function
ODT	Octal debugging technique; on-line debugging technique
od(z)	Output decoupling (zero)
OE	Optical emission
OED	Optoelectric display
OEF	Origin element field
OEM	Original equipment manufacturer
OIC	Only-in-chain
OIS	Office information system
OL	On-line
OLC	Open-loop control
OLHP	Open left-half plane
OLMR	On-line modified refined
OLR	On-line refined
OLRT	On-line, real-time
OLS	Open-loop system
OLTEP	On-line test executive program
OLTS	On-line test system
OLTSEP	On-line, stand-alone executive program
OLTT	On-line terminal test
OMR	Optical mark reading

ONERA Office National d'Etudes et de Recherche Aeronautiques
OPCE Operator control element
OPICT Operator interface control block
OPM Operations per minute
OPS Operations per second
OPTIM Order point technique for inventory management
OPUR Object program utility routines
OQL Outgoing quality level
OR Operational research; operations research
ORHP Open right-half plane
ORRDR Optimal reliable robust decentralized regulator
ORS Optimal real storage
OS Operating system
OSA Open system architecture
OSAF Origin subarea field
OSAM Overflow sequential access method
OSI Open systems interconnection
OSRDR Optimal simultaneous robust decentralized regulator
OST Operation station task
OT Overtime; organ terminal
OTEC Ocean thermal energy conversion
OTRAJ Output trajectory
OUTLM Output limiting facility
PA Product assurance; program attention
PAB Primary application block
PABX Private automatic branch exchange
PAC Program authorized credentials
PAD Packet assembly–disassembly
PAIR Precision approach interferometric radar
PAL Programmable array logic; programmable algorithm machine assembly language
PAM Pulse amplitude modulation
PAP Precision approach radar
PAS Practical asymptotic stability
PAT Peripheral allocation table; programmers aptitude test
PATRIC Pattern recognition and information correlation (system)
PAV Program activation vector
PAX Private automatic exchange
PBA Printed board assembly
PBG Power bond graph
PBX Private branch exchange
PC Personal computer; path control; plug compatible; printed circuit; production control
PCA Primary communication attachment
PCAM Punch card accounting machine
PCB Page control block; printed circuit board; process control block; program communication block
PCD Partition control descriptor; preconfigured definition
PCE Processing and control element
PCI Program check interruption; program controlled interruption
PCL Process control language
PCM Plug compatible manufacturers; pulse coded modulation; punch card machine
PCMI Photochronic microimagery
PCP Primary control program
PCR Peripheral control routine
PCS Print contrast signal; project control system
PCT Partition control table
PCU Peripheral control unit
PD Process descriptor
PDA Physical device address
PDE Partial differential equation; perturbed differential equation; periodic differential equation; position-determining equipment
PDESS Partial differential equation system specification
PDF Probability distribution function
pdf Probability density function
PDM Pulse-duration modulation
PDN Public data network
PDR Processing data rate
PDS Partitioned data set
PE Phase encoding; projective equivalence; processor element
PEARL Process and experiment automation real-time language
PEC Page end character; program exception code
PECOS Project evaluation and cost optimization systems
PEL Picture element
PEM Program execution monitor
PEP Partitioned emulation programming
PER Program event recording
PERT Program evaluation and review technique
PES Practical exponential stability; partial exponential stability
PF Pretreatment filter
PFEP Programmable front-end processor
PFR Power fail recovery; prototype fast reactor
PFT Page frame table
PGT Page table
PHB Program header block
PHIL Programmable algorithm machine high-level language
PI Performance index; program isolation; programmed instruction; proportional-plus-integral
PIA Peripheral interface adaptor

PIC	Position-independent code; program information code
PICS	Production information and control system
PID	Personal identification device; proportional-plus-integral-plus-derivative
PIN	Personal identification number
PIO	Peripheral input–output; programmed input–output
PIOCS	Physical input–output control system
PIRV	Programmed interrupt request vector
PIU	Path information unit; peripheral interface unit
PL	Programming language
PL/1*	Programming language 1
PLA	Programmable logic array
PLBR	Prototype large breeder reactor
PLC	Program level change (tape)
PLCB	Pseudo-line control block; program list control block
PLL	Phase-locked loop
PLO	Phase-locked oscillator
PLPS	Presentation level protocol standard
PLR	Program library release; pseudo-linear regression
PLRS	Position, location and reporting system
PLSS	Precision location strike system
PLUM	Priority low-use minimal
PM	Phase modulation; preventive maintenance; program mode
PMBX	Private manual branch exchange
PMCT	Program management control table
PMF	Performance measurement facility; probability mass function
PMI	Personnel management information
PMM	Perfect model matching
PMOS	*p*-channel MOS
PMR	Proton magnetic resonance
PMS	Project management system; public message service
PMTS	Predetermined motion time system
PMX	Private manual exchange
PN	Part number; Petri net
PNA	Project network analysis
PNR	Passenger name record
POF	Point-of-failure (restart)
POH	Power-on hours
POI	Program operator interface
POL	Problem-oriented language
POLE	Point-of-last-environment (restart)
POM	Printed output microfilm; pseudo-overvaluing matrix
POP	Project optimization procedures
POS	Pascal operating system; point of scale; pseudo-overvaluing system
PPA	Protected partition area
PPBS	Planning, programming and budgeting system
PPDE	Parabolic partial differential equation
PPE	Problem program efficiency; problem program evaluator
PPI	Pulses per inch
PPM	Pulse-position modulation
PPS	Positively Poisson stable; pulses per second; plant protection system
PPSN	Public packet-switched network
PPT	Primary program operator interface task
PQA	Protected queue area
PQEL	Partition queue element
PR	Pattern recognition
PR(F)	Positive real (function)
PRBS	Pseudo-random binary sequence
PRC	Postal rate commission; primary return code; program required credentials
PRF	Pulse repetition frequency
PRI	Pulse repetition interval
PROLOG*	Programming in logic
PROM	Programmable read-only memory
PRR	Pulse repetition rate
PRT	Printer
PS	Practical stability; presentation services
PSA	Peripherals Suppliers' Association; process service area
PSB	Program specification block
PSCB	Presentation services command processor
PSCF	Primary system control facility
PSD	Positive semidefinite
PSE	Packet-switching exchange
PSF	Point spread function
PSG	Planning system generator
PSI	Peripheral subsystem interface
PSK	Phase-shift keying
PSM	Production systems management; proportional spacing machine
PSN	Print sequence number; public switched network
PSP	Packet-switching processor
PSR	Programming support representative
PSRR	Product and support requirements request
PSS	Packet-switching service; process-switching services
PST	Priority selection table; program synchronization table
PSTN	Public switched telephone network
PSU	Port sharing unit
PSV	Program status vector
PSW	Processor status word; program status word
PT	Partition table; punched table
PTC	Positive temperature coefficient
PTERM	Physical terminal
PTF	Program temporary fix
PTM	Programmable terminal multiplexor
PTP	Paper-tape punch; point-to-point

PTR Paper-tape reader
PTT Programmer productivity techniques
PTTC Paper-tape transmission code
PTW Page table word
PU Physical unit; pluggable unit
PUB Physical unit block
PUCP Physical unit control point
PVS Program validation services
PWM Pulse-width modulation
PWR Pressurized water reactor
PWS Private wire service
PZT Piezoelectric transducer
QA Quality assurance
Q/A Question/answer
QBE Query by example
QC Quality control
QCB Queue control block
QCPSK Quaternary coherent phase-shift keying
QECB Queue element control block
QIC Quarter-inch cartridge drive compatibility
QISAM Queued-indexed sequential access method
QLF Quadratic Lyapunov function
QNSM Queuing network stochastic model
QOH Quantity on hand
QPI Quadratic-plus-integral (function)
QSAM Queued sequential access method
QTAM Queued telecommunications access method
QUIP Quad-in-line package
R & D Research and development
R-field Resistive field
R/W Read/write
RA Repair action; repeat to address; reduction in area; right atrium
RAD Random-access device
RAIR Remote access immediate response
RAM Random-access memory; resident access method
RAMP Reliability, availability and maintainability program
RAMS Regional air-monitoring station
RAP Response analysis program; reactive atmosphere processing
RAR Rapid-access recording; reasonably assured resource
RAS Reliability–availability–serviceability; remote-access system
ratan Radar and television aid to navigation
RAX Remote-access computing system
RB Return-to-bias
RBA Relative byte address
RBM Real-time batch monitor
RBP Registered business programmer
RBT Remote batch terminal
RBV Return-beam vidicon
RC Radio control
RCB Request control block; resource control block
RCC Radio common carrier
RCD Receiver-carrier detector
RCF Reader's comment form
RCM Random choice method
RCP Recognition and control processor
RCR Required carrier return character
RCS Reloadable control store
RCT Region control task
RDC Remote data concentrator
RDF Record definition field; radial distribution function; radio direction finder
RDH Remote device handler
RDOS Real-time disk operating system
RES Remote entry services
RES Relative electric strength
RETAIN Remote technical assistance and information network
REU Remote entry unit
REW Read, execute, write
REX Route extension
rf Radio frequency
RFA Register field address
RFP Random fixed point; request for proposal
RFQ Request for quotation
RGB Red, green, blue
RGLS Recursive generalized least squares
RGP Remote graphics processor
RH Right hand; request/reponse header
RHP Right half-plane
RHS Right-hand side
RIDS Reset information data set
RIHANS River and harbor aid to navigation system
RIV Recursive instrumental variable
RJE Remote job entry
RKA Runge–Kutta algorithm
RKM Runge–Kutta method
RL Root locus
RLA Remote loop adapter
RLD Relocation dictionary
RLM Resident load module
RLS Recursive least square (algorithm)
RLT Root-locus technique
RMC Rod memory computer
RMI Radio magnetic indicator
RML Recursive maximum likelihood
RMM Read-mostly memory
RMS Root mean square; recovery management support; remote manipulator system
RNX Restricted numeric exchange
RO Receive only
ROC Receiver operating characteristic
ROI Return on investment
ROM Read-only memory

RON	Run occurence number
ROS	Read-only storage or store; resident operating system
ROTR	Receive-only typing reperforator
ROW	Rest of the world
RP	Roughing pump
RPE	Required page end character
RPG	Report program generator
RPL	Request parameter list
rpm	Rotations per minute; revolutions per minute
RPN	Reverse Polish notation
RPQ	Request for price quotation
RPS	Rotational position sensing
RPT	Repeat character
RPV	Reactor pressure vessel
RPW	Running process word
RQE	Reply queue element
RRDR	Reliable robust decentralized regulator
RRDRP	Reliable robust decentralized regulator problem
RRN	Relative record number
RS	Reciprocal system; record separator
RSAM	Relative sequential access method
RSCS	Remote spooling communications subsystem
RSDS	Relative sequential data set
RSF	Remote support facility
RSID	Resource identification table
RSM	Real storage management
RSP	Required space character
RSPT	Real storage page table
RSR	Reactor safety research
RSS	Resources security system
RT	Reciprocal transformation; real time; receiver–transmitter
RTAM	Remote terminal access method
RTB	Response/throughput bias
RTI	Real-time interface
RTL	Resistor–transistor logic
RTM	Real-time monitor; registered trade mark
RTOS	Real-time operating system
RTS	Reactive terminal service; remote terminal supervisor; request to send; reactor trip system
RU	Request/response unit
RUSS	Robotic ultrasonic scanning system
RV	Random variable
RVA	Recorded-voice announcement
RVI	Reverse interrupt
RVR	Runway visual range
RVT	Resource vector table
RWM	Read/write memory
RWO	Right, wrong, omits
Rx	Receiver
RZ	Return to zero
S & L	Savings & loan
SA	Sinoatrial
SAB	Secondary application block; session awareness block
SAC	Store access control
SACP	Selected area channelling pattern
SADT	Structured analysis and design technique
SAFE	Structural analysis by finite elements
SAGMOS	Self-aligning gate MOS
SAM	Successive approximation method; sequential access method
SAMIS	Structural analysis and matrix interpretive system
SAMOS	Silicon and aluminum MOS
SAP	General structural analysis program
SAS	Stability augmented system
SASI	Shugart Associates system interface
SAT	System access technique
SAVT	Secondary address vector table; save area table
SAW	Surface acoustic wave
SAYE	Save as you earn
SBA	Shared batch area; small business administration
SBC	Single-board computer; small business computer
SBS	Satellite business systems; small business system; subscript character
SBVP	Stochastic boundary-value problem
SC	Session control; sufficient condition
SCA	System control area
SCADA	Supervisory control and data acquisition
SCANIT	Scan-only intelligent terminal
SCB	Station control block; string control byte
SCC	Specialized common carriers
SCD	System contents directory
SCDDS	Switched-circuit digital data service
SCDR	Store controller definition record; subsystem controller definition record
SCF	System control facility
SCL	System control language
SCP	Support control program; system control programming
SCPC	Single channel per carrier
SCR	Silicon-controlled rectifier
SCT	Special characters table; system configuration table
SD	Semidefinite; stability domain
SDA	Source date automation
SDI	Selective dissemination of information; serial data in
SDL	System directory list
SDLC	Synchronous data link control
SDO	Serial data out
SDR	Statistical data recorder; system definition record
SDW	Segment descriptor language

SDX	Satellite data exchange
SE	System equivalence; effort source; systems engineer; systems engineering
SEO	Self-excited oscillations
SEP	Separate element pricing
SERC	Science and Engineering Research Council (UK)
SEREP	System error recording editing program
SESLP	Sequential explicit stochastic linear programming
SF	Flow source
SFC	Sectored file controller
SGHWR	Steam-generated heavy-water reactor
SGJP	Satellite graphic job processor
SGL	System generation language
SGP	Statistics generation program
SGR	Self-generation recycle
SGT	Segment table
SHY	Syllable hyphen character
SI	Système International; shift in character
SIAM	Society for Industrial and Applied Mathematics; system integrated access method
SIB	Session information block
SIC	Standard industrial classification; specific inductive capacitance
SICB	Subinterrupt control block
SIG	Special interest group
SII	Standard individual identifier
SIMD	Single instruction stream, multiple data stream
SIMP	Satellite information message protocol
SIO	Serial input–output
SIP	Single-in-line package; system initialize program
SIR	Shuttle imaging radar
SISD	Single instruction stream, single data stream
SISO	Single-input, single-ouput
SIU	System integration unit
SJCC	Spring Joint Computer Conference
SKU	Stock-keeping unit
SLAP	Subscriber line-access protocol
SLAR	Side-looking airborne radar
SLC	Single line controller
SLD	Straight-line depreciation
SLIB	Subsystem library
SLIC	Subscriber line interface circuit
SLIP	Serviceability level indicator processing
SLR	Side-looking radar
SLSI	Super-large-scale integration
SLSS	System library subscription service
SLT	Solid logic technology
SLU	Secondary logical unit
SM	Stochastic model
SMA	Skeletal muscle activity
SME	Society of Manufacturing Engineers (USA)
SMES	Superconducting magnet energy storage
SMF	System management facility
SML	Spool multileaving
SMM	System management monitor
SMQ	Save/restore message queue
SMS	System measurement software
SN	Scalar norm
S–N	Signal-to-noise
SNA	Systems network architecture
SNBU	Switched-network backup
SNF	Sequence number field
SNR	Signal-to-noise ratio
SO	Shift-out character
SOA	Safe operating area; start of address
SOD	Sum-of-years digit
SOF	Start-of-format control
SOH	Start of header, start of heading character
SOM	Start of message; system operator's manual
SOP	Standard operating procedure; study organization plan
SOR	Signal operator responsibility
SOS	Silicon on sapphire; sophisticated operating system
SOTUS	Sequentially operated teletypewriter universal selector
SP	Singular perturbation; symbolic polynomial; pressure source; single precision; space character; structured programming
SP+R	Singular perturbation and reciprocal transformation
SPA	Scratched area
SPC	Stored-program control
SPDT	Single-pole, double-throw
SPEC	Speech predictive encoding
SPF	Structured programming facility
SPG	Sort program generator
SPI	Single program initiator
SPIE	Society of Photo-Optical Instrumentation Engineers (USA)
SPIN	Strategies and policies in informatics
SPM	Source program maintenance
SPN	Switched public network
SPOOL	Simultaneous peripheral operations on-line
SPPS	Subsystem program preparation support
SPR	System parameter record
SPR(F)	Strict positive real (function)
SPROM	Switched programmable read-only memory
SPS	Superscript character; solar power satellite
SPT	System parameter table
SQ	Flow source

SQA	System queue area
SQD	Signal quality detector
SQL	Structured query language
SQR	Square root
SRB	Service request block
SRDR	Simultaneous robust decentralized regulator
SRF	Software recovery facility
SRM	System resources manager; standard reference material
SRST	System resource and status table
SRT	Segmentation register table; single requestor terminal
SRTS	Special reverse charge service
SS	Solid state; start–stop
SSA	Segment search argument; slave service area; structured systems analysis
SSAM	Slave service area module
SSCF	Secondary system control facility
SSCP	System service control point
SSDR	Supermarket subsystem definition record
SSE	Strict system equivalence; switching system engineer
SSI	Small-scale integration
SSID	Subsystem identification
SSL	Source statement library
SSMVT	State space and multivariable theory
SSP	System service program
SSR	Serially reusable resource; status save area; solid-state relay; secondary surveillance radar
SSRP	Steady-state regulator problem
SSS	Subsystem support services; semistate space
SST	System scheduler table
ST	Straight time; synchronization table; system table
STAIRS	Storage and information retrieval system
STARS 2	Shell theory automated for rotational structures
STC	Self-tuning control
STCB	Subtask control block
STD	Subscriber trunk dialling
STDM	Statistical time-division multiplexing
STE	Segment table entry
STEC	Store exception condition
STL	Schottky transistor logic
STM	Short-term memory
STMT	Statement
STOCS	Small terminal oriented computer systems
STP	Stop character
STR	Self-tuning regulation; synchronous transmit/receive; synchronous transmitter/receiver
STRAJ	State trajectory
STRAM	Synchronous transmit/receive access method
STST	System task-set table
STW	System tape writer
STX	Start of text
SU	Selectable unit; signalling unit
SUB	Substitute character
SUM	Summator
SVA	Shared virtual area
SVC	Supervisor call instruction
SVF	State-variable filter
SVS	Single virtual storage system
SVT	System variable table
SW	Switch character
SWA	Scheduler work area
SWADS	Schedular work area data set
SWAMI	Software-aided multifont input
SYN	Synchronous idle character
SYNC	Synchronizing character
SYSGEN	System generation
SYSLOG	System log
SYSRES	System residence disk
T & M	Time & materials
TA	Transactional analysis
TAB	Tape-automated bonding; tone answer-back
TACT	Transient area control table
TAD	Transient area descriptor
TAF	Thick arrow form
TAG	Time automated grid
TBM	Terabit memory; to be determined
TBO	Time between overhaul
TC	Thermocouple; transmission control
TCAM	Telecommunications access method
TCAS	Terminal control address space; traffic alert and collision avoidance system
TCB	Task control block; thread control block
TCDMS	Telecommunication/data management system
TCF	Terminal configuration facility
TCM	Thermally controlled module
TCP	Topologically close packed; transmission channelling pattern
TCU	Terminal control unit; transmission control unit
TD	Transmitter–distributor
TDF	Transborder data flow
TDL	Terminal display language; transformation definition language
TDM	Time-division multiplexing
TDMA	Time-division multiple access
TDOS	Tape disk operating system
TDRM	Time-domain reflectometry microcomputer
TDRSS	Tracking and delay relay satellite system
TDS	Transaction data set; transaction driven system

TELNET	Telephone network
TELOPS	Telemetry on-line processing system
TES	Thermal energy storage
TF	Transformer
TGID	Transmission group identifier
TH	Transmission header
TICCETT	Time-shared interactive computer-controlled educational television
TIOC	Terminal input–output coordinator
TIOT	Task input–output table; terminal input–output task
TIP	Terminal interface processor
TIQ	Task input queue
TJB	Time-sharing job control block
TJID	Terminal job identification
TL	Thermoluminescence
TLA	Telex line adapter
TLAB	Translation look-aside buffer
TLAM	Tangential linear associated model
TLP	Term lease plan
TLS	Tactical landing system
TLU	Table look-up
TM	Tape mark
TML	Teradyne modelling language
TMP	Terminal monitor program
TMR	Triple modular redundancy
TMS	Telecommunications message switcher
TMU	Transmission message unit
TNS	Transaction network service
TOAF	Telecommunication access method origin address field
TOC	Table of contents; total organic carbon
TOCT–TOU	Time of check to time of use
TOD	Time of day
TOF	Time of flight
TOFFEA	Two-fluid flow equation analysis
TOLTEP	Teleprocessing on-line test executive program
TOM	Transparent office manager
TOS	Tape operating system
TOTE	Teleprocessing online test executive
TP	Teleprocessing; transaction processing
TPBVP	Two-point boundary-value problem
TPI	Tracks per inch
TPL	Terminal programming language
TPLIB	Transient program table
TPNS	Teleprocessing network simulator
TPR	Technical proposal requirements
TR	Tubular reactor
TRACS	Traffic reporting and control system
TRAN	Transmit
TRC	Table reference character
TRE	Time request element
TRI	Transmission interface converter
TRM	Test request message
TS	Total stability; time sharing; transmission services
TSAC	Time slot assignment circuit
TSB	Terminal status block
TSC	Time-sharing control task
TSCB	Task set control block
TSID	Time-sharing input queue control block
TSO	Time-sharing option
TSOS	Time-sharing operating system
TSRT	Task set reference table
TSS	Time-sharing system
TT	Transmitting typewriter
TTD	Temporary text delay
TTF	Terminal transaction facility
TTL	Transistor–transistor logic
TTY	Teletype; teletypewriter
TU	Tape unit
TWA	Time-weighted average; two-way alternate
TWS	Translator writing systems; two-way simultaneous
TWT	Travelling-wave tube
TWTA	Travelling-wave-tube amplifier
TWX	Teletypewriter exchange
TX	Transmit
UA	Uniform attractivity
UADS	User attribute data set
UART	Universal asynchronous receiver–transmitter
UAS	Uniform asymptotic stability
UB	Ultimate boundedness; upper bound
UBHR	User block handling routine
UBS	Unit backspace character
UC	Unit circle
UCF	Utility control facility
UCS	Universal character set
UDC	Universal decimal classification
UDS	Universal data set
UFP	Utility facilities program
UHL	User header label
UID	Universal individual identifier
ULA	Uncommitted logic array
ULC	Universal logic circuit
UN–S	Unstable, Poisson stable
UPAS	Uniform practical asymptotic stability
UPC	Universal product code
UPL	User programming language
UPS	Uniform practical stability; uninterruptible power supply
URC	Unit record controller
US	Uniform stability; unit separator
USART	Universal synchronous–asynchronous receiver–transmitter
USASI	USA Standards Institute
USERID	User identification
USRT	Universal synchronous receiver–transmitter
USS	Unformatted system services
UTL	User trailer label
UTS	Uniform total stability; unbound task set; universal time-share system
UUB	Uniform ultimate boundedness
UUD	Unit under development

UUT	Unit under test
uv	Ultraviolet
uv ROM	Ultraviolet read-only memory
VAB	Voice answer-back
VAC	Value-added carrier
VAI	Video-assisted instruction
VAM	Virtual access method
VAN	Value-added network
VCBA	Variable control block area
VCO	Voltage-controlled oscillator
VCR	Video cassette recorder
VDI	Virtual device interface
VDL*	Vienna definition language
VDP	Video display processor
VDT	Video display terminal
VDU	Video display unit; visual display unit
vf	Voice frequency
VFR	Visual flying rules
VFU	Vertical format unit
VG	Voice grade
VHLL	Very-high-level language
VHM	Very high modulus
VHPIC	Very-high-performance integrated circuit
VHRR	Very-high-resolution radiometer
VHSIC	Very-high-scale integrated circuit
VICC	Visual information control console
VIDAC	Virtual data acquisition and control
VIO	Virtual input–output
VIP	Verifying interpreting punch; visual information project
VIR	Visual and infrared radiometer
VISSR	Visual infrared spin–scan radiometer
VIT	Very intelligent terminal
VLA	Very large array
VLCD	Very-low-cost display
VLF	Vector Lyapunov function
VLSI	Very-large-scale integration
VM	Virtual memory
VMM	Variable mission manufacturing
VMOS	Virtual memory operating system; vertical MOS
VMS	Virtual memory system
VMT	Video matrix terminal
VN	Vector norm
VNL	Via net loss
VOGAD	Voice-operated gain-adjusting device
VOS	Virtual operating system
VP	Verifying punch
VR	Virtual route
VRAM	Video memory RAM
VRC	Vertical redundancy check; visible record computer
VRID	Virtual route identifier
VS	Visual system; virtual storage
VSAM	Virtual storage access method
VSB	Vestigial side band; very small business user
VSC	Variable-structure controller
VSE	Virtual storage extended
VSM	Virtual storage management
VSN	Volume serial number
VSPX	Vehicle scheduling program extended
VT	Vertical tabulation character
VTAM	Virtual telecommunications access method
VTOC	Volume table of contents
VTR	Video tape recorder
VU	Voice unit
VUVM	Voluntary universal marking program
WACK	Wait before receive positive acknowledgement
WADS	Wide area data service
WAP	Wavefront array processor
WATS	Wide area telephone service
WBS	Work breakdown structure
WCS	Writeable control store
WD	Wiring diagram
WDC	World data center
WIP	Work in process; work in progress
WJCC	Western Joint Computer Conference
WM	Word mark
WP	Word processing
WPM	Words per minute
WPS	Words per second
WRU	Who are you?
WT	Wild type
WTO	Write-to-operator
WTOR	Write-to-operator with reply
WWMCCS	World-wide military command and control system
WXTRN	Weak external reference
WYSIWYG	What you see is what you get
XMOS	High-speed MOS
XTC	External transmit clock
YBP	Years before present
YMD	Year, month, day
YTD	Year to date
ZBB	Zero base budgetting
ZCR	Zero crossing rate